Optical Tweezers
Methods and Applications

SERIES IN OPTICS AND OPTOELECTRONICS

Series Editors: **E. Roy Pike**, Kings College, London, UK
Robert G. W. Brown, University of California, Irvine

Recent titles in the series

Optical Tweezers
Methods and Applications

Edited by

Miles J. Padgett
University of Glasgow, Scotland, UK

Justin E. Molloy
MRC National Institute for Medical Research
London, UK

David McGloin
University of Dundee, Scotland, UK

CRC Press
Taylor & Francis Group
Boca Raton London New York

CRC Press is an imprint of the
Taylor & Francis Group, an **informa** business

A TAYLOR & FRANCIS BOOK

Chapman & Hall/CRC
Taylor & Francis Group
6000 Broken Sound Parkway NW, Suite 300
Boca Raton, FL 33487-2742

First issued in paperback 2019

ISBN-13: 978-1-4200-7412-3 (hbk)
ISBN-13: 978-0-367-38393-0 (pbk)

Library of Congress Cataloging-in-Publication Data

Optical tweezers : methods and applications / editors, Miles J. Padgett, Justin Molloy, David McGloin.
    p. cm. -- (Series in optics and optoelectronics)
  Includes bibliographical references and index.
  ISBN 978-1-4200-7412-3 (hardcover : alk. paper)
  1. Optical tweezers. 2. Optoelectronics. 3. Imaging systems in biology. I. Padgett, Miles J. II. Molloy, Justin. III. McGloin, David.

TK8360.O69O68 2010
681'.757--dc22
                                                    2009038146

Visit the Taylor & Francis Web site at
http://www.taylorandfrancis.com

and the CRC Press Web site at
http://www.crcpress.com

Contents

Preface

It is now more than 20 years since Arthur Ashkin and co-workers reported the demonstration of an optical trap for micron-sized particles based on a single, tightly focused laser beam. Since that time, the technical development of these optical tweezers has progressed hand in hand with their application throughout both the biological and physical sciences. This book reproduces approximately 60 landmark papers grouped into chapters covering both the techniques and their applications. Each chapter is introduced by a brief commentary that sets the papers into their historical and contemporary context.

Chapter 1 covers the pioneering work of Ashkin, featuring both his initial work on radiation pressure and the immediate progress following his seminal demonstration. Many of the original uses of optical tweezers were within biological systems; Chapter 2 covers these and features many of the more recent biological applications. Chapter 3 concerns the extensive use of optical tweezers for the measurement of piconewton forces, especially within biological systems. Chapter 4 examines the various approaches that have been applied to the modeling of forces within optical tweezers. These analyses include the mechanisms underpinning the basic tweezers force and also the interaction between many particles held within multiple tweezers systems.

In addition to their use in the biological sciences, optical tweezers have proved an invaluable tool within colloidal science, and various contributions to this field are covered in Chapter 5. Chapter 6 concerns the conversion of optical tweezers into optical spanners, where the trapped particle can be rotated either by transfer of optical angular momentum or by direct control of the trap positions.

Beyond trapping a single particle, optical tweezers can be configured to trap multiple objects simultaneously. Most recently this has centered on the use of spatial light modulators to create holographic tweezers, and these various developments are covered in Chapter 7. One of the new emerging applications of optical tweezers is within microfluidic systems. Chapter 8 covers these developments and their applications across scientific fields.

Our hope is that this book serves as an introduction to the topic by providing the key papers themselves and the additional references within the chapter introductions.

Miles J. Padgett
Justin E. Molloy
David McGloin

Reproduced Papers by Chapters

1

Optical Tweezers: The Early Years

Optical tweezers are a manifestation of the fact that light is able to exert a force on matter. Clearly, this is not an intuitive notion, as everyday experience leaves us with no impression that light carries momentum. However, quantum theory quantitatively states the momentum carried per photon. Indeed, the predication that there exists "radiation pressure" predates quantum mechanics and was suggested as a result of the electromagnetic theory developed by James Clerk Maxwell [1]:

> In a medium in which the waves are propagated there is a pressure in the direction normal to the wave, and numerically equal to the energy contained in unit volume.

Such pressure was also predicted by Bartoli [2] in a thermodynamic context and it became known as the Maxwell–Bartoli force. In fact, suggestions of such light pressure had an even earlier genesis, being hypothesized by Kepler (1619) when trying to determine the nature of comet tails, and had subsequently been studied by luminaries such as Euler (1746) (see mention in Refs. [3–6]). There were even attempts to carry out experiments to observe radiation pressure (see discussion in Ref. [3]) but these were hindered by thermal effects, although they did lead to Crooks's discovery of radiometric forces [7].

It is perhaps surprising to see that early experiments were successfully carried out to observe this radiation pressure before the invention of the laser. Two independent results were published in 1901. The papers by P. N. Lebedev (working in Moscow) [3] and by E. F. Nichols and G. F. Hull [4] (see **PAPER 1.1**), at Dartmouth University in the United States, took similar approaches to looking at the effect that light, produced by an arc lamp, had on thin vanes. Lebedev used very low gas pressures and thin metallic vanes while Nichols and Hull made use of high gas pressures and silvered glass vanes. In the latter case, measurements were carried out to ensure that the results were valid in the high-pressure limit, with the observation that the effects of the presence of the gas were negligible over short exposures of the vanes to the light. Nichols and Hull carried out a subsequent analysis of the experiments [5,6] highlighting both the historical background to the experiments and a comparison of the two experiments. They concluded that both experiments had measurement errors but could both safely be considered to have qualitatively proved the existence of a radiation pressure but without quantitatively following the theory of Maxwell–Bartoli forces. The quantitative demonstration was published in 1903 by Nichols and Hull [5,6].

Optical manipulation in its current form was pioneered by Arthur Ashkin working at Bell Labs in the 1970s and 1980s. By this time, the laser had been discovered and was a well-established tool. The much higher intensities producible by lasers allowed optical forces to be studied more easily. However, radiometric forces dominate optical forces when present and make the observation of the direct effect of optical forces difficult. This issue, coupled with a lack of laser sources, explains the absence of any substantial work in this area between the observation of the radiation pressure effect and Ashkin's work. One of Ashkin's key insights was realizing that by making use of transparent particles in a transparent medium he could still observe radiation pressure due to small amounts of reflection from such particles while avoiding problems associated with heating.

In his first paper [8] (see **PAPER 1.2**), Ashkin laid the foundation for the whole field. Using a weakly focused laser beam, he observed particle guiding: particles are drawn into the beam and travel in the direction of the beam propagation. He also demonstrated that there exists a gradient force, caused by the intensity gradient of the laser beam being used, that pulls high-refractive-index (compared to the surrounding medium) particles toward the beam axis but repels low-refractive-index particles (air bubbles) away from the axis. He hypothesized that light could be used to rotate particles (see Chapter 6) and observed that particles such as droplets can be accelerated. He also discussed how the techniques could be applied to atoms, work that would ultimately lead to the magneto-optical trap and the atomic dipole trap (intricately related to optical forces on microscopic particles). The final major result from the paper is that by using two beams that counterpropagate an optical trap can be created in which the particles are confined through a combination of radiation pressure and gradient force.

The next demonstration of radiation pressure by Ashkin was carried out the following year with J. M. Dziedzic as a coauthor. Here, trapping was achieved using a single beam [9] (see **PAPER 1.3**), simply by directing the beam upward so that the radiation pressure balances the gravitational force on the particle. The particles in this case are glass beads suspended in air. An interesting observation is that a second beam was used, orthogonal to the first, that could be used to drag the bead up or down within the levitating beam: turning this beam off after dragging allowed the bead to move back to the equilibrium position without undergoing any oscillations, indicating that the particle is heavily overdamped. Furthermore, the authors also say that they were able to use laser modes other than the normal TEM00 Gaussian beam, such as TEM01 and TEM01*, a doughnut mode.

Ashkin had clearly seen the possibilities for his new technique and went on to carry out a number of key experiments in the area of optical levitation over the next decade or so, showing that levitation was possible in high vacuum [10], how particle position could be stabilized using feedback mechanisms [11], and observing new types of nonlinear effects [12], as well as examining the effect of radiation pressure on a liquid interface [13] (still topical today with the continued debates about the momentum of light in a dielectric medium [14]).

Of the early optical manipulation work, one of the most abiding applications is that of the optical levitation of droplets [15] (see **PAPER 1.4**). Although most of the current applications that make use of optical manipulation were only really opened up by the invention of optical tweezers, the study of aerosols has until recently remained largely in the domain of optical levitation. Ashkin and Dziedzic's 1975 paper (**PAPER 1.4**) outlines how both solid and liquid airborne particles can be trapped and the types of experiments that can be carried out. These include experiments such as crystallization and simple measurements on particle interactions. It also highlights some of the issues associated with levitation, such as multiple particles being trapped simultaneously and the lack of independent control of these multiple droplets.

Ashkin and Dziedzic [16] show the use of the stabilization method to make sensitive measurements on levitated droplets. In this paper, the effect of wavelength and size is examined to see how the optical forces vary. In particular, the paper makes use of resonances excited within the droplets (whispering-gallery-mode type resonances) to provide highly sensitive size determination. Accuracies of 1 part in 105 or 106 are claimed. This type of technique is now commonly used for size measurement and can be used in combination with other techniques, such as Raman spectroscopy [17], to provide accurate size and composition determination of a particle.

It is clear from the literature that Ashkin and his colleagues were very much working in isolation on optical manipulation during the 1970s. Many of the papers citing Ashkin's original work in the years following it were concerned with atom trapping, while only a few other authors were concerned with microscopic manipulation. The application of radiation pressure-based traps fell out of favor somewhat with the development of the optical tweezers and the number of citations of the original work remained fairly static until the mid 1990s but has slowly picked up until the current day; there has been a significant rediscovery of this early work in recent years as people have become more interested in dual-beam and fiber-based traps.

In 1986, Ashkin, Dziedzic, Bjorkholm, and Chu, working at Bell Labs, demonstrated a new way to trap particles making use of a single laser beam [18] (see **PAPER 1.5**). By focusing a laser through a high-numerical-aperture microscope objective (all the radiation pressure work to date was based on low-numerical-aperture lenses), they found that a particle could be trapped not only in the plane transverse to the laser beam propagation but also in the axial direction. So, in this geometry, particles could be trapped using a beam pointing downward, holding the particle against gravity. The authors demonstrated that their tweezers were able to trap particles in the range 25 nm to 10 μm; they used a second beam to guide particles into the trapping region of the tweezers' beam.

As with the original radiation pressure trapping paper, the authors discussed how they felt their tool would be used: "They also open a new size regime to optical trapping encompassing macromolecules, colloids, small aerosols, and possibly biological particles." This proved to be very prescient, as this is very accurate description of what optical tweezers are currently used for. Indeed, the only surprising thing was that there was some doubt over whether biological samples could be trapped; there appears to have been no work done on biological samples using radiation pressure trapping in the early years of the technique. This was quickly shown to be possible by Ashkin and Dziedzic [19] (see **PAPER 1.6**), in which both viruses and bacteria were trapped using a green laser. The disadvantage of using green wavelengths is that biological samples are easily damaged due to increased absorption of visible light by biological samples, an effect often termed "opticution."

A safer form of trapping was demonstrated by Ashkin and colleagues in late 1987 in which an infrared laser was used to trap cells without damaging them [20]. It is also notable as it makes use of a dual-beam optical tweezers for the first time (although this technique had been used in a levitation experiment previously [21]). The use of infrared beams is of key importance in biological experiments and nearly all experiments in the biological arena make use of such wavelengths.

The development of optical tweezers was far more rapid than that of radiation pressure traps, with the research community realizing fairly quickly the wide range of applications that the technique enabled. Still, it is interesting to note that around 50% of the citations of the original tweezers paper have come in the past 5 years or so. So it seems that there is still much work that can be done using optical tweezers, with new application areas opening up all the time and increasingly sophisticated techniques being developed. The ease with which tweezers can be combined with other techniques in areas such as microscopy, spectroscopy, and microfluidics suggests that the field has a strong future in the coming decades.

Endnotes

1. J.C. Maxwell, *A Treatise on Electricity and Magnetism*, vol. 2, Oxford, Clarendon Press (1873).
2. A. Bartoli, Il calorico raggiante e il secondo principio di termodynamica, *Nuovo Cimento* **15**, 196–202 (1884).
3. P.N. Lebedev, Experimental examination of light pressure, *Ann. Phys.* **6**, 433 (1901).
4. E.F. Nichols and G.F. Hull, A preliminary communication on the pressure of heat and light radiation, *Phys. Rev.* **13**, 307–320 (1901).
5. E.F. Nichols and G.F. Hull, The pressure due to radiation (2nd paper), *Phys. Rev.* **17**, 26 (1903).
6. E.F. Nichols and G.F. Hull, The pressure due to radiation (2nd paper), *Phys. Rev.* **17**, 91 (1903).
7. W. Crooks, On attraction and repulsion resulting from radiation, *Philos. Trans. R. Soc. London* **164**, 501–527 (1874).
8. A. Ashkin, Acceleration and trapping of particles by radiation pressure, *Phys. Rev. Lett.* **24**, 156–159 (1970).
9. A. Ashkin and J.M. Dziedzic, Optical levitation by radiation pressure, *Appl. Phys. Lett.* **19**, 283–285 (1971).
10. A. Ashkin and J.M. Dziedzic, Optical levitation in high-vacuum, *Appl. Phys. Lett.* **28**, 333–335 (1976).
11. A. Ashkin and J.M. Dziedzic, Feedback stabilization of optically levitated particles, *Appl. Phys. Lett.* **30**, 202–204 (1977).
12. A. Ashkin and J.M. Dziedzic, Observation of a new nonlinear photoelectric effect using optical levitation, *Phys. Rev. Lett.* **36**, 267–270 (1976).
13. A. Ashkin and J.M. Dziedzic, Radiation pressure on a free liquid surface, *Phys. Rev. Lett.* **30**, 139–142 (1973).
14. R.N.C. Pfeifer, T.A. Nieminen, N.R. Heckenberg, and H. Rubinsztein-Dunlop, Colloquium: Momentum of an electromagnetic wave in dielectric media, *Rev. Mod. Phys.* **79**, 1197–1216 (2007).
15. A. Ashkin and J.M. Dziedzic, Optical levitation of liquid drops by radiation pressure, *Science* **187**, 1073–1075 (1975).
16. A. Ashkin and J.M. Dziedzic, Observation of resonances in radiation pressure on dielectric spheres, *Phys. Rev. Lett.* **38**, 1351–1354 (1977).
17. R. Symes, R.M. Sayer, and J.P. Reid, Cavity enhanced droplet spectroscopy: Principles, perspectives and prospects, *Phys. Chem. Chem. Phys.* **6**, 474–487 (2004).
18. A. Ashkin, J.M. Dziedzic, J.E. Bjorkholm, and S. Chu, Observation of a single-beam gradient force optical trap for dielectric particles, *Opt. Lett.* **11**, 288–290 (1986).
19. A. Ashkin and J.M. Dziedzic, Optical trapping and manipulation of viruses and bacteria, *Science* **235** 1517–1520 (1987).
20. A. Ashkin, J.M. Dziedzic, and T. Yamane, Optical trapping and manipulation of single cells using infrared-laser beams, *Nature* **330**, 769–771 (1987).
21. A. Ashkin and J.M. Dziedzic, Observation of light-scattering from nonspherical particles using optical levitation, *Appl. Opt.* **19**, 660–668 (1980).

A PRELIMINARY COMMUNICATION ON THE PRESSURE OF HEAT AND LIGHT RADIATION.

By E. F. Nichols and G. F. Hull.

MAXWELL,[1] dealing mathematically with the stresses in an electro-magnetic field, reached the conclusion that "in a medium in which waves are propagated there is a pressure normal to the waves and numerically equal to the energy in unit volume." Bartoli,[2] in 1876, announced that the existence of such a stress was essential to the validity of the second law of thermodynamics. He sought for such a pressure experimentally but failed to get conclusive results. The problem has been discussed theoretically, on the basis of thermodynamics, by Boltzmann[3] and by Galitzine.[4] Lebedew[5] applies such a pressure, opposed to the sun's gravitation, to account for the solar repulsion of comets' tails; and more recently Arrhenius[6] attempts to explain the aurora borealis on similar grounds.

Every approach to the experimental solution of the problem has hitherto been balked by the disturbing action of gases which it is impossible to remove entirely from the space surrounding the body upon which the radiation falls. The forces of attraction or repulsion, due to the action of gas molecules, are functions of the temperature difference between the body and its surroundings, caused by the absorption by the body of a portion of the rays which fall

[1] J. C. Maxwell, Elec. & Mag., 1st Ed., Oxford, 1873, Vol. 2, p. 391; also 2d Ed., Oxford, 1876, Vol. 2, p. 401.

[2] S. Bartoli, Sopra i movimenti prodotti della luce e dal calore: Florence, Le Monnier, 1876; also Exner's Repert d. Phys., 21, p. 198. 1885.

[3] L. Boltzmann, Wied. Ann., 22, p. 31; also p. 291; also Wied. Ann., 31, p. 139.

[4] B. Galitzine, Wied. Ann., 47, p. 479, 1892; also Phil. Mag., 85, p. 113, 1893.

[5] P. Lebedew, Wied. Ann., 45, p. 292, 1892.

[6] S. Arrhenius, Konigl. Vetanskaps-Akademiens, Förhandlingar, 1900; also Phys. Zeitschrift, Nov. 10 and 17, 1900.

upon it, and of the pressure of the gas surrounding the illuminated body. In the particular form of apparatus used in the present study the latter function appears very complicated, and certain peculiarities of the gas action remain inexplicable upon the basis of any simple group of assumptions which the writers have so far been able to make.

Since we can neither do away entirely with the gas nor calculate its effect under varying conditions, the only hopeful approach which remains is to devise apparatus and methods of observation which will reduce the errors due to gas action to a minimum. There are four ways in which the partial elimination of the gas action may be accomplished experimentally :

1. The surfaces which receive the radiation, the pressure of which is to be measured, should be as perfect reflectors as possible. This will reduce the gas action by making the rise of temperature due to absorption small while the radiation pressure will be increased ; the theory requiring that a beam, totally reflected, exert twice the pressure of an equal beam completely absorbed.

2. By studying the action of a beam of constant intensity upon the same surface surrounded by air at different pressures, certain pressures may be found where the gas action is smaller than at others.

3. The apparatus—some sort of torsion balance—should carry two surfaces symmetrically placed with reference to the rotation axis, and the surfaces on the two arms should be as nearly equal as possible in every respect save one—the forces due to radiation and gas action should have the same sign on one side and opposite signs on the other. In this way a mean of the resultant forces on the two sides should be, in part at least, free from gas action.

4. Radiation pressure, from its nature, must reach its maximum value instantly, while observation has shown that the gas action begins at zero and increases with length of exposure, rising rapidly at first, then more slowly to its maximum effect, which, in most of the cases observed, was not reached until the exposure had lasted from $2\frac{1}{2}$ to 3 minutes. For large gas pressures an even longer exposure was necessary to reach stationary conditions. The gas action may be thus still further reduced by a ballistic or semi-ballistic method of measurement.

The form of suspension of the torsion balance, used to measure radiation pressure in the present study, is seen in Fig. 1. The ro-

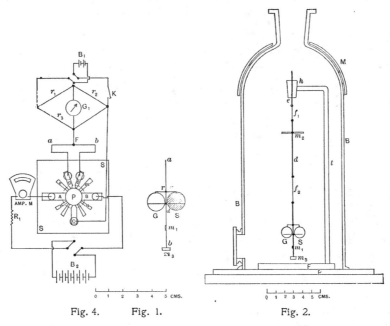

Fig. 4.   Fig. 1.   Fig. 2.

tation axis *ab* was a fine rod of drawn glass. At *r*, a drawn glass cross-arm *c*, bent down at either end into the form of a small hook,

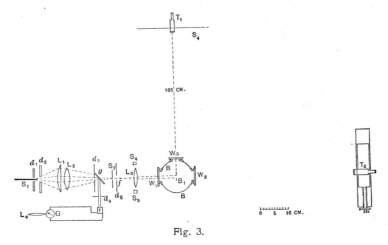

Fig. 3.

was attached. The surfaces G and S, which received the light beam, were circular microscope cover-glasses, 12.8 mm. in diameter and 0.17 mm. thick, weighing approximately 51 mgs. each. Through each glass a hole 0.5 mm. or less in diameter was drilled near the edge, by means of which the glasses could be hung on the hooks on the cross-arm c. Two other drawn-glass cross-arms were attached to the rotation axis at d, one on each side. The cover-glasses slipped easily between these, and were thus held securely in one plane. Further down on ab, a small silvered plane mirror m_1 was made fast at right angles to the plane of G and S. This mirror was polished bright on the silver side so that the scale at S_4 (Fig. 3), could be read in either face. A small brass weight m_3 (Fig. 1), of 452 mgs. mass and of known dimensions, was attached at the lower end of ab. The cover glasses which served as vanes were silvered and brilliantly polished on the silvered sides, and so hung on the small hooks on c that the glass face of one, and the silver face of the other, were presented to the light. A quartz fiber f_2 (Fig. 2), 3 cms. long, was made fast to the upper end of ab, and to the lower end of a fine glass rod d which carried a horizontal magnet m_2. The rod d was in turn suspended by a short fiber to a steel pin e which could be raised or lowered in the bearing h. The whole was carried by a bent glass tube t, firmly fastened to a solid brass foot F, resting on a plane, ground-glass plate P, cemented to a brass platform mounted on three levelling screws not shown in Fig. 2. A bell-jar B, 25 cms. high and 11 cms. in diameter covered the balance. The flange of the bell-jar was ground to fit the plate P. A ground-in hollow glass stopper fitted the neck of the bell-jar, which could thus be put in connection with a system of glass tubes leading to a Geissler mercury pump, a MacLeod pressure gauge, and a vertical glass tube dipping into a mercury cup and serving as a rough manometer for measuring the larger gas pressures employed during the observations. The low pressures were measured on the MacLeod gauge in the usual way. A semicircular magnet M, fitted to the vertical curvature of the bell-jar, was used to direct the suspended magnet m_2, and thus control the zero position of the torsion balance. By turning M through 180°, the opposite faces of the vanes G and S could be presented to the light.

THE ARRANGEMENT OF APPARATUS.

A horizontal section of the apparatus through the axis of the light beam, is shown in Fig. 3, which like the other figures is, in its essential features, drawn to scale. The white-hot end of the horizontal carbon S, of a 90° arc lamp, fed by alternating current, served as a source. The arc played against the end of the horizontal carbon from the vertical carbon which was screened from the lenses L_1 and L_2 by an asbestos diaphragm d_2. The lens L_4 projected an enlarged image of the arc and carbons on a neighboring wall, so that the position of the carbons and the condition of the arc could be seen at all times by both observers. The cone of rays passing through the small diaphragm d_2 fell upon the glass condensing lenses L_1, L_2 which formed an image of the carbon at f. At d_3 a diaphragm, 11.25 mm. in diameter, was interposed, which permitted only the central portion of the cone of rays to pass. Just beyond d_3, a plane plate of glass g was placed in the path of the beam and reflected part of the beam through the diaphragm d_4, to a thermopile T, connected with a galvanometer G. The beam which traversed the plate g passed on to a shutter at S_2. The shutter was worked by a magnetic escapement, operated by the seconds contact of a standard clock. The observer at T_1 might choose the second for opening or closing the shutter, but the shutter's motion always took place at the time of the seconds contact in the clock. Any exposure was thus of some whole number of seconds duration. The opening in the shutter was such as to let through, at the time of exposure, all of the direct beam which passed through d_3, but shut out the stray light. It was made to correspond in every way to the diaphragm d_4, so that for all of the reflected beam which passed d_4, the corresponding transmitted portion passed S_2 and no more. The glass lens L_3 focused a sharp image of the aperture d_3 in the plane of the vanes of the torsion balance B_1 under the bell-jar B. The bell-jar was provided with three plate glass windows W_1, W_2, W_3. The first two gave a circular opening 42 mm. in diameter, and the third was used in connection with the telescope T_1, by which an image of the scale S_4 was seen in the small mirror m_1 (Figs. 1, 2). The lens L_3 was arranged to move horizontally between the stops

S_3 and S_4. The stops were so adjusted that when the lens was against S_3 the sharp image of the aperture d_3 fell centrally upon one vane; and when against S_4 the image fell centrally upon the other. This adjustment, which was a very important one, was made by the aid of a telescope T_2, mounted on the carriage of a dividing engine. This was used to observe and measure the positions of the vanes and the rotation axis, as well as the positions of the images of d_3, when the lens L_3 was against the stops S_3 and S_4. For the latter measurements, the vanes could be moved out of the way, by turning the suspension through 90° by the control magnet M (Fig. 2).

METHODS OF OBSERVATION.

The observations leading to the results given later were of three different kinds: (1) The calibration of the torsion balance; (2) the measurement of the pressure of radiation in terms of the constant of the balance, and (3) the measurement of the energy of the same beam in electrical units by the aid of a bolometer described later.

1. The determination of the constant of the torsion balance was made by removing the vanes G and S and accurately measuring the period of vibration. Its moment of inertia was easily computed from the masses and distribution of the separate parts about the axis of rotation. From the moment of inertia, period, the lever arms of G and S, and the distance of the scale S_4, the constant of the balance for a centimeter deflection was found to be 4.65×10^{-5} dynes.

2. In the measurements of radiation pressure, it was easier to refer the intensity of the beam at each exposure to some arbitrary standard which could be kept constant than to try to hold the lamp as steady as would otherwise have been necessary. For this purpose, the thermopile T (Fig. 3) was introduced, and simultaneous observations were made of the relative intensity of the reflected beam by the deflection of the galvanometer G, and the pressure due to the transmitted beam by the deflection of the torsion balance. The actual deflection of the balance was then reduced to a deflection corresponding to a galvanometer deflection of 100 scale divisions. The galvanometer sensitiveness was carefully tested at

intervals during the work, and was found as nearly constant as the character of the observations required. All observations of pressure were thus reduced to the pressure due to a beam of fixed intensity.

At each gas pressure in the bell-jar, at which radiation pressure measurements were taken, two sets of observations were made. In one of these sets, static conditions were observed, and in the other, the deflections of the balance due to short exposures were measured. In the static observations, each vane of the balance was exposed in turn to the beam from the arc lamp, the exposures lasting in each case until the turning points of the swings showed that static conditions had been reached. The combined pressure, due to radiation and gas action would thus be equal to the product of the angle of deflection and the constant of the balance. The torsion system was then turned through 180° by rotating the outside magnet, and similar observations made on the reverse side of the vanes, after which the system was turned back to its first position, and the earlier set of observations repeated, and so on.

The considerations underlying the above method of procedure, are as follows: The beam from the lamp, before reaching the balance, passed through three thick glass lenses and two glass plates. All wave-lengths destructively absorbed by the glass, were thus sifted out of the beam by the time it reached the glass balance vanes. The silver coatings on the vanes absorbed therefore more than the glass. The radiation pressure was always away from the source irrespective of the way the vanes were turned, while the gas action would be exerted mainly on the silvered side of the vanes. As the face of the vane on one side of the balance was silver, and on the other side glass backed by silver, the two forces should act in the same direction on one vane, and oppositely on the other. If each silver coating reflected equally on its two faces, the average of the deflections on the two arms of the balance, should be free in part, from gas action.

This partial elimination of gas action would be more complete but for two reasons: (1) The reflection at the exposed glass face of one vane diminishes the energy incident upon the silver coating behind it and the resultant absorption in the same proportion. (2) The glass faces of the two vanes, when exposed, will not, in all probability, be equally heated. In the case of the vane which receives the light on its silvered face, the glass face can be warmed only by the conduction of heat from the silver coating through the glass, while in the vane receiving light on its glass face, some heat from absorption of the beam in passing through the glass will doubtless be added to this heat due to conduction. Although the vanes, when freshly silvered and polished, were so brilliant on the silvered sides that it was often difficult to tell which face was silver and which glass, the coatings were found in use to deteriorate rapidly, and the silver faces whitened earlier than the glass. After several hours of observing, vanes freshly silvered at the beginning had to be removed and resilvered. A higher rate of deterioration was noticeable when working in low gas pressures than in high.

By the reversal of suspension and averaging the results gained with the suspension direct and reversed, nearly all errors due to lack of symmetry in the balance, or in the position of the light images with reference to the rotation axis ; or errors due to lack of uniformity in the distribution of radiant intensity in different parts of the image, as well as errors due to certain differences between the mirrors themselves, could be eliminated.

Static observations were made at eight different gas pressures between 0.06 mm. of mercury and 96 mm. Within this range, as was afterwards shown by the series of semiballistic in connection with the static observations, the gas action varied in magnitude between one-tenth, and 6 or 7 times, the radiation pressure. The gas action was, however, at certain pressures so complicated that no rational way of treating the static measurements at the different gas pressures succeeded in establishing the existence of a constant force in the direction of propagation of the light beam. It was plain, therefore, that further elimination of the gas action must be sought in exposures so short that the gas action would not have time to reach more than a small fraction of its stationary value. This led to the method of semiballistic observations.

If the vibrations of any suspension be undamped, a constant deflecting force, acting for one-fourth of the period of oscillation, will cause a throw equal to $\sqrt{2}$ times the permanent deflection for the same force.   A correction for damping when considerable may also be introduced.   The period of the balance as described was 23.1 secs.   This was increased to 24 secs. by adding masses to the system.   Deflections due to six seconds' exposure were then observed in the same manner, with reversals and all, as has been already described under the static observations.   Because of the shorter time involved in making individual observations, a considerably greater number of ballistic measurements were made at each gas pressure.   In all the observations yet made, not a single ballistic deflection contrary to the direction of radiation pressure was obtained, and in the static observations, the first throw on exposing was invariably in the radiation pressure direction, irrespective of the direction of the final deflection.

The average ballistic deflection obtained, reduced to the equivalent permanent deflection for each of eight different gas pressures, is given in the accompanying table.   The column p gives the gas pressure in the bell-jar in millimeters of mercury, and column d the reduced permanent deflection in millimeters for the radiation pressure.

p	d	p	d
96.3	19.7	33.4	21.1
67.7	21.0	1.2	20.9
37.9	21.6	0.13	26.8
36.5	22.1	0.06	23.2
Ave.			22.5

For gas pressures between 45 and 30 mm. the gas action is small and increases toward both higher and lower pressures in the range studied.   Taking the product of the average deflection in cms. by the constant of the torsion balance, the radiation pressure was $2.25 \times 4.65 \times 10^{-5} = 1.05 \times 10^{-4}$ dynes.

THE BOLOMETER.

To test the above value of radiation pressure with the theory, it was necessary to measure the energy of the radiation causing the pressure. This was attempted with the aid of a bolometer of special construction.

On a sheet of platinum 0.001 mm. thick, rolled in silver (by the firm Sy & Wagner, Berlin), a circle P (Fig. 4), 11.3 mm. in diameter, was drawn. The sheet was cut from the edges inward to the circumference of the circle, in such a way as to leave five principal strips A, B, C, D, E, connected to the circle in the manner shown. Other narrower strips, as e, m, n, o, etc., were left to give the disk additional support. The disk, by means of the connecting arms, was mounted with asphalt varnish centrally over a hole 14 mm. in diameter, bored through a slab S cut from a child's school slate. Portions of the silver not to be removed by the acid were carefully covered by the asphalt varnish. Thus on the strips A and B, the silver was protected to the very edge of the circle, while on all the other arms, the silver was left exposed back to the edge of the boring in the slate. The whole system was then plunged into warm nitric acid, and the silver eaten away from all unprotected surfaces, leaving only the thin platinum sheet, which was blackened by elec tric deposition of platinum, by the method used in the manufacture of the Reichsanstalt bolometers. At A, B, C, D, E, holes were bored extending through the slate, and copper washers were soldered to the silver strips, and binding posts were attached.

The torsion balance was removed from under the bell-jar, the bolometer put in the place of one of the vanes, and the bell-jar replaced. Connections to the bolometer were made as schematically shown in Fig. 4. The disk P was the exact size of the light image thrown on the vanes in the pressure measurements, and the intention was to heat the disk by allowing the image to fall on it, and then with the light turned off, to heat it to the *same temperature* by sending a current through it from A to B. If r be the resistance from A to B in ohms, when exposed to the lamp, and i be the current in ampères which gives the same temperature in P as that given by the absorbed radiation, then $i^2 r \times 10^7$ will be the activity of the beam in erg-seconds. The temperature of the disk, whether exposed to the radiation or heated by the current, was shown by the resistance of the disk in the direction C to D and E which was made one arm of a Wheatstone Bridge. The relation of the heating current to the bridge was adjusted as follows: With the key K open, so that no current flowed through the bridge, the heating current from six storage cells B_2 was turned on, and the sliding contact at F so set that the bridge galvanometer zero was not changed by reversing the heating current. The equipotential point to c was found very near the middle of the wire ab, which showed the current distribution of P to be symmetrical with respect to a diameter at right angles to AB. The key K was then closed making the bridge current, and the bridge was balanced. The bolometer was next exposed to the radiation, and simultaneous observations were made on the two galvanometers G (Fig. 3) and G_1 (Fig. 4) and the deflection of the latter, was reduced to that corresponding to a deflection of 100 divisions of galvanometer G, as was done in the radiation pressure observations. During the pressure, as well as the bolometer measurements, the electric lamp was regulated by the observer at the galvanometer G, and the lamp held reasonably close to this standard value. The reductions were, therefore, of the simplest kind. After shutting off the light, the heating current was turned on and regulated by means of the variable resistance R_1 (Fig. 4), so that nearly the same throw was obtained from the galvanometer G, as when the bolometer was exposed to the lamp. All deflections of the galvanometer G_1 were taken with the bridge current both direct and reversed, to eliminate any local disturbances in the bridge. With the heating current on, galvanometer deflections were read with the heating current both direct and reversed. The heating current was read in ampères from a Siemens & Halske direct-reading precision milliampèremeter. From repeated observations the current which gave the same heating effect as the light beam, was $i = 0.865$ amp. The resistance between the binding posts A and B, was measured with the lamp on, and gave $r = 0.278$ ohm. The intensity of the beam in erg-seconds was thus $= 0.278 \times 0.75 \times 10^7$.

Considering the beam as a cylinder containing energy which is transmitted at the velocity v of radiation, the energy contained in a length of 3×10^{10} cms. would, according to the bolometer results be $i^2 r \times 10^7$, and the consequent pressure in dynes, against the totally absorbing end of the cylinder must, according to Maxwell, equal the energy per unit length, or

$$p = \frac{i^2 r \times 10^7}{v}.$$

If the incident beam falls upon a totally reflecting surface the energy per unit volume is doubled and the deduction from Maxwell is that the pressure is equal to $\dfrac{2i^2 r \times 10^7}{v}$.

According to Bartoli and Boltzmann, the pressure due to a beam totally reflected must on thermodynamic grounds also be $\dfrac{2i^2 r \times 10^7}{v}$.

If the beam is but partially reflected, however, the pressure would be

$$p = \frac{(1 + r_1) i^2 r \times 10^7}{v},$$

where r_1 equals the percentage of the incident energy reflected Considering the character of the light from the arc lamp, and the heavy infra-red absorption of the glass masses traversed, the reflecting power of the bright silver coatings was estimated to be nearly that for the wave-lengths of the D lines, or about 92 per cent.[1] Putting $1 + r = 1.92$ we have from the bolometer results

$$\frac{1.92 \times 0.278 \times 0.75 \times 10^7}{3 \times 10^{10}} = 1.34 \times 10^{-4} \text{ dynes}$$

as the pressure which the torsion balance should have shown, instead of the observed value 1.05×10^{-4}. Some of the mirrors used in the pressure observations were already well below this maximum reflecting power, so that the value 1.34×10^{-4} should doubtless be still further reduced, but no quantitative va ue can be assigned to such a correction.

Taking the results as they stand, the measured radiation pressure is to the radiation pressure which the theory applied to the bolometer measurements would require as $1.05 : 1.34$ or as $78 : 100$.

---

[1] E. F. Nichols, Phys. Rev., 4, p. 303 also F. Paschen, Ann. Phys., 4, p. 304, 1901.

Some sources of uncertainty in the pressure measurements have already been mentioned, but certain sources of error in the bolometer measurements in which the nature of the effect upon the result is known, should be more clearly indicated.

1. Errors in the bolometer which would tend to produce a greater divergence between the two results obtained are :

(*a*) The reflection of radiation by the platinum-black coating (this error is small) and (*b*) the heat conductivity of the black coating. In the case of the lamp, the transformation of radiant energy into heat takes place near the outer surface of the platinum-black covering, and the electrical resistance of P is affected only after the heat has passed by conduction through this coating to the sheet of platinum underneath. In the case of heating by the current, on the other hand, the heat is generated in the conducting sheet directly, and is dissipated at the outer surface after conduction through the black layer. This error is probably small also.[1]

2. Errors which would tend to bring the two results into closer agreement are two in number. . (*a*) The measured resistance between the binding posts A and B must be greater than that of the disk P by the resistance of the silver strips leading from the binding posts on both sides to P, and by any contact resistances involved. The magnitude of this error is unknown, but it may be large. (*b*) The distribution of current lines of flow for both heating current and bridge current would be such as to cause an error in the result of the same sign as the resistance error. The maximum heating would occur where the silver strips A and B join the disk, for here the stream lines will be most congested; on the other hand, the stream lines of the bridge current in these regions will be most sparse because nearly all the lines traversing the disk on the two sides will be drawn off into the edges of these same silver strips by the higher conductivity of the latter. This error may also be considerable. In a new bolometer now under construction the errors due to false resistance and unfavorable distribution of flow lines will be greatly reduced. Another series of radiation pressure measure-

[1] See Kurlbaum : Temperaturdifferenzen zwischen der Oberfläche und dem Innern eines strahlendes Korpers. Ann. Phys., 2, p. 554.

ments, made under more favorable conditions than those here reported, is also in progress.

The writers believe that the observations already in hand are sufficient to prove experimentally the existence of a pressure, not due to gas molecules, of the nature and order of magnitude of radiation pressure, but toward a close quantitative measurement of this pressure much remains still to be done.

THE WILDER LABORATORY, DARTMOUTH COLLEGE,
HANOVER, N. H., August, 1901.

ACCELERATION AND TRAPPING OF PARTICLES BY RADIATION PRESSURE

A. Ashkin

Bell Telephone Laboratories, Holmdel, New Jersey 07733

(Received 3 December 1969)

Micron-sized particles have been accelerated and trapped in stable optical potential wells using only the force of radiation pressure from a continuous laser. It is hypothesized that similar accelerations and trapping are possible with atoms and molecules using laser light tuned to specific optical transitions. The implications for isotope separation and other applications of physical interest are discussed.

This Letter reports the first observation of acceleration of freely suspended particles by the forces of radiation pressure from cw visible laser light. The experiments, performed on micron-sized particles in liquids and gas, have yielded new insights into the nature of radiation pressure and have led to the discovery of stable optical potential wells in which particles were trapped by radiation pressure alone. The ideas can be extended to atoms and molecules where one can predict that radiation pressure from tunable lasers will selectively accelerate, trap, or separate the atoms or molecules of gases because of their large effective cross sections at specific resonances. The author's interest in radiation pressure from lasers stems from a realization of the large magnitude of the force, and the observation that it could be utilized in a way which avoids disturbing thermal effects. For instance a power $P = 1$ W of cw argon laser light at $\lambda = 0.5145$ μm focused on a lossless dielectric sphere of radius $r = \lambda$ and density $= 1$ gm/cc gives a radiation pressure force $F_{rad} = 2qP/c = 6.6 \times 10^{-5}$ dyn, where q, the fraction of light effectively reflected back, is assumed to be of order 0.1. The acceleration $= 1.2 \times 10^{8}$ cm/sec$^2 \cong 10^{5}$ times the acceleration of gravity.

Historically,[1,2] the main problem in studying radiation pressure in the laboratory has been the obscuring effects of thermal forces. These are caused by temperature gradients in the medium surrounding an object and, in general, are termed radiometric forces.[3] When the gradients are caused by light, and the entire particle moves, the effect is called photophoresis.[3,4] These forces are usually orders of magnitude larger than radiation pressure. Even with lasers, photophoresis usually completely obscures radiation pressure.[5] In our work, radiometric effects were avoided by suspending relatively transparent particles in relatively transparent media. We operated free of thermal effects at 10^{3} times the power densities of Ref. 5.

The first experiment used transparent latex spheres[6] of 0.59-, 1.31-, and 2.68-μm diam freely suspended in water. A TEM$_{00}$-mode beam of an argon laser of radius $w_0 = 6.2$ μm and $\lambda = 0.5145$ μm was focused horizontally through a glass cell 120 μm thick and manipulated to focus on single particles. See Fig. 1(a). Results were observed with a microscope. If a beam with milliwatts of power hits a 2.68-μm sphere off center, the sphere is simultaneously drawn in to the beam axis and accelerated in the direction of the light. It moves with a limiting velocity of microns per second until it hits the front surface of the glass cell where it remains trapped in the beam. If the beam is blocked, the sphere wanders off by Brownian motion. Similar effects occur with the other sphere sizes but more power is required for comparable velocities. When mixed, one can accelerate 2.68-μm spheres and leave 0.585-μm spheres behind. The particle velocities and the trapping on the beam axis can

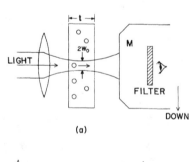

(a)

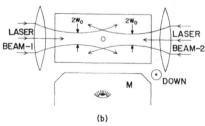

(b)

FIG. 1. (a) Geometry of glass cell, $t = 120$ μm, for observing micron particle motions in a focused laser beam with a microscope M. (b) The trapping of a high-index particle in a stable optical well. Note position of the TEM$_{00}$-mode beam waists.

be understood as follows (see Fig. 2): The sphere of high index $n_H = 1.58$ is situated off the beam axis, in water of lower index $n_L = 1.33$. Consider a typical pair of rays symmetrically situated about the sphere axis B. The stronger ray (a) undergoes Fresnel reflection and refraction (called a deflection here) at the input and output faces. These result in radiation pressure forces $F_R{}^i$, $F_R{}^o$ (the input and output reflection forces), and $F_D{}^i$, $F_D{}^o$ (the input and output deflection forces), directed as shown. Although the magnitudes of the forces vary considerably with angle Φ, qualitatively the results are alike for all Φ. The radial (r) components of $F_D{}^i$, $F_D{}^o$ are much larger than $F_R{}^i$ and $F_R{}^o$ (by ~10 at $\Phi = 25°$). All forces give accelerations in the $+z$ direction. $F_R{}^i$ and $F_R{}^o$ cancel radially to first order. $F_D{}^i$ and $F_D{}^o$ add radially in the $-r$ direction, thus the net radial force for the stronger ray is <u>inward</u> toward higher light intensity. Similarly the symmetrical weak ray (b) gives a net force along $+z$ and a net <u>outward but weaker</u> radial force. Thus the sphere as a whole is accelerated <u>inward</u> and <u>forward</u> as observed. To compute the z component of the force for a sphere on axis, one integrates the perpendicular (s) and parallel (p) components of the plane-polarized beam over the sphere. This yields an effective $q = 0.062$. This geometric optic result (neglecting diffraction) is identical with the asymptotic limit of a wave analysis by Debye[2] for an incident plane wave. He finds $q = 0.06$. From the force we get the limiting velocity v in a viscous medium using Stokes's law. For $r \ll w_0$,

$$v = 2qPr/3c\pi w_0{}^2\eta, \qquad (1)$$

where η is the viscosity. For $P = 19$ mW, $w_0 = 6.2$ μm, and a sphere of $r = 1.34$ μm in water ($\eta = 1 \times 10^{-2}$ P), one computes $v = 29$ μm/sec. We measured $v = 26 \pm 5$ μm/sec which is good agreement. In the above, the sphere acts as a focusing lens. If one reverses the relative magnitudes of the indices of the media, the sphere becomes a diverging lens, the sign of the radial deflection forces reverse, and the sphere should be pushed <u>out</u> of the beam. This prediction was checked experimentally in an extreme case of a low-index sphere in a high-index medium, namely an air bubble. Bubbles, of ~8-μm diam, were generated by shaking a high-viscosity medium consisting of an 80% by weight mixture of glycerol in water. It was found that the bubbles were <u>always pushed out</u> of the light beam as they were accelerated along, as expected. In the same

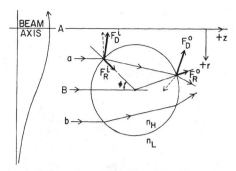

FIG. 2. A dielectric sphere situated off the axis A of a TEM_{00}-mode beam and a pair of symmetric rays a and b. The forces due to a are shown for $n_H > n_L$. The sphere moves toward $+z$ and $-r$.

medium of $n = 1.44$, the 2.68-μm spheres of $n = 1.58$ were still focusing. At higher powers the bubbles are expected to deform. This would result in a deformation contribution to the radiation pressure force as postulated by Askaryon.[7] Our observation of the attraction of high-index spheres into regions of high light intensity is related to the deformation of a liquid surface postulated by Kats and Kantorovich.[8]

The experimentally observed radial inward force on the high-index spheres suggest a means of constructing a true <u>optical potential well</u> or "optical bottle" based on radiation pressure alone. If one has two opposing equal TEM_{00} Gaussian beams with beam waists located as shown in Fig. 1(b), then a sphere of high index will be in stable equilibrium at the symmetry point as shown (i.e., any displacement gives a restoring force). <u>Such trapping was observed experimentally</u> in an open cell filled with 2.68-μm spheres in water as sketched in Fig. 1(b). Here the entire beam is viewed at once. Particles are observed by their brilliant scattered light. With 128 mW in only one of the beams, a maximum particle velocity of ~220 μm/sec was observed as particles traversed the entire near field. The calculated velocity is 195 μm/sec. For trapping, the two opposing beams were introduced. Particles that drift near either beam are drawn in, accelerated to the stable equilibrium point, and stop. To check for stability one can interrupt one beam for a moment. This causes the particle to accelerate rapidly in the remaining beam. When the opposing beam is turned on again the particle returns to its equilibrium point, only more slowly since it is now acted on by the differential force. Interrupting the other beam reverses the behavior. In other experiments, ~5-μm-diam water droplets from an atomizer were accelerated in air with a single beam. At 50 mW, velocities ~0.25 cm/sec were observed. Such motions could be seen with the naked eye. The behavior of the droplets was in qualitative agreement with expectation.

In our experiments it is clear that we have discriminated against radiometric forces. These forces push more strongly on hot surfaces and would push high-index spheres and bubbles out of the beam; whereas our high-index spheres were drawn into the beam. Even the observed direction of acceleration along the beam axis is the opposite of the radiometric prediction. A moderately absorbing focusing sphere concentrates more heat on the downstream side of both the ball and the medium and should move upstream into the light (negative photophoresis).[9] For water drops in air we can invoke the well-confirmed formula of Hettner[10] and compute the temperature gradient needed across a 5-μm droplet to account radiometrically for the observed velocity of 0.25 cm/sec. From Stokes's formula, $F = 2.1 \times 10^{-7}$ dyn. Hettner's formula then requires a gradient of 0.5°C across the droplet. No such gradients are possible with the 50 mW used. For water and glycerol the gradients are also very low.

The extension to vacuum of the present experiments on particle trapping in potential wells would be of interest since then any motions are frictionless. Uniform angular acceleration of trapped particles based on optical absorption of circular polarized light or use of birefringent particles is possible. Only destruction by mechanical failure should limit the rotational speed. In vacuum, particles will heat until they are cooled by thermal radiation or vaporize. With the minimum power needed for levitation, micron spheres will assume temperatures of hundreds to thousands of degrees depending on the loss. The ability to heat in vacuum without contaminating containing vessels is of interest. Acceleration of neutral spheres to velocities ~10^6-10^7 cm/sec is readily possible using powers that avoid vaporization. In this regard one could attempt to observe and use the resonances in radiation pressure predicted by Debye[2] for spheres with specific radii. The separation of micron- or submicron-sized particles by radiation pressure based on radius as demonstrated experimentally could also be useful [see Eq. (1)].

Finally, the extension of the ideas of radiation pressure from laser beams to atoms and molecules opens new possibilities. In general, atoms

and molecules are quite transparent. However, if one uses light tuned to a particular transition, the interaction cross section can be much larger than geometric. For example, an atom of sodium has $\pi r^2 = 1.1 \times 10^{-15}$ cm² whereas, from the absorption coefficient,[11] the cross section σ_T at temperature T for the D_2 resonance line at $\lambda = 0.5890$ μm is $\sigma_T = 1.6 \times 10^{-9}$ cm² $= 0.5\lambda^2$ for $T < 40$°K (region of negligible Doppler broadening). The absorption and isotropic reradiation by spontaneous emission of resonance radiation striking an atom results in an average driving force or pressure in the direction of the incident light. We shall attempt to show that radiation pressure from a laser beam on resonance can work as an actual optical gas pump and operate against significant gas pressures. Figure 3(a) shows a schematic version of such a pump. Imagine two chambers initially filled with sodium vapor, for example. A transparent pump tube of radius w_0 is uniformly filled with laser light tuned to the D_2 line of Na from the left. Let the total optical power P and the pressure p_0 be low enough to neglect light depletion and absorption saturation. Most atoms are in the ground state. The average force on an atom is $P\sigma_T/c\pi w_0^2$ and is constant along the pump. Call x_{cr} the critical distance. It is the distance traveled by an atom in losing its average kinetic energy $\frac{1}{2}mv_{av}^2$. That is, $Fx_{cr} = \frac{1}{2}mv_{av}^2 \cong kT$. The variation of pressure in a gas

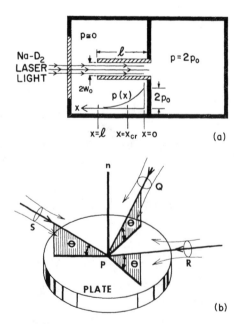

FIG. 3. (a) Schematic optical gas pump and graph of Na pressure $p(x)$. (b) Geometry of gas confinement about point P of a plane surface.

with a constant force is exponential at equilibrium. Thus

$$p(x) = p_0 e^{-Fx/kT} = p_0 e^{-x/x_{cr}}, \qquad (2)$$

$$x_{cr} = \pi w_0^2 ckT/P\sigma_T. \qquad (3)$$

Next, consider higher power. Saturation sets in. Population equalization occurs between upper and lower levels for those atoms of the Doppler-broadened line of width $\Delta\nu_D$, within the natural width $\Delta\nu_n$ of line center. A "hole" is burned in the absorption line and the power penetrates more deeply into the gas. But there is a net absorption, even when saturated, due to the ever-present spontaneous emission from the upper energy level. The average force per atom also saturates and is constant along the tube. Its value is $(h/\tau_n\lambda)(\Delta\nu_n/\Delta\nu_D)$, where τ_n is the upper level natural lifetime.[12] Lastly, we consider the effect of collision broadening due to a buffer gas on the force per atom. With collision one replaces $1/\tau_n$ by $(1/\tau_n + 1/\tau_L)$ and $\Delta\nu_n$ by $(\Delta\nu_n + \Delta\nu_L)$ in the average saturated force, where $\Delta\nu_L = \frac{1}{2}\pi\tau_L$ is the Lorentz width. This enhances the force greatly. Then

$$x_{cr} = \frac{kT\lambda}{h}\left(\frac{\tau_n\tau_L}{\tau_n+\tau_L}\right)\left(\frac{\Delta\nu_D}{\Delta\nu_n+\Delta\nu_L}\right). \qquad (4)$$

As an example, consider Na vapor at $p_0 = 10^{-3}$ Torr ($n_0 = 3.4\times10^{13}$ atoms/cc and $T = 510°K$), buffered by helium at 30 Torr. Take a tube of $l = 20$ cm with diameter $2w_0 = 10^{-2}$ cm. For $\tau_n = 1.48 \times 10^{-8}$ sec, $\Delta\nu_D = 155\Delta\nu_n$ (at $T = 510°K$), and $\Delta\nu_L \cong 30\Delta\nu_n$,[13] one finds $x_{cr} = 1.5$ cm and $l = 20$ cm $= 13.3 x_{cr}$. Thus $p(l) = 2p_0 e^{-13.3} = 2\times10^{-3}\times1.7 \times10^{-6} = 3.4\times10^{-9}$ Torr. Essentially complete separation has occurred. This requires a total number of photons per second of $2\pi\omega_0^2 x_{cr}n_0/(1/\tau_n + 1/\tau_L) \cong 1.7\times10^{19} \cong 6$ W. Under saturated conditions there is little radiation trapping of the scattered light. Almost all the incident energy leaves the gas without generating heat. The technique applies for any combination of gases. Even different isotopes of the same atom or molecule could be separated by virtue of the isotope shift of the resonance lines. The possibilities for

forming atomic or molecular beams with specific energy states and for studying chemical reaction kinetics are clear. The possibility of obtaining significant population inversions by resonant gas pumping remains to be evaluated. One can also show that gas can be optically trapped at the surface of a transparent plate. For example [see Fig. 3(b)], three equal TEM$_{00}$-mode beams with waists at points Q, R, and S directed equilaterally at point P, at some angle θ, result in a restoring force for displacements of an atom about P. Gas trapped about P could serve as a windowless gas target in many experimental situations. The perfection of accurately controlled frequency-tunable lasers is crucial for this work.

It is a pleasure to acknowledge stimulating conversations with many colleagues; in particular, J. G. Bergman, E. P. Ippen, J. E. Bjorkholm, J. P. Gordon, R. Kompfner, and P. A. Wolff. I thank J. M. Dziedzic for making his equipment and skill available.

---

[1] E. F. Nichols and G. F. Hull, Phys. Rev. 17, 26, 91 (1903).

[2] P. Debye, Ann. Physik 30, 57 (1909).

[3] N. A. Fuchs, *The Mechanics of Aerosols* (The Macmillan Company, New York, 1964).

[4] F. Ehrenhaft and E. Reeger, Compt. Rend 232, 1922 (1951).

[5] A. D. May, E. G. Rawson, and E. H. Hara, J Appl. Phys. 38, 5290 (1967); E. G. Rawson and E. H. May, Appl. Phys. Letters 8, 93 (1966).

[6] Available from the Dow Chemical Company.

[7] G. A. Askar'yan, Zh. Eksperim. i Teor. Fiz.—Pis'ma Redakt. 9, 404 (1969) [translation: JETP Letters 9, 241 (1969)].

[8] A. V. Kats and V. M. Kantorovich, Zh. Eksperim. i Teor. Fiz.—Pis'ma Redakt. 9, 192 (1969) [translation: JETP Letters 9, 112 (1969)].

[9] See Ref. 3, p. 60.

[10] G. Z. Hettner, Physics 37, 179 (1926); Ref. 3, p. 57.

[11] A. C. G. Mitchell and M. W. Zemansky, *Resonance Radiation and Excited Atoms* (Cambridge University Press, New York, 1969), p. 100.

[12] J. P. Gordon notes that power broadening occurs at still higher powers. This increases the hole width and the average force $\sim\sqrt{P}$.

[13] See Ref. 11, p. 166.

Optical Levitation by Radiation Pressure

A. Ashkin and J. M. Dziedzic

Bell Telephone Laboratories, Holmdel, New Jersey 07733
(Received 14 June 1971; in final form 13 August 1971)

The stable levitation of small transparent glass spheres by the forces of radiation pressure has been demonstrated experimentally in air and vacuum down to pressures ~1 Torr. A single vertically directed focused TEM_{00}-mode cw laser beam of ~ 250 mW is sufficient to support stably a ~20-μ glass sphere. The restoring forces acting on a particle trapped in an optical potential well were probed optically by a second laser beam. At low pressures, effects arising from residual radiometric forces were seen. Possible applications are mentioned.

This letter reports the observation of stable optical levitation of transparent glass spheres in air and vacuum by the forces of radiation pressure from laser light. The technique used involves properties of radiation pressure previously deduced from experiments on small transparent micron-sized spheres in liquid.[1] It was shown that a light beam striking a sphere of a high index of refraction in a position of transverse gradient of light intensity not only exerts a force directed along the light beam, but also has a transverse component of force which pushes the particle toward the region of maximum light intensity. This fact makes stable optical potential wells possible.[1] In contrast to magnetic[2] and electrostatic[3] feedback levitation, or electrodynamic levitation,[4] optical levitation based on the potential well provided by radiation pressure is truly stable in a dc sense with the particle at rest at the equilibrium point. In this regard, magnetically levitated superconductors[5] are similarly stable at rest.

In our experiment a single vertically directed focused cw laser beam was used to lift a glass sphere off a glass plate and stably levitate it. Figure 1 shows the basic apparatus. About 100–500 mW of 5145-Å laser light in the TEM_{00} mode is focused by a 5-cm lens and directed vertically on a selected sphere of ~15–25 μ in diameter, initially at rest in position A on the glass plate. The power is such that a force of several g is applied at A with the particle at the beam waist ($2w_0 \cong 25$ μ). This force is directly calculable from the index of refraction of the sphere ($n = 1.65$) as indicated in Ref. 1. This, by itself, is insufficient to break the strong van der Waals attraction to the supporting plate, which for a 20-μ sphere is ~$10^4 g$. This bond can, however, be broken acoustically by setting up a vibration with a piezoelectric ceramic cylinder cemented to the glass plate. Tuning the driving frequency rapidly through a mechanical resonance momentarily shakes the particle loose and it begins to rise up into the diverging Gaussian beam. It comes to equilibrium at position B about 1 mm above the beam waist where radiation pressure and gravity balance. A glass enclosure over the plate serves to minimize air currents. By moving the lens, the beam and hence the particle can be moved anywhere within the enclosure. It can even be deposited on the roof for careful subsequent examination. Figure 2 shows a 20-μ particle levitated in air and photographed by

its scattered light. The particle is extremely stable and can remain aloft for hours.

Operation at reduced pressure was accomplished with a simple vacuum cell and an adjustable leak valve. Levitation down to pressures as low as 1 Torr was observed before the particles were lost. It is felt that residual radiometric forces coupled with reduced viscous air damping account for this loss. As the pressure was reduced, the equilibrium position B gradually dropped down toward the beam focus indicating the onset of an additional downward force, thought to be radiometric in origin. Due to residual optical absorption and the lenslike character of the sphere, the top of the sphere is slightly hotter than the bottom. This gives a downward radiometric force (negative photophoresis) which initially increases as the pressure is reduced. The thermal force F_{th} is proportional to $1/p$ down to ~10 Torr, where the mean free path is comparable with the sphere diameter.[6] Also, the viscosity and thermal conductivity of the gas are roughly constant for pressures down to ~10 Torr. Below this pressure, radiometric forces, viscosity, and thermal conductivity begin to decrease. Experimentally, at low pressure the particles begin to become less stable

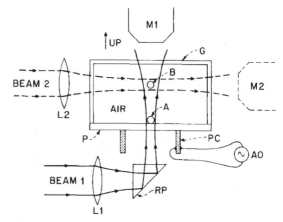

FIG. 1. Levitation apparatus. Particle at A is shaken loose acoustically and lifted to B by TEM_{00}-mode beam 1. TEM_{00}-mode beam 2 is introduced later as a probe beam to study the strength of the trapping forces. L1 and L2 are lenses, P is a glass plate, G is a glass enclosure about 1.5 cm high, RP is a reflecting prism, PC is a piezoelectric ceramic cylinder driven by audio-oscillator AO, and M1 and M2 are microscopes.

FIG. 2. Photograph of a ~20-μ transparent glass particle being levitated about 1 cm above a glass plate by a 250-mW vertically directed laser beam (shown as beam 1 in Fig. 1). The bright spot is the particle (vastly overexposed) photographed by its own scattered light. The ~90° Mie scattering from the particle is seen on a screen placed at the rear and side of the glass enclosure.

horizontally and vertically and even begin to spin at an increasing rate about a vertical axis prior to breaking loose from the optical trapping forces. Thus we see that even radiometric forces which are one or two orders of magnitude less than radiation pressure at atmospheric pressure can cause trouble at reduced pressure. A calculation of optical absorption in the particle, based on the estimated radiometric forces, indicates an absorption loss $\alpha \cong 5 \times 10^{-2}$ cm^{-1} which is quite high. The glass spheres were made commercially by the Flexolite Corp. of St. Louis, Mo. for reflectors. Since a loss of $\alpha \cong 10^{-4}-10^{-5}$ cm^{-1} is possible in glass, this thermal limitation is not fundamental.

It can be estimated that the horizontal restoring forces are much stronger than the vertical restoring forces about the equilibrium point of the optical trap. This is directly observable in an auxiliary experiment in which a particle levitated in a 250-mW vertical beam 1 is illuminated by a horizontal beam 2 of adjustable power (see Fig. 1). Microscopes 1 and 2 are used to project enlarged views of the beams and particles on viewing screens. In photographs 1(a) and 1(b) of Fig. 3, taken with microscope 1, we observe the horizontal displacement of the particle in beam 1 as the power in beam 2 is increased to ~125 mW. This power almost pushes the particle out of beam 1. Thus we find a maximum transverse trapping acceleration of ~$\frac{1}{2}g$ for beam 1. This is very large and accounts for the high degree of horizontal stability observed. The much weaker vertical stability manifests itself in microscope 2 in the sensitivity of the vertical equilibrium level to minor power fluctuations. It is interesting to note that the

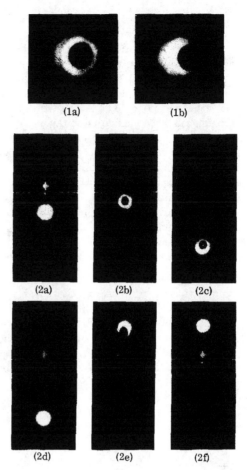

FIG. 3. Probing the horizontal and vertical trapping forces with auxiliary transverse beam 2. Top views 1(a) and 1(b) from microscope 1 show the horizontal displacement of the particle in beam 1. Side views 2(a) and 2(f) from microscope 2 show the vertical position of the particle in beam 1 as it is hit by transverse beam 2.

probe beam 2 itself can be used to stabilize the particle vertically. View 2(a) shows the particle, seen by its own scattered light as two bright sources located near the top and bottom of the particle,[7] sitting at its equilibrium level marked by the white line. Just below it is beam 2, adjusted to about 40 mW. In view 2(b), beam 2 is raised so its fringe field hits the particle, whereupon it draws the particle down into it, close to its axis. In this position the transverse stability of only 40 mW in beam 2 is sufficient to hold the particle essentially fixed vertically as the power fluctuates. In fact, it is now possible to lower beam 2 and drag the particle down many particle diameters into the region of slowly increasing vertical force [view 2(c)] before it breaks free and returns to its equilibrium position [view 2(d)]. The particle can also be lifted up, as shown in view 2(e), before it breaks free and drops back to the equilibrium level [view 2(f)]. In situations 2(d) and 2(f), where the particle returns to equilibrium, we can conveniently observe the effects of viscous damping inasmuch as the particle returns without undergoing any vertical oscillations.

Many other trapping configurations having even greater stability based on several beams are possible. For instance, a third beam, opposing beam 2, and of the same power would keep the equilibrium point on the axis of beam 1 while adding to the vertical stability. Other laser modes such as TEM_{01} and TEM_{01}^* (the doughnut mode) have been successfully used, although the TEM_{00} mode is optimum. Other particle shapes, if not too extreme, are useable, possibly even partly hollow particles. With a different launching technique, it certainly should be possible to levitate very much smaller particles, even perhaps in the submicron range. It should be noted, however, that the principles of optical levitation should not be expected to provide any significant trapping of atoms, even though the pressure of resonance radiation on atoms is quite a significant effect.[8]

The technique of optical levitation will probably have use in applications where the precision micromanipulation of small particles, free from any supports, is important such as in light scattering from single small particles (Mie scattering) or in laser-initiated fusion experiments. If the viscous damping can be further reduced, applications to inertial devices such as gyroscopes and accelerometers become possible. Measurement of low optical absorptions, absolute optical power measurement, and pressure measurement are also likely areas of application. Levitation may also provide an interesting adjunct to Millikan-type experiments on charged particles. The extreme simplicity of the technique and its remarkable stability recommend its use.

[1]A. Ashkin, Phys. Rev. Letters 24, 156 (1970).

[2]J.W. Beams, Science 120, 619 (1954).

[3]C.B. Strang of the Martin-Marietta Corp. (private communication); also Martin-Marietta, Report No. OR9638, 1968 (unpublished).

[4]R.F. Wuerker, H. Shelton, and R.V. Langmuir, J. Appl. Phys. 30, 342 (1959).

[5]I. Simon, J. Appl. Phys. 24, 19 (1953).

[6]N.A. Fuchs, *The Mechanics of Aerosols* (Macmillan, New York, 1964).

[7]The interference observed in the 90° Mie scattering of Fig. 2 can be ascribed to the interference from these two bright sources. One can determine the particle diameter d by measuring its distance from the observing screen and the fringe spacing, since the separation between the two near-field sources is $\cong (1 + \sqrt{2})\,d/2$.

[8]A. Ashkin, Phys. Rev. Letters 25, 1321 (1971).

Reports

Optical Levitation of Liquid Drops by Radiation Pressure

Abstract. Charged and neutral liquid drops in the diameter range from 1 to 40 microns can be stably levitated and manipulated with laser beams. The levitation technique has been extended toward smaller particles (about 1 micron), lower laser power (less than 1 milliwatt), and deeper traps (greater than ten times the particle's weight). The techniques developed here have particular importance in cloud physics, aerosol science, fluid dynamics, and optics. The interactions of the drops with light, the electric field, the surrounding gas, and one another can be observed with high precision.

We report here on a study of the optical levitation of charged and neutral liquid drops with laser beams. The techniques developed here have particular importance in cloud physics, aerosol science, fluid dynamics, and optics. Optical levitation is based on the ability of light to stably trap nonabsorbing particles by the force of radiation pressure (*1*). In this technique a continuous-wave (cw) vertically directed focused TEM$_{00}$ Gaussian-mode laser beam not only supports the particle's weight but pulls the particle transversely into the region of high light intensity on the beam axis. Once the particle is trapped, one can manipulate it by moving the beam about. This levitation technique has been demonstrated for glass spheres and other nonabsorbing particles (*1, 2*). Applications of levitation and radiation pressure have been discussed (*3*).

The proposed application of levitation to cloud physics and aerosol science is a natural extension of the technique, since progress in these fields has hinged crucially on studies of the interactions of single particles (*4*). A comparison of optical levitation with other techniques that have been used in these sciences for manipulating particles, such as electrodynamic suspension, acoustic suspension, the use of wind tunnels, and suspension on fine wires, indicates that only optical levitation has the combined ability to handle the drops (~ 1 to 40 μ) found in natural clouds with such simplicity, high positional stability, ease of manipulation, and ease of observation. The advantage of this technique results from the depth of the optical trap, its highly localized nature approximating the particle size, and the ease with which the beam can

be moved in space. We will discuss specifically how we generate charged and neutral drops in the 1- to 40-μ diameter range, trap them singly and in small clouds in the light beams, move them about, measure their charge, and observe their interactions with the electric field, light, one another, and the surrounding gas.

Figure 1 is a sketch of the basic apparatus. A vertically directed cw laser beam is focused by a lens L and introduced into a glass box B from below. The beam is aligned with a hole H, about 0.5 mm in diameter, located in a sliding roof cover C. A liquid droplet cloud is sprayed into a large storage vessel V with an atomizer A where it can settle slowly by gravity. Some drops fall through the hole and enter the light beam where, if their sizes are in the correct range, they can be trapped and levitated. With microscope M_1 one can view the levitated drops from the side. The drops are not only seen by scattered light from the levitating beam but are visible in silhouette against the projection lamp P. The enlarged view from M_1 is projected on a screen for ease of observation and height measurement. Two plane electrodes E_1 and E_2 with a narrow slit, which allows the beam to pass through, can be used to apply an essentially uniform electric field to the particles. Once drops have been collected by the beam, one can stop the rain of particles by pushing the sliding roof cover C aside with the sliding glass plate G. Thus, without opening the box to significant disturbing air currents, the source can be removed and replaced by a transparent plate for viewing with microscope M_2 from above.

The levitated drops in general arrange themselves in order of size, with the largest closest to the beam focus. Figure 2 shows various situations. With multiple drops the upper drops are arranged in a light intensity distribution modified by the lower drops. Thus in Fig. 2b′ the upper drop is located outside the shadow cast by the lower drop and sits off-axis in the high-intensity light ring which diffracts past the lower drop. With many drops (Fig. 2, c and d) the particles become more closely coupled as a result of their effects on the light distribution and also as a result of electrostatic attractions and repulsions. Up to 20 or so drops can collect in a fixed array which undergoes rearrangement when significantly dis-

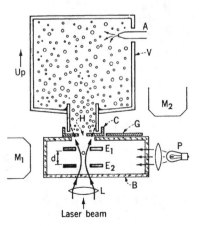

Fig. 1. Sketch of the experimental apparatus. Dimension d is ~0.6 cm.

turbed. Drops were made from various liquids. Drops of pure water in the 10- to 30-μ range evaporate so rapidly in room air that they are held only for about 30 seconds. Instead of adding water vapor to the surrounding air to increase drop life, we made drops from liquids with low vapor pressure such as silicone oil or water-glycerol mixtures with composition ratios varying from ten parts water and one part glycerol to pure glycerol. Silicone oil drops persist in this apparatus almost indefinitely. However, a pure glycerol drop with a diameter of ∼ 12 μ was observed to evaporate to ∼ 1 μ in 3 hours. We gradually reduced the power of the argon ion laser operating at 5145 Å from ∼ 40 to ∼ 0.2 mw to keep the drop from rising too high in the levitating beam as its mass decreased. The impressively low power of 0.2 mw indicates how little power is required for the levitation of particles approaching 1 μ in diameter. The 35-μ particle of Fig. 2a is held by ∼ 400 mw.

The electrical charge carried by drops can be determined from a comparison of the applied electric forces with the light forces. Thus, if we displace a drop upward with an electric field, then the percentage reduction in light needed to restore the drop to its original position represents the fraction of the particle's weight supported by the field. From the weight and the applied field, one can determine the total charge. Water-glycerol mixtures are polar liquids and give drops with various charges (plus, minus, or neutral) as expected. The charges are often high enough so that a force several hundreds of times the weight of the particle, mg (where m is the mass and g is the gravitational acceleration), can be applied with reasonable fields. Nonpolar silicone oil gives drops with much less charge.

The magnitude of the restoring forces acting on levitated particles corresponds to ∼ mg (2). For larger particles (∼ 40 μ) the magnitude of the restoring forces gives rise to a very stable trap. Thus it would take an air current of ∼ 5 cm/sec to eject a 40-μ particle from the beam. Because of increased viscous drag, small particles (∼ 1 μ) are stable up to an air current of only ∼ 3×10^{-3} cm/sec. This makes manipulation of 1-μ drops difficult. However, it is possible to deepen these traps by orders of magnitude. By increasing the strength of the upward supporting light and balancing it with a downward electric or optical force, one can maintain levitation and increase the trap depth in proportion to the increase in light power. Fortunately this method is most useful with small drops where the levitating power is initially quite low. In practice, we have deepened traps by more than an order of magnitude by using a downward electric force. This scheme may permit the levitation of even smaller particles, possibly in the submicron range. It also has the important advantage of operating in the zero-gravity environment of space experiments. The opposite procedure of applying an upward electric force and decreasing the supporting light beam results ultimately in a transition to the Millikan type of support where uniform electric forces alone support a particle in neutral equilibrium. Although easily done for small particles (∼ 1 to 4 μ), this procedure has less practical interest since we lose the advantages of stability. For larger particles where light power is a consideration and damping is not severe, partial electric support has merit. With the field uniformity of our plates we have easily supported 90 percent of a particle's weight electrically.

Potential applications of the levitation technique for single trapped drops include measurements of the rates of evaporation, condensation, charging, and neutralization. The effects of changes in the ion content and flow rate of the surrounding air and also changes in the drop itself, such as the effects of the addition of dissolved impurities (salts) and various solid nuclei, can be studied. In addition to making possible the direct observation of the drops in a high-power microscope while levitated, the levitation technique can also be used to deposit drops elsewhere for measurements.

Concerning solid nuclei, the smallest solids that we have levitated thus far are ∼ 4-μ dried latex spheres (Dow Corning). Because latex is slightly absorbing in 5145-Å light, we used ∼ 1 mw of 6328-Å light to levitate the spheres. We also expect that one can grow a small crystal within a levitated drop by partial evaporation of a drop containing dissolved salts. Possibly we can evaporate the solvent completely and levitate the crystal itself, if its shape is not too extreme. The most irregularly shaped objects that we have levitated thus far are random clumps of two to five glass spheres. These invariably orient themselves in the beam. These objects, incidentally, could be

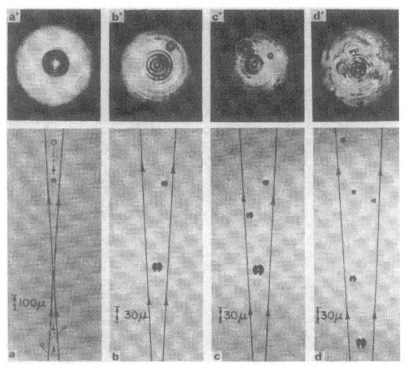

Fig. 2. Photos of optically levitated drops. (a–d) Side views taken with microscope M_1. Beam shapes are shown for reference; (a) also shows the trajectories of drops colliding with the levitated drop. (a'–d') Corresponding top views taken with microscope M_2; (a'–c') focus on the highest drop, and (d') focuses lower in the beam, showing the diffraction rings from the four lowest drops.

used for studying Mie scattering from oriented irregular particles. In addition, our observations of levitation in complex light intensity distributions, such as in Fig. 2d, as well as interesting cases of levitation of small drops in the irregularities of the light beam occurring just above splattered fallen drops lying on the cell floor, indicate that levitation does not require an extreme uniformity of geometry. Finally, concerning ice crystals, we believe that fairly regular crystals such as prisms *should* levitate and that more complex shapes *may* levitate. Possibly an electric field would orient the more plate-like crystals and thus aid in levitation.

Levitation permits various drop interactions to be observed. Thus, if two levitated drops (Fig. 2b) have opposite charges, we can force them closer and closer together by applying an external field until they finally coalesce. The fused drop remains levitated at a new height with the combined mass and charge. We have also directly observed drop-drop collisions. Often a levitated drop is struck from above by a heavier drop that is drawn into the beam as it falls. Alternatively, a levitated drop can be hit from below by lighter drops that wander into the lower regions of the beam where they are drawn in, are driven upward through the beam focus, and then encounter the levitated drop (see Fig. 2a). In these encounters we have often observed drop coalescence when the partners have opposite charges and misses when the drops have like charges. Drops can grow by a factor of 4 or 5 in diameter by successive collisions.

The simple experiments described here indicate the potential of the technique for studying the important problems of droplet growth by accretion in Langmuir type collisions with the use of cloud size droplets. In more sophisticated experiments prepared target and incident drops of known size and charge, each initially held in its own optical trap, could be used. They could subsequently be placed in the same beam and be brought together at varying speeds, with an electric field or light being used as the driving mechanism. The parameters of the collision, the particle trajectories, and the detailed fluid dynamics of the drop coalescence could be observed with high-speed movie cameras viewing from different angles. Finally we have made observations on the question of the coalescence efficiency of drops making physical contact. Occasionally we observed two drops roughly equal in size come side by side in the beam and seemingly touch for seconds prior to coalescence. This result should be checked with the use of movie cameras, as suggested above.

A. Ashkin
J. M. Dziedzic
Bell Laboratories,
Holmdel, New Jersey 07733

References and Notes

1. A. Ashkin, *Phys. Rev. Lett.* **24**, 156 (1970).
2. ―――― and J. M. Dziedzic, *Appl. Phys. Lett.* **19**, 283 (1971); *ibid.* **24**, 586 (1974).
3. A. Ashkin, *Sci. Am.* **226**, 63 (February 1972).
4. B. J. Mason, *The Physics of Clouds* (Oxford Univ. Press, London, 1971); C. N. Davies, Ed., *Aerosol Science* (Academic Press, London, 1966); Proceedings of the International Colloquium on Drops and Bubbles, California Institute of Technology, Pasadena, 28–30 August 1974.

14 November 1974

Observation of a single-beam gradient force optical trap for dielectric particles

A. Ashkin, J. M. Dziedzic, J. E. Bjorkholm, and Steven Chu

AT&T Bell Laboratories, Holmdel, New Jersey 07733

Received December 23, 1985; accepted March 4, 1986

Optical trapping of dielectric particles by a single-beam gradient force trap was demonstrated for the first reported time. This confirms the concept of negative light pressure due to the gradient force. Trapping was observed over the entire range of particle size from 10 μm to ~25 nm in water. Use of the new trap extends the size range of macroscopic particles accessible to optical trapping and manipulation well into the Rayleigh size regime. Application of this trapping principle to atom trapping is considered.

We report the first experimental observation to our knowledge of a single-beam gradient force radiation-pressure particle trap.[1] With such traps dielectric particles in the size range from 10 μm down to ~25 nm were stably trapped in water solution. These results confirm the principles of the single-beam gradient force trap and in essence demonstrate the existence of negative radiation pressure, or a backward force component, that is due to an axial intensity gradient. They also open a new size regime to optical trapping encompassing macromolecules, colloids, small aerosols, and possibly biological particles. The results are of relevance to proposals for the trapping and cooling of atoms by resonance radiation pressure.

A wide variety of optical traps based on the basic scattering and gradient forces of radiation pressure have been demonstrated or proposed for the trapping of neutral dielectric particles and atoms.[2–4] The scattering force is proportional to the optical intensity and points in the direction of the incident light. The gradient force is proportional to the gradient of intensity and points in the direction of the intensity gradient. The single-beam gradient force trap is conceptually and practically one of the simplest radiation-pressure traps. Although it was originally proposed as an atom trap,[1] we show that its uses also cover the full spectrum of Mie and Rayleigh particles.

It is distinguished by the feature that it is the only all-optical single-beam trap. It uses only a single strongly focused beam in which the axial gradient force is so large that it dominates the axial stability. In the only previous single-beam trap, the so-called optical levitation trap, the axial stability relies on the balance of the scattering force and gravity.[5] In that trap the axial gradient force is small, and if one turns off or reverses the direction of gravity the particle is driven out of the trap by the axial scattering force.

There were also relevant experiments using gradient forces on Rayleigh particles that did not strictly involve traps, in which liquid suspensions of submicrometer particles acted as an artificial nonlinear optical Kerr medium.[6]

The physical origin of the backward gradient force in single-beam gradient force traps is most obvious for particles in the Mie size regime, where the diameter is large compared with λ. Here one can use simply ray optics to describe the scattering and optical momentum transfer to the particle.[7,8] In Fig. 1a) we show the scattering of a typical pair of rays A of a highly focused beam incident upon a 10-μm lossless dielectric sphere, for example. The principal part of the momentum transfer from the incident light to the particle is due to the emergent rays A′, which are refracted by the particle. Successive surface reflections, such as R_1 and R_2, contribute a lesser scattering. For a glass particle in water the effective index m, equal to the index of the particle divided by the index of the medium, is about 1.1 to 1.2, and the sphere acts as a weak positive lens. If we consider the direction of the resulting forces F_A on the particle that are due to refraction of rays A in the weak-lens regime, we see as shown in Fig. 1a) that there is a substantial net backward trapping-force component toward the beam focus.

Figure 2 sketches the apparatus used for trapping Mie or Rayleigh particles. Spatially filtered argon-laser light at 514.5 nm is incident upon a high-numerical-aperture (N.A. 1.25) water-immersion microscope objective, which focuses a strongly convergent downward-directed beam into a water-filled glass cell. Glass Mie particles are introduced into the trap by an auxiliary vertically directed holding beam,[5] which lifts particles off the bottom of the cell and manipulates them to the focus. Rayleigh particles are simply dispersed in water solution at reasonable concentrations and enter the trapping volume by Brownian diffusion. A microscope M is used to view the trapped particles visually off a beam splitter S or by recording the 90° scatter with a detector D.

Figure 1b) is a photograph of a 10-μm glass sphere of index about 1.6 trapped and levitated just below the beam focus of a ~100-mW beam. The picture was taken through a green-blocking filter using the red fluorescence of the argon laser beam in water in order to make the trajectories of the incident and scattered

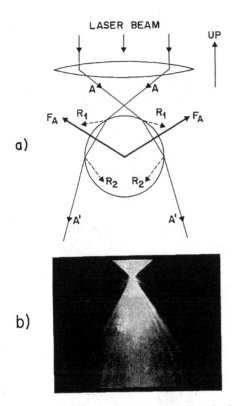

Fig. 1. a) Diagram showing the ray optics of a spherical Mie particle trapped in water by the highly convergent light of a single-beam gradient force trap. b) Photograph, taken in fluorescence, of a 10-μm sphere trapped in water, showing the paths of the incident and scattered light rays.

beams visible. The sizable decrease in beam angle of the scattered light, which gives rise to the backward force, is clearly seen. The stria in the forward-scattered light arise from the usual Mie-scattering ring pattern.

Next consider the possibility of single-beam trapping of submicrometer Rayleigh particles whose diameter $2r$ is much less that λ. Although we are now in the wave-optic regime, we will again see the role of the strong axial gradient in producing a net backward axial force component. For Rayleigh particles in a medium of index n_b the scattering force in the direction of the incident power is $F_{scat} = n_b P_{scat}/c$, where P_{scat} is the power scattered.[9] In terms of the intensity I_0 and effective index m

$$F_{scat} = \frac{I_0}{c} \frac{128\pi^5 r^6}{3\lambda^4} \left(\frac{m^2-1}{m^2+2}\right)^2 n_b. \qquad (1)$$

The gradient force F_{grad} in the direction of the intensity gradient for a spherical Rayleigh particle of polarizability α is[6]

$$F_{grad} = -\frac{n_b}{2}\alpha\nabla E^2 = -\frac{n_b^3 r^3}{2}\left(\frac{m^2-1}{m^2-2}\right)\nabla E^2. \qquad (2)$$

This Rayleigh force component, in analogy with the gradient force for Mie particles, can be related to the lenslike properties of the scatterer.

As for atoms,[1] the criterion for axial stability of a single-beam trap is that R, the ratio of the backward axial gradient force to the forward-scattering force, be greater than unity at the position of maximum axial intensity gradient. For a Gaussian beam of focal spot size w_0 this occurs at an axial position $z = \pi w_0^2/\sqrt{3}\,\lambda$, and we find that

$$R = \frac{F_{grad}}{F_{scat}} = \frac{3\sqrt{3}}{64\pi^5} \frac{n_b^2}{\left(\dfrac{m^2-1}{m^2+2}\right)} \frac{\lambda^5}{r^3 w_0^2} \geq 1, \qquad (3)$$

where λ is the wavelength in the medium. This condition applies only in the Rayleigh regime where the particle diameter $2r \lesssim 0.2\lambda \cong 80$ nm. In practice we require R to be larger than unity. For example, for polystyrene latex spheres in water with $m = 1.65/1.33 = 1.24$ and $2w_0 = 1.5\lambda = 0.58\ \mu$m we find for $R \geq 3$ that $2r \leq 95$ nm. Thus with this choice of spot size we meet the stability criterion over the full Rayleigh regime. The fact that $R < 3$ for $2r > 95$ nm does not necessarily imply a lack of stability for such larger particles since we are beyond the range of validity of the formula. Indeed, as we enter the transition region to Mie scattering we expect the ray-optic forward-scattering picture to be increasingly valid. As will be seen experimentally we have stability from the Rayleigh regime, through the transition region, into the full Mie regime. For silica particles in water with $m = 1.46/1.33 = 1.10$ and $2w_0 = 0.58\ \mu$m we find for $R \geq 3$ that $2r \leq 126$ nm. For high-index particles with $m \equiv 3.0/1.33 = 2.3$ we find that $2r \leq 61$ nm.

The stability condition on the dominance of the backward axial gradient force is independent of power and is therefore a necessary but not sufficient condition for Rayleigh trapping. As an additional sufficient trapping condition we have the requirement[1] that the Boltzmann factor $\exp(-U/kT) \ll 1$, where $U = n_b\alpha E^2/2$ is the potential of the gradient force. As was previously pointed out,[6] this is equivalent to requiring that the time to pull a particle into the trap be much less than the time for the particle to diffuse out of the trap by Brownian motion. If we set $U/kT \geq 10$,

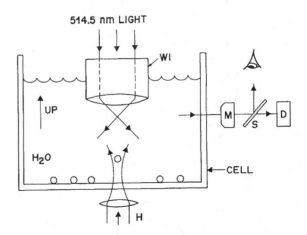

Fig. 2. Sketch of the basic apparatus used for the optical trapping of Mie and Rayleigh particles in water by means of a single-beam gradient force radiation-pressure trap.

for example, and use a power of ~ 1.5 W focused close to the limiting spot diameter of 0.58 μm $\cong 1.5\lambda$, we find for silica that the minimum theoretical particle size that satisfies this condition is $2r = 19$ nm. For polystyrene latex, the minimum particle size that can be trapped under these conditions is $2r = 14$ nm. With a high-index particle of $m = 3/1.33 = 2.3$ the theoretical minimum size is $2r \cong 9$ nm.

Additional experiments were performed on individual colloidal polystyrene latex particles in water. Unfortunately the particles exhibit a form of optical damage at high optical intensities. For 1.0-μm spheres with a trapping power of a fraction of a milliwatt, particles survived for tens of minutes and then shrank in size and disappeared. Spheres of 0.173 μm were trapped for several minutes with a power of a milliwatt before being lost. Particles of 0.109-μm diameter required about 12–15 mW and survived about 25 sec. With 85- and 38-nm latex particles the damage was so rapid that it was difficult to observe the scattering reliably. It was nevertheless clear that trapping occurred over full size range from Mie to Rayleigh particles.

The remarkable uniformity of latex particles was evident from the small variation of $\pm 15\%$ in the 90° scatter of 0.109-μm particles. Since the scattering is closely Rayleigh this corresponds to a diameter variation of $\pm 2.4\%$. Subsequently we determined the size of unknown silica Rayleigh particles by comparing their scatter with the scatter from the 0.109-μm particles taken as a standard, using Eq. (1). Although the 0.109-μm particles are not strictly Rayleigh, one can make a modest theoretical correction[10] of ~ 1.06 to the effective particle size.

Trapping of nominally spherical colloidal silica particles was observed by using commercially available Nalco and Ludox samples[11] diluted with distilled water. With a high concentration we quickly collect many particles in the trap and observe a correspondingly large scattering. At reduced concentration we can observe single particles trapped for extended time. Once a particle is captured at the beam focus we observe an apparent cessation of all Brownian motion and a large increase in particle scattering.

With silica samples we always observe a wide distribution of particle sizes as evidenced by the more than an order-of-magnitude difference in scattering from particles trapped with a given laser power. Particle damage by the light was not a serious problem with silica particles. The smaller particles of the distribution showed only slight changes of scattering over times of minutes. The larger particles would often decay by factors up to 3 in comparable times.

Measurements were made on a Nalco 1060 sample with a nominal size of about 60 nm and an initial concentration of silica of 50% by weight diluted to one part in 10^5–10^6 by volume. Trapping powers of 100–400 mW were used. The absolute size of the Nalco 1060 particles as determined by comparison with the 0.109-μm latex standard varied from ~ 50 to 90 nm with many at ~ 75 nm. We also studied smaller silica particles using a Ludox TM sample with nominal particle diameter of ~ 21 nm and a Nalco 1030 sample with nominal distribution of 11 to 16 nm. Dilutions of $\sim 10^6$–10^7 and powers of ~ 500 mW to 1.4 W were used. With both samples we were limited by laser power in the minimum-size particle that could be trapped. With 1.4 W of power the smallest particle trapped had a scattering that was a factor of $\sim 3 \times 10^4$ less than from the 0.109-μm standard. This gives a minimum particle size of 26 nm assuming a single spherical scatterer. The measured minimum size of 26 nm compares with the theoretically estimated minimum size of ~ 19 nm for this power, based on $U/kT = 10$ and spot size $2w_0 = 0.58$ μm. This difference could be resolved by assuming a spot size $2w_0 = 0.74$ μm $= 1.28$ (0.58 μm) $\cong 1.9$ λ.

Experimentally we found that we could introduce a significant drift of the fluid relative to the trapped particle by moving the entire cell transversely relative to the fixed microscope objective. This technique gives a direct method of measuring the maximum trapping force. It also implies the ability to separate a single trapped particle from surrounding untrapped particles by a simple flushing technique.

Our observation of trapping of a 26-nm silica particle with 1.4 W implies, by simple scaling, the ability to trap a 19.5-nm particle with $m = 1.6/1.33 = 1.20$ and a 12.5-nm particle of $m = 3.0/1.33 = 2.26$ at the same power. These results suggest the use of the single-beam gradient force traps for other colloidal systems, macromolecules, polymers, and biological particles such as viruses. In addition to lossless particles with real m there is the possibility of trapping Rayleigh particles with complex m for which one can in principle achieve resonantly large values of the polarizability α. Finally, we expect that these single-beam traps will work for trapping atoms[1] as well as for macroscopic Rayleigh particles since atoms can be viewed as Rayleigh particles with a different polarizability.

References

1. A. Ashkin, Phys. Rev. Lett. **40,** 729 (1978).
2. A. Ashkin, Science **210,** 1081 (1980); V. S. Letokhov and V. G. Minogin, Phys. Rep. **73,** 1 (1981).
3. A. Ashkin and J. P. Gordon, Opt. Lett. **8,** 511 (1983).
4. A. Ashkin and J. M. Dziedzic, Phys. Rev. Lett. **54,** 1245 (1985).
5. A. Ashkin and J.M. Dziedzic, Appl. Phys. Lett. **19,** 283 (1971).
6. P. W. Smith, A. Ashkin, and W. J. Tomlinson, Opt. Lett. **6,** 284 (1981); and A. Ashkin, J. M. Dziedzic, and P. W. Smith, Opt. Lett. **7,** 276 (1982).
7. A. Ashkin, Phys. Rev. Lett. **24,** 146 (1970).
8. G. Roosen, Can. J. Phys. **57,** 1260 (1979).
9. See, for example, M. Kerker, *The Scattering of Light* (Academic, New York, 1969), p. 37.
10. W. Heller, J. Chem. Phys. **42,** 1609 (1965).
11. Nalco Chemical Company, Chicago, Illinois; Ludox colloidal silica by DuPont Corporation, Wilmington, Delaware.

Optical Trapping and Manipulation of Viruses and Bacteria

A. Ashkin and J. M. Dziedzic

Optical trapping and manipulation of viruses and bacteria by laser radiation pressure were demonstrated with single-beam gradient traps. Individual tobacco mosaic viruses and dense oriented arrays of viruses were trapped in aqueous solution with no apparent damage using ~120 milliwatts of argon laser power. Trapping and manipulation of single live motile bacteria and *Escherichia coli* bacteria were also demonstrated in a high-resolution microscope at powers of a few milliwatts.

W E REPORT THE EXPERIMENTAL demonstration of optical trapping and manipulation of individual viruses and bacteria in aqueous solution by laser light using single-beam gradient force traps. Individual tobacco mosaic viruses (TMV) and oriented arrays of viruses were optically confined within volumes of a few cubic micrometers, without obvious damage, and manipulated in space with the precision of the optical wavelength. The ability of the same basic optical trap to confine and manipulate motile bacteria was also demonstrated. We have used the trap as an "optical tweezers" for moving live single and multiple bacteria while being viewed under a high-resolution optical microscope. These results suggest that the techniques of optical trapping and manipulation, which have been used to advantage with particles in physical systems, are also applicable to biological particles. Optical trapping and manipulation of small dielectric particles and atoms by the forces of radiation pressure have been studied since 1970 (*1–4*). These are forces arising from the momentum of the light itself. Early demonstrations of optical trapping (*1*) and optical levitation (*5*) involved micrometer-size transparent dielectric spheres in the Mie regime (where the dimensions *d* are large compared to the

wavelength λ). More recently optical trapping of submicrometer dielectric particles was demonstrated in the Rayleigh regime (where *d* << λ). Single dielectric particles as small as ~260 Å (*6*) and even individual atoms (*7*) were trapped with single-beam gradient traps (*8*).

Single-beam gradient traps are conceptually and practically the simplest. They consist of only a single strongly focused Gaussian laser beam having a Gaussian transverse intensity profile. In such traps the basic scattering forces and gradient force components of radiation pressure (*1, 3, 4, 8*) are configured to give a point of stable equilibrium located close to the beam focus. The scattering force is proportional to the optical intensity and points in the direction of the incident light. The gradient force is proportional to the gradient of intensity and points in the direction of the intensity gradient. Particles in a single-beam gradient trap are confined transverse to the beam axis by the radial component of the gradient force. Stability in the axial direction is achieved by making the beam focusing so strong that the axial gradient force component, pointing toward the beam focus, dominates over the scattering force trying to push the particle out of the trap. Thus one has a stable trap based solely on optical forces, where gravity plays no essential role as was the case for the levitation trap (*5*). It works over a particle

size range of 10^5, from ~10 μm down to a few angstroms, which includes both Mie- and Rayleigh-size particles.

The sensitivity of laser trap effectiveness to optical absorption and particle shape is of particular importance for the trapping of biological particles. Absorption can cause an excessive temperature rise or additional thermally generated (radiometric) forces as a result of temperature gradients within a particle (*9*). In general, the smaller the particle size the less the temperature rise and the less the thermal gradients for a given absorption coefficient (*10*). Particle shape plays a larger role in the trapping of Mie particles than Rayleigh particles. For Mie particles both the magnitude and direction of the forces depend on the particle shape (*3*). This restricts trapping to fairly simple overall shapes such as spheres, ellipsoids, or particles whose optical scattering varies slowly with orientation in the beam. In the Rayleigh regime, however, the particle acts as a dipole (*6*) and the direction of the force is independent of particle shape; only the magnitude of the force varies with orientation. A significant conclusion of this work is that important types of biological particles in both the Mie and Rayleigh regimes have optical absorptions and shapes that fall within the scope of single-beam gradient traps.

As a first test for trapping of small biological particles we tried TMV, a much studied virus that can be prepared in monodisperse colloidal suspension in water at high concentrations (*11, 12*). Its basic shape is cylindrical with a diameter of 200 Å and a length of 3200 Å (*13*). Although its volume of about 470 Å^3 is typical of a Rayleigh particle, its length is comparable to the wavelength in the medium and we expect some Mie-like behavior in its light scattering. TMV particles have a negative charge in solution (*14*) and an index of refraction (*15*) of about 1.57. Our virus samples were prepared from the same batches used in experi-

AT&T Bell Laboratories, Holmdel, NJ 07733.

From A. Ashkin and J. M. Dziedzic, "Optical Trapping and Manipulation of Viruses and Bacteria," Science. 235: 1517–1520 (1987). Reprinted with permission from AAAS.

ments on the self-alignment of TMV in parallel arrays in dense aqueous suspensions (*13, 16*). These samples, suitably diluted, were studied in essentially the same apparatus (*6*) previously used for trapping of silica and polystyrene latex colloidal suspensions and Mie-size dielectric spheres (see Fig. 1).

When one visually observes the 90° scattering from untrapped viruses in the vicinity of the focus, one sees the random-walk motion and intensity fluctuations, or flickering, characteristic of Brownian diffusion in position and orientation. These rotational intensity fluctuations of about an order of magnitude result from the large changes in polarizability of a cylindrical particle with orientation in the light beam (*17*). At laser power levels of about 100 to 300 mW we begin to see trapping. The capture of a virus manifests itself as a sudden increase in the 90° scattering, as shown in Fig. 2. As more viruses are captured in the trap, we see further abrupt changes in scattering. If we block the beam momentarily (at points B) the trap empties and the trapping sequence repeats. Not only does all apparent positional Brownian motion disappear when viruses are captured but the intensity fluctuations characteristic of a freely rotating TMV particle are also greatly reduced (Fig. 2). This strongly indicates the angular alignment of the individual and multiple viruses within the trap. In previous work on trapped silica colloids we deduced that the successive particles entering the trap do not coalesce into a single particle but form a dense fixed array of separated particles (*18*), presumably held apart by electrostatic repulsion. It is likely therefore that charged viruses in the trap also form a dense array of separate particles. We further suspect that the trapped viruses are oriented parallel to one another as in dense oriented arrays of free TMV suspensions (*13, 16*).

The size of an unknown captured particle can be obtained by comparing the magnitude of its 90° scattering with that from a polystyrene latex calibrating sphere of known size, in the same trap. This comparison technique was used in our previous work on colloidal silica (*6*). It gives high accuracy for spherical Rayleigh particles because of the r^6 dependence of Rayleigh scattering. In the Rayleigh limit, all parts of the particle radiate in phase as a single dipole. Since TMV is not strictly a Rayleigh particle, we might expect interference effects due to optical path differences from different parts of the same particle to reduce the scattering below the full Rayleigh value. However, if we compare the 90° scatter of the average of 72 single particles from one sample, measured over a period of several days, with the 90° scatter from the calibrat-

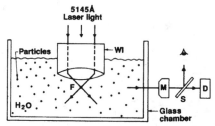

Fig. 1. Apparatus used for optical trapping of TMV particles and mobile bacteria. Spatially filtered argon laser light at 5145 Å is focused to a spot diameter of about 0.6 μm in the water-filled chamber by the high numerical aperture (1.25) water-immersion microscope objective (WI) forming a single-beam gradient trap near the beam focus (F). The 90° scattering from trapped particles can be viewed visually through a beam splitter (S) with a microscope (M) or recorded using a photodetector (D).

ing sphere, using the Rayleigh formula, we find an effective volume of about (450 Å)³. This volume corresponds to a cylinder 200 Å in diameter and about 3100 Å long, which is quite close to the volume of TMV. We conclude from this that we are looking at single TMV particles, and further that the axis of the TMV particle in the trap is oriented closely perpendicular to the beam axis along the optical electric field, since only then can all parts of the cylindrical virus radiate in phase at 90° as a Rayleigh particle. Looking at the size uniformity of our 72 particles we find, assuming that all differences in scatter are due to changes in the length of the viruses, that 75% of all particles lie in a length range within about 20% of the average. Thus we find that this particular sample is quite monodisperse and the measured length is $L = 3100 \pm 700$ Å.

Not much size information can be deduced from the observed changes in 90° scatter as additional virus particles enter a trap and form an aligned array, since the combined scattering field in such cases is the result of interference of the fields from each

particle of the array. Not only are the positions and phases of the fields of the various particles unknown, but they probably change as additional particles enter the trap. Examples of destructive interference and a decrease in total 90° scattering as additional particles enter a trap are seen at times of 7.4 and 10.8 minutes in Fig. 2. In a related experiment (*18*) on the angular distribution of scattered light from trapped colloidal silica particles we found that when a similar decrease in 90° scattering occurred on entry of an additional particle, it was possible to find another direction where the phases added constructively and actually gave an increase in scattering.

In the above discussion we deduced the optical alignment of TMV along the optical field E on purely experimental grounds. This also makes sense energetically since this optical alignment results in the maximum polarizability α of the cylindrical virus. Indeed, the TMV, when aligned, not only feels the deepest trapping potential $\alpha E^2/2$ but also experiences strong realigning torques when rotated, due to the angular dependence of $\alpha E^2/2$. More detailed information on the angular orientation of TMV within the trap could be obtained from scattering experiments with an additional low-power probe laser beam of different wavelength and varying directions of polarization relative to the virus axis.

Another conclusion that can be drawn from the data of Fig. 2 is that it is possible to trap viral material without any gross optical damage. Indeed, single viruses have been trapped for tens of minutes with no apparent changes in size as indicated by the constant magnitude of the scattering. The viability of the virus after trapping was not examined. In previous experiments optical damage was a serious problem, which limited the trapping of small Rayleigh-sized polystyrene latex particles and even to some extent silica particles. For silica this damage was subsequently eliminated (*18*) by *p*H changes or additions of potassium silicate to the solution, which points to a surface photochemical reaction as the damage mechanism. For virus particles the fact that the strong optical absorptions are in the ultraviolet probably contributes to their optical stability in the visible light range. As with silica colloids (*6*) we were able to manipulate captured TMV particles within their environment by moving either the light beam or the entire chamber. This implies the ability to separate trapped viruses by means of a simple flushing technique, for example.

The major problem encountered in the experiment was one of reproducibility. Thus far we have had two batches of dense virus (10 and 50% by weight) and had successful

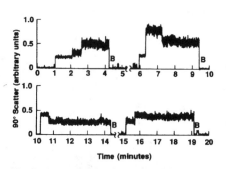

Fig. 2. Scattered light observed at 90° as successive TMV viruses enter the optical trap. At times labeled "B" the trapping beam is momentarily blocked, releasing the viruses. The trap subsequently refills with new virus particles.

trapping runs with samples diluted from each batch which lasted several days. However, with other samples from the same batches we trapped many fewer TMV-sized particles. Instead we trapped mainly larger-sized clumps and smaller-sized single particles. These larger trapped clumps usually decrease in size (damage) in just a few minutes. Large clumps, which also decayed in time, were previously trapped in our silica experiments (6). The smaller-sized trapped particles are stable in size and are probably just smaller pieces of virus. On this assumption, the lengths of some of these smaller viral pieces that we observed at trapping powers of ~1.5 W were ~270 Å. If these small particles are typical of the index of refraction of other proteins, then this observation implies an ability to trap proteins with molecular weight $M \geq 3 \times 10^6$ at a power of 1.5 W.

The lack of consistency in TMV trapping could be the result of either the dilution process or the trap geometry. With TMV samples, proper pH and low ionic content are needed to avoid polydispersity. At $pH > 9$ the TMV falls apart and at $pH < 6$ it aggregates. We attempted to maintain the pH of our diluted samples close to 7. To do this we used deionized or distilled water with small volumes of buffer solution added to adjust the pH, with no improvement in the consistency of our results. Regarding trap geometry, we often found laser beam wander to be a problem and usually checked our overall geometry by looking for good trapping of 600-Å silica test particles. It is possible, however, that the trap is much more tolerant of aberrations for the spherical 600-Å Rayleigh particles than for 3200-Å TMV. If the TMV sits transverse to the beam axis, the virus would extend to the edges of the beam to regions not felt by a silica particle. Although it requires more power, a trap with a larger diameter focal spot might be more favorable for a particle of this shape. Another obvious experiment is to study the trapping of more spherical viruses such as tomato bushy-stunt virus (19) with a diameter of ~450 Å.

In most of our experiments with silica colloids or TMV in water, we noticed the appearance of some strange new particles in diluted samples that had been kept around for several days. They were quite large compared to Rayleigh particles, on the basis of their scattering of light, and were apparently self-propelled. They were clearly observed moving through the distribution of smaller slowly diffusing Rayleigh-sized colloidal particles at speeds as high as hundreds of micrometers per second. They could stop, start up again, and frequently reversed their direction of motion at the boundaries of the

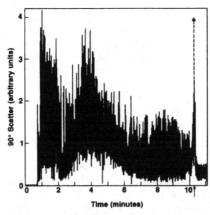

Fig. 3. Scattered light at 90° from a live bacterium trapped by ~5 mW of laser power. At about 10.3 minutes (indicated by arrow) the power was increased to 100 mW. The bacterium was killed and apparently loses much of its cell contents.

trapping beam, when they encountered a dark region, indicating some sort of attraction toward the light. Their numbers increased rapidly as time went by. When examined under 800× magnification in an optical microscope, they were clearly identifiable as rod-like motile bacteria, propelled by rotating tails. There were at least two types of bacteria, about 0.5 and 1.5 μm in length. Optically they resembled small, transparent Mie particles with an index of refraction close to unity.

When one of these bacteria wandered into or was possibly attracted into the trap, it was captured. It was observable through microscope M either by eye or on the photo detector, as a fluctuating signal, as it struggled unsuccessfully to escape from the trap. The far-field forward scatter from the bacteria could also be seen on a screen placed outside the cell. To help capture the bacteria we initially set the laser power at ~50 mW. Once the bacterium had been captured, we quickly lowered the power to ~5 mW to reduce the possibility of optically damaging the bacterium. Figure 3 shows the recorded 90° scatter as a function of time, taken at 5 mW, as a bacterium propels itself about in the trap. After about 10 minutes we raised and maintained the power at ~100 mW. This, as is seen, was sufficient to kill the trapped bacterium. The light scattering stopped fluctuating and decreased to quite a low value as the bacterium apparently vented some of it contents into the surroundings. The remains of the cell could be held in the trap with laser power as low as ~0.5 mW. It was reported that similar venting of a cell's contents occurred in experiments on the puncture of blood cells by pulsed laser beams (20).

In other experiments we illuminated the entire trapping region with a wide low-power auxiliary red laser beam directed transverse to the trap axis. We then viewed the scene through a red-pass filter with microscope M, either visually or with a video camera and recorder. We could observe the capture of free-swimming bacteria and their subsequent release as the trapping beam was blocked. Several bacteria were occasionally trapped simultaneously. We were also able to demonstrate optical micromanipulation of single trapped bacteria, within the liquid, by moving either the trapping beam or the entire chamber and its liquid.

In the above experiment the low resolution of the side-viewing microscope M and the presence of red laser interference rings made it difficult to resolve details of the trapped bacteria. An obvious extension of this viewing technique is to introduce the trapping laser beam directly into a high-resolution microscope through a beam splitter. By using a water-immersion microscope objective with a high numerical aperture for both laser trapping and viewing through a filter, we were able to simultaneously trap, manipulate, and observe bacteria or other particles with high resolution. For convenience we placed our water samples containing bacteria under a cover slip and used a water-immersion objective designed for operation through the cover slip. An additional lens mounted outside the microscope on an *xyz* mechanical micromanipulator was used to focus the laser trapping beam within the field of view of the microscope and move it about transversely without any need to touch the microscope. We observed the scene either by eye or with a video camera. At power levels as low as 3 to 6 mW we were able to move the beam about and capture a free-swimming bacterium anywhere in the field of view. Once the bacterium had been captured, we could rapidly move it transversely and continue to catch more bacteria until we had a half dozen or more within the trap. Rapid transverse motion without loss of bacteria is possible because of the strong transverse gradient trapping forces. The trapping forces in the axial or z direction are stronger in the forward direction of the light rather than in the backward direction (6, 7). Thus any rapid upward motion of the focus can result in escape of particles.

At lower power levels, from 1 to 3 mW, and probably less, we discovered another trapping mode in which bacteria were trapped against the bottom surface of the slide. In this low-power surface mode of operation the mechanical surface provides the backward trapping force needed to prevent the escape of the particle out the bottom of the trap in the direction of the weakest trapping force. It is still possible to move particles about transversely over the surface in this mode because the transverse forces remain quite strong even at the lower power. Bacteria captured by either mode of trapping with powers in the range from 1 to 6 mW have survived for hours in the laser light with no apparent damage.

We performed subsequent experiments using the high-resolution microscope with *Escherichia coli* bacteria. These bacteria are much less motile and could be captured and manipulated rapidly at surfaces and in the bulk fluid with powers as low as a fraction of a milliwatt with no apparent change in behavior or appearance. At powers of 100 mW or more it was possible to observe a shrinkage in the size of the *E. coli* as they become optically damaged in a time of about a minute. With yet another sample of highly motile bacteria we observed a gradual loss in motility of trapped bacteria in about a half-minute with powers as low as 10 to 20 mW. In all cases where optical damage was observed with the 5145-Å green argon laser line it might be advantageous to use other laser wavelengths.

One advantage offered by the high-resolution microscope was the ability to study the trapping forces on bacteria in some detail. For example a bacterium, while being manipulated close to the surface of the slide,

would occasionally manage to attach itself to the surface with its tail and remain tethered. Under these conditions it was still possible to optically manipulate the particle in a circle around its tether and observe the action of the optical forces. Although we do not have a complete description of the trapping forces for complex shaped particles like bacteria, it is clear from these experiments that the same qualitative features, based on simple ray-optics and refraction, that account for trapping of Mie-sized spheres (3, 4, 6) and spheroids (21) still apply here. For example when the beam center is moved toward the edges of a bacterium where refraction is large and asymmetric, we generate large transverse gradient forces which in effect drag the particle with the beam in the direction which recenters the particle on the beam. This is the same basic effect that accounts for the centering of a small Mie particle at the position of maximum light intensity of a large Gaussian beam. However, for the case of a large particle and a small beam, as we have here, it is clearly the local shape of particle that dominates the net force.

REFERENCES AND NOTES

1. A. Ashkin, *Phys. Rev. Lett.* **24**, 156 (1970).
2. _____, *ibid.* **25**, 1321 (1970).
3. _____, *Science* **210**, 1081 (1980).
4. G. Roosen, *Can. J. Phys.* **57**, 1260 (1979).
5. A. Ashkin and J. M. Dziedzic, *Appl. Phys. Lett.* **19**, 283 (1971).
6. _____, J. E. Bjorkholm, S. Chu, *Opt. Lett.* **11**, 288 (1986).
7. S. Chu, J. E. Bjorkholm, A. Ashkin, A. Cable, *Phys. Rev. Lett.* **57**, 314 (1986).
8. A. Ashkin, *ibid.* **40**, 729 (1978).
9. _____ and J. M. Dziedzic, *Appl. Phys. Lett.* **28**, 333 (1976).
10. P. W. Dusel, M. Kerker, D. D. Cooke, *J. Opt. Soc. Am.* **69**, 55 (1979).
11. S. Fraden, A. J. Hurd, R. B. Meyer, M. Cahoon, D. L. D. Caspar, *J. Phys. (Paris) Colloq.* **46**, C3-85 (1985).
12. H. Boedtker and N. S. Simmons, *J. Am. Chem. Soc.* **80**, 2550 (1958).
13. J. A. N. Zasadzinski and R. B. Meyer, *Phys. Rev. Lett.* **56**, 636 (1986).
14. V. A. Parsegian and S. L. Brenner, *Nature (London)* **259**, 632 (1976).
15. M. A. Lauffer, *J. Phys. Chem.* **42**, 935 (1938).
16. J. A. N. Zasadzinski, M. J. Sammon, R. B. Meyer, in *Proceedings of the 43rd Annual Meeting of the Electron Microscopy Society of America*, G. Bailey, Ed. (San Francisco Press, San Francisco, 1985), p. 524.
17. H. C. van de Hulst, *Light Scattering by Small Particles* (Dover, New York, 1981), pp. 70–73.
18. A. Ashkin and J. M. Dziedzic, unpublished data.
19. C. Tanford, *Physical Chemistry of Macromolecules* (Wiley, New York, 1961), chap. 2.
20. K. O. Greulich *et al.*, paper Tu GG-11 presented at the International Quantum Electronics Conference, San Francisco, June 1986.
21. A. Ashkin and J. M. Dziedzic, *Appl. Opt.* **19**, 660 (1980).
22. We thank J. A. N. Zasadzinski and R. B. Meyer for providing samples of TMV colloidal solution, and T. Yamane for *E. coli* samples. Helpful discussions with J. E. Bjorkholm and S. Chu are gratefully acknowledged.

29 September 1986; accepted 30 January 1987

2

Applications in Biology

The importance of optical traps in biology arose from the development of the single-beam gradient trap, or "optical tweezers," in 1986 [1] (see **PAPER 1.5**). Optical tweezers have since been applied in cell biology [2] and in the rapidly expanding field of single-molecule research [3]. Optical forces produced by commonly available lasers lie in the piconewton (pN) range, which is just right to experiment with biological cells and individual molecules. Because most biological materials absorb only weakly in the near infrared region of the electromagnetic spectrum, Nd:YAG ($\lambda = 1064$ nm) lasers are well suited to optical tweezers applications in biology [2,4,5]. Optical forces generated by relatively inexpensive, diode-pumped lasers (>100 mW power) easily overcome the relatively weak forces associated with thermal agitation, gravity, and fluid flow and enable biological cells and microorganisms (including viruses, bacteria, sperms, red blood cells, and other mammalian cells) to be picked up and manipulated at will within aqueous media. However, the biggest impact in biology so far has been the ability to make ultra-high-resolution mechanical measurements on individual biological molecules. Optical tweezers are suitable for many studies in biology because they can readily produce forces of between 0.1 pN to 500 pN.

It is helpful at this point to consider the relative size of the forces within a biological context: 1000 pN (1 nanonewton) will break a covalent bond and is enough to pull apart two mammalian cells; 150 pN will disrupt avidin–biotin and antibody–antigen interactions* and will also break an actin filament; 60 pN is sufficient to unravel the DNA double helix, unfold protein tertiary structure, and distort or extrude biological membranes; 30 pN will halt DNA processing enzymes like helicases and polymerases and is sufficient to overcome the thrust generated by the bacterial flagellar motor; 10 pN is sufficient to stall[†] cytoskeletal motor proteins like myosin, kinesin, and dynein, which are responsible for powering cell motility and muscle contraction.

When Ashkin first started experimenting with optical traps, he realized that an unfocused laser beam would draw objects of high refractive index toward the center of the beam and propel them in the direction of propagation. An arrangement of two counterpropagating beams allowed objects to be confined in three dimensions [6] (see **PAPER 1.2**). He then discovered that a single, tightly focused laser beam could capture small dielectric particles in three dimensions [1] (see **PAPER 1.5**). This discovery revolutionized the field in terms of biological applications as it meant that objects could be picked up and moved at will using a single laser beam within a light microscope—an instrument common to all biology labs. The single-beam gradient trap, "optical tweezers," requires that light is brought to a diffraction-limited focal spot. For a Gaussian laser beam of wavelength $\lambda = 1064$ nm focused by an objective lens of numerical aperture N.A.=1.3, the full width at half maximum (FWHM) diffraction-limited spot size is around 500 nm (given by FWHM = 0.6 λ/N.A.). High-numerical-aperture objective lenses with excellent correction for spatial and chromatic aberrations have been designed by microscope manufacturers with the specific aim of achieving the highest possible image resolution from biological specimens. A survey of lenses indicates that some are well suited to optical tweezers in terms of transmission in the near infrared [7]. The maximum force and stiffness (i.e., restoring force per unit displacement from rest position) of optical tweezers increases as a quadratic function of

---

* The lifetime of a chemical bond, τ_F, subjected to a given force, F, decreases exponentially with force compared to the value measured at zero load, τ_0, according to Boltzmann statistics. The work done in extending the chemical bond (–Fd) depends upon a characteristic distance, d (~ 0.1 nm), and thermal energy is given by k_bT. So lifetime is strongly dependent on load; for example, if $\tau_0 = 10,000$ s, then $\tau_{100pN} = 800$ s, $\tau_{500pN} = 37$ ms.

$$\tau_F = \tau_0 \times e^{\left(\frac{-Fd}{k_bT}\right)}$$

[†] Motor proteins are driven by the available free energy change associated with ATP hydrolysis, ΔG^{ATP} (about 50% of the total free energy change is available, which under physiological conditions gives –60 kJ/mol · 0.5 = 50 pN.nm). Stall force, F_{max}, depends upon the size of the movement produced, d_{uni}, as the motor protein hydrolyzes a single ATP molecule. Since kinesin advances 8 nm for each ATP molecule hydrolyzed, its stall force is about 6 pN.

$$F_{max} = \frac{\Delta G^{ATP}}{d_{uni}}$$

particle size up to 1 μm diameter and is linearly related to optical power [8] (see **PAPER 3.3**). Laser power of 100 mW gives an optical trap of 0.08 pN.nm^{-1} stiffness, and a maximum force of about 20 pN. This is sufficient to hold and manipulate cells, distort membranes, and push and pull on single molecules.

The earliest biological studies were made using material that was large enough to manipulate directly using optical tweezers. Ashkin et al. used optical tweezers to capture bacteria (about 2 μm long) and tobacco mosaic virus (100 nm dia.) [9] (see **PAPER 1.6**), to manipulate single cells [10] and cell organelles [11], and, finally, to measure the force of cell organelle movement inside living cells [12]. An early worry concerning the use of optical tweezers with biological materials (especially living cells or intact organisms) was that the very high flux of laser light (10^7 W/cm^2) might cause cell damage or death (termed "optiction" by Ashkin [13]). Damage might result directly from interaction between light and biological molecules by breaking covalent bonds, causing photochemical reactions to occur, or by heating. Biological molecules are easily damaged by heat and many denature when heated above 45°C. Infrared light is known to cause damage to bacteria and cultured mammalian cells mainly from oxygen-dependent photochemical reactions. The most harmful wavelengths in the near infrared region are between 850 and 950 nm and the safest lie immediately to either side of this [7]. If the material is transparent (i.e., absorbs < 10%), then little heating occurs because the surface-area-to-volume ratio is extremely large and the rate of heat loss by convection is correspondingly rapid [14].

The functionality of optical tweezers can be enhanced by combining the trapping laser beam with additional lasers that are then used either to image or to cut (using UV light) the biological specimen. These more complex tools and optical configurations are reviewed elsewhere [15] but of particular note are the high-precision instruments used to image cell topology by raster scanning an optically trapped particle [16,17]. More recently, high-throughput systems have been created that allow many living cells to be sorted and/or arranged using holographic techniques [18] (see **PAPER 2.1**), [19] (Chapter 7). By combining a counterpropagating optical trap arrangement with a microfluidic system, Guck and colleagues developed a high-throughput optical "stretcher" that causes mechanical extension of cells along the optical axis as they pass through a microfluidic channel (see Chapter 8). It was proposed that this device might be used as a diagnostic tool to assay cell bulk stiffness [20] (see **PAPER 2.2**).

Basic studies of the mechanical properties of red blood cells were initially made using optically trapped microspheres attached to the cell membranes [21], [22] (see **PAPER 2.3**). Knowing the stiffness of the optical

tweezers, the applied forces were estimated by measuring the displacement of the bead from the center of the optical trap. These studies revealed how mechanical properties depend both on the cell membrane and on the underlying cytoskeleton, which consists of the proteins actin and spectrin. It was discovered that the outer cell membrane, known as the plasma membrane, of cultured mammalian cells could be detached from the underlying cytoskeleton and pulled out into a long membrane tether [23,24]. The mechanics of membrane tether formation were modeled by these authors in order to derive contributions from lipid bending and cytoskeletal adhesions.

Later studies investigated the direct effect of load and cell activation, via signaling pathways, on membrane-cytoskeleton and extracellular adhesion molecules [25,26]. The ultimate refinement of this approach has been to study the mechanical properties of individual integral proteins that are embedded in the membrane. Integral membrane proteins have a variety of critical functions in the cell, such as ion channels, receptors for hormones, and cell adhesion proteins, and their interaction with lipids and the underlying cytoskeleton is functionally significant. Optical tweezers have allowed individual membrane proteins to be manipulated using a microsphere coated with antibodies. The optically trapped bead serves as a handle that can be used to grab onto a specific protein and then drag it through the lipid membrane. By blocking the laser beam, the bead is suddenly released from the tweezers and its elastic recovery in position can then be monitored by video microscopy [27] (see **PAPER 2.4**). Such experiments have allowed interactions between specific membrane proteins and other cell constituents to be probed by mechanical means. The authors of one study [28] used this approach to probe how an integral membrane protein interacted with the underlying cytoskeletal network. They found that it acted as a physical barrier, or fence, that tended to corral proteins into localized regions or microdomains.

Because biological molecules (e.g., DNA, proteins, and lipid membranes) are less than 25 nm in diameter, individual molecules are too small to manipulate directly with light as they simply do not interact with a sufficient flux of photons. Instead, workers again couple the molecules to a glass or polystyrene microsphere that acts as a microscopic "handle" with which to manipulate associated molecules or cellular structures [3,4,29,30]. The reproducible size and spherical shape of synthetic microspheres allows easy calibration of the optical tweezers system so that quantitative measurements of force and movement can be made. The use of synthetic microspheres or beads has been adopted for many studies of molecular mechanics and biomolecular interactions [3]. Latex particles (refractive index, $n = 1.6$) are commercially available with a wide variety of surface chemistries and beads of 1-μm diameter are readily visualized by fluorescence or bright-field optical microscopy, making them the "bead of choice." The general approach is to adhere purified molecules to the bead surface by specific coupling chemistries or by nonspecific adsorption. The bead is then readily manipulated using the optical tweezers, and mechanical (e.g., binding) interactions between the surface-bound molecule and a fixed substratum can be probed. At low molecular surface densities (e.g., 1 to 10 molecules per bead), the incidence of binding events is governed by Poissonian statistics.* So by scoring the number of events observed at different molecule-to-bead coupling ratios, the number of molecules required to give an individual response can be deduced. Simple dilution takes us to a regime where single-molecule interactions can be observed and recorded.

Over the past 10 to 15 years, optical tweezers have revolutionized biology by spearheading the new field of single-molecule research and they remain one of the most important tools in the armory of single-molecule techniques. Their use has revealed important information about molecular structure and mechanisms, and their biggest impact to date has been in studies of molecular motors—the biological machines that convert chemical potential energy into mechanical work. These studies include work on bacterial flagellar motors [31–33], kinesin [34] (see **PAPER 2.5**) and [35], myosin [36–39] (see [37] **PAPER 2.6**), DNA processing enzymes [40–42], the mechanical properties of DNA [43] (see **PAPER 2.7**), and the folding of protein tertiary structure [44]. Molecular motors perform highly diverse functions in biology, from packaging RNA into viral capsids; copying, transcribing, and translating DNA; causing intracellular movements in plants, animals, and fungi; and making protozoa, bacteria, and sperm swim or glide and enabling entire animals to move around. Most molecular motors are built of protein but some have RNA as a major structural component (e.g., the ribosome). They come in two varieties: rotary motors and linear motors. Rotary motors are driven by the electrochemical gradients present across biological lipid membranes. They are embedded within the membrane and work like miniature turbines, coupling the flow of ions to drive their rotation. Linear motors produce their driving force from changes in chemical potential arising from the hydrolysis of adenosine triphosphate (ATP) to adenosine diphosphate + inorganic phosphate

(ADP+Pi). They move along filamentous tracks made of either DNA, RNA, or the cytoskeletal proteins, actin, and microtubules. Kinesins move on microtubules, myosins move along actin filaments and there are simply countless different motors that move on DNA and RNA. Linear motors have a force-producing "motor" region that contains the active site responsible for ATP hydrolysis, a "binding site" for the cognate filament track, and a "neck" or "tail" region that can amplify the motor movement and also attach to specific cargos to be transported around the cell.

Optical tweezers have given massive insight into the way different motor proteins work. In particular, by enabling measurement of the minute forces (pN), movements (nm), and torques (pN.nm) that they produce [33], [34] (see **PAPER 2.5**), [37] (see **PAPER 2.6**). Many of these high-resolution mechanical studies have required specialized apparatus to be constructed, and often new experimental findings have been driven by technological advances in tweezers design or implementation. In recent years, computer control of the apparatus has allowed new experimental modalities to be realized simply by modifying computer software. We will now highlight some of the discoveries that have advanced our understanding of molecular mechanisms in biology over the past 20 years.

The critical issue in making calibrated measurements is that the thermal motion of optically trapped particles has an amplitude and bandwidth governed by trap stiffness.[†] At commonly used trap stiffnesses (0.1 to 0.01 pN.nm^{-1}), the bandwidth, f_c, is about 100–500 Hz so standard video imaging frame rates ($f_{sample} = 25$–30 Hz) do not satisfy the Shannon–Nyquist sampling criterion (ideally $f_{sample} \geq 2f_c$). This means that if video images are used to localize particle position, data are undersampled. Also, because cameras integrate optical signals over each frame, consecutive images undergo temporal, low-pass filtering. Therefore, dynamic information is lost and the amplitude of thermal noise is poorly determined. Finally, one should recall that in the early 1990s it was a major technical challenge to capture, store, and analyze large quantities of video-format data using a personal computer. Early studies focused on measuring DC signals, like the maximum force that a motor could generate. To do this, the laser power was gradually turned down until the molecular motor attached to the particle

---

* The number of observed interactions, N_{obs}, between an optically trapped bead that has been coupled to a species of biological molecule at a bead-to-molecule molar ratio, a, depends upon the proportion of active molecules, λ (a fitting parameter), and the number, b, that are required to produce each observed interaction:

$$N_{obs} = 1 - e^{\frac{-(a^b)}{\lambda}}$$

---

† Motion of a spherical, optically trapped particle of radius $r = 500$ nm held in water of viscosity $\eta = 0.001$ N.s.m^{-2} (Stokes drag = β) by an optical tweezer of stiffness $\kappa = 0.02$ pN.nm^{-1} is overdamped. The power spectral density of its thermal motion is Lorenzian with characteristic roll-off frequency, $f_c = 300$ Hz.

$$f_c = \frac{\kappa}{2\pi\beta} \quad \text{where} \quad \beta = 6\pi\eta r$$

generated sufficient force to escape the trap, called the "escape force" [45]. Escape forces could be calibrated by flowing solution past the optically trapped particle at different velocities, v, until the viscous drag force, βv (see footnote 4) pulled it out of the trap. However, there was a pressing need to make time-resolved measurements on motor proteins, and for this reason use of split photodiode detectors revolutionized the field because they enabled high-bandwidth (up to 100 kHz) readout of position and easy logging of x,y coordinates data via analogue-to-digital converters. The mechanical stability of the microscope system could then be easily characterized, optical tweezers stiffness accurately calibrated, and particle position correctly determined over the full bandwidth of its thermal motion.

One of the first quantitative studies using optical tweezers in biology [32] focused on understanding how the bacterial flagellar motor works. Many bacteria bear one or more "whiplike" flagellae, just a few nanometers in diameter but around 5 µm long, that project from the surface of the bacterium out into the surrounding medium. The flagellum is rotated about its base by a rotary motor made of a protein complex that is powered by the flow of ions down a chemiosmotic gradient. The flagellum, composed of the protein flagellin, has a helical shape and when rotated at its base (at up to 1000 Hz) generates thrust like an Archimedes screw. This drives the bacterium through solution at up to 50 µm.s^{-1}. How this biological motor (just 45 nm across) works is of great interest to biologists, physicists, engineers, and nanotechnologists. It is a molecular machine driven by an ionic flux (of H$^+$ or Na$^+$) that somehow produces torque as ions pass through two structural regions of the motor called the rotor and stator. The first calibrated measurements of force and movement made using optical tweezers were of the compliance of the flagellar motor [32]. In this early study, the tweezers were used to grab and forcibly rotate a bacterium that was tethered to a microscope coverslip by its flagellum. Forces were calibrated by measuring the time constant of elastic recoil of the bacterium in the viscous medium. This study paved the way to making better calibrated measurements, using optical tweezers as an ultrasensitive force transducer. Early studies showed that if sufficient force were applied to the motor it would stop and even reverse its motion [31]. Over the years, the greatest single challenge has been to measure the expected stepwise motion of the rotor subunit as individual ions are admitted through the system. It was hoped that the tweezers stiffness might reduce thermal motion to the point where angular steps might be resolved above thermal noise. An exciting recent result was reported in a landmark study [33] that showed discrete, stepwise motions, 26 steps/revolution, with each step driven by the flux of about 100 protons.

Linear motors produce force and movement via the cyclical interaction between the motor and its cognate track coupled to the breakdown of ATP. During the chemical cycle, the motor binds to the track, changes shape (producing a small movement), and then detaches. While detached, motor shape is reset so when it rebinds it can work again. Each interaction between motor and track consumes one ATP molecule by converting it to ADP+Pi. Optical tweezers have allowed us to measure the minute forces and movements produced as kinesin and myosin motors move on their cognate tracks [34] (see **PAPER 2.5**), [37] (see **PAPER 2.6**). These micromechanical studies address important questions such as: How much force and movement are produced as a single ATP molecule is hydrolyzed? How are the chemical and mechanical cycles coupled together? How do two-headed motors move in a coordinated fashion to "walk" processively along their track? How do the kinetics of force production depend upon external load? Which parts of the motor are important for producing movement? And, how do some motors move in the opposite direction to others?

The first motor system to be studied in detail was the kinesin-microtubule motor responsible for transporting vesicles within nerve cells. These carry membrane-bound bags of neurotransmitters from the spine all the way to the tips of the fingers. Long-range transport over a distance of a meter or so would take many thousands of years to accomplish if left to passive diffusion, whereas an ATP-powered kinesin molecule can make the journey in about 2 days. We know that kinesin moves along the microtubules inside the cell, but the question is: how does it work? Does it shuffle along the microtubule using the energy from ATP to rectify Brownian motion, or does it actually "walk" in a deterministic fashion, moving in discrete steps from one microtubule binding site to the next?

Following work that showed kinesin operated as a single molecule [46], Block and others [47], [34] (see **PAPER 2.5**) surmised that optical tweezers might reduce thermal motion to the point where individual kinesin steps (if they existed) should become apparent. They bound kinesin to a single, optically trapped bead and carefully lowered it onto a microtubule visualized using differential interference contrast (DIC) microscopy. The microtubule was bound firmly to the microscope coverslip and immersed in a buffered salt solution containing various ions and the chemical fuel ATP. The laser light was captured after passing though the bead and sent to a photodiode detector. The DIC optics of the microscope used a Wollaston prism to produce two adjacent spots of orthogonally polarized light just a few tens of nanometers apart along the axis of the Wollaston prism shear. The two beams passed through the bead and were scattered to slightly different extents depending upon the position of the bead along the shear axis. By collecting the exiting light and separating it into two orthogonal polarizations, the relative intensity of the two beams could be measured to estimate the displacement along the shear axis. Svoboda and colleagues [34] (see **PAPER 2.5**) found that by reducing the concentration of ATP to just a few micromolars, bead movement became slow and the individual, jerky motions produced by the kinesin became visible. These individual 8-nm steps correspond exactly to the repeat distance of the microtubule lattice and imply that a kinesin moves from one binding site to the next as it advances along the microtubule.

Since this early groundbreaking study, over 50 papers have been published in which the mechanical properties of kinesins have been probed using optical tweezers. One important technical advance was to measure the force–velocity relationship by applying a force-feedback clamp in which displacement of the bead from the trap center was kept constant [48]. A recent publication in the field has pushed the sensitivity of the method to the point where stepping at physiological concentrations of ATP is observed and the time-course of individual steps can be investigated with 50 microsecond time resolution. In this study, variable forces were applied using a fully computer-controlled apparatus so that backward stepping (e.g., reversal of the motor) under extremes of load could be studied [35].

The acto-myosin motor system is responsible for muscle contraction and a host of other cellular motilities. Myosin works by binding to actin, pulling, and then releasing. Each chemical cycle produces a few nanometers of movement and a few piconewtons of force. The major experimental challenge with this system is that interactions between myosin and actin are intermittent. So the flexible actin filament must be held in a fixed position, close to an individual myosin, to enable repeated interactions to be observed. Actin is a flexible polymer so it must be pulled out straight between two optically trapped microspheres using two independently controlled optical tweezers.

Different laboratories working on this problem came up with somewhat different solutions. Finer et al. [36] and Guilford et al. [49] used a circularly polarized laser source that was split into two separate light paths using a polarizing beam splitter. One of the light beams passed through a dual-axis acousto-optic deflector (AOD) so that its angular deflection (hence, *x,y* position at the object plane in the microscope) could be rapidly controlled. The other beam of light was deflected using motorized mirrors. The beams were recombined using a second polarizing beam splitter just before they entered the microscope light path. The other approach, adopted by Molloy et al. [37] (see **PAPER 2.6**), was to multiplex a single laser beam between two different positions using a dual-axis AOD. This group found that by chopping a

single laser beam rapidly between two sets of *x,y* coordinates (at 20 kHz), two separate optical traps could be synthesized and their positions independently controlled by sending suitable command signals to the AOD electronics. In all these experimental setups, a four-quadrant photodiode detector was used to monitor bead position using bright-field illumination. All of the groups also incorporated a position feedback servo loop giving the instruments two basic modes of operation: a free-running, open-loop mode whereby bead position was allowed to fluctuate within the optical trap, and a closed-loop feedback mode in which bead position was clamped. When the feedback loop was switched on, the effective stiffness of the optical tweezers was increased by about 2–3 orders of magnitude and the amplitude of thermal motion was drastically reduced as the laser was moved to compensate for bead motion.

To make the recordings, a single actin filament was adhered at either end to optically trapped beads that had been surface-coated with a chemically modified form of myosin that bound tightly, and almost irreversibly, to actin. The actin filament was then pulled taut by moving the two tweezers apart and positioned immediately above a third surface-attached bead to which active myosin molecules had been attached. This is called the "three-bead geometry." In the presence of ATP, myosin underwent repeated cycles of interaction with the actin filament, producing square wavelike displacements and forces. Analysis of these optical trapping data sets enabled both the size and timing of the molecular motions to be deduced. The maximum movement produced by myosin working under low load was 5 nm and the peak force was around 9 pN [50].

There are now known to be at least 18 different families of myosin and the human genome has 39 different myosin genes. The three-bead geometry described above has been widely adopted in studies of these different myosin types. Some myosins act alone within the cell while others are coupled together and work in teams to generate very large forces. It was discovered that one type, called myosin V (five), walks processively along actin [39,51] taking very large, 36-nm steps. The two neck regions of myosin V are long enough to span the gap between target actin binding sites that are found to arise every 36 nm along the actin helical structure [52]. A single-headed version of this protein was found to produce only 25 nm of movement for each ATP hydrolyzed. It was proposed that the "missing distance" required to reach the next actin binding site was produced by a rectified Brownian search of the actin filament lattice [39]. Several studies have been made in which load has been applied to this motor as it moves on actin [53–55] and these give insights on how the chemical cycles of the two myosin heads might be coupled by internal (intramolecular) strain.

In order to address the issue of which part of the myosin molecule generates movement, recombinant protein technology was used to create myosins with different neck lengths. The movement generated by different constructed molecules was found to be directly proportional to neck length [56,57]. This argues strongly for the idea that this part of the molecule acts like a simple lever to amplify motions deep within the motor domain.

In order to understand the detailed mechanochemistry of the myosin cycle, it became necessary to increase the time resolution of the measurements. Many of the critical chemical events arise on the millisecond timescale. The time response of the optical tweezers mechanical measurements (of force and distance) is limited by the relaxation time of the bead within the optical potential. In simple terms, response time can be made faster either by reducing the bead size (lowering viscous drag) or by increasing the tweezers' stiffness. We can see that simply reducing bead radius, r, does not necessarily improve time resolution because viscous drag, β, scales directly with radius (see footnote 4) whereas tweezers stiffness scales with r^2 [8] (see **PAPER 3.3**). Another technique to improve time resolution relies on applying a high-frequency sinusoidal carrier oscillation to one of the optical tweezers that is picked up by motion at the other bead. Using a digital lock-in detector, very small and rapid movements seen in individual binding events could be synchronized with submillisecond resolution and averaged to give nanometer displacement resolution. Using this technique, it was discovered that the myosin motor produces a small structural rearrangement just before it releases the "spent fuel," ADP, from its catalytic site [38]. This finding has since been confirmed in studies of other myosin types [58,59]. These high-resolution studies give critical information about how chemical and mechanical cycles are coupled and help us build a picture of how biological motors act to generate force using just one molecule of fuel at a time.

Following successes made in measuring forces and movements from cytoskeletal molecular motors, workers then turned their attention to studying the properties of DNA and DNA-processing enzymes. Compared to the cytoskeletal filaments (actin and microtubules), DNA is a very flexible polymer. One measure of polymer flexibility is the so-called persistence length (L_p), defined as the minimum separation between two points along a polymer chain at which thermal motions are uncorrelated. The bulk response of the polymer can be modeled as a wormlike chain with end-to-end length (L_o) and persistence length (L_p). The persistence length of actin is 17 μm and this can readily be estimated by viewing the writhing motion of fluorescently labeled single actin filaments in aqueous solution using fluorescence microscopy. The mechanical properties of DNA and RNA are much more difficult to measure because their persistence length, ~ 50 nm, is much shorter than the wavelength of light. The mechanical properties of RNA and DNA are of interest in terms of basic polymer physics and also in a biological context because many RNA/DNA-processing enzymes need to unwind, bend, straighten, or open the backbone strands in order to function. Furthermore, complicated three-dimensional (3D) structures composed of DNA and RNA are of critical importance in gene expression and in RNA and DNA replication, and can also give rise to highly complex enzymatic functions (e.g., the ribosome). Using optical tweezers, the mechanical properties of DNA and RNA can be studied with high precision by simply pulling on either end of the molecule [60] or by dragging it through solution [61]. By exerting very large forces, the classical double-helix structure can be opened out and converted to an extended ladder. The optical tweezers apparatus used by Baumann et al. [43] (see **PAPER 2.7**) was based on one of Ashkin's early designs [6] (see **PAPER 1.2**) using two counterpropagating laser beams to produce a very stiff optical trap. This format enables large forces to be generated so that very high solution flow rates can be achieved without losing beads from the tweezers. By flowing solution past an optically trapped bead to which DNA has been attached, the DNA streams out in the flow field. The far end of the DNA was grabbed by a stiff glass microneedle that acted as a mechanical ground. Calibrated tensions were applied by moving the tweezers and measuring the bead displacement from its central resting position. This allowed the force-extension property of DNA to be determined all the way from the low-force regime (0–50 pN) where it behaves as a wormlike chain, through the intermediate-force (50–100 pN) regime where the double-helix structure is destroyed, to the high-force regime (> 100 pN) where the DNA ultimately breaks. More recently, the mechanical properties of DNA stem loops or "hairpins" have been studied by pulling them apart using tweezers [62–65]. These studies have enabled the free-energy landscape of DNA/RNA superstructures caused by base-pairing and the effect of base-pair mismatch on hairpin stability to be investigated.

Given our ability to manipulate DNA, it was not long before attention turned to the enzymes that move along it, performing complex functions such as copying it into RNA [40,41,66–69]. One of the most important events in biology is the process by which DNA is transcribed into RNA within the cell nucleus. Transcription is conducted by an enzyme (or enzyme complex) called RNA polymerase (RNApol). RNApol binds onto a specific region of DNA called the promoter and then moves along it, opening the DNA double helix and copying the genetic code into a newly synthesized strand of RNA. When RNApol reaches a sequence called a stop codon, transcription ends and the nascent RNA chain is released. In the living cell, RNA moves out of the nucleus and into the cytoplasm where it is translated into protein by the ribosome [70]. RNApol is a model system for studying the protein–DNA and protein–RNA interactions, and optical tweezers experiments herald a completely new era in understanding molecular biology at a mechanical level.

To date, RNApol has been studied by several groups using various trapping geometries, including the counterpropagating trap plus needle system [67], the single-beam gradient trap [41], and the dual-trap, "three bead" geometry [40]. In these studies, the most dramatic observation is that RNApol has an ATP-driven motor that pulls it along the DNA strand with a force of up to 25 pN. This force is five times greater than that of the cytoskeletal motors kinesin and myosin. More detailed analysis of the process reveals that RNApol makes several abortive binding attempts before it properly engages and moves along DNA. Its velocity depends upon the availability of RNA nucleosides and also on the concentration of ATP fuel for the motor. The most recent optical-tweezers-based studies have been made with remarkable angstrom (0.1 nm) precision [42,71,72]. These investigations showed that the enzyme probably moves along one DNA base at a time, hydrolyzing a new ATP for each 0.34 nm step along the DNA helix. The distance moved per ATP hydrolyzed is 10 times smaller than for kinesin and myosin, and this perhaps explains how it is capable of generating such large forces as it moves (see footnote 2).

Another important challenge in biology is to understand how an amino acid sequence folds into a functional 3D protein structure. If we understand the basic rules of how a polypeptide folds into a functional protein, then one day we might be able to simply compute structure and biological function from genomic data. One approach has been to study the unfolding and refolding pathway of a polypeptide using optical tweezers. The first optical-tweezers-based protein unfolding studies were made on the repetitive folded domains of the giant protein titin (one of the largest known proteins at 2 MDa) [44,73]. Titin is found in muscle fibers, where it spans the length of the sarcomere and is thought to form a scaffold network to aid assembly of myosin filaments. It is responsible for the elasticity of the resting muscle fiber and consists of tandem immunoglobulin-like (Ig) and fibronectin-like domains. The fact that it is a 2-µm long, highly repetitive modular structure makes it ideal for mechanical unfolding studies. From an experimental standpoint, it is ideal because one might expect to see a characteristic mechanical "fingerprint" as each of the domains unfolds. Using antibodies specific to either end of the titin molecule, it was attached by one end to a microscope coverslip and at the other to a plastic microsphere. By capturing the bead in optical tweezers, loads could be applied by moving the tweezers laterally. As the protein was put under increasing tension, the Ig domains sequentially unfolded, leaving behind disordered polypeptides that behaved as a

wormlike chain. The succession of unfolding events gave force-extension diagrams that had a sawtooth appearance. The force required to unfold a single Ig domain could be determined accurately and the end-to-end length of the unfolded polypeptide chain could also be measured. Most excitingly, by returning the polypeptide to its starting length and waiting for a period of time, it was discovered that the domains refolded so that the unfolding path could be studied again [44].

In this chapter, we have described some of the experimental systems that have been studied by optical tweezers. Their major impact in biology so far has been in the area of single-molecule research because optical tweezers allow us to apply and measure forces and displacements at a molecular scale. The fact that we can adjust the mechanical stiffness of optical tweezers over a range that spans that of biological molecules makes them suitable to perform studies that are currently inaccessible to related techniques like atomic force microscopy (AFM). Many processes in biology are driven by energies close to thermal noise, including the action of molecular motors, DNA and RNA folding, protein folding, and protein–protein and protein–ligand interactions. In the future, optical tweezers will undoubtedly continue to make significant contributions to the study of molecular biophysics and we expect that they will be increasingly important in whole cell manipulation.

Endnotes

1. A. Ashkin, J.M. Dziedzic, J.E. Bjorkholm, and S. Chu, Observation of a single-beam gradient force optical trap for dielectric particles, *Opt. Lett.* **11**, 288–290 (1986).
2. M.P. Sheetz, ed. Laser tweezers in cell biology, *Methods in Cell Biology*, ed. L. Wilson and P. Matsudaira, vol. 55, San Diego, Academic Press (1998), 223.
3. A.D. Mehta, M. Rief, J.A. Spudich, D.A. Smith, and R.M. Simmons, Single-molecule biomechanics with optical methods, *Science* **283**, 1689–1695 (1999).
4. J.E. Molloy and M.J. Padgett, Lights, action: Optical tweezers, *Contemp. Phys.* **43**, 241–258 (2002).
5. K. Svoboda and S.M. Block, Biological applications of optical forces, *Annu. Rev. Biophys. Biomolec. Struct.* **23**, 247–285 (1994).
6. A. Ashkin, Acceleration and trapping of particles by radiation pressure, *Phys. Rev. Lett.* **24**, 156–159 (1970).
7. K.C. Neuman, E.H. Chadd, G.F. Liou, K. Bergman, and S.M. Block, Characterization of photodamage to *Escherichia coli* in optical traps, *Biophys. J.* **77**, 2856–2863 (1999).
8. R.M. Simmons, J.T. Finer, S. Chu, and J.A. Spudich, Quantitative measurements of force and displacement using an optical trap, *Biophys. J.* **70**, 1813–1822 (1996).
9. A. Ashkin and J.M. Dziedzic, Optical trapping and manipulation of viruses and bacteria, *Science* **235**, 1517–1520 (1987).

10. A. Ashkin, J.M. Dziedzic, and T. Yamane, Optical trapping and manipulation of single cells using infrared-laser beams, *Nature* **330**, 769–771 (1987).

11. A. Ashkin and J.M. Dziedzic, Internal cell manipulation using infrared laser traps, *Proc. Natl. Acad. Sci. U.S.A.* **86**, 7914–7918 (1989).

12. A. Ashkin, K. Schutze, J.M. Dziedzic, U. Euteneuer, and M. Schliwa, Force generation of organelle transport measured *in vivo* by an infrared laser trap, *Nature* **348**, 346–348 (1990).

13. A. Ashkin and J.M. Dziedzic, Optical trapping and manipulation of single living cells using infrared-laser beams, *Berichte der Bunsen Gesellschaft fur Physikalische Chemie* **93**, 254–260 (1989).

14. E.J.G. Peterman, F. Gittes, and C.F. Schmidt, Laser-induced heating in optical traps, *Biophys. J.* **84**, 1308–1316 (2003).

15. K.O. Greulich, *Micromanipulation by Light in Biology and Medicine: The Laser Microbeam and Optical Tweezers*, Basel, Birkhauser Verlag (1999), 300.

16. L.P. Ghislain and W.W. Webb, Scanning-force microscope based on an optical trap, *Opt. Lett.* **18**, 1678–1680 (1993).

17. E.L. Florin, A. Pralle, J.K.H. Horber, and E.H.K. Stelzer, Photonic force microscope based on optical tweezers and two-photon excitation for biological applications, *J. Struct Biol.* **119**, 202–211 (1997).

18. M.P. MacDonald, G.C. Spalding, and K. Dholakia, Microfluidic sorting in an optical lattice, *Nature* **426**, 421–424 (2003).

19. P. Jordan, J. Leach, M. Padgett, P. Blackburn, N. Isaacs, M. Goksor, D. Hanstorp, A. Wright, J. Girkin, and J. Cooper, Creating permanent 3D arrangements of isolated cells using holographic optical tweezers, *Lab Chip* **5**, 1224–1228 (2005).

20. J. Guck, R. Ananthakrishnan, H. Mahmood, T.J. Moon, C.C. Cunningham, and J. Kas, The optical stretcher: A novel laser tool to micromanipulate cells, *Biophys. J.* **81**, 767–784 (2001).

21. K. Svoboda, C.F. Schmidt, D. Branton, and S.M. Block, Conformation and elasticity of the isolated red blood cell membrane skeleton, *Biophys. J.* **63**, 784–793 (1992).

22. J. Sleep, D. Wilson, R. Simmons, and W. Gratzer, Elasticity of the red cell membrane and its relation to hemolytic disorders: An optical tweezers study, *Biophys. J.* **77**, 3085–3095 (1999).

23. J.W. Dai and M.P. Sheetz, Mechanical properties of neuronal growth cone membranes studied by tether formation with laser optical tweezers, *Biophys. J.* **68**, 988–996 (1995).

24. J.W. Dai and M.P. Sheetz, Membrane tether formation from blebbing cells, *Biophys. J.* **77**, 3363–3370 (1999).

25. D. Raucher, T. Stauffer, W. Chen, K. Shen, S.L. Guo, J.D. York, M.P. Sheetz, and T. Meyer, Phosphatidylinositol 4,5-bisphoshate functions as a second messenger that regulates cytoskeleton-plasma membrane adhesion, *Cell* **100**, 221–228 (2000).

26. D. Choquet, D.P. Felsenfeld, and M.P. Sheetz, Extracellular matrix rigidity causes strengthening of integrin-cytoskeleton linkages, *Cell* **88**, 39–48 (1997).

27. M. Edidin, S.C. Kuo, and M.P. Sheetz, Lateral movements of membrane-glycoproteins restricted by dynamic cytoplasmic barriers, *Science* **254**, 1379–1382 (1991).

28. Y. Sako and A. Kusumi, Barriers for lateral diffusion of transferring receptor in the plasma-membrane as characterized by receptor dragging by laser tweezers—Fence versus tether, *J. Cell Biol.* **129**, 1559–1574 (1995).

29. R.M. Simmons and J.T. Finer, Optical tweezers (Glasperlenspiel-II), *Curr. Biol.* **3**, 309–311 (1993).

30. A.D. Mehta, J.T. Finer, and J.A. Spudich, Reflections of a lucid dreamer: Optical trap design considerations, *Meth. Cell Biol.* **55**, 47–69 (1998).

31. R.M. Berry and H.C. Berg, Absence of a barrier to backwards rotation of the bacterial flagellar motor demonstrated with optical tweezers, *Proc. Natl. Acad. Sci. U.S.A.* **94**, 14433–14437 (1997).

32. S.M. Block, D.F. Blair, and H.C. Berg, Compliance of bacterial flagella measured with optical tweezers, *Nature* **338**, 514–518 (1989).

33. Y. Sowa, A.D. Rowe, M.C. Leake, T. Yakushi, M. Homma, A. Ishijima, and R.M. Berry, Direct observation of steps in rotation of the bacterial flagellar motor, *Nature* **437**, 916–919 (2005).

34. K. Svoboda, C.F. Schmidt, B.J. Schnapp, and S.M. Block, Direct observation of kinesin stepping by optical trapping interferometry, *Nature* **365**, 721–727 (1993).

35. N.J. Carter and R.A. Cross, Mechanics of the kinesin step, *Nature* **435**, 308–312 (2005).

36. J.T. Finer, R.M. Simmons, and J.A. Spudich, Single myosin molecule mechanics: Piconewton forces and nanometre steps, *Nature* **368**, 113–118 (1994).

37. J.E. Molloy, J.E. Burns, J. Kendrick-Jones, R.T. Tregear, and D.C.S. White, Movement and force produced by a single myosin head, *Nature* **378**, 209–212 (1995).

38. C. Veigel, L.M. Coluccio, J.D. Jontes, J.C. Sparrow, R.A. Milligan, and J.E. Molloy, The motor protein myosin-I produces its working stroke in two steps, *Nature* **398**, 530–533 (1999).

39. C. Veigel, F. Wang, M.L. Bartoo, J.R. Sellers, and J.E. Molloy, The gated gait of the processive molecular motor, myosin V, *Nat. Cell Biol.* **4**, 59–65 (2002).

40. G.M. Skinner, C.G. Baumann, D.M. Quinn, J.E. Molloy, and J.G. Hoggett, Promoter binding, initiation, and elongation by bacteriophage T7 RNA polymerase, *J. Biol. Chem.* **279**, 3239–3244 (2004).

41. H. Yin, M.D. Wang, K. Svoboda, R. Landick, S.M. Block, and J. Gelles, Transcription against an applied force, *Science* **270**, 1653–1657 (1995).

42. K.M. Herbert, A. La Porta, B.J. Wong, R.A. Mooney, K.C. Neuman, R. Landick, and S.M. Block, Sequence-resolved detection of pausing by single RNA polymerase molecules, *Cell* **125**, 1083–1094 (2006).

43. C.G. Baumann, S.B. Smith, V.A. Bloomfield, and C. Bustamante, Ionic effects on the elasticity of single DNA molecules, *Proc. Natl. Acad. Sci. U.S.A.* **94**, 6185–6190 (1997).

44. L. Tskhovrebova, J. Trinick, J.A. Sleep, and R.M. Simmons, Elasticity and unfolding of single molecules of the giant muscle protein titin, *Nature* **387**, 308–312 (1997).

45. S.C. Kuo and M.P. Sheetz, Force of single kinesin molecules measured with optical tweezers, *Science* **260**, 232–234 (1993).

46. J. Howard, A.J. Hudspeth, and R.D. Vale, Movement of microtubules by single kinesin molecules, *Nature* **342**, 154–158 (1989).

47. S.M. Block, L.S.B. Goldstein, and B.J. Schnapp, Bead movement by single kinesin molecules studied with optical tweezers, *Nature* **348**, 348–352 (1990).

48. K. Visscher, M.J. Schnitzer, and S.M. Block, Single kinesin molecules studied with a molecular force clamp, *Nature* **400**, 184–189 (1999).

49. W.H. Guilford, D.E. Dupuis, G. Kennedy, J.R. Wu, J.B. Patlak, and D.M. Warshaw, Smooth muscle and skeletal muscle myosins produce similar unitary forces and displacements in the laser trap, *Biophys. J.* **72**, 1006–1021 (1997).

50. T. Nishizaka, H. Miyata, H. Yoshikawa, S. Ishiwata, and K. Kinosita, Unbinding force of a single motor molecule of muscle measured using optical tweezers, *Nature* **377**, 251–254 (1995).

51. A.D. Mehta, R.S. Rock, M. Rief, J.A. Spudich, M.S. Mooseker, and R.E. Cheney, Myosin-V is a processive actin-based motor, *Nature* **400**, 590–593 (1999).

52. W. Steffen, D. Smith, R. Simmons, and J. Sleep, Mapping the actin filament with myosin, *Proc. Natl. Acad. Sci. U.S.A.* **98**, 14949–14954 (2001).

53. M. Rief, R.S. Rock, A.D. Mehta, M.S. Mooseker, R.E. Cheney, and J.A. Spudich, Myosin-V stepping kinetics: A molecular model for processivity, *Proc. Natl. Acad. Sci. U.S.A.* **97**, 9482–9486 (2000).

54. S. Uemura, H. Higuchi, A.O. Olivares, E.M. De La Cruz, and S. Ishiwata, Mechanochemical coupling of two substeps in a single myosin V motor, *Nat. Struct. Mol. Biol.* **11**, 877–883 (2004).

55. C. Veigel, S. Schmitz, F. Wang, and J.R. Sellers, Load-dependent kinetics of myosin-V can explain its high processivity, *Nat. Cell Biol.* **7**, 861–869 (2005).

56. C. Ruff, M. Furch, B. Brenner, D.J. Manstein, and E. Meyhofer, Single-molecule tracking of myosins with genetically engineered amplifier domains, *Nat. Struct. Biol.* **8**, 226–229 (2001).

57. T. Sakamoto, S. Schmitz, J.E. Molloy, C. Veigel, F. Wang, and J.R. Sellers, Effect of myosin V neck length on processivity and working stroke, *Mol. Biol. Cell* **13**, 2572 (2002).

58. C. Veigel, J.E. Molloy, S. Schmitz, and J. Kendrick-Jones, Load-dependent kinetics of force production by smooth muscle myosin measured with optical tweezers, *Nat. Cell Biol.* **5**, 980–986 (2003).

59. M. Capitanio, M. Canepari, P. Cacciafesta, V. Lombardi, R. Cicchi, M. Maffei, F.S. Pavone, and R. Bottinelli, Two independent mechanical events in the interaction cycle of skeletal muscle myosin with actin, *Proc. Natl. Acad. Sci. U.S.A.* **103**, 87–92 (2006).

60. S.B. Smith, Y.J. Cui, and C. Bustamante, Overstretching b-DNA: The elastic response of individual double-stranded and single-stranded DNA molecules, *Science* **271**, 795–799 (1996).

61. T.T. Perkins, S.R. Quake, D.E. Smith, and S. Chu, Relaxation of a single DNA molecule observed by optical microscopy, *Science* **264**, 822–826 (1994).

62. J. Liphardt, B. Onoa, S.B. Smith, I. Tinoco, and C. Bustamante, Reversible unfolding of single RNA molecules by mechanical force, *Science* **292**, 733–737 (2001).

63. B. Onoa, S. Dumont, J. Liphardt, S.B. Smith, I. Tinoco, and C. Bustamante, Identifying kinetic barriers to mechanical unfolding of the *T. thermophila* ribozyme, *Science* **299**, 1892–1895 (2003).

64. W.J. Greenleaf, K.L. Frieda, D.A.N. Foster, M.T. Woodside, and S.M. Block, Direct observation of hierarchical folding in single riboswitch aptamers, *Science* **319**, 630–633 (2008).

65. M.T. Woodside, P.C. Anthony, W.M. Behnke-Parks, K. Larizadeh, D. Herschlag, and S.M. Block, Direct measurement of the full, sequence-dependent folding landscape of a nucleic acid, *Science* **314**, 1001–1004 (2006).

66. M.D. Wang, M.J. Schnitzer, H. Yin, R. Landick, J. Gelles, and S.M. Block, Force and velocity measured for single molecules of RNA polymerase, *Science* **282**, 902–907 (1998).

67. R.J. Davenport, G.J.L. Wuite, R. Landick, and C. Bustamante, Single-molecule study of transcriptional pausing and arrest by e-coli RNA polymerase, *Science* **287**, 2497–2500 (2000).

68. G.J.L. Wuite, S.B. Smith, M. Young, D. Keller, and C. Bustamante, Single-molecule studies of the effect of template tension on T7 DNA polymerase activity, *Nature* **404**, 103–106 (2000).

69. J.W. Shaevitz, E.A. Abbondanzieri, R. Landick, and S.M. Block, Backtracking by single RNA polymerase molecules observed at near-base-pair resolution, *Nature* **426**, 684–687 (2003).

70. F. Vanzi, Y. Takagi, H. Shuman, B.S. Cooperman, and Y.E. Goldman, Mechanical studies of single ribosome/mRNA complexes, *Biophys. J.* **89**, 1909–1919 (2005).

71. E.A. Abbondanzieri, W.J. Greenleaf, J.W. Shaevitz, R. Landick, and S.M. Block, Direct observation of base-pair stepping by RNA polymerase, *Nature* **438**, 460–465 (2005).

72. W.J. Greenleaf and S.M. Block, Single-molecule, motion-based DNA sequencing using RNA polymerase, *Science* **313**, 801 (2006).

73. M.S.Z. Kellermayer, S.B. Smith, H.L. Granzier, and C. Bustamante, Folding-unfolding transitions in single titin molecules characterized with laser tweezers, *Science* **276**, 1112–1116 (1997).

Microfluidic sorting in an optical lattice

M. P. MacDonald[1], **G. C. Spalding**[1,2] **& K. Dholakia**[1]

[1]*School of Physics and Astronomy, University of St Andrews, North Haugh, St Andrews, Fife KY16 9SS, UK*
[2]*Department of Physics, Illinois Wesleyan University, Bloomington, Illinois 61702, USA*

The response of a microscopic dielectric object to an applied light field can profoundly affect its kinetic motion[1]. A classic example of this is an optical trap, which can hold a particle in a tightly focused light beam[2]. Optical fields can also be used to arrange, guide or deflect particles in appropriate light-field geometries[3,4]. Here we demonstrate an optical sorter for microscopic particles that exploits the interaction of particles—biological or otherwise—with an extended, interlinked, dynamically reconfigurable, three-dimensional optical lattice. The strength of this interaction with the lattice sites depends on the optical polarizability of the particles, giving tunable selection criteria. We demonstrate both sorting by size (of protein microcapsule drug delivery agents) and sorting by refractive index (of other colloidal particle streams). The sorting efficiency of this method approaches 100%, with values of 96% or more observed even for concentrated solutions with throughputs exceeding those reported for fluorescence-activated cell sorting[5]. This powerful, non-invasive technique is suited to sorting and fractionation within integrated

letters to nature

('lab-on-a-chip') microfluidic systems, and can be applied in colloidal, molecular and biological research.

In microfluidics, flow is predominantly laminar, creating challenges in the design of actuators such as mixers and sorters. The extremely low Reynolds numbers associated with microfluidic flows (for example, 10^{-3}) mean that inertial effects cannot be used to sort particles, yet the ability to select and sort both colloidal and biological matter—in a fast and efficient manner related to its physical properties—is a key requirement in this environment. Existing techniques[6,7] for separation of macro-molecules typically require batch processing, while the most novel alternatives are reliant upon diffusion[8–10]. Such approaches cannot separate objects as large as cells or other colloidal suspensions. The current standard for separating such suspensions, fluorescence-activated cell sorting (FACS), requires fluorescence to distinguish between cell types[11]. For particles that do not fluoresce, sedimentation[12] can be used but this gives low resolution and is not suitable for small volumes of analyte. Where small volumes are involved the trend is towards integrating devices into micro-total analysis systems, or 'lab-on-a-chip'[5]. Here we introduce an effective, all-optical, dynamically reconfigurable means of sorting in a microfluidic sample cell. This powerful yet simple technique is non-invasive, sorting particles without physical contact. Sole use of optical forces simplifies surface interaction and sterility issues by removing the extremely high surface area associated with any physical sieve or gel.

Recent research into sculpting optical wavefronts and intensity profiles has led to enhanced levels of control over microscopic matter and biological particles[13]. Interferometric patterns of light can create one-, two- or three-dimensional optical lattices, enabling numerous studies in atomic and colloidal science[4,13,14]. Here we exploit the influence of optical lattices on matter to enforce the strong lateral separation required for continuous processing while maintaining compatibility with the full mesoscopic range of particle sizes (up to the scale of ~100 μm). The interaction between optical lattices and matter depends upon a relative polarizability that grows with the size of the particle yet remains effective down to atomic scales[15]. Selected particle types follow defined paths through an optical lattice, providing lateral fractionation. Importantly, the optical lattice is three-dimensional in nature, providing the ability to sort particles throughout a three-dimensional flow.

A two-dimensional array of traps has been used recently[16] to explore the incursion—from all in-plane angles simultaneously—of colloid into the optical array. This led to a study[4] of the angular deviation of single, well-isolated (that is, extremely dilute) particle trajectories as a function of a relative rotation between an externally imposed fluid flow and the [100] direction of the trap array. These studies used solely monodisperse colloidal solutions, and thus did not demonstrate fractionation or separation of mixed particle streams, but there was a suggestion that optical arrays might be used for that purpose[4]. A key observation was that this array of traps could provide only weak angular deflection (no more than 11°). Recognizing the limitation of using discrete (point-like), isotropic traps, we instead utilize an extended, three-dimensional optical lattice of interlinked sites, with linkages established along the desired channelling direction. We show that the optimum lattice type for high-angle separation lies between the two extremes of discrete traps and extended guides, where we have strongly interlinked localized intensity maxima.

Here a five-beam interference pattern is used to sort co-streaming particles in microfluidic flows. Figure 1 describes the overall concept for a sorting machine based on optical forces—interaction with optical fields provides a selective means of removing material matching specific criteria from an otherwise laminar stream. When a flow of mixed particles is passed through the lattice, selected particles are strongly deflected from their original trajectories while others pass straight through largely unhindered, depending upon their sensitivity to the optical potential. In our apparatus, a single

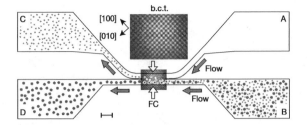

Figure 1 The concept of optical fractionation. Low Reynolds number flows will be laminar: without an actuator all particles from chamber B would flow into chamber D. Chamber A would typically introduce a 'blank' flow stream, although this could be any stream into which the selected particles are to be introduced. By introducing a three-dimensional optical lattice—in this case a body-centred tetragonal (b.c.t.) lattice—into the fractionation chamber (FC), one species of particle is selectively pushed into the upper flow field. The reconfigurability of the optical lattice allows for dynamic updating of selection criteria. For weakly segregated species, the analyte can be either recirculated through the optical lattice or directed through cascaded separation chambers. This latter option also allows the use of multiple selection criteria in a single integrated chip. The flow volume in our current sample cells is 100 μm thick; scale bar, 40 μm.

1,070-nm laser beam (coherence length 1.2 cm) passes through a diffractive beam splitter, producing four beams diverging from the central, undiffracted spot at 4.6°, in a cross shape (Supplementary Fig. A). Collimating optics provide an 'infinity space' in which we have independent control over the phase and amplitude of each of these five beams before co-focusing through an aspheric lens produces a large, three-dimensional optical lattice through multi-beam interference.

First we demonstrate purely chemical separation (Fig. 2); a co-streaming mixture of same-sized (2 μm diameter) silica and polymer spheres passes through our optical lattice. There is 45° angular separation—even at flow speeds in excess of 35 μm s^{-1} (for a laser output of 530 mW and densities as shown). Differences in contrast allow the two, co-flowing species to be separately tracked, via automated particle identification with tenth-pixel resolution[17] (here ~8 nm). We have used this precision to assemble a variety of statistics (such as the size and location distributions of particle displacements[18]) for a number of polydisperse systems passing through the lattice.

One measure of efficiency, appropriate for two-species flows (Fig. 2), is an outcome-based figure of merit (FOM) given by the percentage of species A deflected by the optical lattice minus the percentage of species B deflected. Figure 3 shows the experimental results for this efficiency FOM, as a function of flow speed, taken from statistical analysis of more than 1,200 particle trajectories in a dense flow of mixed 2-μm polymer and silica spheres. As the flow speed increases, we see changes in the particle displacement statistics[18], from an exponential distribution at low speeds (characteristic of hopping transport) to a gaussian distribution (reflecting guided flow, which can be further tuned from pseudo-ballistic to diffusively guided regimes and—finally—to the limit where all particles flow past unhindered). Between the regimes of hopping and free flow, we observe selective dynamics[18] and, accordingly, the value of the FOM rises, approaching 100% in the semi-dilute limit, with values of 96% or more sustained even for concentrated solutions as shown in Fig. 3 (an earlier proposal to use only a single laser beam[19] achieved only a fraction of one per cent in efficiency). Three mechanisms account for reductions from 100% efficiency: if the polydisperse flow is not well covered by the optical lattice, if the flow is beyond the semi-dilute limit (particularly if large aggregates are present), or if the flow rate is not optimized for the laser power (which amounts to a balance between kinetic and optical potential energies[18]). The influence of the first two of these mechanisms is kept to a minimum

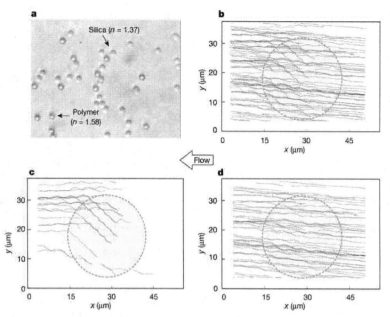

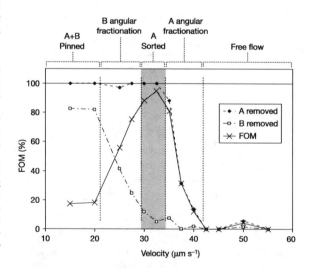

Figure 2 Separation by index of refraction. Same-sized (2 μm diameter) polymer and silica spheres in water co-flow from right to left through the optical lattice at a fluid speed of 30 μm s⁻¹ with a total incident laser power of 530 mW. **a**, An image of the silica/polymer mix, indicating the typical particle density. *n*, index of refraction. **b**, Polymer trajectories are shown in red, and silica trajectories in black, with a green circle indicating the *x–y* range over which the optical lattice is most intense. **c**, The polymer tracks show a deflection in excess of 45° while (**d**) the silica tracks are only slightly modulated by the optical lattice. This very large angular separation is attributed to the linking of intensity maxima along the 100 direction of the optical lattice.

by the use of high-angle deflections in the optical lattice.

Each of these limitations upon efficiency can be rephrased as a limitation upon the overall throughput of the system. Typical throughput of our prototypical system was ~25 particles per second, slightly higher than those reported for separation in a microfabricated FACS. Optical fractionation does not require tagging or endogenous fluorophores. Additionally, optical fractionation will allow highly parallel sorting, with throughputs that will scale with the volume of the sample cell, providing high parallel throughput and ~100% efficiency for deflection of, for example, a selected cell type, with complete exclusion of other types.

Using the set-up described above, and 530 mW of power (measured directly at the laser), a monodisperse (2 μm) collection of protein microcapsules is extracted from a polydisperse particle stream moving at 20 μm s⁻¹ (Fig. 4). These microcapsules are an ultrasound contrast agent that can be used to locally permeate cell membranes (a technique known as sonoporation[20]), providing direct DNA transfection or drug delivery. Using optical fractionation to create a collection of protein microcapsules of a tailored size provides an essential level of control for studies of drug delivery at the cellular level. Practical biological potential (specifically for haematology or immunology) has also been demonstrated by separating erythrocytes (red blood cells) from lymphocytes dispersed in a culture medium (M.P.M., L. Paterson, G.C.S., A. Richies and K.D., manuscript in preparation).

Whereas other technologies have trouble distinguishing non-fluorescent particle sizes that differ by just 20%, optical fractionation offers unparalleled exponential size selectivity: the probability for activation over a potential barrier involves a Boltzmann factor, an exponential dependence upon the ratio of (1) the additional energy required to overcome the barrier to (2) the relevant thermal energy scale, which is set by the temperature. Term (1) is dependent upon particle size, resulting in an exponential sensitivity to size. This result is robustly general, even as we tune the interconnectivity

Figure 3 Typical efficiencies. Our efficiency figure of merit (FOM) is, for the co-flowing mixtures of two species shown in Fig. 2, the percentage of species A (polymer) deflected minus the percentage of species B (silica) deflected, plotted as flow speed increases, or laser power is reduced. At speeds just beyond de-pinning, both species are deflected by the optical lattice. Appropriate choice of particle speeds and laser power yields optimal separation. The data shown are based on more than 1,200 particle trajectories, with a laser output of 530 mW. For each material system, similar statistics are gathered in order to establish acceptable operating parameters (here, for example, the optimal measured flow speed was 32.5 μm s⁻¹). The particles were suspended in water. Use of a more viscous fluid such as a culture medium simply reduces the optimum flow velocity for optical sorting, but does not change the general behaviour of the technique.

letters to nature

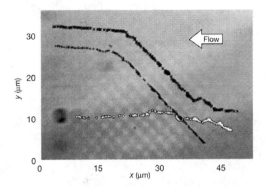

Figure 4 Optical fractionation by size. Black crosses represent the frame-by-frame positions (tracked at standard video rates with ~8 nm precision) of two 2-μm diameter protein microcapsules (drug delivery agents) as they flow from right to left across the optical lattice. The flow speed is 20 μm s^{-1} with a total incident laser power of 530 mW. Again, significant angular deflection is achieved, while a co-flowing 4-μm-diameter capsule of the same sort flows nearly straight through (white dots). The collection of 2-μm capsules allows a controlled study of drug delivery at the cellular level.

of the lattice. Such anisotropy merely changes a prefactor in term (1), for motion along the direction of lowered barrier heights.

We note that it is necessary to move beyond the limit of discrete, isotropic traps to achieve the sorts of concentrated, mixed-flow separations demonstrated here. Rather than using kinetic lock-in of hopping probabilities[4], we provide a more deterministic form of channelling. We use light patterns that contain a tailored degree of 'interconnectedness' (that is, lower potential energy barriers) along the direction of desired channelling[1]. Supplementary Fig. B provides details of the experimental parameters used to tune these linkages, along with detailed mapping of lattice parameters and the linkages themselves. The details of the linkages between lattice sites plays a key role in optical fractionation. Distinct individual lattice points are ideally suited to trapping particles, whereas interlinks between the lattice sites are central to deflecting particles into a desired trajectory. The optimum lattice for fractionation with high angles of displacement lies between a discrete lattice of intensity maxima and continuous guides; we went on to find that the body-centred tetragonal (b.c.t.) lattice, used for the results shown in Figs 2–4, is satisfactory. This is demonstrated clearly in Supplementary Fig. D, where the linked b.c.t. lattice has both the highest efficiency and the highest throughput when fractionating at 45°. More detailed calculations and discussion of this will appear elsewhere[18].

Arrays composed of discrete, isotropic traps tend to 'jam' (which is why studies involving such arrays tend to use extremely dilute flows). The dwell time in strongly localizing traps significantly slows particles with respect to the input stream. Jamming is likely to occur in long arrays of this type, effectively preventing particles that would otherwise pass through un-deflected from doing so, and leading to 'spill-over' contamination of the uptake stream.

Practical fractionating arrays must be long enough to transport the selected particles (transversely) from the most distant portion of the input stream all the way to the uptake stream, where they are extracted. For dense flows, greater length is required, but in general the larger the angle of deflection through the lattice, the shorter the lattice can be, making fractionation at 45° with a linked b.c.t. lattice particularly well suited to practical implementation of optical fractionation. □

Received 1 July; accepted 21 October 2003; doi:10.1038/nature02144.

1. Tatarkova, S. A., Sibbett, W. & Dholakia, K. Brownian particle in an optical potential of the washboard type. *Phys. Rev. Lett.* **91**, 038101 (2003).
2. Ashkin, A., Dziedzic, J. M., Bjorkholm, J. E. & Chu, S. Observation of a single-beam gradient force optical trap for dielectric particles. *Opt. Lett.* **11**, 288–290 (1986).
3. Burns, M. M., Fournier, J. M. & Golovchenko, J. A. Optical matter—crystallization and binding in intense optical-fields. *Science* **249**, 749–754 (1990).
4. Korda, P. T., Taylor, M. B. & Grier, D. G. Kinetically locked-in colloidal transport in an array of optical tweezers. *Phys. Rev. Lett.* **89**, 128301 (2002).
5. Fu, A. Y., Spence, C., Scherer, A., Arnold, F. H. & Quake, S. R. A microfabricated fluorescence-activated cell sorter. *Nature Biotechnol.* **17**, 1109–1111 (1999).
6. Han, J. & Craighead, H. G. Separation of long DNA molecules in a microfabricated entropic trap array. *Science* **288**, 1026–1029 (2000).
7. Nykypanchuk, D., Strey, H. H. & Hoagland, D. A. Brownian motion of DNA confined within a two-dimensional array. *Science* **297**, 987–990 (2002).
8. Ertas, D. Lateral separation of macromolecules and polyelectrolytes in microlithographic arrays. *Phys. Rev. Lett.* **80**, 1548–1551 (1998).
9. Duke, T. A. J. & Austin, R. H. Microfabricated sieve for the continuous sorting of macromolecules. *Phys. Rev. Lett.* **80**, 1552–1555 (1998).
10. Chou, C. F. *et al.* Electrodeless dielectrophoresis of single- and double-stranded DNA. *Biophys. J.* **83**, 2170–2179 (2002).
11. Galbraith, D. W., Anderson, M. T. & Herzenberg, L. A. in *Methods in Cell Biology* Vol. 58 (eds Sullivan, K. F. & Kay, S. A.) 315–341 (Academic, London, 1999).
12. Athanasopoulou, A., Koliadima, A. & Karaiskakis, G. New methodologies of field-flow fractionation for the separation and characterization of dilute colloidal samples. *Instrum. Sci. Technol.* **24**, 79–94 (1996).
13. Dholakia, K., Spalding, G. C. & MacDonald, M. Optical tweezers: The next generation. *Phys. World* **15**, 31–35 (2002).
14. MacDonald, M. P. *et al.* Creation and manipulation of three-dimensional optically trapped structures. *Science* **296**, 1101–1103 (2002).
15. Greiner, M., Mandel, O., Esslinger, T., Hansch, T. W. & Bloch, I. Quantum phase transition from a superfluid to a Mott insulator in a gas of ultracold atoms. *Nature* **415**, 39–44 (2002).
16. Korda, P. T., Spalding, G. C. & Grier, D. G. Evolution of a colloidal critical state in an optical pinning potential landscape. *Phys. Rev. B* **66**, 024504 (2002).
17. Crocker, J. C. & Grier, D. G. Methods of digital video microscopy for colloidal studies. *J. Colloid Interface Sci.* **179**, 298–310 (1996).
18. MacDonald, M. P., Spalding, G. C. & Dholakia, K. Transport and fractionation of brownian particles in an optical lattice. *Phys. Rev. Lett.* (submitted).
19. Imasaka, T., Kawabata, Y., Kaneta, T. & Ishidzu, Y. Optical chromatography. *Anal. Chem.* **67**, 1763–1765 (1995).
20. Marmottant, P. & Hilgenfeldt, S. Controlled vesicle deformation and lysis by single oscillating bubbles. *Nature* **423**, 153–156 (2003).

Supplementary Information accompanies the paper on **www.nature.com/nature**.

Acknowledgements We thank P. Campbell for supplying protein microcapsules, and A. Riches for blood samples. This work was supported by the UK Engineering and Physical Sciences Research Council, the Research Corporation, and the National Science Foundation.

Competing interests statement The authors declare that they have no competing financial interests.

Correspondence and requests for materials should be addressed to M.P.M. (mpm4@st-and.ac.uk).

The Optical Stretcher: A Novel Laser Tool to Micromanipulate Cells

Jochen Guck,* Revathi Ananthakrishnan,* Hamid Mahmood,* Tess J. Moon,[†¶] C. Casey Cunningham,[‡] and Josef Käs[*§¶‖]

*Center for Nonlinear Dynamics, Department of Physics, University of Texas at Austin, Texas 78712, [†]Department of Mechanical Engineering, University of Texas at Austin, Texas 78712, [‡]Baylor University Medical Center, Dallas, Texas 75246, [§]Institute for Molecular and Cellular Biology, University of Texas at Austin, Texas 78712, [¶]Texas Materials Institute, University of Texas at Austin, Texas 78712, [‖]Center for Nano- and Molecular Science and Technology, University of Texas, Austin, Texas 78712 USA

ABSTRACT   When a dielectric object is placed between two opposed, nonfocused laser beams, the total force acting on the object is zero but the surface forces are additive, thus leading to a stretching of the object along the axis of the beams. Using this principle, we have constructed a device, called an optical stretcher, that can be used to measure the viscoelastic properties of dielectric materials, including biologic materials such as cells, with the sensitivity necessary to distinguish even between different individual cytoskeletal phenotypes. We have successfully used the optical stretcher to deform human erythrocytes and mouse fibroblasts. In the optical stretcher, no focusing is required, thus radiation damage is minimized and the surface forces are not limited by the light power. The magnitude of the deforming forces in the optical stretcher thus bridges the gap between optical tweezers and atomic force microscopy for the study of biologic materials.

INTRODUCTION

For almost three decades, laser traps have been used to manipulate objects ranging in size from atoms to cells (Ashkin, 1970; Chu, 1991; Svoboda and Block, 1994). The basic principle of laser traps is that momentum is transferred from the light to the object, which in turn, by Newton's second law, exerts a force on the object. Thus far, these optical forces have solely been used to trap an object. The most common laser trap is a one-beam gradient trap, called optical tweezers (Ashkin et al., 1986). Optical tweezers have been an invaluable tool in cell biological research: for trapping cells (Ashkin et al., 1987; Ashkin and Dziedzic, 1987), measuring forces exerted by molecular motors such as myosin or kinesin (Block et al., 1990; Shepherd et al., 1990; Kuo and Sheetz, 1993; Simmons et al., 1993; Svoboda et al., 1993), or the swimming forces of sperm (Tadir et al., 1990; Colon et al., 1992), and for studying the polymeric properties of single DNA strands (Chu, 1991).

In contrast, the optical stretcher is based on a double-beam trap (Ashkin, 1970; Constable et al., 1993) in which two opposed, slightly divergent, and identical laser beams with Gaussian intensity profile trap an object in the middle. This trapping is stable if the total force on the object is zero and restoring. This condition is fulfilled if the refractive index of the object is larger than the refractive index of the surrounding medium and if the beam sizes are larger than the size of the trapped object. In extended objects such as cells, the momentum transfer primarily occurs at the surface. The total force acting on the center of gravity is zero because the two-beam trap geometry is symmetric and all

the resulting surface forces cancel. Nevertheless, if the object is sufficiently elastic, the surface forces stretch the object along the beam axis (see Fig. 1) (Guck et al., 2000). At first, this optical stretching may seem counterintuitive, but it can be explained in a simple way. It is well known that light carries momentum. Whenever a ray of light is reflected or refracted at an interface between media with different refractive indices, changing direction or velocity, its momentum changes. Because momentum is conserved, some momentum is transferred from the light to the interface and, by Newton's second law, a force is exerted on the interface.

To illustrate, let us consider a ray of light passing through a cube of optically denser material (see Fig. 2). As it enters the dielectric object, the light gains momentum so that the surface gains momentum in the opposite (backward) direction. Similarly, the light loses momentum upon leaving the dielectric object so that the opposite surface gains momentum in the direction of the light propagation. The reflection of light on either surface also leads to momentum transfer on both surfaces in the direction of light propagation. This contribution to the surface forces is smaller than the contribution that stems from the increase of the light's momentum inside the cube. The two resulting surface forces on front and backside are opposite and tend to stretch the object (Guck et al., 2000). However, the asymmetry between the surface forces leads to a total force that acts on the center of the cube. If there is a second, identical ray of light that passes through the cube from the opposite side, there is no total force on the cube, but the forces on the surface generated by the two rays are additive. In contrast to asymmetric trapping geometries, where the total force is the trapping force used in optical traps, the optical stretcher exploits surface forces to stretch objects. Light powers as high as 800 mW in each beam can be used, which lead to surface forces up to hundreds of pico-Newton. There is no problem with radiation damage to the cells examined, which is not surprising because the laser beams in the optical stretcher

Received for publication 15 August 2000 and in final form 6 May 2001.

Address reprint requests to Jochen Guck, University of Texas at Austin, Center for Nonlinear Dyanmics, RLM14.206, 26th and Speedway, Austin, TX 78712. Tel.: 512-475-7647; Fax: 512-471-1558; E-mail: jguck@chaos.ph.utexas.edu.

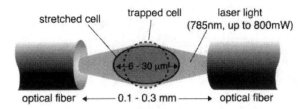

FIGURE 1 Schematic of the stretching of a cell trapped in the optical stretcher. The cell is stably trapped in the middle by the optical forces from the two laser beams. Depending on the elastic strength of the cell, at a certain light power the cell is stretched out along the laser beam axis. The drawing is not to scale; the diameter of the optical fibers is $125 \pm 5\ \mu m$.

are not focused, minimizing the light flux through the cells in comparison to other optical traps (see Viability of Stretched Cells). To demonstrate this concept of optical deformability, we stretched osmotically swollen erythrocytes and BALB 3T3 fibroblasts.

Human erythrocytes, i.e., red blood cells (RBCs), were used as initial test objects. Red blood cells offer several advantages as a model system for this type of experiment in that they lack any internal organelles, are homogeneously filled with hemoglobin, and can be osmotically swollen to a spherical shape. They are thus close to the model of an isotropic, soft, dielectric sphere without internal structure that we used for the calculation of the stress profiles (see below). Furthermore, they are very soft cells and deformations are easily observed. As an additional advantage, RBCs have been studied extensively and their elastic properties are well known (Bennett, 1985, 1990; Mohandas and Evans, 1994). The only elastic component of RBCs is a thin membrane composed of a phospholipid bilayer sandwiched between a triagonal network of spectrin filaments on the inside and glycocalix brushes on the outside (Mohandas and Evans, 1994). The ratio between cell radius ρ and membrane thickness h, $\rho/h \approx 100$. This means that the bending energy is negligibly small compared to the stretching energy (see Deformation of Thin Shells). Thus, linear membrane theory can be used to predict the deformations of RBCs subjected to the surface stresses in the optical stretcher. By comparing the deformations observed in the optical stretcher with the deformations expected, we quantitatively verified the forces predicted from our calculations.

The BALB 3T3 fibroblasts under investigation are an example of typical eukaryotic cells that, in contrast to RBCs, have an extensive three-dimensional (3D) network of protein filaments throughout the cytoplasm as the main elastic component (Lodish et al., 1995). In this network, called the cytoskeleton, semiflexible actin filaments, rod-like microtubules, and flexible intermediate filaments are arranged into an extensive, 3D compound material with the help of accessory proteins (Adelman et al., 1968; Pollard, 1984; Elson, 1988; Janmey, 1991). Classical concepts in polymer physics fail to explain how these filaments provide

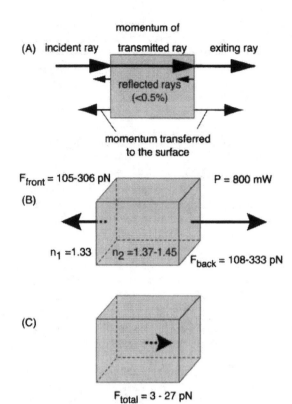

FIGURE 2 Momentum transfer and resulting forces on a dielectric box due to one laser beam incident from the left. (*A*) A small portion of the incident light is reflected at the front surface. The rest enters the box and gains momentum due to the higher refractive index inside. On the back, the same fraction is reflected and the exiting light loses momentum. The lower arrows indicate the momentum transferred to the surface. (*B*) The resulting forces for a light power of 800 mW at the front and the back are $F_{\text{front}} = 105–306$ pN and $F_{\text{back}} = 108–333$ pN, respectively, depending on the refractive index of the material. Note that the force on the back is larger than the force on the front. (*C*) Due to the difference between forces on front and back, there is a total force, $F_{\text{total}} = F_{\text{back}} - F_{\text{front}} = 3 - 27$ pN, acting on the center of gravity of the box. This total force pushes the box away from the light source. An elastic material will be deformed by the forces acting on the surface, which are an order of magnitude larger than the total force.

mechanical stability to cells (MacKintosh et al., 1995), but, in most cells, cytoskeletal actin is certainly a main determinant of mechanical strength and stability (Stossel, 1984, Janmey et al., 1986; Sato et al., 1987; Elson, 1988).

The actin cortex is a thick ($\rho/h \approx 10$) homogeneous layer just beneath the plasma membrane. In cells adhered to the substrate, additional bundles of individual actin filaments, called stress fibers, insert into focal adhesion plaques and span the entire cell interior. Dynamic remodeling of this network of F-actin facilitates such important cell functions as motility and the cytoplasmic cleavage as the last step of mitosis (Pollard, 1986; Carlier, 1998; Stossel et al., 1999). Cells are drastically softened by actin-disrupting cytochalasins (Petersen et al., 1982; Pasternak and Elson, 1985) and gelsolin (Cooper et al., 1987), indicating the importance of actin. More recently, frequency-dependent atomic force microscopy (AFM)-based microrheology showed that fibroblasts exhibit the same viscoelastic signature as homogeneous actin networks in vitro (Mahaffy et al., 2000). Another experiment (Heidemann et al., 1999) investigated the response of rat embryo fibroblasts to mechanical deformation by glass needles. Actin and microtubules were tagged with green fluorescent protein and the role of these two cytoskeletal components in determining cell shape during deformation was directly visualized. Again, actin was found to be almost exclusively responsible for the cell's elastic response, whereas microtubules clearly showed fluid-like behavior.

In nonmitotic cells, microtubules radiate outward from the microtubule-organizing center just outside the cell nucleus (Lodish et al., 1995). They serve as tracks for the motor proteins dynein and kinesin to transport vesicles through the cell. Microtubules are also required for the separation of chromosomes during mitosis (Mitchison et al., 1986; Mitchison, 1992). Intermediate filaments are unique to multicellular organisms and comprise an entire class of flexible polymers that are specific to certain differentiated cell types (Herrmann and Aebi, 1998; Janmey et al., 1998). For example, vimentin is expressed in mesenchymal cells (e.g., fibroblasts). Vimentin fibers terminate at the nuclear membrane and at desmosomes, or adhesion plaques, on the plasma membrane. Another type of intermediate filament is lamin, which makes up the nuclear lamina, a polymer cortex underlying the nuclear membrane (Aebi et al., 1986). Intermediate filaments are often colocalized with microtubules, suggesting a close association between the two filament networks. Both microtubules and intermediate filaments are thought to be less important for the elastic strength and structural response of cells subjected to external stress (Petersen et al., 1982; Pasternak and Elson, 1985; Heidemann et al., 1999; Rotsch and Radmacher, 2000). However, intermediate filaments become more important at large deformations that cannot be achieved with deforming stresses of several Pascal. Intermediate filaments are also more important to elasticity in adhered cells as opposed to suspended cells, where the initially fully extended filaments become slack (Janmey et al., 1991; Wang and Stamenovic, 2000). Despite these experiments, a quantitative description of the cytoskeletal contribution to a cell's viscoelasticity is still missing. The optical stretcher can be used to measure the viscoelastic properties of the entire cytoskeleton and to shed new light on the problem of cellular elasticity.

The ability to withstand deforming stresses is crucial for cells and has motivated the development of several techniques to investigate cell elasticity. Atomic force microscopy (Radmacher et al., 1996), manipulation with microneedles (Felder and Elson, 1990), microplate manipulation (Thoumine and Ott, 1997), and cell poking (Dailey et al., 1984) are not able to detect small variations in cell elasticity

because these detection devices have a very high spring constant compared to the elastic modulus of the material probed. The AFM technique has recently been improved for cell elasticity measurements by attaching micron-sized beads to the scanning tip to reduce the pressure applied to the cell (Mahaffy et al., 2000). Micropipette aspiration of cell segments (Discher et al., 1994) and displacement of surface-attached microspheres (Wang et al., 1993) can provide inaccurate measurements if the plasma membrane becomes detached from the cytoskeleton during deformation. In addition, all of these techniques are very tedious and only probe the elasticity over a relatively small area of a cell's surface. Whole-cell elasticity can be indirectly determined by measurements of the compression and shear moduli of densely packed cell pellets (Elson, 1988; Eichinger et al., 1996), or by using microarray assays (Carlson et al., 1997). However, these measurements only represent an average value rather than a true single-cell measurement, and depend on noncytoskeletal forces such as cell–cell and cell–substrate adhesion. The optical stretcher is a new tool that not only circumvents most of these problems, but also permits the handling of large numbers of individual cells by incorporation of an automated flow chamber, fabricated with modern soft lithography techniques, that guides cells through the detector.

MATERIALS AND METHODS

Erythrocyte preparation

The buffer for the RBCs was derived from Zeman (1989) and Strey et al. (1995) and consisted of 100 mM NaCl, 20 mM Hepes buffer (pH 7.4), 25 mM glucose, 5 mM KCl, 3 mM $CaCl_2$, 2 mM $MgCl_2$, 0.1 mM adenine, 0.1 mM inosine, 1% (volume) antibiotic-antimycotic solution, 0.25–1.5% albumin, and 5 units/ml heparin. All reagents were purchased from Sigma (St. Louis, MO) unless stated otherwise. Red blood cells were obtained by drawing ~10 μl of blood from the earlobe or fingertip. The blood was diluted with 4 ml of the buffer. Because the buffer has a physiological osmolarity (~270 mOsm), the RBCs initially have a flat, biconcave, disc-like shape. However, the buffer was then diluted to lower the osmolarity to 130 mOsm, at which point the RBCs swell to assume a spherical shape. The average radius of the swollen RBCs was measured to be $\rho = 3.13 \pm 0.15$ μm using phase contrast microscopy. The error given is the standard deviation (SD) of 55 cells measured. The refractive index of spherical RBCs, $n = 1.378 \pm 0.005$ (Evans and Fung, 1972), the refractive index of the final buffer was measured to be $n = 1.334 \pm 0.001$, both of which were used for the calculations in the RBC stretching experiments.

Eukaryotic cell preparation

As prototypical eukaryotic cells, BALB 3T3 fibroblasts (CCL-163) were obtained from American Type Culture Collection (Manassas, VA) and maintained in Dulbecco's Modified Eagle's Medium with 10% nonfetal calf serum and 10 mM Hepes at pH 7.4. For cells to be trapped and stretched in the optical stretcher, they must be in suspension. Because these are normally adherent cells, single-cell suspensions for each experiment were obtained by incubating the cells with 0.25% trypsin-EDTA solution at 37°C for 4 min. After detaching, the activity of trypsin-EDTA was inhibited by adding fresh culture medium. This treatment causes the cells

For the fluorescence studies of the actin cytoskeleton of BALB 3T3 cells in suspension, we used TRITC-phalloidin (Molecular Probes, Inc., Eugene, OR), a phallotoxin that binds selectively to filamentous actin (F-actin), increasing the fluorescence quantum yield of the fluorophore rhodamine several-fold over the unbound state (Allen and Janmey, 1994). This assures that predominantly actin filaments are detected rather than actin monomers. Before staining, the cells were spun down and gently resuspended in a 4% formaldehyde solution for 10 min to fix the actin cytoskeleton. They were then washed three times with PBS, permeabilized with a 0.1% Triton-X 100 solution for 2 min, and washed three more times. Then the cells were stained with a 1-μg/ml TRITC-phalloidin solution for 10 min, followed by a final washing step (3× with PBS). Fluorescence images were acquired with an inverted microscope (Axiovert TV100, Carl Zeiss, Inc., Thornwood, NY) and deconvolved using a Jansson–van Cittert algorithm with 100 iterations (Zeiss KS400 software).

Silica and polystyrene beads

The silica and polystyrene beads used for the calibration of the image analysis algorithm and for the shooting experiments were purchased from Bangs Laboratories, Inc. (Fishers, IN). Their radii were $\rho = 2.50 \pm 0.04$ μm (SD) and $\rho = 2.55 \pm 0.04$ μm (SD), respectively, as given in the specifications provided by the manufacturer. Using index matching, their indices of refraction were measured to be $n = 1.430 \pm 0.003$ for silica beads using mixtures of water and glycerol, and $n = 1.610 \pm 0.005$ for polystyrene beads using mixtures of diethyleneglycolbutylether and α-chloronaphthalene. The index of refraction of water, used for the calculation of the forces on the silica and polystyrene beads in the shooting experiments, was measured to be $n = 1.333 \pm 0.001$.

Experimental setup

The setup of the experiment (see Fig. 4) is essentially a two-beam fiber

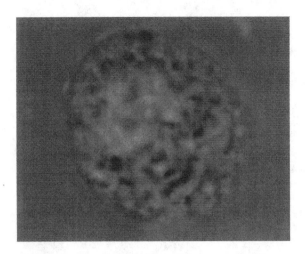

FIGURE 3   Phase contrast image of a BALB 3T3 fibroblast in an albumin solution with refractive index $n = 1.370 \pm 0.005$. At the matching point, the contrast between cell and surrounding is minimal. The lighter parts of the cytoplasm have a slightly lower refractive index, whereas the darker parts have a slightly higher refractive index than the bulk of the cell. The cell's radius is $\rho \approx 8.4$ μm.

to stay suspended as isolated cells for 2–4 h. Once in suspension, the cells assumed a spherical shape. Their average radius was $\rho = 9.2 \pm 2.8$ μm (SD of 20 cells measured), and their average refractive index, $n = 1.370 \pm 0.005$, was measured using index matching in phase contrast microscopy (see Fig. 3) (Barer and Joseph, 1954, 1955a,b). The refractive index of the cell medium, which was used for the calculations of the fibroblast shooting and stretching experiments, was measured to be $n = 1.335 \pm 0.002$.

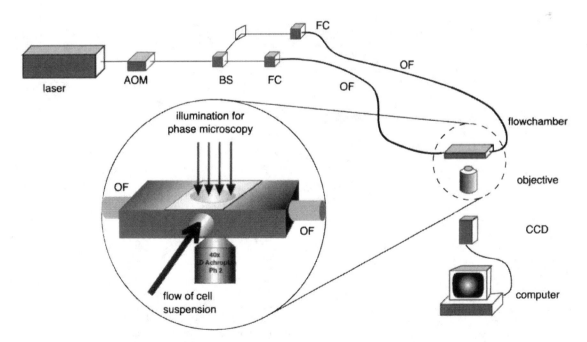

FIGURE 4   Setup of the optical stretcher. The intensity of the laser beam is controlled by the acousto-optic modulator (*AOM*), split in two by a beam splitter (*BS*), and coupled into optical fibers (*OF*) with two fiber couplers (*FC*). The inset shows the flow chamber used to align the fiber tips and to stream a cell suspension through the trapping area. Digital images of the trapping and optical stretching were recorded by a Macintosh computer using a CCD camera.

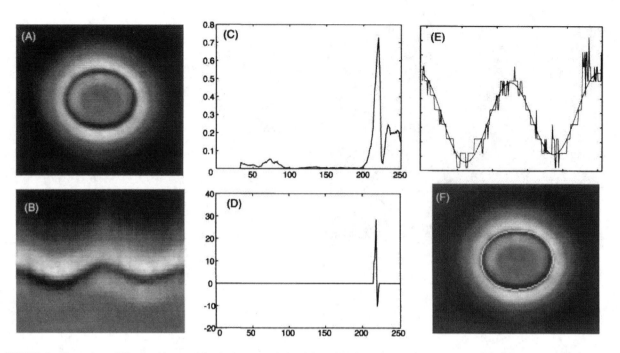

FIGURE 5   Illustration of the algorithm used for the image analysis of the cell deformation in the optical stretcher. (*A*) Original phase contrast image of a stretched spherical RBC. The diameter of the cell is about 6 μm. (*B*) The original image mapped onto a rectangle. The inside of the cell is the lower part of the picture. (*C*) Line scan across the cell boundary from top to bottom after squaring the grayscale values. (*D*) Line scan after thresholding and division by the first spatial derivative. (*E*) The zigzag line is the boundary of the cell as extracted from the binary image. The smooth line is the inverse Fourier transform of the three dominant frequencies in the original data. (*F*) The white line shows the image of this smooth line representing the boundary of the cell as detected by the algorithm converted back into the cell image. The line matches the cell's boundary to a high degree.

trap (Constable et al., 1993). A tunable, cw Ti-Sapphire laser (3900S, Spectra Physics Lasers, Inc., Mountain View, CA) with up to 7W of light power served as light source at a wavelength of $\lambda = 785$ nm (30 GHz bandwidth). An acousto-optic modulator (AOM-802N, IntraAction Corp., Bellwood, IL) was used to control the beam intensity, i.e., the surface forces. This can be done with frequencies between 10^{-2} and 10^3 Hz, thus allowing for time-dependent rheological measurements in the frequency range most relevant for biological samples. The beam was split in two by a nonpolarizing beam-splitting cube (Newport Corp., Irvine, CA) and then coupled into single-mode optical fibers (mode field diameter = 5.4 $\pm$ 0.2 μm, NA 0.11). The fiber couplers were purchased from Oz Optics, Ltd. (Carp, ON, Canada) and the single-mode optical fibers from Newport. The optical fibers not only simplify the setup of the experiment, they also serve as additional spatial filters and guarantee a good spatial mode quality (TEM$_{00}$). The maximum light powers achieved in this setup were 800 mW in each beam at the object trapped. The power exiting the fiber was measured before and after each experimental run to verify the stability of the coupling over the 1–2-h period. All power values given are measured with a relative error of ± 1% (SD).

For trapping and stretching cells in the optical stretcher, the fibers' alignment is crucial. For the RBCs, a solution similar to the one described in Constable et al. (1993) was used: a glass capillary with a diameter between 250 and 400 μm was glued onto a microscope slide, and the fibers were pressed alongside so that they were colinear and facing each other (not shown in Fig. 4). Red blood cells were so light that they sunk very slowly and could be trapped out of a cell suspension placed on top of the fiber ends. For the BALB 3T3 fibroblasts, we used a flow chamber geometry (see inset of Fig. 4) that allowed us to stream a suspension of cells directly through the gap between the optical fibers.

After successfully trapping one cell, the flow was stopped and the cell's elasticity was measured. Then the cell was released and the flow was started again until the next cell was trapped. The microscope slide, or the flow chamber, was mounted on an inverted microscope equipped for phase contrast and fluorescence microscopy. Phase contrast images of the trapping and stretching were obtained with a CCD camera (CCD72S, MTI-Dage, Michigan City, IN). The pixel size for all magnifications used was calibrated with a 100 lines/mm grating, which allowed for absolute distance measurements. The side length of the square image pixels was 118 $\pm$ 2 nm for the 40$\times$ objective with additional 2.5$\times$ magnification lens, used for the cell-size measurements. To measure larger distances, such as the distance between the fiber tip and the trapped object, we used a 20$\times$ objective, which resulted in an image pixel size of 611 $\pm$ 5 nm/pixel. All stretching experiments were done at room temperature.

Image analysis

After completing the experiments, image data were analyzed on a Macintosh computer to quantify the deformation of a cell in the optical stretcher. The algorithm was developed in the scientific programming environment MATLAB (MathWorks, Inc., Natick, MA), which treats bitmap images as matrices. Figure 5 *A* shows a typical phase contrast image of an RBC stretched at moderate light powers ($P \approx 100$ mW). The boundary of the cell in the image is the border between the dark cell and the bright halo. The goal was to extract the shape of the cell from this image to use the quantitative information for further evaluation.

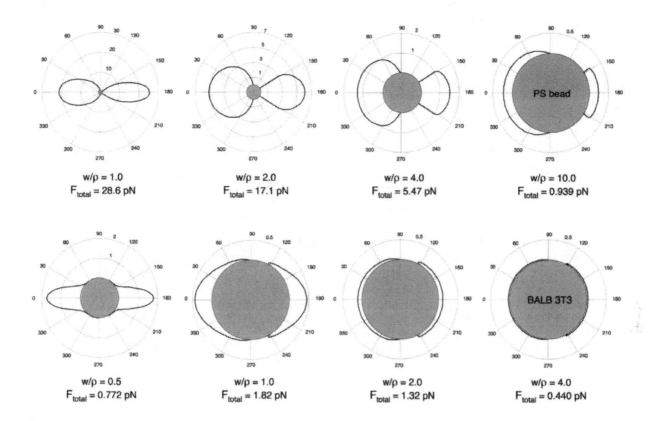

FIGURE 6 Surface stress profiles for one laser beam incident from the left on a polystyrene bead (*top row*) and a BALB 3T3 fibroblast (*bottom row*) for different ratios between the beam radius w and the object's radius ρ. The radii of the polystyrene bead and the fibroblast used for this calculation were $\rho = 2.55\ \mu m$ and $\rho = 7.70\ \mu m$ and the refractive indices were $n_2 = 1.610$ and $n_2 = 1.370$, respectively. The total light power $P = 100$ mW for all profiles. The concentric rings indicate the stress in Nm^{-2} (note the different scales). The resulting total force after integration over the surface acting on the center of gravity of the object is noted in each case.

The algorithm consists of the following steps. First, the geometrical center of the cell is found and used as the origin of a polar coordinate system. The grayscale values along the radii outward from the center are reassigned to Cartesian coordinates (see Fig. 5 B). Essentially, the image is cut along one radius and mapped onto a rectangle. The bottom of Fig. 5 B is the interior of the cell and the top is the outside. The cell boundary is along the wavy line between the black and the white bands. Next, the image is squared to enhance the contrast between cell and background. The effect of this filter can be seen in Fig. 5 C, which shows a line scan across the cell boundary. If we assume that the boundary between inside and outside coincides with the first peak, when moving from the outside to the inside of the cell, where the slope of the line scan is zero, we can drastically enhance the signal-to-noise ratio by dividing the data by their first spatial derivative. Because this would also enhance peaks in the background noise, we first threshold the image at a low value to set the background to zero. The combined effect of this mathematical filter can be seen in Fig. 5 D. The first peak, taken as the cell boundary, is then easy to detect (zigzag line in Fig. 5 E). The resolution up to this point is identical to the pixel resolution of the microscope/CCD system, i.e., 118 ± 2 nm/pixel. To further improve this resolution, we use the physical constraint that the cell boundary has to be smooth on this length scale. This is implemented by Fourier-decomposing the boundary data and by filtering out the high spatial frequency noise, which increases the resolution to an estimated ± 50 nm. The smooth line in Fig. 5 E is the inverse Fourier transform of

the remaining frequencies. The information about the deformation of cells is then extracted from the resulting function. Figure 5 F shows the original cell with the boundary as detected with this algorithm.

In general, this sort of image analysis can yield resolutions down to ± 11 nm (see, for example, Käs et al., 1996), which is well below the optical resolution of the microscope and also below the pixel resolution. The reason, in short, is that we do not want to resolve two close-by objects, which is limited to a distance of about half the wavelength. Instead, the goal is to detect how much an edge, characterized by a large change in intensity, is moving.

The absolute size determination of an object using this algorithm depends somewhat on the exact definition of the boundary between object and surrounding medium in the phase-contrast image. Our choice, as described above, was driven by the investigation of images of silica beads with known size. The estimated resolution of ± 50 nm is in agreement with measurements of these beads, which have a radius of $\rho = 2.50 \pm 0.04\ \mu m$ (SD). This resolution is certainly sufficient to discriminate between stretched and unstretched cells as reported further below. The advantages of this algorithm are its speed, precision, and its ability to detect the shape of any cell.

While the radii of the cells were measured with the algorithm, the distances between the fiber tip and the cell, d, were measured in a simpler way by counting pixels in images. For the 20$\times$ objective, this can be done with a pixel resolution of $\pm 0.6\ \mu m$.

RESULTS AND DISCUSSION

Theory

Total force for one beam

The simplest way to describe the interaction of light with cells is by ray optics (RO). This approach is valid when the size of the object is much larger than the wavelength of the light. The diameter of cells, 2ρ, is on the order of tens of microns. Cell biological experiments, such as the optical stretching of cells, are performed in aqueous solution, and water is sufficiently transparent only for electromagnetic radiation in the near infrared (the laser used was operated at a wavelength of $\lambda = 785$ nm). Thus, the criterion for ray optics, $2\pi\rho/\lambda \approx 25$–$130 \gg 1$, is fulfilled (van de Hulst, 1957).

The idea is to decompose an incident laser beam into individual rays with appropriate intensity, momentum, and direction. These rays propagate in a straight line in uniform, nondispersive media and can be described by geometrical optics. Each ray carries a certain amount of momentum p proportional to its energy E and to the refractive index n of the medium it travels in, $p = nE/c$, where c is the speed of light in vacuum (Ashkin and Dziedzic, 1973; Brevik, 1979). When a ray hits the interface between two dielectric media with refractive indices n_1 and n_2, some of the ray's energy is reflected. Let us assume that $n_2 > n_1$ and $n_2/n_1 \approx 1$, which is the case for biological objects in aqueous media, and that the incidence is normal to the surface. The fraction of the energy reflected is given by the Fresnel formulas (Jackson, 1975), $R \approx 10^{-3}$. The momentum of the reflected ray, $p_r = n_1RE/c$, and the momentum of the transmitted ray, $p_t = n_2(1 - R)E/c$ (Ashkin and Dziedzic, 1973; Brevik, 1979). The incident momentum, $p_i = n_1E/c$, has to be conserved at the interface. The difference in momentum between the incident ray and the reflected and transmitted rays, $\Delta p = p_i + p_r - p_t$, is picked up by the surface, which experiences a force F according to Newton's second law,

$$F = \frac{\Delta p}{\Delta t} = \frac{n_1 \Delta E}{c \Delta t} = \frac{n_1 QP}{c}, \tag{1}$$

where P is the incident light power and Q is a factor that describes the amount of momentum transferred ($Q = 2$ for reflection, $Q = 1$ for absorption). For partial transmission of one laser beam hitting a flat interface at normal incidence as described above, $Q_{\text{front}} = 1 + R - n(1 - R) = -0.086$ ($n_1 = 1.33$, $n_2 = 1.43$, $n = n_2/n_1$). This force acts in the backward direction, away from the denser medium (see also Fig. 2). The transmitted ray eventually hits the backside of the object and again exerts a force on the interface. Here, $Q_{\text{back}} = [n + Rn - (1 - R)](1 - R) = 0.094$, and the force acts in the forward direction, again away from the denser medium. For the total force acting on the object's center of gravity, $Q_{\text{total}} = Q_{\text{front}} + Q_{\text{back}} = 0.008$. The total force is

obviously an order of magnitude smaller than either one of the surface forces.

If the ray hits the interface under an angle $\alpha \neq 0$, it changes direction according to Snell's law, $n_1\sin\alpha = n_2\sin\beta$, where β is the angle of the transmitted ray. In this case, the vector nature of momentum has to be taken into account and R becomes a function of the incident angle α. R is taken to be the average of the coefficients for perpendicular and parallel polarization relative to the plane of incidence. This is a negligible deviation from the true situation (the error in the stress introduced by this simplification is smaller than 2% for $n_2 = 1.45$, and smaller than 0.5% for $n_2 = 1.38$), but it simplifies the calculation and preserves symmetry of the problem with respect to the laser axis. The components of the force in terms of Q on the front side, parallel and perpendicular to the beam axis, are

$$\begin{aligned} Q_{\text{front}}^{\text{parallel}}(\alpha) &= \cos(0) - R\cos(\pi - 2\alpha) \\ &\quad - n(1 - R)\cos(2\pi - \alpha + \beta) \\ &= 1 + R(\alpha)\cos(2\alpha) \\ &\quad - n(1 - R(\alpha))\cos(\alpha - \beta) \\ &= Q_{\text{front}}(\alpha)\cos\phi, \end{aligned} \tag{2a}$$

and

$$\begin{aligned} Q_{\text{front}}^{\text{perpendicular}}(\alpha) &= \sin(0) - R\sin(\pi - 2\alpha) \\ &\quad - n(1 - R)\sin(2\pi - \alpha + \beta) \\ &= R(\alpha)\sin(2\alpha) \\ &\quad + n(1 - R(\alpha))\sin(\alpha - \beta) \\ &= Q_{\text{front}}(\alpha)\sin\phi, \end{aligned} \tag{2b}$$

where ϕ is the angle between the beam axis and the direction of the momentum transferred. Similarly, on the back the components of the surface force are

$$\begin{aligned} Q_{\text{back}}^{\text{parallel}}(\alpha) &= (1 - R(\alpha))[n\cos(\alpha - \beta) \\ &\quad + nR(\beta)\cos(3\beta - \alpha) \\ &\quad - (1 - R(\beta))\cos(2\alpha - 2\beta)] \\ &= Q_{\text{back}}(\alpha)\cos\phi, \end{aligned} \tag{3a}$$

and

$$\begin{aligned} Q_{\text{back}}^{\text{perpendicular}}(\alpha) &= (1 - R(\alpha))[-n\sin(\alpha - \beta) \\ &\quad + nR(\beta)\sin(3\beta - \alpha) \\ &\quad - (1 - R(\beta))\sin(2\alpha - 2\beta)] \\ &= Q_{\text{back}}(\alpha)\sin\phi. \end{aligned} \tag{3b}$$

Subsequent reflected and refracted rays can be neglected because $R < 0.005$ for all incident angles. The magnitude of the force in terms of Q on either the front or the backside is given by

$$Q_{\text{front/back}}(\alpha) = \sqrt{(Q_{\text{front/back}}^{\text{parallel}}(\alpha))^2 + (Q_{\text{front/back}}^{\text{perpendicular}}(\alpha))^2}, \quad (4)$$

which is a function of the incident angle α, and the direction of the force is

$$\phi_{\text{front/back}}(\alpha) = \arctan\left(\frac{Q_{\text{front/back}}^{\text{perpendicular}}(\alpha)}{Q_{\text{front/back}}^{\text{parallel}}(\alpha)}\right). \quad (5)$$

The forces on front and back are always normal to the surface for all incident angles. Thus, the stress σ, i.e., the force per unit area, along the surface where the ray enters and leaves the cell is

$$\sigma_{\text{front/back}}(\alpha) = \frac{F}{\Delta A} = \frac{n_1 Q_{\text{front/back}}(\alpha) I(\alpha)}{c}, \quad (6)$$

where $I(\alpha)$ is the intensity of the light. Figure 6 shows stress profiles calculated for spherical objects with the refractive index of polystyrene beads, and with the average refractive index of RBCs hit by one laser beam with Gaussian intensity distribution. The profiles are rotationally symmetric with respect to the beam axis. The sphere acts as a lens and focuses the rays on the back toward the beam axis, which results in a narrower stress profile. Integrating this asymmetrical stress over the whole surface yields the total force on the object's center of mass, which pushes the object in the direction of the beam propagation. At the same time, the applied stress stretches the object in both directions along the beam axis. Due to the cylindrical symmetry, the total force has only a component in the direction of the light propagation, which is generally called scattering force. If the object is displaced from the beam axis, this symmetry is broken. It experiences a force, called gradient force, perpendicular to the axis, which pulls the object toward the highest laser intensity at the center of the beam if the refractive index of the object is greater than that of the surrounding medium. Because the gradient force is restoring, after the object reaches the axis, it will stay there as long as no other external forces are present and the gradient force is zero.

The stress profile and the total force depend on the ratio between beam radius w and sphere radius ρ, and on the relative index of refraction, $n = n_2/n_1$. If there is only one beam shining on the object, the total force will accelerate it. Because the beam is slightly divergent (the beam radius doubles from $w = 2.7$ μm at the fiber tip to $w = 5.4$ μm over a distance of 70 μm), the beam radius w, and therefore the stress profiles and the total force, are functions of the distance d from the fiber end (see Fig. 8). Smaller beam size with respect to the object results in higher light intensity and thus greater stress on the surface ($w/\rho \ll 1$). As the beam

radius w increases with increasing distance d and w/ρ approaches one, the light intensity and the magnitude of the surface stress decrease. However, the total force increases because the asymmetry between the front and back becomes more pronounced. As the beam size becomes much larger than the object ($w/\rho \gg 1$), the surface stresses and the total force vanish, because less light is actually hitting the object. The highest total forces ($w/\rho = 1$) calculated for polystyrene beads in water and RBCs in their final buffer for a light power of $P = 100$ mW are $F_{\text{total}} = 28.6 \pm 0.9$ pN and $F_{\text{total}} = 1.82 \pm 0.06$ pN, respectively. The relative error in the force calculations, due to uncertainties in the measurement of the relevant quantities (indices of refraction, light power, radius, distance between fiber and object) as given earlier, is 3.2%. In general, the magnitude of the surface stress and the total force increase with higher relative indices of refraction n.

The change in total force as the object is pushed away from the light source can be measured by setting the accelerating total force F_{total} equal to the Stokes drag force acting on the spherical object,

$$F_{\text{total}} = 6\pi\eta\rho v, \quad (7)$$

where η is the viscosity of the surrounding medium and v is the velocity of the object. The viscous drag on a spherical object can depend strongly on the proximity of boundaries. A correction factor a can be found in terms of the ratio between the radius of the sphere ρ and the distance to the closest boundary b (Svoboda and Block, 1994),

$$a = \frac{1}{1 - \dfrac{9}{16}\dfrac{\rho}{b} + \dfrac{1}{8}\left(\dfrac{\rho}{b}\right)^3 - \dfrac{45}{256}\left(\dfrac{\rho}{b}\right)^4 - \dfrac{1}{16}\left(\dfrac{\rho}{b}\right)^5 + \cdots}. $$

$$(8)$$

The Reynolds number is on the order of 10^{-4}, so inertia can be neglected. The measurement of the total force on different objects was used to investigate to what extent cells can be approximated as objects with a homogeneous index of refraction (see Shooting Experiments).

Stress profiles for two beams

A configuration with two opposing, identical laser beams functions as a stable optical trap where the dielectric object is held between the two beams. When the object is trapped, the surface stresses caused by the two incident beams are additive. Fig. 7 shows the resulting stress profiles for RBCs, which are rotationally symmetric with respect to the beam axis. If the object is centered, the surface stresses cancel upon integration, and the total force is zero. Otherwise, restoring gradient and scattering forces will pull the object back into the center of the trap. The trapping force is the minimal force required to pull the object completely out of

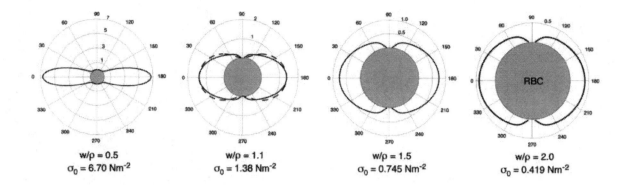

FIGURE 7   Surface stress profiles on an RBC trapped in the optical stretcher for different ratios between the beam radius w and the cell radius ρ. The total power in each beam was $P = 100$ mW for all profiles. The radius of the RBC used for this calculation was $\rho = 3.30$ μm, and the refractive index was $n_2 = 1.378$. The concentric rings indicate the stress in Nm^{-2}. The peak stresses σ_0 along the beam axis (0° and 180° direction) are given below each profile. The trapping of the cell for $w/\rho < 1$ is unstable (see text). The dashed line for $w/\rho = 1.1$ shows the $\sigma_r(\alpha) = \sigma_0\cos^2(\alpha)$ approximation of the true stress profile.

the trap, which is equal to the greatest gradient force encountered as the object is displaced.

Again, the shape of the profile changes depending on n and the ratio w/ρ. The smaller the beam size with respect to the object, the greater the stress in the vicinity of the axis. However, the trap is not stable if the beams are smaller than the object. This has been discussed and experimentally shown by Roosen (1977). For $w/\rho \gg 1$, the surface stresses become very small. The ideal trapping situation is when the beams are only slightly larger than the object ($w/\rho \gtrsim 1$). For example, for $w/\rho = 1.1$, the peak stresses σ_0 along the beam axis for RBCs trapped with two 100-mW beams, $\sigma_0 = 1.38 \pm 0.05$ Nm^{-2}. The relative error in the stress calculations, due to uncertainties in the measurement of the relevant quantities as given earlier, is 3.0%. In this case, the stress profile can be well approximated by $\sigma_r(\alpha) = \sigma_0\cos^2(\alpha)$ (see Fig. 7 for $w/\rho = 1.1$). This functional form of the stress profile makes an analytical solution of the deformation of certain elastic objects tangible.

Deformation of thin shells

Erythrocytes were used initially because they are soft, easy to handle and to obtain, and their deformations are easily observed. They are also much more accessible to theoretical modeling than eukaryotic cells with their highly complex and dynamic internal structures. Thus, RBCs can be considered well-defined elastic objects that can be used to verify the calculated stress profiles. The only elastic component of RBCs is a thin composite shell made of the plasma membrane, the two-dimensional cytoskeleton, and the glycocalix. The ratio between shell radius ρ and shell thickness h, $\rho/h \approx 100$. In this case, membrane theory can

be used to describe deformations due to surface stresses (Mazurkiewicz and Nagorski, 1991; Ugural, 1999).

Membrane theory is the simplification of a more general theory of the deformation of spherical shells in which the bending energy U_b of the shell is neglected and only the membrane (or stretching) energy U_m is considered. It can be shown that the ratio of those two energies for the case of axisymmetric stress, as applied with the optical stretcher, is $U_b/U_m = 4h^2/3\rho^2 \approx 10^{-4}$ for RBCs. The stress σ_r applied to a spherical object in the optical stretcher has the form, $\sigma_r = \sigma_0\cos^2(\alpha)$, as shown above. Spherical coordinates are an obvious choice, where the radial direction is denoted by r, the polar angle by θ, and the azimuthal angle is ϕ. The coordinate system is oriented such that the incident angle of the rays, α, in the previous section is identical to the polar angle θ (the laser beams are traveling along the z-axis). The total energy U of a thin shell consists of the membrane energy and the work done by the stress applied and is given by,

$$U = 2\pi\rho^2 \int \left[\left(\frac{Eh}{2(1-v^2)}\left[\varepsilon_\theta^2 + \varepsilon_\phi^2 + 2v\varepsilon_\theta\varepsilon_\phi\right]\right) \right. \tag{9}$$
$$\left. - \sigma_r u_r \times \sin(\theta) \right] d\theta,$$

where ϵ_θ and ϵ_ϕ are the strains in the polar and meridional direction, respectively, u_r is the radial deformation of the membrane, E is the Young's modulus, and v is the Poisson ratio. The connections between the strains and the deformations are

$$\varepsilon_\theta = \frac{1}{\rho}\left(\frac{du_\theta}{d\theta} - u_r\right) \tag{10a}$$

and

$$\varepsilon_\phi = \frac{1}{\rho}(u_\theta \cot(\theta) - u_r), \qquad (10b)$$

where u_θ is the deformation in meridional direction. The radial and meridional displacements, which describe the experimentally observed deformation of the dielectric object in the optical stretcher, can be found by using Euler's equations,

$$\frac{1}{\rho}\frac{d}{d\theta}\frac{dF}{du'_\theta} - \frac{dF}{du_\theta} = 0 \qquad (11a)$$

and

$$\frac{1}{\rho}\frac{d}{d\theta}\frac{dF}{du'_r} - \frac{dF}{du_r} = 0, \qquad (11b)$$

where F is the integrand of the energy functional,

$$u'_\theta = \frac{1}{\rho}\frac{du_\theta}{d\theta} \qquad (12a)$$

and

$$u'_r = \frac{1}{\rho}\frac{du_r}{d\theta}. \qquad (12b)$$

Using Eqs. 10, 11, and 12 and the explicit form of $\sigma_r(\theta)$, we find the following expressions for the radial and the meridional deformations of the membrane,

$$u_r(\theta) = \frac{\rho^2\sigma_0}{4Eh}[(5 + v)\cos^2(\theta) - 1 - v] \qquad (13a)$$

and

$$u_\theta(\theta) = \frac{\rho^2\sigma_0(1 + v)}{2Eh}\cos(\theta)\sin(\theta). \qquad (13b)$$

As expected from the symmetry of the problem, the deformations u_r and u_θ are independent of the azimuthal angle ϕ, and there is also no displacement u_ϕ in this direction. Figure 11 shows the shapes of thin shells calculated with Eq. 13a for $v = 0.5$, which is normal for biological membranes, and $Eh = (3.9 \pm 1.4) \times 10^{-5}\,\text{Nm}^{-1}$, which was found to be the average for RBCs from the experiments (see below) for increasing stresses σ_0. Because these equations are linear, they only hold for small strains ($<10\%$). In the microscope images, which are cross-sections of the objects because the focal depth of the objective is much smaller than the diameter of the RBCs, only the radial deformations can be observed. The direct comparison between the theoretically expected deformations $u_r(\theta)$ and the experimentally observed radial deformations help to establish the RO model as valid explanation for the optical stretching of soft dielectric objects.

Experiments

Shooting experiments

To test the assumptions underlying the RO calculations as described above, we measured the total force acting on different objects. It was not clear if it was permissible to model living cells with their organelles and other small-scale structures as homogeneous spheres with an isotropic index of refraction. Although this assumption is obvious for RBCs homogeneously filled with hemoglobin, it might be questionable in the case of eukaryotic cells containing organelles and other internal structures (see Fig. 3). In this series of experiments, individual silica beads, polystyrene beads, or fibroblasts were trapped in the optical stretcher. The setup was identical to the one used for the trapping and stretching of RBCs (see Experimental Setup and Fig. 4). After stably trapping the objects, we blocked one of the laser beams. The total force from the other beam accelerated the object away from the light source.

The total force was determined using Eqs. 7 and 8 and the velocities, radii, and distances measured during the experiment. In our setup, the distance b between the moving objects and the coverslip as closest boundary was $b = 62.5 \pm 2.5\,\mu\text{m}$ (half the diameter of the optical fiber). In the case of the silica and polystyrene beads, this distance is about 25 times the radius of the beads and the correction factor $a = 1.023$. For the cells, the distance is ~7–9 times the radius and $a = 1.072$–1.090. The viscosity η used in the calculation was that for water at 25°C, $\eta = 0.001$ Pa s. Figure 8 shows the total force measured as a function of the distance d between the fiber tip and the object, as well as the total force as expected from our RO calculations. The error bars shown are statistical errors in the experimental data. The relative errors introduced by the uncertainties in the measurements of radii, velocities, and distances are negligibly small (1.1–3.1%). It is not surprising that the experimental data points for silica beads and polystyrene beads match the theory because these are truly spherical and have an isotropic index of refraction. The fact that, also, the experimental results for the cells matched the theory was proof that even eukaryotic cells can be treated using this simple RO model. The magnitude and the dependence of the total force on the distance d are in good agreement with the results by (Roosen, 1977).

Stretching of erythrocytes

Single, osmotically swollen RBCs were trapped in the optical stretcher at low light powers ($P \approx 5$–10 mW). The light power was then increased to a higher value between $P = 10$ and $P = 800$ mW for approximately 5 s, an image of the stretched cell was recorded, and the light power was decreased again to the original value. During the short time intervals when stress was applied, we did not observe any creep, i.e., any increase in deformation during the duration

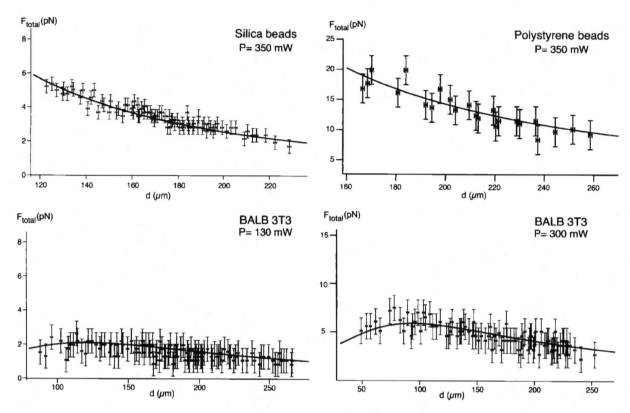

FIGURE 8   The total forces from one laser beam on silica beads, polystyrene beads, and BALB 3T3 fibroblasts as functions of the distance d between the fiber tip and the object. The data points are from measuring the total force on the objects in the optical stretcher after obstructing one of the laser beams. The solid lines represent the total forces calculated using ray optics. The radii of the silica beads, the polystyrene beads, and the fibroblasts were $\rho = 2.50 \pm 0.04$ μm, $\rho = 2.55 \pm 0.04$ μm, and $\rho = 7.70 \pm 0.05$ μm, and the refractive indices were $n_2 = 1.430 \pm 0.003$, $n_2 = 1.610 \pm 0.005$, and $n_2 = 1.370 \pm 0.005$, respectively. The total light power P is indicated in each case. The error bars represent standard deviations; the error in the distance measurement was ± 0.6 μm.

of the stretching. Also, the cells did not show any hysteresis or any kind of plastic deformation up to $P \approx 500$ mW. Figure 9 shows a sequence of RBC images recorded at increasing light powers. It is obvious that the deformation increased with the light power used. The radius along the beam axis increased from 3.13 ± 0.05 μm in the first image to 3.57 ± 0.05 μm in the last image, a relative increase of $14.1 \pm 0.3\%$, whereas the radius in the perpendicular direction decreased from 3.13 ± 0.05 μm to 2.77 ± 0.05 μm (relative change $-11.5 \pm 0.2\%$).

From the radius of the cell, the distance between cell and fiber tips, the refractive indices of cell and medium, and the power measured, we calculated the stress profiles for each cell using the RO model. As mentioned earlier, the relative error in the stress calculation is 3.0%. The peak stress in the last image of Fig. 9 was calculated as $\sigma_0 = 1.47 \pm 0.03$ Nm^{-2}. Figure 10 shows the relative increase in radius along the beam axis ($\theta = 0$) and the relative decrease perpendicular ($\theta = \pi/2$) versus the peak stress σ_0 for 55 RBCs. The error bars shown are statistical errors. The relative errors in the calculation of the relative changes and the peak stresses due to uncertainties in the relevant quantities measured are

comparatively small. The solid line in Fig. 10 shows a fit of $u_r(0)/\rho$ (see Eq. 13a) to the experimental data. For the fit, the errors in the peak stresses were neglected and a singular value decomposition algorithm was used. The data points were weighed with the inverse of the standard deviations. The resulting slope was 0.080 ± 0.011 m^2N^{-1}, which yielded $Eh = (3.9 \pm 1.4) \times 10^{-5}$ Nm^{-1}. The intercept was zero. The correlation coefficient for the fit was, $r = 0.92$, excluding the last data point. This shows that up to $\sigma_0 \approx 2$ Nm^{-2} ($P \approx 350$ mW) and relative deformations of about 10%, the response of the RBCs was linear. In this regime, linear membrane theory can be used to describe the deformation of RBCs.

In the literature, usually the cortical shear modulus Gh is given, rather than the Young's modulus Eh. The quantities are related by $Gh = Eh/2(1 + \nu) = Eh/3$. In our case, the shear modulus, $Gh = (1.3 \pm 0.5) \times 10^{-5}$ Nm^{-1}. This value is in good agreement with values reported previously from micropipette aspiration measurements, which yielded shear moduli in the range $6–9 \times 10^{-6}$ Nm^{-1} (Hochmuth, 1993). Micropipette aspiration is the most established technique for measuring cellular elasticities, and the value for the shear

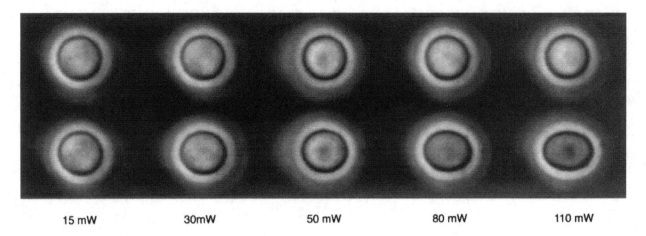

15 mW 30mW 50 mW 80 mW 110 mW

FIGURE 9 Typical sequence of the stretching of one osmotically swollen RBC for increasing light powers. The top row shows the RBC trapped at 5 mW in each beam. The power was then increased to the higher power given below, which lead to the stretching shown underneath, and then reduced again to 5 mW. The stretching clearly increases with increasing light power. The radius of the unstretched cell was $\rho = 3.13 \pm 0.05$ μm, and the distance between the cell and either fiber tip was $d \approx 60$ μm. The images were obtained with phase contrast microscopy. Any laser light was blocked by an appropriate filter.

modulus of the RBC membrane has been confirmed many times and is well accepted.

More recently, optical tweezers were used to measure the shear modulus with very differing results. In these experi-ments, beads were attached to the membrane on opposite sides of the RBC, trapped with optical tweezers, and then displaced. The values found ranged from $(2.5 \pm 0.4) \times 10^{-6}$ Nm^{-1} (Hénon et al., 1999) to 2×10^{-4} Nm^{-1} (Sleep et al.,

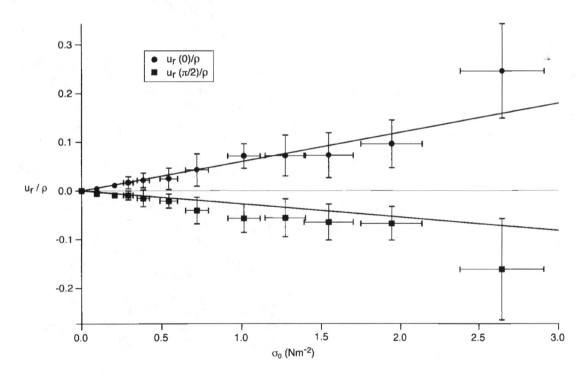

FIGURE 10 Relative deformation u_r/ρ of RBCs along (*positive values*) and perpendicular (*negative values*) to the laser axis in the optical stretcher as a function of the peak stress σ_0. The error bars for the relative deformations and the peak stresses are standard deviations. The solid line shows a fit of Eq. 13a as derived from membrane theory to the data points using a singular value decomposition algorithm. The linear correlation coefficient for this fit, $r = 0.92$, (excluding the last data point) indicates a linear response of the RBCs to the applied stress. Beyond a peak stress $\sigma_0 \approx 2$ Nm^{-2} the deformation starts deviating from linear behavior.

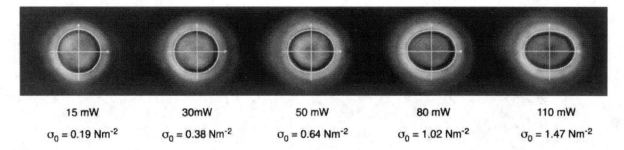

15 mW	30mW	50 mW	80 mW	110 mW
$\sigma_0 = 0.19$ Nm^{-2}	$\sigma_0 = 0.38$ Nm^{-2}	$\sigma_0 = 0.64$ Nm^{-2}	$\sigma_0 = 1.02$ Nm^{-2}	$\sigma_0 = 1.47$ Nm^{-2}

FIGURE 11   Comparison between the deformations of RBCs observed in the optical stretcher and the deformations expected from membrane theory (*white lines*). The peak stresses σ_0 calculated using ray optics, which are shown with each image, were used for the membrane theory calculations. The resulting theoretical shapes, calculated with Eq. 13a, were overlaid on the original images from Fig. 9 to show the excellent agreement between the two for all peak stresses.

1999). Because this technique applies point forces to the membrane, the stress is highly localized and leads to non-linear deformations. The discrepancy between these values and the established values for the shear modulus can probably be attributed to this different load condition.

Furthermore, the theoretically expected and the observed shapes of RBCs in the optical stretcher coincide well (see Fig. 11). The white lines are the shapes of thin shells with RBC material properties as predicted by linear membrane theory subjected to the surface profile calculated by the RO model. These lines were overlaid on the images of the stretched RBCs in Fig. 9. The excellent agreement between the predicted and the observed shaped shows that using RO theory is sufficient to calculate the surface stress on cells in the optical stretcher. An ab inito treatment of the interaction of a spherical dielectric object in an inhomogeneous electromagnetic wave using Maxwell's equations and surface stress tensor would be much more difficult and is also not necessary in this case. The RO model is powerful enough to accurately predict the qualitative and the quantitative aspect of the stretching. Ray optics has the additional benefit of being much more accessible.

Beyond $\sigma_0 \approx 2$ Nm^{-2}, the response of the RBCs became nonlinear and the shapes observed began to diverge from the ones expected from the linear membrane theory. Figure 12 shows the response of such a cell. At a peak stress of $\sigma_0 = 2.55 \pm 0.10$ Nm^{-2}, the cell was stretched from a radius along the beam axis of 3.36 ± 0.05 μm to 6.13 ± 0.05 μm (relative change $82 \pm 3\%$), whereas the perpendicular radius decreased from 3.38 ± 0.05 μm to 2.23 ± 0.05 μm (relative change $-34 \pm 2\%$). For the same cell, this transition from linear to nonlinear response was repeatable several times. Due to the variance in age, size, and elasticity between individual cells, the point of transition differed between different cells. If stretched even further (beyond $\sigma_0 \approx 3$ Nm^{-2}), the cells would rupture, as proven by the visually detectable release of hemoglobin from the cells.

Stretching of eukaryotic cells

Similar to most eukaryotic cells, BALB 3T3 fibroblasts are heavily invested by an extensive 3D cytoskeletal network of polymeric filaments, mainly actin filaments, which is largely preserved even in the suspended state (see below). Because these cells resemble solid spheres made of a non-

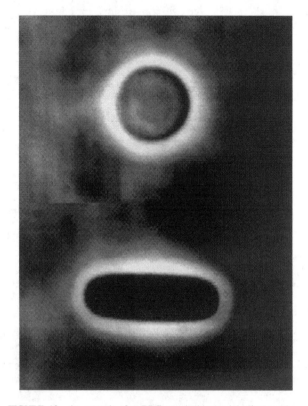

FIGURE 12   An example of an RBC stretched beyond the linear regime. The top part shows the undeformed cell trapped at 5 mW with a radius $\rho = 3.36 \pm 0.05$ μm. In the lower part, the cell is stretched with a peak stress $\sigma_0 \approx 2.55 \pm 0.10$ Nm^{-2} to $\rho = 6.13 \pm 0.05$ μm along the beam axis, an increase of $82 \pm 3\%$. After reducing the power, the cell returned to its original shape.

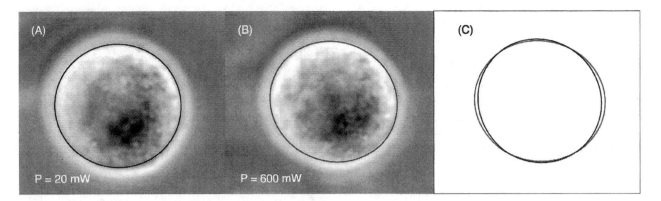

FIGURE 13 The stretching of a BALB 3T3 fibroblast in the optical stretcher. Due to the more extensive cytoskeleton of eukaryotic cells, the deformation is less obvious than for the RBCs. Still, the radius increased from (A) $\rho = 11.23 \pm 0.05$ μm for the cell trapped at 20 mW to (B) $\rho = 11.84 \pm 0.05$ μm along the beam axis at 600 mW, and decreased from 11.53 ± 0.05 μm to 11.27 ± 0.05 μm in the perpendicular direction. This is a relative deformation of $5.43 \pm 0.04\%$ and $-2.25 \pm 0.01\%$, respectively, for a peak stress $\sigma_0 = 5.3 \pm 0.2$ Nm^{-2}. The deformation can be seen much easier in (C) where the outlines of the stretched and unstretched cell, found with the image analysis algorithm, are overlaid.

uniform, complex compound material, we expected that they would not be stretched as significantly as RBCs, which are essentially thin shells. Figure 13 shows a fibroblast (A) trapped at $P = 20$ mW light power and (B) stretched with $P = 600$ mW. The peak stress in this case was, $\sigma_0 = 5.3 \pm 0.2$ Nm^{-2}, and the relative deformation along the beam axis was $5.43 \pm 0.04\%$, and $-2.25 \pm 0.01\%$ in the perpendicular direction. The small degree of deformation of this fibroblast is only possible to detect by using the algorithm for the extraction of the cell boundary (C). Any eukaryotic cell can be stretched this way. Although the applied stress can be calculated and the resulting deformation of cells can be measured, it would be incorrect to calculate a Young's modulus assuming a homogeneous elastic sphere. It would require complex modeling and probably time dependent measurements to extract elastic properties of the different cytoskeletal components.

Viability of stretched cells

The viability of the cells under investigation was an important issue because dead cells do not maintain a representative cytoskeleton. Even though care had been taken to avoid radiation damage to cells in the optical stretcher by selecting a wavelength (785 nm) with low absorption, their viability was checked on a case-by-case basis. Our approach to this was twofold. The appearance of BALB 3T3 cells is significantly different when they are not alive. Living cells in phase contrast microscopy show a characteristic bright rim around their edge. Dead cells usually have no sharp contour and appear diffuse. After a cell had been trapped and stretched for several minutes, it was compared with other cells that had not been irradiated. In all cases, the cells looked alike and normal. A second, more careful approach was the use of the vital stain Trypan Blue. As long as a cell is alive, it is able to prevent the dye from entering the cytoplasm. When the cell is dead or does not maintain its normal function, the dye will penetrate the cell membrane and the whole cell appears blue. After adding 5% (volume) Trypan Blue to the cell suspension, no staining was observed. These tests show that the cells survived the conditions in an optical stretcher without any detectable damage.

It is not obvious that it is possible to use two 800-mW laser beams for the deformation of such delicate objects as cells without causing radiation damage. An important consideration is the careful choice of the least damaging wavelength. For the trapping of inanimate matter, such as glass or silica beads, the choice of wavelength is not critical. Using a short wavelength might be desirable for optical tweezers because it results in higher gradients and better trapping efficiencies because the spot size of a focused beam is about half the wavelength of the light used. However, short wavelengths are not appropriate to preserve biological objects such as cells because the absorption by chromophores in cells is low in the infrared and increases with decreasing wavelength (Svoboda and Block, 1994). The absorption peaks of proteins, for example, are found in the ultraviolet region of the electromagnetic spectrum. Therefore, researchers resorted to the 1064 nm of an Nd-YAG laser and achieved better results (Ashkin et al., 1987). At first sight, this choice seems also less than optimal because 70% of a cell's weight is water (Alberts et al., 1994), which absorbs more strongly with increasing wavelength. Most cells trapped with optical tweezers do not survive light powers greater than 20–250 mW, depending on the specific cell type and the wavelength used (Ashkin et al., 1987; Ashkin and Dziedzic, 1987; Kuo and Sheetz, 1992). However, recent work on optical tweezers shows that local heating of water can be ruled out as limiting factor in optical trapping experiments in the near infrared region. Theoretical calculations predict a temperature increase of less than 3 K/100 mW in the wavelength range 650-1050 nm for durations of up to 10 s (Schönle and Hell, 1998). This is in agreement with earlier experimental work, where an average temperature increase of $<1.00 \pm 0.30°C/100$ mW at powers up to 400 mW was observed in Chinese hamster ovary cells (CHO cells) trapped in optical tweezers at a wavelength of 1064 nm (Liu et al., 1996).

Even though heating due to water absorption seems to be unimportant in trapping experiments, the cells could still be damaged by radiation in other ways. Studies directly monitoring metabolic change and cellular viability of biological samples trapped in optical tweezers addressed this concern. Microfluorometric measurements on CHO cells (Liu et al., 1996) show that up to 400 mW of 1064-nm cw laser light does not change the DNA structure or cellular pH. Still, the right choice of wavelength is important, because rotating *E. coli* assays reveal that there is photodamage with maxima at 870 and 930 nm and minima at 830 and 970 nm (Neuman et al., 1999). There seems to be evidence that the presence of oxygen is involved in the damage pathway but the direct origin of the damage remains unclear. According to this study, damage at 785 nm, the wavelength we used, is at least as small as at 1064 nm, the most commonly used wavelength for biological trapping experiments. Also, the sensitivity to light was found to be linearly related to the intensity, which rules out multi-photon processes.

In the light of these findings, it is not surprising that we did not observe any damage to the cells trapped and stretched in the optical stretcher. This becomes even more plausible if one considers that the beams are not focused and the power densities are lower than in optical tweezers by about two orders of magnitude for the same light power ($\sim 10^5$ rather than 10^7 W/cm^2). Thus, much higher light powers, i.e., higher forces, can obviously be used without the danger of "optication".

Cytoskeleton of cells in suspension

Finally, because there was some concern that eukaryotic cells, apart from leukocytes, might dissolve their actin cytoskeleton when they are in suspension, we examined the actin cytoskeleton of BALB 3T3 fibroblasts in suspension using TRITC-phalloidin labeling and fluorescence microscopy. Figure 14 clearly shows that suspended cells do have an extensive actin network throughout the whole cell. In particular, the peripheral actin cortex can be seen at the plasma membrane. The only features of the cytoskeleton not present in suspension are stress fibers, which is consistent with the absence of focal adhesion plaques in suspended cells. Even without stress fibers, the BALB 3T3 displayed a large resistance to deformation in the optical stretcher. Stress fibers are predominantly seen in cells adhered to a substrate and are less pronounced in cells embedded in a tissue matrix. Thus, the situation with many stress fibers is probably as unphysiologic as the lack thereof in suspended cells.

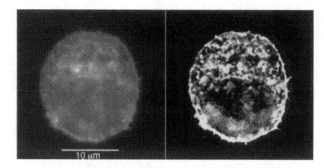

FIGURE 14 Fluorescence image of a BALB 3T3 fibroblast in suspension. The cell had been fixed and the actin cytoskeleton labeled with TRITC-phalloidin. The left panel shows the original image and the right panel shows the same cross-section through the cell after deconvolution. Clearly, the cell maintains a filamentous actin cytoskeleton even in suspension.

Our finding is in agreement with other studies that report the disappearance of stress fibers and a redistribution of actin. Furthermore, these studies show that the influence of the cytoskeleton on the deformability of normally adherent cells such as bovine aortic endothelial cells (Sato et al., 1987), chick embryo fibroblasts (Thoumine and Ott, 1997), and rat embryo fibroblasts (Heidemann et al., 1999) can be investigated when the cells are in suspension.

This view is supported by recent frequency-dependent AFM microrheology experiments, comparing in vitro actin gels and NIH3T3 fibroblasts (Mahaffy et al., 2000). These experiments showed that the viscoelastic signature of the cells resembled that of homogeneous actin gels, so that the elastic strength of cells can be almost entirely attributed to the actin cortex, which is still present in suspended cells. There is actually an advantage to investigating cells in suspension, because most polymer theories are for in vitro, isotropic actin networks. It might be especially interesting to be able to compare these theories with data from living cells in which only the actin cortex is present and anisotropic structures such as stress fibers are absent.

CONCLUSION AND OUTLOOK

Although the possibility of optically trapping biological matter with lasers is well known and commonly used, the optical deformability of dielectric matter had been ignored. The optical stretcher proves to be a nondestructive optical tool for the quantitative deformation of cells. The forces exerted by light due to the momentum transferred are sufficient not only to hold and move objects, but also to directly deform them. The important point is that the momentum is predominantly transferred to the surface of the object. The total force, i.e., gradient and scattering force, which traps the object, arises from the asymmetry of the resulting surface stresses when the object is displaced from its equilibrium position and acts on the center of gravity of the object. These trapping forces are significantly smaller than the forces on the surface because the surface forces almost completely cancel upon integration. This is most obvious in a two-beam trap as used for the optical stretcher. When the dielectric object is in the center of the optical trap, the total force is zero, whereas the forces on the surface can be as high as several hundred pico-Newton. In synopsis, the forces applicable for cell elasticity measurements with the optical stretcher range from those possible with conventional optical tweezers to those achieved with atomic force microscopes.

Somewhat surprising was the finding that the surface forces pull on the surface rather than compressing it. The RO approach readily explains this behavior by the increase of the light's momentum as it enters the denser medium and the resulting stresses on the surface. An equivalent approach would be to think in terms of minimization of energy. It is energetically favorable for a dielectric object to have as much of its volume located in the area with the highest intensity along the laser beam axis. The result is that the cell is pulled toward the axis. Even though this is conceptionally correct, it would be much harder to calculate.

The optical deformability of cells can be used to distinguish between different cells by detecting phenomenological differences in their elastic response as demonstrated for RBCs and BALB 3T3 fibroblasts. At the same time, trapped cells did not show any sign of radiation damage, even when stretched with up to 800 mW of light in each beam, and maintained a representative cytoskeleton in suspension. In the same way, the optical stretcher can be used for quantitative research on the cytoskeleton. For example, it is commonly believed that the actin part of the cytoskeleton, and especially the actin cortex, is most important for the elasticity of the cell. This hypothesis can be easily tested using the optical stretcher to measure the elasticity after genetically altering the relative amount of the three different main cytoskeletal components and their accessory proteins.

There might also be a biomedical application of the optical stretcher. Due to its simple setup and the incorporation of an automated flow chamber, the optical stretcher has the potential to measure the elasticity of large numbers of cells in a short amount of time. We expect to be able to measure one cell per second, which is a large number of cells compared to existing methods for measuring cell elasticity. Given the limited lifespan of a living cell sample in vitro, the measurement frequency is crucial to ensure good statistics. This naturally suggests the optical stretcher for applications in the research and diagnosis of diseases that result in abnormalities of the cytoskeleton. Better knowledge of the basic cell biology of the cytoskeleton contributes to the understanding of these disorders and can affect diagnosis and therapy of these diseases. Cytoskeletal changes are significant in, and are even used to diagnose, certain diseases such as cancer. Existing methods of cancer detection rely on markers and optical inspection (Sidransky,

1996). Using measurements of cytoskeletal elasticity as an indicator for malignancy could be a novel approach in oncology. For example, effects of malignancy on the cytoskeleton that have been reported include increasing disorder of the actin cytoskeleton (Koffer et al., 1985; Takahashi et al., 1986), changes in the absolute amount of total actin and the relative ratio of the various actin isoforms (Wang and Goldberg, 1976; Goldstein et al., 1985; Leavitt et al., 1986; Takahashi et al., 1986; Taniguchi et al., 1986), an overexpression of gelsolin in breast cancer cells (Chaponnier and Gabbiani, 1989), and the lack of filamin in human malignant melanoma cells (Cunningham et al., 1992). All of these changes will likely result in an altered viscoelastic response of these cells that can be detected with the optical stretcher. In fact, models of actin networks (MacKintosh et al., 1995) show that the shear modulus scales with actin concentration raised to 2.2. Even a slight decrease in actin concentration should result in a detectable decrease of the cell's elasticity. In this way, the optical stretcher could advance to a diagnostic tool in clinical laboratories. This novel technique would require minimal tissue samples, which could be obtained using cytobrushes on the surfaces of the lung, esophagus, stomach, or cervix, or by fine-needle aspiration using stereotactic, ultrasonographic, or MRI guidance (Dunphy and Ramos, 1997; Fajardo and DeAngelis, 1997).

The authors would like to thank Alan Chiang, Benton Pahlka, Christian Walker, and Robert Martinez for their help with the experiments. We are grateful to Rebecca Richards-Kortum, John Wright (National Science Foundation Integrative Graduate Education and Research Training program user facility), Carole Moncman, Eric Okerberg, and Kung-Bin Sung for their invaluable suggestions and assistance with the fluorescence microscopy, and to David Humphrey, Martin Forstner, and Douglas Martin for many supporting discussions.

The work was supported by the Whitaker Foundation grant #26-7504-94, and by the National Institutes of Health grant 26-1601-1683.

REFERENCES

Adelman, M. R., G. G. Borisy, M. L. Shelanski, R. C. Weisenberg, and E. W. Taylor. 1968. Cytoplasmic filaments and tubules. *Fed. Proc.* 27:1186–1193.

Aebi, U., J. Cohn, L. Buhle, and L. Gerace. 1986. The nuclear lamina is a meshwork of intermediate-type filaments. *Nature.* 323:560–564.

Alberts, B., D. Bray, J. Lewis, M. Raff, K. Roberts, and J. D. Watson. 1994. Molecular Biology of the Cell. Garland Publishing, New York. 786–861.

Allen, P. G., and P. A. Janmey. 1994. Gelsolin displaces phalloidin from actin filaments. A new fluorescence method shows that both Ca^{2+} and Mg^{2+} affect the rate at which gelsolin severs F-actin. *J. Biol. Chem.* 269:32916–32923.

Ashkin, A. 1970. Acceleration and trapping of particles by radiation pressure. *Phys. Rev. Lett.* 24:156–159.

Ashkin, A., and J. M. Dziedzic. 1973. Radiation pressure on a free liquid surface. *Phys. Rev. Lett.* 30:139–142.

Ashkin, A., and J. M. Dziedzic. 1987. Optical trapping and manipulation of viruses and bacteria. *Science.* 235:1517–1520.

Ashkin, A., J. M. Dziedzic, J. E. Bjorkholm, and S. Chu. 1986. Observation of a single-beam gradient force optical trap for dielectric particles. *Opt. Lett.* 11:288–290.

Ashkin, A., J. M. Dziedzic, and T. Yamane. 1987. Optical trapping and manipulation of single cells using infrared laser beams. *Nature.* 330:769–771.

Barer, R., and S. Joseph. 1954. Refractometry of living cells, part I. Basic principles. *Q. J. Microsc. Sci.* 95:399–423.

Barer, R., and S. Joseph. 1955a. Refractometry of living cells, part II. The immersion medium. *Q. J. Microsc. Sci.* 96:1–26.

Barer, R., and S. Joseph. 1955b. Refractometry of living cells, part III. Technical and optical methods. *Q. J. Microsc. Sci.* 96:423–447.

Bennett, V. 1985. The membrane skeleton of human erythrocytes and its implication for more complex cells. *Annu. Rev. Biochem.* 54:273–304.

Bennett, V. 1990. Spectrin-based membrane skeleton—a multipotential adapter between plasma membrane and cytoplasm. *Physiol. Rev.* 70:1029–1060.

Block, S. M., L. S. B. Goldstein, and B. J. Schnapp. 1990. Bead movement by single kinesin molecules studied with optical tweezers. *Nature.* 348:348–352.

Brevik, I. 1979. Experiments in phenomenological electrodynamics and the electromagnetic energy-momentum tensor. *Phys. Rep.* 52:133–201.

Carlier, M. F. 1998. Control of actin dynamics. *Curr. Opin. Cell Biol.* 10:45–51.

Carlson, R. H., C. V. Gabel, S. S. Chan, R. H. Austin, J. P. Brody, and J. W. Winkelman. 1997. Self-sorting of white blood cells in a lattice. *Phys. Rev. Lett.* 79:2149–2152.

Chaponnier, C., and G. Gabbiani. 1989. Gelsolin modulation in epithelial and stromal cells of mammary carcinoma. *Am. J. Pathol.* 134:597–603.

Chu, S. 1991. Laser manipulation of atoms and particles. *Science.* 253:861–866.

Colon, J. M., P. G. Sarosi, P. G. McGovern, A. Ashkin, and J. M. Dziedzic. 1992. Controlled micromanipulation of human sperm in three dimensions with an infrared laser optical trap: effect on sperm velocity. *Fertil. Steril.* 57:695–698.

Constable, A., J. Kim, J. Mervis, F. Zarinetchi, and M. Prentiss. 1993. Demonstration of a fiber-optical light-force trap. *Opt. Lett.* 18:1867–1869.

Cooper, J. A., J. Bryan, B. Schwab, III, C. Frieden, D. J. Loftus, and E. L. Elson. 1987. Microinjection of gelsolin into living cells. *J. Cell Biol.* 104:491–501.

Cunningham, C. C., J. B. Gorlin, D. J. Kwiatkowski, J. H. Hartwig, P. A. Janmey, H. R. Byers, and T. P. Stossel. 1992. Actin-binding protein requirement for cortical stability and efficient locomotion. *Science.* 255:325–327.

Dailey, B., E. L. Elson, and G. I. Zahalak. 1984. Cell poking. Determination of the elastic area compressibility modulus of the erythrocyte membrane. *Biophys. J.* 45:661–682.

Discher, D. E., N. Mohandas, and E. A. Evans. 1994. Molecular maps of red cell deformation: hidden elasticity and in situ connectivity. *Science.* 266:1032–1035.

Dunphy, C., and R. Ramos. 1997. Combining fine-needle aspiration and flow cytometric immunophenotyping in evaluation of nodal and extranodal sites for possible lymphoma: a retrospective review. *Diagn. Cytopathol.* 16:200–206.

Eichinger, L., B. Köppel, A. A. Noegel, M. Schleicher, M. Schliwa, K. Weijer, W. Wittke, and P. A. Janmey. 1996. Mechanical perturbation elicits a phenotypic difference between dictyostelium wild-type cells and cytoskeletal mutants. *Biophys. J.* 70:1054–1060.

Elson, E. L. 1988. Cellular mechanics as an indicator of cytoskeletal structure and function. *Annu. Rev. Biophys. Biophys. Chem.* 17:397–430.

Evans, E., and Y. C. Fung. 1972. Improved measurements of the erythrocyte geometry. *Microvasc. Res.* 4:335–347.

Fajardo, L. L., and G. A. DeAngelis. 1997. The role of stereotactic biopsy in abnormal mammograms. *Surg. Oncol. Clin. North Am.* 6:285–299.

Felder, S., and E. L. Elson. 1990. Mechanics of fibroblast locomotion: quantitative analysis of forces and motions at the leading lamellas of fibroblasts. *J. Cell Biol.* 111:2513–2526.

Guck, J., R. Ananthakrishnan, T. J. Moon, C. C. Cunningham, and J. Käs. 2000. Optical deformability of soft biological dielectrics. *Phys. Rev. Lett.* 84:5451–5454.

Goldstein, D., J. Djeu, G. Latter, S. Burbeck, and J. Leavitt. 1985. Abundant synthesis of the transformation-induced protein of neoplastic human fibroblasts, plastin, in normal lymphocytes. *Cancer Res.* 45:5643–5647.

Heidemann, S. R., S. Kaech, R. E. Buxbaum, and A. Matus. 1999. Direct observations of the mechanical behaviors of the cytoskeleton in living fibroblasts. *J. Cell Biol.* 145:109–122.

Hénon, S., G. Lenormand, A. Richert, and F. Gallet. 1999. A new determination of the shear modulus of the human erythrocyte membrane using optical tweezers. *Biophys. J.* 76:1145–1151.

Herrmann, H., and U. Aebi. 1998. Structure, assembly, and dynamics of intermediate filaments. *Subcell. Biochem.* 31:319–362.

Hochmuth, R. M. 1993. Measuring the mechanical properties of individual human blood cells. *J. Biomech. Eng.* 115:515–519.

Jackson, J. D. 1975. Classical Electrodynamics. John Wiley and Sons, New York. 281–282.

Janmey, P. A. 1991. Mechanical properties of cytoskeletal polymers. *Curr. Opin. Cell Biol.* 3:4–11.

Janmey, P. A., V. Euteneuer, P. Traub, and M. Schliwa. 1991. Viscoelastic properties of vimentin compared with other filamentous biopolymer networks. *J. Cell Biol.* 113:155–160.

Janmey, P. A., J. Peetermans, K. S. Zaner, T. P. Stossel, and T. Tanaka. 1986. Structure and mobility of actin filaments as measured by quasielastic light scattering, viscometry, and electron microscopy. *J. Biol. Chem.* 261:8357–8362.

Janmey, P. A., J. V. Shah, K. P. Janssen, and M. Schliwa. 1998. Viscoelasticity of intermediate filament networks. *Subcell. Biochem.* 31:381–397.

Käs, J., H. Strey, J. X. Tang, D. Finger, R. Ezzell, E. Sackmann, and P. A. Janmey. 1996. F-actin, a model polymer for semiflexible chains in dilute, semidilute, and liquid crystalline solutions. *Biophys. J.* 70:609–625.

Koffer, A., M. Daridan, and G. Clarke. 1985. Regulation of the microfilament system in normal and polyoma virus transformed cultured (BHK) cells. *Tissue Cell.* 17:147–159.

Kuo, S. C., and M. P. Sheetz. 1992. Optical tweezers in cell biology. *Trends Cell Biol.* 2:116–118.

Kuo, S. C., and M. P. Sheetz. 1993. Force of single kinesin molecules measured with optical tweezers. *Science.* 260:232–234.

Leavitt, J., G. Latter, L. Lutomski, D. Goldstein, and S. Burbeck. 1986. Tropomyosin isoform switching in tumorigenic human fibroblasts. *Mol. Cell. Biol.* 6:2721–2726.

Liu, Y., G. J. Sonek, M. W. Berns, and B. J. Tromberg. 1996. Physiological monitoring of optically trapped cells: assessing the effects of confinement by 1064-nm laser tweezers using microfluorometry. *Biophys. J.* 71:2158–2167.

Lodish, H., D. Baltimore, A. Berk, S. L. Zipurski, P. Matsudaira, and J. Darnell. 1995. Molecular Cell Biology. Scientific American Books, New York. 1051–1059.

MacKintosh, F. C., J. Käs, and P. A. Janmey. 1995. Elasticity of semiflexible biopolymer networks. *Phys. Rev. Lett.* 75:4425–4428.

Mahaffy, R. E., C. K. Shih, F. C. MacKintosh, and J. Käs. 2000. Scanning probe-based frequency-dependent microrheology of polymer gels and biological cells. *Phys. Rev. Lett.* 85:880–883.

Mazurkiewicz, Z. E., and R. T. Nagorski. 1991. Shells of Revolution. Elsevier Publishing, New York. 360–369.

Mitchison, T., L. Evans, E. Schulze, and M. Kirschner. 1986. Sites of microtubule assembly and disassembly in the mitotic spindle. *Cell.* 45:515–527.

Mitchison, T. J. 1992. Compare and contrast actin filaments and microtubules. *Mol. Biol. Cell.* 3:1309–1315.

Mohandas, N., and E. Evans. 1994. Mechanical properties of the red cell membrane in relation to molecular structure and genetic defects. *Annu. Rev. Biophys. Biomol. Struct.* 23:787–818.

Neuman, K. C., E. H. Chadd, G. F. Liou, K. Bergman, and S. M. Block. 1999. Characterization of photodamage to *Escherichia coli* in optical traps. *Biophys. J.* 77:2856–2863.

Pasternak, C., and E. L. Elson. 1985. Lymphocyte mechanical response triggered by cross-linking surface receptors. *J. Cell Biol.* 100:860–872.

Petersen, N. O., W. B. McConnaughey, and E. L. Elson. 1982. Dependence of locally measured cellular deformability on position on the cell, temperature, and cytochalasin B. *Proc. Natl. Acad. Sci. U.S.A.* 79:5327–5331.

Pollard, T. D. 1984. Molecular architecture of the cytoplasmic matrix. *Kroc Found. Ser.* 16:75–86.

Pollard, T. D. 1986. Assembly and dynamics of the actin filament system in nonmuscle cells. *J. Cell. Biochem.* 31:87–95.

Radmacher, M., M. Fritz, C. M. Kacher, J. P. Cleveland, and P. K. Hansma. 1996. Measuring the viscoelastic properties of human platelets with the atomic force microscope. *Biophys. J.* 70:556–567.

Roosen, G. 1977. A theoretical and experimental study of the stable equilibrium positions of spheres levitated by two horizontal laser beams. *Opt. Commun.* 21:189–195.

Rotsch, C., and M. Radmacher. 2000. Drug-induced changes of cytoskeletal structure and mechanics in fibroblasts: an atomic force microscopy study. *Biophys. J.* 78:520–535.

Sato, M., M. J. Levesque, and R. M. Nerem. 1987. An application of the micropipette technique to the measurement of the mechanical properties of cultured bovine aortic endothelial cells. *J. Biomech. Eng.* 109:27–34.

Schönle, A., and S. W. Hell. 1998. Heating by absorption in the focus of an objective lens. *Opt. Lett.* 23:325–327.

Shepherd, G. M., D. P. Corey, and S. M. Block. 1990. Actin cores of hair-cell stereocilia support myosin motility. *Proc. Natl. Acad. Sci. U.S.A.* 87:8627–8631.

Sidransky, D. 1996. Advances in cancer detection. *Sci. Am.* 275:104–109.

Simmons, R. M., J. T. Finer, H. M. Warrick, B. Kralik, S. Chu, and J. A. Spudich. 1993. Force on single actin filaments in a motility assay measured with an optical trap. *In* The Mechanism of Myofilament Sliding in Muscle Contraction. H. Sugi and G. H. Pollack, editors. Plenum, New York. 331–336.

Sleep, J., D. Wilson, R. Simmons, and W. Gratzer. 1999. Elasticity of the red cell membrane and its relation to hemolytic disorders: an optical tweezers study. *Biophys. J.* 77:3085–3095.

Stossel, T. P. 1984. Contribution of actin to the structure of the cytoplasmic matrix. *J. Cell Biol.* 99:15s–21s.

Stossel, T. P., J. H. Hartwig, P. A. Janmey, and D. J. Kwiatkowski. 1999. Cell crawling two decades after Abercrombie. *Biochem. Soc. Symp.* 65:267–280.

Strey, H., M. Peterson, and E. Sackmann. 1995. Measurements of erythrocyte membrane elasticity by flicker eigenmode decomposition. *Biophys. J.* 69:478–488.

Svoboda, K., and S. M. Block. 1994. Biological applications of optical forces. *Annu. Rev. Biophys. Biomol. Struct.* 23:147–285.

Svoboda, K., C. F. Schmidt, B. J. Schnapp, and S. M. Block. 1993. Direct observation of kinesin stepping by optical trapping interferometry. *Nature.* 365:721–727.

Tadir, Y., W. H. Wright, O. Vafa, T. Ord, R. H. Asch, and W. M. Berns. 1990. Force generated by human sperm correlated to velocity and determined using a laser generated optical trap. *Fertil. Steril.* 53:944–947.

Takahashi, K., V. I. Heine, J. L. Junker, N. H. Colburn, and J. M. Rice. 1986. Role of cytoskeleton changes and expression of the H-ras oncogene during promotion of neoplastic transformation in mouse epidermal JB6 cells. *Cancer Res.* 46:5923–5932.

Taniguchi, S., T. Kawano, T. Kakunaga, and T. Baba. 1986. Differences in expression of a variant actin between low and high metastatic B16 melanoma. *J. Biol. Chem.* 261:6100–6106.

Thoumine, O., and A. Ott. 1997. Time scale dependent viscoelastic and contractile regimes in fibroblasts probed by microplate manipulation. *J. Cell Sci.* 110:2109–2116.

Ugural, A. C. 1999. Stresses in Plates and Shells. McGraw-Hill, New York. 339–409.

van de Hulst, H. C. 1957. Light Scattering by Small Particles. John Wiley and Sons, New York. 172–176.

Wang, E., and A. R. Goldberg. 1976. Changes in microfilament organization and surface topography upon transformation of chick embryo fibroblasts with Rous sarcoma virus. *Proc. Natl. Acad. Sci. U.S.A.* 73:4065–4069.

Wang, N., J. P. Butler, and D. E. Ingberg. 1993. Mechanotransduction across the cell surface and through the cytoskeleton. *Science.* 260:1124–1127.

Wang, N., and D. Stamenovic. 2000. Contribution of intermediate filaments to cell stiffness, stiffening, and growth. *Am. J. Physiol. Cell Physiol.* 279:C188–C194.

Zeman, K. 1989. Untersuchung physikalisch und biochemisch induzierter Änderungen der Krümmungselastizität der Erythrozytenmembran mittels Fourierspektroskopie der thermisch angeregten Oberflächenwellen (Flickern). Doktorarbeit. Physikalische Fakultät, Technische Universität München, Germany.

Elasticity of the Red Cell Membrane and Its Relation to Hemolytic Disorders: An Optical Tweezers Study

John Sleep, David Wilson, Robert Simmons, and Walter Gratzer

MRC Unit of Muscle and Cell Motility, Randall Institute, Kings College London, 26–29 Drury Lane, London WC2B 5RL, United Kingdom

ABSTRACT   We have used optical tweezers to study the elasticity of red cell membranes; force was applied to a bead attached to a permeabilized spherical ghost and the force-extension relation was obtained from the response of a second bead bound at a diametrically opposite position. Interruption of the skeletal network by dissociation of spectrin tetramers or extraction of the actin junctions engendered a fourfold reduction in stiffness at low applied force, but only a twofold change at larger extensions. Proteolytic scission of the ankyrin, which links the membrane skeleton to the integral membrane protein, band 3, induced a similar effect. The modified, unlike the native membranes, showed plastic relaxation under a prolonged stretch. Flaccid giant liposomes showed no measurable elasticity. Our observations indicate that the elastic character is at least as much a consequence of the attachment of spectrin as of a continuous membrane-bound network, and they offer a rationale for formation of elliptocytes in genetic conditions associated with membrane-skeletal perturbations. The theory of Parker and Winlove for elastic deformation of axisymmetric shells (accompanying paper) allows us to determine the function BH^2 for the spherical saponin-permeabilized ghost membranes (where B is the bending modulus and H the shear modulus); taking the literature value of 2×10^{-19} Nm for B, H then emerges as 2×10^{-6} Nm^{-1}. This is an order of magnitude higher than the value reported for intact cells from micropipette aspiration. Reasons for the difference are discussed.

INTRODUCTION

The red blood cell membrane has unique viscoelastic properties, which resemble, in some respects, those of a fluid, in others, those of a solid. Their structural basis remains a matter of debate. The high resistance of the membrane to changes in surface area is a characteristic of phospholipid bilayers, whereas its response to shear deformation depends on the cytoskeletal network of proteins that coats its cytoplasmic surface (Evans and Hochmuth, 1978; Berk et al., 1989; Mohandas and Evans, 1994).

The distortion of the cell in response to an applied mechanical stress has been observed in a variety of ways, but nearly all quantitative data on elastic and rheoviscous properties are the outcome of one technique. This is micropipette aspiration (Evans, 1973), in which a protrusion from a flaccid membrane is created and drawn into the pipette. The relation between pressure and extension of the protrusion then delivers a value of the shear elastic modulus. The relaxation rate, when the pressure is released, can also give an estimate of the shear viscosity.

We have explored the scope of the optical tweezers technique for applying a defined linear stretching force to the red cell membrane and measuring the response to fast and slow induced distortions. We have sought to define the structural features of the membrane skeleton that control the elastic properties and to relate them to the effects of hered-

itary cytoskeletal anomalies. The results show several unexpected features and differences from earlier data.

MATERIALS AND METHODS

Preparative and analytical procedures

Red blood cells were obtained from the blood bank and were no more than a week old. To prepare ghosts, cells were three times washed with isotonic phosphate-buffered saline (PBS), pH 7.6, the upper layer, containing leukocytes, being discarded each time. The cells were then suspended at 10% hematocrit and 0.1 volume of 10 mg ml^{-1} saponin was added, together, in some cases, with 1 μl/ml of cell suspension of 20 mg ml^{-1} phenylmethylsulphonyl fluoride in ethanol. The cell suspension was left at room temperature for 20 min and the ghosts and residual unlysed cells were then pelleted at $40,000 \times g$; the pale ghost layer was collected and washed three more times with PBS. Ghosts obtained by hypotonic lysis were also examined. These were prepared by adding the packed, washed cells to 20 volumes of ice-cold 10 mM sodium phosphate, pH 7.6, together with phenylmethylsulphonyl fluoride. Ghosts were collected by pelleting as before and washed 2–3 times with the lysis medium. They were then made isotonic in PBS and 1 mM in magnesium-ATP, by addition of a ten-times concentrated stock solution, and incubated for 1 h at 37°C.

To determine whether saponin lysis had caused loss of cholesterol, total lipid was extracted with 2:1 chloroform-methanol from membranes prepared by hypotonic and saponin lysis. Phospholipid in the extract was assayed by ashing with perchloric acid and determining the orthophosphate concentration by the ammonium molybdate color reaction, whereas cholesterol was assayed by the ferric chloride color reaction. Analytical procedures were as set out by Kates (1972).

To examine how the mechanical properties of the membrane are related to the known protein–protein interactions that maintain the integrity of the membrane-associated network, three types of modification were undertaken. The targets of the three modifying agents are depicted schematically in Fig. 1 A and the results of the modifications, as revealed by gel electrophoresis, are shown in Fig. 1 B.

To effect dissociation of the spectrin tetramers to dimers in situ (Fischer et al., 1978), the saponin-lysed ghosts were suspended at their original concentration in PBS, adjusted to pH 7.0. The suspension was then made 2 mM in N-ethylmaleimide (NEM) and left to react at room temperature for

Received for publication 21 January 1999 and in final form 31 July 1999.

Address reprint requests to Dr. John Sleep, Department of Biophysics, Cell and Molecular Biology, The Randall Institute, King's College, 26–29 Drury Lane, London WC2B 5RL, U.K. Tel.: +44-171-836-8851; Fax: 44-171-497-9078; E-mail: john@muscle.rai.kcl.ac.uk.

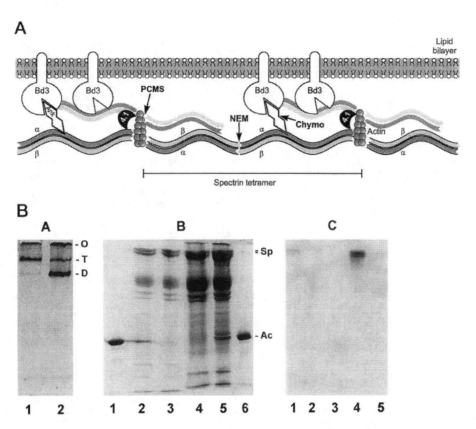

FIGURE 1 (*A*) Schematic representation of the red cell membrane structure and sites of functional modifications. The membrane skeletal network is composed of an approximately hexagonal lattice, in which the structural members are spectrin tetramers, made up of $\alpha\beta$ heterodimers, associated head-to-head. The junctions of the lattice consist of short actin filaments, together with protein 4.1, which binds to spectrin and actin. Each spectrin tetramer is attached by way of an ankyrin molecule to a population of the transmembrane protein, band 3. Arrows show the sites of action of the modifying reagents. NEM dissociates some 70% of the spectrin tetramers to dimers, PCMS causes dissociation of the actin protofilaments, with complete loss of actin from the membrane, and chymotrypsin (Chymo) severs the ankyrin, and thus the link between spectrin and band 3, without significantly degrading other proteins. (*B*) Modifications to membrane structure. *Panel A*, Dissociation of spectrin tetramers in situ by treatment with NEM: spectrin was extracted in the cold from NEM-treated ghosts. Gel electrophoresis shows that spectrin from untreated ghosts (*lane 1*) remained almost entirely tetrameric, whereas that from the modified ghosts was largely (~70%) dissociated into dimers (*lane 2*). O denotes the origin of migration (top of the gel). The staining here represents the oligomeric fraction of spectrin, associated with actin and 4.1. T is spectrin tetramer, and D the dimer. *Panel B*, Dissociation of membrane skeletal actin protofilaments by treatment of ghosts with *p*-chloromercuriphenyl sulphonate. SDS-polyacrylamide gel electrophoresis shows membrane proteins of untreated ghosts applied at two concentrations (*lanes 2 and 5*) and of treated ghosts at the same concentrations as the untreated (*lanes 3 and 4*). Lanes 1 and 6 show the migration of pure actin. Sp denotes the spectrin doublet and Ac actin. The only proteins that can be seen to have been affected by the reagent are actin and the minor actin-binding protein, band 4.9, migrating just above the actin. *Panel C*, Proteolytic cleavage of ankyrin with chymotrypsin: electroblot of SDS-gel, stained with anti-ankyrin followed by second antibody. The undigested ghosts were applied on lane 4 and purified spectrin on lane 3. Lane 1 shows traces of ankyrin remaining after a digestion of 6 min, whereas no ankyrin is detected after a digestion of 180 min (*lane 2*) or after 20 min (*lane 5*), the conditions used to generate the modified ghosts for the optical tweezers experiments. The gel stained with Coomassie Brilliant Blue R revealed no detectable loss of any other zones after the 20-minute treatment.

1 hr. The ghosts were recovered by pelleting, a twofold molar excess of dithiothreitol over the original NEM concentration was added, and the ghosts were washed twice with PBS, pH 7.6. To assay the extent of tetramer dissociation, the spectrin was extracted by washing the ghosts with ice-cold 0.25 mM phosphate, pH 8, and dialyzing against the same buffer in the cold. The spectrin was recovered in the supernatant after centrifugation at $100,000 \times g$ for 15 min and analyzed by electrophoresis in a 4.5% polyacrylamide gel in a 0.1 M Tris-Bicine buffer system, run in the cold. The gel, stained with Coomassie Brilliant Blue R250, was evaluated by densitometry (Fig. 1 *B*, panel A). The spectrin tetramers were also dissociated in situ by incubating ghosts in 10 mM sodium phosphate, pH 8.0, at 37°C for 1 hr (Liu and Palek, 1980). They were then made isotonic and kept cold. The spectrin was extracted and analyzed by gel electrophoresis in the cold, as above.

Membrane-skeletal actin was dissociated from the protein network by the method of Gordon and Ralston (1990): ghosts were washed with 5 mM phosphate, pH 8.2, and pelletted. To the loose pellet 0.1 volume of 10 mM *p*-chloromercuriphenylsulphonic acid (PCMS), adjusted to pH 8.0, was added, and the reaction was allowed to proceed for 1 h on ice. The pellet was then suspended in PBS, pH 7.6, containing a fivefold molar excess of dithiothreitol over PCMS, and the ghosts were washed twice with PBS. To determine the extent of extraction of the actin, an aliquot of the ghost preparation was dissolved in 1% sodium dodecyl sulphate, diluted tenfold with water, followed by 0.1 volume of a ten-times concentrated polyacrylamide gel sample buffer (Laemmli, 1970), containing sucrose, β-mercaptoethanol, and Bromophenol Blue tracker dye, heated at 100°C for 5 min and applied to a 9% sodium dodecylsulphate-polyacrylamide gel. Muscle actin was also applied to the gel to ensure correct identification of the actin zone (Fig. 1 *B, panel B*). To determine whether the reagent had caused dissociation of spectrin tetramers, spectrin was extracted in the cold, as described above, from an aliquot of the treated ghosts and analyzed by polyacrylamide gel electrophoresis in the cold in the absence of denaturant.

Cleavage of ankyrin in unsealed hypotonic ghosts occurs on very mild treatment with trypsin (Jinbu et al., 1984) or chymotrypsin (Pinder et al., 1995), with little or no detectable damage to other proteins. We found that scission of ankyrin in ghosts derived from saponin-lysed cells at physiological ionic strength required considerably greater exposure to the enzyme, but that the other membrane-associated proteins were correspondingly more resistant. Freshly dissolved chymotrypsin was added to ghost pellets on ice to give a concentration of 10 μg ml^{-1}. Aliquots taken at various times were quenched with excess chymostatin. The ghosts were washed twice with PBS and prepared as before for SDS-polyacrylamide gel electrophoresis. To analyze the residual content of intact ankyrin the ghost protein was dissolved in SDS-gel sample buffer as before and separated in 5.6% gels in the buffer system of Fairbanks et al. (1971). Duplicate gel lanes were stained with Coomassie Brilliant Blue R250 or electroblotted onto nitrocellulose membrane. The membrane was blocked with milk powder, then treated with a rabbit polyclonal anti-ankyrin antibody (prepared by J.C. Pinder in this laboratory, Randall Institute, King's College, London) and the blots were developed with second antibody, using the chemiluminescent (ECL) system (Amersham, Little Chalfont, U.K.). Analysis by this method revealed that, under the above conditions, a chymotryptic digestion of 20 min sufficed to eliminate all intact ankyrin, band 2.1 (Fig. 1 *B, panel C*). Electrophoresis revealed no detectable damage to any other proteins; in particular, protein 4.1, which is very sensitive to proteolysis, remained intact. In addition, we extracted spectrin from the chymotrypsin-treated ghosts and examined it electrophoretically as above, to show that there was no conversion of tetramers to dimers.

Giant phospholipid vesicles with the composition of the red cell membrane, prepared by the method of Käs and Sackmann (1991), were given to us by Dr. P. McCauley (Imperial College, London).

Polystyrene latex beads of 1 μm diameter, derivatized with aldehyde groups (Interfacial Dynamics Corp., Portland, Orgeon), were incubated for 2 h at room temperature with 0.1 mg ml^{-1} wheat germ agglutinin (Sigma, Poole, U.K.) in 0.3 M sodium chloride, 40 mM sodium borate, pH 8.2. An excess of neutralized glycine was then added, and the beads were washed by pelleting from suspensions in PBS. For experiments on the phospholipid vesicles, the beads were similarly reacted with annexin V (Alexis Corp., San Diego, California), which had first been dialyzed against the reaction medium to remove the admixture of glycine added by the manufacturer.

Optical tweezers

The apparatus was based on an inverted microscope (Zeiss Axiovert, Oberkochen, Germany). Polarizing prisms were used to split the laser beam (Nd-YLF 1.047 μm TFR, Spectra Physics, Mountain View, California) to give two independently movable single-beam gradient traps, the stiffnesses of the two traps being equalized with a $\lambda/2$ plate. A 63×, infinity-corrected, 1.4 N.A. objective was used for these studies. Trap stiffness was measured from the Brownian motion of a trapped bead, and the values derived from the corner frequency and from the standard deviation of the bead position were in satisfactory agreement (0.08 pN nm^{-1}). In the experiments, one of the traps was left stationary, and the position of the bead in this trap was monitored by focussing an image onto a photodiode quadrant detector. The position of the second bead was controlled by a pair of acousto-optical modulators, which allowed the trap to be moved rapidly over a range of about 4 μm. The method measured the compliance of the link between the two beads but did not exclude the possibility that the two ends of the cell, or more likely the bead cell linkages at each end, were behaving differently. To address this question, the positions of the two beads relative to the center of the cell were analyzed in a series of 25 video frames. On average, the compliance of the link between each of the two beads and the center of the cell differed by 30%. The compliance of the stronger of the two links, which might be regarded as the best estimate, is about 15% less than the average value we report.

RESULTS

Principles of the measurements

The principle of the optical-tweezers approach to the measurement of elastic properties consists in attaching two adhesive beads to the cell at opposite ends of a diameter, and holding one in place with one trap, while moving the other with a second trap, to induce a tension (positive or negative) in the cell. By monitoring the movement of the first bead in response to the controlled displacement of the second bead, a force-extension profile can be generated. Two types of force-extension curves were investigated: in the first, the periodic method, a triangular wave motion (1–2 Hz, which is well below the frequency at which a lag in response, indicative of viscous relaxation, develops) was imposed on the moving trap; in the second, the stepwise method, a series of stepped increases in length were imposed on the cell, the time between steps being such as to allow the tension in the cell to relax back to close to its equilibrium value and thus reveal whether it undergoes plastic yield.

A number of procedures were tried for generating beads that would attach tightly to the cell exterior. The best results were obtained with aldehyde-derivatized beads, to which wheat germ agglutinin was covalently coupled. This lectin has high affinity for the carbohydrate of the exposed sialoglycoproteins, the glycophorins. Beads prepared in this way were used for all studies described here, other than those on synthetic phospholipid vesicles.

Properties of red cell ghosts

The application of the optical-tweezers technique to the native biconcave cell or to the resealed ghost presents a number of problems. The opacity of intact cells reduces the precision with which the position of an attached bead can be determined with the quadrant detector. More fundamental is the difficulty of interpreting force-extension profiles for objects that do not have rotational symmetry about the axis joining the beads. Finally, because of the impermeability of the membrane, intact cells change shape under conditions of constant volume, which means that the response to strain is very dependent upon the pressure difference across the membrane; the mechanical properties of the protein cytoskeleton may then no longer be the dominant influence on the force-extension curve. For these reasons, we have chosen to study ghosts prepared by saponin lysis. Saponin belongs to a group of glycosylated sterols, which includes digitonin, and interacts specifically with membrane cholesterol to cause permeabilization of the membrane (Elias et al., 1978). The holes are too small to be visible in the electron microscope, although they allow the passage of large proteins (Seeman, 1967). The only perceptible structural changes in the membrane are the appearance of small surface pits, 40–50 Å across (Seeman et al., 1973), and corrugations in the freeze-fracture faces (Elias et al., 1978).

We have found that ghosts generated in this manner are mainly spherical, with a smooth contour and a relatively uniform appearance. The membrane composition appears to be unchanged, or at all events, the cholesterol:phospholipid ratio was the same as that in ghosts prepared by hypotonic lysis within the precision (~5%) of our assay. Saponin-lysed cells were previously shown to freely admit proteins, such as G-actin and gelsolin (Pinder et al., 1986). We have further shown that, when spectrin, labeled with fluorescein isothiocyanate (freed of excess reagent by gel filtration), was added to the ghosts, the fluorescent intensity inside the ghost had become indistinguishable from that in the surrounding buffer within the time (a few seconds) between mixing and observation. Resealed hypotonically generated ghosts, by contrast, showed no penetration of fluorescence into the lumen. Considering the difference of two orders of magnitude between the diffusion coefficients of spectrin and of water, we therefore take it that volume equilibration in the saponin-lysed ghosts is effectively instantaneous, and the constant-volume restriction is therefore eliminated. The uniformity of shape is an additional important advantage for optical trap experiments.

Comparison of the saponin-treated ghosts with intact red cells and with smooth resealed hypotonically generated ghosts shows that the force-extension curves are broadly similar, and it is therefore likely that the elastic characteristics of the membrane are little changed by exposure to saponin.

On application of the maximum force used in this study of 25 pN, the axial ratio of the ghost increased from unity to about 1.2. The cells were also compressed for half the cycle, but, in this case, it cannot be ensured that the pressure remains orthogonal to the membrane surface, and lateral displacement of the bead was often observed. No quantitative analysis in the compressive part of the force-extension profile was therefore attempted.

Phosphorylation of membrane proteins

Incubation of the ghosts with magnesium-ATP to phosphorylate spectrin and other membrane skeletal proteins made no detectable difference to the shape of the ghost or to the force-extension profiles. This is consistent with the lack of any such metabolic effect on the elastic properties of red cells, measured by the micropipette technique (Meiselman et al., 1978) (though, of course, in intact cells or resealed ghosts, ATP depletion induces shape changes). The ATP incubation was therefore omitted in most experiments.

Effects of structural perturbations of the membrane skeleton

Modifications of the membrane cytoskeleton in the ghosts were undertaken in an attempt to identify the structural features associated with the elastic properties. The protein–protein interactions responsible for the cohesion of the membrane-cytoskeletal complex can be separated into "horizontal" and "vertical" kinds, that is, those in the plane of the skeletal network and those orthogonal to the plane of the membrane and affecting, therefore, the membrane–network interaction. Many genetic defects in both these categories have been discovered and are associated with characteristic pathological phenotypes (Lux and Palek, 1995). Figure 1 *A* shows, in schematic form, the sites of the modifications that we have carried out.

The most common horizontal defects, which give rise to hereditary elliptocytosis and hemolytic disease, are spectrin mutations in the self-association site of the $\alpha\beta$-dimers that form the structural members of the network (spectrin tetramers) by interacting head-to-head. This condition can be reproduced by treating the ghosts with *N*-ethylmaleimide (Fischer et al., 1978) or by incubating them at low ionic strength (Liu and Palek, 1980). Both treatments dissociate a maximum of about 70% of the spectrin tetramers into dimers (Fig. 1 *B*, panel *A*). Most of our experiments were performed on the *N*-ethylmaleimide-treated ghosts because of the slow reversal of the electrostatically induced dissociation when the ghosts are returned to an isotonic medium at room temperature.

A still more radical disruption of the membrane skeletal network can be achieved by dissociating the short actin filaments that make up the lattice junctions, with complete loss of actin from the membrane. The thiol-specific organomercurial, PCMS, has been shown to react with membrane skeletal actin, as well as with spectrin, and only to a small extent with any other major membrane-associated proteins (Gordon and Ralston, 1990). We found, in agreement with these workers, that, under their conditions of reaction, the actin was almost entirely lost (Fig. 1 *B*, *panel B*). Extraction of these ghosts at low ionic strength in the cold liberated only a minor proportion of the spectrin, but this fraction was tetrameric. It seems likely, therefore, that the action of the reagent is effectively confined to the elimination of actin, together with the minor actin-binding protein, band 4.9, as observed by Gordon and Ralston (1990).

To examine the effects of disturbing vertical interactions in the membrane, we found conditions for cleaving the ankyrin by proteolysis, with no discernible degradation of any other proteins. These were based on earlier observations on proteolysis of open ghosts at low ionic strength (Jinbu et al., 1984; Pinder et al., 1995). We found that all the ankyrin was degraded (Fig. 1 *B*, *panel C*), presumably into its spectrin- and band 3-binding domains (Hall and Bennett, 1987). The resulting ghosts were somewhat irregular in outline and showed some tendency to throw off vesicles.

Force-extension profiles

The response of an unmodified ghost to a periodic applied force and a typical force-extension relationship are shown in

Fig. 2. Two factors that could be imagined to limit the reproducibility of the measurements are the natural variation in cell size within a single blood sample (Jay, 1975) and the precision with which the two beads can be positioned at opposite ends of a diameter. However, the theory of Parker and Winlove (1999) predicts that the stiffness will be dependent on only the cube-root of the radius, and, over the small range of cell sizes in a population, we found no significant correlation between cell size and stiffness. We also found that, when the beads were deliberately positioned off the diameter, there was only a modest difference in the measured response. Averaged force-extension profile for normal and variously modified red cell ghosts are shown in Fig. 3.

Around the point of zero extension, all the modified cells showed a fourfold diminution in stiffness, relative to the untreated controls, but the difference from normals was strikingly reduced to a factor of only about 2 at the highest applied forces (Fig. 3, *B–D*). The observations on the NEM-treated ghosts stand in contrast to the somewhat elevated shear elastic modulus derived from micropipette aspiration (Chabanel et al., 1989: Rangachari et al., 1989).

Theoretical analysis

The accompanying work (Parker and Winlove, 1999) analyses the response to tension applied at opposite poles of a spherical shell. Resistance to polar extension arises from two sources, the out-of-plane bending stiffness, B, and the in-plane shear modulus, H. Their relative contributions are expressed in terms of the nondimensional parameter $C = a^2 H/B$, where a is the radius of the sphere. The analysis shows that, for values of $C > 10$, if the dimensional force F^* is scaled, such that $F_s = F^*(aBH^2)^{-1/3}$, a plot of F_s against the fractional extension, ϵ, is almost independent of C. This leads to the approximation: $F^* \approx 5\epsilon(aBH^2)^{1/3}$, from which BH^2 can be directly determined. For the unmodified ghosts, a fractional extension of 0.1 gives a force of about 15 pN, leading to a value for BH^2 of 9×10^{-27} N^{3}m^{-1}. The modified ghosts all behave in a fairly similar manner, an extension of 0.1 giving forces of 4, 4.9, and 3.6 pN for the chymotrypsin-, PCMS-, and NEM-treated ghosts, respectively. These correspond to values of 1.7×10^{-28}, 3.14×10^{-28}, and 1.3×10^{-28} N^{3}m^{-1} for BH^2.

As will be discussed, B can probably be taken to be a function of the lipid bilayer, and should thus be little affected by modifications of the associated proteins. Literature values of B lie in the range $1.8–7 \times 10^{-19}$ Nm (Evans, 1983; Strey et al., 1995). If we assume Evans's preferred value of 2×10^{-19} Nm, then for unmodified cells, H must be about 2×10^{-4} Nm^{-1}, and for modified cells, about 3×10^{-5} Nm^{-1}, corresponding to values for C of 9000 and 1300: the requirement for $C > 10$ is thus met. The protein modifications reduce the force developed for a given small extension by a factor of about 4, but because of its dependence on $H^{2/3}$ the shear modulus is changed by a factor of about 7.

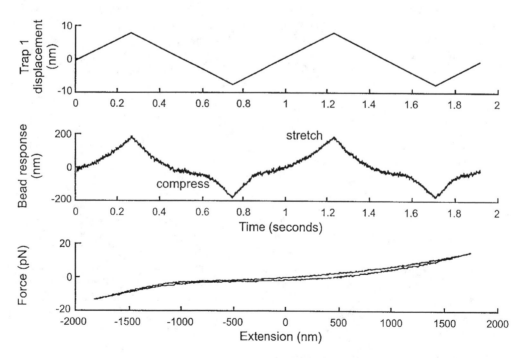

FIGURE 2 (*A*) Movement of the driven bead in response to a periodic tensile force in the form of a triangular wave-form. The response of the indicator bead is shown in *B*, while *C* shows a typical force-extension profile for a normal ghost, the upper curve corresponding to a stretch and the lower to the ensuing relaxation.

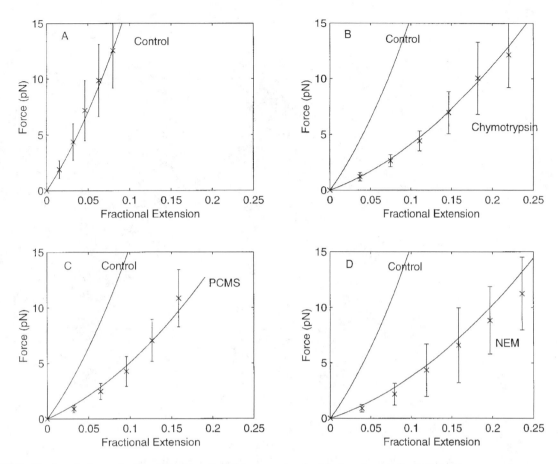

FIGURE 3   The dynamic force-extension relationships obtained from experiments of the type shown in Fig. 2. Each plot represents the average of 7–10 experiments. Error bars correspond to ± 1 SD. (*A*), The results for a set of control cells and an averaged curve for 4 such sets of data is shown in the other panels: Ghosts were treated as described (*B*), with chymotrypsin to cleave the ankyrin bridges between spectrin and the transmembrane protein, band 3; (*C*), with PCMS to eliminate the actin network junctions; and (*D*), with NEM to dissociate spectrin tetramers. The curves correspond to the theory of Parker and Winlove (1999) using a value for *C* of 1000 and 9×10^{-27} Nm^{-1} for *BH*^{2} for the unmodified cells, and 1.7×10^{-28}, 3.14×10^{-28}, and 1.3×10^{-28} for cells treated with chymotrypsin (*B*), PCMS (*C*), and NEM (*D*) (see Fig. 1 *A*).

Plastic yield of modified membranes

The formation of elliptocytes in the circulation when, in consequence of a genetic defect, a significant proportion of the spectrin is in the form of the dimer, implies that shape recovery after exposure to high shear (as when the cell passes through a capillary) may be incomplete. The resulting tendency for the cell to align itself with the long axis in the direction of fluid flow would ensure that this remains the preferred direction of stretching. To determine whether a yield phenomenon of this nature can be induced in cells in which a high proportion of spectrin dimers has been artificially generated, we applied prolonged stretches to the NEM-treated ghosts. The appearance of such a stretched cell is shown in Fig. 4. Whereas untreated control ghosts maintained tension for at least several minutes under a constant applied force of 20 pN, the treated cells exhibited a yield effect; this could be repeated many times on further stretching (Fig. 5). Averaged data for NEM-treated ghosts, compared to controls are shown in Fig. 6 *C*.

Both the chymotrypsin- and PCMS-treated ghosts behaved similarly to those treated with NEM and exhibited plastic yield under constant applied tension (Fig. 6, *A* and *B*).

Protein-free phospholipid vesicles

The extensibility of giant unilamellar lipid vesicles (Needham and Nunn, 1990; Käs and Sackmann, 1991) was tested. The vesicles were prepared from lipid of the same composition as the red cell membrane and are heterogeneous in size and shape, though in some cases approximately biconcave. Vesicles of the approximate size of a red cell were selected, and we found that beads coated with annexin V, which has a high calcium-dependent affinity for phosphatidylserine (Kuypers et al., 1996), would attach satisfactorily to the membrane surface. Sealed spherical vesicles, present in the preparations, could not (by reason of their already minimum surface:volume ratio in the osmotically

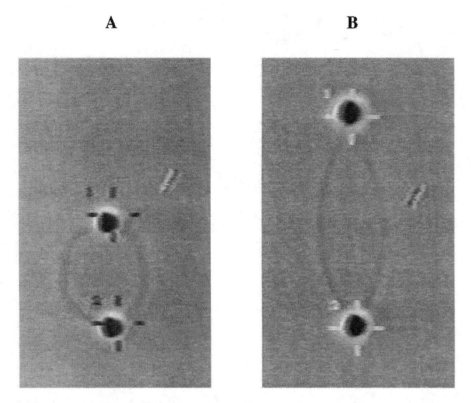

FIGURE 4 Appearance of NEM-modified red cell ghost in the optical trap, (*A*) before and (*B*) after yielding in response to a prolonged stretch.

inflated condition) be deformed at the forces generated by the optical trap. After treatment with saponin, however, the membranes of these vesicles became highly deformable. Because of their extreme softness and also the variation in size, no quantitative measurements of extension as a function of force were attempted, but comparison with ghosts and intact biconcave red cells left no room for doubt that the vesicle bilayer was very compliant and quite unlike the natural membrane. This observation agrees with the conclusions of micropipette analysis (Needham and Nunn, 1990). At the opposite extreme lie saponin ghosts, which have had their cytoskeletal protein networks cross-linked by treatment with glutaraldehyde: such cross-linked ghosts are completely inextensible by the optical tweezers.

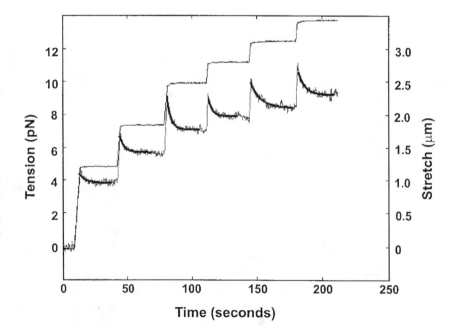

FIGURE 5 Response to prolonged stretches. A series of stretches were imposed on the ghost, which was allowed to relax back to a condition of near-equilibrium tension after each stretch. An exponential was fitted to the relaxation phases, the average rate being 0.1 s^{-1}. The quality of fit to such a single exponential and the rate were both somewhat variable between red cells modified in the three ways depicted in Fig. 1 *A*. Typical results are shown for a chymotrypsin-treated ghost. Untreated cells showed negligible relaxation.

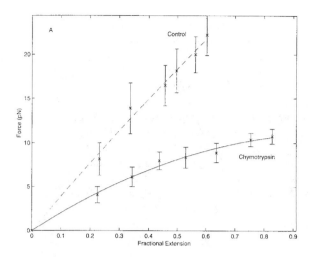

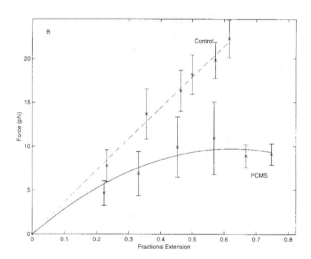

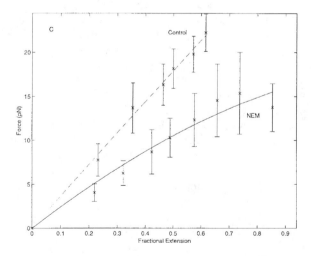

FIGURE 6 Force as a function of extension after relaxation of the cell to a near-equilibrium state, as shown in Fig 5. Profiles for ghosts, treated as indicated, with (*A*) chymotrypsin, (*B*) PCMS, and (*C*) NEM are compared in each case with averaged data for untreated control ghosts. The dashed line represents data for untreated control cells.

DISCUSSION

To allow the quantitative interpretation of force-extension relationships, using the theory for deformation of axisymmetric shells developed by Parker and Winlove (1999), we have studied saponin-permeabilized spherical ghosts. It seems likely that the elastic properties of these membranes differ little from those of intact red cells: we have observed, in the first place, that they exhibit force-extension profiles in much the same force regime (see also Hénon et al., 1999, discussed below). Moreover, saponin is a steroidal compound, analogous to cholesterol and known to act by binding to cholesterol in the membrane and not to lyse membranes that contain no cholesterol. Any saponin remaining in the membrane after lysis and extensive washing is unlikely to alter the membrane properties, since it has been shown that gross changes in the cholesterol content of the red cell membrane have no discernible effect on either the pressure-extension relation or the viscoelasticity, measured in the micropipette (Chabanel et al., 1983).

The theory developed in the accompanying paper (Parker and Winlove, 1999) predicts both the shape of the shell and the force as a function of polar extension for given values of the bending modulus *B* and the shear modulus *H*. Experimental constraints, however, allowed us to extract only the product BH^2 and not the individual moduli. The maximum force available in the optical tweezers was insufficient to engender a large enough distortion to permit modeling of the shape of the cell envelope. Moreover, the most sensitive region for the analysis of cell shape is that around the poles, where precise measurements are vitiated by the bead images. A more fundamental limit to interpretation of data at large strains may be that *H* does not remain invariant with ϵ. Our data, at all events, are confined, for the unmodified cells, to the small extension regime.

To proceed further, we need then to assume a literature value for either *B* or *H,* and we have chosen to put $B = 2 \times 10^{-19}$ Nm because this lies in the range given both by a micropipette method (Evans, 1983), which involves a much greater bending distortion than in our experiments, and by an analysis of the flicker phenomenon (Strey et al., 1995), in which the bending distortions are small. This value of *B* results in an estimate of 2×10^{-4} Nm^{-1} for *H*, which is an order of magnitude greater than the value given by micropipette methods (Evans and Hochmuth, 1978; Hochmuth and Hampel, 1979). A possible explanation emerges from the observation by Discher and Mohandas (1996) that large extensions pulled in the micropipette are accompanied by an effective phase separation, caused by the failure of the integral membrane proteins to follow the membrane flow and, instead, to accumulate at the entrance to the capillary.

Stokke et al. (1986a,b) have represented the red cell membrane as a composite of the lipid bilayer and an attached ionic protein gel, in which the spectrin tetramers function as entropy springs. Their model has the consequence that the pressure-extension relation in the micropipette should be determined by the (unknown) ratio of the elastic shear modulus and the modulus of area compression of the cytoskeletal layer. In these circumstances, the shear modulus cannot be explicitly determined from the pressure-extension plot (Stokke et al., 1986b). The differences between our results and those of micropipette aspiration may then throw light on the origins of the elastic properties.

The modifications of the membrane-associated proteins cause a reduction in BH^2 by a factor of about 100. The literature values of B for cells (Evans, 1983; Strey et al., 1995) are close enough to those for protein-free lipid vesicles of similar size (Schneider et al., 1984; Faucon et al., 1989; Evans and Ravicz, 1990) to justify the assumption that B is unchanged by the modifications. Thus, the low value of BH^2 for the modified ghosts appears to be a consequence of a reduction of the shear modulus by almost an order of magnitude. The modifications do not change the elastic character of the cell, as observed in the optical tweezers, to anything near that of protein-free lipid membranes; we conclude that neither extensive interruptions of the membrane skeletal network nor scission of a primary attachment to the bilayer prevents the associated proteins from substantially increasing the shear elasticity of the membrane. The membrane skeletal constituent, band 4.1, remains after all the modifications and presumably retains its attachments to the membrane by way of the transmembrane protein, glycophorin C, and the peripheral protein p55 (see e.g., Hemming et al., 1995), but the binary association of spectrin with 4.1 is weak (Tyler et al., 1980) and, in any case, in the absence of actin protofilaments, there is nothing to retain the 4.1 molecules in the form of junctional clusters. It thus seems likely that the association per se of spectrin with the membrane is enough to render the material elastic. Our observations on cells in which the ankyrin has been severed imply that other interactions of spectrin, including that with the inner-leaflet anionic phospholipid, phosphatidylserine (see e.g., Maksymiw et al., 1987), may suffice to ensure a strong enough interaction to generate elasticity. It has to be recognized, however, that factors other than those we have considered here may influence the mechanical characteristics of the membrane. It has been suggested, for instance, that the interaction of the transmembrane protein, band 3, with lipid may be one such (Peters et al., 1996).

An obvious qualitative explanation for the elasticity of the red cell membrane is that the primary skeletal network constituent, the spectrin tetramer, functions, like the polymer of rubber, as an entropy spring: its configurational entropy is diminished by the restriction of its ends to a separation of about half the root-mean-square end-to-end distance of the molecule in free solution. The entropy-spring model for the network has been developed quantitatively (Kozlov and Markin, 1987; Boal, 1994), but does not provide the only possible basis of membrane elasticity. From a study of the changes in dimensions of isolated membrane skeletons as a function of temperature and medium composition, Vertessy and Steck (1989) suggested that the elasticity was determined by protein–protein interactions. McGough and Josephs (1990) inferred from the evidence of

electron microscopy that the spectrin tetramers in the cytoskeletal lattice have the form of bihelical springs, which expand or contract without bending, in response to shear. Hansen et al. (1996) have developed a model that predicts the values of the shear modulus and the modulus of area expansion of a membrane with the network geometry of the red cell cytoskeleton, on the assumption that the spectrin tetramers behave in the manner suggested by McGough and Josephs and function as Hookean springs. Hansen et al. (1997a) further concluded that the deduced spring constant could not readily be accounted for by an entropy spring model because of insufficient flexibility of the spectrin molecule. The elastic properties of the model membrane were found (Hansen et al., 1997b) to be dependent on the network functionality (the average number of spectrin tetramers radiating from each network junction), and the analysis allows explicit predictions of how these properties would be expected to change when the functionality is altered, as in genetically abnormal cells.

At least for the case of the modified cells, in which the continuity or functionality of the network or its attachment to the bilayer is grossly disrupted, the elasticity of the membrane varies with applied force, and probably has more than one component. This behavior must also set limits on applicability of the analysis of Hansen et al. (1997b). They showed that their theory can explain the somewhat reduced shear modulus observed (Waugh and Agre, 1988) in hereditary spherocytes, characterized by a deficit of spectrin, if these cells embody a reduction in network functionality; however, this does not accord with the results of electron microscopy, which reveals a network of normal geometry in human (Liu et al., 1990) and mouse (Yi et al., 1997) spherocytes. The replacement of spectrin tetramers by dimers in cases of hereditary elliptocytosis, caused by mutations in the dimer–dimer association sites, was treated similarly by Hansen et al., and, as before, a decreased shear modulus is predicted; yet here again experimental micropipette measurements record an increase (Chabanel et al., 1989). This is also the case for cells treated, as in the present work, with NEM to dissociate 70% of the spectrin tetramers (Chabanel et al., 1989; Rangachari et al., 1989), where the theory predicts a diminution in shear modulus by two orders of magnitude. Hansen et al. (1997b) conclude that their elastic network model is inadequate to explain the nature of the elasticity of these cells. It may be noted that, for the case of an isotropic elastic network governed by entropy springs, the force required for a given extension also increases with increasing network functionality (Treloar, 1975). In our measurements, the stiffness of the membrane does indeed decrease with the loss of connecting lattice elements to about the extent demanded by the theory of Hansen et al. (1997b).

The plastic deformation of the modified membranes under tension affords an explanation for the elliptocytosis that invariably accompanies genetic anomalies associated with interruptions in the membrane skeletal network (Lux and Palek, 1995). In these cases, the asymmetric cell contour must develop during physiological flow, which implies irreversibility of deformation. This may originate in failure of the shape to recover rapidly enough after a transient distortion, so that the asymmetry determines the direction of application of the next induced stress. It should be emphasized that our results do not necessarily bear on the stability of the membrane toward shearing forces, which is a property separable from its elastic characteristics (Chasis and Mohandas, 1986).

There have been three previous applications of optical tweezers to studies on red cells: Svoboda et al. (1992) used an optical trap to immobilize a cell while irrigating it with a nonionic detergent to liberate the membrane skeleton, and examined the contraction of the network with increasing ionic strength. Bronkhorst et al. (1995) used a multiple trap to measure rates of shape recovery of cells subjected to bending deformations. Most recently, and since the present work was submitted for publication, a further study has appeared, showing force-extension profiles for intact red cells (Hénon et al., 1999). Discocytic and osmotically swollen, nearly spherical cells were examined, and the data show that the extension for a given force of these cells is about twice what we observe on the permeabilized spherical ghosts. This is in good agreement with our own observations on unlysed dicocytes. Such a difference between the stretch response of flaccid intact cells and of the spherical membranes that we have studied is in accord with qualitative expectation. Hénon et al. (1999) derive from their results a shear modulus of 2×10^{-6} Nm^{-1}, which is two orders of magnitude lower than the value we have inferred from our data. Their analysis assumes that the biconcave cell can be treated as a planar disc and that their nearly spherical form as a sphere. Their treatment further implicitly assumes that the contribution of the bending stiffness of the membrane is negligible, whereas the exact solution of Parker and Winlove (1999) for an axisymmetric form implies that the force for a given extension is proportional to $B^{1/3}$. It is difficult to assess the validity of the assumptions made by Hénon et al. (1999), but we surmise that these, rather than any differences between the membranes of the intact and lysed cells, are responsible for the discrepancy between the conclusions. Our results suggest that the optical tweezers technique should have considerable advantages for the quantitative study of elasticity of membranes and cells.

We thank Dr. J. C. Pinder for help with the analysis of modified ghost membranes, Dr. P. McCauley for gifts of giant phospholipid vesicles, to Dr. D. N. Fenner for help and discussion of theoretical treatments, and to Dr. G. B. Nash for valuable advice and discussion.

REFERENCES

Berk, D. A., R. M. Hochmuth, and R. E. Waugh. 1989. Viscoelastic properties and rheology. *In* Red Blood Cell Membranes. P. Agre and J. C. Parker, editors. Marcel Dekker Inc, New York and Basel, pp. 423–454.

Boal, D. 1994. Computer simulation of a model network for the erythrocyte cytoskeleton. *Biophys. J.* 67:521–529.

Bronkhorst, P. J. H., G. J. Streeker, J. Grimbergen, E. J. Nijhof, J. J. Sixma, and G. J. Brakenhoff. 1995. A new method to study shape recovery of red blood cells using multiple optical trapping. *Biophys. J.* 69:1666–1673.

Chabanel, A., M. Flamm, K. L. P. Sung, M. M. Lee, D. Schachter, and S. Chien. 1983. Influence of cholesterol content on red cell membrane viscoelasticity and fluidity. *Biophys. J.* 44:171–176.

Chabanel, A., K. L. P. Sung, J. Rapiejko, J. T. Prchal, S.-C. Liu, and S. Chien. 1989. Viscoelastic properties of red cell membrane in hereditary elliptocytosis. *Blood.* 73:592–595.

Chasis, J. A., and N. Mohandas. 1986. Erythrocyte membrane deformability and stability: two distinct membrane properties that are independently regulated by skeletal protein associations. *J. Cell Biol.* 103:343–350.

Discher, D. E., and N. Mohandas. 1996. Kinematics in red-cell aspiration by fluorescence imaged microdeformation. *Biophys. J.* 71:1680–1694.

Elias, P. M., J. Goerke, and D. S. Friend. 1978. Freeze-fracture identification of sterol-digitonin complexes in cell and liposome membranes. *J. Cell Biol.* 78:577–593.

Evans, E. A. 1973. A new membrane concept applied to the analysis of fluid shear- and micropipette-deformed red blood cells. *Biophys. J.* 13:941–954.

Evans, E. A. 1983. Bending elastic modulus of red blood cell membrane derived from buckling instability in micropipet aspiration tests. *Biophys. J.* 43:27–30.

Evans, E. A., and R. M. Hochmuth. 1978. Mechano-chemical properties of membrane. *Curr. Top. Membr. Transp.* 10:1–64.

Evans, E. A., and W. Ravicz. 1990. Entropy-driven tension and bending elasticity in condensed-fluid membranes. *Phys. Rev. Lett.* 64:2094–2097.

Fairbanks G, T. L. Steck, and D. F. H. Wallach. 1971. Electrophoretic analysis of the major polypeptides of the human erythrocyte membrane. *Biochemistry.* 10:2606–2617.

Faucon, J. F., M. D. Mitov, P. Meleard, I. Rivas, and P. Bothorel. 1989. Bending elasticity and thermal fluctuations of lipid membranes. Theoretical and experimental requirements. *J. Physiol. (Paris).* 50:2389–2392.

Fischer, T. M., C. W. M. Haest, M. Stohs, D. Kamp, and B. Deuticke. 1978. Selective alteration of erythrocyte deformability by SH-reagents. Evidence for involvement of spectrin in membrane shear elasticity. *Biochim. Biophys. Acta.* 510:270–282.

Gordon, S., and G. B. Ralston. 1990. Solubilization and denaturation of monomeric actin from erythrocyte membranes by *p*-chloromercuribenzenesulfonate. *Biochim. Biophys. Acta.* 1025:43–48.

Hall, T. G., and V. Bennett. 1987. Regulatory domains of erythrocyte ankyrin. *J. Biol. Chem.* 262:10537–10545.

Hansen, J. C., R. Skalak, S. Chien, and A. Hoger. 1996. An elastic network model based on the structure of the red blood cell membrane skeleton. *Biophys. J.* 70:146–166.

Hansen, J. C., R. Skalak, S. Chien, and A. Hoger. 1997a. Spectrin properties and the elasticity of the red blood cell membrane skeleton. *Biorheology.* 34:327–348.

Hansen, J. C., R. Skalak, S. Chien, and A. Hoger. 1997b. Influence of network topology on the elasticity of the red blood cell membrane skeleton. *Biophys. J.* 72:2369–2381.

Hemming, N. J., D. J. Anstee, M. A. Staricoff, M. J. A. Tanner, and N. Mohandas. 1995. Identification of the membrane attachment sites for protein 4.1 in the human erythrocyte. *J. Biol. Chem.* 270:5360–5366.

Hénon, S., G. Lenormand, A. Richert, and F. Gallet. 1999. A new determination of the shear modulus of the human erythrocyte membrane using optical tweezers. *Biophys. J.* 76:1145–1151.

Hochmuth, R. M., and W. L. Hampel. 1979. Surface elasticity and viscosity of red cell membrane. *J. Rheology* 23:669–680.

Jay, A. W. L. 1975. Geometry of the human erythrocyte. I. Effect of albumin on cell geometry. *Biophys. J.* 15:205–222.

Jinbu, Y., S. Sato, T. Nakao, M. Nakao, S. Tsukita, and S. Tsukita. 1984. The role of ankyrin in shape and deformability change of human erythrocyte ghosts. *Biochim. Biophys. Acta.* 773:237–245.

Käs, J., and E. Sackmann, E. 1991. Shape transitions and shape stability of giant phospholipid vesicles in pure water induced by area-to-volume changes. *Biophys. J.* 60:825–844.

Kates, M. 1972. Techniques in Lipidology. North-Holland/American Elsevier, Amsterdam, London, New York. 352–365.

Kozlov, M. M., and V. S. Markin. 1987. Model of red blood cell membrane skeleton: electrical and mechanical properties. *J. Theor. Biol.* 129: 439–452.

Kuypers, F. A., R. A. Lewis, M. Hua, M. A. Schott, D. Discher, J. D. Ernst, and B. H. Lubin. 1996. Detection of altered membrane phospholipid asymmetry in subpopulations of human red blood cells using fluorescently labeled annexin V. *Blood.* 7:1179–1187.

Laemmli, U. K. 1970. Cleavage of structural proteins during the assembly of bacteriophage T4. *Nature.* 227:680–685.

Liu, S.-C., L. H. Derick, P. Agre, and J. Palek. 1990. Alterations of the erythrocyte membrane skeletal ultrastructure in hereditary spherocytosis, hereditary elliptocytosis, and pyropoikilocytosis. *Blood.* 76: 198–206.

Liu, S.-C., and J. Palek. 1980. Spectrin tetramer–dimer equilibrium and the stability of erythrocyte membrane skeletons. *Nature.* 285:586–588.

Lux, S. E., and J. Palek. 1995. Disorders of the red cell membrane. *In* Blood: Principles and Practice of Hematology. R. I. Handin, S. E. Lux, and T. P. Stossel, editors. J. B. Lippincott Co., Philadelphia. 1701–1818.

McGough, A. M., and R. Josephs. 1990. On the structure of erythrocyte spectrin in partially expanded membrane skeletons. *Proc. Natl. Acad. Sci. USA.* 87:5208–5212.

Maksymiw, R., S. Sui, H. Gaub, and E. Sackmann. 1987. Electrostatic coupling of spectrin dimers to phosphatidylserine containing lipid lamellae. *Biochemistry.* 26:2983–2990.

Meiselman, H. J., E. A. Evans, and R. M. Hochmuth. 1978. Membrane mechanical properties of ATP-depleted human erythrocytes. *Blood.* 52: 499–504.

Mohandas, N., and E. Evans. 1994. Mechanical properties of the red cell membrane in relation to molecular structure and genetic defects. *Ann. Rev. Biophys. Biomol. Struct.* 23:787–818.

Needham, D., and R. S. Nunn. 1990. Elastic deformation and failure of lipid bilayer membranes containing cholesyterol. *Biophys. J.* 58: 997–1009.

Parker, K. H., and C. P. Winlove. 1999. Deformation of spherical vesicles with permeable, constant-area membranes: application to red cells. *Biophys. J.* 77:3096–3107.

Peters, L. L., R. A. Shivadasani, S.-C. Liu, M. Hanspal, K. M. John, J. M. Gonzalez, C. Brugnara, B. Gwynn, N. Mohandas, S. L. Alper, S. H. Orkin, and S. E. Lux. 1996. Anion exchanger 1 (band 3) is required to prevent erythrocyte membrane surface loss but not to form the membrane skeleton. *Cell.* 86:917–927.

Pinder, J. C., A. Pekrun, A. M. Maggs, A. P. R. Brain, and W. B. Gratzer. 1995. Association state of human red blood cell band 3 and its interaction with ankyrin. *Blood.* 85:2951–2961.

Pinder, J. C., A. G. Weeds, and W. B. Gratzer. 1986. Study of actin filament ends in the human red cell membrane. *J. Mol. Biol.* 191: 461–468.

Rangachari, K., G. H. Beaven, G. B. Nash, B. Clough, A. R. Dluzewski, Myint-Oo, R. J. M. Wilson, and W. B. Gratzer. 1989. A study of red cell membrane properties in relation to malarial invasion. *Mol. Biochem. Parasitol.* 34:63–74.

Schneider, M. D., J. T. Jenkins, and W. W. Webb. 1984. Thermal fluctuations of large quasi-spherical bimolecular phospholipid vesicles. *J. Physiol. (Paris).* 45:1457–1462.

Seeman, P. 1967. Transient holes in the erythrocyte membrane during hypotonic hemolysis and stable holes in the membrane after lysis by saponin and lysolecithin. *J. Cell Biol.* 32:55–70.

Seeman, P., D. Cheng, and G. H. Iles. 1973. Structure of membrane holes in osmotic and saponin hemolysis. *J. Cell Biol.* 56:519–527.

Stokke, B. T., A. Mikkelsen, and A. Elgsaeter. 1986a. The human erythrocyte membrane skeleton may be an ionic gel. I. Membrane mechanochemical properties. *Eur. Biophys. J.* 13:203–218.

Stokke, B. T., A. Mikkelsen, and A. Elgsaeter. 1986b. The human erythrocyte membrane skeleton may be an ionic gel. III. Micropipette aspiration of unswollen erythrocytes. *J. Theor. Biol.* 123:205–211.

Strey, H., M. Peterson, and E. Sackmann. 1995. Measurement of erythrocyte membrane elasticity by flicker eigenmode decomposition. *Biophys. J.* 69:478–488.

Svoboda, K., C. F. Schmidt, D. Branton, and S. M. Block. 1992. Conformation and elasticity of the isolated red blood cell membrane skeleton. *Biophys. J.* 63:784–793.

Treloar, L. R. G. 1975. The Physics of Rubber Elasticity. Clarendon Press, Oxford, U.K. 74–79

Tyler, J. M., B. N. Reinhardt, and D. Branton. 1980. Association of erythrocyte membrane proteins. Binding of purified bands 2.1 and 4.1 to spectrin. *J. Biol. Chem.* 255:7034–7049.

Vertessy, B. G., and T. L. Steck. 1989. Elasticity of the human red cell membrane skeleton. Effects of temperature and denaturants. *Biophys. J.* 55:255–262.

Waugh, R. E., and P. Agre. 1988. Reduction of erythrocyte membrane viscoelastic coefficients reflect spectrin deficiencies in hereditary spherocytosis. *J. Clin. Invest.* 81:133–141.

Yi, S. J., S.-C. Liu, L. H. Derick, J. Murray, J. E. Barker, M. R. Cho, J. Palek, and D. E. Golan. 1997. Red cell membranes of ankyrin-deficient mice lack band 3 tetramers but contain normal skeletons. *Biochemistry.* 36:9596–9604.

Lateral Movements of Membrane Glycoproteins Restricted by Dynamic Cytoplasmic Barriers

MICHAEL EDIDIN, SCOT C. KUO, MICHAEL P. SHEETZ*

Cell membranes often are patchy, composed of lateral domains. These domains may be formed by barriers within or on either side of the membrane bilayer. Major histocompatibility complex (MHC) class 1 molecules that were either transmembrane- ($H-2D^b$) or glycosylphosphatidylinositol (GPI)-anchored (Qa2) were labeled with antibody-coated gold particles and moved across the cell surface with a laser optical tweezers until they encountered a barrier, the barrier-free path length (BFP). At room temperature, the BFPs of Qa2 and $H-2D^b$ were 1.7 ± 0.2 and 0.6 ± 0.1 (micrometers $\pm$ SEM), respectively. Barriers persisted at 34°C, although the BFP for both MHC molecules was fivefold greater at 34°C than at 23°C. This indicates that barriers to lateral movement are primarily on the cytoplasmic half of the membrane and are dynamic.

ALTHOUGH THE LATERAL DIFFUsion of a few membrane proteins, notably visual rhodopsin, is rapid and appears to be hindered only by the

M. Edidin, Department of Biology, The Johns Hopkins University, Baltimore, MD 21218.
S. C. Kuo and M. P. Sheetz, Department of Cell Biology, Duke University Medical Center, Durham, NC 27710.

*To whom correspondence should be addressed.

From M. Edidin, S. C. Kuo, and M. P. Sheetz, "Lateral Movements of Membrane Glycoproteins Restricted by Dynamic Cytoplasmic Barriers," Science. 254: 1379–1382 (1991). Reprinted with permission from AAAS.

viscosity of membrane lipids, the lateral diffusion of most proteins is hindered in several ways. Significant fractions of most membrane proteins are immobile, and diffusion coefficients for the mobile fractions are 10- to 100-fold lower than that of rhodopsin (1). In erythrocyte membranes, the spectrin-actin complex limits the lateral diffusion of band 3 (micrometer scale) (2) but has little effect on rotation of band 3

Fig. 1. Planar movement of a gold particle trapped by the laser optical tweezers. The gold particle was bound to the membrane by a monoclonal antibody (MAb) to Qa2, the GPI-linked protein, and was captured and moved laterally by the optical tweezers at a rate of 0.5 to 1 μm s⁻¹. The video overlay in each panel indicates time of day (hh:mm:ss), with the date portion monitoring a fraction of the laser power (in milliwatts). A stationary feature on the cell surface (marked by parentheses) was used to determine the relative distance the particle moved while in the optical trap (marked by carets). Murine

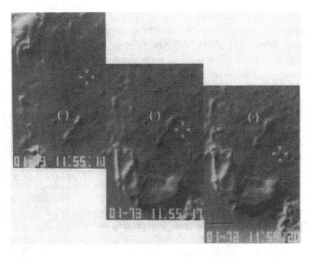

HEPA-OVA cells were stably transfected with the Qa2 gene; the level of Qa2 expression was similar to that of the endogenous MHC class 1 molecule H-2Dᵇ (*13*). Cells were plated on acid-washed cover slips (22 by 22 mm, No. 0) and cultured overnight in DME-FCS [Dulbecco's modified Eagle's medium (Gibco) with 10% fetal calf serum]. Gold particles (40 nm, E-Y Scientific) were mixed with an equal volume of MAb 20-8-4 (0.2 to 0.5 mg ml⁻¹) (*13*) in phosphate-buffered saline (PBS) (10 mM PO₄, 140 mM KCl, pH 7.4) and incubated for 10 min on ice before addition of dry bovine serum albumin (BSA) to a final concentration of 5 mg ml⁻¹. Particles were washed three times by centrifuging for 15 min at 6000*g* and resuspending the pellet in PBS with BSA (5 mg ml⁻¹); the final resuspended volume after washing was 10% of the original particle volume. Bath sonication was used to disrupt aggregates. Cells were washed with phenol red–free MEME (minimal essential medium with Earle's salts, Gibco) buffered by 10 mM Hepes (MEME-Hepes) and assembled in a flow-cell cover slip chamber that allowed exchanging the medium with a 1:9 dilution of the gold particle suspension in MEME-Hepes. Samples were maintained at 34°C by a hot-air incubator (constructed with the use of a plastic enclosure around the microscope optics with hot air supplied by a commercial hair dryer controlled with a Variac; a thermocouple in the enclosure monitored the temperature). The laser optical trap was constructed as described (*9*), but we substituted a 12-W laser (model 116F from Quantronix) to provide greater beam power. In the experiments, 15 to 30 mW of beam power was used at the sample. Scale bar = 2 μm.

(nanometer scale) (*3*). Thus, spectrin and associated proteins may form corrals partitioning the membrane (*4*). In this model, proteins move freely within a corral, on a scale of hundreds of nanometers, but only rarely cross the boundaries of corrals. Although the erythrocyte membrane is specialized and may be unusual, the model of constraints to lateral movement in its membrane may be used to explain anomalies in the lateral diffusion of other molecules, such as MHC class 1 molecules (*5*).

The properties of the MHC class 1 glycoproteins, H-2Dᵇ and Qa2, are ideal for addressing the nature of restricted diffusion. Structurally, these two glycoprotein molecules are 80% homologous and have globular extracellular domains (*6*). However, the two proteins are anchored differently in the membrane: H-2Dᵇ has a transmembrane segment and cytoplasmic tail, whereas Qa2 is anchored by a GPI linkage. Fluorescence photobleaching and recovery (FPR) measurements of lateral diffusion in the plasma membrane show different behaviors for these two molecules. The measurements are consistent with the notion that H-2Dᵇ molecules are confined to domains approximately 1 μm in radius and Qa2 molecules are free to cross these boundaries (*5*). The mobile

fraction, *R*, of H-2Dᵇ molecules decreases with increasing size of the bleached spot, but the mobile fraction of Qa2 molecules remains constant (*5*). Because the diffusion coefficients for the mobile fraction of the two molecules are similar in small bleached areas, only the long-range diffusion of H-2Dᵇ appears restricted. These data suggested that the cytoplasmic tail or transmembrane region, or both, encountered barriers in the cytoplasmic half of the membrane that were absent in the extracellular half. With the use of gold particle "handles" bound to these proteins by specific antibodies, we tested the barrier hypothesis by directly measuring the BFP for the two classes of MHC molecules by dragging them through the membrane with laser optical tweezers.

Membrane proteins tagged with gold or other particles provide a means to track the diffusional and directed movements of membrane glycoproteins. The diffusion coefficients, *D*, derived from single particle tracking have corresponded well with those measured by FPR (*7*). We measured *D* = 1.3 ± 0.2 × 10⁻¹⁰ cm² s⁻¹ (average ± SEM) for 26 particles attached to H-2Dᵇ. For 23 particles attached to Qa2, *D* = 2.1 ± 0.3 × 10⁻¹⁰ cm² s⁻¹. *D* from FPR

measurements of diffusion over comparable areas (a fraction of a square micrometer) are 2 × 10⁻¹⁰ to 4 × 10⁻¹⁰ cm² s⁻¹ for both kinds of MHC molecules (*5*).

The infrared laser tweezers can capture and move these gold-labeled MHC molecules. The mechanism of optically trapping such small particles (Rayleigh size range) has been described elsewhere (*8*). We positioned the beam waist of the laser trap such that gold particles free in solution were held in the plane of focus of the microscope when trapped (*9*). When subjected to viscous force in a flow cell, the trapped particle was always within a 0.5-μm radius until this viscous force exceeded the trapping force. As long as the membrane drag was less than the maximum force of the trap, particles dragged along the plane of the membrane remained in this radius (see Fig. 1). Moving the microscope stage at a rate of 0.5 to 1.0 μm s⁻¹ kept the membrane drag minimal and yet allowed easy visualization of particles escaping the trap (>0.5 μm from center of trap) after encountering an obstacle. Under our experimental conditions (15 to 30 mW of infrared laser power at the sample), the maximum lateral force of the trap on the gold particles was <0.5 pN.

The BFP of each trapped particle was measured from video images, with stationary particles serving as reference points (Fig. 1). At room temperature (23°C), gold particles attached to H-2Dᵇ molecules had a lower BFP than gold particles attached to Qa2 molecules. H-2Dᵇ could be moved only 0.6 ± 0.1 μm (average ± SEM; number of determinations, *n* = 68) in any direction before being pulled from the trap. Particles attached to Qa2 molecules could be moved 1.7 ± 0.2 μm (*n* = 53). The maximum BFP of gold–H-2Dᵇ was 3.5 μm, whereas the maximum BFP of gold-Qa2 was ~8 μm (Fig. 2, A and B). The statistical distribution of these BFP values appears exponential, but rigorous model testing was beyond the scope of this study.

The BFP of gold-labeled MHC molecules increased with increased temperature (34°C). However, the average BFP of the transmembrane-anchored H-2Dᵇ molecules was still one-third the BFP of the GPI-anchored Qa2 molecules, 3.5 ± 0.6 μm (*n* = 29) for H-2Dᵇ compared to 8.5 ± 0.8 μm (*n* = 50) for Qa2 (Fig. 2, C and D). These averages understate the differences between the two molecules. Particles on Qa2 molecules can be moved large distances on cells at 34°C. Seven out of 51 particles remained in the trap when the underlying surface was displaced a distance >15 μm. Ten other particles remained in the trap while the surface was displaced by >10 μm (Fig. 2, C and D).

Our earlier FPR measurements of membrane domains were made at 21°C (*5*). The results above implied that FPR at elevated temperatures should also detect membrane domains and that these domains would appear larger than at room temperature. This was indeed the case (Fig. 3). The mobile fraction of labeled H-2D^b molecules was higher at 37°C ($R = 0.7$) than at room temperature [$R = 0.5$; see (*5*) for the range of values at room temperature] and decreased when the size of the bleached spot was increased (Fig. 3, A and B). The mobile fraction of Qa2, which was also higher at 37°C than at room temperature, did not change when the size of the bleached

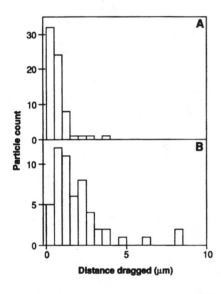

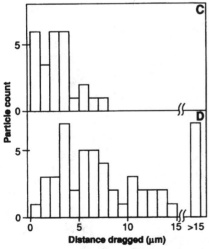

Fig. 2. Distance traveled at 23°C by 40-nm gold particles attached to MHC molecules on HEPA-OVA cells. Measurements were made as described in Fig. 1: (**A**) H-2D^b molecules bound to MAb 28-14-8 (*14*) labeled with gold particles. (**B**) Qa2 molecules detected with gold-labeled MAb 20-8-4 (*15*). Distance traveled at 34°C by 40-nm gold particles attached to MHC molecules: (**C**) H-2D^b molecules. (**D**) Qa2 molecules.

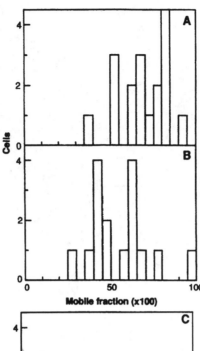

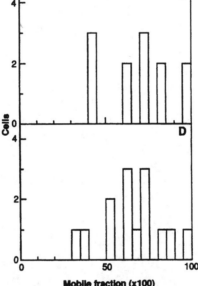

Fig. 3. FPR measurements of the mobile fractions, *R*, of MHC molecules labeled at 37°C with fluorescent adducts of the same Fab fragments that were used for BFP measurements, MAb 28-14-8 for H-2D^b and MAb 20-8-4 for Qa2. On the same day, measurements of *D* were made at, 22°C with a laser spot size of 0.65 μm: $D = 13 \pm 2 \times 10^{-10}$ cm^2 s^{-1} ($n = 15$) for H-2D^b and $D = 13 \pm 2 \times 10^{-10}$ cm^2 s^{-1} ($n = 22$) for Qa2. The difference between these values and previous values (*5*) is most likely due to the smaller number of observations. There was no significant increase in *D* with a 15°C increase in temperature. (**A**) H-2D^b. Radius of the laser spot, 0.65 μm; mobile fraction, 0.70 ± 0.04 (mean ± SEM); $D = 5.6 \pm 1 \times 10^{-10}$ cm^2 s^{-1} ($n = 16$). (**B**) H-2D^b. Radius of the laser spot, 1.1 μm; mobile fraction, 0.55 ± 0.04 (mean ± SEM); $D = 16 \pm 4 \times 10^{-10}$ cm^2 s^{-1} ($n = 15$). (**C**) Qa2. Radius of the laser spot, 0.65 μm; mobile fraction, 0.70 ± 0.06 (mean ± SEM); $D = 11 \pm 2 \times 10^{-10}$ cm^2 s^{-1} ($n = 12$). (**D**) Qa2. Radius of the laser spot, 1.1 μm; mobile fraction, 0.71 ± 0.06 (mean ± SEM); $D = 32 \pm 6 \times 10^{-10}$ cm^2 s^{-1} ($n = 12$).

spot was increased (Fig. 3, C and D).

Both laser trapping and FPR experiments indicate that there are barriers to lateral movements of MHC molecules. Both methods also show that the GPI-anchored Qa2 molecules encounter barriers less frequently than do H-2D^b molecules. The major difference between the two types of molecules is that H-2D^b molecules extend completely through the membrane bilayer and bear a cytoplasmic tail (*6*).

These structural data suggest a model for the more limited lateral mobility of H-2D^b molecules. These molecules collide with a cellular structure underlying or within the cytoplasmic half of the membrane, a membrane-associated cytoskeleton. Interactions with lipid domains or other transmembrane proteins anchored to a cytoskeleton are discounted because they should also affect Qa2 mobility. However, the extent of lateral mobility of Qa2 molecules is likely to be limited by collisions with other membrane glycoproteins that are either corralled or anchored to a cytoskeleton. The barriers to lateral displacement we observe cannot account for the decreased diffusion coefficients of glycoproteins in biological membranes (*1*). This is evident when we compare the fivefold increase in BFP that accompanies an 11°C increase in temperature with the small change in *D* observed for the same increase in temperature (see caption of Fig. 3).

The barriers on the cytoplasmic half of the membrane are likely to contain spectrin and its membrane anchor ankyrin. Spectrin and ankyrin are major constituents of the erythrocyte membrane skeleton and occur in many other cell types as well (*10*). Barriers composed of a perfect spectrin mesh, 0.2 μm on a side, would create smaller domains than implied by our work. However, larger values of BFP do not necessarily imply larger distances between elements of a stable cytoskeletal meshwork. Rather, the BFP could reflect the probability of encountering a stationary element of a dynamic and changing mesh whose characteristic dimensions are smaller than the BFP. Such a metabolically active membrane skeleton, with transient breaks in the meshwork, would be indistinguishable from an intact, stable meshwork of larger dimensions. The increased BFP of particles at 34°C compared to 23°C is not consistent with an inert web of barriers. A dynamic, hence imperfect, matrix of spectrin could regulate both short-range (*11*) and long-range (*12*) molecular interactions on the cell surface.

REFERENCES AND NOTES

1. K. Jacobson, A. Ishihara, R. Inman, *Annu. Rev. Physiol.* 49, 163 (1987); M. Edidin, in *The Structure of Cell Membranes*, P. Yeagle, Ed. (CRC Press, Boca Raton, FL, in press).

2. M. P. Sheetz, M. Schindler, D. E. Koppel, *Nature* **285**, 510 (1980).
3. E. A. Nigg and R. J. Cherry, *Proc. Natl. Acad. Sci. U.S.A.* **77**, 4702 (1980).
4. M. P. Sheetz, *Semin. Hematol.* **20**, 175 (1983).
5. M. Edidin and I. Stroynowski, *J. Cell Biol.* **112**, 1143 (1991).
6. I. Stroynowski, *Annu. Rev. Immunol.* **8**, 501 (1990).
7. M. P. Sheetz, S. Turney, H. Qian, E. L. Elson, *Nature* **340**, 284 (1989); R. N. Ghosh and W. W. Webb, *Biophys. J.* **57**, 286a (1990); M. de Brabander *et al.*, *J. Cell Biol.* **112**, 111 (1991).
8. A. Ashkin, J. M. Dziedzic, J. E. Bjorkholm, S. Chu, *Opt. Lett.* **11**, 288 (1986); S. M. Block, in *Noninvasive Techniques in Cell Biology*, J. K. Foskett and S. Grinstein, Eds. (Wiley-Liss, New York, 1990), pp. 375–402.
9. D. Kucik, S. C. Kuo, E. L. Elson, M. P. Sheetz, *J. Cell Biol.* **114**, 1029 (1991).
10. G. V. Bennett, *Curr. Opin. Cell Biol.* **2**, 51 (1990).
11. R. Peters, *FEBS Lett.* **234**, 1 (1988).
12. M. Saxton, *Biophys. J.* **58**, 1303 (1990).
13. I. Stroynowski, M. Soloski, M. G. Low, L. E. Hood, *Cell* **50**, 759 (1987).
14. K. Ozato and D. Sachs, *J. Immunol.* **125**, 2473 (1980).
15. _____, *ibid.* **126**, 317 (1980).

16. Supported by NIH grants AI14584 (to M.E.) and GM 36277 (to M.P.S.), a grant from the Muscular Dystrophy Association (to M.P.S.), and a grant from the Jane Coffin Childs Memorial Fund for Medical Research (to S.C.K.).

5 August 1991; accepted 7 October 1991

Direct observation of kinesin stepping by optical trapping interferometry

Karel Svoboda[*†], **Christoph F. Schmidt**[*‡], **Bruce J. Schnapp**[§]
& Steven M. Block[*‖]

* Rowland Institute for Science, 100 Edwin Land Boulevard, Cambridge, Massachusetts 02142, USA
† Committee on Biophysics, Harvard University, Cambridge, Massachusetts 02138, USA
§ Department of Cell Biology, Harvard Medical School, Boston, Massachusetts 02115, USA

Do biological motors move with regular steps? To address this question, we constructed instrumentation with the spatial and temporal sensitivity to resolve movement on a molecular scale. We deposited silica beads carrying single molecules of the motor protein kinesin on microtubules using optical tweezers and analysed their motion under controlled loads by interferometry. We find that kinesin moves with 8-nm steps.

ENZYMES such as myosin, kinesin, dynein and their relatives are linear motors converting the energy of ATP hydrolysis into mechanical work, moving along polymer substrates: myosin along actin filaments in muscle and other cells; kinesin and dynein along microtubules. Motion derives from a mechanochemical cycle during which the motor protein binds to successive sites along the substrate, in such a way as to move forward on average[1‑3]. Whether this cycle is accomplished through a swinging crossbridge, and how cycles of advancement are coupled to ATP hydrolysis, have been the subject of considerable debate[4‑6]. *In vitro* assays for motility[7,8], using purified components interacting in well-defined experimental geometries, permit, in principle, measurement of speeds, forces, displacements, cycle timing and other physical properties of individual molecules, using native or mutant proteins[9‑14].

Are there steps?

Do motor proteins make characteristic steps? That is, do they move forward in a discontinuous fashion, dwelling for times at

‡ Present address: Department of Physics, University of Michigan, Ann Arbor, Michigan 48109, USA.
‖ To whom correspondence should be addressed.

ARTICLES

well-defined positions on the substrate, interspersed with periods of advancement? We define the step size, which may be invariant or represent a distribution of values, as the distance moved forwards between dwell states. (Here we use 'step size' in its physical sense, and not to mean the average distance moved per molecule of ATP hydrolysed, also termed the 'sliding distance'[4,9,11,15-18]. The latter corresponds to the physical step size, or an integral multiple thereof, in models in which ATP hydrolysis and stepping are tightly coupled.) If stepping occurs, the distributions of step sizes and dwell times will place constraints on possible mechanisms for movement.

The motor protein kinesin has advantages over myosin for physical studies, despite a comparative dearth of biochemical data. It can be made to move slowly, permitting better time-averaging of position, and it remains attached to the substrate for a substantial fraction of the kinetic cycle[12,13,19], reducing the magnitude of brownian excursions. Movement by single molecules of kinesin has been demonstrated[12,13], and kinesin can transport small beads, which provide high-contrast markers for motor position in the microscope[8]. Unlike actin filaments, microtubules are relatively rigid[20], and can be visualized by video-enhanced differential interference-contrast microscopy[21] (DIC). Finally, recombinant kinesin expressed in bacteria has been shown to move *in vitro*, paving the way for future study[14].

Significant technical difficulties nevertheless exist in measuring movements of single molecules. The motions occur on length scales of ångströms to nanometres and on timescales of milliseconds and less. To obtain the high spatial and temporal sensitivity required, we combined optical tweezers[22,23] with a dual-beam interferometer[24] to produce an 'optical trapping interferometer'. A laser, focused through a microscope objective of high numerical aperture, provides position detection and trapping functions simultaneously, and can produce controlled, calibratable forces in the piconewton range. Using this device, we captured silica beads with kinesin molecules bound to their surface out of a suspension and deposited them onto microtubules immobilized on a coverslip. We then observed the fine structure of the motion as beads developed load by moving away from the centre of the trap.

Our data provide evidence for steps under three sets of conditions. At moderate levels of ATP and low loads, kinesin movement was load-independent, and the relatively high speed, in combination with brownian motion, precluded direct visualization of steps. A statistical analysis of the trajectories, however, revealed a stepwise character to the motion. At saturating levels of ATP and high loads, or at low levels of ATP and low loads, (when movement is slowed mechanically or chemically), it was possible to see the abrupt transitions directly. We estimate the step size to be 8 nm, a distance that corresponds closely to the spacing between adjacent α-β tubulin dimers in the protofilament of the microtubule[25], suggesting that the elementary step spans one dimer.

Optical trapping interferometer

Polarized laser light is introduced at a point just below the objective Wollaston prism into a microscope equipped with DIC optics (Fig. 1a). The prism splits the light into two beams with orthogonal polarization: these are focused to overlapping, diffraction-limited spots at the specimen plane, and together they function as a single optical trap[23]. A phase object in the specimen plane inside the region illuminated by the two spots (the detector zone) introduces a relative retardation between the beams, so that when they recombine and interfere in the condenser Wollaston prism, elliptically polarized light results (Fig. 1a, left). The degree of ellipticity is measured by additional optics[24] and provides a sensitive measure of retardation, which changes during movement, passing through zero when the object exactly straddles the two spots (for example, a bead moved along a microtubule, as shown in Fig. 1a, right). For small excursions

(out to ~150 nm), the output of the detector system is linear with displacement (Fig. 1a, inset).

Over most of its bandwidth, detector noise is at or below 1 Å/$\sqrt{Hz}$ (Fig. 1b). The response to a 100 Hz sinusoidal calibration signal of 1 nm amplitude (Fig. 1b, inset) shows this ångström-level noise. An optical trapping interferometer has advantages over conventional split-photodiode systems[26,27]. Because it is a non-imaging device, it is relatively insensitive to vibrations of the photodetector. Laser light levels ensure that detectors do not become shot-noise-limited. The detector zone can be repositioned rapidly within the microscope field of view. Because trapping and position-sensing functions are provided by the same laser beam, the two are intrinsically aligned. Finally, the arrangement does not interfere with the simultaneous use of conventional DIC imaging.

Trapping force can be calibrated in a number of ways[28-31]. Two independent methods were used here. A rapid, convenient method applicable for small excursions from the trap centre is to measure the thermal motion of an unbound, trapped bead (Fig. 1c). The optical trap behaves like a linear spring, so that dynamics correspond to brownian motion in a harmonic potential, which has a lorentzian power spectrum. Experimental spectra are well fitted by lorentzians, and the corner frequency provides the ratio of the trap spring constant to the frictional drag coefficient of the bead[32]. The latter is obtained from Stokes' law, after correction for the proximity of the coverslip surface[33]. This approach permits forces on individual beads to be characterized *in situ*, just before or after their use in motility assays. A second approach, useful for larger excursions, is to move the stage in a sinusoidal motion while recording bead displacement from the fixed trap position (details are given in Fig. 1 legend). Both methods give results agreeing within 10%.

Low-load regime

Silica beads were incubated with small amounts of kinesin, such that they carried fewer than one active molecule, on average, and were deposited on microtubules with the optical trap. The power was set to ~17 mW at the specimen plane, providing a nominal trapping force that varied linearly from 0 pN at the centre to ~1.5 pN at the edge of the detector zone (~200 nm). The ATP concentration was fixed at 10 μM. Mean bead velocity in the outermost region of the trap, corresponding to the greatest force, was within 10% of that measured near the trap centre (51 nm s^{-1} versus 54 nm s^{-1}). As observed previously[13], beads frequently released from the microtubule after a variable period of progress (runs) and were drawn rapidly back to the trap centre (within $\leqslant$ 2 ms), whereupon they rebound and began to move again (Fig. 2a). Multiple cycles of attachment, movement and release were observed, with the same bead passing repeatedly through the detector region (as many as 20 times). The points of release, hence the corresponding force levels, were not the same at each pass. Measurements were terminated when a bead travelled out of the trap altogether or became stuck.

Bead displacement during one run is shown in Fig. 2b. The thermal noise in the displacement signal decreased as the bead developed load and walked toward the edge of the trap (Fig. 2c), indicating a nonlinear elasticity. This can be explained as follows. The bead's position is determined by its linkage to two springs. The first is the optical trap, with spring constant k_{trap}. The second is the linkage connecting the bead to the microtubule, acting through the motor, with spring constant k_{motor}. Both springs are extended as the motor pulls forward. If these springs were linear, thermal motion would be independent of the equilibrium position, with a mean-square displacement given by $\langle x^2 \rangle = k_B T/(k_{trap} + k_{motor})$, where k_B is Boltzmann's constant and T is the temperature. Therefore, the data of Fig. 2c imply that k_{motor} gets stiffer with increased extension. It is this diminution of noise, in part, that makes it possible to see steps: when the amplitude of the thermal noise is substantially larger than the step size, steps cannot be detected by any of the methods used here (our unpublished computer simulations). This underscores the need to keep the bead-to-motor linkage stiff in experiments designed to look for molecular-scale motion, and also for imposing a stiff external tension. Considerable variability was observed in the mean noise level from bead to bead in these

experiments, as well as in the degree of noise reduction under tension (by factors of 1–3). For this reason, and because of the relatively large extension (tens of nanometres), we consider it unlikely that stretching of the kinesin molecule itself was chiefly responsible for nonlinear behaviour. One reasonable explanation

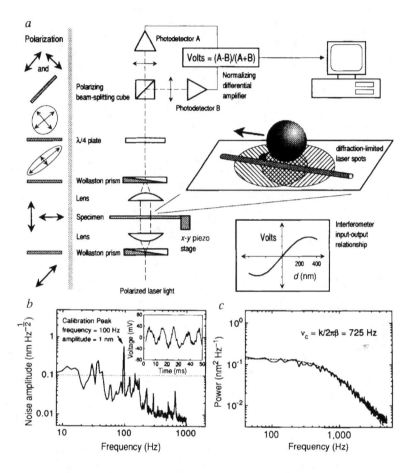

FIG. 1 The optical trapping interferometer. *a* (left), The diagram illustrates the polarization state of light as it passes through elements of the system, viewed along the optical axis. *a* (right), A schematic of the instrument. Polarized laser light passes through a Wollaston prism and is focused to two overlapping diffraction-limited spots ($\sim 1\,\mu$m diameter) with orthogonal polarization, separated by roughly ~ 250 nm. After passage through the specimen (a bead propelled along a microtubule by a kinesin molecule), light recombines in the upper Wollaston and develops slightly elliptical polarization (see text). Ellipticity is measured by a quarter waveplate, which produces nearly circularly polarized light, followed by a polarizing beam-splitting cube, which splits the light into two nearly equal components. The difference in intensity is detected by photodiodes and a normalizing differential amplifier. Signals were analysed offline (LabView, National Instruments). *b*, Sensitivity of the interferometer. The graph shows the spectral noise density of the interferometer responding to a 100-Hz calibration signal of 1-nm amplitude. The large peak (arrow) corresponds to the signal. The detector voltage output (inset) shows both signal and noise. *c*, Force calibration of the optical trap. The thermal noise spectrum of a bead trapped using 58 mW of laser power is shown (solid line). The spectrum is fitted by a lorentzian (dashed line).

METHODS. We used a modified inverted microscope (Axiovert 35, Carl Zeiss) equipped with Nomarski DIC optics (Plan Neofluar 100×/1.3NA oil objective) fixed to a vibration isolation table (TMC Corp.). Light from a Nd:YLF laser (CW, 3W TEM$_{00}$, $\lambda=1,047$ nm; Spectra Physics) was coupled by an optical fibre. Beam-steering was accomplished with a telescope arrangement[13,23]. An *x–y* piezo stage (Physik Instrumente) allowed positioning of the specimen under computer control. To measure instrument noise, a bead was embedded in polyacrylamide to suppress brownian motion and introduced into the detector. The stage was moved in a sinusoidal motion along the *x* axis to provide a calibration signal while recording voltage, and the power spectrum was computed. The power spectrum was scaled by computing the integral under the calibration peak and setting this equal to the mean-square value of the sinusoidal displacement amplitude. To calibrate the trap stiffness, and thereby the force, a bead in solution was trapped (typically $\sim 2\,\mu$m above the coverslip) and its brownian motion recorded, from which a power spectrum was computed. The corner frequency provides the ratio of trap stiffness to the viscous drag of the bead; drag was calculated from the bead's diameter and corrected for proximity to the coverslip. One-sided power spectra were normalized such that $\langle x^2 \rangle = \int_0^\infty S(f)\,df$, where $\langle x^2 \rangle$ is the mean-square displacement and $S(f)$ is the power at a given frequency. Trapping force changed by less than 10% as the distance from the coverslip was reduced from 2 to 1 μm (data not shown). The force profile of the trap was also mapped over a larger range of displacements by moving the stage in sinusoidal fashion (amplitude $A=2.5\,\mu$m at frequency *f*) while monitoring the peak displacement, *y*, of a bead. At low frequencies, the force exerted by the fluid is $F=2\pi A\beta f=k_{\text{trap}}y$, where $\beta=5.7\times10^{-6}$ pN s nm^{-1} is the drag coefficient of the bead. The proportionality between *y* and *f* gave the mean stiffness, $(4.3\pm0.3)\times10^{-4}$ pN nm^{-1} mW^{-1}, which was approximately constant out to displacements of ± 200 nm (data not shown). The foregoing stiffness was used to compute mean forces in the outermost regions of the trap. Power at the specimen plane was estimated by measuring the external power with a meter and applying an attenuation factor of

58%, corresponding to the transmittance of the objective at 1,064 nm, determined separately (data not shown). Kinesin was purified from squid optic lobe by microtubule affinity[47] and tubulin from bovine brain[48]. Experiments were done at room temperature. Silica beads (0.6 μm diameter, 6×10^6 w/v final concentration; Bangs Labs) were incubated for at least 5 min in buffer (80 mM PIPES, 1 mM MgCl$_2$, 1 mM EGTA, pH 6.9, 50 mM KCl, 0.5 mM dithiothreitol, 50 μg ml^{-1} filtered casein, 1–500 μM ATP, 0.5 μg ml^{-1} phosphocreatine kinase, 2 mM phosphocreatine; in some experiments, the last three reagents were replaced by 2 mM AMP-PNP). Beads were incubated for >1 h with kinesin diluted $\sim 1:10,000$ from stock ($\sim 50\,\mu$g ml^{-1} kinesin heavy chain). Taxol-stabilized microtubules were introduced into a flow chamber in which the coverslip had been treated with 4-aminobutyl-dimethylmethoxysilane (Huls America): microtubules bound tightly to this surface. The chamber was then incubated with 1 mg ml^{-1} casein (in 80 mM PIPES, 1 mM MgCl$_2$, 1 mM EGTA, pH 6.9) for 10 min, rinsed with buffer, and kinesin-coated beads introduced. Beads were captured from solution with the trap and deposited on a microtubule selected with its long axis parallel to the Wollaston shear direction, the direction of detector sensitivity. Fewer than half the beads bound and moved when placed on a microtubule. Under these conditions, beads carry Poisson-distributed numbers of functional motors[13]. The bead size in these studies was such that the chance that any bead carried two motors in sufficient proximity to interact simultaneously with the microtubule was less than 2%, assuming, generously, that kinesin heads can reach 100 nm from their points of attachment.

ARTICLES

for the nonlinearity is that the bead–motor linkage behaves like an entropic spring, with the swivelling motion of the bead at the end of its tether contributing degrees of freedom. This linkage becomes taut under load, such that subsequent displacements of the motor are communicated sharply to the bead.

Beads under the conditions shown in Fig. 2a, b moved close to 50 nm s^{-1}. The speed and the thermal noise in the data made a statistical analysis necessary to detect periodic structure. We used the following approach. First, reduced-noise segments of all runs in an experiment were selected, corresponding to movements under load, and filtered to further reduce noise (see below). Then, a histogram of all pairwise differences in displacement was computed for each record[34]. This 'pairwise distance distribution function' (PDF) shows spatial periodicities in stepping motion, independent of times at which steps occur. Next, PDFs were averaged. Finally, the power spectrum of the average PDF was computed: this produces a peak at the mean spatial frequency of the stepper.

For a noiseless, stochastic stepper (equidistant advances after exponentially distributed time intervals), the PDF gives a set of evenly spaced peaks at multiples of the step distance. For noisy steppers, peaks become broadened, disappearing altogether when their peak widths (r.m.s. amplitudes of brownian motion) exceed the step size. The detection of steps in the presence of noise can be improved by low-pass filtering the records with an appropriate filter frequency so as to preserve stepwise character. For this purpose, we used a nonlinear median filter[35]. Computer simulations showed that the peak stepping signal recovered from a background of gaussian noise by an appropriately chosen median filter was approximately twice the amplitude of an equivalent, linear Bessel filter (our unpublished data).

The PDF and associated power spectrum for a single run along a microtubule are shown in Fig. 3a, b. Even in individual runs, there was often an indication of periodicity in the PDF, with a spacing from 6–8 nm, although there was considerable

variation from run to run. Averaged data (Fig. 4a, b), gave peaks corresponding to an uncorrected spacing of 6.7 ± 0.2 nm (mean ± s.e.). Essentially identical data were obtained for beads moving on axonemes (data not shown). Note that backward movements, corresponding to negative distances in the PDF, were practically non-existent. Can our instrument, in combination with the analysis, reliably detect nanometre-sized steps in a noisy background? To test this, we attached beads carrying single motors to microtubules with AMP-PNP, a non-hydrolysable ATP analogue. Beads bound in rigor by AMP-PNP exhibited brownian motion but did not translocate. The piezo stage of the microscope was then stepped stochastically to move the entire specimen (bead and microtubule) through the detector zone, parallel to the long axis of the microtubule, in a series of 8-nm increments, at the same speed as the kinesin-based movement of Fig. 4a. The PDF and power spectrum for this experiment gave peaks at nearly the same positions (Fig. 4c, d). (Similarly, experiments with a stage moved in 4-nm increments gave peaks at half this spacing; data not shown.) Smaller, secondary peaks (measured relative to the sloping background) at subharmonics of the main spatial frequency are usually seen: these arise mainly from the variation in peak heights in the PDF. The measured periodicity was 6.5 ± 0.2 nm, a distance that is 19% smaller than the anticipated value (8 nm). The discrepancy is a consequence of stretching in the elastic bead–motor linkage under load, causing beads to move only a fraction of the motor displacement. (The fraction is $k_{motor}/(k_{trap}+k_{motor})$ for linear springs.) This interpretation was confirmed by an experiment with beads bound directly to the coverslip that had dramatically reduced brownian motion. When the stage was again stepped stochastically by 8 nm, we obtained a strong peak at 8.0 ± 0.2 nm (Fig. 4e, f). Therefore, applying a 19% correction to the spectrum of Fig. 4b yields an adjusted estimate of 8.3 ± 0.2 nm for the mean periodicity of kinesin-based movement.

Could the filtering and/or statistical analysis produce

FIG. 2 *a*, Distance of a bead from the trap centre over 25 s. Multiple cycles of movement, release (solid arrows) and reattachment can be seen. The bead stuck briefly (<1 s) at 23 s (dashed arrow), then continued. The apparent peak between seconds 8 and 9 (third arrow) is a consequence of interferometer nonlinearity: the bead actually passed the turnover point in voltage at ~280 nm (Fig. 1a, inset) and continued its forward movement briefly before releasing. Portions of records beyong 200 nm were not analysed. *b*, Detail from the dashed box in *a*. The raw data (lower trace) were median-filtered for subsequent analysis (upper trace). *c*, The mean-square noise of the track in *b* as a function of distance from the trap centre. The noise decreases as the bead is placed under tension by the trap.
METHODS. Data were acquired at 1 kHz. A cubic polynomial was fitted to the response function (Fig. 1a, inset), and used to convert voltage to distance, a procedure that extended the usable instrument range to ±200 nm with a ±5% error (data not shown). The algorithm computes the absolute value of displacement, so that the distance from the centre of the trap is rectified. Record segments corresponding to movement near the trap centre were not used for analysis. The response function was determined by tracking the motion of beads stuck tightly to the coverslip surface and moved with computer-controlled voltage waveforms supplied to the piezo stage. The piezo stage was calibrated with nanometre-scale precision using video-based methods[38] against a 10-μm diamond-ruled grating (Donsanto Corp.). A calibration procedure was implemented that allowed beads of slightly varying size to be used, even though these scatter different amounts of light. This was made possible by the observation that the turnover point in the response function (Fig. 1a, inset) occurs at a fixed distance from the trap centre, independent of the bead size, and that the response function scales with voltage at that point. By measuring the turnover voltage, it was possible to establish an absolute correspondence between voltage and distance. Only runs longer than 100 nm were analysed for steps, as follows. A line was

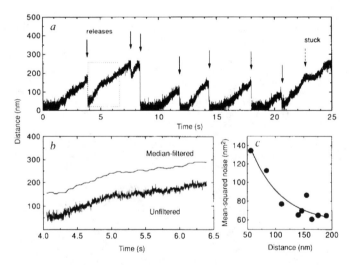

fitted to the run, and the slope used as a first estimate of velocity, v'. A set of non-overlapping segments of duration $2d/v'$ was then fitted to the run, where d is an assumed step size, and an improved estimate of velocity, $\langle v \rangle$, was computed from these segments. The filter frequency was chosen such that $f_c \approx 2\langle v \rangle/d$. Data were filtered with a Bessel filter at $1.2f_c$ and then by a median filter[35] of rank r, chosen such that $f_c = 1/(2r+1)$. This procedure was determined by computer simulations to provide reliable results with stochastic steppers subject to gaussian white noise. The exact choice of d is not critical: for the final data analysis, $d=8$ nm was assumed, but other values (from 4 to 16 nm) gave essentially identical results. The noise in *c* is the mean-square deviation of unfiltered data from successive line segments.

artefactual results? To answer this question, we again bound beads to microtubules with AMP-PNP, but moved the specimen smoothly through the trap at the same mean speed. The PDF and associated power spectrum (Fig. 4g, h) had similar baselines, but no peaks corresponding to spatial periodicities. The variation in several such spectra provided a means of estimating the statistical significance of the peak in Fig. 4b: its likelihood of random occurrence is $P < 0.00001$.

High-load regime

The laser power was set to ~58 mW at the specimen plane, a level that provided a nominal force that varied linearly from 0 pN at the centre to ~5 pN at the edge of the detector zone. The ATP concentration was raised to 500 µM (saturation level). When beads carrying single kinesin molecules were placed on microtubules, their motion was more erratic than at lower powers. Beads near the trap centre, experiencing low loads, moved rapidly at 300–500 nm s^{-1}. Under higher loads towards the edge of the trap, beads markedly slowed (or became stuck), although some were still able to escape altogether, opposing forces up to 5 pN. Other beads slowed or stuck before reaching the trap's edge, then continued forward at increased speed. Shuttering the trap briefly (<0.5 s) during a slow (or stuck) episode enabled some beads, but not all, to regain forward motion. Multiple transitions between fast- and slow-moving phases were seen in individual records, with beads entering the slow phase over a range of positions (loads) in the trap. Beads still underwent

multiple cycles of movement, release and reattachment, similar to the behaviour observed at lower loads.

In the slow-moving phase, the reduced speed permitted visualization of steps. Figure 5a, b shows two records at high load, each containing roughly ten abrupt displacements forwards. The heights of these steps varied from ~5–18 nm, with the smaller transitions averaging 8 ± 2 nm (mean $\pm$ s.d.; $n = 16$; Fig. 5c-f). Beads also advanced through what appeared to be 'double steps' (~17 nm), but it was not possible, given the bandwidth and noise in these signals, to determine whether such jumps represented two steps in rapid succession or one single step. Assuming that motors are stochastic steppers, one expects an exponential distribution of dwell times with numerous short steps. Records

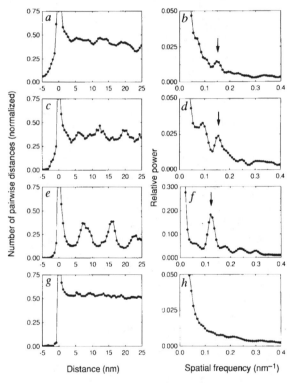

FIG. 4 *a*, The averaged pairwise distance distribution function (PDF) for kinesin movement, taken from 17 records of 10 different beads. *b*, The power spectrum of the data in *a*, with a peak at 0.149 ± 0.004 nm^{-1} (arrow), corresponding to a periodicity of 6.7 ± 0.2 nm. *c*, The averaged PDF for beads attached to the microtubule using AMP-PNP, with an 8-nm stochastically stepped stage. *d*, The power spectrum of the data in *c*, with a main peak at 0.154 ± 0.004 nm^{-1} (arrow), corresponding to a periodicity of 6.5 ± 0.2 nm. *e*, The averaged PDF for beads attached directly to the coverslip, with an 8-nm stochastically stepped stage. *f*, The spectrum of the data in *e*, with a peak at 0.125 ± 0.003 nm^{-1} (arrow), corresponding to a periodicity of 8.0 ± 0.2 nm. *g*, The averaged PDF for beads attached to the microtubule using AMP-PNP with a smoothly moved stage. *h*, The power spectrum of *g*. The average amplitude fluctuation about the mean for records of smooth movement was 0.002 units over the range of spatial frequencies 0.1–0.2; that is, there are no statistically significant peaks in this trace. All PDFs and power spectra were normalized to unity. Note different scale in *e*.
METHODS. PDFs in each panel represent averages of 17 records. Peak positions in power spectra and corresponding error estimates were determined by fitting gaussians. To estimate the statistical significance of the peak in *b*, the standard deviation per point, σ_k, for 17 spectra of individual runs of a smoothly stepped stage was computed. The likelihood of the peak was $\langle (S_k - \bar{N}_k)^2 \rangle^{1/2} / \langle \sigma_k \rangle$, where S_k is the average spectrum of kinesin-driven movement in *b*, and $\bar{N}_k$ is the average spectrum of smooth movement in *h*; averages were taken over the four spatial frequencies comprising the peak.

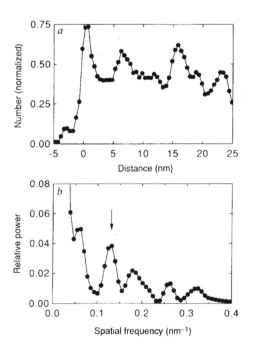

FIG. 3 *a*, The pairwise distance distribution function (PDF) for a single run. The density of interpoint distances in a median-filtered record is shown: multiple peaks with regular spacing can be identified. *b*, The normalized power spectrum of the data in *a*, showing a prominent peak at ~0.130 nm^{-1} (arrow), corresponding to periodicity of 7.7 nm.
METHODS. PDFs were computed by binning distance differences $(x_j - x_i)$ for all $(j > i)$ in a histogram, with bin width 0.5 nm. The PDF was normalized to unity at the first bin and smoothed with a 3-point moving window. The one-sided power spectrum, $S(k)$, was computed from the PDF by FFT and normalized to unity at $S(0)$. The PDF is identical to the autocorrelation function of the density of distances from the trap centre. $S(k)$ is analogous to the square of the structure factor used in X-ray scattering, with the density of distances being analogous to electron density[49].

ARTICLES

of the type shown in Fig. 5*a*, *b* are relatively hard to obtain: additional records must be collected before a more complete statistical analysis, including measurements of dwell-time distributions, will be possible. In parts of the record, the distance from the centre of the trap appeared to shorten briefly (~ 10 ms) just before a step (Fig. 5*e*, *f*), but this was not seen in all cases (Fig. 5*c*, *d*). We do not know if this behaviour reflects an intrinsic property of the motor, but the shortening is not due to kinesin unbinding because substrate release for periods as short as a millisecond would result in the bead being pulled off completely (Fig. 2).

Low-ATP regime

Reasoning that steps might also be resolved under low loads if beads moved slowly enough we restored the power to 17 mW and lowered the ATP concentration to 1 or 2 μM. When beads carrying single kinesin molecules were placed on a microtubule, their speeds were slow, about 5–15 nm s^{-1}, and single steps could again be seen in records that were selected for low noise (and therefore, presumably, stiffer bead–motor linkages) (Fig. 6*a–e*). A statistical analysis of these, like that performed in the low-load

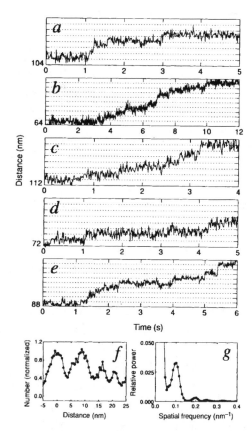

FIG. 6 *a, b*, Multiple steps in the displacement records of two different beads moving at an ATP concentration of 2 μM. *c–e*, Multiple steps in the displacement records of three different beads moving at an ATP concentration of 1 μM. Horizontal gridlines (dotted) have been drawn at a spacing of 8 nm. A 10-point jumping average of data taken at 1 kHz is plotted for an effective rate of 100 Hz. *f, g*, The PDF and associated power spectrum of the record in *e*, showing a periodicity of 8.8 ± 0.5 nm. Mean step size and s.d. for records in *a–e* were computed from peak positions determined from PDFs. A correction for the extension in k_{motor} has not been applied to these data, which were selected for their low noise and which presumably reflect stiffer linkages.

regime, showed that steps measured 8.2 ± 1.1 nm (mean ± s.d., as determined from peak positions in power spectra for all five panels), the same size as those seen in the high-load regime. Figure 6*f*, *g* shows an example of the PDF and associated power spectrum for the data of Fig. 6*e*, where the mean periodicity was 8.8 ± 0.5 nm.

Is it possible that kinesin motors travel a longer, possibly variable, distance with each ATP hydrolysis, moving by some integral multiple of this spacing? That is, does a single hydrolysis result in a sequence of physical steps, rather than just one, as has been proposed for myosin[4,5,16,36]? If this were the case for kinesin, then for extremely low concentrations of ATP (the diffusion limit), one would expect a motor to advance in a rapid burst of steps (through $\gg 8$ nm) before stopping to wait for the next ATP molecule. This would produce motion characterized by periods of zero advancement interspersed with clusters of steps. The mean rate of advance during a step series ought to be high, close to the speed for saturating levels of ATP (~ 500 nm s^{-1}). We saw no evidence for step clusters in any of our records at 1 or 2 μM ATP: overall rates of advancement were quite uniform.

Discussion

The finding of discrete stepping behaviour should allow direct measurement of kinetic parameters of the mechanochemical cycle, by determining the timing of various phases of the motion. For example, releases of the type shown in Fig. 2*a* permit an estimate of the cycle off-time. Once a bead near the edge of the trap detaches under low load conditions, it is returned rapidly to the centre through a distance, *x*, of ~ 200 nm within an average time $\tau = 1.8 \pm 0.4$ ms (mean ± s.e.). To make net forward progress, the kinesin off-time, τ_{off}, must be less than the time required for the trap to pull the bead back by one step, *d*, that is, $\tau_{off} < d/v = d\tau/x$, where *v* is the return speed at the edge of the trap. Using 8 nm for *d* gives $\tau_{off} \leqslant 72$ μs. This time is an upper limit, because velocity is practically unaffected by trapping forces. Such a short off-time signifies that kinesin somehow remains bound, sustaining load throughout most of its activity, perhaps moving hand-over-hand[12,37]. Estimates of other kinetic parameters are ongoing subjects of investigation.

It is known that kinesin motors do not wander over the surface lattice of a microtubule, but move instead along straight paths parallel to a protofilament[38–40]. Kinesin head fragments cloned from squid[41] or *Drosophila*[42] and expressed in bacteria have been used to decorate microtubules. Such fragments are non-functional in motility assays, but they bind microtubules and have microtubule-activated ATPase activity. Saturation binding experiments indicate a stoichiometry of one kinesin heavy-chain fragment to each tubulin dimer[42], and crosslinking studies demonstrate that it is the β-subunit of tubulin that is bound[41]. Electron microscopy of decorated microtubules shows an 8-nm repeat arising from the head spacing[41,42]. Taken together with our observations, this suggests that the two heads of a kinesin molecule walk along a single protofilament—or walk side-by-side on two adjacent protofilaments—stepping ~ 8 nm at a time, making one step per hydrolysis (or perhaps fewer, requiring multiple hydrolyses per step). If the foregoing model is correct, then during movement of single molecules, the ATP hydrolysis rate, r_{ATP}, should be related to the step size, *d*, and speed, *v*, through $r_{ATP} = v/d$. Movement at saturating levels of ATP ($v \approx 500$ nm s^{-1} and even higher[20]) implies $r_{ATP} \geqslant 60$ s^{-1} mol^{-1}, 6–10-fold higher than reported[43–45]. The hydrolysis rates for motility assays *in vitro* must therefore be substantially higher than values inferred from solution biochemistry, or motors must make several steps per ATP hydrolysed. The latter is difficult to reconcile with our data, particularly at low concentrations of ATP.

Occasionally jumps have been seen[38] in selected records of kinesin movement during video tracking, measuring 3.7 ± 1.7 nm, although most of the movement was subjectively

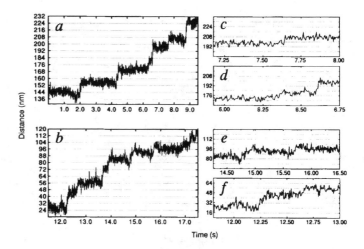

FIG. 5 *a, b*, Multiple steps in the displacement records of two different beads in the high-load regime. *c, d*, Details of selected steps in *a*. *e, f*, Details of selected steps in *b*. Horizontal gridlines (dotted) have been drawn at a spacing of 8 nm. A 5-point jumping average of data taken at 1 kHz is plotted for an effective rate of 200 Hz. The mean step size for these records was computed from averaged values of distance determined during segments of traces before and after transitions, identified by eye. Records at high loads were not corrected for stretching in k_{motor} because the load presumably extends that linkage to be taut.

smooth. Our instrument resolves distances of this size, but we did not find evidence for significant periodicities near ~4 nm. The earlier work, however, was done before the development of single-motor assays, and beads carried unknown numbers of motors, probably several. It is possible that the shorter spacing, if real, might reflect an effect of multiple heads stepping along neighbouring protofilaments.

Our data from moving beads subjected to trap-induced tensions suggest that speeds are largely unaffected by forces of 1.5 pN, and that single molecules of kinesin can still transport beads against loads up to ~5 pN. These findings seem incompatible with the conclusions of Kuo and Sheetz[30], who reported measuring the 'isometric' force generated by single kinesin molecules at 1.9 ± 0.4 pN. In their experiments, nominally arrested beads continued to move; that is, the isometric condition was not truly fulfilled, but this difficulty alone is probably insufficient to explain the discrepancy.

Finding displacement steps in the movement of kinesin molecules is in many respects analogous to detecting current steps in single-channel recordings of neurons. In both cases, steps reflect underlying motions of individual proteins, and allow one to make meaningful measurements at the level of single molecules. Before the advent of single-channel recording, Johnson (thermal) noise due to the leakage resistance of electrodes prevented the resolution of picoampere-sized current steps. Improved methods for increasing this resistance provided conditions that led directly to step detection[46]. As a result, single-channel recording has produced many insights in molecular neuroscience during the past decade. For mechanoenzymes, brownian (thermal) noise of objects pulled by motors prevented the resolution of nanometre-sized displacement steps. Additional stiffness, coming in our case from the optical trap and its effect on the bead linkage, provides the required reduction in noise, permitting visualization of steps. ☐

Received 16 July; accepted 27 September 1993.

1. Huxley, H. E. *Science* **164**, 1356–1366 (1969).
2. Squire, J. *The Structural Basis of Muscle Contraction* (Plenum, London, 1981).
3. Bagshaw, C. R. *Muscle Contraction* (Chapman & Hall, London, 1982).
4. Burton, K. J. *Musc. Res. Cell Motil.* **13**, 590–607 (1992).
5. Simmons, R. M. *Curr. Biol.* **2**, 373–375 (1992).
6. Cooke, R. *CRC Crit. Rev. Biochem.* **21**, 53–117 (1986).
7. Kron, S. J. & Spudich, J. A. *Proc. natn. Acad. Sci. U.S.A.* **83**, 6262–6276 (1986).
8. Vale, R. D., Schnapp, B. J., Reese, T. S. & Sheetz, M. P. *Cell* **40**, 559–569 (1985).
9. Ishijima, A., Doi, T., Sakurada, K. & Yanagida, T. *Nature* **352**, 301–306 (1991).
10. Uyeda, T. Q. P., Kron, S. J. & Spudich, J. A. *J. molec. Biol.* **214**, 699–714 (1990).
11. Uyeda, T. Q. P., Warrick, H. M., Kron, S. J. & Spudich, J. A. *Nature* **352**, 307–311 (1991).
12. Howard, J., Hudspeth, A. J. & Vale, R. D. *Nature* **342**, 154–158 (1989).
13. Block, S. M., Goldstein, L. S. B. & Schnapp, B. J. *Nature* **348**, 348–352 (1990).
14. Yang, J. T., Saxton, W. M., Stewart, R. J., Raff, E. C. & Goldstein, L. S. B. *Science* **249**, 42–47 (1990).
15. Toyoshima, Y., Kron, S. J. & Spudich, J. A. *Proc. natn. Acad. Sci. U.S.A.* **87**, 7130–7134 (1990).
16. Yanagida, T., Arata, T. & Oosawa, F. *Nature* **316**, 366–369 (1985).
17. Harada, Y., Sakurada, K., Aoki, T., Thomas, D. & Yanagida, T. *J. molec. Biol.* **216**, 49–68 (1990).
18. Higuchi, H. & Goldman, Y. *Nature* **353**, 352–354 (1991).
19. Spudich, J. A. *Nature* **348**, 284–285 (1990).
20. Gittes, F., Mickey, B., Nettleton, J. & Howard, J. *J. Cell Biol.* **120**, 923–934 (1993).
21. Schnapp, B. J. *Meth. Enzym.* **134**, 561–573 (1986).
22. Ashkin, A., Dziedzic, J. M., Bjorkholm, J. E. & Chu, S. *Optics Lett.* **11**, 288–290 (1986).
23. Block, S. M. in *Noninvasive Techniques in Cell Biology* 375–401 (Wiley-Liss, New York, 1990).
24. Denk, W. & Webb, W. W. *Appl. Optics* **29**, 2382–2390 (1990).
25. Amos, L. & Klug, A. *J. Cell Sci.* **14**, 523–549 (1974).
26. Kamimura, S. *Appl. Optics* **26**, 3425–3427 (1987).
27. Kamimura, S. & Kamiya, R. *J. Cell Biol.* **116**, 1443–1454 (1992).
28. Block, S. M., Blair, D. F. & Berg, H. C. *Nature* **338**, 514–517 (1989).
29. Ashkin, A., Schuetze, K., Dziedzic, J. M., Euteneuer, U. & Schliwa, M. *Nature* **348**, 346–348 (1990).
30. Kuo, S. C. & Sheetz, M. P. *Science* **260**, 232–234 (1993).
31. Simmons, R. M. et al. in *Mechanisms of Myofilament Sliding in Muscle* (eds Sugi, H. & Pollack, G.) 331–336 (Plenum, New York, 1993).
32. Wang, C. W. & Uhlenbeck, G. E. in *Selected Papers on Noise and Stochastic Processes* (ed. Wax, N.) 113–132 (Dover, New York, 1954).
33. Happel, J. & Brenner, H. *Low Reynolds Number Hydrodynamics* 322–331 (Kluwer Academic, Dordecht, 1991).
34. Kuo, S. C., Gelles, J., Steuer, E. & Sheetz, M. P. *J. Cell Sci. suppl.* **14**, 135–138 (1991).
35. Gallagher, N. C. Jr & Wise, G. L. *IEEE Trans. Acoust. Speech Sign. Proc.* **29**, 1136–1141 (1981).
36. Oosawa, F. & Hayashi, S. *Adv. Biophys.* **22**, 151–183 (1986).
37. Schnapp, B. J., Crise, B., Sheetz, M. P., Reese, T. S. & Khan, S. *Proc. natn. Acad. Sci. U.S.A.* **87**, 10053–10057 (1990).
38. Gelles, J., Schnapp, B. J. & Sheetz, B. J. *Nature* **331**, 450–453 (1988).
39. Kamimura, S. & Mandelkow, E. *J. Cell Biol.* **188**, 865–875 (1992).
40. Ray, S., Meyhoefer, E., Milligan, R. A. & Howard, J. *J. Cell Biol.* **121**, 1083–1093 (1993).
41. Song, Y.-H. & Mandelkow, E. *Proc. natn. Acad. Sci. U.S.A.* **90**, 1671–1675 (1993).
42. Harrison, B. C. et al. *Nature* **362**, 73–75 (1993).
43. Kusnetsov, S. A. & Gelfand, V. I. *Proc. natn. Acad. Sci. U.S.A.* **83**, 8530–8534 (1986).
44. Hackney, D. *Proc. natn. Acad. Sci. U.S.A.* **85**, 6314–6318 (1988).
45. Gilbert, S. P. & Johnson, K. A. *Biochemistry* **32**, 4677–4684 (1993).
46. Sakmann, B. & Neher, E. *Single-Channel Recording* (Plenum, New York, 1983).
47. Weingarten, M. D., Suter, M. M., Littman, D. R. & Kirschner, M. W. *Biochemistry* **13**, 5529–5537 (1974).
48. Schnapp, B. J. & Reese, T. S. *Proc. natn. Acad. Sci. U.S.A.* **86**, 1548–1552 (1989).
49. Glatter, O. & Kratky, H. C. *Small Angle X-Ray Scattering* (Academic, New York, 1982).

ACKNOWLEDGEMENTS. We thank W. Denk, C. Godek, M. Meister, P. Mitra and R. Stewart for discussions and advice, W. Hill for electronic design, and R. Stewart for help with motility assays and protein purification. This work was supported by the Rowland Institute for Science (K.S., C.F.S., S.M.B.), and partial support from the NIH (K.S. and B.J.S.), the Lucille P. Markey Charitable Trust (B.J.S.), and the University of Michigan (C.F.S.).

LETTERS TO NATURE

Movement and force produced by a single myosin head

J. E. Molloy, J. E. Burns, J. Kendrick-Jones[*]**, R. T. Tregear**[*] **& D. C. S. White**

Department of Biology, University of York, York YO1 5DW, UK
[*] Laboratory of Molecular Biology, Hills Road, Cambridge CB2 2QH, UK

MUSCLE contraction is driven by the cyclical interaction of myosin with actin, coupled to the breakdown of ATP. Studies of the interaction of filamentous myosin[1] and of a double-headed proteolytic fragment, heavy meromyosin (HMM)[2,3], with actin have demonstrated discrete mechanical events, arising from stochastic interaction of single myosin molecules with actin. Here we show, using an optical-tweezers transducer[2,4], that a single myosin subfragment-1 (S1), which is a single myosin head, can act as an independent generator of force and movement. Our analysis accounts for the broad distribution of displacement amplitudes observed, and indicates that the underlying movement (working stroke) produced by a single acto-S1 interaction is ~4 nm, considerably shorter than previous estimates[1-3,5] but consistent with structural data[6]. We measure the average force generated by S1 or HMM to be at least 1.7 pN under isometric conditions.

Previous studies[7-9], under conditions in which several molecules are interacting with the actin filament, indicated that both S1 and single-headed myosin are capable of generating movement and force. To demonstrate that such events occur between a single myosin head and an actin monomer, we have measured interactions between S1 and actin using an optical-tweezers apparatus[5]; its means of operation is similar to that used previously[2], except that it synthesizes a double trap by rapid scanning of a single laser beam, and has a position detector with a frequency response high enough (2.2 kHz) to allow observation of the full amplitude of brownian motion.

An actin filament was suspended between two polystyrene beads held in optical traps, and its intermittent interaction with S1 or HMM bound to a stationary glass sphere at low surface density was monitored by measuring the motion of one of the beads (see Fig. 1 legend). The surface concentration of S1 or HMM and the bulk concentration of ATP were chosen so that mechanical interactions between a single S1 head and the actin filament could be readily observed.

When low concentrations of either S1 or HMM were used, fast-rising displacements could be clearly distinguished from thermal vibration of the actin filament (Fig. 1a-c). Most of the displacements were in one direction, but negative displacements were also observed (down arrows). By rotating the two trap positions through 180° about the myosin contact, so that the polarity of the actin filament was reversed, we found that the preferred direction of the displacements was also reversed and that ~80% of displacements occurred in the preferred direction. In 10 μM ATP, the average duration of positive and negative displacements was 90 ms and was independent of the amplitude of the observed attachment (Fig. 1g), suggesting that the presence of a preferred direction for attachment is not a kinetic effect arising from differences in probability of detachment, but is probably due to a myosin work stroke with its polarity determined by that of the actin filament. Lowering the ATP concentration from 10 μM (Fig. 1b) to 2 μM (Fig. 1c) increased the mean duration. Those amplitudes of positive and negative displacements which could be distinguished from thermal noise were measured from the mean resting position. Histograms of the distribution of measured events are shown for both HMM and S1 in Fig. 2a, b. The striking feature of our data is the significant number of very large displacements (up to 30 nm) and negative displacements (down to −25 nm). Large-amplitude

positive displacements have previously been observed[2,3], especially at low optical trap stiffness, but negative displacements had not.

Noise of the displacement records is due to the thermal motion of the bead. During both positive and negative displacement events the amplitude of brownian noise was reduced, indicating that system stiffness had increased (Fig. 1a-c). Particularly in traces obtained at 5 and 2 μM ATP, periods of low noise were also observed without significant changes in mean displacement (Fig. 1c). S1 stiffness, κ_m, predicted from noise analysis during these periods, was similar irrespective of displacement ($\kappa_{xm} - \kappa_x = 0.48$ pN nm^{-1}).

Using low variance of noise (Fig. 1d) as the criterion for attachment, we again measured the size of displacements produced by S1 and HMM relative to the mean rest position. In contrast to the method above, this technique enabled us to identify work strokes that drove the bead to near its mean position. The resulting histograms are shown in Fig. 2c, d.

In feedback (force) mode (Fig. 1 legend) the bead was held in a fixed position, the forces exerted by the molecular interactions being compensated by movement of the laser trap, allowing estimates of the isometric forces produced by single actomyosin interactions (Fig. 3a). The mean force of the events was 1.7 pN and their duration was similar to that of the displacements. We found that HMM produced approximately the same amount of force as S1 (Fig. 2e, f).

Controlled, large-amplitude length changes were applied by imposing movement to the laser beam holding the bead imaged on the photodetector. The forces were measured from the movement of the laser beam required to produce the correct movement of the bead (Fig. 3b). The minimum stiffness of a single acto-myosin interaction was determined from the change in force required to produce a given displacement. Mean stiffness, measured from the gradient of length–tension diagrams, was 0.17 pN nm^{-1} (Fig. 3c), much less than the above estimates from noise analysis, but stiffer than the trap stiffness used for our estimates of the working stroke, and similar to estimates of crossbridge stiffness in muscle fibres.

The similarity between the displacements and forces produced by S1 and HMM suggests that the same molecular process is generating the mechanical events found in both preparations. The possibility that the S1 measurements might arise from contaminant HMM is excluded by the absence of HMM in gels of the S1 preparation (Fig. 1 legend) and the fact that only three times more S1 was needed than HMM to produce these mechanical results. We conclude that a single myosin head can generate force and movement.

To extract the size of the working stroke from our data, the randomizing effect of thermal filament displacement has to be considered. We believe that crossbridge attachment will occur with equal intrinsic probability over the full range of the brownian motion of the actin filament. The start position of the myosin molecule relative to the bead rest position is arbitrary, certainly between different bead/actin filament preparations, and probably to some extent between records using any particular actin filament. The actin filament is free to rotate in an unconstrained fashion, so on average every actin monomer must be equivalent. The beads and actin filament would be expected to rotate over several seconds[10] (that is, the period of a single recording). Thus dependence of attachment upon azimuthal orientation[5] will be detectable only in recordings made with sufficient speed. Different recordings will have different azimuthal start points. Although summation of mechanical events will be very rare at the dilutions of protein used, there is likely to be more than one S1 head able to interact with the actin filament over the time scale of these experiments.

Our interpretation of the distribution of displacements in Fig. 2a, b is that crossbridge attachments occur at any axial position of the actin filament, and that the working stroke produces a displacement from that arbitrary starting point. The distribution

LETTERS TO NATURE

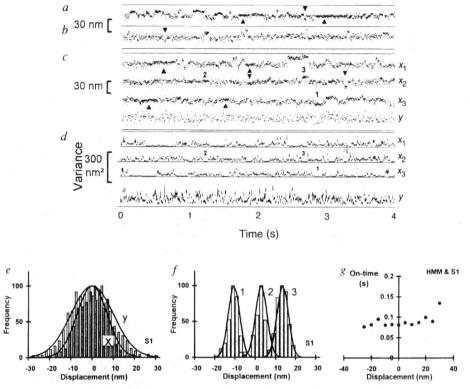

FIG. 1 Displacement of actin filament under near-zero load conditions resulting from interactions with low densities of *a*, HMM and *b–f*, S1. *a*, Record obtained on contact of HMM (bound to surface at 2 μg ml⁻¹) with an actin filament in 10 μM ATP; κ_{trap}, 0.022 pN nm⁻¹. *b*, Similar record obtained with S1, at density of 5 μg ml⁻¹, in 10 μM ATP; κ_{trap}, 0.021 pN nm⁻¹. *c*, Three sequential records $(x_1–x_3)$ obtained on contact of S1 heads, at 6 μg ml⁻¹, with an actin filament in 2 μM ATP; κ_{trap}, 0.042 pN nm⁻¹. The 'noise' content of these records is due to residual brownian motion of the trapped bead. Detector noise was small by comparison (0.8 nm r.m.s.). Note the reduction in thermal vibration during the displacements. Negative displacements marked with down-pointing arrows. Periods showing long low thermal noise and little change of mean position are marked with up-pointing arrows. The fourth trace (y) is the record obtained in the y-axis during the sampling of x_1. Sampling frequency 10 kHz, 5-point average. *d*, Running 13-point variance of the traces in *c*. The gridlines are the zero-axis for the trace immediately above that line. *e,f*, Histograms of the distribution of displacements during the records shown in *c*. *e*, The distribution in the y-axis and in the x-axis during a 'high-noise' section of the trace. The lines are Boltzmann distributions with zero mean and stiffness $\kappa_y = 0.04$ pN nm⁻¹ and $\kappa_x = 0.08$ pN nm⁻¹. The data indicate that the compliance in the x-direction is half that in the y-direction ($\kappa_x = 2 \kappa_y$ (see Methods)). This y-axis stiffness is very similar to that determined by other means for a single bead. *f*, The distribution during the 'low-noise' sections marked 1, 2 and 3 in *c*. The fitted curves assume a stiffness κ_{xm} of 0.56 pN nm⁻¹. To estimate the amplitude of brownian motion the detector bandwidth must be greater than the brownian noise bandwidth (~600 Hz; controlled by viscous drag on the bead and trap stiffness¹²). Our detector bandwidth (2.2 kHz) was sufficient to estimate the amplitude of brownian noise but gave poor phase information. We were therefore unable to determine accurately the start position of a bead immediately before a crossbridge interaction. *g*, Plot of attached lifetime (on-time) versus displacement during attachment. On-time is nearly independent of displacement polarity and magnitude.

METHODS. S1 was prepared by digestion of rabbit psoas muscle myosin with papain under conditions of low salt and high magnesium; the reaction was stopped with E64 and the supernatant column-purified. The S1 ran as a single peak on ion-exchange and gel filtration chromatography. All three light chains were present, and there was no sign of HMM in SDS–PAGE. F-actin, HMM and NEM-myosin were prepared by standard methods. To make the 1.1-μm polystyrene beads bind to F-actin irreversibly, they were coated with 1 μg ml⁻¹ BSA-TRITC and 10 μg ml⁻¹ NEM-treated monomeric myosin. Glass microspheres (1.7 μm diameter) were applied to a coverslip as a suspension in 0.1% nitrocellulose; after drying, the coverslip was made part of

a 0.2-mm deep flow chamber and S1 was bound in a low ionic strength solution (in mM: 25 KCl, 10 Tris, pH7, 4 MgCl₂, 1 EGTA). Rhodamine–phalloidin labelled F-actin and NEM-myosin labelled polystyrene beads were introduced in an ATP-containing *in vitro* motility assay buffer¹³ at 23 °C. Two NEM-coated beads were captured by optical traps and an actin filament strung between them. The position of one bead was monitored by brightfield microscopy, using a four-quadrant photodetector². A glass microsphere (coated with S1 or HMM) was brought up close to the actin filament and the interaction with S1 or HMM observed. Each optical trap imposes a trap-stiffness κ_{trap} to its bead, the value of which can be set by varying the laser power. The actin filament stiffness in tension (as is the case in these experiments) is sufficiently great for the 10-μm lengths used¹⁴ that it can be considered rigid. The axial (x-axis) system stiffness κ_x is thus double the individual trap stiffness ($\kappa_x = 2 \kappa_{trap}$). When a myosin head attaches, its stiffness, κ_m, adds to provide a total stiffness $\kappa_{xm} = \kappa_x + \kappa_m$. The y-axis stiffness κ_y will, under all conditions, be approximately equal to κ_{trap}. The mean square deviation from the rest position in the absence of acto-myosin interaction is related to the system stiffness ($\langle x^2 \rangle = kT/\kappa_x$; $\langle y^2 \rangle = kT/\kappa_{trap}$)¹². The noise seen on the traces is close to that predicted from the independently derived value of κ_{trap} (Fig. 1*e,f*) and displays a Boltzmann distribution with half-width at half-height (ln 2) $2kT/\kappa_x)^{1/2} \approx 12$ nm for the trap stiffness we have used (see below; Fig. 2*a–d*). κ_{trap} was calibrated using a bead with no actin filament attached (when $\kappa_x = \kappa_y = \kappa_{trap}$) by three methods. First, with the trapped bead held 5 μm from the coverslip a large amplitude triangle wave motion was applied to the microscope slide and Stokes' force² (6πηav, with viscosity η, bead radius a, stage velocity v) produced a bead movement proportional to trap compliance. Second, the mean square brownian motion was measured ($\langle x^2 \rangle$) over several seconds at 2.5 kHz bandwidth and the equipartition principle applied ($\frac{1}{2}\kappa_{trap}\langle x^2 \rangle = \frac{1}{2}kT$) (ref. 12). Third, the most convenient check of trap stiffness was by on-line power-spectrum analysis using the 3-dB roll-off in brownian noise spectrum ($v_c = \kappa/2\pi\beta$; ($\beta = 6\pi\eta a$)¹²; this was measured 5 μm from the coverslip surface. Using this method, laser power was adjusted until data matched a predetermined lorenzian curve. Four-quadrant detector sensitivity was determined by applying a known displacement to the laser tweezer position in the x- and y-axes. When feedback was applied between the quadrant detector and the optical trap position, stiffness was raised to 5 pN nm⁻¹, measured by applying large Stokes' force. With an actin filament held taut between two trapped beads, κ_x was usually set to be between 0.03 and 0.05 pN nm⁻¹. This stiffness is sufficiently lower than that of the acto-S1 interaction (at least 0.17 pN nm⁻¹; see Figs 1*e, f* and 3*c*) to allow the magnitude of the elementary movement to be estimated.

FIG. 2 *a, b,* Frequency distribution of the amplitude of displacements, A_x, scored by measuring displacements that were evident above background of residual brownian noise (*a,* 39 traces from 18 filaments; *b,* 73 traces from 21 filaments). The fits are Boltzmann distributions with $A_x = A_0.\exp\left\{-\frac{1}{2}\kappa(x-x_0)^2/kT\right\}$, in which x_0 is the displacement caused by crossbridge interaction. The arbitrary κ and amplitudes were fitted by assuming that all of the data belong to a single distribution in which values are missing for zero and small displacements (for fitting method see *e* below). The crossbridge work stroke, x_0, of these fits is 3.5 nm for S1 and 5.0 nm for HMM. *c, d,* Frequency distribution of the amplitude of displacements measured for a more restricted data set (*c,* 9 traces from 4 filaments; *d,* 5 traces from 2 filaments) in which the system was sufficiently stable to estimate crossbridge attachments by the increase in system stiffness (fall in local variance). Using this method, zero and near-zero displacements were quantified (see Fig. 1*c, d*). The solid lines have the same values of κ and x_0 as in *a* and *b,* with A_0 scaled to fit the frequency of observations. The broken lines are the best-fit Boltzmann distributions to the data. For S1 the value of x_0 predicted from these data is 3.3 ± 0.73 (s.e.m.) nm and the value of κ is 0.052 pN nm^{-1}. For HMM, $x_0 = 7.9 \pm 0.91$ nm, with $\kappa = 0.031$ pN nm^{-1}. Note that there are many observations with small or zero displacement. *e,* The data of *a* and *b* were recalculated as log(frequency) versus $(x - x_0)^2$, where x is the observed displacement and x_0 is a constant, thereby causing both 'tails' of the distributions to be positive, and making normal distributions linear, giving a gradient proportional to κ, and the *y*-intercept of log(A_0). For this calculation, frequencies of 0 or 1 were omitted, as were those for displacements between -7.5 and $+10$ nm (so only the tails of the histograms were used). The small graphs on the left-hand side show plots of log(frequency) versus $(x - x_0)^2$ for three values of x_0. The main figure shows the value of the regression coefficient of the best linear fit to this relationship versus x_0. The parameters with the highest regression coefficients gave, for S1, $x_0 = 3.5$ nm, $\kappa = 0.046$ pN nm^{-1}, $A_0 = 131$; and for HMM, $x_0 = 5.0$ nm, $\kappa = 0.047$ pN nm^{-1}, $A_0 = 126$. Note that, in all cases, the fitted value of κ is very close to the value of κ_x for these experiments. *f, g,* Isometric forces from S1 and HMM; amplitude of each event was estimated by eye.

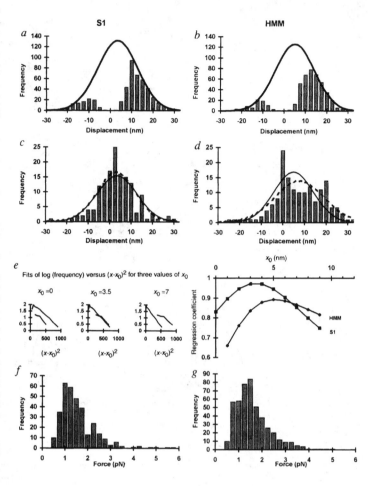

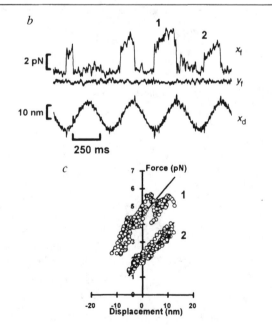

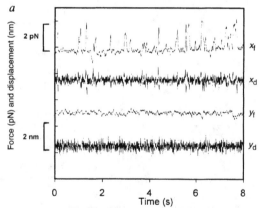

FIG. 3 Force resulting from near-isometric interaction of S1 with actin filaments. 10 µM ATP. *a,* Record of isometric forces. Forces were generated predominantly in one direction, but a few negative forces were detected. Because of the geometry of the optical tweezers, and the fact that only one bead has feedback applied to its position, it is not possible to quantify negative forces. X_f, Y_f; X_d, Y_d: forces and displacements in *x* and *y* directions, respectively. *b,* Record of forces with 2.5 Hz, 20 nm forced axial sinusoidal displacement. System stiffness 5 pN nm^{-1}. *c,* Force–extension plot during the marked regions in *b*; fitted lines have a gradient of 0.17 pN nm^{-1} (mean value was 0.17 pN nm^{-1}, $N = 7$ filaments; 25 cyles total).

LETTERS TO NATURE

of displacements will thus have the same half-width as that of the brownian motion in the absence of attachments, but will be shifted by the size of the working stroke (x_0) from the mean rest position. The observed data can be fitted by such a distribution shifted by about 4–5 nm (Fig. 2a, b). However, observations at or close to zero displacement are missing because, in this and previous analyses[2,3,5], only displacements that deviate clearly from the mean rest position can be scored. The bimodal distribution is simply the two tails of a single broad distribution of crossbridge interactions (Fig. 2c, d). For S1 the data obtained by scoring attachments (Fig. 2c) fit closely to that predicted (solid curve of Fig. 2a). For HMM the fit is less good, and it is possible that both heads are contributing to the data observed.

We do not know the extent to which compliance of attachment between S1 and the nitrocellulose substrate acts as an elastic element during force generation; a consequence of this is that our estimates of force and stiffness are lower limits. However, unknown external compliance does not contribute to our estimate of displacement because this is determined from measurements at low force. In addition, the orientation of S1 or HMM on the substrate will presumably be random in these experiments. If the attached head orientates itself with sufficient flexibility that it 'points' in the direction of the actin filament then the above values are correct; however, if the movement produced

is determined by its orientation with the substrate then the true value is determined by averaging a cosine term through 180 °, which means that our values of both force and displacement will be underestimates by a factor of $\pi/2$ compared to those measured in muscle-fibre experiments[11]. The finding that the myosin work-stroke size is close to the F-actin monomer repeat is intriguing. Similar findings have been made for kinesin, for which the work stroke seems to be close to the tubulin monomer repeat[12]. □

Received 2 June; accepted 14 September 1995.

1. Ishijima, A. et al. *Biochem. biophys. Res. Commun.* **199,** 1057–1063 (1994).
2. Finer, J. T., Simmons, R. M. & Spudich, J. A. *Nature* **368,** 113–118 (1994).
3. Miyata, H. et al. *J. Biochem., Tokyo* **115,** 644–647 (1994).
4. Ashkin, A., Dziedzic, J. M., Bjorkholm, J. E. & Chu, S. *Optics Lett.* **11,** 288–290 (1986).
5. Molloy, J. E. et al. *Biophys. J.* **68,** 298s–305s (1995).
6. Rayment, I. et al. *Science* **261,** 50–58 (1993).
7. Harada, Y., Noguchi, A., Kishino, A. & Yanagida, T. *Nature* **326,** 805–808 (1987).
8. Toyoshima, Y. Y. et al. *Nature* **328,** 536–539 (1987).
9. Kishino, A. & Yanagida, T. *Nature* **334,** 74–76 (1988).
10. Einstein, A. *Theory of the Brownian Movement* (ed. Furth, R.) (Dover, 1956).
11. Huxley, A. F. & Simmons, R. M. *Nature* **233,** 533–538 (1971).
12. Svoboda, J. K., Schmidt, C. F., Schnapp, B. J. & Block, S. M. *Nature* **365,** 721–727 (1993).
13. Kron, S. J., Toyoshima, Y. Y., Uyeda, T. Q. P. & Spudich, J. A. *Meth. Enzym.* **196,** 399–416 (1991).
14. Kojima, H., Ishijima, A. & Yanagida, T. *Proc. natn. Acad. Sci. U.S.A.* **91,** 12962–12966 (1994).

ACKNOWLEDGEMENTS. We thank J. C. Sparrow, J. D. Currey, S. M. Block and J. Howard for discussions. This work was supported by the Royal Society and BBSRC.

Ionic effects on the elasticity of single DNA molecules

CHRISTOPH G. BAUMANN*, STEVEN B. SMITH†, VICTOR A. BLOOMFIELD*, AND CARLOS BUSTAMANTE†‡§

*Department of Biochemistry, University of Minnesota, St. Paul, MN 55108; and †Institute of Molecular Biology and Department of Chemistry and ‡Howard Hughes Medical Institute, University of Oregon, Eugene, OR 97403

Communicated by Brian W. Matthews, University of Oregon, Eugene, OR, April 7, 1997 (received for review February 4, 1997)

ABSTRACT We used a force-measuring laser tweezers apparatus to determine the elastic properties of λ-bacteriophage DNA as a function of ionic strength and in the presence of multivalent cations. The electrostatic contribution to the persistence length P varied as the inverse of the ionic strength in monovalent salt, as predicted by the standard worm-like polyelectrolyte model. However, ionic strength is not always the dominant variable in determining the elastic properties of DNA. Monovalent and multivalent ions have quite different effects even when present at the same ionic strength. Multivalent ions lead to P values as low as 250–300 Å, well below the high-salt "fully neutralized" value of 450–500 Å characteristic of DNA in monovalent salt. The ions Mg^{2+} and $Co(NH_3)_6^{3+}$, in which the charge is centrally concentrated, yield lower P values than the polyamines putrescine^{2+} and spermidine^{3+}, in which the charge is linearly distributed. The elastic stretch modulus, S, and P display opposite trends with ionic strength, in contradiction to predictions of macroscopic elasticity theory. DNA is well described as a worm-like chain at concentrations of trivalent cations capable of inducing condensation, if condensation is prevented by keeping the molecule stretched. A retractile force appears in the presence of multivalent cations at molecular extensions that allow intramolecular contacts, suggesting condensation in stretched DNA occurs by a "thermal ratchet" mechanism.

Ions strongly affect such biologically significant behavior of DNA as wrapping around nucleosomes, packaging inside bacteriophage capsids, and binding to proteins involved in transcriptional initiation and elongation. While such influence is often explained by relatively simple ion exchange equilibria (1, 2), in many cases ions appear to exert their effect by modifying the structure and mechanical properties of DNA—its bending and torsional rigidity. The study of the ionic dependence of the elastic properties of DNA is therefore essential to understand the energetics of these key biological processes.

Recent technical advances in nanomanipulation have allowed the mechanical behavior of single DNA molecules to be studied. Magnetic beads (3, 4), micro fibers (5), optical traps (6), and hydrodynamic flow (7) have been used to apply a wide range of forces to individual bacteriophage DNA molecules. Smith *et al.* (6) have recently described three regimes in the elastic response of λ-bacteriophage DNA (λ DNA) molecules. Between 0.01 and 10 pN, the molecule behaves as an entropic spring and is well described by the worm-like chain (WLC) model (8–10), from which a persistence length, P, and a contour length, L_o, can be obtained. Between 10 and 65 pN, the molecule deviates from the predictions of the *inextensible* WLC as it extends beyond its B-form contour length. From this "enthalpic elasticity" regime an elastic stretch modulus, S, can be obtained. Finally, at about 65 pN, the molecule suddenly

yields in a highly cooperative fashion and overstretches ≈1.7 times. A more recent study using an optical trapping interoferometer (11) has probed the entropic and enthalpic elasticity of single plasmid-length DNA molecules.

In this paper we used a force-measuring optical trap instrument (6) to determine the elastic properties of λ DNA as a function of ionic strength and in the presence of multivalent cations. Our results show the following effects, some of them unexpected, of ion concentration, valence, and structure on DNA elasticity: (*i*) The electrostatic contribution to P varies as the inverse of the ionic strength, I, in monovalent salt, as predicted by the standard uniformly charged cylinder model (12, 13). However, ionic strength is not always the dominant variable in determining the elastic properties of DNA; monovalent and multivalent ions have quite different effects even when present at the same ionic strength. Multivalent ions lead to persistence lengths as low as 250–300 Å, well below the high-salt "fully neutralized" value of 450–500 Å characteristic of DNA in monovalent salt. (*ii*) Ions in which the charge is centrally concentrated [Mg^{2+}, $Co(NH_3)_6^{3+}$] lead to lower P values than those in which the same charge is linearly distributed (putrescine^{2+}, spermidine^{3+}). (*iii*) The elastic stretch modulus and persistence length display opposite trends with ionic strength—a direct contradiction of macroscopic elasticity theory. (*iv*) DNA is still well described by the WLC model at concentrations of trivalent cations capable of inducing DNA collapse or condensation, if collapse is prevented by keeping the molecule stretched. (*v*) No evidence of the buckling transition that has been postulated to underlie condensation (14) is observed. Instead, a retractile force appears in the presence of multivalent cations at molecular extensions that allow intramolecular contacts, suggesting the existence of a "thermal ratchet" condensation mechanism in stretched DNA.

WLC Elastic Behavior

The WLC model describes the behavior of a DNA molecule as intermediate between a rigid rod and a flexible coil, accounting for both local stiffness and long-range flexibility (15). The flexibility of the chain is described by the persistence length P, the distance over which two segments of the chain remain directionally correlated. An interpolation formula that describes the extension x of a WLC with contour length L_o in response to a stretching force F is (8, 9)

$$\frac{FP}{kT} = \frac{1}{4}\left(1 - \frac{x}{L_o}\right)^{-2} - \frac{1}{4} + \frac{x}{L_o}, \qquad [1]$$

where k = the Boltzmann constant and T = absolute temperature. This equation describes the entropic elasticity of the WLC, arising from the reduced entropy of the stretched chain ($x \leq 16$ μm for λ DNA), and assumes that the DNA is *inextensible*.

Abbreviations: BB, background buffer; WLC, worm-like chain.
§To whom reprint requests should be addressed at: Institute of Molecular Biology, University of Oregon, Eugene, OR 97403-1229. e-mail: carlos@alice.uoregon.edu.

Near full extension, x approaches L_o as $F^{-1/2}$ (8, 9, 16). In this regime DNA can also be enthalpically stretched beyond the contour length defined by B-DNA geometry (6). An equation that describes this *extensible* WLC regime ($14 \le x \le 17 \ \mu m$ for λ DNA) is (10)

$$\frac{x}{L_o} = 1 - \frac{1}{2}\left(\frac{kT}{FP}\right)^{1/2} + \frac{F}{S},$$ [2]

where S is the elastic stretch modulus. The first two terms on the right hand side of Eq. **2** describe the extension of the molecule as $x \to L_o$ and correspond to the strong-stretching limit of Eq. **1**. The third term describes the extension of the molecule beyond its canonical B-DNA length; i.e., it is the enthalpic component of the elasticity. The range of applicability for Eq. **1** may be extended by including this enthalpic component (11); however, nearly identical values of P, L_o, and S are obtained.

At extensions beyond these two elastic regimes, DNA undergoes a reversible overstretch transition to a form $\approx 70\%$ longer than canonical B-DNA (5, 6).

Behavior and Analysis of Force–Extension Curves

λ DNA molecules were tethered between two streptavidin-coated latex beads (diameter = 3.54 μm). One bead was held by a micropipette while the other was optically trapped by force-measuring laser tweezers (Fig. 1). The extension of the DNA molecule was determined from the distance between beads. The force acting on the molecule was inferred from the displacement of the laser beams on position-sensitive photodetectors, and it was calibrated against the viscous drag on a bead by using Stokes law (ref. 6; C.B., D. Keller, and S.B.S., unpublished work).

The 5′-overhangs of λ DNA (methylated *c*1857*ind* 1 *Sam* 7; New England Biolabs) were biotinylated with the Klenow fragment of DNA polymerase using biotin-11-dCTP (Sigma), dATP, dGTP, and dUTP as described previously (6). Single-strand nicks were repaired with DNA ligase. After biotinylation and nick ligation, DNA stocks were stored in an EDTA-containing buffer. Monovalent salt solutions were prepared by diluting 100 mM cacodylate pH 7 buffer stocks (86.2 mM sodium cacodylate, 13.8 mM cacodylic acid) supplemented with either 100 or 500 mM NaCl (total Na$^+$ concentration ca.

186 and 586 mM, respectively). MgCl$_2$·2H$_2$O (Baker), putrescine·2HCl (Sigma), spermidine·3HCl (Sigma), and hexaammine cobalt(III) trichloride (Eastman), hereafter referred to as Co(NH$_3$)$_6^{3+}$, were utilized without further purification and prepared as 0.1 M stocks in deionized water ($\rho \ge$ 12 MΩ·cm). Final concentrations of the di- and trivalent cations were prepared in background buffer (BB): 1 mM NaCl/1 mM cacodylate, pH 7 (total Na$^+$ concentration $\approx$1.86 mM). Complete buffer exchange between experiments was ensured by monitoring the conductivity of the fluid chamber eluant. Experimental buffers containing Mg^{2+} were purged from the chamber by extensive washing with an EDTA-containing buffer.

Dependence of *F–x* Curves on Monovalent Salt Concentration. Force vs. extension curves for single λ DNA molecules at NaCl concentrations of 1, 50, and 500 mM are plotted in Fig. 2*A*. The entropic and enthalpic stretching regimes are enlarged in Fig. 3*A*, distinctly showing that P decreases and S increases as the salt concentration is raised. A lower value of P causes the force curve to rise more abruptly at low extensions, since the more flexible chain exists in a less extended conformation than the stiffer chain with higher P. Increases in S are readily apparent as increased slopes of the *F–x* curves at extensions approaching the contour length. As observed previously (6), decreasing monovalent ionic strength lowers the force at which λ DNA cooperatively overstretches to a conformation $\approx$1.7 times longer than B-DNA. Individual molecules often display hysteresis (Fig. 2*A*, arrows) upon relaxation of this form.

The persistence lengths, contour lengths, and elastic stretch moduli were extracted by fitting the entropic and enthalpic stretching data with the appropriate WLC models: *inextensible* WLC (Eq. **1**), strong-stretching limit (Eq. **2** without F/S term), and *extensible* WLC (complete Eq. **2**). P and S values are shown in Table 1. L_o was 16.5 ± 0.2 μm, and it showed no trend with

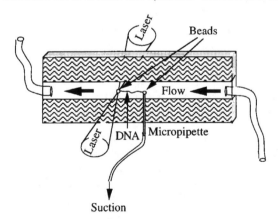

FIG. 1. Schematic diagram showing the geometry of the experiment. Two counter propagating laser beams "trap" a bead in the middle of a specially designed fluid chamber, far (>100 μm) from the chamber walls. The other bead is held by a micropipette introduced inside the chamber. The micropipette is moved to increase or decrease the tension in the DNA molecule that bridges the two beads. The environment around the molecule can be altered simply by exchanging the solution inside the chamber.

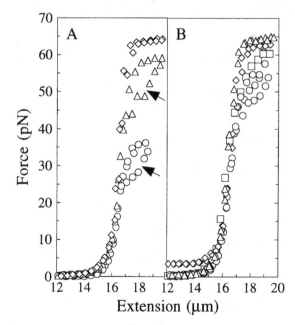

FIG. 2. Ionic effects on the elastic response of single λ DNA molecules. A portion of the overstretch transition is shown for all molecules. (*A*) Stretching λ DNA in 1 ($\bigcirc$), 50 ($\triangle$), and 500 ($\diamondsuit$) mM NaCl. Individual molecules often display melting hysteresis (arrows) upon relaxation of the overstretched form. (*B*) Stretching λ DNA in the presence of di- and trivalent cations with BB: 100 μM MgCl$_2$ ($\square$), 100 μM putrescine^{2+} ($\bigcirc$), 100 μM spermidine^{3+} ($\triangle$), or 25 μM Co(NH$_3$)$_6^{3+}$ ($\diamondsuit$). To compare behavior in BB alone refer to the *F–x* curve for 1 mM NaCl.

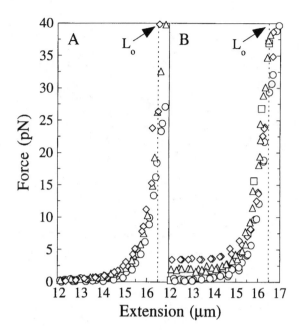

FIG. 3. Ionic effects on the entropic and enthalpic elasticity of single λ DNA molecules. The B-form contour length (L_o) is 16.5 μm (broken line). (*A*) Stretching λ DNA in 1 (○), 50 (△), and 500 (◇) mM NaCl. (*B*) Stretching λ DNA in the presence of di- and trivalent cations with BB: 100 μM $MgCl_2$ (□), 100 μM putrescine^{2+} (○), 100 μM spermidine^{3+} (△), or 25 μM $Co(NH_3)_6^{3+}$ (◇). To compare behavior in BB alone refer to the *F–x* curve for 1 mM NaCl. The nonzero force (1–5 pN) observed at $x < 14$ μm with spermidine^{3+} and $Co(NH_3)_6^{3+}$ was reproducible and is thought to represent intramolecular condensation.

ionic strength. This corresponds to 3.40 ± 0.04 Å/base pair (bp), within the range determined from crystallography for B-DNA (3.32 ± 0.19 Å) (18, 19). At the strong-stretching limit where *P* dominates the elasticity ($x \approx 14$–16 μm), λ DNA follows the WLC behavior predicted by Eq. **2** without the enthalpic stretching term *F/S*. A plot of data for 500 mM NaCl and the predicted WLC behavior (solid line) are shown in Fig. 4*A*.

Dependence of *F–x* Curves on Multivalent Cations. The elastic response of λ DNA molecules in BB with Mg^{2+}, putrescine, spermidine, and $Co(NH_3)_6^{3+}$ is shown in Fig. 2*B*. *P*, L_o, and *S* were derived from the entropic and enthalpic elastic behavior of λ DNA (Fig. 3*B*) as explained above. *P* and *S* are given in Table 2. L_o values did not differ significantly from those observed in monovalent salt, showing that multivalent cations do not provoke observable secondary structure tran-

Table 1. Effect of monovalent ionic strength on the persistence length and elastic modulus

Ionic strength, mM	Inextensible WLC *P*, Å	Strong stretching limit *P*, Å	Extensible WLC	
			P, Å	*S*, pN
1.86	963 ± 48	862 ± 49	949 ± 59	649 ± 82
3.72	704 ± 36	668 ± 32	757 ± 25	745 ± 100
5.58	605 ± 34	629 ± 20	767 ± 54	476 ± 142
7.44	479 ± 30	494 ± 83	622 ± 37	686 ± 65
9.30	621 ± 83	656 ± 97	652 ± 27	452 ± 35
18.6	523 ± 54	521 ± 56	529 ± 95	532 ± 67
93.0	438 ± 14	521 ± 31	511 ± 18	1,006 ± 2
186	561 ± 31	541 ± 33	525 ± 124	1,401 ± 313
586	451 ± 21	472 ± 21	559 ± 32	1,435 ± 160

Results are reported as the mean ± standard error, $n \geq 3$.

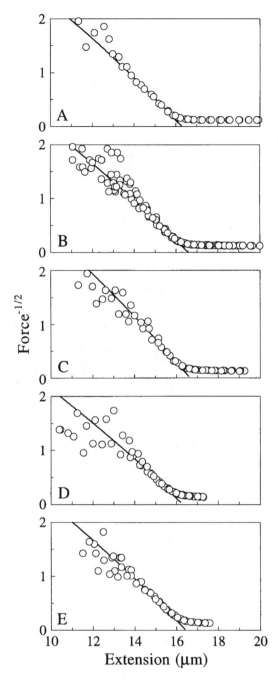

FIG. 4. Ionic effects on the high-force entropic elastic response of single λ DNA molecules. These molecules display strong-stretching WLC behavior for $x \approx 14$–16 μm as denoted by the linear relation between force$^{-1/2}$ and *x*. A linear fit of these data yields the contour (L_o) and persistence (*P*) lengths: (*A*) 500 mM NaCl, $L_o = 16.5$ μm, and $P = 446$ Å; (*B*) 100 μM $MgCl_2$ + BB, $L_o = 16.6$ μm, and $P = 419$ Å; (*C*) 100 μM putrescine^{2+} + BB, $L_o = 16.6$ μm, and $P = 560$ Å; (*D*) 5 μM $Co(NH_3)_6^{3+}$ + BB, $L_o = 16.3$ μm, and $P = 379$ Å; and (*E*) 25 μM spermidine^{3+} + BB, $L_o = 16.5$ μm, and $P = 443$ Å. The data for $x < 13$ μm were not included in these fits.

sitions in λ DNA under the conditions employed here. The di- and trivalent cations strongly reduce *P* and increase *S* of λ DNA. As shown in Fig. 4 *B–E*, strong-stretching WLC behavior is still followed (solid lines) despite the large reductions in *P*.

Table 2. Effect of multivalent cations on the persistence length and elastic modulus

Solution condition	Inextensible WLC P, Å	Strong-stretching limit P, Å	Extensible WLC	
			P, Å	S, pN
BB	963 ± 48	862 ± 49	949 ± 59	649 ± 82
100 μM Mg^{2+} + BB	409 ± 37	455 ± 74	508 ± 103	957 ± 203
100 μM Put^{2+} + BB	581 ± 48	560 ± 19	684 ± 31	945 ± 68
200 μM Put^{2+} + BB	329 ± 20	341 ± 6	309 ± 31	1,094 ± 360
25 μM Spd^{3+} + BB	320 ± 18	368 ± 75	419 ± 47	1,006 ± 193
100 μM Spd^{3+} + BB	440 ± 13	375 ± 44	463 ± 65	1,175 ± 141
2 μM Co(NH$_3$)$_6^{3+}$ + BB	407 ± 48	476 ± 126	422 ± 1	696 ± 183
5 μM Co(NH$_3$)$_6^{3+}$ + BB	243 ± 18	382 ± 101	352 ± 32	1,144 ± 134
25 μM Co(NH$_3$)$_6^{3+}$ + BB	299 ± 21	177 ± 30	265 ± 46	1,010 ± 229

Results are reported as the mean ± standard error, $n \geq 3$. Put, putrescine; Spd, spermidine.

Multivalent cations also increase the force at which the molecule overstretches, yielding values characteristic of high monovalent salt (Fig. 2B).

The nonzero force (1–5 pN) observed with 25 μM Co(NH$_3$)$_6^{3+}$ and 100 μM spermidine at $x < 14$ μm (Fig. 3B) is likely to be due to side-by-side association of the DNA which, if unconstrained, would lead to intramolecular collapse. This phenomenon was frequently observed with concentrations of spermidine and Co(NH$_3$)$_6^{3+}$ sufficient to condense DNA (20, 21); it was reversible, and could be avoided by keeping the molecule stretched. A nonzero force baseline was added to Eqs. 1 and 2 when fitting these F–x curves. The advantages of single-molecule manipulations are obvious here, as we are able to determine the elastic properties of DNA at concentrations of spermidine and Co(NH$_3$)$_6^{3+}$ that would cause condensation in bulk.

Persistence Length

Monovalent Salt Dependence of P. P values tabulated in Table 1 as a function of monovalent ionic strength are plotted in Fig. 5. The three WLC fits (data points in Fig. 5) yield very similar values of P (normally within 10%). Models which incorporate chain extensibility yield essentially the same P as those that do not, probably because the entropic and enthalpic components dominate the elastic behavior of the molecule in different parts of the F–x curve. The nonelectrostatic contribution dominates P at $I > 20$ mM, in accord with previous experimental determinations using light scattering (22) and flow birefringence (23) with phage DNA, and electro-optical measurements (24) with short (41–256 bp) DNA fragments.

Throughout the range of I investigated here, the data are well fit with a nonlinear Poisson–Boltzmann theory for uniformly charged cylinders (12, 13)

$$P = P_o + P_{el} = P_o + \frac{1}{4\kappa^2 l_B} = P_o + 0.324 I^{-1} \text{ Å}, \quad [3]$$

where P_o and P_{el} are the nonelectrostatic and electrostatic contributions to P, $1/\kappa$ is the Debye–Hückel screening length, and l_B is the Bjerrum length (7.14 Å in water at 25°C for double-stranded DNA). The last equality gives P in Å with I in molar units. The curve in Fig. 5 is drawn with $P_o = 500$ Å; values between 450 and 500 Å fit equally well, in agreement with most previous determinations (25).

Values of P with Multivalent Cations. The values of P with Mg^{2+}, putrescine^{2+}, spermidine^{3+}, or Co(NH$_3$)$_6^{3+}$ are strikingly lower than those at essentially the same ionic strength but with Na$^+$ as the only cation. Twenty-five micromolar spermidine and 5 μM Co(NH$_3$)$_6^{3+}$ are the critical concentrations required to condense DNA under these buffer conditions (20, 21). At these concentrations, P was reduced 2- to 4-fold relative to low monovalent salt buffer. Higher concentrations did not produce significant further reductions. The effects of Co(NH$_3$)$_6^{3+}$ on the entropic elasticity are dramatic (Fig. 6): as little as 2 μM Co(NH$_3$)$_6^{3+}$ reduced P to roughly the same extent as 100 μM spermidine^{3+}, and the plateau value of P at higher Co(NH$_3$)$_6^{3+}$ concentrations is near 250 Å, half the limiting value for monovalent salt. This fact may have some bearing on the efficiency with which Co(NH$_3$)$_6^{3+}$ condenses DNA (21) (see below). Mg^{2+} and putrescine^{2+} cannot condense DNA in aqueous solution, yet also reduce P by 2- to 3-fold at low ionic strength. We note that Co(NH$_3$)$_6^{3+}$ reduces P more strongly

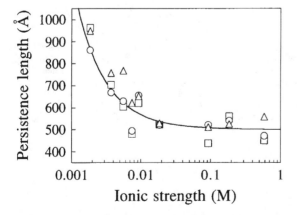

FIG. 5. Dependence of persistence length (P) on monovalent (Na$^+$) ionic strength. Points are from Table 1: □, *inextensible* WLC; ○, strong-stretching limit; △, *extensible* WLC. Line calculated from Eq. 3 with $P_o = 500$ Å.

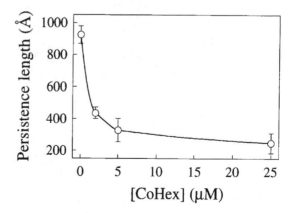

FIG. 6. Dependence of persistence length (P) on Co(NH$_3$)$_6^{3+}$ (CoHex) concentration for λ DNA in BB. Points and error bars are mean and standard deviations of the P values in Table 2 obtained from fits to the three different WLC models.

than isovalent spermidine, and Mg^{2+} more than putrescine^{2+}. This is probably due to the central concentration of charge in $Co(NH_3)_6^{3+}$ and Mg^{2+} compared with the linear distribution of charges in the polyamines.

These effects of Mg^{2+}, spermidine^{3+}, and $Co(NH_3)_6^{3+}$ on the entropic elasticity of DNA agree well with those found previously. Electro-optical studies of short DNA fragments showed that 100 μM Mg^{2+} and 10 μM $Co(NH_3)_6^{3+}$ reduced P to $\approx$350 Å and 200 Å, respectively, in low monovalent salt (26). Linear dichroism and viscometry studies of T7 DNA showed that a precondensation spermidine concentration of 20 μM decreased P to $\approx$445 Å (27). Our result for 100 μM spermidine, $\approx$430 Å, shows this to be a limiting value, very near the accepted value of P_o.

Despite the large alteration of the entropic elasticity by multivalent cations, λ DNA still follows the linear behavior predicted by the strong-stretching limit of the WLC model (Fig. 4 *B–E*). Changes in the persistence length of a WLC could be due to changes in permanent static bending, thermally induced dynamic bending, or both (28). Although, strictly speaking, our data cannot differentiate between ion-induced changes in static vs. dynamic bending, large permanent bends would probably cause deviations from the behavior predicted by the strong-stretching limit (Eq. 2), and this is not observed (C.B., D. Keller, and S.B.S., unpublished work). Furthermore, site-binding is usually invoked to explain the formation of static bends, yet NMR studies of di- and multivalent cation mobility show most ions are loosely associated with random DNA sequences rather than tightly bound (29–31). We suggest that decreases in P arise mainly from enhanced bending fluctuations skewed toward the concentrated positive charge at sites of transiently associated di- or multivalent counterions. Theory (I. Rouzina and V.A.B., unpublished work) predicts a few transiently associated cations per persistence length, each inducing a bend of $\approx$20° distributed over 6 bp, would be sufficient to cause the observed decrease in P. Thermal fluctuations result in an $\approx$4° bend per bp, thus the postulated transient bending is similar in magnitude to thermal bending and should not perturb the B-form helix.

The reduced P caused by multivalent cations will decrease the volume occupied by the DNA coil, enhancing the close packing of DNA and increasing the probability of intrachain contacts through a reduction in search space. However, an increase in DNA flexibility alone will not drive the collapse of DNA, since divalent cations do not produce this effect. Rather, the driving force for DNA collapse must also reside in the ability of a cation to reduce most, but not all, of the DNA phosphate charge while influencing helix hydration (32), helix secondary structure (33), and/or electrostatic attractions through ion correlation (34, 35).

Stretch Elasticity

Dependence of S on Monovalent Salt. The elastic stretch modulus of single λ DNA molecules in monovalent salt buffer was determined by fitting F–x data to the *extensible* WLC model, Eq. 2, yielding the S values shown in Table 1. The scatter for individual S values is high (mean SD $\approx$ 28%), but S clearly decreases as the ionic strength is lowered, a trend opposite to that of P. This presents a surprising contradiction to the predictions of macroscopic elasticity theory for a homogeneous elastic rod. According to that theory (36), if a uniform rod of cross-sectional area A has a stretch modulus of S, then the Young's modulus E of the material constituting that rod is given by

$$E = S/A. \qquad [4]$$

The bending rigidity of the rod, B, varies with the Young's modulus and the cross-sectional area moment of inertia M as

$$B = EM. \qquad [5]$$

The persistence length of a WLC is given by $P = B/kT$ (37) so that

$$P = EM/kT. \qquad [6]$$

For a cylindrical molecule, $A = \pi r^2$ and $M = \pi r^4/4$ and therefore

$$P = Sr^2/4kT. \qquad [7]$$

Eq. 7 indicates that P and S should vary in the same way with I if r remains approximately constant, while Table 1 shows that they change in opposite directions. Taking a high-salt value for S of 1400 pN from Table 1 and using $r = 10$ Å gives $E = 4.5 \times 10^8$ Pa from Eq. 4 and $P = 845$ Å from Eq. 6, clearly too large for the high-salt case. A typical low-salt value for S of 500 pN would give $P = 400$ Å by Eq. 7, whereas the geometric persistence length in low salt has increased by electrostatic effects to beyond 800 Å.

Relaxing the assumption that the molecular radius remains constant cannot account for these variations. To make Eq. 7 consistent with the measured S and P values from Table 1 requires that the radius increase from 8 Å at high salt to 16 Å at low salt, a prediction totally at variance with experiment. Fluorescence polarization anisotropy measurements of intercalated ethidium showed that the radius of 48-bp DNA fragments increased by only 0.3 Å between 100 and 20 mM Na^+ (38).

The decrease in S with decreasing I implies that, as the ionic strength is lowered, DNA becomes more susceptible to enthalpic elongation. This effect is more complex than a simple increase in electrostatic repulsion among the DNA phosphates. A charged Hookean spring would adopt a longer equilibrium contour length if the ionic strength were reduced, and its spring constant (S value) would also increase. A possible explanation for the decrease of S with I is that at low I there is localized melting in A+T-rich regions. Increases in I would reduce both local melting and intramolecular electrostatic repulsion, thus simultaneously increasing S and decreasing P. Thus it appears that ionic strength effects on S and P reflect different molecular mechanisms, and therefore are not governed by macroscopic elasticity theory.

This is not the first indication that DNA does not behave like a classical macroscopic cylinder. Other evidence comes from a comparison of the bending rigidity and the constant of torsional rigidity C (39). These are related as

$$B/C = 1 + \sigma, \qquad [8]$$

where σ is the Poisson ratio—i.e., the negative of the ratio of radial compression to axial elongation (36). For an incompressible rod, $\sigma = \frac{1}{2}$, and thermodynamic stability requires $-1 < \sigma < \frac{1}{2}$. Values of $\sigma < 0$ correspond to a thickening of the rod as it is stretched, a behavior unknown in homogeneous, isotropic elastic material. From Eq. 5, $B = 2 \times 10^{-19}$ erg·cm at 25°C if $P = 500$ Å. Values of C vary considerably from one investigation to the next, but are generally in the range 2×10^{-19} to 3.4×10^{-19} erg·cm (for reviews see refs. 25 and 40). These values give σ from 0 to -0.4.

Of course, there is no reason to assume that double-stranded DNA, with its external phosphates, internal stacked base pairs, and major and minor grooves, should behave like an isotropic, homogeneous, cylindrical rod. Indeed, its volume and mass are not strictly fixed, since the presence of a spine of hydration may depend on mechanical stress, temperature, or type of salt. But the simple rod assumption is commonly made in modeling DNA behavior, and the results we discuss here are a dramatic illustration of the limitations of this point of view.

S with Multivalent Cations. Fits to the *extensible* WLC model show that di- and trivalent cations stabilize the DNA duplex against enthalpic elongation (Table 2). Condensing concentrations of spermidine^{3+} (≥ 25 μM) and Co(NH$_3$)$_6^{3+}$ (≥ 5 μM) stabilize λ DNA against elongation as if it were in 50 mM NaCl. Mg^{2+} and putrescine^{2+} also increase S to similar levels. These observations parallel results with di- and trivalent cations in DNA melting temperature experiments (41–44), and they support localized melting as a possible explanation for decreases in S as discussed above. Mg^{2+} and the naturally occurring polyamines may act *in vivo* to oppose forces which destabilize or melt duplex DNA during supercoiling, replication, and transcription.

Concluding Remarks

Single-molecule methods offer several advantages over bulk studies: (*i*) The three elastic regimes can be investigated separately, making it possible to distinguish their possibly different molecular origins. (*ii*) Molecules of different lengths can be studied in a variety of solution conditions at molecular extensions where excluded volume effects, which often complicate the interpretation of bulk experiments (ref. 22 and references therein), may be neglected. (*iii*) DNA is condensed into compact particles by multivalent cations. In single-molecule experiments the extension of the molecule can be controlled so as to prevent its condensation, making it possible to separate the effects of condensation from changes in the elasticity of DNA induced by these cations. (*iv*) Manipulations of DNA at molecular extensions that preclude intramolecular contacts can be used, as reported here, to test the competing hypotheses that cation-induced condensation is due to attractive intersegmental free energy arising from electrostatics, hydration, and helix perturbation (17), or from abrupt buckling transitions (14). (*v*) Condensation of single molecules probably involves loop formation to yield a stable toroidal nucleus. Since this mechanism requires slack in the molecule, it can be prevented by applying external tension. The retractile force seen in Fig. 3B may have been generated when thermal motion created temporary slack, thus enabling loops to form and side-by-side association to occur. Such association acted like a pawl in a ratchet to take up any slack and prevent the reversal of thermal motion. Since the permanent tension generated in the molecule could not have occurred without thermal motion, the constrained condensation behaves as a "thermal ratchet."

We thank Dr. Steven Block for many helpful discussions. This work was supported in part by National Science Foundation Grant MBC 9118482 and National Institutes of Health Grant GM 32543 to C.B., National Institutes of Health Grant GM 28093 to V.A.B., and a National Institutes of Health Traineeship (GM 08277) to C.G.B.

1. Manning, G. S. (1978) *Q. Rev. Biophys.* **11**, 179–246.
2. Record, M. T., Jr., Anderson, C. F. & Lohman, T. M. (1978) *Q. Rev. Biophys.* **11**, 103–178.
3. Smith, S. B., Finzi, L. & Bustamante, C. (1992) *Science* **258**, 1122–1126.
4. Strick, T. R., Allemand, J.-F., Bensimon, D., Bensimon, A. & Croquette, V. (1996) *Science* **271**, 1835–1837.
5. Cluzel, P., Lebrun, A., Heller, C., Lavery, R., Viovy, J.-L., Chatenay, D. & Caron, F. (1996) *Science* **271**, 792–794.
6. Smith, S. B., Cui, Y. & Bustamante, C. (1996) *Science* **271**, 795–799.
7. Perkins, T. T., Smith, D. E., Larson, R. G. & Chu, S. (1995) *Science* **268**, 83–87.
8. Bustamante, C., Marko, J. F., Siggia, E. D. & Smith, S. (1994) *Science* **265**, 1599–1600.
9. Marko, J. F. & Siggia, E. D. (1995) *Macromolecules* **28**, 8759–8770.
10. Odijk, T. (1995) *Macromolecules* **28**, 7016–7018.
11. Wang, M. D., Yin, H., Landick, R., Gelles, J. & Block, S. M. (1997) *Biophys. J.* **72**, 1335–1346.
12. Odijk, T. (1977) *J. Polym. Sci. Polym. Phys. Ed.* **15**, 477–483.
13. Skolnick, J. & Fixman, M. (1977) *Macromolecules* **10**, 944–948.
14. Manning, G. S. (1980) *Biopolymers* **19**, 37–59.
15. Grosberg, A. Y. & Khokhlov, A. R. (1994) *Statistical Physics of Macromolecules* (Am. Inst. Phys. Press, New York), pp. 3, 5, 217–220.
16. Kovac, J. & Crabb, C. C. (1982) *Macromolecules* **15**, 537–541.
17. Bloomfield, V. A. (1996) *Curr. Opin. Struct. Biol.* **6**, 334–341.
18. Dickerson, R. E., Drew, H. R., Conner, B. N., Wing, R. M., Fratini, A. V. & Kopka, M. L. (1982) *Science* **216**, 475–485.
19. Saenger, W. (1984) *Principles of Nucleic Acid Structure* (Springer, New York).
20. Wilson, R. W. & Bloomfield, V. A. (1979) *Biochemistry* **18**, 2192–2196.
21. Widom, J. & Baldwin, R. L. (1980) *J. Mol. Biol.* **144**, 431–453.
22. Sobel, E. S. & Harpst, J. A. (1991) *Biopolymers* **31**, 1559–1564.
23. Cairney, K. L. & Harrington, R. E. (1982) *Biopolymers* **21**, 923–934.
24. Porschke, D. (1991) *Biophys. Chem.* **40**, 169–179.
25. Hagerman, P. J. (1988) *Annu. Rev. Biophys. Biophys. Chem.* **17**, 265–286.
26. Porschke, D. (1986) *J. Biomolec. Struct. Dyn.* **4**, 373–389.
27. Baase, W. A., Staskus, P. W. & Allison, S. A. (1984) *Biopolymers* **23**, 2835–2851.
28. Schellman, J. A. & Harvey, S. C. (1995) *Biophys. Chem.* **55**, 95–114.
29. Rose, D. M., Polnaszek, C. F. & Bryant, R. G. (1982) *Biopolymers* **21**, 653–664.
30. Berggren, E., Nordenskiöld, L. & Braunlin, W. H. (1992) *Biopolymers* **32**, 1339–1350.
31. Wemmer, D. E., Srivenugopal, K. S., Reid, B. R. & Morris, D. R. (1985) *J. Mol. Biol.* **185**, 457–459.
32. Rau, D. C. & Parsegian, V. A. (1992) *Biophys. J.* **61**, 246–259.
33. Reich, Z., Ghirlando, R. & Minsky, A. (1991) *Biochemistry* **30**, 7828–7836.
34. Oosawa, F. (1971) *Polyelectrolytes* (Dekker, New York).
35. Rouzina, I. & Bloomfield, V. A. (1996) *J. Phys. Chem.* **100**, 9977–9989.
36. Landau, L. D. & Lifshitz, E. M. (1986) *Theory of Elasticity* (Pergamon, Oxford).
37. Landau, L. D. & Lifshitz, E. M. (1970) *Statistical Physics* (Pergamon, Oxford).
38. Fujimoto, B. S., Miller, J. M., Ribeiro, N. S. & Schurr, J. M. (1994) *Biophys. J.* **67**, 304–308.
39. Manning, G. S. (1985) *Cell. Biophys.* **7**, 57–89.
40. Crothers, D. M., Drak, J., Kahn, J. D. & Levene, S. D. (1992) *Methods Enzymol.* **212**, 3–29.
41. Dove, W. F. & Davidson, N. (1962) *J. Mol. Biol.* **5**, 467–478.
42. Eichhorn, G. L. & Shin, Y. A. (1968) *J. Am. Chem. Soc.* **90**, 7323–7328.
43. Thomas, T. J. & Bloomfield, V. A. (1984) *Biopolymers* **23**, 1295–1306.
44. Thomas, T. J., Kulkarni, G. D., Greenfield, N. J., Shirahata, A. & Thomas, T. (1996) *Biochem. J.* **319**, 591–599.

3

Measuring Forces and Motion

If one wishes to build a physical structure or a chemical process, then an understanding of the mechanical properties of the materials is essential. Biologists and clinicians want to understand how natural materials are made, how they work, and why they sometimes fail. While an engineer might use a macroscopic tool to study hard building materials, biologists require a microscopic device to study soft biological cells and molecules. The reason that optical tweezers have found so many applications in biology is that they are just right for grabbing and manipulating soft biological materials [1,2].

At moderate laser powers (~ 100 mW), optical tweezers have a stiffness that is similar to biological molecules like proteins, DNA, and lipid membranes. We know that some biological materials are hard—like bones, mollusk shells, insect cuticles, and woody plants—but the living material of all organisms is mechanically "soft." Biology has evolved to work best at the ambient temperature on Earth and we find that thermal energy ($k_b T$; Boltzmann's constant $\cdot$ 300 K = 4×10^{-21} J or 4 pN.nm) does sufficient mechanical work (force · distance) to bend or distort biological molecules and make them diffuse within aqueous and lipid media. The molecules and chemical reactions of life exploit weak forces and small energy differences, and in order to investigate them we need a delicate tool and a sensitive measuring device.

This chapter describes the progress made in adapting optical tweezers as an engineering tool to measure minute forces (piconewtons), distances (nanometers), and energy differences that are close to the thermal limit [3] (see **PAPER 3.1**); (see also Refs. [4,5] and open-access Web resource [6]). The mechanical properties that one might wish to measure using optical tweezers are the steady-state, force-extension relationship (elastic modulus) and the time-dependent viscous properties that are a function of the rate of change in length or force (viscous modulus). The general approach to studying hard and soft biological materials has firm roots in macroscopic mechanical studies [7]. We find that, for small amplitude perturbations of force and length, most biological materials have a linear response in terms of viscous and elastic moduli. However, if loads or extensions are large or applied very rapidly, then interesting nonlinearities or discontinuities are often observed. The critical point is that mechanical studies can give important insights into the function of biological molecules (e.g., Ref. [8]). For instance, imagine being able to pick up a single molecule from one of your muscles and then sense how it moves as it breaks down the chemical fuel that makes your muscles contract.

In the early 1990s, several laboratories adopted optical tweezers in order to record the mechanical properties of individual molecules [9] (see **PAPER 2.5**), [10–13], [14] (see **PAPER 2.6**). The critical issue in working with single molecules is that the amount of force, movement, and total energy involved is very small—usually around 10 $k_b T$, that is, the same order of magnitude as thermal energy.

For small displacements, the force produced by optical tweezers along a particular axis depends upon the change in momentum of light passing through each hemisphere of a spherical particle along that axis [15] (see **PAPER 3.2**). The greater the imbalance in momentum change, the greater the restoring force. To compute the force as a function of displacement over large displacements, we find that numerical integration is required (see Chapter 4 and references therein). Experiment shows that for small displacements (< ±250 nm) of a spherical particle, restoring force is directly proportional to displacement [16] (see **PAPER 3.3**). In other words, the optical tweezers behaves as a Hookean spring. Stiffness can be increased either by increasing the gradient in light intensity (by reducing the spot size relative to the bead diameter) or simply by increasing the total flux. The total distance over which the restoring force acts depends upon the sizes of the bead and the spot of light. At extreme bead deflection, the restoring force declines and eventually the direction of the force is reversed so that it tends to push the bead away from the light spot. For small bead displacements, the equation of motion governing the behavior of a bead of mass m in a tweezers of stiffness κ within a medium that gives a viscous damping β results from the balance between inertial, viscous, and elastic forces (Equation 3.1):

$$m \frac{\partial^2 x}{\partial t^2} + \beta \frac{\partial x}{\partial t} + \kappa x = 0 \qquad (3.1)$$

In the absence of damping (e.g., $\beta = 0$), the result would be an oscillator with resonant frequency f_{res}, given by Equation (3.2):

$$f_{res} = \frac{1}{2\pi}\sqrt{\frac{\kappa}{m}} \qquad (3.2)$$

In typical applications, the stiffness of the optical tweezers is around 0.05 pN nm^{-1} (5×10^{-5} N.m^{-1}) and the trapped objects are around 1 μm diameter (corresponding to a mass of 5×10^{-16} kg). Hence, the resonant frequency is approximately 50 kHz. However, because most experiments are performed in an aqueous medium, significant damping force arises. For micron-sized particles of radius r moving in a fluid of viscosity η, the Stokes drag constant, β, is given by Equation (3.3):

$$\beta = 6\pi r\eta \qquad (3.3)$$

For a 1-μm diameter sphere in water, $\beta = 1 \times 10^{-8}$ N.s.m^{-1}. The combination of viscous damping and the springlike stiffness of the optical tweezers gives rise to a single-pole, low-pass filter with −3 db frequency, f_0, of around 800 Hz, given by Equation (3.4):

$$f_0 = \frac{\kappa}{2\pi\beta} \qquad (3.4)$$

Since this is much lower than the resonant frequency, bead motion is very overdamped. Because most experiments are performed at room temperature, the bead held in the tweezers undergoes thermal fluctuations of mean squared amplitude $\langle x^2 \rangle$ given by the principle of equipartition of energy. When the tweezers stiffness is 0.04 pN.nm^{-1} and the temperature = 300 K, the root mean squared amplitude is around 10 nm (Equation 3.5):

$$\frac{1}{2}\kappa <x^2> = \frac{1}{2}k_b T \qquad (3.5)$$

Note that this also means that it is very unlikely that a trapped particle will spontaneously diffuse from the grasp of the optical tweezers, which have a capture range of about 300 nm.

Given that the mechanical system is overdamped, we find that thermal noise exhibits a Lorenzian power density spectrum. The amplitude of motion, A_f, over each frequency interval, f, is given by Equation (3.6):

$$A_f = \frac{4k_b T\beta}{\kappa^2(1+(f/f_0)^2)} \qquad (3.6)$$

Most optical-tweezers-based mechanical testing apparatus are constructed using a light microscope combined with a number of external optical components mounted on a vibration isolation table (or optical table) [3] (see **PAPER 3.1**), [17]. To make measurements with nanometer precision, great care must be taken to isolate the system from mechanical and acoustic noise and thermal drift. One of the worst sources of noise is the beam-pointing instability as the laser beam traverses the optical table. Several mechanical design strategies minimize this source of noise:

- Use a high-stability laser source.
- Expand the laser beam early on in the light path.
- Minimize the total path length.
- Enclose the light path using tubes that are just large enough to accommodate the optical elements.
- If multiple traps are used then the different beam paths should be matched and should run as close as possible to each other so as to reject common-mode movements ([18–22]; see [20] **PAPER 3.4**).

The main cause of pointing instability is due to light passing through regions of air at different temperatures and hence differing refractive indexes. It was reported that stability (particularly drift < 0.1 Hz) can be improved dramatically by enclosing the optical components and filling the enclosure with low-refractive-index gas (helium) at atmospheric pressure [23]. Careful design is required to prevent air leaking into the system, as this might be expected to cause unwanted local refractive index changes.

To make mechanical measurements, the material under test must be held at a minimum of two points. One of these points is considered a mechanical ground, and the position of the other is referenced to that fixed point. Most optical-tweezers-based mechanical testing apparatus usually have three different mechanical ground points: the laser mounting position (fixed to the optical table); the microscope coverslip (fixed to the microscope stage); and the position detector (fixed to the microscope camera port). These three reference points are located up to 1 m away from one another. High-resolution measurements are only possible because unwanted movement (drift) of the laser and the position detector are demagnified by the optical system. The 100X objective lens and various other accessory lenses required to generate the optical tweezers give about one-thousand-fold magnification. This means that one micrometer (or one microradian of arc) movement measured at the detector or laser mounting points is equivalent to one nanometer at the microscope coverslip. One way to maximize positional stability and resolution is to use two optical tweezers as the mechanical handles. One tweezers can be considered as a mechanical ground point and the other as the moveable point. If the two tweezers have a common origin, then most sources of noise will cancel because they will show up as a common mode signal that is easily rejected. Using multiple traps also enables more complex mechanical properties to be studied (e.g., three-point loading). In order to produce dual optical tweezers, two different methods have been developed [16] (see **PAPER 3.3**), [24] (see **PAPER 3.5**). One of the most exciting recent developments in the field is to synthesize multiple optical tweezers using holographic elements [25,26] (see Chapter 7). Multiple tweezers created in this way can be readily adapted to make quantitative measurements of force and displacement [27,28].

For high-resolution studies (especially those made on single molecules or colloidal particles), we need to consider the frequency bandwidth of our measurements and that of any possible mechanical noise (see Equation 3.6). For most studies, the mechanical response of interest lies in the region 0.01 Hz to about 10,000 Hz. Once measurement bandwidth has been established, the experimenter must make efforts to characterize and reduce noise within that bandwidth. Generally, thermal drift dominates at low frequencies while acoustic and building noise dominate at higher frequencies [3] (see **PAPER 3.1**), [5].

In many early optical tweezers experiments, bead position was measured using video microscopy [29–32], [33] (see **PAPER 2.4**). Video images were analyzed using various methods to determine the centroid of the bead image. One limitation of these studies, which was soon realized by all experimenters, is that the bandwidth of video recording undersampled the thermal noise of the optically trapped particle and this led to major problems with data interpretation. A major advantage of video is that a large field of view is recorded by pixilated detectors that have excellent spatial homogeneity. Furthermore, the information content of video is very high, and digital image analysis enables microscope stage drift to be cancelled by monitoring movement of fiducial markers. In the 1990s it was a major challenge to capture video frames and analyze and store the data at full video frame rate using a personal computer. However, high-speed modern computers now mean that video is being used very successfully for position detection in optical tweezers applications [27]. Modern cameras permit readout speeds of up to 1000 Hz (giving useful data up to 500 Hz) and this is fast enough to sample thermal motion of optically trapped particles and to measure fast dynamics in biology and colloid science.

In order to make calibrated measurements of bead displacement and the corresponding forces, split photodiode detectors are the simplest devices to build and use. They produce electrical signals that can be transferred to computers using analogue-to-digital converters. In principle, they can give up to 100 kHz bandwidth and subangstrom sensitivity [34].

One approach has been to measure the scattering of the laser light used for the optical tweezers after it has passed though the particle [12], [9] (see **PAPER 2.5**), [35] (see **PAPER 3.6**). Svoboda developed an interferometric detector that measured the relative scattering of two orthogonally polarized beams positioned just a few tens of nanometers apart at the optical tweezers [9] (see **PAPER 2.5**). The beams were separated using a polarizing beam splitter after they had passed though the bead, and their relative intensities were measured using two photodiodes. The detector system is sensitive to displacement along the axis of separation of the two light beams. An alternative system, developed by Smith [12] and Allersma [35] (see **PAPER 3.6**), simply used a single-laser trapping beam and then monitored the scattering of the beam using a four-quadrant detector. In both systems the signal is proportional to the bead displacement relative to the trap position so it can be calibrated to read out force directly. The signal can also be used to provide a closed-loop, force-feedback control by maintaining bead displacement at a fixed value by servoing either the trap or stage positions [36–38].

Another approach has been to measure bead position by casting its image onto a split photodiode detector. This requires the use of another light source to give either bright-field or dark-field illumination. This sort of detector can be calibrated to give bead position relative to the microscope axis and can be considered as a displacement signal relative to that ground point. Forces can be calculated if the trap position is known. This sort of signal is ideally suited to position feedback control [16] (see **PAPER 3.3**), [24] (see **PAPER 3.5**).

Closed-loop, force-feedback control works by controlling bead position relative to the trap center position. An electrical signal from the position sensor is fed back to move the position of an optical element (e.g., acousto-optic modulator or galvanometer mirror) such that the displacement of the bead within the tweezers is kept constant. This means that the load applied to the material under test is fixed at a known value and therefore the extension of any compliant linkages in series with the specimen is constant [36,38–41]. The linking material might consist of an antibody or streptavidin-biotin or the tail region of kinesin or myosin. The linker is sometimes more compliant than the specimen under test so if an overall length change is imposed, part of it is taken up in extending the linker. In their experiments with kinesin, a motor protein that moves along microtubules, Visscher et al. [38] were able measure the force–velocity relationship for the motor by fixing the load at various predetermined values.

Position feedback locks the absolute position of the bead relative to the microscope axis and the force (e.g., relative laser position) changes to compensate for any changes in external load, which allows the absolute extension of a material to be controlled. This type of feedback permits absolute bead position to be maintained, and therefore Brownian motion is suppressed. The stiffness of the tweezers is increased theoretically to infinity but practically is limited by the bandwidth and sensitivity of the detector and optical control system, and also by the maximum linear range of the optical tweezers [16] (see **PAPER 3.3**), [24] (see **PAPER 3.5**).

One ingenious version of the feedback-control loop approach used a dual optical tweezers to hold an individual actin filament while the position of each bead was monitored independently. The measured position of one bead was used to control the tweezers holding the other bead, and because the two were connected by the actin the bead position could be held fixed. In this way the compliance of the actin and both of its end connections was effectively cancelled. This was important because the experimenters wished to apply a known load, rapidly (within < 1 ms), to a myosin molecule that was bound at the middle of the actin filament [42] (see **PAPER 3.7**).

Closed-loop control of the optical tweezers mechanical apparatus can also be applied via the microscope stage position. This requires that the microscope stage is an electromechanical device (usually a piezoelectric substage is used). One group took the video image of a fiducial marker and sent the centroid position signal to a negative feedback loop in order to clamp to the microscope stage position and compensate for any thermal drift. Meanwhile, the position of two beads held in a dual-trap arrangement was measured using two four-quadrant photodiode detectors. By holding a single actin filament between both of the optically trapped beads, it could then be moved slowly past a single myosin molecule so that interactions at each actin binding site could be mapped with nanometer precision [43].

The use of optical tweezers as a mechanical testing apparatus requires that the tweezers can be manipulated with nanometer precision (using a galvanometer or motorized mirrors, acousto- or electro-optic deflectors, or holographic gratings). Furthermore, the microscope stage, which often acts as a mechanical ground, must be held fixed relative to the tweezers position and the associated position sensor must also be stably mounted.

Because the objective lens used to produce the optical tweezers gives a highly magnified image of the trapped particle, nanometer-scale detection systems can be fabricated using millimeter-sized sensors. Modern charge-coupled device (CCD) or complementary metal-oxide semiconductor (CMOS) cameras give excellent spatial linearity over a huge measurement range. A camera with a field of view of 100 μm can give 1-nm position sensitivity over the entire field of view (e.g., 1 part in 100,000). Early experiments used split photodiode detectors because they are inexpensive, easy to make, and simple to interface to a computer. They give high-bandwidth and high-sensitivity signals that are proportional either to bead position or to force.

With careful construction and good environment control, an optical tweezers mechanical test apparatus can have stability in the nanometer range. The apparatus can be calibrated either using thermal motion of the trapped particle (Equations 3.5 and 3.6) or by applying viscous loads (Equations 3.3 and 3.4). This enables the transfer function between bead displacement and force to be determined. Optical tweezers mechanical testing apparatus have enabled the properties of individual molecules to be measured with pN-nm resolution over a bandwidth of 10 mHz to 10 kHz. Finally, computer control allows forcing functions and feedback loops to be used so that optically trapped particles can be used to apply controlled forces and displacements to specimens. These sophisticated testing protocols can then be used to probe detailed mechanistic questions about how biological molecules work (Chapter 2) and increase our understanding of interactions between colloidal particles (Chapter 5). Over the past few years, the advent of holographic tweezers and ultrafast camera systems has opened exciting new possibilities in terms of high-speed nanomechanical measurements of force and movement. The ability to manipulate multiple particles and measure forces and displacements at several positions simultaneously makes many new experimental formats possible.

Endnotes

1. K. Svoboda and S.M. Block, Biological applications of optical forces, *Annu. Rev. Biophys. Biomolec. Struct.* **23**, 247–285 (1994).
2. A.D. Mehta, M. Rief, J.A. Spudich, D.A. Smith, and R.M. Simmons, Single-molecule biomechanics with optical methods, *Science* **283**, 1689–1695 (1999).
3. K.C. Neuman and S.M. Block, Optical trapping, *Rev. Sci. Instrum.* **75**, 2787–2809 (2004).
4. J.R. Moffitt, Y.R. Chemla, S.B. Smith, and C. Bustamante, Recent advances in optical tweezers, *Annu. Rev. Biochem.* **77**, 19.1–19.24 (2008).
5. F. Gittes and C.F. Schmidt, Signals and noise in micromechanical measurements, *Meth. Cell Biol.* **55** 129–156 (1998).

6. A.E. Knight, G.I. Mashanov, and J.E. Molloy, Single molecule measurements and biological motors, *Eur. Biophys. J.* **35**, 89 (2005).

7. S.A. Wainwright, W.D. Biggs, J.D. Currey, and J.M. Gosline, *Mechanical Designs in Organisms*, Princeton, NJ, Princeton University Press (1976).

8. A.F. Huxley and R.M. Simmons, Proposed mechanism of force generation in muscle, *Nature* **233**, 533–538 (1971).

9. K. Svoboda, C.F. Schmidt, B.J. Schnapp, and S.M. Block, Direct observation of kinesin stepping by optical trapping interferometry, *Nature* **365**, 721–727 (1993).

10. J.T. Finer, R.M. Simmons, and J.A. Spudich, Single myosin molecule mechanics: Piconewton forces and nanometre steps, *Nature* **368**, 113–118 (1994).

11. T.T. Perkins, S.R. Quake, D.E. Smith, and S. Chu, Relaxation of a single DNA molecule observed by optical microscopy, *Science* **264**, 822–826 (1994).

12. S.B. Smith, Y.J. Cui, and C. Bustamante, Overstretching b-DNA: The elastic response of individual double-stranded and single-stranded DNA molecules, *Science* **271**, 795–799 (1996).

13. W.H. Guilford, D.E. Dupuis, G. Kennedy, J.R. Wu, J.B. Patlak, and D.M. Warshaw, Smooth muscle and skeletal muscle myosins produce similar unitary forces and displacements in the laser trap, *Biophys. J.* **72**, 1006–1021 (1997).

14. J.E. Molloy, J.E. Burns, J. Kendrick-Jones, R.T. Tregear, and D.C.S. White, Movement and force produced by a single myosin head, *Nature* **378**, 209–212 (1995).

15. A. Ashkin, Forces of a single-beam gradient laser trap on a dielectric sphere in the ray optics regime, *Biophys. J.* **61**, 569–582 (1992).

16. R.M. Simmons, J.T. Finer, S. Chu, and J.A. Spudich, Quantitative measurements of force and displacement using an optical trap, *Biophys. J.* **70**, 1813–1822 (1996).

17. M.J. Lang and S.M. Block, Resource letter: LBOT-1: Laser-based optical tweezers, *Am. J. Phys.* **71**, 201–215 (2003).

18. K. Visscher, S.P. Gross, and S.M. Block, Construction of multiple-beam optical traps with nanometer-resolution position sensing, *IEEE J. Sel. Top. Quant. Electron.* **2**, 1066–1076 (1996).

19. J.E. Molloy, Optical chopsticks: Digital synthesis of multiple optical traps, *Meth. Cell Biol.* **55**, 205–216 (1998).

20. C. Veigel, M.L. Bartoo, D.C.S. White, J.C. Sparrow, and J.E. Molloy, The stiffness of rabbit skeletal actomyosin cross-bridges determined with an optical tweezers transducer, *Biophys. J.* **75**, 1424–1438 (1998).

21. J.R. Moffitt, Y.R. Chemla, D. Izhaky, and C. Bustamante, Differential detection of dual traps improves the spatial resolution of optical tweezers, *Proc. Natl. Acad. Sci. U.S.A.* **103**, 9006–9011 (2006).

22. K.M. Herbert, A. La Porta, B.J. Wong, R.A. Mooney, K.C. Neuman, R. Landick, and S.M. Block, Sequence-resolved detection of pausing by single RNA polymerase molecules, *Cell* **125** 1083–1094 (2006).

23. E.A. Abbondanzieri, W.J. Greenleaf, J.W. Shaevitz, R. Landick, and S.M. Block, Direct observation of base-pair stepping by RNA polymerase, *Nature* **438**, 460–465 (2005).

24. J.E. Molloy, J.E. Burns, J.C. Sparrow, R.T. Tregear, J. Kendrick-Jones, and D.C. White, Single-molecule mechanics of heavy meromyosin and S1 interacting with rabbit or *Drosophila* actins using optical tweezers, *Biophys. J.* **68**, 298S–303S; 303S–305S (1995).

25. P. Jordan, J. Leach, M. Padgett, P. Blackburn, N. Isaacs, M. Göksor, D. Hanstorp, A. Wright, J. Girkin, and J. Cooper, Creating permanent 3D arrangements of isolated cells using holographic optical tweezers, *Lab Chip* **5**, 1224–1228 (2005).

26. J. Leach, G. Sinclair, P. Jordan, J. Courtial, M.J. Padgett, J. Cooper, and Z.J. Laczik, 3D manipulation of particles into crystal structures using holographic optical tweezers, *Opt. Express* **12**, 220–226 (2004).

27. S. Keen, J. Leach, G. Gibson, and M.J. Padgett, Comparison of a high-speed camera and a quadrant detector for measuring displacements in optical tweezers, *J. Opt. A—Pure Appl. Opt.* **9**, S264–S266 (2007).

28. R. Di Leonardo, J. Leach, H. Mushfique, J.M. Cooper, G. Ruocco, and M.J. Padgett, Multipoint holographic optical velocimetry in microfluidic systems, *Phys. Rev. Lett.* **96** 134502 (2006).

29. T. Nishizaka, H. Miyata, H. Yoshikawa, S. Ishiwata, and K. Kinosita, Unbinding force of a single motor molecule of muscle measured using optical tweezers, *Nature* **377**, 251–254 (1995).

30. H. Miyata, H. Hakozaki, H. Yoshikawa, N. Suzuki, K. Kinosita Jr., T. Nishizaka, and S. Ishiwata, Stepwise motion of an actin filament over a small number of heavy meromyosin molecules is revealed in an *in vitro* motility assay, *J. Biochem. Tokyo* **115**, 644–647 (1994).

31. S.C. Kuo and M.P. Sheetz, Force of single kinesin molecules measured with optical tweezers, *Science* **260**, 232–234 (1993).

32. S.M. Block, D.F. Blair, and H.C. Berg, Compliance of bacterial flagella measured with optical tweezers, *Nature* **338**, 514–518 (1989).

33. M. Edidin, S.C. Kuo, and M.P. Sheetz, Lateral movements of membrane-glycoproteins restricted by dynamic cytoplasmic barriers, *Science* **254**, 1379–1382 (1991).

34. N.J. Carter and R.A. Cross, Mechanics of the kinesin step, *Nature* **435**, 308–312 (2005).

35. M.W. Allersma, F. Gittes, M.J. deCastro, R.J. Stewart, and C.F. Schmidt, Two-dimensional tracking of ncd motility by back focal plane interferometry, *Biophys. J.* **74**, 1074–1085 (1998).

36. K. Visscher and S.M. Block, Versatile optical traps with feedback control, in *Molecular Motors and the Cytoskeleton, Part B*, San Diego, Academic Press (1998), 460–489.

37. M.J. Schnitzer, K. Visscher, and S.M. Block, Force production by single kinesin motors, *Nat. Cell Biol.* **2**, 718–723 (2000).

38. K. Visscher, M.J. Schnitzer, and S.M. Block, Single kinesin molecules studied with a molecular force clamp, *Nature* **400**, 184–189 (1999).

39. M.D. Wang, M.J. Schnitzer, H. Yin, R. Landick, J. Gelles, and S.M. Block, Force and velocity measured for single molecules of RNA polymerase, *Science* **282**, 902–907 (1998).

40. D.E. Smith, S.J. Tans, S.B. Smith, S. Grimes, D.L. Anderson, and C. Bustamante, The bacteriophage phi 29 portal motor can package DNA against a large internal force, *Nature* **413**, 748–752 (2001).

41. M. Rief, R.S. Rock, A.D. Mehta, M.S. Mooseker, R.E. Cheney, and J.A. Spudich, Myosin-V stepping kinetics: A molecular model for processivity, *Proc. Natl. Acad. Sci. U.S.A.* **97**, 9482–9486 (2000).

42. Y. Takagi, E.E. Homsher, Y.E. Goldman, and H. Shuman, Force generation in single conventional actomyosin complexes under high dynamic load, *Biophys. J.* **90**, 1295–1307 (2006).

43. W. Steffen, D. Smith, R. Simmons, and J. Sleep, Mapping the actin filament with myosin, *Proc. Natl. Acad. Sci. U.S.A.* **98**, 14949–14954 (2001).

REVIEW ARTICLE

Optical trapping

Keir C. Neuman and Steven M. Block[a)]

Department of Biological Sciences, and Department of Applied Physics, Stanford University, Stanford, California 94305

(Received 15 January 2004; accepted 14 June 2004; published 2 September 2004)

Since their invention just over 20 years ago, optical traps have emerged as a powerful tool with broad-reaching applications in biology and physics. Capabilities have evolved from simple manipulation to the application of calibrated forces on—and the measurement of nanometer-level displacements of—optically trapped objects. We review progress in the development of optical trapping apparatus, including instrument design considerations, position detection schemes and calibration techniques, with an emphasis on recent advances. We conclude with a brief summary of innovative optical trapping configurations and applications.
© *2004 American Institute of Physics.* [DOI: 10.1063/1.1785844]

I. INTRODUCTION

Arthur Ashkin pioneered the field of laser-based optical trapping in the early 1970s. In a series of seminal papers, he demonstrated that optical forces could displace and levitate micron-sized dielectric particles in both water and air,[1] and he developed a stable, three-dimensional trap based on counterpropagating laser beams.[2] This seminal work eventually led to the development of the single-beam gradient force optical trap,[3] or "optical tweezers," as it has come to be known.[4] Ashkin and co-workers employed optical trapping in a wide-ranging series of experiments from the cooling and trapping of neutral atoms[5] to manipulating live bacteria and viruses.[6,7] Today, optical traps continue to find applications in both physics and biology. For a recent survey of the literature on optical tweezers see Ref. 8. The ability to apply picoNewton-level forces to micron-sized particles while simultaneously measuring displacement with nanometer-level precision (or better) is now routinely applied to the study of molecular motors at the single-molecule level,[9–19] the physics of colloids and mesoscopic systems,[20–29] and the mechanical properties of polymers and biopolymers.[18,20,30–43] In parallel with the widespread use of optical trapping, theoretical and experimental work on fundamental aspects of optical trapping is being actively pursued.[4,20,44–48] In addition to the many excellent reviews of optical trapping[9,49–53] and specialized applications of optical traps, several comprehensive guides for building optical traps are now available.[54–60] For the purpose of this review, we will concentrate on the fundamental aspects of optical trapping with particular emphasis on recent advances.

Just as the early work on optical trapping was made possible by advances in laser technology,[4] much of the recent progress in optical trapping can be attributed to further technological development. The advent of commercially available, three-dimensional (3D) piezoelectric stages with ca-

pacitive sensors has afforded unprecedented control of the position of a trapped object. Incorporation of such stages into optical trapping instruments has resulted in higher spatial precision and improved calibration of both forces and displacements. In addition, stage-based force clamping techniques have been developed that can confer certain advantages over other approaches of maintaining the force, such as dynamically adjusting the position or stiffness of the optical trap. The use of high-bandwidth position detectors[61] improves force calibration, particularly for very stiff traps, and extends the detection bandwidth of optical trapping measurements. In parallel with these technological improvements, recent theoretical work has led to a better understanding of 3D position detection[62–64] and progress has been made in calculating the optical forces on spherical objects with a range of sizes.[65,66]

II. PRINCIPLES OF OPTICAL TRAPPING

An optical trap is formed by tightly focusing a laser beam with an objective lens of high numerical aperture (NA). A dielectric particle near the focus will experience a force due to the transfer of momentum from the scattering of incident photons. The resulting optical force has traditionally been decomposed into two components: (1) a scattering force, in the direction of light propagation and (2) a gradient force, in the direction of the spatial light gradient. This decomposition is merely a convenient and intuitive means of discussing the overall optical force. Following tradition, we present the optical force in terms of these two components, but we stress that both components arise from the very same underlying physics (see theoretical progress, below for a unified expression). The scattering component of the force is the more familiar of the two, which can be thought of as a photon "fire hose" pushing the bead in the direction of light propagation. Incident light impinges on the particle from one direction, but is scattered in a variety of directions, while some of the incident light may be absorbed. As a result, there

[a)]Electronic mail: sblock@stanford.edu

is a net momentum transfer to the particle from the incident photons. For an isotropic scatter, the resulting forces cancel in all but the forward direction, and an effective scattering cross section can be calculated for the object. For most conventional situations, the scattering force dominates. However, if there is a steep intensity gradient (i.e., near the focus of a laser), the second component of the optical force, the gradient force, must be considered. The gradient force, as the name suggests, arises from the fact that a dipole in an inhomogeneous electric field experiences a force in the direction of the field gradient.[67] In an optical trap, the laser induces fluctuating dipoles in the dielectric particle, and it is the interaction of these dipoles with the inhomogeneous electric field at the focus that gives rise to the gradient trapping force. The gradient force is proportional to both the polarizability of the dielectric and the optical intensity gradient at the focus.

For stable trapping in all three dimensions, the axial gradient component of the force pulling the particle towards the focal region must exceed the scattering component of the force pushing it away from that region. This condition necessitates a very steep gradient in the light, produced by sharply focusing the trapping laser beam to a diffraction-limited spot using an objective of high NA. As a result of this balance between the gradient force and the scattering force, the axial equilibrium position of a trapped particle is located slightly beyond (i.e., down-beam from) the focal point. For small displacements (~150 nm), the gradient restoring force is simply proportional to the offset from the equilibrium position, i.e., the optical trap acts as Hookean spring whose characteristic stiffness is proportional to the light intensity.

In developing a theoretical treatment of optical trapping, there are two limiting cases for which the force on a sphere can be readily calculated. When the trapped sphere is much larger than the wavelength of the trapping laser, i.e., the radius $(a) \gg \lambda$, the conditions for Mie scattering are satisfied, and optical forces can be computed from simple ray optics (Fig. 1). Refraction of the incident light by the sphere corresponds to a change in the momentum carried by the light. By Newton's third law, an equal and opposite momentum change is imparted to the sphere. The force on the sphere, given by the rate of momentum change, is proportional to the light intensity. When the index of refraction of the particle is greater than that of the surrounding medium, the optical force arising from refraction is in the direction of the intensity gradient. Conversely, for an index lower than that of the medium, the force is in the opposite direction of the intensity gradient. The scattering component of the force arises from both the absorption and specular reflection by the trapped object. In the case of a uniform sphere, optical forces can be directly calculated in the ray-optics regime.[68,69] The extremal rays contribute disproportionally to the axial gradient force, whereas the central rays are primarily responsible for the scattering force. Thus, expanding a Gaussian laser beam to slightly overfill the objective entrance pupil can increase the ratio of trapping to scattering force, resulting in improved trapping efficiency.[69,70] In practice, the beam is typically expanded such that the $1/e^2$ intensity points match the objective aperture, resulting in ~87% of the incident power enter-

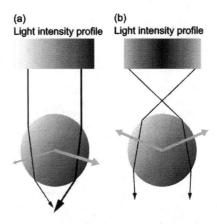

FIG. 1. Ray optics description of the gradient force. (A) A transparent bead is illuminated by a parallel beam of light with an intensity gradient increasing from left to right. Two representative rays of light of different intensities (represented by black lines of different thickness) from the beam are shown. The refraction of the rays by the bead changes the momentum of the photons, equal to the change in the direction of the input and output rays. Conservation of momentum dictates that the momentum of the bead changes by an equal but opposite amount, which results in the forces depicted by gray arrows. The net force on the bead is to the right, in the direction of the intensity gradient, and slightly down. (B) To form a stable trap, the light must be focused, producing a three-dimensional intensity gradient. In this case, the bead is illuminated by a focused beam of light with a radial intensity gradient. Two representative rays are again refracted by the bead but the change in momentum in this instance leads to a net force towards the focus. Gray arrows represent the forces. The lateral forces balance each other out and the axial force is balanced by the scattering force (not shown), which decreases away from the focus. If the bead moves in the focused beam, the imbalance of optical forces will draw it back to the equilibrium position.

ing the objective. Care should be exercised when overfilling the objective. Absorption of the excess light by the blocking aperture can cause heating and thermal expansion of the objective, resulting in comparatively large (~μm) axial motions when the intensity is changed. Axial trapping efficiency can also be improved through the use of "donut" mode trapping beams, such as the TEM_{01}^* mode or Laguerre-Gaussian beams, which have intensity minima on the optical propagation axis.[69,71–73]

When the trapped sphere is much smaller than the wavelength of the trapping laser, i.e., $a \ll \lambda$, the conditions for Raleigh scattering are satisfied and optical forces can be calculated by treating the particle as a point dipole. In this approximation, the scattering and gradient force components are readily separated. The scattering force is due to absorption and reradiation of light by the dipole. For a sphere of radius a, this force is

$$F_{\text{scatt}} = \frac{I_0 \sigma n_m}{c}, \tag{1}$$

$$\sigma = \frac{128 \pi^5 a^6}{3 \lambda^4} \left(\frac{m^2 - 1}{m^2 + 2} \right)^2, \tag{2}$$

where I_0 is the intensity of the incident light, σ is the scattering cross section of the sphere, n_m is the index of refraction of the medium, c is the speed of light in vacuum, m is the ratio of the index of refraction of the particle to the index of the medium (n_p/n_m), and λ is the wavelength of the trapping laser. The scattering force is in the direction of propagation of the incident light and is proportional the intensity. The time-averaged gradient force arises from the interaction of the induced dipole with the inhomogeneous field

$$F_{\text{grad}} = \frac{2\pi\alpha}{cn_m^2} \nabla I_0, \qquad (3)$$

where

$$\alpha = n_m^2 a^3 \left(\frac{m^2 - 1}{m^2 + 2} \right) \qquad (4)$$

is the polarizability of the sphere. The gradient force is proportional to the intensity gradient, and points up the gradient when $m > 1$.

When the dimensions of the trapped particle are comparable to the wavelength of the trapping laser ($a \sim \lambda$), neither the ray optic nor the point-dipole approach is valid. Instead, more complete electromagnetic theories are required to supply an accurate description.[74–80] Unfortunately, the majority of objects that are useful or interesting to trap, in practice, tend to fall into this intermediate size range ($0.1-10\lambda$). As a practical matter, it can be difficult to work with objects smaller than can be readily observed by video microscopy (~ 0.1 μm), although particles as small as ~ 35 nm in diameter have been successfully trapped. Dielectric microspheres used alone or as handles to manipulate other objects are typically in the range of $\sim 0.2-5$ μm, which is the same size range as biological specimens that can be trapped directly, e.g., bacteria, yeast, and organelles of larger cells. Whereas some theoretical progress in calculating the force on a sphere in this intermediate size range has been made recently,[65,66] the more general description does not provide further insight into the physics of optical trapping. For this reason we postpone discussion of recent theoretical work until the end of the review.

III. DESIGN CONSIDERATIONS

Implementing a basic optical trap is a relatively straightforward exercise (Fig. 2).[55,58] The essential elements are a trapping laser, beam expansion and steering optics, a high NA objective, a trapping chamber holder, and some means of observing the trapped specimen. Optical traps are most often built by modifying an inverted microscope so that a laser beam can be introduced into the optical path before the objective: the microscope then provides the imaging, trapping chamber manipulation, and objective focus functions. For anything beyond simply trapping and manually manipulating objects, however, additional elements become necessary. Dynamic control of trap position and stiffness can be achieved through beam steering and amplitude modulation elements incorporated in the optical path before the laser beam enters the objective. Dynamic control over position and stiffness of the optical trap has been exploited to implement position- and force-clamp systems. Position clamps, in which the position of a trapped object is held constant by varying the force, are well suited for stall force measurements of molecular motors.[39,49,81–83] Force clamps, in which the force on a trapped object is fixed by varying the position of the trap, are

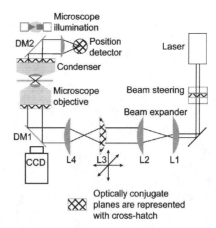

FIG. 2. Layout of a generic optical trap. The laser output beam usually requires expansion to overfill the back aperture of the objective. For a Gaussian beam, the beam waist is chosen to roughly match the objective back aperture. A simple Keplerian telescope is sufficient to expand the beam (lenses $L1$ and $L2$). A second telescope, typically in a $1:1$ configuration, is used for manually steering the position of the optical trap in the specimen plane. If the telescope is built such that the second lens, $L4$, images the first lens, $L3$, onto the back aperture of the objective, then movement of $L3$ moves the optical trap in the specimen plane with minimal perturbation of the beam. Because lens $L3$ is optically conjugate (conjugate planes are indicated by a cross-hatched fill) to the back aperture of the objective, motion of $L3$ rotates the beam at the aperture, which results in translation in the specimen plane with minimal beam clipping. If lens $L3$ is not conjugate to the back aperture, then translating it leads to a combination of rotation *and* translation at the aperture, thereby clipping the beam. Additionally, changing the spacing between $L3$ and $L4$ changes the divergence of the light that enters the objective, and the axial location of the laser focus. Thus, $L3$ provides manual three-dimensional control over the trap position. The laser light is coupled into the objective by means of a dichroic mirror ($DM1$), which reflects the laser wavelength, while transmitting the illumination wavelength. The laser beam is brought to a focus by the objective, forming the optical trap. For back focal plane position detection, the position detector is placed in a conjugate plane of the condenser back aperture (condenser iris plane). Forward scattered light is collected by the condenser and coupled onto the position detector by a second dichroic mirror ($DM2$). Trapped objects are imaged with the objective onto a camera. Dynamic control over the trap position is achieved by placing beam-steering optics in a conjugate plane to the objective back aperture, analogous to the placement of the trap steering lens. For the case of beam-steering optics, the point about which the beam is rotated should be imaged onto the back aperture of the objective.

well suited for displacement measurements.[49,56,81,84,85] Incorporation of a piezoelectric stage affords dynamic positioning of the sample chamber relative to the trap, and greatly facilitates calibration. Furthermore, for the commonly employed geometry in which the molecule of interest is attached between the surface of the trapping cell and a trapped bead "handle," piezoelectric stages can be used to generate a force clamp.[86–88] The measurement of force and displacement within the optical trap requires a position detector, and, in some configurations, a second, low power laser for detection. We consider each of these elements in detail.

A. Commercial systems

Commercial optical trapping systems with some limited capabilities are available. Cell Robotics[89] manufactures a laser-trapping module that can be added to a number of inverted microscopes. The module consists of a 1.5 W diode pumped Nd:YVO$_4$ laser ($\lambda = 1064$ nm) with electronic intensity control, and all of the optics needed to both couple the laser into the microscope and manually control the position of the trap in the specimen plane. The same module is incorporated into the optical tweezers workstation, which includes a microscope, a motorized stage and objective focus, video imaging, and a computer interface. Arryx Incorporated[90] manufactures a complete optical trapping workstation that includes a 2 W diode pumped solid-state laser ($\lambda = 532$ nm), holographic beam shaping and steering, an inverted microscope, a motorized stage, and computer control. Holographic beam shaping provides control over the phase of the trapping laser,[91,92] which allows multiple, individually addressable, optical traps in addition to high order, complex trapping beams. An integrated optical trap is also available from PALM Microlaser Technologies,[93] either alone or incorporated with their microdissection system. The PALM system employs an infrared trapping laser and computer control of the stage, similar to the other optical trapping systems. The commercial systems tend to be expensive, but they offer turnkey convenience at the price of flexibility and control. None of the systems currently comes equipped with position detection capabilities beyond video imaging, and only one (Arryx) provides dynamic control over the trap position, but with an unknown update rate (~ 5 Hz or less). Overall, these systems are adequate for positioning and manipulating objects but are incapable, without further modifications, of ultrasensitive position or force measurements. As commercial systems become increasingly sophisticated and versatile, they may eventually offer an "off-the-shelf" option for some optical trapping applications. In deciding between a commercial or custom-built optical trap, or among commercial systems, several factors should be considered. Basic considerations include cost, maximum trap force and stiffness, choice of laser wavelength (important for biological samples), specimen or trap positioning capability, optical imaging modes, position-detection capabilities, and sample geometry. In addition, flexibility and the possibility to upgrade or improve aspects of the system should also be considered. How easily can the optical system be modified or adapted? Can the functionality be upgraded? Perhaps the most fundamental question concerns the decision to buy or to build. Whereas building a basic optical trap is now standard practice in many labs, it requires a certain familiarity with optics and optical components (in relation to the complexity of the optical trap), as well as a significant time investment for the design, construction, and debugging phases. These factors should be weighed against the potential benefits of reduced cost, increased flexibility and greater control of home-built optical traps.

B. Trapping laser

The basic requirement of a trapping laser is that it delivers a single mode output (typically, Gaussian TEM$_{00}$ mode) with excellent pointing stability and low power fluctuations. A Gaussian mode focuses to the smallest diameter beam waist and will therefore produce the most efficient, harmonic trap. Pointing instabilities lead to unwanted displacements of the optical trap position in the specimen plane, whereas power fluctuations lead to temporal variations in the optical

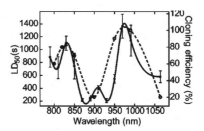

FIG. 3. The wavelength dependence of photodamage in *E. coli* compared to Chinese hamster ovary (CHO) cells. (Solid circles and solid line, left axis, half lethal dose time for *E. coli* cells (LD$_{50}$); open circles and dashed line, right axis, cloning efficiency in CHO cells determined by Liang *et al.* (Ref. 96) (used with permission). Lines represent cubic spline fits to the data). The cloning efficiency in CHO cells was determined after 5 min of trapping at 88 mW in the specimen plane (error bars unavailable), selected to closely match to our experimental conditions (100 mW in the specimen plane, errors shown as ± standard error in the mean). Optical damage is minimized at 830 and 970 nm for both *E. coli* and CHO cells, whereas it is most severe in the region between 870 and 930 nm (reprinted from Ref. 95).

trap stiffness. Pointing instability can be remedied by coupling the trapping laser to the optical trap via an optical fiber, or by imaging the effective pivot point of the laser pointing instability into the front focal plane of the objective. Both of these solutions however, trade reduced pointing stability against additional amplitude fluctuations, as the fiber coupling and the clipping by the back aperture of the microscope objective depend on beam pointing. Thus, both power and pointing fluctuations introduce unwanted noise into any trapping system. The choice of a suitable trapping laser therefore depends on several interdependent figures of merit (power, power stability, pointing stability, thermal drift, wavelength, mode quality, etc.).

Output power of the trapping laser and the throughput of the optical system will determine the maximum attainable stiffness and force. As discussed above, trapping forces depend on multiple parameters and are difficult to calculate for most conditions of practical interest. Generally speaking, maximum trapping forces on the order of 1 pN per 10 mW of power delivered to the specimen plane can be achieved with micron-scale beads.[9] As a specific example, trapping a 0.5 μm polystyrene ($n = 1.57$) sphere in water with a TEM$_{00}$ 1064 nm laser that overfills a 1.2 NA objective by $\sim 10\%$ ($1/e^2$ intensity points matched to the aperture radius), gives a stiffness of 0.16 pN/nm per W of power in the specimen plane. In practice, laser power levels can range from a few mW to a Watt or more in the specimen plane, depending on details of the laser and setup, objective transmittance, and the desired stiffness.

Wavelength is an important consideration when biological material is trapped, particularly for *in vivo* trapping of cells or small organisms.[94] There is a window of relative transparency in the near infrared portion of the spectrum (~ 750–1200 nm), located in the region between the absorption of proteins in the visible and the increasing absorption of water towards the infrared.[9] Substantial variation with wavelength of optical damage to biological specimens is observed even within the near infrared region (Fig. 3), with damage minima occurring at 970 and 830 nm[95–97] for bacterial cells of *Escherichia coli*. If damage or "opticution"[98] of biological specimens is not a concern, then the choice of wavelength becomes less critical, but the potential effects of heating resulting from light absorption by the medium or the trapped particle should certainly be considered.[99-101] The optimal choice of trapping wavelength will also depend on the transmission of the objective used for optical trapping (discussed below), as well as the output power available at a given wavelength.

In practice, a variety of lasers has been employed for optical trapping. The factors discussed above, along with the cost, will determine the final selection of a trapping laser. The laser of choice for working with biological samples is currently the neodymium:yttrium–aluminum–garnet (Nd:YAG) laser and its close cousins, neodymium: yttrium–lithium–fluoride (Nd:YLF), and neodymium: yttrium–orthovanadate (Nd:YVO$_4$). These lasers operate in the near infrared region of the spectrum at 1.047, 1.053, or 1.064 μm, which helps to limit optical damage. Diode pumped versions of these lasers offer high power (up to 10 W or even more) and superior amplitude and pointing stability. An additional advantage of diode-pumped solid-state (DPSS) lasers is that the noise and heat of the laser power supply can be physically isolated from the laser itself and the immediate region of the optical trap. The output of the pump diodes can be delivered to the laser head via an optical fiber bundle, in some cases up to 10 m in length. The main drawback of such DPSS lasers is their cost, currently on the order of $5–10 K per W of output power. Diode lasers afford a lower-cost alternative in a compact package and are available at several wavelengths in the near infrared, but these devices are typically limited to less than ~250 mW in a single-transverse mode, the mode required for efficient trapping. Diode lasers also suffer significantly from mode instabilities and noncircular beams, which necessitates precise temperature control instrumentation and additional corrective optics. By far the most expensive laser option is a tunable cw titanium:sapphire (Ti:sapphire) laser pumped by a DPSS laser, a system that delivers high power (~1 W) over a large portion of the near infrared spectrum (~750–950 nm), but at a current cost in excess of $100 K. The large tuning range is useful for parametric studies of optical trapping, to optimize the trapping wavelength, or to investigate the wavelength-dependence of optical damage.[95] A Ti:sapphire laser is also employed for optical trapping *in vivo*[94] since it is the only laser currently available that can deliver over ~250 mW at the most benign wavelengths (830 and 970 nm).[95]

In optical trapping applications where no biological materials will be trapped, any laser source that meets the basic criteria of adequate power in the specimen plane, sufficient pointing and amplitude stability, and a Gaussian intensity profile, may be suitable. Optical traps have been built based on argon ion,[3] helium-neon,[102] and diode laser sources,[103,104] to name a few. The DPSS lasers employed in our lab for biological work supply ~4 W of power at 1064 nm with power fluctuations below 1%–2% and a long-term pointing stability of ±50 μrad.

C. Microscope

Most optical traps are built around a conventional light microscope, requiring only minor modifications. This approach reduces the construction of an optical trap to that of coupling the light from a suitable trapping laser into the optical path before the objective without compromising the original imaging capabilities of the microscope. In practice, this is most often achieved by inserting a dichroic mirror, which reflects the trapping laser light into the optical path of the microscope but transmits the light used for microscope illumination. Inverted, rather than upright, microscopes are often preferred for optical trapping because their stage is fixed and the objective moves, making it easier to couple the trapping light stably. The use of a conventional microscope also makes it easier to use a variety of available imaging modalities, such as differential interference contrast and epifluorescence.

With more extensive modifications, a position detector can be incorporated into the trapping system. This involves adding a second dichroic mirror on the condenser side of the microscope, which reflects the laser light while transmitting the illuminating light. In order to achieve the mechanical stability and rigidity required for nanometer scale position measurements, more extensive modifications of the microscope are generally required.[50,59] In the current generation of optical traps, the rotating, multiobjective turret is conventionally replaced with a custom-built single objective holder, along with a mount for the dichroic mirror. The original stage is removed and the microscope is modified to accommodate a more substantial stage platform, holding a crossed-roller bearing stage (for coarse movement) mounted to a piezoelectric stage with feedback (for fine movement). Finally, the condenser assembly is attached to a fine focus transport (similar to that used for the objective) that is then mounted to the illumination column by a rigid aluminum beam.[59]

An alternative to the redesign and retrofitting of a commercial microscope is to build the entire optical trap from individual optical components.[57,103,104] This approach is slightly more involved, as the entirety of the imaging and trapping optical paths have to be designed and built. The increase in complexity, however, can be offset by increased flexibility in the design and a wider choice of components, greater access to the optical paths, and reduced cost.

D. Objective

The single most important element of an optical trap is the objective used to focus the trapping laser. The choice of objective determines the overall efficiency of the optical trapping system (stiffness versus input power), which is a function of both the NA and the transmittance of the objective. Additionally, the working distance and the immersion medium of the objective (oil, water, or glycerol) will set practical limits on the depth to which objects can be trapped. Spherical aberrations, which degrade trap performance, are proportional to the refractive index mismatch between the immersion medium and the aqueous trapping medium. The deleterious effect of these aberrations increases with focal depth. The working distance of most high NA oil immersion

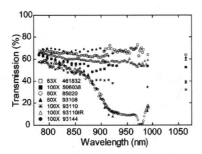

FIG. 4. Microscope objective transmission curves. Transmission measurements were made by means of the dual-objective method. Part numbers are cross-referenced in Table I. The uncertainty associated with a measurement at any wavelength is ~5% (reprinted from Ref. 95).

objectives is quite short (~0.1 mm), and the large refractive index mismatch between the immersion oil ($n = 1.512$) and the aqueous trapping medium ($n \sim 1.32$) leads to significant spherical aberrations. In practice, this limits the maximum axial range of the optical trap to somewhere between 5 and 20 μm from the coverglass surface of the trapping chamber.[104] Trapping deeper into solution can be achieved with water immersion objectives that minimize spherical aberration[105] and which are available with longer working distances. A high NA objective (typically, 1.2–1.4 NA) is required to produce an intensity gradient sufficient to overcome the scattering force and produce a stable optical trap for microscopic objects, such as polystyrene beads. The vast majority of high NA objectives are complex, multielement optical assemblies specifically designed for imaging visible light, not for focusing an infrared laser beam. For this reason, the optical properties of different objectives can vary widely over the near infrared region (Fig. 4).[9,95] Generally speaking, objectives designed for general fluorescence microscopy display superior transmission over the near infrared compared to most general-purpose objectives, as do infrared-rated objectives specifically produced for use with visible and near infrared light (Table I). Given the wide variation in transmission characteristics for different objectives, an objective being considered for optical trapping should be characterized at the wavelength of the trapping light. Manufacturers rarely supply the transmission characteristics of objectives outside the visible portion of the spectrum. When transmission characteristics in the near infrared are provided, the figures may represent an overestimate, since the throughput of the objective is often measured using an integrating sphere, which

also registers scattered light that is not well focused, and hence does not contribute to trapping. To measure the effective transmission of a high NA objective accurately, the dual objective method is preferred,[9,95,106] in which two identical, matched objectives are used to focus and then recollimate the laser beam (the transmission of a single objective is the square root of the transmission for the objective pair). Furthermore, because the transmission may depend on the degree to which light is bent, the laser beam should be expanded to fill the objective rear aperture. It should be noted that the extremely steep focusing produced by high NA objectives can lead to specular reflection from surfaces at the specimen plane, so simply measuring the throughput of an objective by placing the probe of a power meter directly in front of the objective lens results in an underestimation of its transmission. This approach is not recommended.

E. Position detection

Sensitive position detection lies at the heart of quantitative optical trapping, since nanoscale measurements of both force and displacement rely on a well-calibrated system for determining position. Position tracking of irregularly shaped objects is feasible, but precise position *and force calibration* are currently only practical with spherical objects. For this purpose, microscopic beads are either used alone, or attached to objects of interest as "handles," to apply calibrated forces. The position detection schemes presented here were primarily developed to track microscopic silica or polystyrene beads. However, the same techniques may be applied to track other objects, such as bacterial cells.[107–109]

1. Video based position detection

For simple imaging of a trapped particle, a video camera mounted to the camera port of the microscope (or elsewhere) often suffices. By digitally processing the signal acquired from the camera, and knowing the size subtended by a single pixel (e.g., by calibrating the video picture against a distance standard, such as a ruled objective micrometer), the position of a trapped object can be determined with subpixel accuracy (typically, to within ~5 nm or better), using any of several centroid-finding algorithms.[110–112] Video tracking of trapped objects using such algorithms has been implemented in real time,[113,114] but this approach is restricted to video acquisition rates (typically ~25–120 Hz), and the precision is ultimately limited by video timing jitter (associated with frame

TABLE I. Transmission of microscope objectives, cross-referenced with Fig. 2.

Part No.	Manufacturer	Magnification/ Tube length (mm)/ Numerical aperture	Type designation	Transmission (±5%)			
				830 (nm)	850 (nm)	990 (nm)	1064 (nm)
461832	Zeiss	63/160/1.2 Water	Plan NeoFluar	66	65	64	64
506038	Leica	100/∞/1.4-0.7 Oil	Plan Apo	58	56	54	53
85020	Nikon	60/160/1.4 Oil	Plan Apo	54	51	17	40
93108	Nikon	60/∞/1.4 Oil	Plan Apo CFI	59	54	13	39
93110	Nikon	100/∞/1.4 Oil	Plan Apo CFI	50	47	35	32
93110IR	Nikon	100/∞/1.4 Oil	Plan Apo IR CFI	61	60	59	59
93144	Nikon	100/∞/1.3 Oil	Plan Fluor CFI	67	68	—	61

acquisition) or variations in illumination. In principle, temporal resolution could be improved through the use of high speed video cameras. Burst frame rates in excess of 40 kHz can be achieved with specialized complementary metal oxide semiconductor (CMOS) cameras, for example. However, the usefulness of high speed cameras can be limited by computer speed or memory capacity. Current CPU speed limits real-time position tracking to $\sim$500 Hz,[115] while practical storage considerations limit the number of high-resolution frames that can be stored to $\sim 10^5$, which corresponds to less than 2 min of high-speed video at 1 kHz. Even if these technological hurdles are overcome, high-speed video tracking is ultimately limited by the number of recorded photons (since shorter exposures require more illumination), so spatial resolution decreases as the frame rate increases. Generally speaking, the signal-to-noise ratio is expected to decrease as the square root of the frame rate. The discrepancy between the low video bandwidth ($\sim$100 Hz) and the much higher intrinsic bandwidth of even a relatively weak optical trap ($\sim$kHz) results in aliasing artifacts, and these preclude the implementation of many of the most effective calibration methods. Furthermore, video-based methods are not well suited to the measurement of the *relative* position of an object with respect to the trap center, further complicating force determination.

2. Imaging position detector

Several alternative (nonvideo) methods have been developed that offer precise, high-bandwidth position detection of trapped objects. The simplest of these is to image directly the trapped object onto a quadrant photodiode (QPD).[56,116,117] The diode quadrants are then summed pairwise, and differential signals are derived from the pairs for both x and y dimensions. If desired, the differential signals can be normalized by the sum signal from the four quadrants to reduce the dependence of the output on the total light intensity. Direct imaging of a trapped particle is typically restricted to a small zone within the specimen plane, and requires careful coalignment of the trap with the region viewed by the detector. Moreover, the high magnification required to achieve good spatial resolution results in comparatively low light levels at the QPD, ultimately limiting bandwidth and noise performance.[49,50] The latter limitation has been addressed by the use of a diode laser operating just below its lasing threshold, acting as a superbright, incoherent illumination source.[56] Imaging using laser illumination is considered impractical because of the speckle and interference that arise from coherent illumination over an extended region. Various laser phase-randomization approaches may relieve this restriction, but these typically carry additional disadvantages, most often reduced temporal bandwidth.

3. Laser-based position detection

Laser-based position detection is appealing, because it is possible to use a single laser for both trapping and position detection. Unlike the imaging detector scheme described above, laser-based detection requires the incorporation of a dichroic mirror on the condenser side of the microscope to couple out the laser light scattered by the specimen. Further-more, the detector and its associated optics (lens, filters) must be stably mounted on (or next to) the condenser to collect the output light. Two different laser-based position detection schemes have been developed. The first relies on polarization interferometry.[9,49,50,118,119] This method is quite analogous to differential interference contrast (DIC) microscopy, and it relies on a subset of the DIC imaging components within the microscope. Incoming plane polarized laser light is split by a Wollaston prism into two orthogonal polarizations that are physically displaced from one another. After passing through the specimen plane, the beams are recombined in a second Wollaston prism and the polarization state of the recombined light is measured. A simple polarimeter consists of a quarter wave plate (adjusted so that plane-polarized light is transformed into circularly polarized light) followed by a polarizing beam splitter. The intensity in each branch of the beam splitter is recorded by a photodiode, and the normalized differential diode signal supplies the polarization state of the light. A bead centered in the trap introduces an equal phase delay in both beams, and the recombined light is therefore plane polarized. When the bead is displaced from its equilibrium position, it introduces a relative phase delay between the two beams, leading to a slight elliptical polarization after the beams are recombined. The ellipticity of the recombined light can be calibrated against physical displacement by moving a bead a known distance through the optical trap. This technique is extraordinarily sensitive[118] and is, in theory, independent of the position of the trapped object within the specimen plane, because the trapping and detection laser beams are one and the same, and therefore intrinsically aligned. In practice, however, there is a limited range over which the position signal is truly independent of the trap position. A further limitation of this technique is that it is one dimensional: it is sensitive to displacement along the Wollaston shear axis, providing position detection in a single lateral direction.

A second type of laser-based position detection scheme—back focal plane detection—relies on the interference between forward-scattered light from the bead and unscattered light.[59,64,120–122] The interference signal is monitored with a QPD positioned along the optical axis at a plane conjugate to the back focal plane of the condenser (rather than at an imaging plane conjugate to the specimen). The light pattern impinging on the QPD is then converted to a normalized differential output in both lateral dimensions as described above. By imaging the back focal plane of the condenser, the position signal becomes insensitive to absolute bead position in the specimen plane, and sensitive instead to the relative displacement of the bead from the laser beam axis.[120] As with the polarization interferometer, the detection beam and the optical trap are intrinsically aligned, however the QPD detection scheme can supply position information in both lateral dimensions.

Laser-based position detection schemes have also been implemented with a second, low-power detection laser.[49,50,59,81] The experimental complication of having to combine, spatially overlap, and then separate the trapping and detection beams is frequently outweighed by the advantages conferred by having an independent detection laser.

Uncoupling trapping and detection may become necessary, for example, when there are multiple traps produced in the specimen plane, or if the absolute position of a bead is the relevant measure, rather than the relative position of a bead from the optical trap. When dynamic position control of the optical trap is implemented (see below), a separate detection laser permits rapid position calibration of each trapped particle, and greatly simplifies position measurements in situations in which the trap is being moved.[50] The choice of a laser for position sensing is less constrained than that of a trapping laser, and only a few mW of output power suffice for most detection schemes. The total power should be kept as low as feasible to prevent the detection light from generating significant optical forces itself, thereby perturbing the trap. A detection laser wavelength chosen to match the peak sensitivity of the photodetector will minimize the amount of power required in the specimen plane. Separating the detection and trapping wavelengths facilitates combining and separating the two beams, but increases the constraints on the dichroic mirror that couples the laser beams into the microscope. We have found that combining two beams of similar wavelength is most easily accomplished with a polarizing beamsplitter, i.e., the beams are orthogonally plane polarized and combined in the polarizer before entering the microscope. Since the trapping and detection wavelengths are closely spaced, a single reflection band on the coupling dichroic mirror suffices to couple both beams into and out of the microscope. A holographic notch filter in front of the position detector provides ~6 orders of magnitude of rejection at the trapping wavelength, permitting isolation and measurement of the much less intense detection beam.

4. Axial position detection

The detection schemes described above were developed to measure lateral displacement of objects within the specimen plane, a major focus of most optical trapping work. Detecting axial motion within the optical trap has rarely been implemented and has not been as well characterized until recently. Axial motion has been determined by: measuring the intensity of scattered laser light on an overfilled photodiode;[123–126] through two-photon fluorescence generated by the trapping laser;[127–130] and by evanescent-wave fluorescence at the surface of a coverglass.[131,132] Although these various approaches all supply a signal related to axial position, they require the integration of additional detectors and, in some cases, fluorescence capability, into the optical trapping instrument. This can be somewhat cumbersome, consequently the techniques have not been widely adopted. The axial position of a trapped particle can also be determined from the total laser intensity in the back focal plane of the condenser.[62,64] The axial position signal derives from the interference between light scattered by the trapped particle and the unscattered beam. On passing through a focus, the laser light accumulates a phase shift of π, known as the Gouy phase.[133] The axial phase shift is given by $\psi(z)$ $=\tan^{-1}(z/z_0)$, where z_0 is the Rayleigh range ($z_0 = \pi w_0^2/\lambda$, where w_0 is the beam waist and λ is the wavelength of light), and z is the axial displacement from the focus.[133] Light scattered by a particle located near the focus will preserve the

phase that it acquired prior to being scattered, whereas unscattered light will accumulate the full Gouy phase shift of π. The far-field interference between the scattered and unscattered light gives rise to an axial position-dependent intensity, which can be measured, for example, at the back focal plane of the condenser (see below and Fig. 8). This is the axial counterpart, in fact, of the lateral interference signal described above. Axial position detection can be achieved through a simple variant of quadrant photodiode-based lateral position detection. Recording the total incident intensity on the position detector supplies the axial position of trapped particle relative to the laser focus.[63,64] In contrast to lateral position detection, axial position detection is inversely proportional to the NA of the detector.[62,63] When a single detector is used to measure both lateral and axial position simultaneously, an intermediate detector NA should be used to obtain reasonable sensitivity in all three dimensions.

5. Detector bandwidth limitations

Position detection based on lasers facilitates high bandwidth recording because of the high intensity of light incident on the photodetector. However, the optical absorption of silicon decreases significantly beyond ~850 nm, therefore position sensing by silicon-based photodetectors is intrinsically bandwidth limited in the near infrared.[61,134] Berg-Sørenson and co-workers[134] demonstrated that the electrical response of a typical silicon photodiode to infrared light consists of both a fast and a slow component. The fast component results from optical absorption in the depletion region of the diode, where the optically generated electron hole pairs are rapidly swept to the electrodes. This represents the intended behavior of the diode, and is valid at wavelengths that are readily absorbed by the active material, i.e., $\lambda <$ ~1 μm. At longer wavelengths, however, a slow component also appears in the diode response, due to absorption of light beyond the depletion region. Electron–hole pairs generated in this zone must diffuse into the depletion region before flowing on to the electrodes, a much slower process. Infrared light is poorly absorbed by silicon, resulting in a greater proportion of the incident light being absorbed beyond the depletion region, increasing the relative contribution of the slow component. Thus, the output of the diode effectively becomes lowpass filtered (f_{3dB} ~8–9 kHz at 1064 nm) in an intensity-, wavelength-, and reverse bias-dependent manner.[134] In principle, the effect of this lowpass filtering could be calculated and compensated, but in practice, this approach is complicated by the intensity dependence of the parasitic filtering. One workaround would be to employ a detection laser at a wavelength closer to the absorption maximum of silicon, i.e., shorter than ~850 nm. Two other solutions include using nonsilicon-based detectors employing different photoactive materials, or using silicon-based photodetectors with architectures that minimize the parasitic filtering. Peterman and co-workers measured the wavelength dependence of parasitic filtering in a standard silicon-based detector. They also reported an increased bandwidth at wavelengths up to 1064 nm for an InGaAs diode as well as for a specialized, fully depleted silicon detector.[61] We have found that one commercial position sensitive detector (PSD) (Pa-

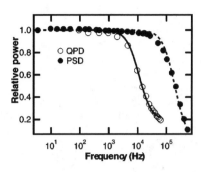

FIG. 5. Comparison of position detector frequency response at 1064 nm. Normalized frequency dependent response for a silicon quadrant photodiode (QPD) (QP50–6SD, Pacific Silicon Sensor) (open circles), and a position sensitive detector (PSD), (DL100–7PCBA, Pacific Silicon Sensor) (solid circles). 1064 nm laser light was modulated with an acousto-optic modulator and the detector output was recorded with a digital sampling scope. The response of the QPD was fit with the function: $\gamma^2+(1-\gamma^2)[1+(f/f_0)^2]^{-1}$, which describes the effects of diffusion of electron-hole pairs created outside the depletion layer (Ref. 134), where γ is the fraction of light absorbed in the diode depletion layer and f_0 is the characteristic frequency associated with light absorbed beyond the depletion layer. The fit returned an f_0 value of 11.1 kHz and a γ parameter of 0.44, which give an effective f_{3dB} of 14.1 kHz, similar to values found in Ref. 134 for silicon detectors. The QPD response was not well fit by a single pole filter response curve. The PSD response, in contrast, was fit by a single pole filter function, returning a rolloff frequency of 196 kHz. Extended frequency response at 1064 nm has also been reported for InGaAs and fully depleted silicon photodiodes (Ref. 61).

cific Silicon Detectors, which supplies output signals similar to those from a QPD, although operating on a different principle), does not suffer from parasitic filtering below ~150 kHz with 1064 nm illumination (Fig. 5).

F. Dynamic position control

Precise, calibrated lateral motion of the optical trap in the specimen plane allows objects to be manipulated and moved relative to the surface of the trapping chamber. More significantly, dynamic computer control over the position and stiffness of the optical trap allows the force on a trapped object to be varied in real time, which has been exploited to generate both force and position clamp measurement conditions.[50,81] Additionally, if the position of the optical trap is scanned at a rate faster than the Brownian relaxation time of a trapped object, multiple traps can be created by time sharing a single laser beam.[49] We consider below the different beam-steering strategies.

1. Scanning mirrors

Traditional galvanometer scanning mirrors benefited from the incorporation of feedback to improve stability and precision. Current commercial systems operate at 1–2 kHz with step response times as short as 100 μs, and with 8 μrad repeatability. The comparatively slow temporal response limits their usefulness for fast-scanning applications, but their low insertion loss and large deflection angles make them a low-cost option for slow-scanning and feedback applications. Recent advances in feedback-stabilized piezoelectric (PZ) systems have resulted in the introduction of PZ scanning mirrors. For the time being, PZ mirrors represent only a slight improvement over galvanometers, with effective op-

eration up to 1 kHz, but just 50 mrad deflection range, and only slightly better resolution and linearity than galvanometers.

2. Acousto-optic deflectors

An acousto-optic deflector (AOD) consists of a transparent crystal inside which an optical diffraction grating is generated by the density changes associated with an acoustic traveling wave of ultrasound. The grating period is given by the wavelength of the acoustic wave in the crystal, and the first-order diffracted light is deflected through an angle that depends on the acoustic frequency through $\Delta\theta=\lambda f/\nu$, where λ is the optical wavelength, and ν and f are the velocity and frequency of the acoustic wave, respectively (ν/f is the ultrasound wavelength). The diffraction efficiency is proportional to depth of the grating, and therefore to the amplitude of the acoustic wave that produced it. AODs are thereby able to control both the trap position (through deflection) and stiffness (through light level). The maximum deflection of an AOD is linearly related to its operating frequency range, and maximum deflections of somewhat over 1° are possible at 1064 nm. AODs are fast: their response times are limited, in principle, by the ratio of the laser spot diameter to the speed of sound within the crystal (~1.5 μs/mm laser diameter for TeO_2 crystals, slightly less for Li_6NbO_3 crystals). In practice, however, the response time of an optical trapping instrument is often limited by other components in the system. A pair of AODs can be combined in an orthogonal configuration to provide both x and y deflections of the optical trap. Due to optical losses in the AODs (an ~80% diffraction efficiency is typical), however, this scheme results in an almost 40% power loss. In addition to mediocre transmission, the diffraction efficiency of an AOD will often vary slightly as a function of its deflection. The resulting position-dependent stiffness variation of the optical trap can either be tolerated (if within acceptable margins for error), calibrated out,[53] or minimized by the selection of a particular range of operating deflections over which the diffraction efficiency is more nearly constant. In practice, however, every AOD needs to be characterized carefully before use for deflection-dependent changes in throughput.

3. Electro-optic deflectors

An electro-optic deflector (EOD) consists of a crystal in which the refractive index can be changed through the application of an external electric field. A gradient in refractive index is established in one plane along the crystal, which deflects the input light through an angle $\theta\propto lV/w^2$, where V is the applied voltage, l is the crystal length, and w is the aperture diameter. Deflections on the order of 20 mrad can be achieved with a switching time as short as 100 ns, sufficient for some optical trapping applications. Despite low insertion loss (~1%) and straightforward alignment, EODs have not been widely employed in optical trapping systems. High cost and a limited deflection range may contribute to this.

G. Piezoelectric stage

Piezoelectric stage technology has been improved dramatically through the introduction of high-precision controllers and sensitive capacitive position sensing. Stable, linear, reproducible, ultrafine positioning in three dimensions is now readily achievable with the latest generation of PZ stages. The traditional problems of hysteresis and drift in PZ devices have been largely eliminated through the use of capacitive position sensors in a feedback loop. With the feedback enabled, an absolute positional uncertainty of 1 nm has been achieved commercially. PZ stages have had an impact on practically every aspect of optical trapping. They can provide an absolute, NIST-traceable displacement measurement, from which all other position calibrations can be derived. Furthermore, these stages permit three-dimensional control of the position of the trap relative to the trapping chamber, which has previously proved difficult or inaccurate.[39] The ability to move precisely in the axial dimension, in particular, permits characterization of the longitudinal properties of the optical trap and can be used to eliminate the creep and backlash typically associated with the mechanical (gear based) focusing mechanism of the microscope. Position and force calibration routines employing the PZ stage are faster, more reproducible, and more precise than previously attainable. Finally, a piezoelectric stage can be incorporated into a force feedback loop[86,135–137] permitting constant-force records of essentially arbitrary displacement, ultimately limited by the stage travel ($\sim$100 μm) rather than the working range of the position detector ($\sim$0.3 μm), the latter being the limiting factor in feedback based on moving the optical trap.[50,59] Stage-based force-feedback permits clamping not only the transverse force, but also the axial force, and hence the polar angle through which the force is applied. Despite these advantages, PZ stages are not without their attendant drawbacks. They are comparatively expensive: a 3D stage with capacitive feedback position sensing plus a digital controller costs roughly $25,000. Furthermore, communication with the stage controller can be slower than for other methods of dynamically controlling trap position (e.g., AODs or EODs), with a maximum rate of $\sim$50 Hz.[59]

H. Environmental isolation

To achieve the greatest possible sensitivity, stability, and signal-to-noise ratio in optical trapping experiments, the environment in which the optical trapping is performed must be carefully controlled. Four environmental factors affect optical trapping measurements: temperature changes, acoustic noise, mechanical vibrations, and air convection. Thermal fluctuations can lead to slow, large-scale drifts in the optical trapping instrument. For typical optical trapping configurations, a 1 K temperature gradient easily leads to micrometers of drift over a time span of minutes. In addition, acoustic noise can shake the optics that couple the laser into the objective, the objective itself, or the detection optics that lie downstream of the objective. Mechanical vibrations typically arise from heavy building equipment, e.g., compressors or pumps operating nearby, or from passing trucks on a roadway. Air currents can induce low-frequency mechanical vi-

brations and also various optical perturbations (e.g., beam deflections from gradients in refractive index produced by density fluctuations in the convected air, or light scattering by airborne dust particles), particularly near optical planes where the laser is focused.

The amount of effort and resources dedicated to reducing ambient sources of noise should be commensurate with the desired precision in the length and time scale of the measurements. Slow thermal drift may not affect a rapid or transient measurement, but could render meaningless the measurement of a slower process. Several methods of reducing noise and drift have been employed in the current generation of optical traps.

The vast majority of optical trapping instruments have been built on top of passive air tables that offer mechanical isolation (typically, -20 dB) at frequencies above $\sim$2–10 Hz. For rejection of lower frequencies, actively servoed air tables are now commercially available, although we are not yet aware of their use in this field. Acoustic noise isolation can be achieved by ensuring that all optical mounts are mechanically rigid, and placing these as close to the optical table as feasible, thereby reducing resonance and vibration. Enclosing all the free-space optics will further improve both mechanical and optical stability by reducing ambient air currents. Thermal effects and both acoustical and mechanical vibration can be reduced by isolating the optical trapping instrument from noisy power supplies and heat sources. Diode pumped solid state lasers are well suited to this approach: since the laser head is fiber coupled to the pump diodes, the power supply can be situated outside of the experimental room. A similar isolation approach can be pursued with noisy computers or power supplies, and even illumination sources, whose outputs can be brought to the instrument via an optical fiber. Further improvements in noise performance and stability may require more substantial modifications, such as acoustically isolated and temperature controlled experimental rooms situated in low-vibration areas. The current generation of optical trapping instruments in our lab[59,138] are housed in acoustically quiet cleanrooms with background noise less than the NC30 (OSHA) rating, a noise level roughly equivalent to a quiet bedroom. In addition, these rooms are temperature stabilized to better than $\pm$0.5 K. The stability and noise suppression afforded by this arrangement has paved the way for high-resolution recording of molecular motor movement, down to the subnanometer level.[85–87]

IV. CALIBRATION

A. Position calibration

Accurate position calibration lies at the heart of quantitative optical trapping. Precise determination of the displacement of a trapped object from its equilibrium position is required to compute the applied force ($F = -\alpha x$, where F is the force, α is the optical trap stiffness, and x is the displacement from the equilibrium trapping position), and permits direct measurement of nanometer-scale motion. Several methods of calibrating the response of a position detector have been developed. The choice of method will depend on

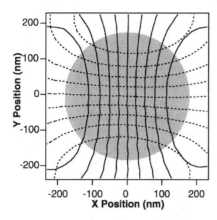

FIG. 6. Lateral two-dimensional detector calibration (adapted from Ref. 59). Contour plot of the x (solid lines) and y (dashed lines) detector response as a function of position for a 0.6 μm polystyrene bead raster scanned through the detector laser focus by deflecting the trapping laser with acousto-optic deflectors. The bead is moved in 20 nm steps with a dwell time of 50 ms per point while the position signals are recorded at 50 kHz and averaged over the dwell time at each point. The x contour lines are spaced at 2 V intervals, from 8 V (leftmost contour) to -8 V (rightmost contour). The y contour lines are spaced at 2 V intervals, from 8 V (bottom contour) to -8 V (top contour). The detector response surfaces in both the x and y dimensions are fit to fifth order two-dimensional polynomials over the shaded region, with less than 2 nm residual root mean square (rms) error. Measurements are confined to the shaded region, where the detector response is single valued.

the position detection scheme, the ability to move the trap and/or the stage, the desired accuracy, and the expected direction and magnitude of motion in the optical trap during an experiment. The most straightforward position calibration method relies on moving a bead through a known displacement across the detector region while simultaneously recording the output signal. This operation can be performed either with a stuck bead moved by a calibrated displacement of the stage, or with a trapped bead moved with a calibrated displacement of a steerable trap.

Position determination using a movable trap relies on initial calibration of the motion of the trap itself in the specimen plane against beam deflection, using AODs or deflecting mirrors. This is readily achieved by video tracking a trapped bead as the beam is moved.[49] Video tracking records can be converted to absolute distance by calibrating the charge coupled device (CCD) camera pixels with a ruled stage micrometer (10 μm divisions or finer),[49,50] or by video tracking the motion of a stuck bead with a fully calibrated piezoelectric stage.[59] Once the relationship between beam deflection and trap position is established, the detector can then be calibrated in one or both lateral dimensions by simply moving a trapped object through the detector active area and recording the position signal.[50,59,81] Adequate two-dimensional calibration may often be obtained by moving the bead along two orthogonal axes in an "X" pattern. However, a more complete calibration requires raster scanning the trapped bead to cover the entire active region of the sensor.[59] Figure 6 displays the two-dimensional detector calibration for a 0.6 μm bead, raster scanned over the detector region using an AOD-driven optical trap. A movable optical trap is typically used with either an imaging position detector, or a second low-power laser for laser-based detection (described above). Cali-

brating by moving the trap, however, offers several advantages. Position calibration can be performed individually for each object trapped, which eliminates errors arising from differences among nominally identical particles, such as uniform polystyrene beads, which may exhibit up to a 5% coefficient of variation in diameter. Furthermore, nonspherical or nonidentical objects, such as bacteria or irregularly shaped particles, can be calibrated on an individual basis prior to (or after) an experimental measurement. Because the object is trapped when it is calibrated, the calibration and detection necessarily take place in the same axial plane, which precludes calibration errors arising from the slight axial dependence of the lateral position signals.

Laser-based detection used in conjunction with a movable trap affords additional advantages. Because the trapping and detection lasers are separate, the focal position of the two can be moved relative to one another in the axial dimension. The maximum lateral sensitivity and minimum variation of lateral sensitivity with axial position occur at the focus of the detection laser. The axial equilibrium position of a trapped object, however, lies above the focus due to the scattering force. Since the detection and trapping lasers are uncoupled, the focus of the detection laser can be made coincident with the axial position of the trapped object, thereby maximizing the detector sensitivity while minimizing the axial dependence of the lateral sensitivity.[59] An additional benefit to using an independent detection laser is that it can be more weakly focused to a larger spot size, since it does not need to trap, thereby increasing the usable detection range. Beyond the added complication and cost of building a movable trap, calibrating with a movable trap has some important limitations. The calibration is limited to the two lateral dimensions, which may be inadequate for experiments where the trapped bead is displaced significantly in the axial dimension.[39,82] Due to the ~4–6-fold lower trap stiffness in the axial dimension, a primarily lateral force pulling an object out of the trapping zone may result in a significant axial displacement. In practice, this situation arises when the trapped object is tethered to the surface of the trapping chamber, e.g., when a bead is attached by a strand of DNA bound at its distal end to the coverglass.[39,82,88,135–137] Accurate determinations of displacement and trapping force in such experiments require axial, as well as lateral, position calibration.

Position calibration is most commonly accomplished by moving a bead fixed to the surface through the detection region and recording the detector output as a function of position. Traditionally, such calibrations were performed in one or two lateral dimensions. The advent of servo-stabilized, 3D piezoelectric (pz) stages has made such calibrations more accurate, easier to perform and—in conjunction with an improved theoretical understanding of the axial position signal—has permitted a full 3D position calibration of an optical trap.[62–64] Whereas full 3D calibration is useful for tracking the complete motion of an object, it is cumbersome and unnecessary when applying forces within a plane defined by one lateral direction and the optical axis. When the trapped object is tethered to the surface of the trapping chamber, for example, it is sufficient to calibrate the axial

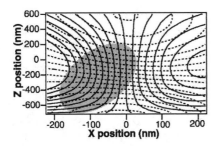

FIG. 7. Axial two-dimensional detector calibration. Contour plot of the lateral (solid lines) and axial (dashed lines) detector response as a function of x (lateral displacement) and z (axial displacement) of a stuck 0.5 μm polystyrene bead moving through the laser focus. A stuck bead was raster scanned in 20 nm steps in x and z. The detector signals were recorded at 4 kHz and averaged over 100 ms at each point. The lateral contour lines are spaced at 1 V intervals, from -9 V (leftmost contour) to 7 V (rightmost contour). The axial contour lines are spaced at 0.02 intervals (normalized units). Measurements are confined to the region of the calibration shaded in gray, over which the surfaces of x and z positions as a function of lateral and axial detector signals were fit to seventh order two-dimensional polynomial functions with less than 5 nm residual rms error.

and the single lateral dimension in which the force is applied. Figure 7 displays the results of such a two-dimensional ("x–z") position calibration for a 0.5 μm bead stuck to the surface of the trapping chamber. The bead was stepped through a raster scan pattern in x (lateral dimension) and z (axial dimension) while the position signals were recorded. Using a stuck bead to calibrate the position detector has some limitations and potential pitfalls. Because it is difficult, in general, to completely immobilize an initially trapped particle on the surface, it is not feasible to calibrate every particle. Instead, an average calibration derived from an ensemble of stuck beads must be measured. Furthermore, the stuck-bead calibration technique precludes calibrating nonspherical or heterogeneous objects, unless these can be attached to the surface (and stereospecifically so) prior to, or after, the experimental measurements. Due to the axial dependence of the lateral position signals ("x–z crosstalk"), using a stuck bead to calibrate only the lateral dimension is prone to systematic error. Without axial position information, it is difficult to precisely match the axial position of a stuck bead with the axial position of a trapped bead. Optically focusing on a bead cannot be accomplished with an accuracy better than $\sim$100 nm, which introduces uncertainty and error in lateral position calibrations for which the axial position is set by focusing. Therefore, even when only the lateral dimensions are being calibrated, it is useful to measure the axial position signal to ensure that the calibration is carried out in the appropriate axial plane.

1. Absolute axial position and measurement of the focal shift

The absolute axial position of a trapped object above the surface of the trapping chamber is an important experimental parameter, because the hydrodynamic drag on an object varies nonlinearly with its height above the surface, due to proximal wall effects (see below and Ref. 9). Absolute axial position measurements may be especially important in situations where the system under investigation is attached to the

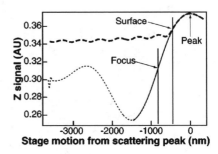

FIG. 8. Axial position signals for a free (heavy dashed line) and stuck (light dashed line) bead as the stage was scanned in the axial direction. All stage motion is relative to the scattering peak, which is indicated on the right of the figure. The positions of the surface (measured) and the focus [calculated from Eq. (5)] are indicated by vertical lines. The axial detection fit [Eq. (5)] to the stuck bead trace is shown in the region around the focus as a heavy solid line.

surface and to a trapped object, as is often the case in biological applications. Force–extension relationships, for example, depend on the end-to-end extension of the molecule, which can only be determined accurately when the axial position of the trapped object with respect to the surface is known. Axial positioning of a trapped object depends on finding the location of the surface of the chamber and moving the object relative to this surface by a known amount. The problem is complicated by the focal shift that arises when focusing through a planar interface between two mismatched indices of refraction e.g., between the coverglass ($n_{glass} \sim 1.5$) and the aqueous medium ($n_{water} \sim 1.3$).[139–144] This shift introduces a fixed scaling factor between a vertical motion of the chamber surface and the axial position of the optical trap within the trapping chamber. The focal shift is easily computed from Snell's law for the case of paraxial rays, but it is neither straightforward to compute nor to measure experimentally when high NA objectives are involved.[144] Absolute axial position determination has previously been assessed using fluorescence induced by an evanescent wave,[131] by the analysis of interference or diffraction patterns captured with video,[113,145] or through the change in hydrodynamic drag on a trapped particle as it approaches the surface.[39] These techniques suffer from the limited range of detectable motion for fluorescence-based methods, and by the slow temporal response of video and drag-force-based measurements.

The position detector sum signal (QPD or PSD output), which is proportional to the total incident intensity at the back focal plane of the condenser, provides a convenient means of both accurately locating the surface of the trapping chamber and measuring the focal shift. In conjunction, these measurements permit absolute positioning of a trapped object with respect to the trapping chamber surface. The detector sum signal as a function of axial stage position for both a stuck bead and a trapped bead are shown in Fig. 8. The stuck bead trace represents the axial position signal of a bead moving relative to the trap. As the bead moves through the focus of the laser (marked on the figure), the phase of light scattered from the bead changes by 180° relative to the unscattered light, modulating the intensity distribution at the back focal plane of the condenser. The region between the extrema of the stuck-bead curve is well described by the expression for axial sensitivity derived by Pralle and co-workers:[62]

$$\frac{I_z}{I}(z) \propto \left(1 + \left(\frac{z}{z_0}\right)^2\right)^{1/2} \sin[\tan^{-1}(z/z_0)], \qquad (5)$$

where an overall scaling factor has been ignored, z is the axial displacement from the beam waist, and $z_0 = \pi w_0^2 / \lambda$ is the Raleigh length of the focus, with beam waist w_0 at wavelength λ. The phase difference in the scattered light is described by the arctangent term, while the prefactor describes the axial position dependent intensity of the scattered light. The fit returns a value for the beam waist, $w_0 = 0.436 \ \mu$m. The equilibrium axial position of a trapped bead corresponds to a displacement of 0.379 μm from the laser focus. A stuck bead scan can also be useful for determining when a free bead is forced onto the surface of the cover slip.

As a trapped bead is forced into contact with the surface of the chamber by the upward stage motion, the free and stuck bead signals merge and eventually become indistinguishable (Fig. 8). The approximate location of the surface with respect to the position of a trapped bead can be determined by finding the point at which both curves coincide. Brownian motion of the trapped bead, however, will shift this point slightly, in a stiffness-dependent manner that will introduce a small uncertainty in the measured position of the surface. The scattering peak in Fig. 8, however, serves as an easily identifiable fiducial reference from which the trapped bead can be moved an absolute distance by subsequent stage motion. In this manner, trapped particles can be reproducibly positioned at a fixed (but uncertain) distance relative to the surface. In order to obtain a precise location of the trapped particle above the surface, both the position of the scattering peak with respect to the surface and the focal shift must be determined. This may be accomplished, for example, by a one-time measurement of the drag on a trapped bead at a series of positions above the scattering peak. The interaction of a sphere with the boundary layer of water near a surface leads to an increase in the hydrodynamic drag β, which can be estimated by Faxen's law for the approximate drag on a sphere near a surface:[9]

$$\beta = \frac{6\pi\eta a}{1 - \frac{9}{16}\left(\frac{a}{h}\right) + \frac{1}{8}\left(\frac{a}{h}\right)^3 - \frac{45}{256}\left(\frac{a}{h}\right)^4 - \frac{1}{16}\left(\frac{a}{h}\right)^5}, \qquad (6)$$

which depends only on the bead radius a, the distance above the surface h, and the viscosity of the liquid η. By measuring the rolloff frequency or the displacement of the trapped bead as the stage is oscillated (see below), the drag force can be determined at different axial stage positions relative to the scattering peak and normalized to the calculated asymptotic value, the Stokes drag coefficient, $6\pi\eta a$. The resulting curve (Fig. 9) is described by a two parameter fit to Eq. (6): a scaling parameter that represents the fractional focal shift and an offset parameter related to the distance between the scattering peak and the coverglass surface. The fit parameters from the curve in Fig. 9 allow absolute positioning of a trapped particle with respect to the surface. The uncertainty in the axial position amounts to roughly 3% of the bead-surface separation, with the residual uncertainty largely due

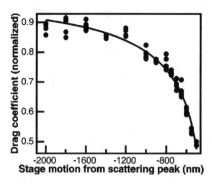

FIG. 9. Normalized drag coefficient (β_0/β, where β_0 is the Stokes drag on the sphere: $6\pi\eta a$) as a function of distance from the scattering peak. The normalized inverse drag coefficient (solid circles) was determined through rolloff measurements and from the displacement of a trapped bead as the stage was oscillated. The normalized inverse drag coefficient was fit to Faxen's law [Eq. (6)] with a height offset ε and scaling parameter δ, which is the fractional focal shift, as the only free parameters: $\beta_0/\beta = 1 - (9/16) \times [a\delta^{-1}(z-\varepsilon)^{-1}] + \frac{1}{8}[a\delta^{-1}(z-\varepsilon)^{-1}]^3 - (45/256)[a\delta^{-1}(z-\varepsilon)^{-1}]^4 - (1/16)[a\delta^{-1}(z-\varepsilon)^{-1}]^5$, where a is the bead radius, z is the motion of the stage relative to the scattering peak, β_0 is the Stoke's drag on the bead, $(6\pi\eta a)$, and β is the measured drag coefficient. The fit returned a fractional focal shift δ of 0.82 ± 0.02 and an offset ε of 161 nm. The position of the surface relative to the scattering peak is obtained by setting the position of the bead center, $\delta(z-\varepsilon)$ equal to the bead radius a, which returns a stage position of 466 nm above the scattering peak, as indicated in Fig. 8.

to the estimate of the focal shift (which leads to a relative rather than an absolute uncertainty). The position of the surface, calculated from the fit parameters of Fig. 9, is indicated in Fig. 8. The focal shift was 0.82 ± 0.02, i.e., the vertical location of the laser focus changed by 82% of the vertical stage motion.

The periodic modulation of the axial position signal as a trapped bead is displaced from the surface (Fig. 8) can be understood in terms of an étalon picture.[146] Backscattered light from the trapped bead reflects from the surface and interferes with forward-scattered and unscattered light in the back focal plane of the condenser. The phase difference between these two fields includes a constant term that arises because of the Gouy phase and another term that depends on the separation between the bead and the surface. The spatial frequency of the intensity modulation is given by $d = \lambda/(2n_m)$, where d is the separation between the bead and the cover slip, λ is the vacuum wavelength of the laser, and n_m is the index of refraction of the medium. This interference signal supplies a second and much more sensitive means to determine the focal shift. The motion of the stage (d_s) and motion of the focus (d_f) are related through a scaling parameter f_s equal to the focal shift $d_f = f_s d_s$. The interference signal is observed experimentally by stage translations. The measured spatial frequency will be given by $d_s = \lambda/(2n_m f_s)$, which can be rearranged to solve for the focal shift $f_s = \lambda/(2n_m d_s)$. The focal shift determined in this manner was 0.799 ± 0.002, which is within the uncertainty of the focal shift determined by hydrodynamic drag measurements (Fig. 9). The true focal shift with a high NA lens is more pronounced than the focal shift computed in the simple paraxial limit, given (from Snell's law) by the ratio of the indices of refraction: $n_m/n_{imm} = 0.878$ for the experimental conditions,

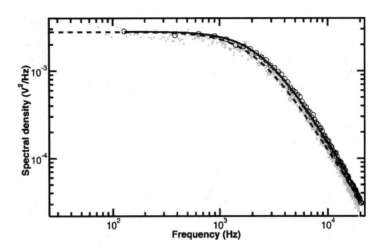

FIG. 10. Power spectrum of a trapped bead. Power spectrum of a 0.5 μm polystyrene bead trapped 1.2 μm above the surface of the trapping chamber recorded with a PSD (gray dots). The raw power spectrum was averaged over 256 Hz windows on the frequency axis (black circles) and fit (black line) to a Lorentzian [Eq. (7)] corrected for the effects of the antialiasing filter, frequency dependent hydrodynamic effects, and finite sampling frequency, as described by Berg–Sørensen and Flyvbjerg (Ref. 148). The rolloff frequency is 2.43 kHz, corresponding to a stiffness of 0.08 pN/nm. For comparison the raw power spectrum was fit to an uncorrected Lorentzian (dashed line), which returns a rolloff frequency of 2.17 kHz. Whereas the discrepancies are on the order 10% for a relatively weak trap, they generally become more important at higher rolloff frequencies.

where n_m is the index of the aqueous medium (1.33) and n_{imm} is the index of the objective immersion oil (1.515). The discrepancy should not be surprising, because the paraxial ray approximation does not hold for the objectives used for optical trapping.[146] The interference method employed to measure the focal shift is both easier and more accurate than the drag-force method presented earlier.

2. Position calibration based on thermal motion

A simple method of calibrating the position detector relies on the thermal motion of a bead of known size in the optical trap.[122] The one-sided power spectrum for a trapped bead is[9]

$$S_{xx}(f) = \frac{k_B T}{\pi^2 \beta (f_0^2 + f^2)}, \qquad (7)$$

where $S_{xx}(f)$ is in units of displacement^2/Hz, k_B is Boltzmann's constant, T the absolute temperature, β is the hydrodynamic drag coefficient of the object (e.g., $\beta = 6\pi\eta a$ for Stokes drag on a sphere of radius a in a medium with viscosity η), and f_0 is the rolloff frequency, related to the trap stiffness through $f_0 = \alpha(2\pi\beta)^{-1}$ for a stiffness α (see below). The detector, however, measures the uncalibrated power spectrum $S_{vv}(f)$, which is related to the true power spectrum by $S_{vv}(f) = \rho^2 \cdot S_{xx}(f)$, where ρ represents the linear sensitivity of the detector (in volts/unit distance). The sensitivity can be found by considering the product of the power spectrum and the frequency squared $S_{xx}(f) \cdot f^2$, which asymptotically approaches the limit $k_B T(\pi^2\beta)^{-1}$ for $f \gg f_0$. Inserting the relationship between the displacement power spectrum and the uncalibrated detector spectrum in this expression and rearranging gives

$$\rho = [S_{vv}(f)\pi^2\beta/k_B T]^{1/2}. \qquad (8)$$

This calibration method has been shown to agree to within ~20% of the sensitivity measured by more direct means, such as those discussed above.[122] An advantage to the

method is that it does not require any means of precisely moving a bead to calibrate the optical trap. However, the calibration obtained by this method is valid only for small displacements, for which a linear approximation to the position signal is valid. In addition, the system detection bandwidth must be adequate to record accurately the complete power spectrum without distortion, particularly in the high frequency regime. System bandwidth considerations are treated more fully in conjunction with stiffness determination, discussed below.

B. Force calibration–stiffness determination

Forces in optical traps are rarely measured directly. Instead, the stiffness of the trap is first determined, then used in conjunction with the measured displacement from the equilibrium trap position to supply the force on an object through Hooke's law: $F = -\alpha x$, where F is the applied force, α is the stiffness, and x is the displacement. Force calibration is thus reduced to calibrating the trap stiffness and separately measuring the relative displacement of a trapped object. A number of different methods of measuring trap stiffness, each with its attendant strengths and drawbacks, have been implemented. We discuss several of these.

1. Power spectrum

When beads of known radius are trapped, the physics of Brownian motion in a harmonic potential can be exploited to find the stiffness of the optical trap. The one-sided power spectrum for the thermal fluctuations of a trapped object is given by Eq. (7), which describes a Lorentzian. This power spectrum can be fit with an overall scaling factor and a rolloff frequency, $f_0 = \alpha(2\pi\beta)^{-1}$ from which the trap stiffness (α) can be calculated if the drag (β) on the particle is known (Fig. 10). For a free sphere of radius a in solution far from any surfaces, the drag is given by the usual Stokes relation $\beta = 6\pi\eta a$, where η is the viscosity of the medium. For a bead trapped nearer the surface of the trapping chamber, additional drag arises from wall effects and must be considered: Faxen's law [Eq. (6)] is appropriate for estimating the drag due to lateral motion. Axial stiffness is also measured via the power spectrum of the axial position signal, but the corrections to the axial drag due to wall effects are larger than those for the lateral drag. The drag on a sphere moving normal to a surface is[147]

$$\beta = \beta_0 \frac{4}{3} \sinh \alpha \sum_{n=1}^{\infty} \frac{n(n+1)}{(2n-1)(2n+3)}$$

$$\times \left[\frac{2 \sinh(2n+1)\alpha + (2n+1)\sinh 2\alpha}{4\sinh^2\left(n+\frac{1}{2}\right)\alpha - (2n+1)^2\sinh^2\alpha} - 1 \right], \quad (9)$$

where

$$\alpha = \cosh^{-1}\left(\frac{h}{a}\right) = \ln\left\{ \frac{h}{a} + \left[\left(\frac{h}{a}\right)^2 - 1 \right]^{1/2} \right\},$$

h is the height of the center of the sphere above the surface, and $\beta_0 = 6\pi\eta a$ is the Stokes drag. The sum converges fairly quickly and ~ 10 terms are required to achieve accurate results. Whereas it is tempting to measure trap stiffness well away from surfaces to minimize hydrodynamic effects, spherical aberrations in the focused light will tend to degrade the optical trap deeper in solution, particularly in the axial dimension. Spherical aberrations lead to both a reduction in peak intensity and a smearing-out of the focal light distribution in the axial dimension.

Determining the stiffness of the optical trap by the power spectrum method requires a detector system with sufficient bandwidth to record faithfully the power spectrum well beyond the rolloff frequency (typically, by more than 1 order of magnitude). Lowpass filtering of the detector output signal, even at frequencies beyond the apparent rolloff leads directly to a numerical underestimate of the rolloff frequency and thereby to the stiffness of the optical trap. Errors introduced by low pass filtering become more severe as the rolloff frequency of the trap approaches the rolloff frequency of the electrical filter. Since the trap stiffness is determined solely from the rolloff of the Lorentzian power spectrum, this method is independent of the position calibration, per se. In addition to determining the stiffness, the power spectrum of a trapped bead serves as a powerful diagnostic tool for optical trapping instruments: alignment errors of either the optical trap or the position detection system lead to non-Lorentzian power spectra, which are easily scored, and extraneous sources of instrument noise can generate additional peaks in the power spectrum.

The measurement and accurate fitting of power spectra to characterize trap stiffness was recently investigated by Berg-Sørensen and Flyvbjerg,[148,149] who developed an improved expression for the power spectrum that incorporates several previously ignored corrections, including the frequency dependence of the drag on the sphere, based on an extension of Faxen's law for an oscillating sphere [Faxen's law, Eq. (6), only holds strictly in the limit of *constant* velocity]. These extra terms encapsulate the relevant physics for a sphere moving in a harmonic potential with viscous damping. In addition to this correction, the effects of finite sampling frequency and signal filtering during data acquisition (due to electronic filters or parasitic filtering by the photosensor) were included in fitting the experimental power spectrum. The resulting fits determine the trap stiffness with an uncertainty of $\sim 1\%$ and accurately describe the shape of the measured spectra. This work underscores the importance of characterizing and correcting the frequency response of the position detection system to obtain accurate stiffness measurements. Figure 10 illustrates a comparison between the fit obtained with the improved fitting routine and an uncorrected fit.

The power spectrum of a trapped bead can also be used to monitor the sample heating due to partial absorption of the trapping laser light. Heating of the trapping medium explicitly changes the thermal kinetic energy term ($k_B T$) in the power spectrum [Eq. (7)] and implicitly changes the drag term as well, $\beta = 6\pi\eta(T)a$, through its dependence on viscosity, which is highly temperature dependent. Peterman and co-workers were able to assess the temperature increase as a function of trapping laser power by determining the dependence of the Lorentzian fit parameters on laser power.[100]

2. Equipartition

The thermal fluctuations of a trapped object can also be used to obtain the trap stiffness through the Equipartition theorem. For an object in a harmonic potential with stiffness α:

$$\frac{1}{2}k_B T = \frac{1}{2}\alpha\langle x^2 \rangle, \quad (10)$$

where k_B is Boltzmann's constant, T is absolute temperature, and x is the displacement of the particle from its trapped equilibrium position. Thus, by measuring the positional variance of a trapped object, the stiffness can be determined. The variance $\langle x^2 \rangle$ is intimately connected to the power spectrum, of course: it equals the integral of the position power spectrum, i.e., the spectrum recorded by a calibrated detector. Besides its simplicity, a primary advantage of the Equipartition method is that it does not depend explicitly on the viscous drag of the trapped particle. Thus, the shape of the particle, its height above the surface, and the viscosity of the medium need not be known to measure the trap stiffness (although, in fairness, both the particle shape and the optical properties of the medium will influence the position calibration itself). The bandwidth requirements of the position detection system are the same as for the power spectral approach, with the additional requirement that the detector must be calibrated. Unlike the power spectral method however, the variance method does not provide additional information about the optical trap or the detection system. For this reason, care should be taken when measuring the stiffness with the Equipartition method. Because variance is an intrinsically *biased estimator* (it is derived from the square of a quantity, and is therefore always positive), any added noise and drift in position measurements serve only to increase the overall variance, thereby decreasing the apparent stiffness estimate. In contrast, low pass filtering of the position signal results in a lower variance and an apparent increase in stiffness.

3. Optical potential analysis

A straightforward extension of the Equipartition method involves determining the complete distribution of particle positions visited due to thermal motions, rather than simply the variance of that distribution. The probability for the displacement of a trapped object in a potential well will be given by a Boltzmann distribution

$$P(x) \propto \exp\left(\frac{-U(x)}{k_B T}\right) = \exp\left(\frac{-\alpha x^2}{2k_B T}\right), \qquad (11)$$

where $U(x)$ is the potential energy and $k_B T$ is the thermal energy. When the potential is harmonic, this distribution is a simple Gaussian parametrized by the trap stiffness α. When the potential is anharmonic, the position histogram can be used, in principle, to characterize the shape of the trapping potential by taking the logarithm and solving for $U(x)$. In practice, this approach is not especially useful without a considerable body of low-noise/low-drift position data, since the wings of the position histogram—which carry the most revealing information about the potential—hold the fewest counts and therefore have the highest relative uncertainty.

4. Drag force method

The most direct method of determining trap stiffness is to measure the displacement of a trapped bead from its equilibrium position in response to viscous forces produced by the medium, generated by moving the stage in a regular triangle wave or sinusoidal pattern. Since forces arise from the hydrodynamics of the trapped object, the drag coefficient, including any surface proximity corrections, must be known. For the case of a sinusoidal driving force of amplitude A_0 and frequency f, the motion of the bead is

$$x(t) = \frac{A_0 f}{\sqrt{f_0^2 + f^2}} \exp[-i(2\pi f t - \varphi)],$$

$$\varphi = -\tan^{-1}(f_0/f), \qquad (12)$$

where f_0 is the characteristic rolloff frequency (above), and φ is the phase delay. Both the amplitude and the phase of the bead motion can be used to provide a measure of trap stiffness.

A triangular driving force of amplitude A_0 and frequency f results in a square wave of force being applied to the bead. For each period of the motion the bead trajectory is

$$x(t) = \frac{\beta A_0 f}{2\alpha}\left[1 - \exp\left(-\frac{\alpha}{\beta}t\right)\right], \qquad (13)$$

where α is the trap stiffness and β is the drag coefficient of the bead, including Faxen's law corrections. Due to the finite response time of the stage, the exponential damping term is convolved with the response time of the stage. Therefore, only the asymptotic value ($\beta A_0 f/2\alpha$) should be used to obtain a reliable estimate of trap stiffness. Drag-force measurements are slow compared with the thermal motion of the particle, so the bandwidth requirements of the detection system are significantly relaxed. Increasing the amplitude or the frequency of the stage motion generates larger displacements of the trapped bead. By measuring the stiffness as a function of bead displacement, the linear region of the trap over

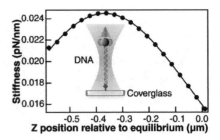

FIG. 11. Axial dependence of lateral stiffness. The experimental geometry for these measurements is depicted in the inset. A polystyrene bead is tethered to the surface of the cover glass through a long DNA tether. The stage was moved in the negative z direction (axial), which pulls the bead towards the laser focus, and the lateral stiffness was determined by measuring the lateral variance of the bead. The data (solid circles) are fit with the expression for a simple dipole [Eq. (14)], with the power in the specimen plane, the beam waist, and an axial offset as free parameters.

which the stiffness is constant can be easily determined.

A variation on the drag force method of stiffness calibration, sometimes called step response calibration, involves rapidly displacing the trap by a small, fixed offset and recording the subsequent trajectory of the bead. The bead will return to its equilibrium position in an exponentially damped manner, with a time constant of α/β as in Eq. (13).

5. Direct measurement of optical force

The lateral trapping force arises from the momentum transfer from the incident laser light to the trapped object, which leads to a change in the direction of the scattered light (Fig. 1). Measuring the deflection of the scattered laser beam with a QPD or other position sensitive detector therefore permits direct measurement of the momentum transfer, and hence the force, applied to the trapped object—assuming that all the scattered light can be collected.[38,57,104] An expression relating the applied force to the beam deflection was presented by Smith *et al.*:[38] $F = I/c \cdot (NA) \cdot X/R_{ba}$ where F is the force, I is the intensity of the laser beam, c is the speed of light, NA is the numerical aperture, X is the deflection of the light, and R_{ba} is the radius of the back aperture of the microscope objective. In principle, this approach is applicable to any optical trapping configuration. However, because it necessitates measuring the total intensity of scattered light, it has only been implemented for relatively low NA, counter-propagating optical traps, where the microscope objective entrance pupils are underfilled. In single-beam optical traps, it is impractical to collect the entirety of the scattered light, owing to the higher objective NA combined with an optical design that overfills the objective entrance pupil.

6. Axial dependence of lateral stiffness

Three-dimensional position detection facilitates measurement of the axial stiffness and mapping of the lateral stiffness as a function of axial position in the trap. Due to the high refractive index of polystyrene beads typically used in optical trapping studies, there is a correspondingly large scattering force in the axial direction. Consequently, the axial equilibrium position of a trapped polystyrene bead tends to lie well beyond the focus, where the lateral intensity gradient—and hence the lateral stiffness—are significantly

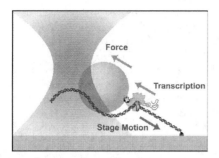

FIG. 12. Cartoon of the experimental geometry (not to scale) for single-molecule transcription experiment. Transcribing RNA polymerase with nascent RNA (gray strand) is attached to a polystyrene bead. The upstream end of the duplex DNA (black strands) is attached to the surface of a flowchamber mounted on a piezoelectric stage. The bead is held in the optical trap at a predetermined position from the trap center, which results in a restoring force exerted on the bead. During transcription, the position of the bead in the optical trap and hence the applied force is maintained by moving the stage both horizontally and vertically to compensate for motion of the polymerase molecule along the DNA (adapted from Ref. 87).

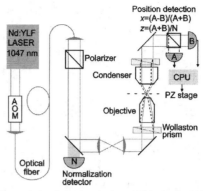

FIG. 13. The optical trapping interferometer. Light from a Nd:YLF laser passes through an acoustic optical modulator (AOM), used to adjust the intensity, and is then coupled into a single-mode polarization-maintaining optical fiber. Output from the fiber passes through a polarizer to ensure a single polarization, through a 1:1 telescope and into the microscope where it passes through the Wollaston prism and is focused in the specimen plane. The scattered and unscattered light is collected by the condenser, is recombined in the second Wollaston prism, then the two polarizations are split in a polarizing beamsplitter and detected by photodiodes A and B. The bleedthrough on a turning mirror is measured by a photodiode (N) to record the instantaneous intensity of the laser. The signals from the detector photodiodes and the normalization diode are digitized and saved to disk. The normalized difference between the two detectors (A and B) gives the lateral, x displacement, while the sum signal (A+B) normalized by the total intensity (N) gives the axial, z displacement.

reduced from their values at the focus. In experiments in which beads are displaced from the axial equilibrium position, the change in lateral trapping strength can be significant. The variation of lateral stiffness as a function of axial position was explored using beads tethered by DNA (1.6 μm) to the surface of the flow chamber (Fig. 11, inset). Tethered beads were trapped and the attachment point of the tether was determined and centered on the optical axis.[39] The bead was then pulled vertically through the trap, i.e., along the axial dimension, by lowering the stage in 20 nm increments. At each position, the lateral stiffness of the trap was ascertained by recording its variance, using the Equipartition method. The axial force applied to the bead tether can increase the apparent lateral stiffness, and this effect can be computed by treating the tethered bead as a simple inverted pendulum.[150,151] In practice, the measured increase in lateral stiffness (given by $\alpha_x = F_a/l$, where α_x is the lateral stiffness, F_a is the axial force on the bead, and l is the length of the tether) resulted in less than a 3% correction to the stiffness and was thereafter ignored in the analysis. An average of 12 measurements is shown in Fig. 11, along with a fit to the lateral stiffness based on a simple dipole and zero-order Gaussian beam model.[152]

$$\alpha_x(z) = \frac{8n_m p}{c w_0}\left(\frac{a}{w_0}\right)^3\left(\frac{m^2-1}{m^2+2}\right)\left(1+\left(\frac{z}{z_0}\right)^2\right)^{-2}, \qquad (14)$$

where n_m is the index of refraction of the medium, p is the laser power in the specimen plane, c is the speed of light, m is the ratio of the indices of refraction of the bead and the medium, and w_0, z and z_0 are the beam diameter at the waist, the axial displacement of the particle relative to the focus, and the Raleigh range, respectively (as previously defined). The data are well fit by this model with the exception of the laser power, which was sixfold lower than the actual power estimated in the specimen plane. A significant discrepancy was anticipated since it had been previously shown that for particle sizes on the order of the beam waist, the dipole approximation greatly overestimates the trap stiffness.[152] The other two parameters of interest are the beam waist and the

equilibrium axial position of the bead in the trap. The fit returned distances of 0.433 μm for the beam waist and 0.368 μm for the offset of the bead center from the focal point. These values compare well with the values determined from the fit to the axial position signal, which were 0.436 and 0.379 μm, respectively (see above). The variation in lateral stiffness between the optical equilibrium position and the laser focus was substantial: a factor of 1.5 for the configuration studied.

V. TRANSCRIPTION STUDIED WITH A TWO-DIMENSIONAL STAGE-BASED FORCE CLAMP

Our interest in extending position detection techniques to include the measurement of force and displacement in the axial dimension arose from the study of processive nucleic acid enzymes moving along DNA (Fig. 12). The experimental geometry, in which the enzyme moving along the DNA pulls on a trapped bead, results in motion of the bead in a plane defined by the direction of the lateral force and the axial dimension. In previous experiments, the effects of axial motion had been calculated and estimated, but not directly measured or otherwise calibrated.[39,82] Improvements afforded by three-dimensional piezoelectric stages permitted the direct measurement of, and control over, the separate axial and lateral motions of the trapped bead. We briefly describe this instrument and the implementation of a two-dimensional force clamp to measure transcription by a single molecule of RNA polymerase.[87]

The optical layout and detection scheme are illustrated in Fig. 13. An existing optical trap[39,153] was modified by adding a normalizing photodetector to monitor the bleedthrough of the trapping laser after a 45° dielectric mirror and a

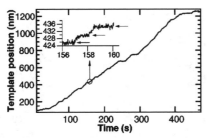

FIG. 14. Two-dimensional, stage based force clamp. Position record of a single RNA polymerase molecule transcribing a 3.5 kbp (1183 nm) DNA template under 18 pN of load. The x and z position signals were low pass filtered at 1 kHz, digitized at 2 kHz, and boxcar averaged over 40 points to generate the 50 Hz feedback signals that controlled the motion of the piezo-electric stage. Motion of the stage was corrected for the elastic compliance of the DNA (Ref. 39) to recover the time-dependent contour length, which reflects the position of the RNA polymerase on the template. Periods of roughly constant velocity are interrupted by pauses on multiple timescales. Distinct pauses can be seen in the trace, while shorter pauses (~1 s) can be discerned in the expanded region of the trace (inset: arrows).

feedback-stabilized three-axis piezoelectric stage (Physik Instrumente P-517.3CD and E710.3CD digital controller) to which the trapping chamber was affixed. The optical trap was built around an inverted microscope (Axiovert 35, Carl Zeiss) equipped with a polarized Nd:YLF laser (TFR, Spectra Physics, $\lambda = 1047$ nm, TEM$_{00}$, 2.5 W) that is focused to a diffraction-limited spot through an objective (Plan Neofluar 100×, 1.3 NA oil immersion). Lateral position detection based on polarization interferometry was implemented. The trapping laser passes through a Wollaston prism below the objective producing two orthogonally polarized and slightly spatially separated spots in the specimen plane; these act as a single trap. The light is recombined by a second Wollaston prism in the condenser, after which it passes through a quarter-wave plate and a polarizing beamsplitter. Two photo-detectors measure the power in each polarization, and the difference between them, normalized by their sum, supplies the lateral position signal. The sum of the detector signals normalized by the incident laser power (from the normalizing detector) provides the axial position signal.[62,64] The axial position signal is a small fraction of the total intensity and is roughly comparable to the intensity noise of the laser. Normalizing the axial position signal with reference to the instantaneous incident laser power, therefore, provides a significant improvement in the signal-to-noise ratio. The two-dimensional position calibration of the instrument, obtained by raster scanning a stuck bead, is shown in Fig. 7. Stiffness in the lateral dimension was measured by a combination of rolloff, triangle-wave drag force, and variance measurements. Stiffness in the axial dimension was measured using the rolloff method and was found to be ~eightfold less than the lateral stiffness.

Single-molecule transcription experiments were carried out with an RNA polymerase specifically attached to the beads and tethered to the surface of the trapping chamber via one end of the template DNA (Fig. 12). Tethered beads were trapped, the surface position was determined as described above, and the bead was centered over the attachment point of the DNA tether, at a predetermined height. Once these initial conditions were established, the two-dimensional

force-clamp routine was begun. The stage was moved in both the axial and lateral directions until the trapped bead was displaced by a predetermined distance from its equilibrium position. Position signals were recorded at 2 kHz and boxcar averaged over 40 points to generate a 50 Hz signal that was used to control the motion of the stage. In this fashion, the displacement of the bead in the trap, and hence the optical force, was held constant as the tether length changed by micron-scale distances during RNA polymerase movement over the DNA template. The motion of RNA polymerase on the DNA can be calculated from the motion of the stage (Fig. 14). Periods of constant motion interrupted by pauses of variable duration are readily observed in the single-molecule transcription trace shown in Fig. 14. Pauses as short as 1 s are readily detected (Fig. 14, inset). Positional noise is on the order of 2 nm, while drift is less than 0.2 nm/s.

Two-dimensional stage based force clamping affords a unique advantage. Since the stiffness in both dimensions is known, the force vector on the bead is defined and constant during an experiment. Tension in the DNA tether opposes the force on the bead, therefore the angle of the DNA with respect to the surface of the trapping chamber is similarly defined and constant. More importantly, the change in the DNA tether length can be calculated from the motion of the stage in one dimension and the angle calculated from the force in both dimensions. As a result, such measurements are insensitive to drift in the axial dimension, which is otherwise a significant source of instrumental error.

VI. PROGRESS AND OVERVIEW OF OPTICAL TRAPPING THEORY

Optical trapping of dielectric particles is sufficiently complex and influenced by subtle, difficult-to-quantify optical properties that theoretical calculations may never replace direct calibration. That said, recent theoretical work has made significant progress towards a more complete description of optical trapping and three-dimensional position detection based on scattered light. Refined theories permit a more realistic assessment of both the capabilities and the limitations of an optical trapping instrument, and may help to guide future designs and optimizations.

Theoretical expressions for optical forces in the extreme cases of Mie particles ($a \gg \lambda$, a is the sphere radius) and Raleigh particle ($a \ll \lambda$) have been available for some time. Ashkin calculated the forces on a dielectric sphere in the ray-optic regime for both the TEM$_{00}$ and the TEM$_{01}^*$ ("donut mode") intensity profiles.[69] Ray-optics calculations are valid for sphere diameters greater than ~10λ, where optical forces become independent of the size of the sphere. At the other extreme, Chaumet and Nieto-Vesperinas obtained an expression for the total time averaged force on a sphere in the Rayleigh regime[154]

$$\langle F^i \rangle = \left(\frac{1}{2}\right) \text{Re}[\alpha E_{0j} \partial^i (E_0^j)^*], \qquad (15)$$

where $\alpha = \alpha_0 (1 - \frac{2}{3} i k^3 \alpha_0)^{-1}$ is a generalized polarizability that includes a damping term, E_0 is the complex magnitude of the electric field, α_0 is the polarizability of a sphere given by Eq.

(4), and k is the wave number of the trapping laser. This expression encapsulates the separate expressions for the scattering and gradient components of the optical force [Eqs. (1) and (3)] and can be applied to the description of optical forces on larger particles through the use of the coupled dipole method.[155] In earlier work, Harada and Asakura calculated the forces on a dielectric sphere illuminated by a moderately focused Gaussian laser beam in the Rayleigh regime by treating the sphere as a simple dipole.[152] The Raleigh theory predicts forces comparable to those calculated with the more complete generalized Lorenz–Mie theory (GLMT) for spheres of diameter up to $\sim w_0$ (the laser beam waist) in the lateral dimension, but only up to $\sim 0.4\lambda$ in the axial dimension.[152] More general electrodynamic theories have been applied to solve for the case of spheres of diameter $\sim\lambda$ trapped with tightly focused beams. One approach has been to generalize the Lorenz–Mie theory describing the scattering of a plane wave by a sphere to the case of Gaussian beams. Barton and co-workers applied fifth-order corrections to the fundamental Gaussian beam to derive the incident and scattered fields from a sphere, which enabled the force to be calculated by means of the Maxwell stress tensor.[76,77] An equivalent approach, implemented by Gouesbet and co-workers, expands the incident beam in an infinite series of beam shape parameters from which radiation pressure cross sections can be computed.[80,156] Trapping forces and efficiencies predicted by these theories are found to be in reasonable agreement with experimental values.[157–159] More recently, Rohrbach and co-workers extended the Raleigh theory to larger particles through the inclusion of second-order scattering terms, valid for spheres that introduce a phase shift, $k_0(\Delta n)D$, less than $\pi/3$, where $k_0 = 2\pi/\lambda_0$ is the vacuum wave number, $\Delta n = (n_p - n_m)$ is the difference in refractive index between the particle and the medium, and D is the diameter of the sphere.[65,66] For polystyrene beads ($n_p = 1.57$) in water ($n_m = 1.33$), this amounts to a maximum particle size of $\sim 0.7\lambda$. In this approach, the incident field is expanded in plane waves, which permits the inclusion of apodization and aberration transformations, and the forces are calculated directly from the scattering of the field by the dipole without resorting to the stress tensor approach. Computed forces and trapping efficiencies compare well with those predicted by GLMT,[66] and the effects of spherical aberration have been explored.[65] Since the second-order Raleigh theory calculates the scattered and unscattered waves, the far field interference pattern, which is the basis of the three-dimensional position detection described above, is readily calculated.[63,64]

VII. NOVEL OPTICAL TRAPPING APPROACHES

Optical trapping (OT) has now developed into an active and diverse field of study. Space constraints preclude a complete survey the field, so we have chosen to focus on a small number of recent developments that seem particularly promising for future applications of the technology.

A. Combined optical trapping and single-molecule fluorescence

Combining the complementary techniques of OT and single-molecule fluorescence (SMF) presents significant technological challenges. Difficulties arise from the roughly 15 orders of magnitude difference between the enormous flux of infrared light associated with a typical trapping laser (sufficient to bleach many varieties of fluorescent dye through multiphoton excitation) compared to the miniscule flux of visible light emitted by a single excited fluorophore. These challenges have been met in a number of different ways. Funatsu and co-workers built an apparatus in which the two techniques were employed sequentially, but not simultaneously.[160] In a separate development, Ishijima and co-workers were able to trap beads attached to the ends of a long (5–10 μm) actin filament while simultaneously monitoring the binding of fluorescent Adenosine triphosphate (ATP) molecules to a myosin motor interacting with the actin filament.[161] In this way, the coordination between the binding of ATP to myosin and the mechanical motion of the actin filament (detected via the optical trap) was determined. This experiment demonstrated the possibility of simultaneous—but not spatially coincident—OT and SMF in the same microscope field of view. In a more recent development, both simultaneous and spatially coincident OT and SMF have been achieved, and used to measure the mechanical forces required to unzip short duplex regions [15 base pair (bp)] of double-stranded DNA.[138] Dye-labeled hybrids were attached via a long (~ 1000 bp) DNA "handle" to a polystyrene bead at one end (using the 3′ end of one strand) and to the coverglass surface at the other (using the 5′ end of the complementary stand). In one experiment, the adjacent terminal ends of the two strands of the DNA hybrid were each labeled with tetramethylrhodamine (TAMRA) molecules. Due to their physical proximity, these dyes self-quenched (the quenching range for TAMRA is ~ 1 nm). The DNA hybrid was then mechanically disrupted ("unzipped") by applying a force ramp to the bead while the fluorescence signal was monitored. The point of mechanical rupture detected with the optical trap was coincident with a stepwise increase in the fluorescence signal, as the two dyes separated, leaving behind a dye attached by one DNA strand to the coverglass surface, as the partner dye was removed with the DNA strand attached to the bead. Control experiments with fluorescent dyes attached to either, but not both, DNA strands verified that the abrupt mechanical transition was specific for the rupture of the DNA hybrid and not, for example, due to breakage of the linkages holding the DNA to the bead or the coverglass surface.

B. Optical rotation and torque

Trapping transparent microspheres with a focused Gaussian laser beam in TEM_{00} mode produces a rotationally symmetric trap that does not exert torque. However, several methods have been developed to induce the rotation of trapped objects.[20,52,162] Just as the change of linear momentum due to refraction of light leads to the production of force, a change in angular momentum leads to torque. Circularly polarized light carries spin angular momentum, of course, and propagating optical beams can also be produced that carry significant amounts of orbital angular momentum, e.g., Laguerre–Gaussian modes.[163] Each photon in such a mode carries $(\sigma+l)\hbar$ of angular momentum, where σ represents the spin angular momentum arising from the polarization state of the light and l is the orbital angular momentum carried by the light pattern. The angular momentum conveyed by the circular polarization alone, estimated at ~10 pN nm/s per mW of 1064 nm light, can be significantly augmented through the use of modes that carry even larger amounts of orbital angular momentum.[164] Transfer of both orbital and spin angular momentum to trapped objects has been demonstrated for absorbing particles.[102,165] Transfer of spin angular momentum has been observed for birefringent particles of crushed calcite,[166] and for more uniform microfabricated birefringent objects.[167,168] Friese and co-workers derived the following expression for the torque on a birefringent particle:[166]

$$\tau = \frac{\varepsilon}{2\omega}E_0^2[\{1-\cos(kd(n_0-n_e))\}\sin 2\varphi$$
$$- \sin(kd(n_0-n_e))\cos 2\varphi \sin 2\theta], \qquad (16)$$

where ε is the permittivity, E_0 is the amplitude of the electric field, ω is the angular frequency of the light, φ describes the ellipticity of the light (plane polarized, $\varphi=0$; circularly polarized, $\varphi=\pi/4$), θ represents the angle between the fast axis of the quarter-wave plate producing the elliptically polarized light and the optic axis of the birefringent particle, k is the vacuum wave number $(2\pi/\lambda)$, and n_0 and n_e are the ordinary and extraordinary indices of refraction of the birefringent material, respectively. Theoretically, all the spin angular momentum carried in a circularly polarized laser beam can be transferred to a trapped object when it acts as a perfect half-wave plate, i.e., $\varphi=\pi/4$ and $kd(n_0-n_e)=\pi$. For the case of plane polarized light, there is a restoring torque on the birefringent particle that aligns the fast axis of the particle with the plane of polarization.[166] Rotation of the plane of polarization will induce rotation in a trapped birefringent particle.

Whereas the transfer of optical angular momentum is a conceptually attractive means of applying torque to optically trapped objects, several other techniques have been employed towards the same end. In one scheme, a high order asymmetric mode, created by placing an aperture in the far field of a laser beam, was used to trap red blood cells: these could be made to spin by rotating the aperture.[169] A more sophisticated version of this same technique involves interfering a Laguerre–Gaussian beam with a plane wave beam to produce a spiral beam pattern.[170] By changing the relative phase of the two beams, the pattern can be made to rotate, leading to rotation in an asymmetric trapped object.[48] Alternatively, the interference of two Laguerre–Gaussian beams of opposite helicity (l and $-l$) creates $2l$ beams surrounding the optical axis, which can be rotated by adjusting the polarization of one of the interfering beams.[46] Additionally, a variety of small chiral objects, such as microfabricated "optical propellers," can be trapped and made to rotate in a symmet-

ric Gaussian beam due to the optical forces generated on asymmetrically oriented surfaces.[171–174]

Rotation of trapped particles is most commonly monitored by video tracking, which is effectively limited by frame rates to rotation speeds below ~15 Hz, and to visibly asymmetric particles (i.e., microscopic objects of sufficient size and contrast to appear asymmetric in the imaging modality used). Rotation rates up to 1 kHz have been measured by back focal plane detection of trapped 0.83 μm beads sparsely labeled with 0.22 μm beads to make these optically anisotropic.[175] Backscattered light from trapped, asymmetric particles has also been used to measure rotation rates in excess of 300 Hz.[102,166]

C. Holographic optical traps

Holograms and other types of diffractive optics have been used extensively for generating complex, high-order optical trapping beams,[20,52,162,165,176] such as the Laguerre–Gaussian modes discussed above. Diffractive optical devices may also be used to synthesize multiple optical traps with arbitrary intensity profiles.[20,91,177–179] A diffractive element placed in a plane optically conjugate to the back aperture of the microscope objective produces an intensity distribution in the specimen plane that is the Fourier transform of the pattern imposed by the element,[177] and several computational methods have been developed to derive the holographic pattern required for any given intensity distribution in the specimen plane.[91,92,180] Generally speaking, diffractive elements modulate both the amplitude and the phase of the incident light. Optical throughput can be maximized by employing diffractive optics that primarily modify the phase but not the amplitude of the incident light, termed kinoforms.[91] Computer-generated phase masks can also be etched onto a glass substrate using standard photolithographic techniques, producing arbitrary, but fixed, optical traps.

Reicherter and co-workers extended the usefulness of holographic optical trapping techniques by generating three independently movable donut-mode trapping beams with an addressable liquid crystal spatial light modulator (SLM).[181] Improvements in SLM technology and real-time hologram calculation algorithms have been implemented, allowing the creation of an array of up to 400 optical traps, in addition to the creation and three-dimensional manipulation of multiple, high order, trapping beams.[92,182,183] Multiple optical traps can also be generated by time sharing, using rapid-scanning techniques based on AODs or galvo mirrors,[49,50] but these are typically formed in just one or two axial planes,[184] and they are limited in number. Dynamic holographic optical tweezers can produce still more varied patterns, limited only by the optical characteristics of the SLM and the computational time required to generate the hologram. Currently, the practical update rate of a typical SLM is around 5 Hz, which limits how quickly objects can be translated.[92] Furthermore, the number and size of the pixels in the SLM restrict the complexity and the range of motion of generated optical traps,[92] while the pixelation and discrete phase steps of the SLM result in diffractive losses. Faster refresh rates (>30 Hz) in a holographic optical trap have recently been reported with a SLM based on ferroelectric, as opposed to

VIII. PROSPECTS

The nearly 2 decades that have passed since Ashkin and co-workers invented the single beam, gradient force optical trap have borne witness to a proliferation of innovations and applications. The full potential of most of the more recent optical developments has yet to be realized. On the biological front, the marriage of optical trapping with single-molecule fluorescence methods[138] represents an exciting frontier with enormous potential. Thanks to steady improvements in optical trap stability and photodetector sensitivity, the practical limit for position measurements is now comparable to the distance subtended by a single base pair along DNA, 3.4 Å. Improved spatiotemporal resolution is now permitting direct observations of molecular-scale motions in individual nucleic acid enzymes, such as polymerases, helicases, and nucleases.[86,87,186] The application of optical torque offers the ability to study rotary motors, such as F_1F_0 ATPase,[187] using rotational analogs of many of the same techniques already applied to the study of linear motors, i.e., torque clamps and rotation clamps.[50] Moving up in scale, the ability to generate and manipulate a myriad of optical traps dynamically using holographic tweezers[20,92] opens up many potential applications, including cell sorting and other types of high-throughput manipulation. More generally, as the field matures, optical trapping instruments should no longer be confined to labs that build their own custom apparatus, a change that should be driven by the increasing availability of sophisticated, versatile commercial systems. The physics of optical trapping will continue to be explored in its own right, and optical traps will be increasingly employed to study physical, as well as biological, phenomena. In one groundbreaking example from the field of nonequilibrium statistical mechanics, Jarzynski's equality[188]—which relates the value of the equilibrium free energy for a transition in a system to a *nonequilibrium* measure of the work performed—was put to experimental test by mechanically unfolding RNA structures using optical forces.[189] Optical trapping techniques are increasingly being used in condensed matter physics to study the behavior (including anomalous diffusive properties and excluded volume effects) of colloids and suspensions,[21] and dynamic optical tweezers are particularly well suited for the creation and evolution of large arrays of colloids in well-defined potentials.[20] As optical trapping techniques continue to improve and become better established, these should pave the way for some great new science in the 21st century, and we will be further indebted to the genius of Ashkin.[3]

ACKNOWLEDGMENTS

The authors thank members of the Block Lab for advice, suggestions, and helpful discussions. In particular, Elio Abbondanzieri helped with instrument construction and all aspects of data collection, Joshua Shaevitz supplied Fig. 6, and Megan Valentine and Michael Woodside supplied valuable comments on the manuscript. We also thank Henrik Flyvbjerg, Kirstine Berg-Sørensen, and the members of their labs for sharing results in advance of publication, for critical reading of the manuscript, and for help in preparing Fig. 10. Finally, we thank Megan Valentine, Grace Liou, and Richard Neuman for critical reading of the manuscript.

[1] A. Ashkin, Phys. Rev. Lett. **24**, 156 (1970).
[2] A. Ashkin and J. M. Dziedzic, Appl. Phys. Lett. **19**, 283 (1971).
[3] A. Ashkin *et al.*, Opt. Lett. **11**, 288 (1986).
[4] A. Ashkin, IEEE J. Sel. Top. Quantum Electron. **6**, 841 (2000).
[5] S. Chu *et al.*, Phys. Rev. Lett. **57**, 314 (1986).
[6] A. Ashkin, J. M. Dziedzic, and T. Yamane, Nature (London) **330**, 769 (1987).
[7] A. Ashkin and J. M. Dziedzic, Science **235**, 1517 (1987).
[8] M. J. Lang and S. M. Block, Am. J. Phys. **71**, 201 (2003).
[9] K. Svoboda and S. M. Block, Annu. Rev. Biophys. Biomol. Struct. **23**, 247 (1994).
[10] A. D. Mehta, J. T. Finer, and J. A. Spudich, Methods Enzymol. **298**, 436 (1998).
[11] A. D. Mehta *et al.*, Science **283**, 1689 (1999).
[12] S. C. Kuo, Traffic **2**, 757 (2001).
[13] A. Ishijima and T. Yanagida, Trends Biochem. Sci. **26**, 438 (2001).
[14] Y. Ishii, A. Ishijima, and T. Yanagida, Trends Biotechnol. **19**, 211 (2001).
[15] S. Jeney, E. L. Florin, and J. K. Horber, Methods Mol. Biol. **164**, 91 (2001).
[16] S. Khan and M. P. Sheetz, Annu. Rev. Biochem. **66**, 785 (1997).
[17] R. Simmons, Curr. Biol. **6**, 392 (1996).
[18] C. Bustamante, J. C. Macosko, and G. J. Wuite, Nat. Rev. Mol. Cell Biol. **1**, 130 (2000).
[19] M. D. Wang, Curr. Opin. Biotechnol. **10**, 81 (1999).
[20] D. G. Grier, Nature (London) **424**, 810 (2003).
[21] D. G. Grier, Curr. Opin. Colloid Interface Sci. **2**, 264 (1997).
[22] J. K. H. Horber, *Atomic Force Microscopy in Cell Biology* (Academic, San Diego, 2002), pp. 1–31.
[23] H. Lowen, J. Phys.: Condens. Matter **13**, R415 (2001).
[24] R. Bar-Ziv, E. Moses, and P. Nelson, Biophys. J. **75**, 294 (1998).
[25] P. T. Korda, M. B. Taylor, and D. G. Grier, Phys. Rev. Lett. **89**, 128301 (2002).
[26] L. A. Hough and H. D. Ou-Yang, Phys. Rev. E **65**, 021906 (2002).
[27] B. H. Lin, J. Yu, and S. A. Rice, Phys. Rev. E **62**, 3909 (2000).
[28] J. C. Crocker and D. G. Grier, Phys. Rev. Lett. **77**, 1897 (1996).
[29] J. C. Crocker and D. G. Grier, Phys. Rev. Lett. **73**, 352 (1994).
[30] C. Bustamante, Z. Bryant, and S. B. Smith, Nature (London) **421**, 423 (2003).
[31] C. Bustamante *et al.*, Curr. Opin. Struct. Biol. **10**, 279 (2000).
[32] L. H. Pope, M. L. Bennink, and J. Greve, J. Muscle Res. Cell Motil. **23**, 397 (2002).
[33] J. F. Allemand, D. Bensimon, and V. Croquette, Curr. Opin. Struct. Biol. **13**, 266 (2003).
[34] A. Janshoff *et al.*, Angew. Chem., Int. Ed. **39**, 3213 (2000).
[35] B. Onoa *et al.*, Science **299**, 1892 (2003).
[36] J. Liphardt *et al.*, Science **292**, 733 (2001).
[37] Z. Bryant *et al.*, Nature (London) **424**, 338 (2003).
[38] S. B. Smith, Y. J. Cui, and C. Bustamante, Science **271**, 795 (1996).
[39] M. D. Wang *et al.*, Biophys. J. **72**, 1335 (1997).
[40] T. T. Perkins *et al.*, Science **264**, 822 (1994).
[41] T. T. Perkins, D. E. Smith, and S. Chu, Science **264**, 819 (1994).
[42] T. T. Perkins *et al.*, Science **268**, 83 (1995).
[43] K. Wang, J. G. Forbes, and A. J. Jin, Prog. Biophys. Mol. Biol. **77**, 1 (2001).
[44] A. Ashkin, Proc. Natl. Acad. Sci. U.S.A. **94**, 4853 (1997).
[45] D. McGloin, V. Garces-Chavez, and K. Dholakia, Opt. Lett. **28**, 657 (2003).
[46] M. P. MacDonald *et al.*, Science **296**, 1101 (2002).
[47] V. Garces-Chavez *et al.*, Nature (London) **419**, 145 (2002).
[48] L. Paterson *et al.*, Science **292**, 912 (2001).
[49] K. Visscher, S. P. Gross, and S. M. Block, IEEE J. Sel. Top. Quantum Electron. **2**, 1066 (1996).
[50] K. Visscher and S. M. Block, Methods Enzymol. **298**, 460 (1998).
[51] J. E. Molloy, Methods Cell Biol. **55**, 205 (1998).
[52] J. E. Molloy and M. J. Padgett, Contemp. Phys. **43**, 241 (2002).
[53] G. J. Brouhard, H. T. Schek III, and A. J. Hunt, IEEE Trans. Biomed. Eng. **50**, 121 (2003).

[54] M. P. Sheetz, *in Laser Tweezers in Cell Biology. Methods in Cell Biology*, edited by L. Wilson and P. Matsudaira (Academic, San Diego, 1998), Vol. 55.

[55] S. M. Block, in *Noninvasive Techniques in Cell Biology*, edited by J. K. Foskett and S. Grinstein (Wiley-Liss, New York, 1990), pp. 375–402.

[56] S. E. Rice, T. J. Purcell, and J. A. Spudich, Methods Enzymol. **361**, 112 (2003).

[57] S. B. Smith, Y. Cui, and C. Bustamante, Methods Enzymol. **361**, 134 (2003).

[58] S. M. Block, in *Constructing Optical Tweezers, Cell Biology: A Laboratory Manual*, edited by D. Spector, R. Goldman, and L. Leinward (Cold Spring Harbor Press, Cold Spring Harbor, NY, 1998).

[59] M. J. Lang *et al.*, Biophys. J. **83**, 491 (2002).

[60] E. Fallman and O. Axner, Appl. Opt. **36**, 2107 (1997).

[61] E. J. G. Peterman *et al.*, Rev. Sci. Instrum. **74**, 3246 (2003).

[62] A. Pralle *et al.*, Microsc. Res. Tech. **44**, 378 (1999).

[63] A. Rohrbach, H. Kress, and E. H. Stelzer, Opt. Lett. **28**, 411 (2003).

[64] A. Rohrbach and E. H. K. Stelzer, J. Appl. Phys. **91**, 5474 (2002).

[65] A. Rohrbach and E. H. Stelzer, Appl. Opt. **41**, 2494 (2002).

[66] A. Rohrbach and E. H. Stelzer, J. Opt. Soc. Am. A Opt. Image Sci. Vis **18**, 839, (2001).

[67] J. D. Jackson, *Classical Electrodynamics*, 2nd ed. (Wiley, New York, 1975).

[68] A. Ashkin, *Methods in Cell Biology* (Academic, San Diego, 1998), Vol. 55, pp. 1–27.

[69] A. Ashkin, Biophys. J. **61**, 569 (1992).

[70] E. Fallman and O. Axner, Appl. Opt. **42**, 3915 (2003).

[71] A. T. O'Neill and M. J. Padgett, Opt. Commun. **193**, 45 (2001).

[72] N. B. Simpson *et al.*, J. Mod. Opt. **45**, 1943 (1998).

[73] M. E. J. Friese *et al.*, Appl. Opt. **35**, 7112 (1996).

[74] E. Almaas and I. Brevik, J. Opt. Soc. Am. B **12**, 2429 (1995).

[75] J. P. Barton, J. Appl. Phys. **64**, 1632 (1988).

[76] J. P. Barton and D. R. Alexander, J. Appl. Phys. **66**, 2800 (1989).

[77] J. P. Barton, D. R. Alexander, and S. A. Schaub, J. Appl. Phys. **66**, 4594 (1989).

[78] P. Zemanek, A. Jonas, and M. Liska, J. Opt. Soc. Am. A Opt. Image Sci. Vis **19**, 1025 (2002).

[79] G. Gouesbet and G. Grehan, Atomization Sprays **10**, 277 (2000).

[80] K. F. Ren, G. Greha, and G. Gouesbet, Opt. Commun. **108**, 343 (1994).

[81] K. Visscher, M. J. Schnitzer, and S. M. Block, Nature (London) **400**, 184 (1999).

[82] M. D. Wang *et al.*, Science **282**, 902 (1998).

[83] G. J. Wuite *et al.*, Nature (London) **404**, 103 (2000).

[84] R. S. Rock *et al.*, Proc. Natl. Acad. Sci. U.S.A. **98**, 13655 (2001).

[85] S. M. Block *et al.*, Proc. Natl. Acad. Sci. U.S.A. **100**, 2351 (2003).

[86] T. T. Perkins *et al.*, Science **301**, 1914 (2003).

[87] K. C. Neuman *et al.*, Cell **115**, 437 (2003).

[88] K. Adelman *et al.*, Proc. Natl. Acad. Sci. U.S.A. **99**, 13538 (2002).

[89] <http://www.cellrobotics.com>

[90] <http://www.arryx.com/home1.html>

[91] E. R. Dufresne *et al.*, Rev. Sci. Instrum. **72**, 1810 (2001).

[92] J. E. Curtis, B. A. Koss, and D. G. Grier, Opt. Commun. **207**, 169 (2002).

[93] <http://www.palm-mikrolaser.com>

[94] S. P. Gross, Methods Enzymol. **361**, 162 (2003).

[95] K. C. Neuman *et al.*, Biophys. J. **77**, 2856 (1999).

[96] H. Liang *et al.*, Biophys. J. **70**, 1529 (1996).

[97] I. A. Vorobjev *et al.*, Biophys. J. **64**, 533 (1993).

[98] A. Ashkin and J. M. Dziedzic, Ber. Bunsenges. Phys. Chem. Chem. Phys. **93**, 254 (1989).

[99] Y. Liu *et al.*, Biophys. J. **68**, 2137 (1995).

[100] E. J. G. Peterman, F. Gittes, and C. F. Schmidt, Biophys. J. **84**, 1308 (2003).

[101] A. Schonle and S. W. Hell, Opt. Lett. **23**, 325 (1998).

[102] M. E. J. Friese *et al.*, Phys. Rev. A **54**, 1593 (1996).

[103] G. J. Wuite *et al.*, Biophys. J. **79**, 1155 (2000).

[104] W. Grange *et al.*, Rev. Sci. Instrum. **73**, 2308 (2002).

[105] S. Inoue and K. R. Spring, *Video Microscopy: The Fundamentals*, 2nd ed. (Plenum, New York, 1997).

[106] H. Misawa *et al.*, J. Appl. Phys. **70**, 3829 (1991).

[107] S. C. Kuo and J. L. McGrath, Nature (London) **407**, 1026 (2000).

[108] Y. L. Kuo *et al.*, Arch. Androl **44**, 29 (2000).

[109] S. Yamada, D. Wirtz, and S. C. Kuo, Biophys. J. **78**, 1736 (2000).

[110] M. K. Cheezum, W. F. Walker, and W. H. Guilford, Biophys. J. **81**, 2378 (2001).

[111] R. E. Thompson, D. R. Larson, and W. W. Webb, Biophys. J. **82**, 2775 (2002).

[112] J. C. Crocker and D. G. Grier, J. Colloid Interface Sci. **179**, 298 (1996).

[113] C. Gosse and V. Croquette, Biophys. J. **82**, 3314 (2002).

[114] M. Keller, J. Schilling, and E. Sackmann, Rev. Sci. Instrum. **72**, 3626 (2001).

[115] V. Croquette (personal communication).

[116] J. T. Finer, R. M. Simmons, and J. A. Spudich, Nature (London) **368**, 113 (1994).

[117] J. E. Molloy *et al.*, Biophys. J. **68**, S298 (1995).

[118] W. Denk and W. W. Webb, Appl. Opt. **29**, 2382 (1990).

[119] K. Svoboda *et al.*, Nature (London) **365**, 721 (1993).

[120] F. Gittes and C. F. Schmidt, Opt. Lett. **23**, 7 (1998).

[121] F. Gittes and C. F. Schmidt, Biophys. J. **74**, A183 (1998).

[122] M. W. Allersma *et al.*, Biophys. J. **74**, 1074 (1998).

[123] L. P. Ghislain and W. W. Webb, Opt. Lett. **18**, 1678 (1993).

[124] L. P. Ghislain, N. A. Switz, and W. W. Webb, Rev. Sci. Instrum. **65**, 2762 (1994).

[125] I. M. Peters *et al.*, Rev. Sci. Instrum. **69**, 2762 (1998).

[126] M. E. J. Friese *et al.*, Appl. Opt. **38**, 6597 (1999).

[127] E. L. Florin *et al.*, J. Struct. Biol. **119**, 202 (1997).

[128] E. L. Florin, J. K. H. Horber, and E. H. K. Stelzer, Appl. Phys. Lett. **69**, 446 (1996).

[129] Z. X. Zhang *et al.*, Appl. Opt. **37**, 2766 (1998).

[130] A. Jonas, P. Zemanek, and E. L. Florin, Opt. Lett. **26**, 1466 (2001).

[131] K. Sasaki, M. Tsukima, and H. Masuhara, Appl. Phys. Lett. **71**, 37 (1997).

[132] A. R. Clapp, A. G. Ruta, and R. B. Dickinson, Rev. Sci. Instrum. **70**, 2627 (1999).

[133] A. E. Siegman, Lasers University Science Books (Sausalito, CA, 1986), p. 1283.

[134] K. Berg-Sorensen *et al.*, J. Appl. Phys. **93**, 3167 (2003).

[135] J. Koch and M. D. Wang, Phys. Rev. Lett. **91**, 028103 (2003).

[136] B. D. Brower-Toland *et al.*, Proc. Natl. Acad. Sci. U.S.A. **99**, 1960 (2002).

[137] S. J. Koch *et al.*, Biophys. J. **83**, 1098 (2002).

[138] M. J. Lang, P. M. Fordyce, and S. M. Block, J. Biol. **2**, 6 (2003).

[139] T. D. Visser and S. H. Wiersma, J. Opt. Soc. Am. A Opt. Image Sci. Vis **9**, 2034 (1992).

[140] S. H. Wiersma *et al.*, J. Opt. Soc. Am. A Opt. Image Sci. Vis **14**, 1482 (1997).

[141] S. H. Wiersma and T. D. Visser, J. Opt. Soc. Am. A Opt. Image Sci. Vis **13**, 320 (1996).

[142] K. Carlsson, J. Microsc. **163**, 167 (1991).

[143] P. Bartlett, S. I. Henderson, and S. J. Mitchell, Philos. Trans. R. Soc. London, Ser. A **359**, 883 (2001).

[144] S. Hell *et al.*, J. Microsc. **169**, 391 (1993).

[145] J. Radler and E. Sackmann, Langmuir **8**, 848 (1992).

[146] K. C. Neuman, E. A. Abbondanzieri, and S. M. Block (unpublished).

[147] H. Brenner, Chem. Eng. Sci. **16**, 242 (1961).

[148] K. Berg-Sørensen and H. Flyvbjerg, Rev. Sci. Instrum. **75**, 594 (2004).

[149] I. M. Toliæ-Nørrelykke, K. Berg-Sørensen, and H. Flyvbjerg, Comput. Phys. Commun. **159**, 225 (2004).

[150] T. R. Strick *et al.*, Biophys. J. **74**, 2016 (1998).

[151] T. R. Strick *et al.*, Science **271**, 1835 (1996).

[152] Y. Harada and T. Asakura, Opt. Commun. **124**, 529 (1996).

[153] K. Svoboda and S. M. Block, Cell **77**, 773 (1994).

[154] P. C. Chaumet and M. Nieto-Vesperinas, Opt. Lett. **25**, 1065 (2000).

[155] P. C. Chaumet and M. Nieto–Vesperinas, Phys. Rev. B **61**, 4119 (2000).

[156] G. Gouesbet, B. Maheu, and G. Grehan, J. Opt. Soc. Am. A Opt. Image Sci. Vis **5**, 1427 (1988).

[157] Y. K. Nahmias and D. J. Odde, Int. J. Quantum Chem. **38**, 131 (2002).

[158] W. H. Wright, G. J. Sonek, and M. W. Berns, Appl. Opt. **33**, 1735 (1994).

[159] W. H. Wright, G. J. Sonek, and M. W. Berns, Appl. Phys. Lett. **63**, 715 (1993).

[160] T. Funatsu *et al.*, Biophys. Chem. **68**, 63 (1997).

[161] A. Ishijima *et al.*, Cell **92**, 161 (1998).

[162] K. Dholakia, G. Spalding, and M. MacDonald, Phys. World **15**, 31 (2002).

[163] S. M. Barnett and L. Allen, Opt. Commun. **110**, 670 (1994).

[164] L. Allen *et al.*, Phys. Rev. A **45**, 8185 (1992).

[165] H. He *et al.*, Phys. Rev. Lett. **75**, 826 (1995).

[166] M. E. J. Friese *et al.*, Nature (London) **394**, 348 (1998).

[167] E. Higurashi, R. Sawada, and T. Ito, J. Micromech. Microeng. **11**, 140 (2001).

[168] E. Higurashi, R. Sawada, and T. Ito, Phys. Rev. E **59**, 3676 (1999).

[169] S. Sato, M. Ishigure, and H. Inaba, Electron. Lett. **27**, 1831 (1991).

[170] M. Padgett *et al.*, Am. J. Phys. **64**, 77 (1996).

[171] E. Higurashi *et al.*, Appl. Phys. Lett. **64**, 2209 (1994).

[172] E. Higurashi *et al.*, J. Appl. Phys. **82**, 2773 (1997).

[173] R. C. Gauthier, Appl. Phys. Lett. **67**, 2269 (1995).

[174] R. C. Gauthier, Appl. Opt. **40**, 1961 (2001).

[175] A. D. Rowe *et al.*, J. Mod. Opt. **50**, 1539 (2003).

[176] H. He, N. R. Heckenberg, and H. Rubinsztein-Dunlop, J. Mod. Opt. **42**, 217 (1995).

[177] E. R. Dufresne and D. G. Grier, Rev. Sci. Instrum. **69**, 1974 (1998).

[178] P. Korda *et al.*, Rev. Sci. Instrum. **73**, 1956 (2002).

[179] D. M. Mueth *et al.*, Phys. Rev. Lett. **77**, 578 (1996).

[180] J. Liesener *et al.*, Opt. Commun. **185**, 77 (2000).

[181] M. Reicherter *et al.*, Opt. Lett. **24**, 608 (1999).

[182] J. E. Curtis and D. G. Grier, Opt. Lett. **28**, 872 (2003).

[183] Y. Igasaki *et al.*, Opt. Rev. **6**, 339 (1999).

[184] A. van Blaaderen *et al.*, Faraday Discuss. **123**, 107 (2003).

[185] W. J. Hossack *et al.*, Opt. Express **11**, 2053 (2003).

[186] J. W. Shaevitz *et al.*, Nature (London) **426**, 684 (2003).

[187] H. Noji *et al.*, Nature (London) **386**, 299 (1997).

[188] C. Jarzynski, Phys. Rev. Lett. **78**, 2690 (1997).

[189] J. Liphardt *et al.*, Science **296**, 1832 (2002).

Forces of a single-beam gradient laser trap on a dielectric sphere in the ray optics regime

A. Ashkin

AT&T Bell Laboratories, Holmdel, New Jersey 07733

ABSTRACT   We calculate the forces of single-beam gradient radiation pressure laser traps, also called "optical tweezers," on micron-sized dielectric spheres in the ray optics regime. This serves as a simple model system for describing laser trapping and manipulation of living cells and organelles within cells. The gradient and scattering forces are defined for beams of complex shape in the ray-optics limit. Forces are calculated over the entire cross-section of the sphere using TEM_{00} and TEM_{01}^* mode input intensity profiles and spheres of varying index of refraction. Strong uniform traps are possible with force variations less than a factor of 2 over the sphere cross-section. For a laser power of 10 mW and a relative index of refraction of 1.2 we compute trapping forces as high as $\sim 1.2 \times 10^{-6}$ dynes in the weakest (backward) direction of the gradient trap. It is shown that good trapping requires high convergence beams from a high numerical aperture objective. A comparison is given of traps made using bright field or differential interference contrast optics and phase contrast optics.

INTRODUCTION

This paper gives a detailed description of the trapping of micron-sized dielectric spheres by a so-called single-beam gradient optical trap. Such dielectric spheres can serve as first simple models of living cells in biological trapping experiments and also as basic particles in physical trapping experiments. Optical trapping of small particles by the forces of laser radiation pressure has been used for about 20 years in the physical sciences for the manipulation and study of micron and submicron dielectric particles and even individual atoms (1–7). These techniques have also been extended more recently to biological particles (8–18).

The basic forces of radiation pressure acting on dielectric particles and atoms are known (1, 2, 19–21). For dielectric spheres large compared with the wavelength, one is in the geometric optics regime and can thus use simple ray optics in the derivation of the radiation pressure force from the scattering of incident light momentum. This approach was used to calculate the forces for the original trapping experiments on micron-sized dielectric spheres (1, 22). These early traps were either all optical two-beam traps (1) or single beam levitation traps which required gravity or electrostatic forces for their stability (23, 24). For particles in the Rayleigh regime where the size is much less than the wavelength λ the particle acts as a simple dipole. The force on a dipole divides itself naturally into two components: a so-called scattering force component pointing in the direction of the incident light and a gradient component pointing in the direction of the intensity gradient of the light (19, 21).

The single-beam gradient trap, sometimes referred to as "optical tweezers," was originally designed for Rayleigh particles (20). It consists of a single strongly focused laser beam. Conceptually and practically it is one of the simplest laser traps. Its stability in the Rayleigh regime is the result of the dominance of the gradient force pulling particles toward the high focus of the beam over the scattering force trying to push particles away from the focus in the direction of the incident light. Subsequently it was found experimentally that single-beam gradient traps could also trap and manipulate micron-sized (25) and a variety of biological particles, including living cells and organelles within living cells (8, 10). Best results were obtained using infrared trapping beams to reduced optical damage. The trap in these biological applications was built into a standard high resolution microscope in which one uses the same high numerical aperture (NA) microscope objective for both trapping and viewing. The micromanipulative abilities of single-beam gradient traps are finding use in a variety of experiments in the biological sciences. Experiments have been performed in the trapping of viruses and bacteria (8); the manipulation of yeast cells, blood cells, protozoa, and various algae and plant cells (10); the measurement of the compliance of bacterial flagella (11); internal cell surgery (13); manipulation of chromosomes (12); trapping and force measurement on sperm cells (14, 15); and recently, observations on the force of motor molecules driving mitochondrion and latex spheres along microtubules (16, 17). Optical techniques have also been used for cell sorting (9).

Qualitative descriptions of the operation of the single-beam gradient trap in the ray optics regime have already been given (25, 26). In Fig. 1 taken from reference 26, the action of the trap on a dielectric sphere is described in terms of the total force due to a typical pair of rays a and b of the converging beam, under the simplifying assumption of zero surface reflection. In this approximation the forces F_a and F_b are entirely due to refraction and are shown pointing in the direction of the momentum change. One sees that for arbitrary displacements of the sphere origin O from the focus f that the vector sum of F_a and F_b gives a net restoring force F directed back to the focus, and the trap is stable. In this paper we quantify the above qualitative picture of the trap. We show how to define the gradient and scttering force on a sphere $\gg \lambda$ in a natural way for beams of arbitrary shape. One can then describe trapping in the ray optics regime in the same terms as in the Rayleigh regime.

Results are given for the trapping forces over the entire cross-section of the sphere. The forces are calculated for input beams with various TEM_{00} and TEM_{01}^* mode intensity profiles at the input aperture of a high numerical aperture trapping objective of NA = 1.25. The results confirm the qualitative observation that good trapping requires the input aperture to be well enough filled by the incident beam to give rise to a trapping beam with high convergence angle. One can design traps in which the trapping forces vary at most by a factor of ~ 1.8 over the cross-section of the sphere with trapping forces as high as $Q = 0.30$ where the force F is given in terms of the dimensionless factor Q in the expression $F = Q(n_1 P/c)$. P is the incident power and $n_1 P/c$ is the incident momentum per second in a medium of index of refraction n_1. There has been a previous calculation of single-beam gradient trapping forces on spheres in the geometrical optics limit by Wright et al. (27), over a limited portion of the sphere, which gives much poorer results. They find trapping forces of $Q = 0.055$ in the above units which vary over the sphere cross-section by more than an order of magnitude.

LIGHT FORCES IN THE RAY OPTICS REGIME

In the ray optics or geometrical optics regime one decomposes the total light beam into individual rays, each with appropriate intensity, direction, and state of polarization, which propagate in straight lines in media of uniform refractive index. Each ray has the characteristics of a plane wave of zero wavelength which can change directions when it reflects, refracts, and changes polarization at dielectric interfaces according to the usual Fresnel formulas. In this regime diffractive effects are neglected (see Chapter III of reference 28).

The simple ray optics model of the single-beam gradient trap used here for calculating the trapping forces on a sphere of diameter $\gg \lambda$ is illustrated in Fig. 2. The trap consists of an incident parallel beam of arbitrary mode structure and polarization which enters a high NA microscope objective and is focused ray-by-ray to a dimensionless focal point f. Fig. 2 shows the case where f is located along the Z axis of the sphere. The maximum convergence angle for rays at the edge of the input aperture of a high NA objective lens such as the Leitz PL APO 1.25W (E. Leitz, Inc., Wetzlar, Germany) or the Zeiss PLAN NEOFLUAR 63/1.2W water immersion objectives (Carl Zeiss, Inc., Thornwood, NY), for example, is $\phi_{max} \cong 70°$. Computation of the total force on the sphere consists of summing the contributions of each beam ray entering the aperture at radius r with respect to the beam axis and angle β with respect the Y axis. The effect of neglecting the finite size of the actual beam focus, which can approach the limit of $\lambda/2n_1$ (see reference 29), is negligible for spheres much larger than λ. The point focus description of the convergent beam in which the ray directions and momentum continue in straight lines through the focus gives the correct incident polarization and momentum for each ray. The rays then reflect and refract at the surface of the sphere giving rise to the light forces.

The model of Wright et al. (27) tries to describe the single-beam gradient trap in terms of both wave and ray optics. It uses the TEM_{00} Gaussian mode beam propagation formula to describe the focused trapping beam and takes the ray directions of the individual rays to be

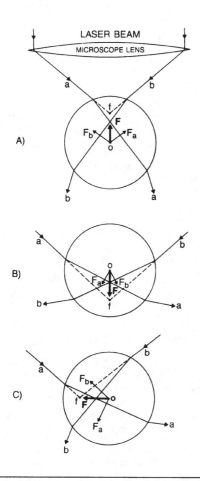

FIGURE 1 Qualitative view of the trapping of dielectric spheres. The refraction of a typical pair of rays a and b of the trapping beam gives forces F_a and F_b whose vector sum F is always restoring for axial and transverse displacements of the sphere from the trap focus f.

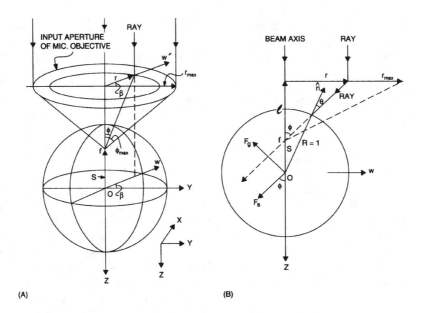

FIGURE 2   (*A*) Single beam gradient force trap in the ray optics model with beam focus *f* located along the Z axis of the sphere. (*B*) Geometry of an incident ray giving rise to gradient and scattering force contributions F_g and F_s.

perpendicular to the Gaussian beam phase fronts. Since the curvature of the phase fronts vary considerably along the beam, the ray directions also change, from values as high as 30° or more with respect to the beam axis in the far-field, to 0° at the beam focus. This is physically incorrect. It implies that rays can change their direction in a uniform medium, which is contrary to geometrical optics. It also implies that the momentum of the beam can change in a uniform medium without interacting with a material object, which violates the conservation of light momentum. The constancy of the light momentum and ray direction for a Gaussian beam can be seen in another way. If one resolves a Gaussian beam into an equivalent angular distribution of plane waves (see Section 11.4.2 of reference 28) one sees that these plane waves can propagate with no momentum or direction changes right through the focus. Another important point is that the Gaussian beam propagation formula is strictly correct only for transversely polarized beams in the limit of small far-field diffraction angles θ', where $\theta' = \lambda/\pi w_0$ (w_0 being the focal spot radius). This formula therefore provides a poor description of the high convergence beams used in good traps. The proper wave description of a highly convergent beam is much more complex than the Gaussian beam formula. It involves strong axial electric field components at the focus (from the edge rays) and requires use of the vector wave equation as opposed to the scalar wave equation used for Gaussian beams (30).

Apart from the major differences near the focus, the model of Wright et al. (27) should be fairly close to the ray optics model used here in the far-field of the trapping beam. The principal distinction between the two calculations, however, is the use by Wright et al. of beams with relatively small convergence angle. They calculate forces for beams with spot sizes w_0 = 0.5, 0.6, and 0.7 μm, which implies values of θ' of ~29, 24, and 21°, respectively. Therefore, these are beams having relatively small convergence angles compared with convergence angles of $\phi_{max} \cong 70°$ which are available from a high NA objective.

Consider first the force due to a single ray of power P hitting a dielectric sphere at an angle of incidence θ with incident momentum per second of n_1P/c (see Fig. 3). The total force on the sphere is the sum of contributions due to the reflected ray of power PR and the infinite number of emergent refracted rays of successively decreasing power PT^2, PT^2R, ... PT^2R^n, ... The quantities R and T are the Fresnel reflection and transmission coefficients of the surface at θ. The net force acting through the origin O can be broken into F_Z and F_Y components as given by Roosen and co-workers (3, 22) (see Appendix I for a sketch of the derivation).

$$F_Z = F_s = \frac{n_1 P}{c} \left\{ 1 + R \cos 2\theta - \frac{T^2[\cos(2\theta - 2r) + R \cos 2\theta]}{1 + R^2 + 2R \cos 2r} \right\} \quad (1)$$

$$F_Y = F_g = \frac{n_1 P}{c} \left\{ R \sin 2\theta - \frac{T^2[\sin(2\theta - 2r) + R \sin 2\theta]}{1 + R^2 + 2R \cos 2r} \right\} \quad (2)$$

where θ and r are the angles of incidence and refraction. These formulas sum over all scattered rays and are therefore exact. The forces are polarization dependent since R and T are different for rays polarized perpendicular or parallel to the plane of incidence.

In Eq. 1 we denote the F_Z component pointing in the direction of the incident ray as the scattering force component F_s for this single ray. Similarly, in Eq. 2 we denote the F_Y component pointing in the direction perpendicular to the ray as the gradient force component F_g for the ray. For beams of complex shape such as the highly convergent beams used in the single-beam gradient trap, we define the scattering and gradient forces of the beam as the vector sums of the scattering and gradient force contributions of the individual rays of the beam. Fig. 2 *B* depicts the direction of the scattering force component and gradient force component of a single ray of the convergent beam

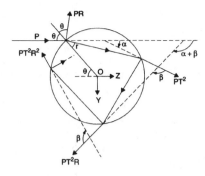

FIGURE 3 Geometry for calculating the force due to the scattering of a single incident ray of power P by a dielectric sphere, showing the reflected ray PR and infinite set of refracted rays PT^2R^n.

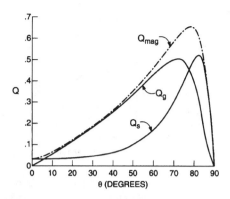

FIGURE 4 Values of the scattering force Q_s, gradient force Q_g, and magnitude of the total force Q_{mag} for a single ray hitting a dielectric sphere of index of refraction $n = 1.2$ at an angle θ.

striking the sphere at angle θ. One can show that the gradient force as defined above is conservative. This follows from the fact that F_g, the gradient force for a ray, can be expressed solely as a function of ρ, the radial distance from the ray to the particle. This implies that the integral of the work done on a particle in going around an arbitrary closed path can be expressed as an integral of $F_g(\rho)d\rho$ which is clearly zero. If the gradient force for a single ray is conservative, then the gradient force for an arbitrary collection of rays is conservative. Thus the conservative property of the gradient force as defined in the geometric optics regime is the same as in the Rayleigh regime. The work done by the scattering force, however, is always path dependent and is not conservative in any regime. As will be seen, these new definitions of gradient and scattering force for beams of more complex shape allow us to describe the operation of the gradient trap in the same manner in both the geometrical optics and Rayleigh regimes.

To get a feeling for the magnitudes of the forces, we calculate the scattering force F_S, the gradient force F_g, and the absolute magnitude of the total force $F_{mag} = (F_S^2 + F_g^2)^{1/2}$ as a function of the angle of incidence θ using Eqs. 1 and 2. We consider as a typical example the case of a circularly polarized ray hitting a sphere of effective index of refraction $n = 1.2$. The force for such a circularly polarized ray is the average of the forces for rays polarized perpendicular and parallel to the plane of incidence. The effective index of a particle is defined as the index of the particle n_2 divided by the index of the surrounding medium n_1; that is, $n = n_2/n_1$. A polystyrene sphere in water has $n = 1.6/1.33 \cong 1.2$. Fig. 4 shows the results for the forces F_S, F_g, and F_{mag} versus θ expressed in terms of the dimensionless factors Q_s, Q_g, and $Q_{mag} = (Q_s^2 + Q_g^2)^{1/2}$, where

$$F = Q\frac{n_1 P}{c}. \qquad (3)$$

The quantity $n_1 P/c$ is the incident momentum per second of a ray of power P in a medium of index of refraction n_1 (19, 31). Recall that the maximum radiation pressure force derivable from a ray of momentum per second $n_1 P/c$ corresponds to $Q = 2$ for the case of a ray reflected perpendicularly from a totally reflecting mirror. One sees that for $n = 1.2$ a maximum gradient force of Q_{gmax} as high as ~ 0.5 is generated for rays at angles of $\theta \cong 70°$. Table I shows the effect of an index of refraction n on the maximum value of gradient force Q_{gmax} occurring at angle of incidence θ_{gmax}. The corresponding value of scattering force Q_s at θ_{gmax} is also listed. The fact that Q_s continues to grow relative to Q_{gmax} as n increases indicates potential difficulties in achieving good gradient traps at high n.

FORCE OF THE GRADIENT TRAP ON SPHERES

Trap focus along Z axis

Consider the computation of the force of a gradient trap on a sphere when the focus f of the trapping beam is located along the Z axis at a distance S above the center of the sphere at O, as shown in Fig. 2. The total force on the sphere, for an axially-symmetric plane-polarized input trapping beam, is clearly independent of the direction of polarization by symmetry considerations. It can therefore be assumed for convenience that the input beam is circularly polarized with half the power in each of two orthogonally oriented polarization components. We find the force for a ray entering the input aperture of the microscope objective at an arbitrary radius r and angle β and then integrate numerically over the distribution of input rays using an AT&T 1600 PLUS personal computer. As seen in Fig. 2, the vertical plane ZW which is rotated by β from the ZY plane contains both the incident ray and the normal to the sphere $\hat{n}$. It is thus the plane of incidence. We can compute the angle of

TABLE 1 For a single ray. Effect of index of refraction n on maximum gradient force Q_{gmax} and scattering force Q_s occurring at angle of incidence θ_{gmax}

n	Q_{gmax}	Q_s	θ_{gmax}
1.1	−0.429	0.262	79°
1.2	−0.506	0.341	72°
1.4	−0.566	0.448	64°
1.6	−0.570	0.535	60°
1.8	−0.547	0.625	59°
2.0	−0.510	0.698	59°
2.5	−0.405	0.837	64°

incidence θ from the geometric relation $R \sin \theta = S \sin \phi$, where R is the radius of the sphere. We take $R = 1$ since the resultant forces in the geometric optics limit are independent of R. Knowing θ we can find F_g and F_s for the circularly polarized ray by first computing F_g and F_s for each of the two polarization components parallel and perpendicular to the plane of incidence using Eqs. 1 and 2 and adding the results. It is obvious by symmetry that the net force is axial. Thus for S above the origin O the contribution of each ray to the net force consists of a negative Z component $F_{gz} = -F_g \sin \phi$ and a positive Z component $F_{sz} = F_s \cos \phi$ as seen from Fig. 2 B. For S below O the gradient force component changes sign and the scattering force component remains positive. We integrate out to a maximum radius r_{max} for which $\phi = \phi_{max} = 70°$, the maximum convergence angle for a water immersion objective of NA = 1.25, for example. Consider first the case of a sphere of index of refraction $n = 1.2$ and an input beam which uniformly fills the input aperture. Fig. 5 shows the magnitude of the antisymmetric gradient force component, the symmetric scattering force component, and the total force, expressed as Q_g, Q_s, and Q_t, for values of S above and $(-S)$ below the center of the sphere. The sphere outline is shown in Fig. 5 for reference. It is seen that the trapping forces are largely confined within the spherical particle. The stable equilibrium point S_E of the trap is located just above the

center of the sphere at $S \cong 0.06$, where the backward gradient force just balances the weak forward scattering force. Away from the equilibrium point the gradient force dominates over the scattering force and Q_t reaches its maximum value very close to the sphere edges at $S \cong 1.01$ and $(-S) \cong 1.02$. The large values of net restoring force near the sphere edges are due to the significant fraction of all incident rays which have both large values of θ, near the optimum value of 70°, and large convergence angle ϕ. This assures a large backward gradient force contribution from the component $F_g \sin \phi$ and also a much-reduced scattering force contribution from the component $F_s \cos \phi$.

Trap along Y axis

We next examine the trapping forces for the case where the focus f of the trapping beam is located transversely along the $-Y$ axis of the sphere as shown in Fig. 6. The details of the force computation are discussed in Appendix II. Fig. 7 plots the gradient force, scattering force, and total force in terms of Q_g, Q_s, and Q_t as a function of the distance S' of the trap focus from the origin along the $-Y$ axis for the same conditions as in III A. For this case the gradient force has only a $-Y$ component. The scattering force is orthogonal to it along the $+Z$ axis. The total force again maximizes at a value $Q_t \cong 0.31$ near the sphere edge at $S' \cong 0.98$ and makes a small angle $\phi = \arctan F_g/F_s \cong 18.5°$ with respect to the Y axis. The Y force is, of course, symmetric about the center of the sphere at O.

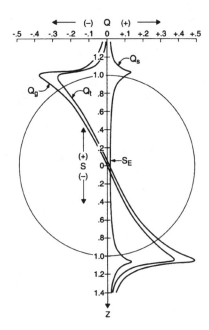

FIGURE 5 Values of the scattering force, gradient force, and total force Q_s, Q_g, and Q_t exerted on a sphere of index of refraction $n = 1.2$ by a trap with a uniformly filled input aperture which is focused along the Z axis at positions $+s$ above and $-s$ below the center of the sphere.

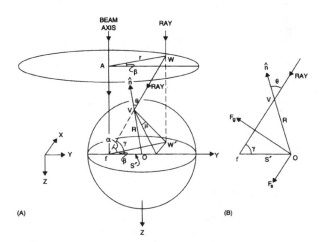

FIGURE 6 (A) Trap geometry with the beam focus f located transversely along the $-Y$ axis at a distance S' from the origin. (B) Geometry of the plane of incidence showing the directions of the gradient and scattering forces F_g and F_s for the input ray.

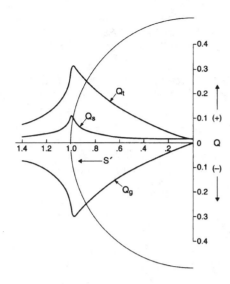

FIGURE 7 Plot of the gradient force, scattering force, and total force Q_g, Q_s, and Q_t as a function of the distance S' of trap focus from the origin along the $-Y$ axis for a circularly polarized trapping beam uniformly filling the aperture and a sphere of index of refraction $n = 1.2$.

General case: arbitrary trap location

Consider finally the most general case where the focus f is situated arbitrarily in the vertical plane through the Z axis at the distance S' from the sphere origin O in the direction of the $-Y$ axis and a distance S'' in the direction of the $-Z$ axis as shown in Fig. 8. Appendix III summarizes the method of force computation for this case.

Fig. 10 shows the magnitude and direction of the gradient force Q_g, the scattering force Q_s, and the total force Q_t as functions of the position of the focus f over the left half of the YZ plane, and by mirror image symmetry about the Y axis, over the entire cross-section of the sphere. This is again calculated for a circularly polarized beam uniformly filling the aperture and for $n = 1.2$. Although the force vectors are drawn at the point of focus f, it must be understood that the actual forces always act through the center of the sphere. This is true for all rays and therefore also for the full beam. It is an indication that no radiation pressure torques are possible on a sphere from the linear momentum of light. We see in Fig. 10 A that the gradient force which is exactly radial along the Z and Y axes is also very closely radial (within an average of $\sim 2°$ over the rest of the sphere. This stems from the closely radially uniform distribution of the incident light in the upper hemisphere. The considerably smaller scattering force is shown in Fig. 10 B (note the change in scale). It is strictly

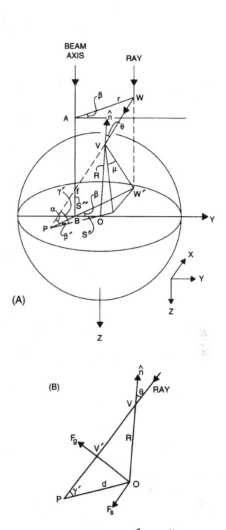

FIGURE 8 (*A*) Trap geometry with the beam focus located at a distance S' from the origin in the $-Y$ direction and a distance S'' in the $-Z$ direction. (*B*) Geometry of the plane of incidence POV showing the direction of gradient and scattering forces F_g and F_s for the ray. Geometry of triangle POB in the XY plane for finding β' and d.

axial only along the Z and Y axes and remains predominantly axial elsewhere except for the regions farthest from the Z and Y axes. It is the dominance of the gradient force over the scattering force that accounts for the overall radial character of the total force in Fig. 10 C. The rapid changes in direction of the force that occur when the focus is well outside the sphere are mostly due to the rapid changes in effective beam direction as parts of the input beam start to miss the

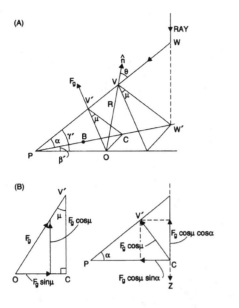

FIGURE 9   Another view of Fig. 8 *A* containing the angle μ between the plane of incidence POV' and the vertical plane WW'P for resolving force components along the coordinate axis.

sphere. We note that the magnitude of the total force Q_t maximizes very close to the edge of the sphere as we proceed radially outward in all directions, as does the gradient and scattering forces. The value of maximum restoring forces varies smoothly around the edge of the sphere from a minimum of $Q_t = 0.28$ in the axially backward direction to a maximum of $Q_t = 0.49$ in the forward direction. Thus, for these conditions the maximum trapping force achieved varies quite moderately over the sphere by a factor of $0.49/0.28 = 1.78$ and conforms closely to the edges of the sphere.

The line EE' marked on Fig. 10 *C* represents the locus of points for which the Z component of the force is zero; i.e., the net force is purely horizontal. If one starts initially at point E, the equilibrium point of the trap with no externally applied forces, and then applies a +Y-directed Stokes force by flowing liquid past the sphere to the right, for example, the equilibrium position will shift to a new equilibrium point along EE' where the horizontal light force just balances the viscous force. With increasing viscous force the focus finally moves to E', the point of maximum transverse force, after which the sphere escapes the trap. Notice that there is a net z displacement of the sphere as the equilibrium point moves from E to E'. We have observed this effect in experiments with micron-sized polystyrene spheres. Sato et al. (18) have recently reported also seeing this displacement.

EFFECT OF MODE PROFILES AND INDEX OF REFRACTION ON TRAPPING FORCES

To achieve a uniformly filled aperture in practice requires an input TEM_{00} mode Gaussian beam with very large spot size, which is wasteful of laser power. We therefore consider the behavior of the trap for other cases of TEM_{00} mode input beam profiles with smaller spot sizes, as well as $TEM_{01}*$ "do-nut" mode beam profiles which preferentially concentrate input light intensity at large input angles ϕ.

TEM_{00} mode profile

Table II compares the performance of traps with $n = 1.2$ having different TEM_{00} mode intensity profiles of the form $I(r) = I_o \exp(-2r^2/w_o^2)$ at the input aperture of the microscope objective. The quantity a is the ratio of the TEM_{00} mode beam radius w_o to the full lens aperture r_{max}. A is the fraction of total beam power that enters the lens aperture. A decreases as a increases. In the limit of a uniform input intensity distribution $A = 0$ and $a = \infty$. For $w_o \leq r_{max}$ we define the convergence angle of the input beam as θ' where $\tan \theta' = w_o/\ell$. ℓ is the distance from the lens to the focus f as shown in Fig. 2 *B*. For $w_o > r_{max}$ the convergence angle is set by the full lens aperture and we use $\theta' = \phi_{max}$, where $\tan \phi_{max} = r_{max}/\ell$. For a NA = 1.25 water immersion objective $\phi_{max} = 70°$. The quality of the trap can be characterized by the maximum strength of the restoring forces as one proceeds radially outward for the sphere origin O in three representative directions taken along the Z and Y axes. We thus list Q_{1max}, the value of the maximum restoring force along the −Z axis, and S_{max}, the radial distance from the origin at which it occurs. Similarly listed are Q_{2max} occurring at S'_{max} along the −Y axis and Q_{3max} occurring at $(-S)_{max}$ along the +Z axis (see Figs. 2, 5, and 6 for a reminder on the definitions of S, $-S$, and S'). S_E in Table II gives the location of the equilibrium point of the trap along the −Z axis as noted in Fig. 5.

One sees from Table II that the weakest of the three representative maximum restoring forces is Q_{1max} occurring in the −Z direction. Furthermore, of all the traps the $a = \infty$ trap with a uniformly filled aperture has the largest Q_{1max} force and is therefore the strongest of all the TEM_{00} mode traps. One can also define the "escape force" of a given trap as the lowest force that can pull the particle free of the trap in any direction. In this context the $a = \infty$ trap has the largest magnitude of escape force of $Q_{1max} = 0.276$. One also sees that the $a = \infty$ trap is the most uniform trap since it has the smallest fractional variation in the extreme values of the restoring forces Q_{1max} and Q_{3max}. If, however, we reduce a to 1.7

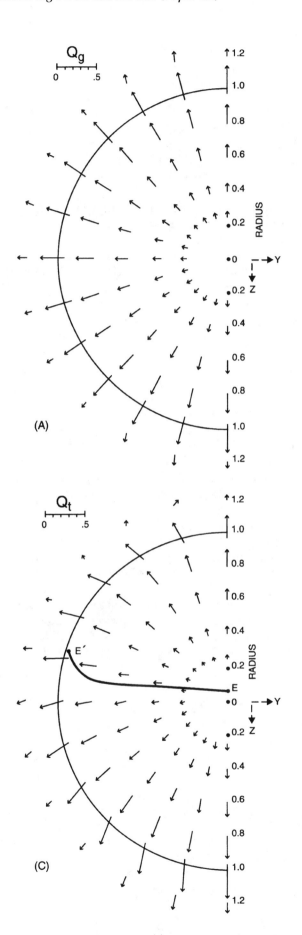

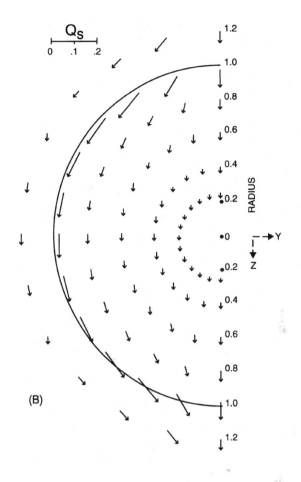

FIGURE 10 *A*, *B*, and *C* show the magnitude and direction of gradient, scattering, and total force vectors $\bar{Q}_g$, $\bar{Q}_s$, and $\bar{Q}_t$ as a function of position of the focus over the YZ plane, for a circularly polarized trapping beam uniformly filling the aperture and a sphere of $n = 1.2$. $\bar{Q}_s$ is the vector sum of $\bar{Q}_g$ and $\bar{Q}_s$. EE' in *C* indicates the line along which $\bar{Q}_t$ is purely horizontal.

TABLE 2   Performance of TEM$_{\infty}$ mode tapes with $n = 1.2$ having different intensity profiles at the input of the microscope objective

a	A	$[Q_{1max},$	$S_{max}]$	$[Q_{2max},$	$S'_{max}]$	$[Q_{3max},$	$(-S)_{max}]$	S_E	θ'
∞	0	−0.276	1.01	0.313	0.98	0.490	1.05	0.06	70°
1.7	0.5	−0.259	1.01	0.326	0.98	0.464	1.05	0.08	70°
1.0	0.87	−0.225	1.02	0.349	0.98	0.412	1.05	0.10	70°
0.727	0.98	−0.184	1.03	0.383	0.98	0.350	1.06	0.13	63°
0.364	1.0	−0.077	1.15	0.498	0.98	0.214	1.3	0.32	45°
0.202	1.0	−0.019	1.4	0.604	0.98	0.147	1.9	0.80	29°

or even 1.0, where the fraction of input power entering the aperture is reasonably high (~ 0.50 or 0.87), one can still get performance close to that of the uniformly filled aperture. Trap performance, however, rapidly degrades for cases of underfilled input aperture and decreasing beam convergence angle. For example, in the trap with $a = 0.202$ and $\theta' \cong 29°$ the value of Q_{1max} has dropped more than an order of magnitude to $Q_{1max} = -0.019$. The maximum restoring forces Q_{1max} and Q_{2max} occur well outside the sphere and the equilibrium position has moved away from the origin to $S_E = 0.8$. This trap with $\theta' \cong 29°$ roughly corresponds to the best of the traps described by Wright et al. (27) (for the case of $w_o = 0.5$ μm). They find for $w_o = 0.5$ μm that the trap has an equilibrium position outside of the sphere and a maximum trapping force equivalent to $Q_{1max} = -0.055$. Any more direct comparision of our results with those of Wright et al. is not possible since they use an approximate force calculation which overestimates the forces somewhat. They do not calculate forces for the beam focus inside the sphere and there are other artifacts associated with their use of Gaussian beam phase fronts to give the incident ray directions near the beam focus.

TEM$_{01}^{*}$ "do-nut" mode profile

Table III compares the performance of several traps based on the TEM$_{01}^{*}$ mode, the so-called "do-nut" mode, which has an intensity distribution of the form $I(r) = I_o (r/w'_o)^2 \exp(-2r^2/w'^2_o)$. The quantity a is now the ratio

of w'_o, the spot size of the do-nut mode, to the full lens aperture r_{max}. All other items in the table are the same as in Table II. For $a = 0.76 \sim 87\%$ of the total beam power enters the input aperture r_{max} and one obtains performance that is almost identical to that of the trap with uniformly filled aperture as listed in Table II. For larger values of a the absolute magnitude of Q_{1max} increases, the magnitude of Q_{2max} decreases, and the fraction of power entering the aperture decreases. Optimal trapping, corresponding to the highest value of escape force, is achieved at values of $a \cong 1.0$ where the magnitudes $Q_{1max} \cong Q_{2max} \cong 0.30$. This performance is somewhat better than achieved with TEM$_{00}$ mode traps.

It is informative to compare the performance of do-nut mode traps with that of a so-called "ring trap" having all its power concentrated in a ring 95–100% of the full beam aperture, for which $\phi \cong \phi_{max} = 70°$. When the ring trap is focused at $S \cong 1.0$ essentially all of the rays hit the sphere at an angle of incidence very close to $\theta_{gmax} = 72°$, the angle that makes Q_g a maximum for $n = 1.2$ (see Table I). Thus the resulting backward total force of $Q_{1max} = 0.366$ at $S = 0.99$, as listed in Table III, closely represents the highest possible backward force on a sphere of $n = 1.2$. The ring trap, however, has a reduced force $Q_{2max} = 0.254$ at $S'_{max} = 0.95$ in the $-Y$ direction since many rays at this point are far from optimal. If we imagine adding an axial beam to the ring beam then we optimally increase the gradient contribution to the force in the $-Y$ direction near $S' = 1.0$ and decrease the

TABLE 3   Performance of TEM*$_{01}$ mode traps with $n = 1.2$ having different intensity profiles at the input of the microscope objective

a	A	$[Q_{1max},$	$S_{max}]$	$[Q_{2max},$	$S'_{max}]$	$[Q_{3max},$	$(-S)_{max}]$	S_E
				TEM*$_{01}$ do-nut mode traps				
1.21	0.40	−0.310	1.0	0.290	0.98	0.544	1.05	0.06
1.0	0.59	−0.300	1.01	0.296	0.98	0.531	1.05	0.06
0.938	0.66	−0.296	1.01	0.298	0.98	0.525	1.05	0.07
0.756	0.87	−0.275	1.01	0.311	0.98	0.494	1.06	0.10
				Ring beam with $\phi = 70°$				
		−0.366	0.99	0.254	0.95	0.601	1.03	
				Ring beam plus axial beam				
		−0.31	0.99	0.31	0.95	0.51	1.03	

Comparison data on a Ring Beam having $\phi = 70°$ and a Ring Beam plus an Axial Beam containing 18% of the power.

overall force in the $-Z$ direction. With 18% of the power in the axial beam one gets $Q_{1max} = Q_{2max} \cong 0.31$. This performance is now close to that of the optimal do-nut mode trap. It is possible to design gradient traps that approximate the performance of a ring trap using a finite number of individual beams (for example, four, three or two beams) located symmetrically about the circumference of the ring and converging to a common focal point at angles of $\phi \cong 70°$. Recent reports (32, 33) at the CLEO-'91 conference presented observations on a trap with two individual beams converging to a focus with $\phi \cong 65°$ and also on a single beam gradient trap using the TEM$_{01}^*$ mode.

Knowledge of the forces produced by ring beams allows one to compare the forces generated by bright field microscope objectives, as have thus far been considered, with the forces from phase contrast objectives of the same NA. For example, assume a phase contrast objective having an 80% absorbing phase ring located between radii of 0.35 and 0.55 of the full input lens aperture. For the case of an input beam uniformly filling the aperture with $n = 1.2$, one finds that the bright field escape force of $Q_{1max} = 0.276$ (see Table II) increases by $\sim 4\%$ to $Q_{1max} = 0.287$ in going to the phase contrast objective. With a TEM$_{00}$ mode Gaussian beam input having $A = 0.87$ and $n = 1.2$, the bright field escape force magnitude of $Q_{1max} = 0.225$ increases by $\sim 2\%$ to $Q_{1max} = 0.230$ for a phase contrast objective. The reason for these slight improvements is that the force contribution of rays at the ring corresponds to $Q_{1max} \cong 0.204$, which is less than the average force for bright field. Thus any removal of power at the ring radius improves the overall force per unit transmitted power. Differential interference contrast optics can make use of the full input lens aperture and thus gives equivalent trapping forces to bright field optics.

Index of refraction effects

Consider, finally, the role of the effective index of refraction of the particle $n = n_1/n_2$ on the forces of a single-beam gradient trap. In Table IV we vary n for two types of trap, one with a uniformly filled input aperture, and the other having a do-nut input beam with $a = 1.0$, for which the fraction of total power feeding the input aperture is 59%. For the case of the uniformly filled aperture we get good performance over the range $n = 1.05$ to $n \cong 1.5$, which covers the regime of interest for most biological samples. At higher index Q_{1max} falls to a value of -0.097 at $n = 2$. This poorer performance is due to the increasing scattering force relative to the maximum gradient force as n increases (see Table I). Also the angle of incidence for maximum gradient force falls for higher n. At $n = 2$ (which corresponds roughly to a particle of index ~ 2.7 in water of index 1.33), the do-nut mode trap is clearly better than the uniform beam trap.

CONCLUDING REMARKS

It has been shown how to define the gradient and scattering forces acting on dielectric spheres in the ray optics regime for beams of complex shape. One can then describe the operation of single beam gradient force traps for spheres of diameter $\gg \lambda$ in terms of the dominance of an essentially radial gradient force over the predominantly axial scattering force. This is analogous to the previous description of the operation of this

TABLE 4   **Effect of index of refraction n on the performance of a trap with a uniformly filled aperture ($a = \infty$) and a do-nut trap with $a = 1.0$**

n	[Q_{1max},	S_{max}]	[Q_{2max},	S'_{max}]	[Q_{3max},	$(-S)_{max}$]	S_E
			Trap with uniformly filled aperture				
1.05	−0.171	1.06	0.137	1.00	0.219	1.06	0.02
1.1	−0.231	1.05	0.221	0.99	0.347	1.06	0.04
1.2	−0.276	1.01	0.313	0.98	0.490	1.05	0.06
1.3	−0.288	0.96	0.368	0.97	0.573	1.04	0.11
1.4	−0.282	0.93	0.403	0.96	0.628	1.02	0.15
1.6	−0.237	0.89	0.443	0.94	0.693	1.00	0.25
1.8	−0.171	0.88	0.461	0.94	0.723	0.99	0.37
2.0	−0.097	0.88	0.469	0.94	0.733	0.99	0.53
			TEM$_{01}^*$ do-nut mode trap with $a = 1.0$				
1.05	−0.185	1.06	0.134	1.00	0.238	1.06	0.02
1.1	−0.250	1.05	0.208	0.99	0.379	1.06	0.03
1.2	−0.300	1.01	0.296	0.98	0.531	1.05	0.06
1.4	−0.309	0.93	0.382	0.95	0.667	1.02	0.13
1.8	−0.204	0.88	0.434	0.94	0.748	0.99	0.32
2.0	−0.132	0.88	0.439	0.94	0.752	0.99	0.42

trap in the Rayleigh regime, where the diameter $\ll \lambda$. Quite strong uniform traps are possible for $n = 1.2$ using the TEM_{01}^* do-nut mode in which the trapping forces vary over the sphere cross-section from a Q value of -0.30 in the $-Z$ direction to 0.53 in the $+Z$ direction. The magnitude of trapping force of 0.30 in the weakest trapping direction gives the escape force which a spherically shaped motile living organism, for example, must exert in order to escape the trap. For a laser power of 10 mW the minimum trapping force or escape force of $Q = 0.30$ is equivalent to 1.2×10^{-6} dynes. This implies that a motile organism 10 μm in diameter which is capable of propelling itself through water at a speed of 128 μm/s will be just able to escape the trap in its weakest direction along the $-Z$ axis. The only possible drawback to using the do-nut mode in practice is the difficulty of generating that mode in the laser. With the simpler TEM_{00} mode beams one can achieve traps with Q's as high as 0.23, for example, with 87% of the laser power entering the aperture of the microscope objective.

The calculation confirms the importance of using beams with large convergence angles θ' as high as $\sim 70°$ for achieving strong traps, especially with particles having lower indices of refraction typical of biological samples. At small convergence angles, less than $\sim 30°$, the scattering force dominates over the gradient force and single beam trapping is either marginal or not possible. One can, however, make a two-beam gradient force trap using smaller convergence angles based on two confocal, oppositely directed beams of equal power in which each ray of the converging beam is exactly matched by an oppositely directed ray. Then the scattering forces cancel and the gradient forces add, giving quite a good trap. Gradient traps of this type have been previously observed in experiments on alternating beam traps (34). The advantage of lower beam convergence is the ability to use longer working distances.

This work using ray optics extends the quantitative description of the single beam gradient trap for spheres to the size regime where the diameter is $\gg \lambda$. In this regime the force is independent of particle radius r. In the Rayleigh regime the force varies as r^3. At present there is no quantitative calculation for the intermediate size regime where the diameter is $\approx \lambda$, in which we expect force variations between r^0 and r^3. This is a more difficult scattering problem and involves an extension of Mie theory (35) or vector methods (36) to the case of highly convergent beams. Experimentally, however, this intermediate regime presents no problems. One can often directly calibrate the magnitude of the trapping force using Stokes dragging forces and thus successfully perform experiments with biological particles of size $\approx \lambda$ (16).

One can get a good idea of the range of validity of the trapping forces as computed in the ray optics regime from a comparison of the scattering of a plane wave by a large dielectric sphere in the ray optics regime with the exact scattering, including all diffraction effects, as given by Mie theory. It suffices to consider plane waves since complex beams can be decomposed into a sum of plane waves. It was shown by van de Hulst in Chapter 12 of his book (35) that ray optics gives a reasonable approximation to the exact angular intensity distribution of Mie theory (except in a few special directions) for sphere size parameters $2\pi r/\lambda = 10$ or 20. The special directions are the forward direction, where a large diffraction peak appears which contributes nothing to the radiation pressure, and the so-called glory and rainbow directions, where ray optics never works. Since these directions contribute only slightly to the total force, we expect ray optics to give fair results down to diameters of approximately six wavelengths or ~ 5 μm for a 1.06-μm laser beam in water. The validity of the approximation should improve rapidly at larger sphere diameters. A similar result was also derived by van de Hulst (35) using Fresnel zones to estimate diffractive effects.

One of the advantages of a reliable theoretical value for the trapping force is that it can serve as a reference for comparison with experiment. If discrepancies appear in such a comparison one can then look for the presence of other forces. For traps using infrared beams there could be significant thermal (radiometric) force contributions due to absorptive heating of the particle or surrounding medium whose magnitude could then be inferred. Detailed knowledge of the variation of trapping force with position within the sphere is also proving useful in measurements of the force of swimming sperm (15).

Received for publication 19 June 1991 and in final form 16 August 1991.

APPENDIX I

Force of a ray on a dielectric sphere

A ray of power P hits a sphere at an angle θ where it partially reflects and partially refracts, giving rise to a series of scattered rays of power $PR, PT^2, PT^2R, \ldots, PT^2R^n, \ldots$. As seen in Fig. 3, these scattered rays make angles relative to the incident forward ray direction of $\pi + 2\theta$, α, $\alpha + \beta, \ldots, \alpha + n\beta \ldots$, respectively. The total force in the Z direction is the net change in momentum per second in the Z direction due to the scattered rays. Thus:

$$F_z = \frac{n_1 P}{c}$$

$$- \left[\frac{n_1 PR}{c} \cos(\pi + 2\theta) + \sum_{n=0}^{\infty} \frac{n_1 P}{c} T^2 R^n \cos(\alpha + n\beta) \right], \quad \text{(A1)}$$

where $n_1 P/c$ is the incident momentum per second in the Z direction. Similarly for the Y direction, where the incident momentum per second is zero, one has:

$$F_Y = 0$$

$$- \left[\frac{n_1 PR}{c} \sin(\pi + 2\theta) - \sum_{n=0}^{\infty} \frac{n_1 P}{c} T^2 R^n \sin(\alpha + \beta) \right]. \quad (A2)$$

As pointed out by van de Hulst in Chapter 12 of reference 35 and by Roosen (22), one can sum over the rays scattered by a sphere by considering the total force in the complex plane, $F_{tot} = F_Z + i F_Y$. Thus:

$$F_{tot} = \frac{n_1 P}{c} [1 + R \cos 2\theta] + i \frac{n_1 P}{c} R \sin 2\theta$$

$$- \frac{n_1 P}{c} T^2 \sum_{n=0}^{\infty} R^n e^{i(\alpha + n\beta)}. \quad (A3)$$

The sum over n is a simple geometric series which can be summed to give:

$$F_{tot} = \frac{n_1 P}{c} [1 + R \cos 2\theta]$$

$$+ i \frac{n_1 P}{c} R \sin 2\theta - \frac{n_1 P}{c} T^2 e^{i\alpha} \left[\frac{1}{1 - Re^{i\beta}} \right]. \quad (A4)$$

If one rationalizes the complex denominator and takes the real and imaginary parts of F_{tot}, one gets the force expressions A1 and A2 for F_Z and F_Y using the geometric relations $\alpha = 2\theta - 2r$ and $\beta = \pi - 2r$, where θ and r are the angles of incidence and refraction of the ray.

APPENDIX II

Force on a sphere for trap focus along Y axis

We treat the case of the beam focus located along the $-Y$ axis at a distance S' from the origin O (see Fig. 6). We first calculate the angle of incidence θ for an arbitrary ray entering the input lens aperture vertically at a radius r and azimuthal angle β in the first quadrant. On leaving the lens the ray stays in the vertical plane AWW'f and heads in the direction towards f, striking the sphere at V. The forward projection of the ray makes an angle α with respect to the horizontal (X, Y) plane. The plane of incidence, containing both the input ray and the normal to the sphere OV, is the so-called γ plane fOV which meets the horizontal and vertical planes at f. Knowing α and β, we find γ from the geometrical relation $\cos \gamma = \cos \alpha \cos \beta$. Referring to the γ plane we can now find the angle of incidence θ from $R \sin \theta = S' \sin \gamma$ putting $R = 1$.

In contrast to the focus along the Z axis, the net force now depends on the choice of input polarization. For the case of an incident beam polarized perpendicular to the Y axis, for example, one first resolves the polarized electric field E into components $E \cos \beta$ and $E \sin \beta$ perpendicular and parallel to the vertical plane containing the ray. Each of these components can be further resolved into the so-called p and s components parallel and perpendicular to the plane of incidence in terms of these angle μ between the vertical plane and the plane of incidence. By geometry, $\cos \mu = \tan \alpha / \tan \gamma$. This resolution yields fractions of the input power in the p and s components given by:

$$f_p = (\cos \beta \sin \mu - \sin \beta \cos \mu)^2 \quad (A5)$$

$$f_s = (\cos \beta \cos \mu + \sin \beta \sin \mu)^2. \quad (A6)$$

If the incident polarization is parallel to the Y axis, then f_p and f_s reverse. Knowing $\theta, f_p,$ and f_s, one computes the gradient and scattering force components for p and s separately using Eqs. A5 and A6 and adds the results.

The net gradient and scattering force contribution of the ray thus computed must now be resolved into components along the coordinate axes (see Fig. 6 B). However, comparing the force contributions of the quartet of rays made up of the ray in the first quadrant and its mirror image rays in the other quadrants we see that the magnitudes of the forces are identical for each of the rays of the quartet. Furthermore, the scattering and gradient forces of the quartet are directly symmetrically about the Z and Y axes, respectively. This symmetry implies that the entire beam can only give rise to a net Z scattering force coming from the integral of the $F_s \cos \phi$ component and a net Y gradient force coming from the $F_g \sin \gamma$ component. In practice we need only integrate these components over the first quadrant and multiply the results by 4 to get the net force. The differences in force that result from the choice of input polarization perpendicular or parallel to the Y axis are not large. For the conditions of Fig. 7 the maximum force difference is $\sim 14\%$ near $S' \cong 1.0$. We have therefore made calculations using a circularly polarized input beam with $f_p = f_s = \frac{1}{2}$, which yields values of net force that are close to the average of the forces for the two orthogonally polarized beams.

APPENDIX III

Force on a sphere for an arbitrarily located trap focus

We now treat the case where the trapping beam is focused arbitrarily in the XY plane at a point f located at a distance S' from the origin in the $-Y$ direction and a distance S'' in the $-Z$ direction (see Fig. 8). To calculate the force for a given ray we again need to find the angle of incidence θ and the fraction of the ray's power incident on the sphere in the s and p polarizations. Consider a ray of the incident beam entering the input aperture of the lens vertically at a radius r and azimuthal angle β in the first quadrant. The ray on leaving the lens stays in the vertical plane AWW'B and heads toward f, hitting the sphere at V. The extension of the incident ray to f and beyond intersects the XY plane at point P at an angle α. The plane of incidence for this ray is the so-called γ' plane POV which contains both the incident ray and the normal to the sphere OV. Referring to the planar figure in Fig. 8 B one can find the angle β' by simple geometry in terms of $S', S'',$ and the known angles α and β from the relation

$$\tan \beta' = \frac{S' \sin \beta}{S' \cos \beta + S''/\tan \alpha}. \quad (A7)$$

We get γ' from $\cos \gamma' = \cos \alpha \cos \beta'$. Referring to the γ' plane in Fig. 8 B we get the angle of incidence θ for the ray from $R \sin \theta = d \sin \gamma'$, putting $R = 1$. The distance d is deduced from the geometric relation:

$$d = \frac{S'' \cos \beta'}{\tan \alpha} + S' \cos(\beta - \beta'). \quad (A8)$$

As in Appendix II, we compute f_p and f_s, the fraction of the ray's power in the p and s polarizations, in terms of the angle μ between the vertical plane W'VP and the plane of incidence POV. We use Eqs. A5 and A6 for the case of a ray polarized perpendicular to the Y axis and the same expressions with f_p and f_s reversed for a ray polarized parallel to the Y axis. To find μ we use $\cos \mu = \tan \alpha / \tan \gamma'$. As in Appendix II we can put $f_p = f_s = \frac{1}{2}$ and get the force for a circularly polarized ray, which is the average of the force for the cases of two orthogonally polarized rays.

The geometry for resolving the net gradient and scattering force contribution of each ray of the beam into components along the axes is now more complex. The scattering force F_s is directed parallel to the incident ray in the VP direction of Fig. 8. It has components $F_s \sin \alpha$ in the $+Z$ direction and $F_s \cos \alpha$ pointing in the BP direction in the XY plane. $F_s \cos \alpha$ is then resolved with the help of Fig. 8 *B* into $F_s \cos \alpha \cos \beta$ in the $-Y$ direction and $F_s \cos \alpha \sin \beta$ in the $-X$ direction. The gradient force F_g points in the direction OV' perpendicular to the incident ray direction VP in the plane of incidence OPV. This is shown in Fig. 8 and also Fig. 9, which gives yet another view of the geometry. In Fig. 9 we consider the plane V'OC, which is taken perpendicular to the γ' plane POV and the vertical plane WW'P. This defines the angle OV'C as μ, the angle between the planes, and also makes the angles OCV', OCP, and CV'P right angles. As an aid to visualization one can construct a true three-dimensional model out of cardboard of the geometric figure for the general case as shown in Figs. 8 and 9. Such a model will make it easy to verify that the above stated angles are indeed right angles, and to see other details of the geometry. We can now resolve F_g into components along the X, Y, and Z axes with the help of right triangles OV'C and CV'P as shown in Fig. 9 *B*. In summary, the net contribution of a ray in the first quadrant to the force is:

$$F(Z) = F_s \sin \alpha + F_g \cos \mu \cos \alpha \qquad (A9)$$

$$F(Y) = -F_s \cos \alpha \cos \beta$$
$$+ F_g \cos \mu \sin \alpha \cos \beta + F_g \sin \mu \sin \beta \qquad (A10)$$

$$F(X) = -F_s \cos \alpha \sin \beta$$
$$+ F_g \cos \mu \sin \alpha \sin \beta - F_g \sin \mu \cos \beta. \qquad (A11)$$

The force equations A9–A11 are seen to have the correct signs since F_s and F_g are, respectively, positive and negative as calculated from Eqs. 1 and 2.

For the general case under consideration we lose all symmetry between first and second quadrant forces and we must extend the force integrals into the second quadrant. All the above formulas which were derived for rays of the first quadrant are equally correct in the second quadrant using the appropriate values of the angles β, β', γ', and μ. For example, in the second quadrant β' can be obtuse. This gives obtuse γ' and obtuse μ. Obtuse μ implies that the γ' plane has rotated its position beyond the perpendicular to the vertical plane AWW'. In this orientation the gradient force direction tips below the XY plane and reverses its Z component as indicated by the sign change in the $F_g \cos \mu \cos \alpha$ term.

There are, however, some symmetry relations in the force contributions of rays of the input beam which still apply. For example, there is symmetry about the Y axis; i.e., rays of the third and fourth quadrants give the same contribution to the Z and Y forces as rays of the first and second quadrants, whereas their X contributions exactly cancel. To find the net force we need only integrate the Y and Z components of first and second quadrants and double the result.

If we make S'' negative in all formulas, we obtain the correct magnitudes and directions of the forces for the case of the focus below the XY plane. Although we find different total force values for S'' positive and S'' negative, i.e., symmetrical beam focus points above and below the XY plane, there still are symmetry relations that apply to the scattering and gradient forces separately. Thus we find that the

Z components of the scattering force are the same above and below but the Y component reverse. For the gradient force the Z components reverse above and below and the Y components are the same. This is seen to be true in Fig. 10. It is also consistent with Fig. 5 showing the forces along the Z axis. This type of symmetry behavior arises from the fact that the angle of incidence for rays entering the first quadrant from above the XY plane (S'' positive) is the same as for symmetrical rays entering in the second quadrant below the XY plane (S'' negative). Likewise the angles of incidence are the same for the second quadrant above and the first quadrant below. These results permit one to directly deduce the force below the XY plane from the values computed above the XY plane. The results derived here for the focus placed at an arbitrary point within the YZ plane are perfectly general since one can always choose to calculate the force in the cross-sectional plane through the Z axis that contains the focus f.

As a check on the calculations one can show that the results putting $S'' = 0$ in the general case are identical with those from the simpler Y axis integrals derived in Appendix II. Also in the limit $S' \rightarrow 0$ one gets the same results as are given by the simpler Z axis integral discussed above.

REFERENCES

1. Ashkin, A. 1970. Acceleration and trapping of particles by radiation pressure. *Phys. Rev. Lett.* 24:156–159.

2. Ashkin, A. 1970. Atomic-beam deflection by resonance-radiation pressure. *Phys. Rev. Lett.* 24:1321–1324.

3. Roosen, G. 1979. Optical levitation of spheres. *Can. J. Phys.* 57:1260–1279.

4. Ashkin, A. 1980. Applications of laser radiation pressure. *Science (Wash. DC)* 210:1081–1088.

5. Chu, S., J. E. Bjorkholm, A. Ashkin, and A. Cable. 1986. Experimental observation of optically trapped atoms. *Phys. Rev. Lett.* 57:314–317.

6. Chu, S., and C. Wieman. 1989. Feature editors, special edition, laser cooling and trapping of atoms. *J. Opt. Soc. Am.* B6:2020–2278.

7. Misawa, H., M. Koshioka, K. Sasaki, N. Kitamura, and H. Masuhara. 1990. Laser trapping, spectroscopy, and ablation of a single latex particle in water. *Chem. Lett.* 8:1479–1482.

8. Ashkin, A., and J. M. Dziedzic. 1987. Optical trapping and manipulation of viruses and bacteria. *Science (Wash. DC)* 235:1517–1520.

9. Buican, T., M. J. Smith, H. A. Crissman, G. C. Salzman, C. C. Stewart, and J. C. Martin. 1987. Automated single-cell manipulation and sorting by light trapping. *Appl. Opt.* 26:5311–5316.

10. Ashkin, A., J. M. Dziedzic, and T. Yamane. 1987. Optical trapping and manipulation of single cells using infrared laser beams. *Nature (Lond.).* 330:769–771.

11. Block, S. M., D. F. Blair, and H. C. Berg. 1989. Compliance of bacterial flagella measured with optical tweezers. *Nature (Lond.).* 338:514–518.

12. Berns, M. W., W. H. Wright, B. J. Tromberg, G. A. Profeta, J. J. Andrews, and R. J. Walter. 1989. Use of a laser-induced force trap to study chromosome movement on the mitotic spindle. *Proc. Natl. Acad. Sci. USA.* 86:4539–4543.

13. Ashkin, A., and J. M. Dziedzic. 1989. Internal call manipulation using infrared laser traps. *Proc. Natl. Acad. Sci. USA.* 86:7914–7918.

14. Tadir, Y., W. H. Wright, O. Vafa, T. Ord, R. H. Asch, and M. W.

Berns. 1989. Micromanipulation of sperm by a laser generated optical trap. *Fertil Steril.* 52:870–873.

15. Bonder, E. M., J. Colon, J. M. Dziedzic, and A. Ashkin. 1990. Force production by swimming sperm-analysis using optical tweezers. *J. Cell Biol.* 111:421A.

16. Ashkin, A., K. Schütze, J. M. Dziedzic, U. Euteneuer, and M. Schliwa. 1990. Force generation of organelle transport measured in vivo by an infrared laser trap. *Nature (Lond.).* 348:346–352.

17. Block, S. M., L. S. B. Goldstein, and B. J. Schnapp. 1990. Bead movement by single kinesin molecules studied with optical tweezers. *Nature (Lond.).* 348:348–352.

18. Sato, S., M. Ohyumi, H. Shibata, and H. Inaba. 1991. Optical trapping of small particles using 1.3 μm compact InGaAsP diode laser. *Optics Lett.* 16:282–284.

19. Gordon, J. P. 1973. Radiation forces and momenta in dielectric media. *Phys. Rev. A.* 8:14–21.

20. Ashkin, A. 1978. Trapping of atoms by resonance radiation pressure. *Phys. Rev Lett.* 40:729–732.

21. Gordon, J. P., and A. Ashkin. 1980. Motion of atoms in a radiation trap. *Phys. Rev. A.* 21:1606–1617.

22. Roosen, G., and C. Imbert. 1976. Optical levitation by means of 2 horizontal laser beams–theoretical and experimental study. *Physics. Lett.* 59A:6–8.

23. Ashkin, A., and J. M. Dziedzic. 1971. Optical levitation by radiation pressure. *Appl. Phys. Lett.* 19:283–285.

24. Ashkin, A., and J. M. Dziedzic. 1975. Optical levitation of liquid drops by radiation pressure. *Science (Wash. DC).* 187:1073–1075.

25. Ashkin, A., J. M. Dziedzic, J. E. Bjorkholm and S. Chu. 1986. Observation of a single-beam gradient force optical trap for dielectric particles. *Optics Lett.* 11:288–290.

26. Ashkin, A., and J. M. Dziedzic. 1989. Optical trapping and manipulation of single living cells using infra-red laser beams. *Ber. Bunsen-Ges. Phys. Chem.* 98:254–260.

27. Wright, W. H., G. J. Sonek, Y. Tadir, and M. W. Berns. 1990. Laser trapping in cell biology. *IEEE (Inst. Electr. Electron. Eng.) J. Quant. Elect.* 26:2148–2157.

28. Born, M., and E. Wolf. 1975. Principles of Optics. 5th ed. Pergamon Press, Oxford. 109–132.

29. Mansfield, S. M., and G. Kino. 1990. Solid immersion microscope. *Appl. Phys. Lett.* 57:2615–2616.

30. Richards, B., and E. Wolf. 1959. Electromagnetic diffraction in optical systems. II. Structure of the image field in an aplanatic system. *Proc. R. Soc. London. A.* 253:358–379.

31. Ashkin, A., and J. M. Dziedzic. 1973. Radiation pressure on a free liquid surface. *Phys. Rev. Lett.* 30:139–142.

32. Hori, M., S. Sato, S. Yamaguchi, and H. Inaba. 1991. Two-crossing laser beam trapping of dielectric particles using compact laser diodes. Conference on Lasers and Electro-Optics, 1991 (Optical Society of America, Washington, D.C.). *Technical Digest.* 10:280–282.

33. Sato, S., M. Ishigure, and H. Inaba. 1991. Application of higher-order-mode Nd:YAG laser beam for manipulation and rotation of biological cells. Conference on Lasers and Electro-Optics, 1991 (Optical Society of America, Washington, D.C.). *Technical Digest.* 10:280–281.

34. Ashkin, A., and J. M. Dziedzic. 1985. Observation of radiation pressure trapping of particles using alternating light beams. *Phys. Rev. Lett.* 54:1245–1248.

35. van de Hulst, H. C. 1981. Light Scattering by Small Particles. Dover Press, New York. 114–227.

36. Kim, J. S., and S. S. Lee. 1983. Scattering of laser beams and the optical potential well for a homogeneous sphere. *J. Opt. Soc. Am.* 73:303–312.

Quantitative Measurements of Force and Displacement Using an Optical Trap

Robert M. Simmons,* Jeffrey T. Finer,[‡] Steven Chu,[§] and James A. Spudich[‡]

*MRC Muscle and Cell Motility Unit, Randall Institute, King's College London, London WC2B 5RL, England; [‡]Departments of Biochemistry and Developmental Biology, Beckman Center, Stanford University Medical Center, and [§]Department of Physics, Stanford University, Stanford, California 94305 USA

ABSTRACT   We combined a single-beam gradient optical trap with a high-resolution photodiode position detector to show that an optical trap can be used to make quantitative measurements of nanometer displacements and piconewton forces with millisecond resolution. When an external force is applied to a micron-sized bead held by an optical trap, the bead is displaced from the center of the trap by an amount proportional to the applied force. When the applied force is changed rapidly, the rise time of the displacement is on the millisecond time scale, and thus a trapped bead can be used as a force transducer. The performance can be enhanced by a feedback circuit so that the position of the trap moves by means of acousto-optic modulators to exert a force equal and opposite to the external force applied to the bead. In this case the position of the trap can be used to measure the applied force. We consider parameters of the trapped bead such as stiffness and response time as a function of bead diameter and laser beam power and compare the results with recent ray-optic calculations.

INTRODUCTION

In a single-beam gradient optical trap, or optical tweezers, a laser beam is brought to a focus in an aqueous solution by a high numerical aperture microscope objective (Ashkin et al., 1986). Any refractile particle near the focus is attracted to it and becomes trapped. The method has been used in a number of biological applications both as a micromanipulator and as a force transducer. Cells, organelles, and chromosomes can be directly manipulated, whereas measurements of the magnitudes of the forces between proteins such as molecular motors require the proteins to be attached to trapped silica or polystyrene beads (see reviews: Block, 1990; Berns et al., 1991; Kuo and Sheetz, 1992; Simmons and Finer, 1993). The trapping force is of the order of 1–100 pN, depending upon the size and refractive index of the particle and the power and wavelength of the laser beam.

Initial attempts to use optical traps as force transducers involved measurement of the applied force necessary to cause the particle to escape the optical trap. This method gives an estimate of the peak force, but it is unable to measure fluctuations in the applied force. More recently, we and others have detected the position of a trapped particle and used it to measure force (Simmons et al., 1993; Kuo and Sheetz, 1993; Svoboda et al., 1993; Finer et al., 1994). In this paper we describe the full details and justification of our method, combining an optical trap with a high-resolution position detector to show that when a force is applied to a trapped bead, the bead is displaced from the center of the trap by an distance that is proportional to the applied force.

When the applied force is changed rapidly, the rise time of the bead displacement is on the millisecond time scale, and thus a trapped bead can be used as a transducer of force fluctuations. We consider parameters of the trapped bead such as stiffness and response time as a function of bead diameter, bead position within the trap, and laser beam power, and compare the results with recent ray-optic calculations of the forces on trapped beads (Ashkin, 1992).

Finally, we show that the performance of an optical trap as a force transducer can be enhanced by means of a feedback circuit. As an external force is applied to a trapped bead, the position of the trap moves to exert an equal and opposite force on the bead. Ashkin and Dziedzic (1977) used feedback to stabilize the vertical (z) position of a particle levitated by a laser beam in a vacuum by modulating the laser beam intensity. In contrast, in our system the feedback is applied to a gradient optical trap in the plane perpendicular to the beam axis using acousto-optic deflectors to shift the trap position rapidly. As a result, the stiffness of the trap is greatly increased, so that the bead is held stationary with nanometer precision, and the position of the trap can be used to measure applied force fluctuations.

MATERIALS AND METHODS

Optics

The apparatus is a modification and extension of an optical trap microscope designed by Chu, Kron, and Sunderman (unpublished). In Fig. 1, the original apparatus consisted of the Nd-YAG laser, mirrors M_1 and M_2 used for steering the laser beam manually, the collimating lens L_2, and a microscope (Zeiss Axioplan). The microscope was modified to accept the laser beam by cutting a hole in the side and inserting a dichroic mirror in place of the eyepiece prisms. The parallel laser beam entered the back aperture of the microscope objective (Zeiss Plan-Neofluar 63×, 1.25 NA, infinity corrected, oil immersion) and was brought to a focus at the specimen, forming the optical trap. A major principle of the optical system

Received for publication 29 December 1995 and in final form 10 January 1996.

Address reprint requests to Dr. James A. Spudich, Department of Biochemistry, Stanford University School of Medicine, Beckman Center, Stanford, CA 94305-5307.

FIGURE 1 Schematic diagram of the optical trap and detection system. The solid lines represent the path of the Nd-YAG laser. The dashed lines represent the illuminating light from a tungsten lamp. Optics include lenses (L_1, L_2), mirrors (M_1, M_2), a dichroic filter (D), and a microscope objective (L_3). The laser beam can be shifted in the plane of the specimen (O) by deflecting it with two orthogonally mounted acousto-optic modulators (AO) and by moving the mirrors M_1, M_2. The illuminating light is split between a video camera (VC) and a quadrant photodiode detector (QD), which indicates the position of a trapped bead. The feedback loops can be closed by feeding the output signals from the quadrant detector into the driver circuits for the voltage-controlled oscillators (VCO), which control the acousto-optic modulators.

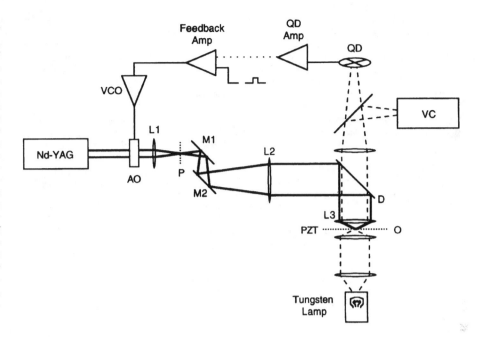

was that the lens L_2 collimated an expanding laser beam at its front focal point, turning it into a parallel beam that just filled the back aperture of the microscope objective, L_3, at its other focal point. A lateral movement of one of the mirrors M_1, M_2 shifted the laser beam parallel to the axis, so that the beam entered the objective at an altered angle to the optical axis. This resulted in a lateral shift in the position of the focal spot. The beam still filled the back aperture of the objective, so the strength of the trap remained roughly constant for displacements of up to 50 μm off-axis. The plane, P, containing the real or apparent point of divergence of the laser beam, was conjugate to the objective focal plane, O. A lateral movement of one of the mirrors M_1, M_2 produced an equal shift of the apparent point source in plane P, and the corresponding movement in plane O was a factor of f_2/f_3 smaller. f_2 was 400 mm and f_3 about 3 mm, so f_2/f_3 was about 150. In experiments in which two beads were to be trapped, the beam was split before it reached mirror M_1, using a half-waveplate followed by a polarizing beam splitter (not shown in Fig. 1). The second pathway also had a pair of mirrors, and the two beams were brought together with a second polarizing beam splitter before lens L_2.

We made the following additions to the apparatus. In some of our experiments (e.g., Finer et al., 1994) it is necessary to move the positions of the traps rapidly under manual control, so we motorized the mirror movements and operated them with a joystick control (Newport positioners 860A-1-HS, controller 860-C2). For fast, electronically controlled movements of the traps, we added two acousto-optic modulators, placed orthogonally (AO_1, AO_2; Isomet 1206C with D323B drivers). An additional lens, L_1, was placed so that the midpoint of AO_1 and AO_2 lay at one focus, and the focal point of L_2 at the other. AO_1 and AO_2 acted as diffraction gratings whose spacings were determined by an acoustical wave of variable frequency. They were set relative to the incident beam at the Bragg angle for the first-order diffracted beam to maximize the intensity in the first-order beam. A change of frequency, resulting from a change in the input voltage to the driver, produced a change in the diffraction angle. The useful range (over which the intensity of the diffracted beam did not vary by more than $\pm 10\%$) was about 12 mradians. The deflection response of an acousto-optic modulator was measured separately; for this measurement the output diffracted laser beam was focused onto a quadrant detector. The response to a square wave input showed a lag of about 4 μs followed by a rise that was 90% complete in 9 μs. Peak-to-peak positional noise was about 10^{-4} of the maximum deflection.

The function of L_1 was to convert an angular deflection from the acousto-optic modulators into a lateral shift of the laser beam parallel to the

optical axis. For a deflection θ, the lateral shift of the beam at plane P was $f_1 \cdot \theta$, and at plane O it was $(f_1 \cdot f_3/f_2) \cdot \theta$. f_1 was determined by the magnification required for the width of the laser beam to fill the back aperture of the objective, as follows. The beam waist at the acousto-optic modulators was about 1 mm; this was necessary for optimal beam quality and was produced using a telescope at the laser. The objective back aperture was 7 mm wide, so the magnification had to be 7. The magnification is given by f_2/f_1, so f_1 had to be 60 mm. The factor ($f_1 \cdot f_3/f_2$) was then about 400 nm $\cdot$ mradian^{-1}. The total useful movement of the trap using an acousto-optic modulator was about 4 μm. The acousto-optic modulators noticeably degraded the quality of the beam and required careful adjustment to optimize the strength of the laser trap. Slightly different settings could produce changes of trap strength differing by a factor of up to 4. At optimal adjustment the efficiency of each modulator was only about 30–40%. Coupled with a restriction on laser power at entry to the acousto-optic modulators of 1 W, the maximum power available at the objective was about 0.2 W.

The laser used in these experiments (Quantronix 416, 10 W Nd-YAG, 1.064 μm) was stable only at laser powers in excess of 5 W. We attenuated the beam immediately after the laser with a half-waveplate and polarizing beam splitter. For finer variations of intensity we used a variable attenuator (Newport 925B) inserted beyond the acousto-optic modulators. An alternative method was to vary the power input to the acousto-optic modulators, but in practice this method was not used much, as it caused an appreciable movement of the beam. The power of the beam entering the microscope was measured from the intensity of the radiation reflected from an optical surface and focused onto a power meter (Coherent Fieldmaster). About half the power would be expected to be lost in the objective (Svoboda and Block, 1994).

Illumination of the specimen was by conventional transmitted light, using a 100-W tungsten lamp with a DC power supply. Illumination for epifluorescence was by a 100-W mercury arc lamp. Light emerging from the microscope was split between a videocamera (VC) and a quadrant photodiode detector (QD) to monitor the bead position, using a beam splitter with selectable ratios (Zeiss 473051). Images of the specimen were projected onto the front surfaces of the video camera and the quadrant detector using eyepieces, 10× in the case of the videocamera and between 7× and 20× with auxiliary singlet lenses as necessary for the quadrant detector, depending on the size of bead in use. The image of the bead filled approximately half the area of the quadrant detector. The distances of the camera and the quadrant detector from the eyepieces could be adjusted to give satisfactory magnifications.

The specimen was viewed using a SIT camera (Hamamatsu C2400–08) with either brightfield or fluorescent illumination, followed by an image-processing system (Hamamatsu Argus 10). The videocamera output was digitized with a frame grabber (Data Translation DT2855) interfaced to a personal computer, and the positions of the mirrors M_1–M_4 were monitored and superimposed on the video image as cursors showing the positions of the traps. The mirror positions were derived using voltage dividers consisting of spring-loaded potentiometers attached to the stage movements. The voltages were read into the computer by analog-to-digital converters (Data Translation DT2814). The focusing of the video camera image could be adjusted by a motorized movement based on a 35-mm camera lens focusing movement, so as to have an independent control over focus. This was necessary because the normal focusing controls of the microscope altered the z-position of the trap as well as the optical focus.

The brightfield image of a bead projected onto the quadrant detector was deliberately defocused so as to enhance the contrast between the bead and the background. This was done by moving the eyepiece away from its normal focus position, so that the image appeared dark on a bright background. Contrast was further enhanced by closing the condenser diaphragm. Adjustments were in general made to optimize the signal while the position of the trap was moved with a square wave (see below).

The whole apparatus was mounted on an antivibration table (Newport). However, there were a number of sources of instrumental noise that affected the measurements. The attachments to the microscope, which was of upright design, introduced a number of low-frequency vibrations, which were excited by the flow of cooling water to the laser. These were substantially reduced, but not eliminated, by surrounding the vertical column extending from the body of the microscope to the quadrant detector and video camera with a metal box filled with lead shot. Pointing instability in the laser and in the laser beam path, which in these experiments was not enclosed, also contributed noise, chiefly drift. Signal averaging was used to estimate the time constant of trapped beads and movement in response to external forces, and this greatly reduced the effect of noise from these sources and from Brownian motion.

Electronics

The x and y coordinates of a trapped bead (x_B, y_B) were determined using a quadrant photodiode detector (Hamamatsu S4349), chosen for its small active area. This minimized dark noise (which is in part proportional to photodiode area) and reduced the amount of magnification needed, leading to a more compact design that was less prone to vibration. The amplifier circuit (Fig. 2 *B*) used low-noise current-to-voltage converters and had switched feedback resistors (omitted from Fig. 2 *B*), which could be selected to maximize the output. x_B and y_B were obtained by appropriate subtractions and additions of the signals. The quadrant detector was mounted on a manual XY movement controlled by micrometers. There was some variation in background illumination over the field at the quadrant detector, which was therefore moved to a point where the background was flat. For most of the measurements in this paper we used 10-MΩ feedback resistors in the quadrant detector, which gave a bandwidth of 2 kHz, measured from the response of the quadrant detector to a light from an LED with a square wave voltage input. Signals were averaged to reduce the noise from Brownian motion and other sources. For measurements where signal averaging was not possible (e.g., when recording Brownian motion) and where it was necessary to maximize the signal relative to instrumental noise, we used 200-MΩ resistors, giving a bandwidth of 100 Hz.

A feedback circuit was designed and built to increase the stiffness of the trap (Fig. 2 *C*). This consisted of integrating and direct pathways for the x_B and y_B signals, and the outputs of the circuit were fed to the drivers for AO_1 and AO_2, with appropriate offsets to produce steady deflections of the laser beam. The operation of the feedback circuit was controlled by analog switches S_1, S_2. In some circumstances, such as when an external force was acting on a trapped bead, the bead position could be far from zero, and when the feedback was switched on, there was a rapid transient as the servo

loop acted to return the bead to its central position. This was minimized by the use of sample-and-hold circuits, as shown in Fig. 2 *A*, to zero the x_B and y_B signals just before the feedback was switched on, so that the bead remained at its offset position.

To a good approximation, movements of the trap position could be regarded as being linear and instantaneous on the time scale of the measurements made in this work. Thus, the trap position (x_T, y_T) could be obtained from the input voltages to the drivers of the acousto-optic modulators. For the present study, the force acting on a bead could be obtained from the x_B, y_B or x_T, y_T signals as appropriate. However, for more general applications we designed further electronic circuits to calculate F_x and F_y, the x and y components of external force acting on a bead, from the difference between the bead position and the position of the trap. An allowance was made for the viscous drag of the solution acting on the bead when it moves, as follows:

$$F_x = k_T(x_T - x_B) - b\frac{dx_B}{dt}$$

$$F_y = k_T(y_T - y_B) - b\frac{dy_B}{dt},$$

where k_T is the stiffness of the trap and b is the damping factor.

Force calibration

Calibration for force was performed by flowing solution past a trapped bead at a known velocity and calculating the force from Stokes' law,

$$F = 6\pi\eta rv,$$

where v is the velocity, r is the radius of the bead, and η is the viscosity of the solution. This was done by moving a microscope substage, consisting of a carrier for a microscope slide, by means of two piezoelectric transducers (PZTs) (Physik Instrumente P771) attached to the stage of the microscope. The PZTs had a natural frequency of about 500 Hz and total range of about 50 μm. They were also used in certain motility assay experiments to track a moving bead by moving the substage and were controlled for this purpose by a joystick. The movement of the stage (usually 5 μm) in response to the triangular wave input used in the calibration procedure was recorded separately, using a low-power objective to project the image of an opaque object onto the quadrant detector. Measurements and calibrations were all done with the trapped bead about 5 μm below the surface of the coverslip, where surface perturbations of Stokes' law should be negligible (Happel and Brenner, 1991; Svoboda and Block, 1994). The trapping force was only slightly lower than at the surface.

Data recording

The x_B, y_B, x_T, and y_T signals were digitized using an analog-digital converter (RC Electronics, ISC-16) interfaced to a personal computer, sampling at rates up to 250 kHz, depending on the bandwidth of the data, and stored on magnetic disk.

Inverted microscope

The records in this paper, with the exception of those in Fig. 12, were obtained using the modified vertical microscope system described in the sections above. We subsequently improved this system in two stages. In the first stage, we built an inverted microscope configuration with improvements in positional stability and with decreased instrumental noise, mainly resulting from substituting a xenon lamp for the tungsten lamp (Finer et al., 1994). Otherwise the optics and electronics were similar to those described above. In the second stage we improved the bandwidth of the detector when using a 100-MΩ resistor to 5 kHz, by the use of a correction circuit

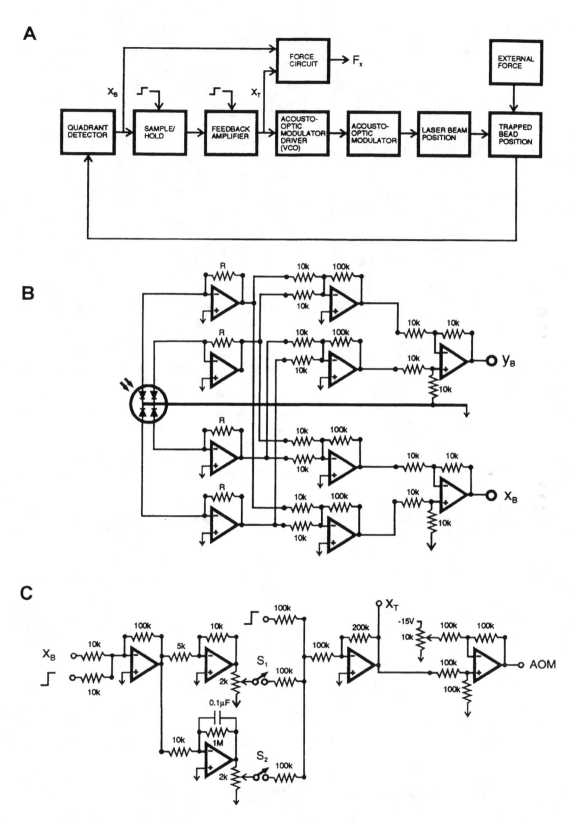

FIGURE 2 (*A*) Block diagram of one channel of electronic circuits showing feedback arrangements. The pulse inputs were used to open and close the feedback loop. (*B*) Quadrant detector circuit. The feedback resistors *R* had a switched range, 10 kΩ to 200 MΩ. Operational amplifiers: first stage, OPA121; second stage, OP270; final stage, AMP03. (*C*) Feedback amplifier circuit (one channel). Operational amplifiers: OP27. S_1, S_2; analog switches used to open and close the feedback loop. Pulse inputs were used to deflect the laser beam.

A

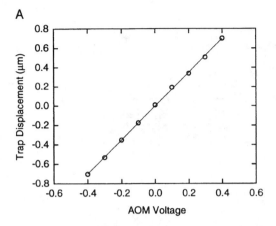

B

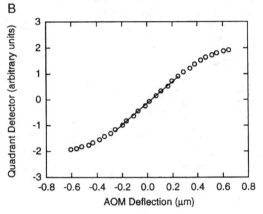

FIGURE 3   (*A*) Acousto-optic modulator (AOM) calibration. There is a linear relation between the input voltage to the AOM and the trap position over the 1.5-μm range shown. (*B*) Quadrant photodiode detector calibration. The AOM deflection represents the bead position. The line shows the linear range of the quadrant detector for bead displacements of 0.2 μm away from the center in each direction.

incorporating a differentiator. The performance of the feedback system was also improved, by incorporating a differential pathway in parallel with the direct and integrating pathways shown in Fig. 2 *C*. The records in Fig. 12 were obtained using the final modifications.

RESULTS

Measurements on 1-μm diameter beads

The majority of our detailed measurements were made on 1-μm beads, which is the diameter used in our experiments on the actomyosin motility assay (Simmons et al., 1993; Finer et al., 1994). We first calibrated the microscope magnification using a stage micrometer, whose image was projected onto the quadrant detector. The distance between successive 10-μm rulings was found by moving the micrometers controlling the position of the detector. Using a trapped bead of 1 μm diameter, we next calibrated AO_1 and AO_2 by applying a square wave to the output stage amplifiers in the feedback circuit (Fig. 2) to move the position of the trap, and measuring on an oscilloscope the movement of the micrometers necessary to null the resulting signal from the quadrant detector (Fig. 3 *A*). Applying a slow triangular wave, we then calibrated the quadrant detector (Fig. 3 *B*). The output was roughly sinusoidal with respect to bead displacement, and for the measurements reported here we kept within the range of movements for which the output of the detector did not depart from linearity by greater than 5%. We refer to this as the linear range of the quadrant detector. In some cases we corrected the responses for the nonlinearity of the detector, but the effect on the results was negligible.

When the position of the trap was moved by applying a square wave input to one of the acousto-optic modulators, a trapped bead followed the movement of the beam with a lag that was nearly first-order for small steps (Fig. 4 *A*). This result shows that a trapped bead behaves as a damped mass in a parabolic energy well, i.e., as if it were attached to the center of the trap by a spring. Calculations using the observed stiffness confirm that the trapped bead can be described as a harmonic oscillator that is so highly damped

FIGURE 4   (*A*) Response of a trapped bead (1-μm diameter) to rapid trap displacements of 300 nm at laser powers ranging from 3 to 150 mW. For laser powers of 52 to 150 mW the response is shown on an expanded time scale. Detector bandwidth was 20 kHz (single-pole, low-pass filter). Each trace is an average of 100 data sweeps. (*B*) The rate constant of the bead response as a function of laser power.

A

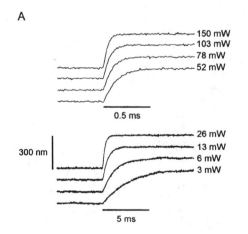

B

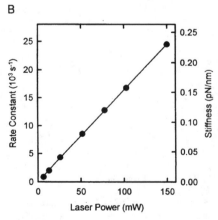

A

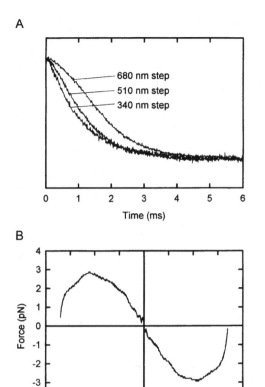

where k is the stiffness of the trap and b is the damping factor. The stiffness depends on the laser power, the size of bead, and its refractive index, and on other factors such as the numerical aperture of the objective (Ashkin, 1992). The dependence of the rate constant, k/b, on laser beam power for 1-μm-diameter beads is shown in Fig. 4 B for the range 3–150 mW. The rate constant varied linearly from 800 s^{-1} for 5 mW to 25,000 s^{-1} for 150 mW. The value of k can be derived by calculating b from Stokes' law (see Materials and Methods); k is shown in Fig. 4 B. It varied from 0.007 to 0.23 pN/nm over the range measured.

The time course of bead movement was independent of the amplitude up to a certain limit, but beyond this there was an increasing lag of the response with increase of step size (Fig. 5 A). This was shown to result from a fall in the force as the trap was offset. The profile of force across the trap was calculated from a record for a large step, by taking the derivative of the response (Fig. 5 B) and plotting it against displacement. The curve shows that, as the trap is displaced from the center of the bead, force changes at first approximately linearly and then reaches a broad peak at about 450 nm, after which it falls off more steeply, so that the total half-width of the trap is about 675 nm. This confirms the calculation by Ashkin (1992) that the maximum force should be reached when the trap is at about the radius of the bead, although his calculation is strictly applicable only to much larger beads.

Another method of exploring the profile of the energy well and obtaining the value of the trap stiffness is to apply an external force to a trapped bead and measure the resulting displacement. This has the advantage that external forces can be applied over a much longer time, and the true steady-state values of displacement can be measured. We did this by applying triangular waves of varying velocity and amplitude to the PZT-controlled microscope substage (Fig. 6 A). The movement of the substage was first recorded by projecting onto the quadrant detector the image of a small opaque object on a microscope slide attached to the substage, using a low-power objective. The movement of the PZTs was slightly nonlinear, but it was found that the average velocity of movement was negligibly different from the velocity calculated from the total excursion and the

FIGURE 5 (*A*) Time course of the response of a trapped 1-μm-diameter bead for rapid trap displacements of 340, 510, and 680 nm. The responses are normalized to fit on the same scale to illustrate the lag in the response for large trap displacements. Detector bandwidth was 20 kHz (single-pole, low-pass filter). Each trace is an average of 100 data sweeps. (*B*) Force profile of an optical trap as a function of displacement of a bead from the center of the trap. The force was calculated by differentiating the bead response to a large trap displacement and calculating the viscous (Stokes') force.

that the inertial term in the equation of motion can be neglected, so that the movement of the bead after a sudden trap displacement is given by

$$x_\mathrm{B} = x_\mathrm{T}(1 - e^{-(k/b)t}),$$

FIGURE 6 (*A*) The response of a trapped bead to a triangular wave input to the microscope stage position. The bead shows a square wave response corresponding to the viscous force. Detector bandwidth was 100 Hz (single-pole, low-pass filter). Each trace is an average of 100 data sweeps. (*B*) Displacement of a bead from the center of the trap as a function of the applied viscous force for 13, 52, and 140 mW laser powers.

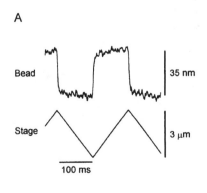

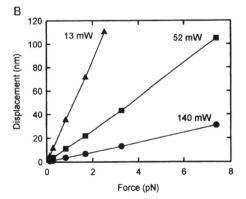

duration. We then measured the average displacement of a trapped bead as a function of Stokes' force. Fig. 6 *B* shows that the steady displacement was linearly related to the applied force, making the important point that the movement of the bead in the trap can be used as a force transducer in the piconewton and sub-piconewton force range. Using long periods of signal averaging to remove noise arising from Brownian motion and other sources, we found that the relationship holds down to the smallest force applied, 0.14 pN.

These measurements gave direct values of trap stiffness that were in good agreement with those estimated from the time constant of displacements produced by the acousto-optic modulators. The force applied in these tests approximated to a square wave, and the resulting displacement had a time constant that is similar to that measured using the acousto-optic modulators under the same conditions. However, the movement produced by the piezoelectric transducers was much slower than that produced by the acousto-optic modulators, and we did not attempt to make any quantitative measurements of rise time using this method.

A method used to measure the force on trapped particles or beads in biological experiments is to estimate the "escape force." As already shown in Fig. 5 *B*, force rises to a maximum with displacement of a trapped bead and then falls off again. So if a force is applied that just exceeds the maximum that the trap can exert, the bead is carried away from the trap, and depending on how long the force is applied, the bead is either attracted back into the trap or escapes. The point at which the escape force was reached was measured by increasing the velocity of a triangular wave applied to one of the substage PZTs and noting when the bead position no longer reached a plateau, but instead continued to move away from the center of the trap. Escape force was found to increase linearly with laser beam power (Fig. 7), as would be expected from the previous measurements (cf. Figs. 4 *B* and 5 *B*).

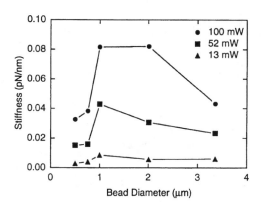

FIGURE 8 Optical trap stiffness as a function of bead diameter for laser powers of 13, 52, and 100 mW.

Measurements as a function of bead diameter

We measured the rate constants and obtained the corresponding values for the trap stiffness, k, for beads of different diameter, ranging from 0.5 to 3.36 μm. The values of k for three laser powers are plotted in Fig. 8. k increased at first with increasing bead diameter, but then reached a broad peak, probably with a maximum between 1 and 2 μm, and then declined. For beads of small diameter relative to the wavelength of the laser ($r \ll \lambda$), stiffness is expected to rise with increasing diameter, as the amount of material in the beam increases and so does the trapping power (Ashkin, 1992). On the other hand, for large particles ($r \gg \lambda$), where the ray optic regime holds, particles intercept all the converging rays at the laser focus, so the trapping force is the same, irrespective of diameter. However, the trapping rays are spread out over a larger distance the bigger the particle, so the stiffness should decrease with increasing diameter: a particle with twice the diameter has to move twice as far for the same force to be produced on it by the trap. Our results,

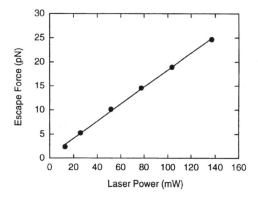

FIGURE 7 Escape force for a trapped bead (1-μm diameter) as a function of laser power. The escape force represents the maximum force an optical trap can exert to keep a bead trapped.

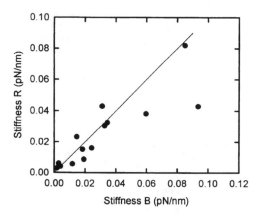

FIGURE 9 Comparison of optical trap stiffness as measured by Brownian motion (stiffness B) and by responses to rapid trap displacements (stiffness R). The line represents the ideal case, where the stiffness measurements are the same by both methods. Bead diameters ranged from 0.5 to 3.36 μm and the laser powers ranged from 13 to 100 mW.

for particles of intermediate diameter, show a peak as expected.

We also estimated stiffness by measuring the Brownian motion on the same samples, recording 50 s of data for each bead size and laser power. From the equipartition of energy, the root mean square Brownian motion along one axis, $\sqrt{x_n^2}$, is given by

$$\frac{1}{2} kx_n^2 = \frac{1}{2} k_B T,$$

where k_B is Boltzman's constant and T is the absolute temperature. Thus the stiffness, k, can be estimated; it is plotted in Fig. 9 against the stiffness calculated from the step changes of trap position. There is a good deal of scatter in the data as a result of four factors: i) for points derived from traps with a high frequency response (large rate constant), there was undoubtedly an underestimate of $\sqrt{x_n^2}$ because of the limited bandwidth of the quadrant detector (resulting from the use of large value feedback resistors needed to minimize the effects of dark noise); ii) points at a high trap stiffness had a higher proportion of instrumental noise; iii) small beads had a finite proportion of dark noise; iv) there was a substantial and variable component of low-frequency noise derived from pointing instability. We did not attempt to correct for these factors, and given the errors, Fig. 9 indicates a reasonable agreement between the two estimates of stiffness.

Escape force was measured for beads of diameter between 0.28 and 3.36 μm. Because the escape force can be measured by visual observation of the video monitor, it was possible by using fluorescent beads to include those of diameter too small to make useful quantitative measurement of trap stiffness. Results for three laser powers are shown in Fig. 10 A. For a given laser power, escape force increases steeply at first with increase of bead diameter and then flattens off toward a plateau. According to Ashkin (1992), for Rayleigh particles ($r \ll \lambda$) trapping force depends on r^3, and for large particles ($r \gg \lambda$) it is independent of r. Our results, which are for sizes of particles in between these regimes, are consistent with these predictions.

For particles large enough to give displacement measurements on the quadrant detector, it was also possible to measure the escape distance, i.e., the distance the bead is displaced from the center of the trap when it escapes the force of the trap, roughly where the maximum force was observed. The values for bead diameters of 0.5–3.36 μm are plotted in Fig. 10 B, which shows that the escape distance lies roughly at half the radius of the bead. For 1-μm beads it would be expected that the escape distance would lie at 450 nm from the center of the trap, the value found for the maximum force in Fig. 5 B, but the value actually obtained was 238 nm. The reason for this discrepancy is most likely that the method used to detect the escape point results in an underestimate of the value. As the velocity of the triangular wave approaches the escape force, the bead displacement record shows a "tail," which we took at the time to be the first sign of the escape process, and noted the force and displacement just as this tail developed. It is more likely that it signifies the start of the broad peak of the profile of Fig. 5 B. The appearance of a tail was probably exacerbated by a slight nonlinearity of the velocity produced by the piezo-electric transducers, resulting in the Stokes' force continuing to rise somewhat during steady motion. A further factor tending to underestimate the position of the maximum of force from the escape point measurements is that the maximum represents an unstable equilibrium, and Brownian motion could help to take the bead beyond the maximum.

A linear relation between escape distance and bead size is to be expected for large particles ($r \gg \lambda$). The escape force is independent of diameter for large particles, but it is produced when a bead is displaced by approximately one diameter from the center of the trap (Ashkin, 1992). For small particles ($r \ll \lambda$), the distance from the center of the trap for the maximum force should be independent of bead size and occur where the square of the field gradient is a maximum. In this case, a rough calculation yields an escape distance on the order of 200 nm. Thus, it would be expected that as bead diameter is increased from zero, the escape distance should at first be independent of bead size, with a value of about 200 nm, and then increase as the size of bead

A

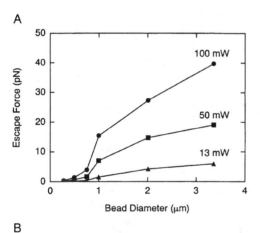

B

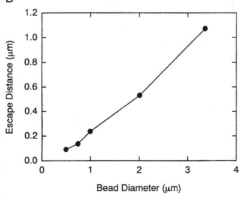

FIGURE 10 (*A*) Escape force for a trapped bead as a function of bead diameter for laser powers of 13, 52, and 100 mW. (*B*) Displacement of a trapped bead at the point of escape as a function of bead diameter for laser power of 100 mW.

FIGURE 11 Optical trap stiffness calibration with and without feedback control. (*A*) Without feedback: the response of a trapped bead to a triangular wave input to the microscope stage position (see Fig. 6). (*B*) With feedback: the laser trap position now shows the response. (*C*) Displacement measured from bead position trace (without feedback) and from trap position trace (with feedback), as a function of applied force. Detector bandwidth was 100 Hz (single-pole, low-pass filter). Each trace is an average of 100 data sweeps.

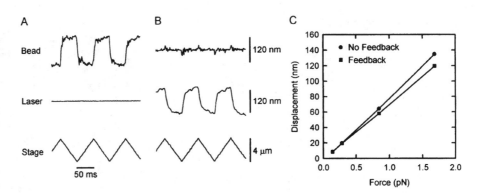

becomes significant compared to λ, finally becoming nearly equal to the diameter (Ashkin, 1992). We were not able to measure the escape distance for beads smaller than 500 nm, but our results show the expected linear relationship as diameter increases.

Feedback control

The results so far show that the displacement of a trapped bead can be used to make time-resolved measurements of force. For measurements of small forces the trap has to be made weak enough to give detectable bead movements above the noise from Brownian motion and instrumental noise. However, this leads to a compliant measuring system, and in some applications it is desirable to make the trap at least an order of magnitude stiffer than the system under investigation to avoid perturbations. We therefore investigated the use of feedback to increase the stiffness of the trap. In feedback mode, the x_B, y_B signals from the quadrant detector are fed through amplifiers to the driver circuits for the acousto-optic modulators (Fig. 2). Ideally, when an external force is applied to a trapped bead, the position of the trap (x_T, y_T) is shifted so that the trap applies an equal and opposite force to the external force acting on the bead. As a result, the bead remains stationary and the trap movement is a measure of the external force. We made some measurements to compare the performance of the system, with and without feedback control. Fig. 11, *A* and *B*, shows records of bead and trap movement to a force applied via a triangular wave input to one of the substage piezoelectric transducers (cf. Fig. 6 *A*). The two measurements give closely similar results, showing that the feedback mode can be used to measure isometric force. Quantitative measurements of x_B, y_B without feedback and x_T, y_T with feedback confirm this (Fig. 11 *C*). We made some measurements of escape force with feedback and found that this was approximately 70% of the escape force without feedback.

It should be noted that in principle feedback can be used to stiffen a trap along a single axis; however, we found single-axis feedback to be unstable in experiments on the actomyosin motility assay (Simmons et al., 1993). Presumably the reason for the instability is that external forces on

a trapped bead that pull it even slightly off axis cause bead movement at right angles, where the trap is less stiff. As a result, the two-dimensional feedback system using two orthogonal acousto-optic modulators described here offers improved stability, even when one-dimensional forces are being measured (Finer et al., 1994).

The feedback system improves the stiffness of the trap (by a factor of about 400 in the example in Fig. 11) without the need to increase the laser power. It is also highly effective at reducing the Brownian motion of a particle (compare Fig. 12, *B* and *C*), albeit at the cost of transferring the noise to the force signal. In all of the figures so far presented in this paper, the experimental data were signal averaged, and this could lead to an overoptimistic impression of the signal-to-noise ratio in single traces. Fig. 12 is included to show the noise levels at typical trap strengths used in experiments on single motor proteins, and we used an improved detector and feedback system to show the

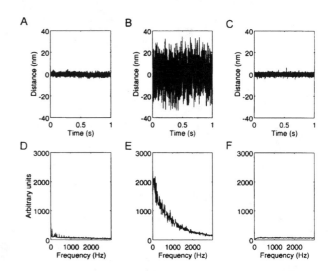

FIGURE 12 Positional noise on a 1-μm diameter bead (*A*) immobilized on a coverglass surface, (*B*) held without feedback in a laser trap with a stiffness of 0.04 pN · nm^{-1}, and (*C*) held in the same laser trap with feedback control. (*D–F*) Fourier transform spectra of the positional noise corresponding to *A–C*, respectively. Detector bandwidth was 5 kHz.

signals at full bandwidth (Inverted microscope, Materials and Methods). In motor protein experiments, displacement has to be measured at a trap stiffness that is as low as possible, typically 0.04 pN $\cdot$ nm^{-1} in our experiments (Finer et al., 1994). This presents a load of 0.4 pN (about 10% of the average isometric force of 3–4 pN) at a typical myosin unitary displacement of 10 nm. Fig. 12 *B* shows that the noise level at this trap stiffness is about 60 nm peak to peak (9.5 nm rms). The noise is primarily due to Brownian motion; the instrumental noise (Fig. 12 *A*) is about 3 nm peak to peak (<1 nm rms). Fig. 12 also shows the noise level when the feedback system is in operation (Fig. 12 *C*), giving a peak-to-peak noise level of about 3 nm (<1 nm rms). The corresponding noise on force was about 2.5 pN peak to peak (0.4 pN rms).

DISCUSSION

Many improvements can be made to the system described here. In preliminary work we have shown that the pointing stability of the laser can be improved by using an optical fiber, which also makes it possible to mount the laser on a separate table, thus reducing vibration. Solid-state lasers and solid-state pumped lasers also present fewer problems from vibration. Inverted microscopes offer improved stage stability compared with the upright design for most of the measurements described here. Improved efficiency and beam quality can be achieved with TeO$_2$ acousto-optic modulators, rather than the PbMbO$_4$ type used here. PZT-operated mirrors can also be used to deflect the laser beam provided they are servo-controlled, with adequate speed for most purposes. Detection noise can be much reduced by using a xenon arc lamp (Fig. 12), and an interferometric design offers improved performance (Denk and Webb, 1990; Svoboda et al., 1993).

Our results show that measurements of the position of a particle in an optical trap can be used to monitor external forces acting on the particle. Of particular interest to studies of biological processes is the observation that a 1-μm bead in a weak trap with feedback control can give simultaneous measurements of force and displacement with subpiconewton and nanometer resolutions, respectively, and with a response time of about 1 ms. These are in the range needed to make measurements on single molecule interactions in actomyosin and kinesin-microtubule motility assays. The main limitation on achieving these resolutions in practice lies in the noise from Brownian motion, at least if signal averaging cannot be used. In making measurements of unitary displacements, the trap stiffness must be sufficiently low that the motor protein movement is unimpeded, and this results typically in a noise of about 60 nm peak to peak (Fig. 12 *B*). Fortunately, the stiffness of the motor protein enhances the trap stiffness, and the Brownian motion noise is

lowered correspondingly (Svoboda et al., 1993; Finer et al., 1994). Alternatively, it may be possible to use a somewhat higher trap stiffness (Fig. 12 *C*), if the relation between displacement and load is not steep in the region of interest (Finer et al., 1994) or the amount of displacement is constrained (Svoboda et al., 1993).

In biological experiments using trapped beads, the nature of the experiment may put limits on the size of bead to be used. However, other things being equal, the data in this paper should be useful in choosing a bead diameter to optimize stiffness, frequency response, or trap strength. In many respects, not the least being ease of detection, bigger is better, but it should be noted that both stiffness (k) and rate constant (k/b) have optimum values.

This work was supported in part by grants from the NSF and AFOSR (SC); a grant from the NIH (JAS); support from the MRC (RMS); travel grants from the Wellcome Trust and Fulbright Commission (RMS); a NATO Collaborative Research Grant (JAS, RMS); and a Human Frontier Science Program grant (JAS, SC, RMS). JTF was a trainee of the Medical Scientist Training Program at Stanford University.

REFERENCES

Ashkin, A. 1992. Forces of a single-beam gradient laser trap on a dielectric sphere in the ray optics regime. *Biophys. J.* 61:569–582.

Ashkin, A., and J. M. Dziedzic. 1977. Feedback stabilization of optically levitated particles. *Appl. Phys. Lett.* 30:202–204.

Ashkin, A., J. M. Dziedzic, J. E. Bjorkholm, and S. Chu. 1986. Observation of a single-beam gradient force optical trap for dielectric particles. *Opt. Lett.* 11:288–290.

Berns, M. W., W. H. Wright, and R. W. Steubing. 1991. Laser microbeam as a tool in cell biology. *Int. Rev. Cytol.* 129:1–44.

Block, S. M. 1990. Optical tweezers: a new tool for biophysics. *In* Noninvasive Techniques in Cell Biology. J. K. Foskett and S. Grinstein, editors. Wiley-Liss, New York. 375–402.

Denk, W., and W. W. Webb. 1990. Optical measurement of picometer displacements of transparent microscopic objects. *Appl. Opt.* 29:2382–2391.

Finer, J. T., R. M. Simmons, and J. A. Spudich. 1994. Single myosin molecule mechanics: piconewton forces and nanometre steps. *Nature.* 368:113–119.

Happel, J., and H. Brenner. 1991. Low Reynolds Number Hydrodynamics, 2nd ed. Kluwer Academic Publishers, Dordecht, the Netherlands.

Kuo, S. C., and M. P. Sheetz. 1992. Optical tweezers in cell biology. *Trends Cell. Biol.* 2:116–118.

Kuo, S. C., and M. P. Sheetz. 1993. Force of single kinesin molecules measured with optical tweezers. *Science.* 260:232–234.

Simmons, R. M., and J. T. Finer. 1993. Glasperlenspiel II. *Curr. Biol.* 3:309–311.

Simmons, R. M., J. T. Finer, H. M. Warrick, B. Kralik, S. Chu, and J. A. Spudich. 1993. Force on single actin filaments in a motility assay measured with an optical trap. *In* Mechanism of Myofilament Sliding in Muscle. H. Sugi and G. H. J. Pollock, editors. Plenum, New York, London. 331–336.

Svoboda, K., and S. M. Block. 1994. Biological applications of optical forces. *Annu. Rev. Biophys. Biomol. Struct.* 23:247–285.

Svoboda, K., C. F. Schmidt, B. J. Schnapp, and S. M. Block. 1993. Direct observation of kinesin stepping by optical trapping interferometry. *Nature.* 365:721–727.

The Stiffness of Rabbit Skeletal Actomyosin Cross-Bridges Determined with an Optical Tweezers Transducer

Claudia Veigel, Marc L. Bartoo, David C. S. White, John C. Sparrow, and Justin E. Molloy

Department of Biology, University of York, York YO1 5YW, England

ABSTRACT   Muscle contraction is brought about by the cyclical interaction of myosin with actin coupled to the breakdown of ATP. The current view of the mechanism is that the bound actomyosin complex (or "cross-bridge") produces force and movement by a change in conformation. This process is known as the "working stroke." We have measured the stiffness and working stroke of a single cross-bridge (κ_{xb}, d_{xb}, respectively) with an optical tweezers transducer. Measurements were made with the "three bead" geometry devised by Finer et al. (1994), in which two beads, supported in optical traps, are used to hold an actin filament in the vicinity of a myosin molecule, which is immobilized on the surface of a third bead. The movements and forces produced by actomyosin interactions were measured by detecting the position of both trapped beads. We measured, and corrected for, series compliance in the system, which otherwise introduces large errors. First, we used video image analysis to measure the long-range, force-extension property of the actin-to-bead connection (κ_{con}), which is the main source of "end compliance." We found that force-extension diagrams were nonlinear and rather variable between preparations, i.e., end compliance depended not only upon the starting tension, but also upon the F-actin-bead pair used. Second, we measured κ_{xb} and κ_{con} during a single cross-bridge attachment by driving one optical tweezer with a sinusoidal oscillation while measuring the position of both beads. In this way, the bead held in the driven optical tweezer applied force to the cross-bridge, and the motion of the other bead measured cross-bridge movement. Under our experimental conditions (at ~2 pN of pretension), connection stiffness (κ_{con}) was 0.26 ± 0.16 pN nm^{-1}. We found that rabbit heavy meromyosin produced a working stroke of 5.5 nm, and cross-bridge stiffness (κ_{xb}) was 0.69 ± 0.47 pN nm^{-1}.

INTRODUCTION

Many types of cellular motility, including muscle contraction, are driven by the cyclical interaction of myosin with actin, coupled to the breakdown of ATP. The current view of the mechanism is that myosin binds to actin with the products of ATP hydrolysis (ADP and phosphate) bound in the catalytic site (cross-bridge attachment). Then, as the products are released, myosin changes conformation to produce a movement or "working stroke" (Huxley, 1969). In the absence of nucleotide, actin and myosin form a tightly bound "rigor" complex. Binding of a new ATP molecule to myosin causes the rigor complex to dissociate (cross-bridge detachment), and subsequent ATP hydrolysis resets the original myosin conformation so that the cycle can be repeated (Lymn and Taylor, 1971). During the cycle, part of the cross-bridge becomes distorted by the working stroke, and mechanical work is stored in this elastic deformation. In this way, the cross-bridge captures the sudden changes in chemical potential associated with steps in the biochemical cycle and is able to do external work on a much slower time scale, e.g., as muscle shortens or vesicles are transported (Huxley and Simmons, 1971). There are three key features to this mechanism; 1) one ATP molecule is broken down per mechanical cycle; 2) the size of the working stroke is determined by the conformation of myosin at the start and

end of the attached period; 3) the cross-bridge is elastic (see review by Cooke, 1998).

Important issues in our understanding of the cross-bridge mechanism are the size of its working stroke and the force that it can produce. Recently, several laboratories have developed single-molecule mechanical transducers that are based on either "optical tweezers" (laser traps) or glass microneedles (e.g., Finer et al. 1994; Saito et al. 1994; Molloy et al., 1995; Ishijima et al., 1996; Dupuis et al., 1997). An advantage of single-molecule experiments over muscle fiber experiments is that the force, working stroke, and kinetics of a single cross-bridge interaction can be measured directly. Measurements of such individual interactions allow critical tests to be made of how actomyosin converts chemical energy to mechanical work. For example, the myosin working stroke should be smaller than the span of a single cross-bridge (a myosin head or S1 is ~16 nm long; Rayment et al., 1993), and the mechanical work done by each cross-bridge interaction must be less than that produced by the breakdown of one ATP molecule. This requires that the elastic element (κ_{xb}) should be of the correct stiffness to store a suitable fraction of the free energy available from ATP breakdown measured under physiological chemical conditions.

The mechanical arrangement used to study actomyosin interactions in most optical tweezer transducers is based on the "three bead" geometry devised by Finer et al. (1994) (see Fig. 1). A single actin filament is suspended between two beads, each held in an independent optical trap (we define these, arbitrarily, as "left" and "right" traps). The filament is positioned above a third bead, on which myosin

Received for publication 30 July 1997 and in final form 19 June 1998.

Address reprint requests to Dr. Justin E. Molloy, Department of Biology, University of York, P.O. Box 373, York YO1 5YW, England.

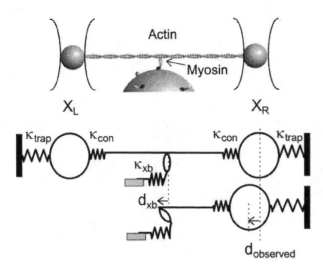

FIGURE 1   The upper panel is a cartoon showing the "three-bead geometry" devised by Finer et al. (1994), used to make single-molecule mechanical measurements from actomyosin. Two latex beads holding an actin filament are manipulated in two independent optical traps. This makes it possible to bring the filament into the vicinity of a third, larger bead that is fixed to the surface of the experimental chamber. The "third bead" is coated with myosin molecules at a low surface density. Actomyosin interactions are monitored by observing the position of the trapped beads with a photodetector (giving the bead positions, X_L and X_R). The lower panel represents the mechanical elements of the system. Upon binding to actin, the myosin cross-bridge forms a mechanical pathway between the beads that are suspended in the optical traps and "ground." The cross-bridge stiffness, κ_{xb}, is linked in series with the actin-to-bead connection stiffness, κ_{con}, and these are combined in parallel with the optical trap stiffness, κ_{trap}. This combination of "springs" gives the overall mounting stiffness κ_x. A fraction of the cross-bridge working stroke, d_{xb}, is taken up by the compliance of the connection, and therefore observed bead displacements ($d_{observed}$) measured with the photodetector need to be corrected for the effects of series compliance in the system.

is deposited at low surface density. This geometry is required to study skeletal muscle myosin (and other motors that spend only a small proportion of their time attached), as it prevents actin from diffusing away from myosin during the detached period of the cross-bridge cycle. Movement and force produced by single cross-bridge interactions are inferred from the position of at least one bead, which is monitored by imaging it onto a quadrant photodiode (4QD). The detector determines the position of the centroid of the image to a resolution of better than 0.5 nm.

Published estimates of the working stroke vary between 5 nm and 25 nm, estimates of stiffness range from 0.16 to 0.6 pN nm^{-1}, and maximum force ranges from 1 to 5 pN (Finer et al., 1994; Guilford et al., 1997; Mehta et al., 1997; Molloy et al., 1995; Nishizaka et al., 1995; Saito et al., 1994; Simmons et al., 1996). Variability in these data between laboratories may, in part, be explained by differences in the type of myosin or subfragment used. However, there are also systematic complications in the measurements, and these have been dealt with in different ways.

1. To determine the cross-bridge working stroke, the stiffness of the apparatus (here the optical traps) must be

much less than that of the cross-bridge (e.g., $\kappa_{trap} \ll \kappa_{xb}$; Fig. 1). This makes it possible for the cross-bridge to undergo its full working stroke unhindered. At such low trap stiffness the beads and associated actin filament necessarily exhibit large amounts of Brownian motion. Determinations of the working stroke depend upon how Brownian motion is accounted for in the analysis (Molloy et al., 1995; Ishijima et al., 1996; Guilford et al., 1997; Mehta et al., 1997).

2. Determinations of the maximum force developed by a cross-bridge under isometric conditions (i.e., no net movement of the molecule) require that the apparatus stiffness is much greater than that of the cross-bridge. This means that the cross-bridge is prevented from moving, allowing its maximum force to be developed. Although the stiffness of the trap can be made sufficiently high by applying feedback ($\kappa_{trap} \approx 10$ pN nm^{-1}; Simmons et al., 1996), the stiffness of the attachments of actin to the two beads is likely to be much smaller (Dupuis et al., 1997; Veigel et al., 1997). Low "connection stiffness" (κ_{con}; Fig. 1) allows movement of the cross-bridge and thus reduces the size of the observed movement and force that it can produce.

To obtain a good estimate of cross-bridge stiffness, κ_{xb}, we have measured and corrected for sources of series elasticity in the system. We have lumped the series elasticity into one component, termed κ_{link}. This consists of the series combination of the actin filament stiffness, κ_{actin}, and the stiffness of its connection to the trapped bead, κ_{con}. We have also refined our measurement of cross-bridge working stroke, d_{xb}, by using position information obtained from both beads. To make these measurements we have developed our apparatus to monitor the positions of both trapped beads simultaneously. One method uses analysis of video images to obtain a linear position signal over a long range (up to 12 μm); and another uses two four-quadrant photodetectors to make high-speed measurements of each trapped bead over a fairly short distance (up to 1 μm). These developments have enabled us to:

1. Determine the compliance of the connection between the NEM-modified myosin-coated beads and the actin filament used in our experiments, e.g., κ_{con}.

2. Determine cross-bridge stiffness (κ_{xb}) by measuring its movement when subjected to an applied load during a single binding event.

3. Test if the length of actin remains constant during its interaction with myosin under low-load conditions.

We discuss the implications of our single-molecule mechanical study in the context of current ideas of the mechanism of force production by actomyosin, which derive mainly from work with muscle fibers.

MATERIALS AND METHODS

Optical tweezers transducer

Our optical tweezers transducer is based around an inverted, fluorescence microscope (Fig. 2). A key feature is that most of the apparatus is computer controlled. This allows different kinds of experiments to be performed simply by running different software.

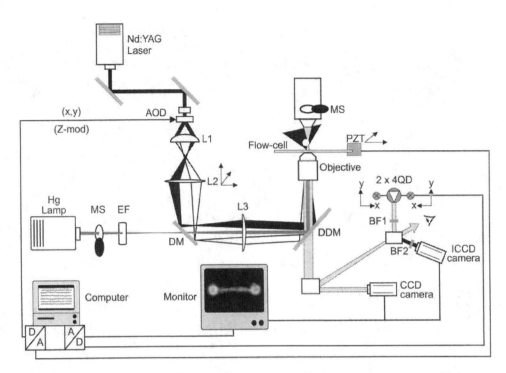

FIGURE 2   The optical trap is built around an inverted microscope (Axiovert 135; Zeiss, Germany). Infrared (1064 nm) laser light, from a diode-pumped Nd:YAG laser (Adlas Model 321, 1064 nm; Adlas, Lubeck, Germany) is combined with green light (EF = 546FS10.25, excitation filter; Andover Corp., Salem, NH) from a mercury arc lamp by use of a "hot mirror" (DM = 820DCSP; Omega Optical, Brattleboro, VT). Both light beams enter the microscope epifluorescence port via a custom-built housing. A dual dichroic mirror (DDM = 570DCLP, Omega; reflects 546 nm and 1064 nm; transmits >570 nm), mounted in the microscope filter block, allows us to use optical tweezers and view rhodamine fluorescence simultaneously. Laser beam alignment is via two mirrors, and the trap position is controlled with two orthogonally mounted acoustooptic deflectors (AODs) (synthesiser/driver, N64010–100 2ASDFS-2, TeO$_2$ crystals N45035-3-6.5 DEG-1.06; NEOS Technologies) controlled by a custom-built computer interface card. To produce two optical traps, we chop between two sets of *x, y* coordinates (to simplify computation, these coordinates are chopped in hardware at 10 kHz). The laser light path is completely enclosed with cardboard tubing to prevent air currents from entering the system at any point. Coarse control of the stage position is by mechanical drives, and a custom-built piezoelectric substage (PZT) allows small range computer-controlled movements of the microscope slide. High-speed position measurements are made with four-quadrant photodiode detectors (2 × 4QD = S1557, Hamamatsu Photonics, Hamamatsu City, Japan; and custom-built electronics). The image is split in half with a 90°, front-surface mirrored, Amici prism, and images of the left and right beads are projected onto the two detectors. Scattered laser light is excluded with a barrier filter (BF1 = short-pass barrier filter). Actin fluorescence was visualized with an intensified CCD camera (Photon-P46036A; EEV, Chelmsford, UK) coupled to a barrier filter (BF2 = LP590; Zeiss). Bright-field illumination (100-W halogen lamp) is used to produce a high-magnification video image (CCD camera, P46310; EEV). An Acroplan 100×, 1.3 N.A. objective and an Optovar 2.5× insert are used to obtain the desired image magnification. Video images from the half-inch format CCD camera attached to the camera port are captured at 512 × 512 pixel resolution, giving 1 pixel = 26 nm. A "slotted-opto switch" detects the position of the microscope prisms used to select different TV/camera ports and permits computer control of mechanical shutters (MS) used to switch between bright-field and fluorescence illumination and of the video source. The 4QDs, AODs, PZT, etc. were cross-calibrated with the video "frame-grabber."

In overview, a single actin filament was attached at either end to two 1.1-μm latex beads that were held and manipulated by two independently controlled optical tweezers. The filament was positioned over a third (glass) bead that was fixed to the surface of the experimental chamber. This bead had been sparsely coated with rabbit heavy meromyosin (HMM) molecules (see Fig. 1). Mechanical interactions between a single HMM molecule and the actin filament were detected from the motion of a bright-field image of the two latex beads cast onto two four-quadrant photodiode detectors (4QD).

Optical tweezers and beam steering

Two independent optical tweezers were synthesized by chopping a single laser beam between two sets of *x, y* coordinates (see Molloy, 1997, for details) with acoustooptical devices (NEOS Technologies, Melbourne, FL). These were controlled by a custom computer interface card with two sets of digital output registers (loaded with the two *x, y* coordinates) that were multiplexed at 10 kHz.

In this paper, stiffness measurements were made by applying a large-amplitude sinusoidal forcing function to one optical tweezer. For this method to work there were three requirements: 1) fast and stable control of the trap position; 2) linearity of the detector signal over the range ± 200 nm; 3) good stage stability over the time course of the experiment. We deal with these issues below.

Acoustooptical deflectors

Acoustooptical deflector (AOD) control of laser position is extremely rapid (response time ≈ 2 μs). However, we found that drift and noise in laser position arose from a variety of sources, including laser pointing stability, AOD noise, and mechanical drift (between the axis of the microscope and the rest of the optical path). We measured the sum of all of these sources of noise (Fig. 3 *A*) and found that stability depended critically upon excluding air currents from the light path; drift was <0.5 nm s^{-1}; noise was <1 nm rms (Fig. 3 *A*).

A AOD controlled laser positioning

B Detector frequency response

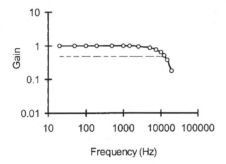

C Detector linearity

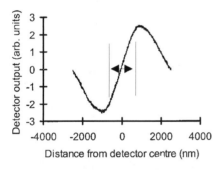

FIGURE 3 (*A*) The speed and stability of AOD-controlled laser positioning were tested by moving the laser beam in a square wave function (30 Hz). The laser beam was reflected from the surface of a silvered coverslip at the microscope object plane and projected onto one of the two 4QDs. The illumination intensity was made the same as that obtained from the bead image. This measures tweezer stability relative to the microscope axis (data sampled at 50 kHz). (*B*) The detector bandwidth was determined over the range of 10 Hz to 20 kHz with the reflected laser beam (as in *A*). The beam position was varied sinusoidally using the AODs, and the input/ output response was determined from the discrete Fourier transform of the data (data sampled at 50 kHz). (*C*) The linearity of the detectors was determined by capturing a latex bead and then moving the bead back and forth with a large-amplitude triangular wave form (± 3 μm). Twenty cycles were averaged to obtain the graph; residual noise is attributable to Brownian motion of the trapped bead.

Four-quadrant photodetectors

We used two four-quadrant photodetectors (4QD) so that the positions of both beads holding the actin filament could be measured. Beads were positioned such that their images lay on either side of the midline of the field of view. The microscope image was split using a mirrored Amici prism (Fig. 2), and each half-image was projected onto a detector. The

detectors could be translated along the *x* (or *y*) axis to accommodate different lengths of actin filament. Changes in the illumination of the four quadrants were used to measure the bead translation parallel (*x*) and perpendicular (*y*) to the actin filament (using electronics similar to those of Simmons et al., 1996). The gain and frequency response of both detectors were matched by adjusting their signal for the same trapped bead.

The response of the 4QDs was found to be linear over a range of ± 300 nm from the detector center (see Fig. 3 *C* for details). This exceeded the maximum range used in our experiments. Detector gain was flat to 10 kHz ($f_c \approx 12.5$ kHz; Fig. 3 *B*), i.e., much greater than the bandwidth of Brownian motion (≈ 600 Hz). Detector noise has both electrical ("dark" noise) and optical (shot noise) sources. We measured the sum of these by evenly illuminating the 4QD (at about the same intensity of light obtained when the bead image is cast on the detector) and recording the output signal (Fig. 4 *A*). The power density spectrum of this noise is shown in Fig. 4 *D*. The bandwidth of the noise is governed by the electronics of the detector circuit. We chose feedback resistors of 100 MΩ ($= R$) in our current-to-voltage "head-stage" circuit. This gave the best compromise between bandwidth (proportional to $1/R$), gain (proportional to R), and resistor noise (Johnson noise, proportional to $R^{0.5}$). Detector noise was $\sim$100-fold smaller than the Brownian motion of the bead, and so minimal correction for detector response was required.

Microscope substage

The final positioning of the HMM-coated bead beneath the suspended actin filament was made using a computer-controlled piezoelectric substage. The range of movement was 25×25 μm^2. Control was by two 12-bit D/A converters (Data Translation; DT2812A) producing 6 nm of displacement per digital bit. Stage position noise (>1 Hz and <10 kHz) was 0.9 nm root mean square (r.m.s.), but long-term stability was poor ($\sim$0.5 nm $\cdot$ s^{-1}; data not shown). This meant that measurements of individual cross-bridge events lasting between 1 ms and 1 s were essentially free of stage positional noise. However, during the course of a single experiment lasting up to 2 h, the stage had to be repositioned several times.

Preparation of proteins, coated beads, experimental chambers, and solutions

F-actin, whole myosin, and HMM were prepared from rabbit skeletal muscle by standard methods (Pardee and Spudich, 1982; Margossian and Lowey, 1982).

Preparation of NEM-myosin

Rabbit skeletal muscle myosin was precipitated from 50 μl of stock solution (25 mg ml^{-1} myosin, stored at $-20°$C in buffered salt solution containing 50% glycerol, prepared as described by Margossian and Lowey, 1982) by the addition of 500 μl of deionized water. The pellet was resuspended in "high salt" buffer (HiS) (500 mM KCl, 4 mM MgCl$_2$, 1 mM EGTA, 20 mM K phosphate buffer, pH 7.2) to give a final concentration of 12 mg ml^{-1}. To this, *N*-ethyl-maleimide (from freshly made 100 mM stock; Sigma Chemical Co.) was added to a final concentration of 4 mM (Meeusen and Cande, 1979). The solution was incubated at 20°C for 30 min, and the reaction was stopped by the addition of 500 μl of 20 mM dithiothreitol in deionized water. The NEM-modified myosin pellet was resuspended in HiS to give a final protein concentration of 80 μg ml^{-1}.

Preparation of NEM-myosin-coated "C-beads"

To make the 1.1-μm polystyrene beads (plain latex beads; LB11 Sigma Chemical Co.) bind to F-actin irreversibly, they were coated with NEM-modified myosin. Ten microliters of beads (10% by mass) was washed twice in 100 μl of deionized water and collected by spinning at 8000 $\times$ *g*

FIGURE 4 Analysis of noise sources in the system in the time and frequency domain. (*A*) Detector noise at full illumination: 1.4 nm rms. (*B*) Brownian noise of a single trapped bead. The r.m.s. deviation in position measured on one axis, over a period of 2.4 s at 2.5 kHz bandwidth, was 14 nm, giving a trap stiffness of 0.02 pN nm^{-1} (5 kHz sample rate). (*C*) Brownian noise of the bead-actin-bead system. *n* indicates intervals of high noise (in the absence of cross-bridge attachment, 11 nm rms. *e* indicates intervals of reduced noise (in the presence of an attached cross-bridge, 3 nm r.m.s. (measured at 3 μM ATP)). (*D*) Spectral analysis of traces in *A–C*. *Trace a*: Detector noise, corner frequency $f_c \approx 12.5$ kHz. *Trace b*: Spectrum for a trapped bead, heavily damped with $f_c \approx 300$ Hz; trap stiffness = 0.02 pN nm^{-1}. *Trace c*: Spectrum of Fig. 3 *C* (*n + e* periods); note that there are two corner frequencies, $f_{c1} \approx 3$ Hz and $f_{c2} \approx 200$ Hz. (*E*) Calibration of trap stiffness using viscous force. *Upper trace*: 5 Hz triangular waveform motion applied to the piezo substage (4.8 μm peak to peak). *Lower trace*: Bead displacement in the *x* direction (average of 10 cycles). The bead moved ~50 nm peak to peak, giving a trap stiffness of 0.02 pN nm^{-1}.

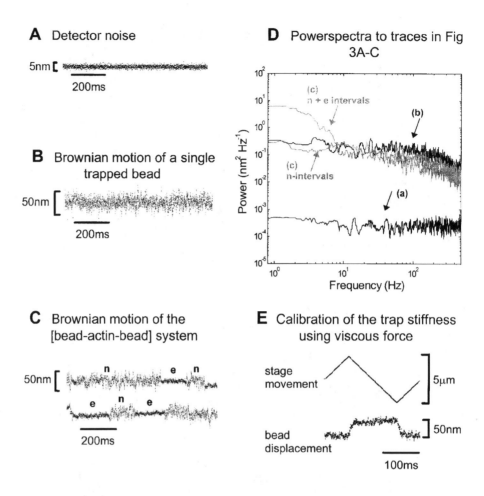

for 1–2 min. This procedure removed most of the surfactant present in the proprietary buffer. The washed beads were resuspended in 50 μl of deionized water (now ~2% by mass). Ten microliters of washed beads was added to 10 μl of NEM-myosin solution (above), and to this 2.5 μl of 0.1 mg ml^{-1} bovine serum albumin–tetramethylrhodamine isothiocyanate (BSA-TRITC) (Sigma) was added. The solution was finally made up to 100 μl with HiS (final solution contains 0.2% beads (2.7 $\times$ 10^6 beads μl^{-1}), 8 μg ml^{-1} NEM-myosin, 2.5 μg ml^{-1} BSA-TRITC) and incubated overnight at 4°C. The resulting C-beads were washed twice and resuspended in assay buffer (AB) (25 mM KCl, 4 mM MgCl$_2$, 1 mM EGTA, 25 mM imidazolium-chloride, pH 7.4; Kron et al., 1991). Storage of C-beads in low-salt solution as opposed to HiS gave much better stability (they behaved well in our experiments for ~4 days), and the stiffness of the actin-bead connection was improved. Short NEM-myosin filaments might project from the surface of such beads. However, the beads tended to aggregate and required dispersion by bath ultrasonication (duration 1–2 s) just before use.

Flow cell construction

The microscope flow cell was constructed from a precleaned 22 $\times$ 50 mm^2 glass microscope slide across which two 3 $\times$ 22 mm^2 strips of (no. 1) coverglass were fixed 15 mm apart with 2 μl of UV-curing epoxy adhesive. A 22 $\times$ 40 mm^2 precleaned coverslip was coated on one surface with 2 μl of 0.1% nitrocellulose dissolved in amyl acetate (Kron et al., 1991). This solution also contained a suspension of 1.7 μm glass microspheres (Bangs Labs, Carmel, IN); ~2 mg ml^{-1} gave a surface density of about one microsphere per 10 μm^2. The precoated coverslip was glued to the cover-

glass strips, orthogonally to the slide, leaving ~10 mm of coverslip projecting from either side. The flow cell was exposed to UV light until the glue was completely cured. UV-curing epoxy adhesive was superior to grease for *z* axis (focus) stability during the experiments.

Solutions

Rabbit skeletal HMM (Margossian and Lowey, 1982) was bound to the nitrocellulose-coated coverglass by allowing 100 μl of 1 μg ml^{-1} HMM dissolved in AB to flow into the flow cell and incubating for 1 min. The coverslip surface was then "blocked" by allowing 100 μl of 1 mg ml^{-1} BSA in AB to flow into the flow cell and leaving for 2 min. Finally, 100 μl of AB-GOC solution was added. AB-GOC solution was prepared as follows. AB was degassed with a vacuum pump, and then an oxygen scavenger system, to reduce photobleaching, of 20 mM DTT, 0.2 mg ml^{-1} glucose oxidase, 0.05 mg ml^{-1} catalase, and 3 mg ml^{-1} glucose was added, together with an ATP backup system (2 mM creatine phosphate, 0.1 mg ml^{-1} creatine phosphokinase). This solution was stored in a disposable 1-ml hypodermic syringe fitted with a narrow-bore needle to reduce oxygen diffusion into the buffer. Before the experiment, 2 μl of rhodam-ine-phalloidin-labeled actin (Molecular Probes; actin concentration 5 μg ml^{-1}), 2 μl of C-beads (above), and ATP at final concentrations between 1 μM and 10 μM were added to 100 μl of AB-GOC. It was important to proceed fairly quickly after the final solution had been added to the flow cell, because the C-beads tended to stick to the coverslip surface. The laboratory was air-conditioned, and the experimental temperature was kept at 23°C.

Optical tweezers procedure

Using fluorescence microscopy, we captured two NEM-myosin-coated beads in the optical tweezers and suspended an actin filament between them. Most of this procedure was performed with the traps held in fixed positions and by steering the stage with the *x-y* mechanical controls. To capture a bead without having to switch the laser tweezer off and on, the bead image was defocused (so the bead lay in a plane between objective and tweezer) just before capture, so that optical scattering forces then acted to push the bead into the stable trapping position. A pretension of 2 pN was applied to the suspended actin filament by moving one of the traps in the *x* direction (along the actin filament axis) and monitoring the motion of the bead held in the stationary trap. The bead-actin-bead assembly was then moved such that each image was cast near or on its respective 4QD. Final alignment of the photodetectors was made using two *x-y* mechanical translators (Fig. 2).

Next, a suitable surface-bound, HMM-coated "third" bead had to be found and positioned beneath the taut actin filament. Surface beads were visualized by bright-field microscopy. To prevent the bead-actin-bead assembly from touching the coverslip surface, the condenser aperture was "stopped down" with the iris diaphragm to give a good depth of focus. The coverslip surface was then surveyed for surface-bound microspheres, at a defocus of ~5 μm. A suitable "third" bead was positioned under the actin filament with the computer-controlled piezo-substage. The condenser aperture was fully opened to allow accurate *z* axis (focus) control, and the focus was adjusted until actomyosin interactions were observed on the highly magnified, bright-field video image of the beads.

Calibration of trap stiffness (κ_{trap})

κ_{trap} was determined by three methods, using a single bead with no actin filament attached (as described by Svoboda and Block, 1994):

1. Stoke's force ($F = \beta v$; v = velocity; $\beta = 6\pi\eta\alpha \approx 10^{-5}$ pN s nm^{-1}; η = viscosity of the solution, α = bead radius) was generated by applying a large-amplitude triangular waveform to the microscope substage, with the trapped bead held 5 μm from the coverslip surface (Fig. 4 *E*). Large forces and motions are produced by this technique, so effects of instrumental noise and calibration errors are minimized.

2. Mean squared Brownian motion ($\langle x^2 \rangle$) was measured and the equipartition principle applied ($\kappa_{trap}\langle x^2 \rangle/2 = \frac{1}{2}kT$) (Fig. 4 *B* and Eq. A3). Data were recorded over a period of several seconds at a bandwidth greater than 2 kHz. Corrections were made for the instrumental noise, which adds to this signal. Because the trap compliance is proportional to x^2, detector noise and calibration are critical to accuracy and errors are worst at high stiffness.

3. With the trapped bead held 5 μm from the coverslip, the power spectrum of the Brownian noise ($f_c = \kappa_{trap}/2\pi\beta$) was determined (Fig. 4 *D*). This method cannot be used to estimate stiffnesses during experiments because the beads are then at an uncertain distance from the coverslip and therefore the viscosity is unknown.

We found that trap stiffness was directly proportional to laser power (data not shown). Experiments were performed using a trap stiffness, κ_{trap}, of 0.02 pN nm^{-1} by adjusting the laser output power and using methods 1–3 (above) to calibrate the stiffness. This was sufficiently lower than the cross-bridge stiffness, to allow good estimates of the working stroke to be made.

RESULTS

Series elasticity and cross-bridge mechanical properties were determined from three different types of measurement.

1. The long-range force-extension property of the bead-actin-bead assembly was measured by moving one laser trap to apply an increasing force and measuring the resulting extension by video image analysis to determine both bead positions.

2. Cross-bridge stiffness and series elasticity of the bead-actin-bead assembly were measured during individual cross-bridge interactions. We used the two four-quadrant photodetectors to measure both bead positions with high time resolution. We determined the stiffnesses either by analysis of the Brownian motion of the beads or by application of a sinusoidal oscillation to the position of one of the laser traps while simultaneously measuring the positions of both of the trapped beads.

3. We measured the working stroke, using the two four-quadrant detectors to monitor both bead positions. By doing this we addressed some of the uncertainties and possible artifacts surrounding our earlier measurements (e.g., Molloy et al., 1995) made with a single detector.

Long-range force-extension property of the actin-to-bead connection, κ_{con}

We measured the long-range force-extension property of the bead-actin-bead assembly by holding one optical trap fixed and applying force by moving the other trap in stepwise increments of 50 nm. The positions of both beads were measured by capturing bright-field video images of the beads and calculating their centers of mass. The applied force was derived from the displacement of the bead held in the fixed trap (from the trap center) and the extension from the distance between the two bead images (see Fig. 5, *A–C*, for details). The gradient of the force-extension diagrams (e.g., Fig. 5 *D*) gave the lumped stiffness of the bead-actin-bead assembly; $\kappa_{link} = \kappa_{actin}\kappa_{con}/(2\kappa_{con} + \kappa_{actin})$ (see Fig. 11 and Eq. A4). These plots were nonlinear, and connection stiffness was greatest at high tension. It would be advantageous to apply large pretensions to the bead-actin-bead assembly and thereby minimize the series compliance. However, to measure the unhindered cross-bridge working stroke, a low-stiffness optical trap is required, and because optical traps only work over a short range (~250 nm), the maximum stable tension that could be applied was ~2 pN. At this tension (Fig. 5 *D*, *dashed box*) the average value of κ_{link} was 0.13 $\pm$ 0.06 pN nm^{-1} (n = 18 different actin filaments).

The shape of the force-extension plots (increase in stiffness with increasing force) was variable between preparations, even though actin filament lengths were similar (4.3 $\pm$ 0.4 μm). This implies that variability of the plots was due mainly to differences in the connection stiffness κ_{con} and not to differences in actin filament stiffness κ_{actin}. If we make the simplifying assumption that κ_{con} is the same at both ends and that κ_{actin} is much larger (~8 pN nm^{-1} for a 5-μm filament; Kojima et al., 1994), then the stiffness at each end $\kappa_{con} \approx 2 \times \kappa_{link} = 0.26$ pN nm^{-1}. This is more than 10 times larger than κ_{trap} and is consistent with the observation that the amplitude of Brownian motion of the bead-actin-bead assembly corresponds to the sum of both

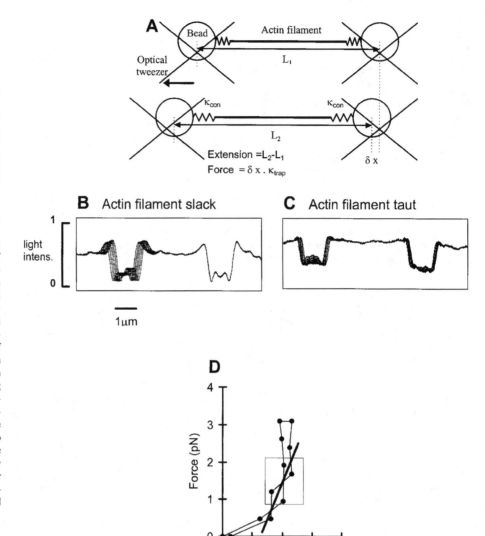

FIGURE 5 (*A*) Static measurements of the bead-actin-bead stiffness using video imaging. Force was applied by stepping the left trap to the left. To determine the position of the beads on the video image, five lines of video data, taken from the central part of the bead image, were averaged over 10 video frames. Bead movements were determined from the position of the center of mass calculated from the video data. Forces were determined from the movement of the right bead. (*B* and *C*) Superposition of six averaged video images. In *B* the actin filament is slack, and when the left bead was displaced the right bead did not move (we found the resolution of this method to be ~2 nm). In *C* the actin filament is held taut between the two beads. Movement of the left trap caused movement of both beads. (*D*) Force-extension diagram of one bead pair for one stretching cycle, which consists of one stretching phase in five steps and a subsequent releasing phase, again in five steps. The least-squares line fitted to the linear part of the curve gives the stiffness κ_{link} in the region of steady tension (1–2 pN) over which other mechanical experiments were performed (in this example 0.13 pN nm^{-1}).

trap stiffnesses (e.g., Fig. 4 *C*, *n* segments r.m.s. deviation = 10–11 nm ≡ 0.33–0.4 pN nm^{-1}; see Eqs. A4 and A5).

κ_{con} and κ_{xb} determined from analysis of Brownian motion

κ_{con} can also be estimated from Brownian motion by extracting the uncorrelated portion of motion of the linked beads (similar in principle to the approach of Mehta et al., 1997). Uncorrelated motion occurs as the beads move either toward or away from each other (rather than motion in the same direction). This releases or extends the series elasticity, κ_{con}. We recorded the motion of both of the beads with the two four-quadrant detectors. The two upper traces in Fig. 6 *B* show the displacement of the left and right beads from a small part of Fig. 6 *A*, but at higher time resolution. The lowest trace shows the difference between the two bead positions, i.e., the uncorrelated motion. The r.m.s. deviation

of the uncorrelated motion, when no HMM was attached, is 6 nm. This gives an estimate for κ_{link} of 0.1 pN nm^{-1}, which is similar to the result obtained in the previous section. This method is inaccurate because the small amplitude of the uncorrelated motion suffers from significant contamination from other noise sources. The problem is exacerbated by the parabolic dependence of stiffness upon displacement amplitude. We would require a much more sensitive and lower-noise detector for this method to work well for larger values of κ_{con} (see Fig. 6 *C*).

Knowing κ_{link} and κ_{trap}, κ_{xb} can be derived from the total stiffness measured during cross-bridge attachment (the total stiffness is then given by the series combination of κ_{xb} and κ_{link} in parallel with κ_{trap}; see Appendix, Fig. 11 and Eqs. A6–A8). We found that the Brownian motion during attached intervals was 3–4 nm (r.m.s.; e.g., Fig. 4 *C*, *e*-segments). For the reasons given in the previous paragraph, this gives a poor estimate for κ_{xb} that is in the range of 1–2 pN nm^{-1}.

FIGURE 6 (*A*) Simultaneous traces of the displacements of both beads holding an actin filament due to interactions with low densities of HMM (bound to the surface at 1 mg ml^{-1}). Three sequential pairs of records (i–iii) are shown. X_R and X_L show the simultaneous positions of the right and left beads, respectively (ATP concentration 3 μM, 23°C). (*B*) (*a, b*). Part of *A*, but at higher time resolution to show displacements of the bead before, during, and after a single HMM attachment. X_R and X_L show displacements of the right- and left-hand beads holding an actin filament. The mean displacements during the attachments, d_R and d_L, are determined from the mean position during the attachment minus the mean position of the baseline measured before and after the attachment. (*c*) Difference between the traces, $X_R - X_L$. (*C*) Graph of the theoretical value of r.m.s. Brownian motion of a trapped bead, calculated for increasing system stiffnesses (see Appendix). The r.m.s. background noise of the detector was ~1.4 nm (Fig. 4). Analysis of the Brownian noise is of use in determining system stiffness only below values of ~0.1 pN nm^{-1}. (*D*) (*a, b*) Distributions of mean displacements for right-hand and left-hand beads during 666 attachments from four actin-filament preparations. For any bead-actin-bead preparation there is a strong bias in one or another direction (determined by the polarity of the actin filament); this direction was made positive in the histograms. The solid curves are Gaussian distributions. The means are equal to the mean value of the events, the amplitudes were determined from the total counts, and the standard deviations were determined from the thermal motion of the bead position in the absence of attachments. For the left- and the right-beads the mean value was 5.04 nm. (*c*) The difference ($d_R - d_L$) was determined on an event-by-event basis and plotted for the 666 events shown above. The solid curve is the Gaussian curve fitting best to the data, with a midposition at 0.005 ± 3.76 nm (S.D.), e.g., centered close to zero.

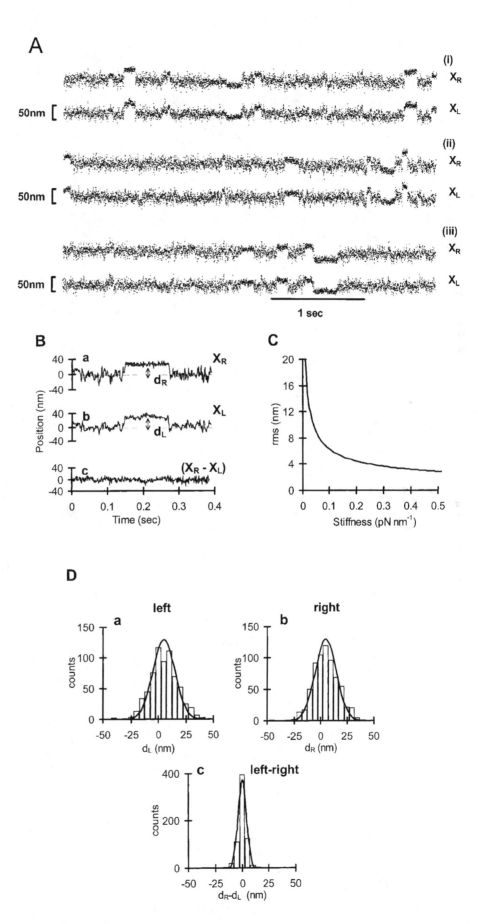

In summary, we found that 1) κ_{con} is nonlinear and variable between preparations; 2) at 1–2 pN pretension of the actin filament, $\kappa_{actin} \gg \kappa_{con} \gg \kappa_{trap}$ (but κ_{con} is probably on the same order of magnitude as κ_{xb}); and 3) because of the manner in which the stiffnesses are combined in series and parallel, analysis of Brownian motion gives a poor estimate of κ_{xb} and κ_{con}.

κ_{con} and κ_{xb} determined from forced oscillations

We measured κ_{con} and κ_{xb} separately by driving one optical trap back and forth with a large-amplitude, sinusoidal oscillation while measuring the position of both trapped beads with the two four-quadrant detectors. The position of the bead in the driven trap was used to derive the applied force, and motion of the other bead gave the extension of the cross-bridge and/or the extension of the actin-to-bead connections (Fig. 7).

At an oscillation frequency of 105 Hz and in the absence of cross-bridge attachment, the amplitude of the driven bead motion was approximately half that of the laser motion, and there was a phase shift of ~30°. The phase shift was due to the viscous drag on the beads (Fig. 8 *B, loop b*). The reduced amplitude of bead motion was caused mainly by the stiffness of the other, stationary trap. We made the peak-to-peak motion of the driven (right-hand side) bead (X_R) ~200 nm. In the absence of an attached cross-bridge, the applied forcing function caused both beads to move sinusoidally. The difference in their movement (the gradient of Fig. 8 *C* is greater than 1; see Eq. A9) arises from length changes in the bead-actin-bead assembly as κ_{link} is subjected to the varying load.

During periods of cross-bridge attachment (Fig. 7, labeled a_1-a_3), motion of the driven bead (X_R) was used to calculate the force applied to the cross-bridge, and motion of the passive bead (X_L) measured the cross-bridge exten-

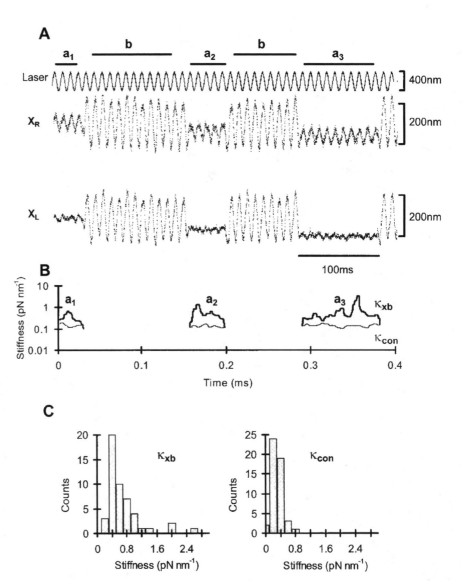

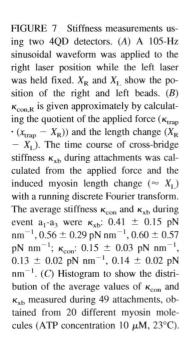

FIGURE 7 Stiffness measurements using two 4QD detectors. (*A*) A 105-Hz sinusoidal waveform was applied to the right laser position while the left laser was held fixed. X_R and X_L show the position of the right and left beads. (*B*) $\kappa_{con,R}$ is given approximately by calculating the quotient of the applied force ($\kappa_{trap} \cdot (x_{trap} - X_R)$) and the length change ($X_R - X_L$). The time course of cross-bridge stiffness κ_{xb} during attachments was calculated from the applied force and the induced myosin length change ($\approx X_L$) with a running discrete Fourier transform. The average stiffness κ_{con} and κ_{xb} during event a_1-a_3 were κ_{xb}: 0.41 ± 0.15 pN nm^{-1}, 0.56 ± 0.29 pN nm^{-1}, 0.60 ± 0.57 pN nm^{-1}; κ_{con}: 0.15 ± 0.03 pN nm^{-1}, 0.13 ± 0.02 pN nm^{-1}, 0.14 ± 0.02 pN nm^{-1}. (*C*) Histogram to show the distribution of the average values of κ_{con} and κ_{xb} measured during 49 attachments, obtained from 20 different myosin molecules (ATP concentration 10 μM, 23°C).

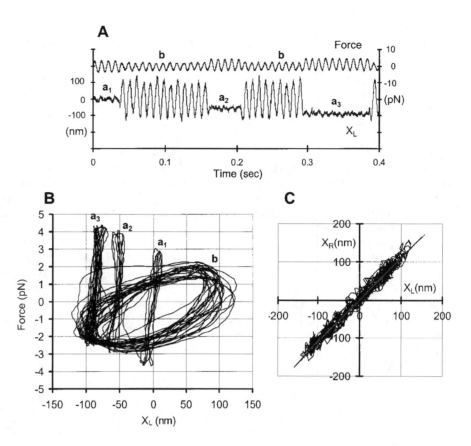

FIGURE 8 (*A*) Applied force (*upper trace*) and movement of the left bead (*lower trace*) from the experimental data shown in Fig. 7 *A*. (*B*) Plotting force against the left bead position gives the stiffness during each attached interval (loops a_1-a_3; 0.38, 0.71, 0.48 pN nm^{-1}, respectively); a correction factor of ~10% should be applied to these slopes to account for κ_{con}. Loop b arises from motion during "detached" intervals. The hysteresis shown by this loop is caused by viscous drag on both beads. (*C*) Left bead versus right bead position during "detached" intervals. There is little phase shift between the motion of the driven (*right*) and the passive (*left*) bead because κ_{link} is large compared to the drag on a single bead (at this forcing frequency); the slope of the curve is given by $\kappa_{link}/(\kappa_{link} + \kappa_{trap})$.

sion. To extract the amplitudes we performed a running discrete Fourier transform at the forcing frequency over a single oscillation period on both the X_R and X_L data. κ_{con} was calculated from the quotient of net extension between the beads and the applied force (Eq. A10). By substituting κ_{con} into Eqs. A12 and A13, κ_{xb} was obtained from Eq. A11. In this way we obtained a running estimate of both κ_{con} and κ_{xb} during each attached interval (Fig. 7 *B*). We could not resolve any systematic change in κ_{xb} or κ_{con} over the attached period. Hence, κ_{xb} and κ_{con} were averaged over each attachment interval and plotted as a histogram (Fig. 7 *C*). The overall mean was 0.69 ± 0.47 pN nm^{-1} for κ_{xb} and 0.31 ± 0.16 pN nm^{-1} for κ_{con}.

κ_{xb} was also determined from a plot of force versus passive bead movement (Fig. 8 *B*, loops a_1, a_2, a_3). In these diagrams the positions of the loops on the *x* axis indicate the starting positions of cross-bridge attachment along the actin filament. The gradient of these diagrams gives κ_{xb}, because the passive bead motion is nearly the same as that of the cross-bridge (average gradient of curves a_1, a_2, a_3 in Fig. 8 *B* = 0.52 pN nm^{-1}).

Displacement produced by a single cross-bridge interaction

By monitoring both bead positions with the two four-quadrant detectors, we were able to address some of the doubts

and problems associated with previous estimates of the cross-bridge working stroke, which were made with an apparatus with only a single detector. For example:

1. Does the motion of a single bead correctly measure the motion of the entire bead-actin-bead assembly, i.e., does the system behave as a nearly "rigid dumbbell"?

2. Does the actin filament change length significantly during cross-bridge attachments?

3. Do artifactual displacements arise if the actin filament binds to HMM that is situated with an orthogonal displacement to the filament center line?

1) Inspection of Fig. 6 shows that the motion of the two beads is well correlated. This means that the bead-actin-bead assembly translates as a nearly rigid body under the influence of Brownian motion. Therefore the actin monomers in the vicinity of the HMM (near the center of the filament) must move in a manner similar to that of the beads.

Attachments were detected from the increase in stiffness (Molloy et al., 1995), and the amplitude of each was measured relative to the mean rest positions for the two beads (Fig. 6 *D*). Attachments were detected automatically by calculating the running variance of the position data (five points), applying a median filter (31 points) to the calculated variance, and then thresholding this data. The data associated with attached intervals were removed, and the remaining data were corrected for baseline drift. Attachment event amplitudes were obtained from the median point in the data relative to corrected baseline at the middle time point of the event. The distributions of event amplitudes, measured separately from data obtained for left and right 4QDs, are shown in Fig. 6 *D*, *a*, *b*. These histograms were fitted well by a single Gaussian distribution, indicating that the data consists of a single population of events. Therefore the mean amplitude of the events was used to measure the average movement produced by the cross-bridge. The spread of the data is explained by the randomizing effect of Brownian motion (Molloy et al., 1995). Both distributions were centered at 5 nm from mean rest position.

2) To determine whether the actin filament changes length during the cross-bridge cycle, the distribution of the difference of left and right bead positions determined on an event-by-event basis was plotted as a histogram (see Fig. 6 *D*, *a*–*c*). A consistent difference in position would indicate that the actin filament changed length when it interacted with the HMM. The mean value of this difference gives the average extension of the bead-actin-bead assembly during events. We found the difference to be 0.06 ± 3.06 nm (SD). The spread of this distribution is explained by the uncorrelated Brownian motion of the beads, which results from compliance of the bead-actin-bead assembly (κ_{link}; see above).

3) If the actin filament binds to an HMM that is displaced laterally from the filament midline (in the *y* axis), this would produce an artifactual observed displacement. This displacement arises because the midpoint of the actin filament would be pulled laterally, and the two beads would therefore move closer together. For example, if the center of a 5-μm-long actin filament bound to an HMM that was 100 nm from the average midline position of the filament, each bead would move $\sim$2 nm inward. If the myosin then underwent a working stroke, the effect would be to increase the observed displacement of one bead and reduce that of the other. However, we found that the average displacement for left and right beads was 5.042 nm and 5.037 nm (respectively), so this potential source of artifact did not seem to affect our results.

DISCUSSION

The aim of this study was to obtain an estimate of the stiffness and working stroke produced by a single acto-HMM interaction under in vitro conditions. The objective was to test if the working stroke and stiffness are consistent with current ideas of how actomyosin functions to produce force and movement. In summary, we found that it was necessary to measure series elastic components in the system and that extension of these components during cross-bridge interactions produced a small measurement error in the working stroke, but a large error in cross-bridge stiffness. Furthermore, by measuring the position of both beads, we found that actin filament length remained constant during cross-bridge interactions, and the segment of the actin

filament that is able to interact with the HMM molecule moved by approximately the same amount as the beads held in the optical traps.

Series elasticity and cross-bridge stiffness

Dupuis et al. (1997) discovered that the bead-actin-bead assembly used in these studies has considerable "end compliance" (or connection stiffness κ_{con}). Their explanation was that extensibility arose from a combination of actin filament flexure at the point of attachment of actin to the trapped bead and rotation of the bead within the optical trap. We measured the long-range series elasticity of our bead-actin-bead assemblies by video microscopy. At the trap stiffness used in our experiments, the maximum pretension that could reasonably be applied ($\sim$2 pN) was insufficient to extend this nonlinear series compliance to a suitably high stiffness; i.e., κ_{con} (0.2 pN nm^{-1}) was lower than the expected value of cross-bridge stiffness (2 pN nm^{-1}; Huxley and Tideswell, 1997). Analysis of the mechanical system (see Appendix, Fig. 9, and Fig. 6 *C*) indicated that estimates obtained by measurement of Brownian motion would give inaccurate estimates of cross-bridge stiffness. Therefore, we developed a novel technique for measuring the cross-bridge stiffness more directly. A force was applied to the cross-bridge by driving one optical trap back and forth with a sinusoidal motion. Cross-bridge distortion was measured from the motion of the other bead held in the fixed optical trap. Using this technique, we found cross-bridge stiffness to be $\sim$0.7 pN nm^{-1}. Recently, Mehta et al. (1997) obtained a similar value of 0.65 pN nm^{-1} for the stiffness of a single rabbit HMM cross-bridge. They used a trapping geometry identical to that employed here and determined cross-bridge stiffness by a method based on analysis of Brownian motion. Nishizaka et al. (1995) measured HMM cross-bridge stiffness by using a single bead held in an optical tweezer that was attached to the end of an actin filament by gelsolin. Use of gelsolin should reduce actin filament flexure, because it binds to the end of the filament. However, they still found a nonlinear length-tension diagram at low force that extended over 30 nm. Their estimate of cross-bridge stiffness, 0.58 pN nm^{-1}, obtained from the steepest region of the curve, is similar to the value we report here.

Several recent studies (e.g., Irving et al., 1995) indicate that the regulatory domain of the myosin head tilts during the working stroke and is distorted by load (Lombardi et al., 1995). If the elasticity resides within the head of the myosin molecule, it is important to ask whether a "cross-bridge" consists of one or both heads of myosin. In this study, we used two-headed HMM, so potentially both heads might bind to actin. If the regulatory domain of each head contributes equally to stiffness, then we might expect the stiffness to be double that of a single head. However, it may be that only one of the two HMM heads can form a stiff connection to actin, as has been suggested for muscle fibers (Offer and Elliott, 1978; Huxley and Tideswell, 1997).

Series elasticity and the working stroke

We have shown that because $\kappa_{con} \gg \kappa_{trap}$, the bead-actin-bead assembly used in these experiments oscillates back and forth as a nearly rigid body, under the influence of Brownian motion. On average, the HMM molecule could interact with any actin monomer that comes within range. This means that the starting point of any individual displacement is unknown and simply cannot be measured for a single observation. Instead, a large number of displacements must be measured and averaged. The expected distribution of displacement amplitudes will have the same r.m.s. deviation as the Brownian noise of the bead-actin-bead assembly. Consequently, if we know the standard deviation of the expected distribution from the overall system stiffness, κ_x, we can calculate the accuracy of our estimate of the displacement from the standard error of the mean. We measured 666 attachments, which had an average observed displacement of 5.0 nm and a standard error of the mean of ± 0.4 nm ((r.m.s. of Brownian noise)/$(n)^{0.5}$). Because κ_{con} is in series with κ_{trap}, it will be extended by the cross-bridge working stroke, and so the observed bead displacement is smaller than the working stroke. From our measurements of κ_{con} and κ_{trap}, the working stroke will be 10% larger than the displacement measured directly from the bead motion. The working stroke is therefore 5.5 ± 0.4 nm (SEM).

We found no change in length of the actin filament caused by its interaction with HMM. During each interaction the entire bead-actin-bead assembly is translated by ~5 nm by the cross-bridge. Therefore, we have shown that length changes in the actin filament neither cause nor contribute significantly to the movement produced by actomyosin interactions at low load.

Cross-bridge working-stroke, stiffness, and energy transduction

The best estimates of mechanical work done per ATP hydrolyzed in muscle fibers come from experiments that were performed to determine the efficiency of muscle contraction. Frog sartorius muscle is the best studied muscle type. Kushmerick and Davies (1969) found that frog sartorius produced 38 pN nm per ATP hydrolyzed (average of their three highest estimates, multiplied by 115% to account for extra ATP usage by Ca^{2+} pumping, as suggested by Woledge et al., 1985). Huxley and Simmons (1971) suggested a similar value of 30 pN nm ($7.3kT$) per interaction. The best recent estimate, based purely on fiber mechanical properties (Linari et al., 1998), produces a value of 27 pN nm per working stroke.

To summarize our results: We have found that κ_{xb} is 0.7 pN nm^{-1}, and the cross-bridge working stroke, d_{xb}, is 5.5 nm. Using these values, it is straightforward to calculate the mechanical work done per interaction, e.g., $\frac{1}{2}d_{xb}^2 \times \kappa_{xb} = 11$ pN nm. This is only one-third of the value obtained from frog muscle fibers. There are several possible explanations and we list their pros and cons below:

1. The mechanical work performed by each actomyosin interaction might be lower in rabbit back muscle than in frog sartorius muscle. We know of no good estimates of work done per ATP hydrolyzed by rabbit fast muscle. There is known to be variability between muscle type and species; higher values have been reported for tortoise muscle (Woledge, 1968) and much lower values for insect flight muscle (Ellington, 1985).

2. The force produced by myosin may be constant over the working stroke. This would make our estimate twice as large as those given above, e.g., work $= d_{xb}^2 \times \kappa_{xb}$. The data of Fig. 8 B (curves a_1, a_2, and a_3) are too noisy to determine whether cross-bridge stiffness is strictly linear, so we cannot rule out the possibility that myosin exerts a nearly constant force during its working stroke. Highly nonlinear elasticity is not easily compatible with most mechanistic schemes for cross-bridge behavior (e.g., Pate and Cooke, 1989).

3. One or both of our measurements (e.g., d_{xb} or κ_{xb}) may underestimate values obtained in the well-ordered filament lattice of muscle fibers.

d_{xb}: Previously (Molloy et al., 1995) we noted that myosin head orientation may affect the size of the observed movement produced by the cross-bridge. HMM molecules that are randomly oriented with respect to the actin filament might produce a mean estimate of the working stroke that is determined by averaging a cosine term through 180°. This would lead to an underestimation of the "true" working stroke produced by correctly oriented HMM molecules by a factor of $\pi/2$. The highly ordered thick and thin filament arrays found in muscle sarcomeres ensure that all of the myosin molecules are aligned parallel to the axis of the actin filament. Ishijima et al. (1996) measured d_{xb} using synthetic myosin rod cofilaments and report a much longer working stroke (17 nm). This value is so large that it is not easily compatible with the idea of a change in cross-bridge conformation causing the movement.

κ_{xb}: Our in vitro measurement of cross-bridge stiffness might not reflect the stiffness of a cross-bridge in a muscle fiber. HMM bound to a nitrocellulose surface may be either stiffer or more compliant than that of a myosin embedded in a thick filament in muscle. If we take our highest estimates of κ_{xb} (2 pN nm^{-1}), then the mechanical work done per working stroke would be 30 pN nm. It is interesting to note that κ_{xb} measured using synthetic myosin rod cofilaments, with correctly oriented myosin heads (Ishijima et al., 1996; 0.14 pN nm^{-1}), was smaller than ours obtained with HMM fixed to nitrocellulose. Such low values of stiffness require either long or multiple working strokes per ATP.

4. There may be more than one cross-bridge interaction per ATP hydrolyzed, and hence the mechanical work done per ATP might be higher. This idea has been proposed for the cross-bridge cycle occurring in muscle fibers that are allowed to shorten rapidly under low load (Piazzesi and Lombardi, 1995). However, we find that the lifetime of the attachments observed here under low load show a first-order dependence upon ATP concentration (data not shown). So it

is most likely that each of the mechanical interactions observed is terminated by the binding of one ATP molecule.

The calculated basic free energy change for ATP breakdown under the in vitro conditions used here is ~ -60 kJ mol^{-1} ($\equiv 100$ pN nm). This is almost twice the energy available in a muscle fiber under physiological conditions (-35 kJ mol^{-1} for creatine phosphate; Woledge and Reilly, 1988). We think it is incorrect to multiply the in vitro basic free energy by an efficiency factor obtained under physiological chemical conditions to obtain a value for expected work output for a single molecule in vitro. To do so would imply that the working stroke or cross-bridge stiffness depends upon ligand concentration. So far we find no evidence for this.

The amount of mechanical work done per ATP hydrolyzed by different myosins under different mechanical and chemical conditions remains an open question. Our measurement of the maximum work obtainable per ATP hydrolyzed under these in vitro conditions is only one-third of that measured in intact frog muscle fibers. However, our results are not inconsistent with current ideas of how actomyosin works.

APPENDIX

Force transducer consisting of a single trapped bead

The properties of this kind of transducer have been described in detail elsewhere (e.g., Svoboda and Block, 1994). Examination of this system reveals the basic properties of the optical tweezers transducer. The equation of motion of the trapped bead in solution is

$$m\frac{\partial^2 x}{\partial t^2} + \beta\frac{\partial x}{\partial t} + \kappa x = 0 \qquad (A1)$$

where m is the bead mass, x is the displacement of the bead from the trap center, β is the viscous drag, and κ_{trap} is the trap stiffness (Fig. 9). For most optical trap experiments, κ_{trap} is adjusted to be ~ 0.02 pN nm^{-1}. For a 1-μm-diameter bead suspended in water, inertial forces are negligible compared to the viscous damping and elastic trapping force, so the first term of Eq. A1 can be ignored. Thermally driven bead motion (Brownian motion) is characterized by a Lorentzian power spectrum with a cutoff frequency (f_c) determined by the ratio of trap stiffness to viscous drag coefficient:

$$f_c = \kappa_{\text{trap}}/(2\pi\beta) \qquad (A2)$$

(e.g., for a 1-μm bead suspended in water; $\beta = 6\pi\eta\alpha \cong 10^{-5}$ pN s nm^{-1}; $f_c \cong 330$ Hz).

If the viscous drag coefficient is known, the measured cutoff frequency can be used to calibrate trap stiffness. However, β depends critically on the proximity of the bead to the glass surface, doubling when a 1-μm-diameter

bead is moved from 2 μm above to 1 μm above the surface (Svoboda and Block, 1994). Hence calibration of trap stiffness by this method is unreliable during actomyosin interactions because the viscous drag coefficient is hard to measure.

A second method for determining trap stiffness is to measure the mean squared deviation in bead position ($\langle x^2 \rangle$). Trap stiffness can be calculated by applying the equipartition theorem:

$$\kappa_{\text{trap}} = \frac{kT}{\langle x \rangle^2} \qquad (A3)$$

(e.g., if $\kappa_{\text{trap}} = 0.02$ pN nm^{-1}, kT (thermal energy) ≈ 4 pN nm; $\langle x \rangle^2 = 200$ nm^2, hence the r.m.s. deviation $= 14$ nm).

Stiffness calculations from this relationship are independent of viscous drag and can be used to measure compliant, spring-like elements. However, the sensitivity of this method is lower at high stiffness because of the quadratic dependence of stiffness upon x position (Fig. 6 C). Calibration of the position detector and measurement and correction for system noise are crucial when high stiffnesses are measured.

Transducers based on a two-bead system

For two beads connected by a rigid filament, the total axial stiffness (κ_x) measured parallel to the filament will be the sum of the two trap stiffnesses. The two trap stiffnesses combine in parallel, not in series, as might first appear from Fig. 10.

If the linkage is compliant (κ_{link}, Fig. 10), the axial stiffness, κ_x (when both trap stiffnesses are the same), is given by

$$\kappa_x = \frac{\kappa_{\text{trap}} \times \kappa_{\text{link}}}{\kappa_{\text{trap}} + \kappa_{\text{link}}} + \kappa_{\text{trap}} \qquad (A4)$$

where

$$\frac{1}{\kappa_{\text{link}}} = \frac{1}{\kappa_{\text{con,R}}} + \frac{1}{\kappa_{\text{actin}}} + \frac{1}{\kappa_{\text{con,L}}}$$

(see Fig. 11).

If $\langle x^2 \rangle$ and κ_{trap} are known, κ_{link} can be calculated from Eqs. A3 and A4:

$$\langle x^2 \rangle = kT\left(\frac{1}{\kappa_x}\right) = \frac{kT}{2}\left(\frac{1}{\kappa_{\text{trap}}} + \frac{1}{(\kappa_{\text{trap}} + 2 \times \kappa_{\text{link}})}\right) \qquad (A5)$$

(e.g., κ_x varies between κ_{trap} and $2\kappa_{\text{trap}}$ for values of κ_{link} between 0 and ∞, and analysis of Brownian motion gives a good estimate of κ_{link} only if κ_{link} is similar to κ_{trap}).

During attachment, cross-bridge stiffness, κ_{xb}, changes the mechanical properties of the system:

$$\kappa_A = \frac{\kappa_{\text{trap}} \times \kappa_{\text{con,L}}}{\kappa_{\text{trap}} + \kappa_{\text{con,L}}} \approx \kappa_{\text{trap}} \qquad \text{for } \kappa_{\text{con}} \gg \kappa_{\text{trap}} \qquad (A6)$$

$$\kappa_B = \kappa_A + \kappa_{\text{xb}} \approx \kappa_{\text{xb}} \qquad \text{for } \kappa_{\text{xb}} \gg \kappa_A \qquad (A7)$$

$$\kappa_X = \frac{\kappa_{\text{con,R}} \times \kappa_B}{\kappa_{\text{con,R}} + \kappa_B} + \kappa_{\text{trap}} \approx \frac{\kappa_{\text{con,R}} \times \kappa_{\text{xb}}}{\kappa_{\text{con,R}} + \kappa_{\text{xb}}} \qquad (A8)$$

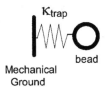

FIGURE 9   Force transducer consisting of a single trapped bead.

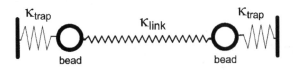

FIGURE 10   Transducers based on a two-bead system.

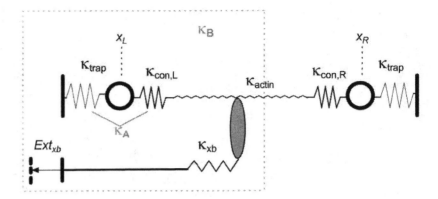

FIGURE 11 Components of κ_{link}.

where κ_{xb} and $\kappa_{\text{con}} \gg \kappa_{\text{trap}}$; κ_{x} is composite stiffness parallel to the actin filament (x direction); $X_{\text{L,R}}$ is the position of the left, right bead; κ_{trap} is trap stiffness; κ_{actin} is actin filament stiffness; $\kappa_{\text{con L,R}}$ is the actin-to-bead connection left, right; κ_{xb} is the cross-bridge stiffness; and Ext_{xb} is the cross-bridge extension.

During cross-bridge attachment, κ_{x} is dominated by κ_{con} and κ_{xb} in series. κ_{xb} can be found by analyzing Brownian motion if $\kappa_{\text{con}} \gg \kappa_{\text{xb}}$. Otherwise κ_{con} and κ_{xb} must be determined independently. To do this we applied a sinusoidal forcing function to the position of one trap (x_{trap}) and observed the positions of both beads (x_{L} and x_{R}). The analysis below assumes that 1) the right trap is moved and the left trap is fixed (e.g., x_{trap} is the position of the right trap); 2) κ_{actin} is much greater than other stiffnesses in the system; 3) $\kappa_{\text{xb}} \gg \kappa_{\text{trap}}$; 4) $\kappa_{\text{con}} = \kappa_{\text{con,R}} = \kappa_{\text{con,L}}$.

During periods when no cross-bridge is attached, κ_{con} is given by

$$\kappa_{\text{con}} = 2 \times \kappa_{\text{trap}} \times \frac{x_{\text{trap}} - x_{\text{R}}}{x_{\text{R}} - x_{\text{L}}} \tag{A9}$$

During periods of cross-bridge attachment, κ_{con} is given by

$$\kappa_{\text{con}} = \kappa_{\text{trap}} \times \frac{x_{\text{trap}} - x_{\text{R}}}{x_{\text{R}} - x_{\text{L}}} \tag{A10}$$

where ($x_{\text{trap}} - x_{\text{R}}$) and ($x_{\text{R}} - x_{\text{L}}$) are the in-phase amplitudes determined by discrete Fourier transform of the data (either as a running window of one cycle or averaged over the entire observation interval). κ_{xb} is given by the quotient of force applied to the cross-bridge and its extension:

$$\kappa_{\text{xb}} = \frac{F_{\text{xb}}}{\text{Ext}_{\text{xb}}} \tag{A11}$$

where

$$\text{Ext}_{\text{xb}} = X_{\text{L}} \times \frac{\kappa_{\text{con}} + \kappa_{\text{trap}}}{\kappa_{\text{con}}} \tag{A12}$$

$$F_{\text{xb}} = \kappa_{\text{trap}} \times (x_{\text{trap}} - x_{\text{R}}) - \kappa_{\text{A}} \times \text{Ext}_{\text{xb}} \tag{A13}$$

Note that if $\kappa_{\text{con}} \gg \kappa_{\text{trap}}$, then left bead movement $x_{\text{L}} \approx \text{Ext}_{\text{xb}}$.

Finally, estimates of the cross-bridge working stroke from the mean of the histogram of observed bead displacements (d_{observed}) should be corrected for extension of the series elastic elements:

xb work stroke

$$= \frac{\kappa_{\text{con}} + \kappa_{\text{trap}} + (2 \times \kappa_{\text{con}} \times \kappa_{\text{trap}})/\kappa_{\text{xb}}}{\kappa_{\text{con}}} \times d_{\text{observed}} \tag{A14}$$

We thank Prof. A. F. Huxley for helpful comments on an earlier version of the manuscript and Dr. A. E. Knight, Dr. R. Thieleczek, and Dr. R. T. Tregear for many helpful discussions.

This work was funded by the Biotechnology and Biological Sciences Research Council, the Royal Society (of London), and the British Heart Foundation.

REFERENCES

Cooke, R. 1998. Actomyosin interaction in striated muscle. *Physiol. Rev.* 77:671–697.

Dupuis, D. E., W. H. Guilford, J. Wu, and D. M. Warshaw. 1997. Actin filament mechanics in the laser trap. *J. Muscle Res. Cell Motil.* 18: 17–30.

Ellington, C. P. 1985. Power and efficiency of insect flight muscle. *J. Exp. Biol.* 115:293–304.

Finer, J. T., R. M. Simmons, and J. A. Spudich. 1994. Single myosin molecule mechanics: piconewton forces and nanometre steps. *Nature.* 386:113–119.

Guilford, W. H., D. H. Dupuis, G. Kennedy, J. Wu, J. B. Patlak, and D. M. Warshaw. 1997. Smooth and skeletal muscle myosin produce similar forces and displacements in the laser trap. *Biophys. J.* 72:1006–1021.

Huxley, A. F., and R. M. Simmons. 1971. Proposed mechanism of force generation in striated muscle. *Nature.* 233:533–538.

Huxley, A. F., and S. Tideswell. 1997. Filament compliance and tension transients in muscle. *J. Muscle Res. Cell Motil.* 17:507–511.

Huxley, H. E. 1969. The mechanism of muscular contraction. *Science.* 164:1356–1366.

Irving, M., T. S. Allen, C. Sabidodavid, J. S. Craik, B. Brandmeier, J. Kendrick-Jones, J. E. T. Corrie, D. R. Trentham, and Y. E. Goldman. 1995. Tilting of the light-chain region of myosin during step length changes and active force generation in skeletal muscle. *Nature.* 375: 688–691.

Ishijima, A., H. Kojima, H. Higuchi, Y. Harada, T. Funatsu, and T. Yanagida. 1996. Multiple and single molecule analysis of the actomyosin motor by nanometer piconewton manipulation with a microneedle: unitary steps and force. *Biophys. J.* 70:383–400.

Kojima, H., A. Ishijima, and T. Yanagida. 1994. Direct measurements of stiffness of single actin filaments with and without tropomyosin by in vitro nanomanipulation. *Proc. Natl. Acad. Sci. USA.* 91:12962–12966.

Kron, S. J., Y. Y. Toyoshima, T. Q. P. Uyeda, and J. A. Spudich. 1991. Assays for actin sliding movement over myosin-coated surfaces. *Methods Enzymol.* 196:399–416.

Kushmerick, M. J., and R. E. Davies. 1969. The chemical energetics of muscle contraction. II. The chemistry, efficiency and power of maximally working sartorius muscles. *Proc. R. Soc. Lond. Biol.* 174:315–353.

Linari, M., I. Dobbie, M. Reconditi, N. Koubassova, M. Irving, G. Piazzesi, and V. Lombardi. 1998. The stiffness of skeletal muscle in isometric contraction and rigor: the fraction of myosin heads bound to actin. *Biophys. J.* 74:2459–2473.

Lombardi, V., G. Piazzesi, M. A. Ferenczi, H. Thirlwell, I. Dobie, and M. Irving. 1995. Elastic distortion of myosin heads and repriming of the working stroke in muscle. *Nature.* 374:553–555.

Lymn, R. W., and E. W. Taylor. 1971. Mechanism of adenosine triphosphate hydrolysis by actomyosin. *Biochemistry.* 21:1925–1928.

Margossian, S. S., and S. Lowey. 1982. Preparation of myosin and its subfragments from rabbit skeletal muscle. *Methods Enzymol.* 85:55–71.

Meeusen, R., and Z. Cande. 1979. *N*-Ethylmaleimide-modified heavy meromyosin, a probe for actomyosin interactions. *J. Cell Biol.* 82:57–65.

Mehta, A. D., J. T. Finer, and J. A. Spudich. 1997. Detection of single molecule interactions using correlated thermal diffusion. *Proc. Natl. Acad. Sci. USA.* 94:7927–7931.

Molloy, J. E. 1997. Optical chopsticks: digital synthesis of multiple optical traps. *Methods Cell Biol.* 55:205–216.

Molloy, J. E., J. E. Burns, J. Kendrick-Jones, R. T. Tregear, and D. C. S. White. 1995. Movement and force produced by single myosin head. *Nature.* 378:209–212.

Nishizaka, T., H. Miyata, H. Yoshikawa, S. Ishiwata, and K. Kinosita, Jr. 1995. Unbinding force of a single motor molecule of muscle measured using optical tweezers. *Nature.* 377:251–254.

Offer, G., and A. Elliott. 1978. Can a myosin molecule bind to two actin filaments? *Nature.* 271:325–329.

Pardee, J. D., and J. A. Spudich. 1982. Purification of muscle actin. *Methods Cell Biol.* 24:274–289.

Pate, E., and R. Cooke. 1989. A model of cross-bridge action: the effects of ATP, ADP and P_i. *J. Muscle Res. Cell Motil.* 10:181–196.

Piazzesi, G., and V. Lombardi. 1995. A cross-bridge model that is able to explain mechanical and energetic properties of shortening muscle. *Biophys. J.* 68:1966–1979.

Rayment, I., H. M. Holden, M. Whittaker, C. B. Yohn, M. Lorenz, K. C. Holmes, and R. A. Milligan. 1993. Structure of the actin-myosin complex and its implications for muscle contraction. *Science.* 261:58–65.

Saito, K., T. Aoki, and T. Yanagida. 1994. Movement of single myosin filaments and myosin step size on an actin filament suspended in solution by a laser trap. *Biophys. J.* 66:769–777.

Simmons, R. M., J. T. Finer, S. Chu, and J. A. Spudich. 1996. Quantitative measurements of forces and displacement using an optical trap. *Biophys. J.* 70:1813–1822.

Svoboda, K., and S. M. Block. 1994. Biological applications of optical forces. *Annu. Rev. Biophys. Biomol. Struct.* 23:247–85.

Veigel, C., D. C. S. White, and J. E. Molloy. 1997. Single molecule energy transduction. *Biophys. J.* 72:A56.

Woledge, R. C. 1968. The energetics of tortoise muscle. *J. Physiol. (Lond.).* 197:685–707.

Woledge, R. C., N. A. Curtin, and E. Homsher. 1985. Energetic aspects of muscle contraction. *Monogr. Physiol. Soc.* 41:167–275.

Woledge, R. C., and P. C. Reilly. 1988. Molar enthalpy change for hydrolysis of phosphorylcreatine under conditions in muscle cells. *Biophys. J.* 54:97–104.

Single-Molecule Mechanics of Heavy Meromyosin and S1 Interacting with Rabbit or *Drosophila* Actins Using Optical Tweezers

Justin E. Molloy,* Julie E. Burns,* John C. Sparrow,* Richard T. Tregear,‡
John Kendrick-Jones‡ and David C. S. White*

*Department of Biology, University of York, United Kingdom, and ‡Laboratory of Molecular Biology, Cambridge, United Kingdom

ABSTRACT   Single-molecule mechanical interactions between rabbit heavy meromyosin (HMM) or subfragment 1 (S1) and rabbit actin were measured with an optical tweezers piconewton, nanometer transducer. Similar intermittent interactions were observed with HMM and S1. The mean magnitude of the single interaction isotonic displacements was 20 nm for HMM and 15 nm with S1. The mean value of the force of single-molecule interactions was 1.8 pN for HMM and 1.7 pN with S1. The stiffness of myosin S1 was determined by applying a sinusoidal length change to the thin filament and measuring the corresponding force; the mean stiffness was 0.13 pN nm^{-1}. By moving an actin filament over a long distance past an isolated S1 head, we found that cross-bridge attachment occurred preferentially at a periodicity of about 40 nm, similar to that of the actin helical repeat. Rate constants for the probability of detachment of HMM from actin were determined from histograms of the lifetime of the attached state. This gave a value of 8 s^{-1} or 0.8 × 10^6 M^{-1} s^{-1} for binding of ATP to the rigor complex. We conclude (1) that our HMM-actin interactions involve just one head, (2) that compliance of the cross-bridge is not in myosin subfragment 2, although we cannot say to what extent contributions arise from myosin S1 or actin, and (3) that the elemental movement can be caused by a change of shape of the S1 head, but that this would have to be much greater than the movements suggested from structural studies of S1 (Rayment et al., 1993).

INTRODUCTION

Optical tweezers (Ashkin et al., 1986) have been used previously to control position and to measure the mechanics of interaction between kinesin and microtubules (Svoboda and Block, 1993) and between myosin and actin (Finer et al., 1994). We have designed and built a transducer based upon optical tweezers as outlined in Fig. 1, similar in principle to that reported by Finer et al. (1994). In our apparatus, several beads can be suspended simultaneously by chopping a single laser beam between different positions, enabling a single actin filament to be attached to a bead at each end and held taut in a specific position and orientation.

We have performed some preliminary experiments to address the following questions:

1) Does single-headed S1 produce the same force and displacement as HMM?

2) What is the force and displacement produced by HMM interacting with actins of different structure?

3) Is the slower F-actin sliding velocity, measured in in vitro assays, of wild-type *Drosophila* and even slower mutant ACT88F^{E93K} due to a foreshortened work stroke or because of lower cross-bridge force?

4) Does S1 interact preferentially with actin monomers that have favorable azimuthal orientation?

5) Is cross-bridge detachment distortion-dependent?

Address reprint requests to Dr. Justin E. Molloy, Department of Biology, University of York, Heslington, York YO1 5DD, U.K. Tel.: 44-904-4324399; Fax: 44-904-432860; E-mail: jem1@york.ac.uk.

Abbreviations used: BSA-TRITC, bovine serum albumin-tetrarhodamine isothiocyanate; NEM, *N*-ethyl-maleimide; HMM, heavy meromyosin; WT, wild type; SDS-PAGE, sodium dodecyl sulphate, polyacrylamide gel electrophoresis; E93K, glutamic acid to lysine at amino acid position 93.

MATERIALS AND METHODS

Protein purification

Actin and HMM

Rabbit F-actin was prepared using a standard method (Pardee and Spudich, 1982). The flight muscle-specific isoform of actin (Act88F) from *Drosophila* was isolated by a scaled-down purification of dissected indirect flight muscles. From 10 flies, we obtained a yield of 5 μg of actin.

Rabbit myosin and chymotryptic HMM were prepared essentially as described by Margossian and Lowey (1982), with the exception that TLCK-treated, α-chymotrypsin was used and its activity stopped with Bowman-Birk inhibitor.

Production of S1

Rabbit skeletal myosin was digested under conditions of low salt and high free magnesium in the presence of activated papain (12 mg/ml myosin in 100 mM NaCl, 4 mM MgCl$_2$, 1 mM DTT, 10 mM imidazole, pH 7.3, 10 μg/ml activated papain, 22°C, 15 min). The reaction was stopped by a twofold excess of E-64 (*trans*-epoxysuccinyl-L-leucylamido(4-guanidino)--butane). Digest products were dialyzed to low ionic strength, and precipitated material was pelleted by centrifugation. The supernatant was column purified (DE52, Whatman, Kent, U.K.), and the central peak of the eluate was concentrated by ultrafiltration (Amicon Ltd. Stonehouse, Gloucs, U.K.) in the presence of 30% sucrose. S1 aliquots were frozen to −80°C. The S1 ran as a single, symmetrical, peak on ion-exchange (DEAE, Whatman) and gel filtration (Sephacryl S200, Pharmacia, Uppsala, Sweden) chromatography. However, SDS-PAGE (gels not shown) indicated that in addition to the intact 95-kDa heavy chain (about 50% of total product), fragments of 75- and 20-kDa were also present in this preparation. All three types of light chain were present.

Production of fluorescently labeled, NEM-myosin-coated latex beads

1.1 μm latex (polystyrene) beads (Sigma Chemical Co., St. Louis, MO) were coated with BSA-TRITC (1 μg/ml) and NEM-treated monomeric myosin (10 μg/ml) (prepared as Sekine and Keilley, 1964) by incubating 0.1% beads, by mass, at 4°C for 1 h in a 0.5 M KCl, buffered (pH 7) solution.

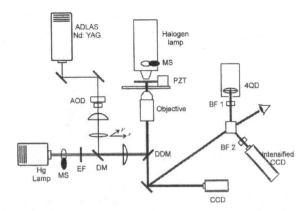

FIGURE 1   Apparatus block diagram. The optical trap is built around an inverted microscope (Axiovert 135, Zeiss, Oberkochen, Germany). Infrared (1064 nm) laser light, from a diode-pumped Nd:YAG laser (Adlas 321, Adlas, Lubeck, Germany) is combined with green light (EF, 546FS10.25, excitation filter, Andover Corp., Salem, NH) from a mercury arc lamp using a "hot mirror" (DM = 820DCSP, Omega Optical, Brattleboro, VT). Both light beams enter the microscope epifluorescence port via a custom-built housing. A dual dichroic mirror (DDM, 570DCLP, Omega Optical; reflects 546 and 1064 nm; transmits >570 nm), mounted in the microscope filter block, allows us to use optical tweezers and view rhodamine fluorescence simultaneously. Laser beam alignment is via two mirrors and control of trap position is made with two orthogonally mounted acousto-optic deflectors (AODs) (NEOS, Melbourne, FL); these are controlled by a custom-built computer interface card. To produce two optical traps, we chop between two sets of *x, y* coordinates (to simplify computation, these coordinates are chopped in hardware at 10 kHz). The laser light path is completely enclosed using cardboard tubing and low density foam rubber to prevent air currents from entering the system at any point. Coarse control of stage position is by mechanical drives and a custom-built piezoelectric substage (PZT) allows small range, computer-controlled movements of the microscope slide. High speed position measurements are made with a 4-quadrant photodiode (4QD, S1557, Hamamatsu Photonics, Hamamatsu City, Japan) and custom-built electronics. Scattered laser light is excluded from this detector (BF1, short-pass barrier filter). Fluorescently labeled actin is viewed with an intensified CCD camera ("Photon - P46036A", EEV, Chelmsford, U.K.) coupled to a barrier filter (BF2 = LP590, Zeiss). Bright-field illumination (100 W halogen lamp) is used to produce a high magnification video image (CCD camera, P46310, EEV). An Acroplan 100X, 2.5 N.A. objective and an Optovar 2.5x insert are used to obtain the desired image magnification. Computer-controlled mechanical shutters (MS) are used to switch between bright-field and fluorescence illumination.

Bead agglomeration was reduced by final addition of 0.5 mg/ml BSA and dispersion by brief ultrasonication at 4°C.

In vitro assays

In vitro motility assays were performed according to Kron et al. (1991). All motility assays were performed at 23°C.

Mechanical measurements made by optical tweezers

Single actin filaments were attached between two NEM-myosin-coated, 1.1-μm diameter latex beads, held in two independently controlled laser traps. The position of one of these beads was monitored by imaging the bead on a quadrant photodetector (Finer et al., 1994). HMM or S1 was supported on a 1.7-μm-diameter glass microsphere attached to a coverslip, and the interactions with the actin filament were enabled by moving the actin filament close to the HMM- or S1-coated glass microsphere. To measure forces and displacements produced by single-motor molecules, the concentration of HMM or S1 applied to the flow-cell was reduced until summation of

events no longer occurred. For both S1 and HMM, about 1–2 μg protein/ml was required. This concentration is reasonable if one assumes that all myosin molecules bind to the nitrocellulose surface, and that the "reach" of an actin filament covers a 30×300 nm^2 area of the 1.7-μm glass microsphere. This area would then contain about 2–5 molecules of motor protein.

In free-run (quasi-isotonic) mode, the stiffness of the laser traps was 0.02 pN nm^{-1} each, giving an overall system stiffness of 0.04 pN nm^{-1}; this stiffness, much less than that of the myosin subfragments, allowed estimates of the elemental movement produced by the interactions between the myosin fragments and the actin. Thermal motion of the bead, while held in the trap, limits the minimum observable stroke size. Because trap and cross-bridge stiffnesses are in series, bead displacement produced by the cross-bridge stroke depends on the relative value of the two stiffnesses.

In feedback (force) mode, the bead was held in a fixed position, forces exerted by the molecular interactions being compensated by movement of the laser trap, allowing estimates of the isometric force of interaction between the molecules. Controlled length changes were applied by imposing movement onto the laser beam holding the bead imaged on the photodetector, the force being measured from the movement of the laser beam required to produce the correct movement of the bead. By this means, stiffness- and distortion-dependent kinetics of the myosin/actin interactions were determined.

RESULTS

Myosin structure

Comparison of responses obtained with rabbit HMM and S1

Fig. 2 shows the interactions obtained between single molecules of rabbit heavy meromyosin and S1 with rabbit actin filaments measured at 10 μM ATP. Responses are shown with the apparatus in (*A*) force and (*B*) displacement modes.

Forces seen with S1 are similar to those observed with HMM, from which we conclude that only one HMM head is producing force under the conditions we have used. The mean value of the forces seen, about 1.8 pN (SD = 0.86 pN, *n* = 724), is less than that, 3.5 pN, reported by Finer et al. (1994) for HMM, as are the maximum values of our 4.5 pN, compared with their 7 pN.

The mean value of displacements seen with S1 (14.8 nm; SD = 5.8; *n* = 264) is significantly less (Student's *t*-test) than seen with HMM (19.7 nm; SD = 7.8; *n* = 261). However, it is possible that the HMM data contain some double strokes and histograms of both data sets (not shown) indicating that the distribution is not normal. So we cannot conclude that the elemental step size of S1 is shorter than that of HMM. The important finding is that for both S1 and HMM displacements of 15 nm are frequently observed.

Actin structure

Comparison of responses seen with different actins

We have measured the force and displacements seen between actin filaments prepared from different sources and rabbit HMM. Fig. 3 shows typical force traces obtained with actin

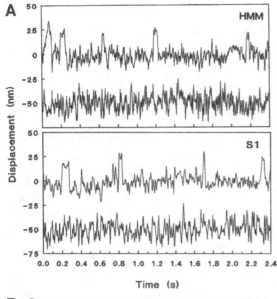

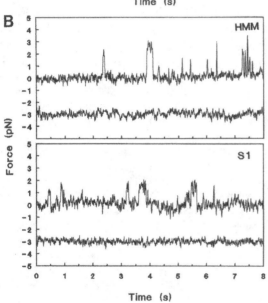

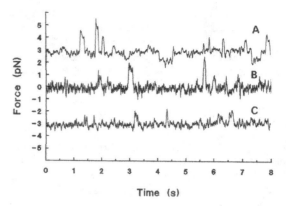

FIGURE 3 Force records of rabbit HMM with: (*A*) rabbit actin (note the negative forces seen in this record; (*B*) *Drosophila* WT actin; (*C*) *Drosophila* ACT88F^{E93K} actin.

Lifetime of attachment: probability of detachment

Fig. 4 shows data of force records for rabbit HMM with rabbit and *Drosophila* WT actin recorded on slow timebases. Also shown are the distribution of the times of attachment, in which attachment is defined as a transition in the force records greater than a threshold (indicated by the horizontal

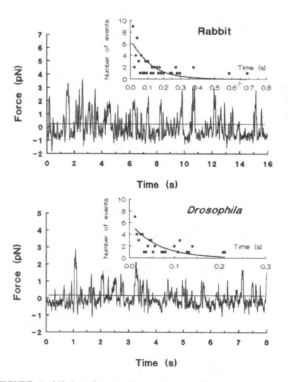

FIGURE 2 Single mechanical interactions between rabbit actin and rabbit HMM or S1 measured at 10 μM ATP (all data were filtered using a 9-point running average). The major noise contribution in all of these records is thermal motion of the bead. The *x* axis stiffness in these measurements is 0.04 pN/nm, and the *y* axis stiffness is 0.02 pN/nm. (*B*) Force records. Traces of force against time. The upper traces are the AOD output parallel to the thin filament (*x*), and the lower traces that perpendicular (*y*) (data sample rate 500 Hz). (*A*) Displacements. Traces of displacement against time. The upper traces are *x* data, and the lower traces are *y* data (data sample rate 2 kHz). Because the traps are aligned in the *x* axis and the actin filament is pulled taut, Brownian noise is less in the *x* than in the *y* axis. During attachments, cross-bridge stiffness adds to the overall stiffness, which constrains the beads' thermal motion, so noise is reduced.

from rabbit, *Drosophila* wild type (WT) and *Drosophila* Act88F^{E93K}. The force pulses were significantly smaller in the *Drosophila* actin interactions (Molloy et al., 1995). However, work stroke sizes were the same.

FIGURE 4 Lifetime of attached states. Records of force against time obtained for rabbit HMM interacting with Rabbit and *Drosophila* actins. There is probably more than one head interacting with the actin filaments in these traces. The horizontal lines indicate the threshold used to determine whether a myosin HMM was attached for the analysis of the distribution of attached lifetimes (histograms, *inset*). The exponential fit for the data obtained using rabbit actin has a rate constant of 8.7 s^{-1}. The exponential fit for that using *Drosophila* actin has a rate constant of 16.0 s^{-1}.

lines), lasting for at least 10 ms. The best-fit exponential curves to this data are also indicated; the rate constants from these fits gives the value of the rate constant for detachment. This preliminary result implies that the attached lifetime of rabbit HMM with *Drosophila* WT actin is shorter than with rabbit actin.

Distortion-dependent responses

What is the stiffness of a single cross-bridge? We have measured the stiffness of the interaction between a single S1 and an actin filament by applying a sinusoidal length change of 20-nm peak-to-peak amplitude at a frequency of 2.5 Hz to a thin filament and after the tension changes induced during the attached phase of the cross-bridge. Fig. 5 *A* shows the results from one such experiment. In Fig. 5 *B* length is plotted against tension for the two attached phases indicated in Fig. 5 *A*. The displacement between the traces is due to a greater tension being exerted by the cross-bridge during attachment 1 than attachment 2. The

stiffnesses of the two plots are about the same, 0.13 pN nm^{-1}. This estimate is greater than that calculated from the mean force and displacements (1.8 pN/15 nm = 0.12 pN nm^{-1}) but less than measurements made using muscle fibers (Huxley and Simmons, 1971).

Distortion dependence of cross-bridge detachment

Fig. 6 is a longer time recording including the same trace as Fig. 5. It is apparent from this trace that the lifetime of the attached state during the stretch phase of the cycle is far greater than that during the release phase (*arrowed*). This is a highly qualitative demonstration of a strong distortion dependence of attachment, and the direction of the effect, that pulling on a cross-bridge tends to keep it attached, as is expected from the Fenn effect.

Distortion dependence of attachment (geometric constraints imposed by stereospecific binding)

Fig. 7 *A* shows the results of a preliminary experiment in which a larger periodic length change was applied to the actin filament to move it slowly past a single myosin S1, and the force of interaction between the S1 and the myosin recorded with the trap in force mode (baseline force arising from the stiffness of the other trap has been subtracted away). The length change was triangular, i.e., backward and forward at a constant velocity as indicated in the figure, at a frequency of 0.25 Hz and peak-to-peak amplitude of 100 nm. The velocity of movement was thus 50 nm s^{-1}, which is much less than the free-sliding velocity of actin filaments measured in these assays (0.5–2 μm s^{-1}). Cross-bridge interactions are seen as vertical displacements of the force trace.

To show the positions at which attachment occurred, the data of Fig. 7 *A* are replotted as a trace of force against position in Fig. 7 *B*. It can be seen that attachments occur preferentially at discrete positions (*arrowed*). The separation of the arrows is about 40 nm, which is very close to the period

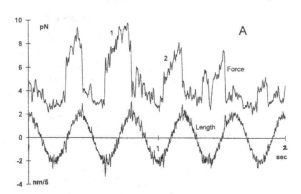

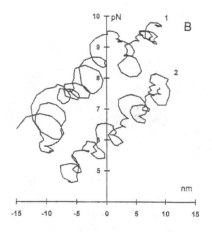

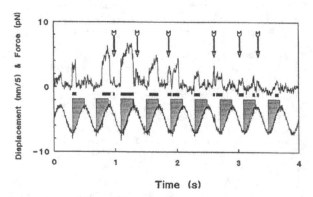

FIGURE 5 Stiffness of S1. (*A*) The lower trace shows the length change applied to the actin filament as a function of time, and the upper trace shows the corresponding trace of force. (*B*) Length against tension for the two attachments labeled 1 and 2 in *A*.

FIGURE 6 Distortion dependence of detachment. A longer record of Fig. 5. The period when the length is increasing is shown shaded. Attachments during the release phase are arrowed. Their duration is much less than that obtained during the stretch phase of the cycle.

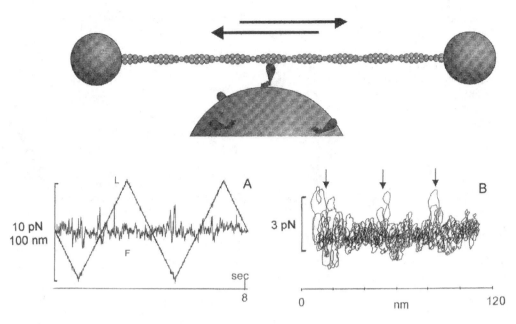

FIGURE 7 Preferred attachment positions on actin filament. Force records obtained during the application of a 100-nm periodic length change applied to a thin filament (shown schematically above). (*A*) Traces of applied length (*L*) and force (*F*) against time. (*B*) Force (*F*) plotted against position (*L*) from *A*.

of the twist of the thin filament of 38.5 nm. Azimuthal attachment selection has been seen in rigor myofibrils (Reedy, 1968), and physiological significance in active muscle has been proposed (Wray, 1979).

DISCUSSION

The results that we have shown are a series of preliminary experiments that demonstrate the capabilities of the optical tweezers apparatus for measuring mechanical and kinetic properties of single-molecule interactions. Finer et al. (1994) showed single molecule interactions between rabbit HMM and rabbit actin filaments. Our findings are in broad agreement with theirs, although we measure slightly longer working strokes and lower mean forces. Lower forces could arise from larger compliance between our myosin subfragments and the nitrocellulose surface, and longer displacements could arise because both their and our data events were analyzed by eye; this may have introduced observer bias.

We show that S1 has very similar forces and displacements to HMM, suggesting that only one HMM head is acting at any one time under isometric conditions. For displacements on the order of 15 nm to be obtained from rotation of a myosin head whose length is also about 15 nm, a change in angle of 1 radian (60°) is required. For an angle change of this magnitude, it is much more straightforward mechanically for this to occur from an angle of less than 90° to one of greater than 90° (see Irving et al., this volume).

The similarity of the results with HMM and S1 suggests that only one head of myosin is required for force production and that myosin S2 is not required per se for force production. The implication is that cross-bridge compliance resides mainly in acto-S1 and not in S2. However, the nature of

attachment between myosin subfragments and the nitrocellulose substrate is not clear, and it is possible that myosin-substrate attachment compliance is required for force production.

Frequently, we observe negative force and displacement events in our data (see, for example, Fig. 3 *A*). We believe these events are real and that they could occur in two possible ways. Either a) cross-bridges attach with negative distortion (i.e., pushing rather than pulling) because of thermal motion of the actin filament and myosin head, or b) at low ATP concentrations cross-bridges sometimes undergo futile work strokes, before attachment to actin, and the negative steps are work stroke reversals after attachment. These possibilities can be tested by performing experiments at different ligand concentrations.

Rabbit versus *Drosophila* wild-type and mutant, E93K, actins

Rabbit HMM interacting with *Drosophila* wild-type and E93K mutant actins produced the same displacement as with rabbit actin, but with lower force in wild-type and lower still in E93K. In this case, the work stroke is independent of actin structure, but cross-bridge force or cross-bridge stiffness is greatly affected by the molecular structure of the acto-myosin binding surface. This implies that the acto-myosin binding site is a major source of cross-bridge compliance.

Shorter attached lifetimes found with *Drosophila* wild-type compared with rabbit actin can be explained by lower cross-bridge stiffness (consistent with the lower forces), which would reduce the free-energy barrier for biochemical transitions and, therefore, increase the detachment rate constant.

In other experiments (in collaboration with Azam Razzaq and Roisean Ferguson), we found that the free-sliding velocity of wild-type *Drosophila* actin was slower than rabbit, whereas E93K was slower still and moved only at very low ionic strength. Together with our findings here, this implies that filament sliding velocities can be controlled not by the size of the working stroke but by the amount of force produced by myosin.

This work was supported by Biotechnology and Biological Sciences Research Council and The Royal Society.

REFERENCES

Ashkin A., J. M. Dziedzic, J. E. Bjorkholm, and S. Chu. 1986. Observation of a single-beam gradient force optical trap for dielectric particles. *Opt. Lett.* 11:288–290.

Drummond, D. R., M. Peckham, J. C. Sparrow, and D. C. S. White. 1990. Alteration in crossbridge kinetics caused by mutations in actin. *Nature.* 348:440–442.

Finer, J. T., R. M. Simmons, and J. A. Spudich. 1994. Single myosin molecule mechanics: piconewton forces and nanometre steps. *Nature.* 368:113–119.

Huxley, A. F., and R. M. Simmons. 1971. Proposed mechanism of force generation in muscle. *Nature.* 223:533–538.

Kron, S. J., Y. Y. Toyoshima, T. Q. P. Uyeda, and J. A. Spudich. 1991. Assays for actin filament sliding movement over myosin coated surfaces. *Methods Enzymol.* 196:399–416.

Margossian, S. S., and S. Lowey. 1982. Preparation of myosin and its subfragments from rabbit skeletal muscle. *Methods Enzymol.* 85:55–71.

Molloy, J. E., J. E. Burns, J. C. Sparrow, and D. C. S. White. 1995. Forces developed by *Drosophila* wild-type and mutant actin interacting with rabbit heavy meromyosin measured by an optical tweezers transducer. *Biophys. J.* 68:238a. (Abstr.)

Pardee, J. D., and J. A. Spudich. 1982. Purification of muscle actin. *Methods Cell Biol.* 24:271–289.

Rayment, I., H. M. Holden, M. Whittaker, C. B. Yohn, M. Lorenz, K. C. Holmes, and R. A. Milligan. 1993. Structure of the actin-myosin complex and its implications for muscle contraction. *Science.* 261:58–65.

Reedy, M. K. 1967. Ultrastructure of insect flight muscle. *J. Mol. Biol.* 31:155–176.

Sekine, T., and W. W. Keilley. 1964. The enzymic properties of *N*-ethylmaleimide modified myosin. *Biochim. Biophys. Acta.* 81:336–345.

Svoboda, K., and S. M. Block. 1993. Direct observation of kinesin stepping by optical trapping interferometry. *Nature.* 365:721–727.

Svoboda, K., and S. M. Block. 1994. Biological applications of optical forces. *Annu. Rev. Biophys. Biomol. Struct.* 23:247–85.

Wray, J. 1979. Filament geometry and the activation of insect flight muscles. *Nature.* 280:325–326.

DISCUSSION

Session Chairperson: Kenneth A. Johnson
Scribe: Seth Hopkins

DAVID WARSHAW: I think you have an elegant approach here of getting the stiffness of a single cross-bridge, and the unique aspect is that you are also doing it under low ATP conditions and so therefore you really have rigor bridges because the ADP would have been off and its waiting for an ATP to come along. So you can get stiffness of a rigor bridge, and I would suggest an important experiment. That would be to try to do this under higher ATP conditions to see if you get the same stiffness to finally answer the question: does a rigor bridge stiffness equal that of an active cross-bridge?

JUSTIN MOLLOY: Right, that would be a nice experiment to do.

AMIT MEHTA: Justin, in the sliding filament in vitro assay, S1 supports movement at a greatly reduced speed with respect to HMM by more than a factor of three. I don't recall the numbers, but that doesn't seem like it can be accounted for by the subtle differences in step and almost same isometric force. Effects such as attachment artifact you would think would affect both assays equally. So I wondered if you had any ideas as to why that would be the case.

MOLLOY: I guess that the attachment artifact is probably the answer to that. In this assay, you are selecting a single head and you are selecting on the basis that it is a working head. In the free-running assay, you might have heads stuck all over the place, some of them may be interferring and slowing the filament down. So with the S1, maybe it just doesn't stick down in the right orientation as nicely as the HMM or something like that.

MEHTA: But wouldn't you expect to get alot more small events if that were the case?

MOLLOY: I guess that's right, but when you are hunting around over a bead, basically you have to do a little hunt for events because there is very low density. You are always looking for events and not nonevents. That would be an experiment worth doing: literally probing for a lack of events and scoring them.

HIDETAKE MIYATA: In your measurement of the S1 stiffness, the amplitude of the response curve becomes smaller—why is that? Another question is how do you decide the cutoff level of the duration time?

MOLLOY: The parameters I chose for the forcing function were chosen based on my knowing the half-time of the on time, so I put the sine wave at 2.5 Hz and that seemed to work out quite nicely, and I set the amplitude at 20 nm to give a reasonable distortion while it was bound. But you are quite right in that record you do see a slight tailing off in the effect and maybe that's just drift, that the actin filament is drifting away from the S1. Because if you have 10 nm worth of drift over 8 s, you would see something just like that.

Two-Dimensional Tracking of ncd Motility by Back Focal Plane Interferometry

Miriam W. Allersma,* Frederick Gittes,* Michael J. deCastro,# Russell J. Stewart,# and Christoph F. Schmidt*

*Department of Physics and Biophysics Research Division, University of Michigan, Ann Arbor, Michigan 48109, and #Department of Bioengineering, University of Utah, Salt Lake City, Utah 84105 USA

ABSTRACT   A technique for detecting the displacement of micron-sized optically trapped probes using far-field interference is introduced, theoretically explained, and used to study the motility of the ncd motor protein. Bead motions in the focal plane relative to the optical trap were detected by measuring laser intensity shifts in the back-focal plane of the microscope condenser by projection on a quadrant diode. This detection method is two-dimensional, largely independent of the position of the trap in the field of view and has ~10-μs time resolution. The high resolution makes it possible to apply spectral analysis to measure dynamic parameters such as local viscosity and attachment compliance. A simple quantitative theory for back-focal-plane detection was derived that shows that the laser intensity shifts are caused primarily by a far-field interference effect. The theory predicts the detector response to bead displacement, without adjustable parameters, with good accuracy. To demonstrate the potential of the method, the ATP-dependent motility of ncd, a kinesin-related motor protein, was observed with an in vitro bead assay. A fusion protein consisting of truncated ncd (amino acids 195–685) fused with glutathione-S-transferase was adsorbed to silica beads, and the axial and lateral motions of the beads along the microtubule surface were observed with high spatial and temporal resolution. The average axial velocity of the ncd-coated beads was 230 ± 30 nm/s (average ± SD). Spectral analysis of bead motion showed the increase in viscous drag near the surface; we also found that any elastic constraints of the moving motors are much smaller than the constraints due to binding in the presence of the nonhydrolyzable nucleotide adenylylimidodiphosphate.

INTRODUCTION

The kinesins are motor proteins responsible for intracellular transport along microtubules of the cytoskeleton. Many active motion processes in eukaryotic cells have been found to be associated with a unique kinesin, or kinesins (for a recent review see Moore and Endow, 1996), prompting a great deal of research focusing on the functional principles of these proteins. Most members of the kinesin family move toward the microtubule plus-end, but some have been found to move toward the minus-end, for example, the ncd and kar3 proteins (McDonald and Goldstein, 1990; Walker et al., 1990; Endow et al., 1994; Kuriyama et al., 1995). Although they move in opposite directions, the motor domains of ncd and kinesin are 43% identical in amino acid sequence (Endow et al., 1990; McDonald and Goldstein, 1990) and have strikingly similar three-dimensional structures (Kull et al., 1996; Sablin et al., 1996). The directionality of kinesin and ncd is likely to be determined within the head domains including the neck region (Stewart et al., 1993; Case et al., 1997; Henningsen and Schliwa, 1997), but the specific microscopic determinants of directionality are not yet known.

Kinesin, originally isolated from squid giant axons (Brady, 1985; Vale et al., 1985), is highly processive; a single dimeric kinesin motor can transport microtubules over many microns, representing hundreds of enzymatic cycles (Howard et al., 1989; Block et al., 1990; Svoboda et al., 1993; Vale et al., 1996). A feature of kinesin's processive mechanism is its path on the surface of microtubules. Video tracking of kinesin-coated beads (Gelles et al., 1988; Wang et al., 1995) and observation of helical tracking of kinesin along supertwisted microtubules (Ray et al., 1993) demonstrated that kinesin dimers follow a single protofilament. Single-headed kinesin, in contrast, supports motility only in large numbers. Beads driven by multiple single-headed kinesins did not appear to follow a straight path along the microtubule surface in video tracking experiments (Berliner et al., 1995), which suggested that kinesin's processivity is due to interactions between the two heads of a dimer. Similar experiments to investigate the path of dimeric ncd have not been reported.

A variety of optical techniques capable of detecting nanometer displacements with high temporal resolution have been developed and applied to studying motor proteins and other biological molecules. Fine glass needles, imaged onto split photodiodes, have been used to observe motions and mechanical responses of myosin (VanBuren et al., 1994; Ishijima et al., 1996) and kinesin (Meyhöfer and Howard, 1995) with nanometer displacement resolution. Similarly, imaging micron-sized beads onto split photodiodes has been used to detect unitary translocation events of myosin (Finer et al., 1994; Simmons et al., 1996) and kinesin molecules (Coppin et al., 1996; Higuchi et al., 1997).

In addition to direct imaging techniques, several phase-sensitive, interferometric detection methods have been de-

Received for publication 14 July 1997 and in final form 7 November 1997.

Address reprint requests to Dr. Christoph Schmidt, Biophysics Research Division, University of Michigan, 930 North University, Chemistry Bldg., Ann Arbor, MI 48109-1055. Tel.: 313-763-9139; Fax: 313-764-3323; E-mail: cfs@umich.edu.

veloped. A spot-focused form of differential interference contrast (DIC) microscopy (Nomarski, 1955; Smith, 1955; Allen et al., 1969) was introduced (Denk and Webb, 1990) that detects the relative phase lag caused by a refractive object between two overlapping, orthogonally polarized laser foci. The method provides uniaxial subnanometer displacement resolution along the direction of the offset between the foci, while measuring the displacement relative to the foci. The detecting laser focus can simultaneously serve as an optical trap (Ashkin et al., 1986; Svoboda and Block, 1994a). The combination of optical trapping and interferometric detection has been used to study kinesin (Svoboda et al., 1993; Svoboda and Block, 1994b; Yin et al., 1995), the mechanical properties of DNA (Smith et al., 1996; Wang et al., 1997), and properties of F-actin gels (Gittes et al., 1997; Schnurr et al., 1997).

Other nonimaging displacement detection techniques generally share the use of split photodiodes to detect diffraction or interference patterns in the trapping laser light. Spatial detection of interference in scanned beam microscopes was first applied to generate phase contrast in scanning transmission electron microscopy (STEM) (Dekkers and de Lang, 1974; Dekkers and de Lang, 1977) and later in laser scanning microscopy (Wilson, 1988). Ghislain and colleagues (Ghislain and Webb, 1993; Ghislain et al., 1994) reported a technique to measure axial and lateral displacements in an optical trap by placing a photodiode between the back-focal plane of the microscope condenser and the intermediate image plane. Smith et al. (1996) reported a related displacement detection scheme in a laser trap formed by two counter-propagating laser beams of low numerical aperture, and Visscher et al. (1996) reported a further variation of the scheme using separate lasers for trapping and detection. When one laser is used for both trapping and detection, displacement is always measured with respect to the center of the trap. With independent lasers, displacement can be measured independent of trap position.

The major advantage of the detection methods based on photodiodes and laser illumination is that nanometer displacements can be measured with bandwidths up to 100 kHz quite easily. Although photodiodes have a potential detection bandwidth of many MHz, practical limits are set by the shot noise due to the finite number of photons detected per sample interval at reasonable laser powers and by electronic noise in the preamplifier (Denk and Webb, 1990; Gittes and Schmidt, 1997a,b). Fast detection avoids the low-pass filtering that is inherent in video imaging and that conceals details of molecular motions.

The approach we present here is to monitor, with a quadrant photodiode, shifts in the pattern of laser intensity in a plane exactly conjugate to the back-focal plane (BFP) of the condenser of our microscope. Laser light intensity shifts are caused by a refractive particle near the laser focus in the specimen plane. The technique works in parallel with optical trapping and has several unique properties: the displacement detection is biaxial, independent of the position of the laser focus in the specimen plane, and compatible

with multiple-trap configurations. We derived a simple but quantitative model explaining the detection scheme as a far-field interference effect between the illuminating light and the light scattered by the trapped object. The model predicts the measured spatial response characteristics of the detector well. The detection method was used to investigate, with high temporal and spatial resolution, the path of beads driven by ncd-glutathione-S-transferase (GST) fusion proteins moving over the microtubule surface. Both axial and lateral motions were observed, and spectral analysis was used to probe the dynamics of the beads.

MATERIALS AND METHODS

Microscope design

Our custom-designed inverted microscope (Fig. 1 *A*) is constructed on an optical rail system (Newport Corp., Irvine, CA), mounted on a vibration isolating optical table (Technical Manufacturing Corp., Peabody, MA). The optical components are from Zeiss (Carl Zeiss, Thornwood, NY). The microscope is designed for trans-illumination DIC microscopy and epi-illumination fluorescence microscopy with simultaneous use of a laser as optical tweezers. For video imaging we use an Ultricon tube camera (VT1000, Dage-MTI, Michigan City, IN). For optical tweezers we use a CW diode pumped Nd:VO$_4$ laser (1.064 μm, 2W, Topaz 106C, Spectra Physics Lasers, Mountain View, CA). It was found to be important to guide the laser beam through tubes (Plexiglas) to prevent pointing fluctuations due to air currents and dust. An optical isolator (Optics for Research, Caldwell, NJ), providing 37 dB isolation against back reflections was necessary to keep the mode pattern of the laser stable. Mode jumps change the beam direction and intensity pattern sufficiently to contribute to noise in the displacement signal. Residual thermal beam pointing fluctuations

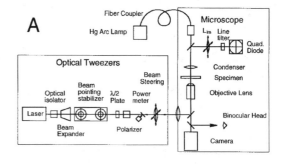

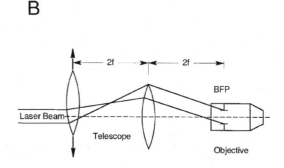

FIGURE 1 (*A*) Schematic diagram of custom microscope as described in the text. (*B*) Arrangement of two equal focal length lenses as a 1:1 telescope for beam steering as described in the text.

were a serious problem at frequencies below ~1 Hz, and therefore we have developed an active feedback controlled beam pointing stabilizer, which will be described elsewhere (unpublished). A 3× beam expander (CVI Laser Corp., Albuquerque, NM) was used to increase the width of the laser beam, reducing the size of the focus in the specimen plane as necessary to achieve trapping. The laser power in the specimen is controlled by a half-wave plate/polarizer combination, so that the laser itself can be run at relatively high power for stable operation.

A 1:1 telescope allows movement of the trap in the specimen plane (Fig. 1 *B*). The beam profile in the plane of the first lens is imaged by the second lens into the BFP of the objective. The image at the BFP remains stationary, but the angle of rays through that plane changes as the first lens is moved laterally to move the focus in the specimen plane. Consequently, the intensity distribution in the condenser BFP does not change with trap position. An angular change of 0.4° in the incoming beam causes a 10-μm displacement in the specimen plane. Laser light is coupled into the microscope optics via a dichroic mirror below the objective (×100, 1.3 NA; Neofluar, Zeiss). After passing through the sample, the beam is recollimated by an oil immersion condenser (1.4 NA; Zeiss), and a second dichroic mirror reflects the laser light out of the imaging path. An additional lens ($F = 50$ mm) images the condenser BFP through a laser line filter onto a quadrant photodiode (10-mm diameter, 10-μm gap; UDT Sensors, Hawthorne, CA) (Fig. 2 *A*). Proper imaging is controlled by replacing the quadrant diode with a CCD camera for alignment.

The two dichroic mirrors allow the simultaneous use of trans-illumination DIC microscopy using the 546-nm line from a mercury arc lamp. For DIC, a pair of Wollaston prisms are mounted below the objective and above the condenser. The laser beam is polarized such that it is not split by the Wollaston prisms and such that the laser light is S-polarized for the dichroics.

The signals from the quadrant diode are amplified by low-noise preamplifiers, networked so that two and two quadrants are grouped for X and Y detection, respectively. The normalized differences are calculated by analog electronics for the X and Y signal, respectively, anti-alias filtered if

desired, and read into an IBM-compatible PC via an A/D board (AT-MIO16X, National Instruments, Austin, TX) (Fig. 2 *B*). These differences are then a measure for the lateral displacement of the trapped particle in the plane normal to the optical axis with respect to the center of the trap. In the figures we give the relative difference between the light intensities I_+ and I_- on the respective halves of the quadrant diode, $(I_+ - I_-)/(I_+ + I_-)$, in percent, ranging from $+100\%$ to -100%. Motor motility data were usually not filtered against aliasing so as to obtain true samples of position (up to the 100-kHz bandwidth of the A/D converter) even when using a low sampling rate. Spectra calculated from these data must be understood to contain aliasing distortions, but the integral over the power spectral density will be correct (which is not the case for filtered data). The signals are displayed and stored using software written in Labview (National Instruments).

Calibration of position detection

As our primary calibration method, beads were immobilized on a coverglass and positioned in the laser focus, and the entire stage was moved with piezoelectric actuators (P-775.00, Polytec PI, Auburn, MA) while the detector signal was recorded together with the driving voltage (Fig. 3 *A*). Producing an accurate triangular motion was complicated by the inherent hysteresis of the piezoelectric actuators. To compensate for this and to calibrate the piezos, video-recorded large-scale bead motion (viewed under

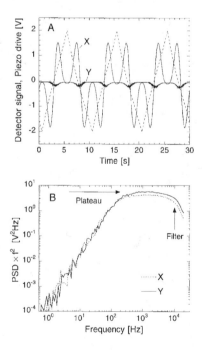

FIGURE 3 Quadrant diode response calibration. (*A*) Detector signals (X and Y, drawn lines) in response to the motion of a silica bead through the laser focus in the X direction. A 0.5-μm silica bead is fixed to a coverslip surface in a water-filled sample chamber. The microscope stage is moved by piezo electric actuators driven with a synthesized signal (- - -) to produce linear motion as described in the text. The amplitude of the motion was 2 μm, and the frequency was 0.1 Hz. (*B*) Power spectral density of the thermal fluctuations of a trapped 0.5-μm silica bead ~4 μm above the substrate surface, multiplied with the square of the frequency. The laser power in the specimen was ~14 mW, the beam diameter in the BFP of the objective ~2 mm. With the assumption of Stokes drag, and the knowledge of the bead size, the detector response can be calculated from the height of the plateau as described in the text. The effect of anti-alias filtering (at 20 kHz) is visible as a steep drop in the spectrum at the high-frequency end.

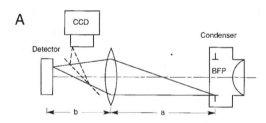

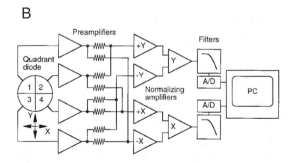

FIGURE 2 (*A*) Schematic diagram of displacement detection by imaging the back-focal plane (BFP) of the microscope condenser onto a quadrant photodiode. A lens of 50 mm focal length is placed such that the BFP of the condenser (~20 mm diameter) is imaged onto the 10-mm-diameter, circular quadrant photodiode. A mirror can be inserted so that the laser diffraction pattern in the BFP can be viewed with a video camera. (*B*) Analog and digital data processing as explained in the text.

DIC) was analyzed by centroid tracking (Metamorph software, Universal Imaging Corp., West Chester, PA) and calibrated against a stage micrometer with 10-μm rulings. Hysteresis of the actuators was fitted to third-order polynomials and found to scale approximately with voltage. To drive the actuators in an accurate triangular motion, we digitally synthesized a proper waveform using the inverse to the hysteresis function and used this to produce up to ~5-μm peak-to-peak bead motion in each direction, typically at a frequency of 0.1 Hz.

Calibration curves, such as in Fig. 3 *A*, have an approximately linear range that increases with bead size; for 0.5-μm beads the response is roughly linear over ±150 nm. From the slope of the calibration curve (in volts/second) and the known stage velocity, the detector response β in volts/nanometer can be calculated. The response is a function of bead size. For small beads, the response should be proportional to d^3; after passing through a maximum, the response should then decrease as $1/d$ for beads much larger than the laser wavelength (see Theory of detection, below). This behavior was experimentally observed (unpublished results).

As a convenient but indirect method of measuring detector response we analyze displacement fluctuations of optically trapped beads of known size. The power spectrum $S(f)$ of these fluctuations near the center of an optical trap is approximately Lorentzian (Svoboda and Block, 1994a; Gittes and Schmidt, 1997). Thus, if the bead has a diameter d in a solvent of viscosity η and κ is the trap stiffness, then the spectrum in units of distance^2/frequency is

$$S(f) = \frac{S_0 f_0^2}{f_0^2 + f^2}, \tag{1}$$

where f is the frequency, $S_0 = 4\gamma k_B T/\kappa^2$ is the zero-frequency intercept of $S(f)$, $f_0 = \kappa/2\pi\gamma$ is the corner frequency of the spectrum, and $\gamma = 3\pi\eta d$ is the Stokes drag coefficient of the bead. One can easily estimate detector sensitivity β from an uncalibrated voltage power spectrum, $S^V(f) = \beta^2 S(f)$, without knowing the stiffness of the trap, by multiplying the voltage power spectrum by f^2. The multiplied spectrum $f^2 S^V(f)$ reaches a plateau for $f \gg f_0$, as shown in Fig. 3 *B*, where the plateau value P^V can be measured and used to estimate the sensitivity β:

$$P^V = \beta^2 S_0 f_0^2 = \beta^2 \frac{k_B T}{3\pi^3 \eta d}.$$

$$\beta = (3\pi^3 \eta P^V d/k_B T)^{1/2} \approx (P^V d/5.0 \times 10^{-20} m^3/s)^{1/2} \tag{2}$$

This last formula applies to water at room temperature and gives β in units of volts/meter. We find this calibration method to agree with the active-displacement calibration, using piezo-actuated stage movement, typically within ~20% for the same size bead. The discrepancy may in part reflect a trapping position that is displaced (due to scattering forces) from the optical focus of the laser. The trap stiffness can be obtained independently of Eq. 2 by measuring the corner frequency of the spectrum, $f_0 = \kappa/2\pi\gamma$.

Proteins

Construction of pGEX-N195

Construction of pGEX-N195 was described previously (Stewart et al., 1993). The fusion protein encoded by pGEX-N195 contains GST, the pGEX-2T thrombin cleavage site, the sequence GPI from the polylinker, and amino acids 195–685 of ncd. This fusion protein, referred to as GST-N195, is similar to the GST-MC1 fusion protein described by Chandra et al. (1993). The GST-MC1 protein contains GST fused to amino acid 210 of ncd. The discrepancy in the numbering of amino acids is due to the choice of the start codon in the cDNA. In our numbering scheme, using the second ATG in the ncd cDNA as the initiator codon, amino acid 195 is the lysine residue encoded by the unique *Afl*II restriction site of ncd. In the numbering scheme used by Chandra et al. (1993), which uses the first ATG in the ncd cDNA as the initiator, amino acid 210 is the lysine residue encoded by the unique *Afl*II site.

Purification of motor proteins and tubulin

GST-N195 was affinity purified from *Escherichia coli* using glutathione-agarose, as described previously (Stewart et al., 1993). Briefly, GST-N195 expression was induced by the addition of 0.1 mM isopropyl β-D-thiogalactopyranoside in approximately mid-log cultures of *E. coli*, which were then shaken at 20–22°C for an additional 4–6 h. After harvesting the cells by centrifugation, they were resuspended in lysis buffer (20 mM PO$_4$, 1 mM EDTA, 150 mM NaCl, pH 7.2) and lysed by sonication. The lysate was clarified by centrifugation (17,000 rpm, Beckman JA-17 rotor). Glutathione-agarose beads (Sigma Chemical Co., St. Louis, MO) were added to the lysate. After mixing at 4°C for 30–60 min, the beads were washed extensively with PEM80 buffer (80 mM Pipes, 1 mM EGTA, 4 mM Mg^{2+}, pH 6.9). GST-N195 was then eluted with 10 mM glutathione in PEM80. GST-N195 was frozen in liquid N$_2$ in small aliquots and stored at −80°C. The protein concentration was ~0.7 mg/ml.

Tubulin was purified from bovine brain following standard recipes (Williams and Lee, 1982). Aliquots were frozen in liquid N$_2$ and stored at −80°C. The protein concentration was ~3.0 mg/ml (Bio-Rad Protein Assay, Bio-Rad Laboratories, Hercules, CA).

Motility assays

Preparation of proteins

Unless noted, all reagents were from Sigma Chemical Co. or Fisher Scientific (Pittsburgh, PA). Tubulin was thawed, polymerized with 1 mM GTP at 37°C for 30 min, and diluted 10-fold into PEM40 (40 mM Pipes, 0.5 mM EGTA, 1 mM Mg^{2+}, 1 mM NaN$_3$) containing 10 μM taxol. For some experiments, the polymerized microtubules were further centrifuged (135,000 × g; Airfuge Ultracentrifuge, Beckman Instruments, Fullerton, CA) to separate microtubules from unpolymerized tubulin. The supernatant was discarded and the pellet was gently resuspended in the original volume of PEM40 with 10 μM taxol. In both cases, microtubules were kept at room temperature, protected from light, and used in experiments for up to 3 days. GST-N195 aliquots were rapidly thawed and brought to 50 mM dithiothreitol. For some experiments, the protein was further centrifuged (135,000 × g; Airfuge Ultracentrifuge) to remove protein aggregates, and the top 80% of the supernatant was recovered. In both cases, the GST-N195 was kept on ice and used in experiments for up to 10 h.

Preparation of GST-N195-coated beads

Silica beads of 0.5 μm diameter (a kind gift from E. Matijewic) were transferred by repeated centrifugation (four times at 16,000 × g; Eppendorf Microcentrifuge, Brinkmann Instruments, Westbury, NY) into anhydrous ethanol and coated by incubation in a 2% v/v solution of Sigmacote (Sigma) in anhydrous ethanol. After another centrifugation step to remove unreacted Sigmacote, the pellet was resuspended in anhydrous ethanol and probe sonicated (Sonifier Cell Disruptor, Branson Ultrasonics Corp., Danbury, CT) to break up bead aggregates. The Sigmacote beads were incubated with GST-N195 on ice for ~5 min at final concentrations of 1.4 × 10^{10} ml^{-1} beads and 7, 1.4, and 0.7 μg/ml protein. The motility assay buffer (AB) consisted of vacuum-degassed PEM40 with an added oxygen-scavenging system, blocking protein and nucleotide (0.018 mg/ml catalase, 3 mg/ml glucose, 0.1 mg/ml glucose oxidase, 0.1 mg/ml casein, 10 μM taxol, 50 mM dithiothreitol, and 1 mM of either MgATP or Mg Adenylylimidodiphosphate (AMP-PNP)). At the protein dilutions we used, multiple motors were expected to be bound to each bead. Assuming, as an upper limit, that all motors in solution adsorbed to the beads, we estimate 2200, 450, and 220 motor dimers per bead for the three dilutions. However, only a fraction of this number may be bound to a bead, and some of the bound motors are possibly inactive.

Assembly of the sample chamber

Motility assays were performed in flow chambers assembled from two coverglasses and narrow strips of double-stick tape (Scotchbrand, 3M, St.

Paul, MN) that were then attached to stainless steel holders for mounting on the microscope stage. KOH/ethanol-cleaned coverglasses were silanized with trimethoxysilylpropyl-diethylenetriamine (DETA, United Chemical Technologies, Bristol, PA) as described previously (Fritz et al., 1995) to provide a positively charged surface to which microtubules strongly attach. Precleaned coverslips were immersed for 5 min in a 1% solution of DETA in 1 mM acetic acid, cleaned by bath sonication for 2 min in a large volume of deionized water, and then dried and cured in an oven at 150°C for 15 min. The chamber was filled with 20 μl of a 1:100 dilution of the microtubule stock solution. After allowing the microtubules to adsorb to the surface for ~2 min, the chamber was flushed with AB buffer containing 1 mg/ml filtered casein. Finally, GST-N195 beads, diluted 1:1000 in AB buffer, were pipetted into the chamber.

Individual beads were captured in solution with the optical tweezers and brought in contact with immobilized microtubules while their motion was monitored with the quadrant diode detector. In the case of bound beads (with AMP-PNP) the motion was purely thermal; with ATP the motion is partly thermal and partly motor driven. Data were digitized at a scan rate of 2 kHz, without any filtering so that the full range of displacements was sampled. Binding and motility of beads on microtubules were also observed with video-enhanced DIC microscopy and recorded on videotape. When the microtubule was not parallel to a detection axis, the X and Y data were rotated (by an angle measured from video) to resolve axial and lateral motion. Such a rotation is meaningful only near the focus center, however, where the detector response is approximately linear.

Video tracks of moving beads were constructed by centroid tracking (Metamorph software, Universal Imaging) to compare with the detector data. The 30-Hz video track was interpolated and the 2-kHz detector data were filtered to a common sample interval; the data were then aligned so as to minimize their mean squared difference.

Theory of detection

Laser trapping has been modeled quantitatively for large particles in a ray optics regime (Ashkin, 1992; Gauthier and Wallace, 1995). When we observe the laser intensity in the condenser BFP with a smaller-than-wave length bead in the vicinity of the laser focus, diffraction and interference, which can not be described by ray optics, appear to be the most prominent effects. Theoretical treatments using rather complicated series expansions have been developed to obtain precise estimates of forces in relatively low-angle Gaussian beams (Gouesbet and Lock, 1994; Lock and Gouesbet, 1994; Harada and Asakura, 1996); however, these do not provide a simple explanation of the far-field intensity. An accessible physical picture, describing how the laser intensity distribution in the BFP depends on experimental parameters, and how it can be used for displacement detection, is highly desirable.

We find that intensity changes in the BFP are quantitatively explicable as an effect of simple interference that occurs throughout the angular range of the focus (in contrast to plane-wave scattering problems (Hulst, 1957) where interference occurs only at zero angle).

The quadrant photodiode in our setup monitors the pattern of laser intensity in the back focal plane (BFP) of the condenser. As the laser enters the sample through the objective, the condenser acts as the imaging lens for the laser focus (see Fig. 1 A). Within ray optics, positional information about the specimen plane is hidden in the BFP. Even within wave optics, the BFP intensity is independent of lateral motion of the focus in the specimen plane, so that one can reposition the trap containing a bead without changing the BFP intensity pattern, apart from nonideality of the optics. The pattern in the BFP represents the angular distribution of light that has passed through the specimen, and thus measuring the BFP intensity distribution is equivalent to an angular scattering experiment. Furthermore, the momentum carried by the scattered light provides the lateral trapping force; in particular, the detection theory discussed here is closely connected with the gradient force exerted by the trap (Gittes and Schmidt, 1998).

For a well corrected lens of focal length f, light at a radius R in the BFP must have emerged from the focus at an angle θ, where $R = f \sin \theta$ (the sine condition (Born and Wolf, 1989)). We describe the direction of light in the

far field by the angle θ from the optical axis and the azimuthal angle ϕ about this axis. This light will be understood to be detected in the BFP at a radius $R = f \sin \theta$ (and at the angle ϕ).

To predict the BFP signal caused by a bead in the vicinity of the laser focus, we need to derive the change in angular intensity distribution that occurs for a given displacement of a bead away from the focus. To attribute this intensity change to interference of scattered light with the original beam, we need a description of the scatterer and the incoming light field. We make the following simplifications.

1) We are concerned with detecting beads that are smaller than or comparable to the wavelength of light. We replace our bead by a Rayleigh scatterer, i.e., a point-like polarizable object. This approximation is not too drastic because we use the uniform-field polarizability α corresponding to our actual sphere radius (Eq. 5, below). In effect, we take the incoming laser field to be uniform over the volume of this sphere.

2) We describe the focus itself in a small-angle (paraxial) approximation, where $\sin \theta \approx \tan \theta \approx \theta$. This is a substantial simplification because the polarization of the light drops out of the problem (Born and Wolf, 1989). If the optical axis is defined as the z direction, we can, for example, take the electric field of the light to point in the x direction everywhere, $E = E_x$. The laser intensity profile is approximately Gaussian with an estimated $1/e^2$ radius of $w_{in} \approx 1.0$ mm at the objective BFP, corresponding to a half-angle of ~30° at the focus (assuming an objective focal length of ~2.1 mm, and $n = 1.33$ in the specimen). At 30°, $\tan \theta / \theta \approx 1.1$ and a paraxial approximation is reasonable. The paraxial approximation also allows for simple and well known expressions for small-angle Gaussian beams (Siegman, 1986). Larger-angle electromagnetic vector solutions to focused Gaussian beams are also available (Yoshida and Asakura, 1974; Axelrod, 1989); however, these are not necessary for the present discussion.

Our model is sketched in Fig. 4 A. Light is focused at the origin ($\vec{r} = 0$). The customary formalism is that the actual electric field E of the light is the real part of a complex wave: $E = \text{Re}\{\hat{E}\}$; time dependence is implicit, in a factor $e^{-i\omega t}$ that is not shown. We observe outgoing light far from the focus, at an angle θ from the optical axis. Using SI electromagnetic units and manipulating formulas for a simple Gaussian beam (Siegman, 1986) the diverging wave far from the focus is

$$\hat{E}(\vec{r}) \approx \frac{-ikw}{r\sqrt{\pi\epsilon_s c_s}} \exp\left\{ikr - \frac{1}{4}k^2w^2\theta^2\right\}, \quad (r \gg w). \quad (3)$$

For the same beam, the field in the focal plane, at a lateral displacement x, is

$$\hat{E}(\vec{r}_s) = \hat{E}(x) = \frac{2}{w\sqrt{\pi\epsilon_s c_s}} e^{-x^2/w^2} \quad \text{(focal plane).} \quad (4)$$

Here $k = 2\pi n_s/\lambda$, where $\lambda = 1.064$ μm is the vacuum wavelength of the light, ϵ_s and n_s are the permittivity and refractive index of the solvent, respectively, c_s is the speed of light in the solvent, and w is the beam half-waist in the focus (the $1/e$ radius of field, not intensity). Equations 3 and 4 are normalized to give the beam a total dimensionless power of 1. Suppose a small sphere of diameter d is displaced laterally by x in the focal plane, which we denote also by the vector $\vec{r}_s$. The particle will have an induced dipole moment $p = 4\pi\epsilon\alpha E$ where α is the electric susceptibility (this particular definition is convenient here). For a sphere with an index of refraction relative to the solvent of $n_r = n/n_s$, the susceptibility is (Jackson, 1975)

$$\alpha = (d/2)^3 \frac{n_r^2 - 1}{n_r^2 + 2}. \quad (5)$$

This comes out to $\alpha \cong 0.0074$ d^3 if we assume $n = 1.45$ for silica beads (Lide, 1992) and $n = 1.33$ for an aqueous solution (Hale and Querry, 1973), both at 1.064 μm wavelength. In the Rayleigh approximation the

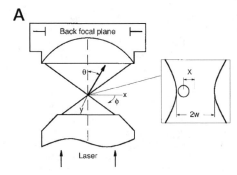

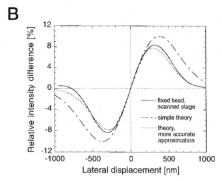

FIGURE 4   (*A*) Diagram indicating the geometry discussed in the text, for interference observed in the back focal plane of the condenser lens between an unscattered Gaussian beam and scattered light from a refractive object located in the focal plane of the beam. For emerging light far from the focus, θ is the angle from the optical axis and ϕ the azimuthal angle about the optical axis. The angular size of the Gaussian beam, $\sim 30°$ in our experiments, is here exaggerated. The inset shows the focal region with a laterally displaced bead. (*B*) The experimental calibration (drawn line) is compared with the simple theory presented in this paper, as well as with an improved calculation presented elsewhere (Gittes and Schmidt, 1998). Both theoretical curves contain no adjustable parameters but depend strongly on the nominal bead diameter of $d = 0.5$ μm and a waist radius $w = 0.53$ μm estimated from the radius of the input beam before the objective.

local electric field $\hat{E}(x)$ acting upon this dipole gives, at large r, a scattered field

$$\hat{E}'(\vec{r}) \approx \frac{k^2 \alpha}{r} \hat{E}(x) \exp(ik|\vec{r} - \vec{r}_s|)$$

$$\approx \frac{k^2 \alpha}{r} \hat{E}(x) \exp ik[r - x \sin \theta \cos \phi], \qquad (6)$$

where θ and ϕ are the angles of observation. The total field far away is $(\hat{E} + \hat{E}')e^{-i\omega t}$ (including the implicit time factor), the sum of the unscattered beam in Eq. 3 and the scattered field in Eq. 6. The time-averaged squared real part of this is $|\hat{E} + \hat{E}'|^2/2$. Therefore, the change in time-averaged light intensity I, due to the interference between the laser beam and the scattered light is

$$\delta I = \frac{\epsilon_s c_s}{2} \{|\hat{E} + \hat{E}'|^2 - |\hat{E}|^2\} \approx \epsilon_s c_s \text{Re}\{\hat{E}\hat{E}'^*\}. \qquad (7)$$

A term $|\hat{E}'|^2$ was discarded; as a result, Eq. 7 represents simple first-order interference. Substituting Eqs. 3, 6, and 4 into Eq. 7 and making a small-θ approximation ($\sin[kx \sin \theta \cos \phi] \cong kx\theta \cos \phi$) gives for the intensity

change

$$\frac{\delta I(x)}{I_{\text{tot}}} \cong \frac{2k^4 \alpha}{\pi r^2} x e^{-x^2/w^2} \theta \cos \phi \, e^{-k^2 w^2 \theta^2/4}. \qquad (8)$$

Equation 8 describes the pattern of intensity modulation caused by a particle displaced by x from the optical axis in the focal plane, observed in the direction (θ, ϕ). Because of the $\cos \phi$ factor, this intensity change consists of a negative lobe on one side of the BFP and a positive lobe on the other, both proportional to x. This means that detection is efficiently performed by simply taking the difference in total power between the two halves of the BFP. Suppose a split diode is oriented with its plus and minus halves along the $\pm x$ axis. The signal change on the plus half (i.e., $-\pi/2 < \phi < \pi/2$) is

$$\frac{\delta I_+}{I_{\text{tot}}} \cong r^2 \int_{-\pi/2}^{\pi/2} d\phi \int_0^{\theta_{\text{max}}} d\theta \sin \theta \, \delta I(x) \qquad (9)$$

The signal change on the plus and minus halves are equal and opposite: $\delta I_- = -\delta I_+$. We can take $\theta_{\text{max}} = \infty$ because $\delta I(x)$ in Eq. 8 is cut off strongly in θ by the exponential factor. Putting Eq. 8 into Eq. 9 (again with $\sin \theta \cong \theta$) and integrating gives

$$\frac{(I_+ - I_-)}{(I_+ + I_-)} \cong \frac{16}{\sqrt{\pi}} \frac{k\alpha}{w^2} (x/w) e^{-(x/w)^2}. \qquad (10)$$

This formula predicts the absolute detector response, as a function of the diameters of the particle and the focus. It is remarkably simple. In Fig. 4 *B*, the prediction of Eq. 10 is compared with an actual calibration (the same one as shown in Fig. 3 *A*). We assume simply the nominal sphere size ($d = 0.5$ μm) and a focus size roughly estimated from the input beam width. For the latter, we assume that the beam enters the microscope with an intensity profile of $\exp(-2R^2/w_{\text{in}}^2)$ where, from beam-width measurements, $w_{\text{in}} \approx 1.0$ mm and. Using $f \approx 2.1$ mm, a basic relationship between angular divergence and focal waist diameters for Gaussian beams (Siegman, 1986) is $kw \approx 2f/w_{\text{in}}$, implying $w = 0.53$ μm. The resulting theoretical curve is similar in range and shape to the measured response curve, notably including its slope in the linear region, despite the lack of adjustable parameters. It is also possible to evaluate a more complicated integral for the intensity imbalance without making the approximation $\sin[kx \sin \theta \cos \phi] \cong kx\theta\cos\phi$ above Eq. 8 (Gittes and Schmidt, 1998), and we include this curve in Fig. 4 *B* to give an idea of the uncertainty in Eq. 10.

As the waist radius w is of the order of λ, the slope of the linear region of Eq. 10 is proportional to $d^3/\lambda w^3 \approx d^3/\lambda^4$. A cubic dependence will arise in all theories based on Rayleigh scattering; it applies to beads somewhat smaller than the wavelength of light, but not necessarily much smaller, as our model works well for spheres a half-wavelength in diameter. For beads much larger than the wavelength of the laser, ray optics become applicable and the wavelength drops out of the problem. Then the bead size is the only relevant length scale and detection can only be a function of x/d. Consequently, the response must decrease as $1/d$ for very large beads.

Control experiments

Fig. 5 demonstrates the sensitivity of the displacement detection. The detector signal was balanced to zero in the X and Y directions by laterally adjusting the lens that images the condenser BFP (L_i in Fig. 1 *A*), without an object in the trap. A silica bead of 0.5 μm diameter, fixed on a glass slide, was then centered in the laser trap by moving the microscope stage with piezo actuators until the detector signal was zeroed again. Then the stage with the bead was moved in a sinusoidal fashion. At a peak-to-peak amplitude of 30 Å the detector signal clearly monitored the bead displacement, both in the X and in the Y direction. At a laser power of ~ 10 mW in the specimen plane, the noise limitation did not stem from photon shot noise but mainly from acoustic noise in the microscope and from beam instabilities. A narrow-band signal like the sinusoidal stage motion could of

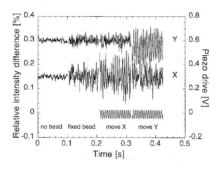

FIGURE 5 Sensitivity of quadrant diode displacement detection. The laser power is 12 mW, the objective BFP is overfilled by inserting an additional beam expander immediately in front of the first telescope lens, and data were digitized at 4 kHz and anti-alias filtered at 2 kHz. The response is plotted as a relative difference between the signals on the two respective halves of the quadrant diode $(I_+ - I_-)/(I_+ + I_-)$, for comparison with Figs. 4 *B*, 8, and 12 (left ordinate). The piezo driving voltage is labeled on the right ordinate). From left to right is shown 1) noise signal without a bead in the laser focus, 2) noise signal with 0.5-μm silica bead laterally centered in the laser focus and adjusted in the axial direction for maximal detector sensitivity, 3) response in X and Y channel to a 30-Å peak-to-peak sinusoidal stage motion in the X direction, and 4) same with 30-Å motion in the Y direction.

course be extracted from the power spectrum of the signal, even if it is not visible in the direct time series data. However, when detecting nonperiodic motion, a common experimental situation with motor proteins, the broad-band noise floor evident in Fig. 5 is more relevant. The signal-to-noise ratio can in principle be improved by low-pass filtering the data above the highest frequency of interest, thereby rejecting high frequency noise contributions (Gittes and Schmidt, 1997a,b), but in our case a substantial part of the noise is from acoustic vibrations of the instrument at relatively low frequencies.

Ideally, displacement detection, being relative to the trap, would be insensitive to repositioning of the trap in the field of view. In practice, a DC offset is generated in the displacement signal by moving the trap in the field of view, as shown in Fig. 6. Without a bead in the laser focus, the detector signal is zeroed with the trap in the center of the field of view and the focus close to the substrate surface. The trap is then moved in increments of ~1 μm for a total of ±10 μm in the X and in the Y direction by moving the first telescope lens as explained in Fig. 1 *B*. The resulting detector signal is given as relative change in the signals from the two respective halves of the quadrant detector compared with the average intensity. For the roughly 20-μm-large field of view, the relative change in the signal is less than 3%. This remaining variation is probably due to less than perfect alignment of the beam-steering lenses and the detector, as well as to contaminations on the optical elements, especially the first telescope lens, which is imaged onto the detector. In practice, the trap can therefore be moved in the specimen without losing AC sensitivity, but a substantial change in the DC offset of the signal has to be taken into account. To give a basis for comparison, the signal of 3% would correspond to a 90-nm displacement of a 0.5-μm bead.

RESULTS

In ATP buffer, at the three concentrations of ncd motor used, a variable fraction of the beads in the sample moved along microtubules. Nonspecific adhesion to the substrate was a persistent phenomenon, even in the presence of 1 mg/ml casein blocking protein. Fig. 7 shows three video frames of a bead moving through the stationary laser focus. For these experiments the number of motors per bead interacting simultaneously with a microtubule was not deter-

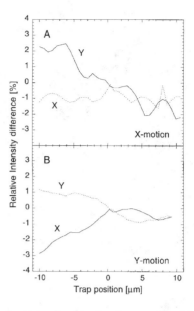

FIGURE 6 Sensitivity of position detection to motion of the trap across the microscope field of view within ±10 μm in X and Y directions (signal balanced in the center of the field of view). For this experiment, no bead is trapped in the laser focus. (*A*) Relative change of the light intensity balance between the X and Y halves of the quadrant diode when moving the trap in the X direction. (*B*) Same as *A* but moving in the Y direction.

mined. The stoichiometry is 200-2000 ncd dimers per bead (see Materials and Methods), but only a fraction of these are expected to be functional. At the lowest motor density, we observed occasional motility, but beads were contained within the relatively weak trap (~4.5×10^{-3} pN/nm). At <450 and <2000 motors per bead, motility was more consistent and the velocity of the beads was determined from the video recordings as 230 ± 30 nm/s. Fig. 8 shows axial displacements measured at high resolution with the quadrant detector for two events such as the one shown in Fig. 7, for a weak (~4.5×10^{-3} pN/nm) and a very weak (~2×10^{-3} pN/nm) trap. Microtubules were not generally aligned with an axis of the detector; therefore, data sets were rotated as described in Materials and Methods. Entering the focus, beads were first pulled forward, which constitutes a negative load; near the center the load decreases to zero and then becomes positive. A zero-load interval is evident as a pause in the track for the stronger trap: attachments apparently remain slack as the motors move from a lagging to an advanced position. Although even the strong trap is relatively weak (maximal force ≤ 0.7 pN) the bead velocity decreased slightly under positive load (163 ± 56 nm/s) as clearly seen in Fig. 8.

Two-dimensional high-resolution detection reveals both axial and lateral displacements of the bead over the microtubule. Lateral thermal displacements might be constrained by multiple attachment points as well as by intrinsic rigidity of the attachments themselves (Wang et al. 1995), and if constrained enough, the path the motors take over the microtubule surface should become visible. Optical trapping, on the other hand, also decreases lateral excursions as can be seen

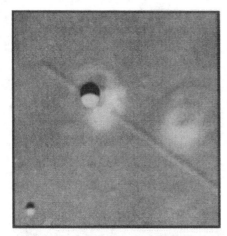

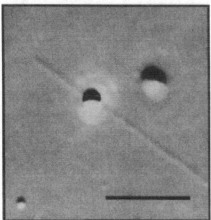

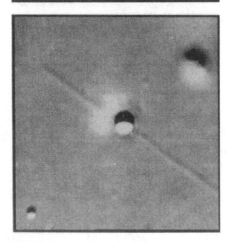

FIGURE 7   Video frames (0, 7.5, and 15 s) of a 0.5-μm silica bead driven by multiple ncd motors through a stationary trap/detector along a microtubule immobilized on a silanized glass substrate. The laser power was 12 mW in the specimen. Scale bar, 5 μm.

in the video track of Fig. 9. Video tracking alone does not provide sufficient information, because the effective low-pass filtering conceals the true excursions of the bead. Fig. 10 *A* shows a video track together with the simultaneous quadrant detector signal, sampled at 2 kHz but unfiltered out to ~100 kHz (a limit due to the sample and hold circuitry of the A/D

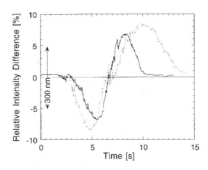

FIGURE 8   Quadrant detector signals from ncd-driven beads traveling through the laser focus, taken at different laser powers and trap stiffnesses. Plotted are the relative intensity differences on the detector versus time. For the gray line, laser power was 12 mW and trap stiffness was 4.5×10^{-3} pN/nm. For the black line, laser power was 4.4 mW and trap stiffness was 2×10^{-3} pN/nm. Data for both tracks were rotated as the microtubules were oriented at 54.7° (high power) and 87° (low power) to the *x* axis, respectively. Therefore, the fringe regions (outside of the marked interval) of the displacement data are distorted (especially for the high-power track). These data (unlike the ones in Fig. 10) were smoothed using a Savitsky-Goley filter (Press et al., 1992) with a window size of 50 points.

conversion board). The full lateral range of motion estimated from the time series data was ~210 nm in the weakest trap and ~200 nm for the stronger trap. The track width outside the focus, estimated from video (Fig. 9) appears as ~260 nm but, because of low-pass filtering, may be underestimated by a factor of 2 (Fig. 10 *A*).

Control beads were prepared identically but with ATP replaced by AMP-PNP. Beads bound to microtubules but sometimes also to the substrate surface at all the motor dilutions used. This made the binding eventually irreversible, even after adding an excess of ATP. Therefore, the results are not reliable with respect to the motor attachment compliance.

A strongly limited lateral motion might have been expected if the motors were protofilament tracking, such as kinesins (Berliner et al., 1995). However, the lateral displacement shown in Fig. 9 from video analysis (low-pass filtered) and in Fig. 10 *A* from interferometric detection

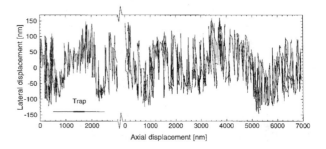

FIGURE 9   Spatial track of ncd-driven 0.5 μm silica bead detected by video centroid tracking, before, while and after passing through the laser focus (laser power 4.4 mW in the specimen, 2×10^{-3} pN/nm trap stiffness). Note the expanded scale in the lateral direction. The horizontal line denotes the extension of the trap between the force maxima, the bar denotes the linear regime. Shown in continuation is the track followed by the same bead while moving along the microtubule with the laser switched off.

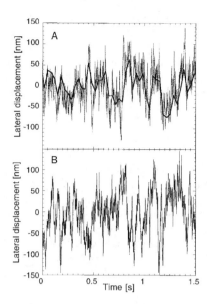

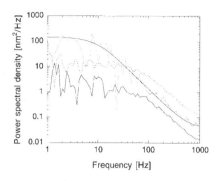

FIGURE 11 Spectral analysis of the lateral motion of an ncd-driven, a stationary (bound with AMP-PNP), and an unbound 0.5-μm silica bead. Single-sided power spectra were calculated from time series data of 1.5 s length, windowed with a Welch window, and smoothed by averaging in equally sized bins in the log domain. The data were not anti-alias filtered during recording to preserve high bandwidth displacement sampling. - - -, unbound bead trapped 1.4 μm above the substrate surface; ——, ncd-coated bead immobilized on a microtubule by AMP-PNP; gray line, ncd-coated moving bead. The drawn smooth line is the theoretical spectrum from which the simulated data track in Fig. 10 B is derived. Aliasing is included, producing the slight upward trend at the end of the spectrum.

FIGURE 10 (*A*) Lateral motion of an ncd-driven 0.5-μm silica bead with respect to the microtubule axis, from quadrant diode detection and video tracking (laser power, 4.4 mW in the specimen; trap stiffness, 2×10^{-3} pN/nm; digitization rate, 2 kHz). The part of the track within the linear range of the detector is shown. Video and quadrant diode tracks were aligned along the time axis first roughly by eye and then more finely by searching for the minimum of the variance of the differences between the video track and a low-pass-filtered version of the quadrant diode signal. (*B*) Simulated track using a program based on a Gaussian random number generator (Press et al., 1992) but with correlations introduced. The two parameters (variance and correlation time) of the correlated Brownian motion were taken from the fit to the data in Fig. 11.

shows no obvious constraint due to motor attachments. From visual inspection it is tempting to interpret the tracks as a relatively slow meandering of the motors over the accessible microtubule surface. On the other hand, this apparent lateral freedom could be a result of very flexible coupling of the beads. To demonstrate that it is possible to quantitatively test the motion by using the high bandwidth displacement data, we compared AMP-PNP-bound stationary beads with moving beads, and we analyzed the power spectra of the lateral displacement signals for both cases and compared with unbound beads away from the surface. We first visually compared the signal measured from the moving beads with simulated pure Brownian motion. Fig. 10 B shows a simulated displacement time series for Brownian motion in a harmonic potential with parameters chosen to produce data with similar spectral properties (amplitude and corner frequency) as observed for the ncd-driven bead (Fig. 10 A). The correlation time of this synthetic track $1/(2\pi f_0)$ is 0.023 s. Naive inspection of the data can be deceptive, as the eye suggests much longer dwell times than this correlation time. Fig. 11 compares power spectral densities calculated from time series positional data for unbound, ncd-driven and stationary beads (bound with AMP-PNP to the microtubule). The unbound bead has a Lorentzian power spectrum as expected (Eq. 1) with a corner frequency of $f_0 \approx 60$ Hz determined by the laser trap stiffness ($\sim 2 \times$

10^{-3} pN/nm). The spectrum of the ncd-driven bead is also well fit by an aliased Lorentzian (taking into account the lack of anti-alias filtering) with a corner frequency $f_0 \approx 7$ Hz. The parameters used to produce the synthetic track in Fig. 10 B were obtained from this fit. The decreasing, high-frequency part of both the ncd-driven and stationary bead spectra has a log-log slope of -2. This reflects purely diffusive motion in the local viscous environment of the beads (e.g., the $\kappa = 0$ limit of Eq. 1 is $S(f) = k_B T/(\gamma \pi^2 f^2)$, independent of binding or motor interaction. The amplitudes of the ncd-driven and stationary bead spectra are both strongly decreased at high frequencies by factors of ~ 3.5 and 10, respectively. This implies that the ncd-driven and stationary beads experience a viscous drag 3.5 and 10 times larger than that of the unbound bead. With the viscous drag known from the high-frequency part of the spectra, the constraint compliances can be deduced from the corner frequency of the power spectral density, as $f_0 \propto \kappa/\gamma$. For the ncd-driven bead the corner frequency decreased by roughly the same factor as the viscous drag increased, although the time series for the moving beads were not long enough to get a precise value for the corner frequency. The results nevertheless imply that there is no large effect of the motor attachments on the elastic constraint of the bead. In contrast, the AMP-PNP-bound bead shows an increased attachment stiffness.

In several cases a motile bead could not escape from the trap before releasing from the microtubule, and repeated escape attempts could be observed as shown in Fig. 12. This behavior may have been due to an especially small number of functional motors on a bead. The linear range of the detector for the 0.5-μm beads is approximately ± 150 nm. Around the maximum in the response signal, the signal becomes ambiguous and the resolution goes to zero. Therefore, instead of calculating the actual displacement with the

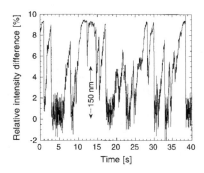

FIGURE 12   Repeated attempts of an ncd-coated 0.5-μm silica bead to escape from the trap (laser power, 12 mW; trap stiffness, 4.5×10^{-3} pN/nm). Plotted is the relative intensity difference on the detector versus time. The microtubule was oriented horizontally, and only the axial signal is shown. For the linear range, the scale in nanometers is given on the right.

help of the calibration curves (Fig. 3), in Fig. 12 we label only the linear part of the position detector range with a displacement scale.

DISCUSSION

We present a high-resolution displacement detection technique for molecular-scale motion, in conjunction with optical trapping, in a light microscope. We provide a simple but quantitative physical model of how a small refractive object near the focus of a laser affects through interference the intensity pattern in the back focal plane (BFP) of the condenser lens collecting the laser light. This picture complements earlier models treating the ray optics regime of large trapped particles (Ashkin, 1992) and is relevant to a practical understanding of laser trapping in general. We draw upon a paraxial Gaussian-beam approximation and well known scattering theory; the unusual feature is that, over the full range of outgoing angles, intensity change is due to a first-order interference effect. More rigorous electromagnetic treatments of spheres near foci (Barton and Alexander, 1989; Barton et al., 1989; Gouesbet and Lock, 1994; Lock and Gouesbet, 1994; Harada and Asakura, 1996) could in principle be extended to this problem, but our simpler model evidently captures the physical mechanism of detection and its dependence on experimental parameters. The model is applicable to related detection techniques that have been developed recently (Ghislain and Webb, 1993; Ghislain et al., 1994; Smith et al., 1996; Visscher et al., 1996).

Experimental detection accuracy was found to be on the order of a nanometer, using 0.5-μm silica beads as probes. Such beads are often used in optical trapping experiments, because they can be coupled to the molecules of interest, such as motor proteins, for micromanipulation. Technically, the BFP method is convenient to implement; it uses less complicated optics than split-beam interferometry and it works well for pure phase objects. Detection is biaxial, and minimal detector realignment is necessary when the laser

focus is moved in the specimen plane. It can be used with double-trap arrangements where two beams are separated by polarization. In comparison with split-beam interferometry (Denk and Webb, 1990; Svoboda et al., 1993), BFP detection is of comparable sensitivity, but it is more susceptible to acoustic vibrations and thermal creep in the instrument as absolute shifts in the BFP intensity pattern are measured; the laser beam cannot be allowed to wander either in the specimen plane or in the BFP.

In exploratory experiments, we have applied BFP detection to the movement of ncd-coated beads along microtubules, where we find several experimental advantages compared with video tracking or uniaxial detection. 1) The high time resolution allows us to detect the position of a tethered bead typically faster than its thermal correlation time, which is the characteristic time it takes to explore its full range of motion. The effective low-pass filtering of video detection of motor tethered beads usually averages over many correlation times (Fig. 10 *A*). 2) The high time resolution makes spectral analysis possible. 3) With the bead bound to an immobilized track, measurement and control of axial and off-axis load is possible; experiments with kinesin (Gittes et al., 1996) have provided evidence for the importance of the load direction and have left open questions. 4) Off-axis motion can be detected and correlated with axial motion. Several studies using video microscopy have focused on the meandering track of motor proteins, dyneins (Wang et al., 1995), and single-headed kinesins (Berliner et al., 1995). These results have been difficult to interpret due especially to the low time resolution of video.

The motility experiments with ncd-coated beads show several conspicuous differences from the kinesin results. Even with multiple motors attached to a single bead, at our lowest motor concentrations, beads detached before reaching the edge of the trap. At the higher concentrations of motors, while moving through and out of the trap, the bead velocity was dependent on load. Both of these findings may be a consequence of a low degree of processivity for ncd, which is consistent with earlier findings from conventional video-based motility assays (our unpublished results). If reconfirmed, this will have important implications for modeling the function of ncd motility and its directionality.

The lateral excursions of a ncd-driven beads were quite large, at least ± 100 nm. The maximal lateral excursions allowed by the geometrical constraint of the substrate surface are approximately ± 120 nm for a 0.5-μm bead, assuming a zero-length linkage and total freedom of motion of the attachment point of the bead on the microtubule surface (Fig. 13). The data are therefore in sharp contrast to the apparently strong lateral constraints found in tracking of even single kinesin molecules (Gelles et al., 1988; Ray et al., 1993; Wang et al., 1995). However, the bead behavior in our experiments could be a consequence of a flexible and extended connection between bead and microtubule, whereas the motors are faithfully tracking protofilaments.

Spectral analysis of the high-resolution quadrant detector data can shed some light on this problem. The power

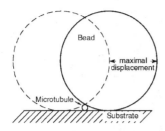

FIGURE 13 Sketch, approximately to scale, of a 0.5-μm bead bound to a microtubule attached to a surface. The maximal possible lateral displacement is $\sim\pm120$ nm.

spectrum contains information about the viscous drag experienced by the bead at high frequencies and, independently of that, information about the elastic constraints in the low-frequency region of the spectrum. Increased drag lowers the corner frequency and decreases the high-frequency amplitudes of the spectrum by the same factor (Eqs. 1 and 2). An increase in horizontal viscous drag is expected in the close vicinity of the surface. If lateral motion is coupled to some degree of vertical displacement (see Fig. 13), the drag coefficient can become very much higher still (Happel and Brenner, 1983).

For both ncd-driven and AMP-PNP-bound stationary beads, the high-frequency regions of the spectra were strongly reduced in amplitude compared with that of an unbound bead (see Fig. 11). This is consistent with a strongly increased drag force very near the surface. The increase in drag by a factor of ~3.5 and 10, respectively, should lower the corner frequency by the same factor. This was not the case for the stationary beads; they thus display an increased stiffness due to attachment through motors and possibly nonspecific binding.

The moving beads, on the other hand, show a decrease in corner frequency by a factor of ~6, which is roughly consistent with the drag increase alone. The fit of the spectrum (Fig. 11) and the simulated data track (Fig. 10 *B*) demonstrate that the lateral motion of the ncd-driven bead looked like simple Brownian motion in a harmonic potential with a drag coefficient close to that of the immobilized beads. The simulated track also demonstrates that it is easy to wrongly identify apparent dwell times that suggest stepping from protofilament to protofilament. The fact that no apparent increase in constraint stiffness is observed (relative to the trap stiffness experienced by the unbound bead) could be due to long and flexible tethers between bead and motors. Such tethers could be formed by aggregates of motors; however, multiple attachments through several motors would tend to limit the lateral excursions strongly if they all stayed processively bound to their protofilaments. Alternatively, if ncd were a low-processivity motor, the bead could perform a fast random walk over the surface. The characteristic sideways stepping time would be related to the cycle time of the motor. This stepping time would have to be fast enough at the saturating ATP concentrations used here to

make the motion resemble unhindered thermal motion (constrained to the microtubule surface).

Key experiments for the future, using our or a similar detection method will be to use specific attachments with well characterized compliances (which can be measured by spectral analysis), variation of ATP concentrations to vary the time scale of the motor-driven diffusion process, and variations in trapping stiffness or feedback-controlled beam steering to keep the load exactly zero.

We acknowledge generous technical support from the Rowland Institute for Science, particularly Winfield Hill. We thank Winfried Denk, Ernst Keller, Karel Svoboda, and Steven Block for helpful discussions.

This work has been supported in part by the Whitaker Foundation, the National Science Foundation (grant BIR 95–12699), and by the donors of the Petroleum Research Fund, administered by the American Chemical Society.

REFERENCES

Allen, R. D., G. B. David, and G. Nomarski. 1969. The Zeiss-Nomarski differential interference equipment for transmitted light microscopy. *Z. Wiss. Mikroskopie.* 69:193–221.

Ashkin, A. 1992. Forces of a single-beam gradient laser trap on a dielectric sphere in the ray optics regime. *Biophys. J.* 61:569–582.

Ashkin, A., J. M. Dziedzic, J. E. Bjorkholm, and S. Chu. 1986. Observation of a single-beam gradient force optical trap for dielectric particles. *Opt. Lett.* 11:288–290.

Axelrod, D. 1989. Fluorescence polarization microscopy. *Methods Cell Biol.* 30:333–352.

Barton, J. P., and D. R. Alexander. 1989. Fifth-order corrected electromagnetic field components for a fundamental Gaussian beam. *J. Appl. Phys.* 66:2800–2802.

Barton, J. P., D. R. Alexander, and S. A. Schaub. 1989. Theoretical determination of net radiation force and torque for a spherical particle illuminated by a focused laser beam. *J. Appl. Phys.* 66:4594–4602.

Berliner, E., E. C. Young, K. Anderson, H. K. Mahtani, and J. Gelles. 1995. Failure of a single-headed kinesin to track parallel to microtubule protofilaments. *Nature.* 373:718–721.

Block, S. M., L. S. Goldstein, and B. J. Schnapp. 1990. Bead movement by single kinesin molecules studied with optical tweezers. *Nature.* 348:348–352.

Born, M., and E. Wolf. 1989. Principles of Optics. Pergamon Press, Oxford.

Brady, S. T. 1985. A novel brain ATPase with properties expected for the fast axonal transport motor. *Nature.* 317:73–75.

Case, R. B., D. W. Pierce, N. HomBooher, C. L. Hart, and R. D. Vale. 1997. The directional preference of kinesin motors is specified by an element outside of the motor catalytic domain. *Cell.* 90:959–966.

Chandra, R., E. D. Salmon, H. P. Erickson, A. Lockhart, and S. A. Endow. 1993. Structural and functional domains of the Drosophila ncd microtubule motor protein. *J. Biol. Chem.* 268:9005–9013.

Coppin, C. M., J. T. Finer, J. A. Spudich, and R. D. Vale. 1996. Detection of sub-8-nm movements of kinesin by high-resolution optical-trap microscopy. *Proc. Natl. Acad. Sci. USA.* 93:1913–1917.

Dekkers, N. H., and H. de Lang. 1974. Differential phase contrast in a STEM. *Optik.* 41:452–456.

Dekkers, N. H., and H. de Lang. 1977. A detection method for producing phase and amplitude images simultaneously in a scanning transmission electron microscope. *Philips Tech. Rev.* 37:1–9.

Denk, W., and W. W. Webb. 1990. Optical measurement of picometer displacements of transparent microscopic objects. *Appl. Opt.* 29:2382–2391.

Endow, S. A., S. Henikoff, and L. Soler-Niedziela. 1990. Mediation of meiotic and early mitotic chromosome segregation in *Drosophila* by a protein related to kinesin. *Nature.* 345:81–83.

Endow, S. A., S. J. Kang, L. L. Satterwhite, M. D. Rose, V. P. Skeen, and E. D. Salmon. 1994. Yeast Kar3 is a minus-end microtubule motor protein that destabilizes microtubules preferentially at the minus ends. *EMBO J.* 13:2708–2713.

Finer, J. T., R. M. Simmons, and J. A. Spudich. 1994. Single myosin molecule mechanics: piconewton forces and nanometre steps. *Nature.* 368:113–119.

Fritz, M., M. Radmacher, J. P. Cleveland, P. K. Hansma, M. W. Allersma, R. J. Stewart, and C. F. Schmidt. 1995. Imaging microtubules under physiological conditions with atomic force microscopy. *Biophys. J.* 68:A288.

Gauthier, R. C., and S. Wallace. 1995. Optical levitation of spheres: analytical development and numerical computations of the force equations. *J. Opt. Soc. Am. B.* 12:1680–1686.

Gelles, J., B. J. Schnapp, and M. P. Sheetz. 1988. Tracking kinesin-driven movements with nanometre-scale precision. *Nature.* 331:450–453.

Ghislain, L. P., N. A. Switz, and W. W. Webb. 1994. Measurement of small forces using an optical trap. *Rev. Sci. Instrum.* 65:2762–2768.

Ghislain, L. P., and W. W. Webb. 1993. Scanning-force microscope based on an optical trap. *Opt. Lett.* 18:1678–1680.

Gittes, F., E. Meyhöfer, B. Sung, and J. Howard. 1996. Directional loading of the kinesin motor molecule as it buckles a microtubule. *Biophys. J.* 70:418–429.

Gittes, F., and C. F. Schmidt. 1997. Signals and noise in micromechanical measurements. *Methods Cell Biol.* 55:129–156.

Gittes, F., and C. F. Schmidt. 1998. Interference model for back focal plane displacement detection in optical tweezers. *Opt. Lett.* In press.

Gittes, F., B. Schnurr, P. D. Olmsted, F. C. MacKintosh, and C. F. Schmidt. 1997. Microscopic viscoelasticity: shear moduli of soft materials determined from thermal fluctuations. *Phys. Rev. Lett.* 79:3286–3289.

Gouesbet, G., and J. A. Lock. 1994. Rigorous justification of the localized approximation to the beam-shape coefficients in generalized Lorenz-Mie theory. II. Off-axis beams. *J. Opt. Soc. Am. A.* 11:2516–2525.

Hale, G. M., and M. R. Querry. 1973. Optical constants of water in the 200-nm to 200-μm wavelength region. *Appl. Opt.* 12:555–563.

Happel, J., and H. Brenner. 1983. Low Reynolds Number Hydrodynamics: With Special Applications to Particulate Media. Kluwer Academic, Dordrecht, The Netherlands.

Harada, Y., and T. Asakura. 1996. Radiation forces on a dielectric sphere in the Rayleigh scattering regime. *Opt. Commun.* 124:529–541.

Henningsen, U., and M. Schliwa. 1997. Reversal in the direction of movement of a molecular motor. *Nature.* 389:93–96.

Higuchi, H., E. Muto, Y. Inoue, and T. Yanagida. 1997. Kinetics of force generation by single kinesin molecules activated by laser photolysis of caged ATP. *Proc. Natl. Acad. Sci. USA.* 94:4395–4400.

Howard, J., A. J. Hudspeth, and R. D. Vale. 1989. Movement of microtubules by single kinesin molecules. *Nature.* 342:154–158.

Hulst, H. C. 1957. Light Scattering by Small Particles. Wiley, New York.

Ishijima, A., H. Kojima, H. Higuchi, Y. Harada, T. Funatsu, and T. Yanagida. 1996. Multiple- and single-molecule analysis of the actomyosin motor by nanometer-piconewton manipulation with a microneedle: unitary steps and forces. *Biophys. J.* 70:383–400.

Jackson, J. D. 1975. Classical Electrodynamics. Wiley, New York.

Kull, F. J., E. P. Sablin, R. Lau, R. J. Fletterick, and R. D. Vale. 1996. Crystal structure of the kinesin motor domain reveals a structural similarity to myosin. *Nature.* 380:550–555.

Kuriyama, R., M. Kofron, R. Essner, T. Kato, S. Dragas-Granoic, C. K. Omoto, and A. Khodjakov. 1995. Characterization of a minus end-directed kinesin-like motor protein from cultured mammalian cells. *J. Cell Biol.* 129:1049–1059.

Lide, D. R. 1992. CRC Handbook of Chemistry and Physics. CRC Press, Boca Raton, FL.

Lock, J. A., and G. Gouesbet. 1994. Rigorous justification of the localized approximation to the beam-shape coefficients in generalized Lorenz-Mie theory. I. On-axis beams. *J. Opt. Soc. Am. A.* 11:2503–2515.

McDonald, H. B., and L. S. Goldstein. 1990. Identification and characterization of a gene encoding a kinesin-like protein in *Drosophila*. *Cell.* 61:991–1000.

Meyhöfer, E., and J. Howard. 1995. The force generated by a single kinesin molecule against an elastic load. *Proc. Natl. Acad. Sci. USA.* 92:574–578.

Moore, J. D., and S. A. Endow. 1996. Kinesin proteins: a phylum of motors for microtubule-based motility. *Bioessays.* 18:207–219.

Nomarski, G. 1955. Microinterféromètre différentiel à ondes polarisées. *J. Phys. Radium.* 16:S9–S13.

Press, W. H., B. P. Flannery, S. A. Teukolsky, and W. T. Vetterling. 1992. Numerical Recipes in C: The Art of Scientific Computing. Cambridge University Press, Cambridge, UK.

Ray, S., E. Meyhöfer, R. A. Milligan, and J. Howard. 1993. Kinesin follows the microtubule's protofilament axis. *J. Cell Biol.* 121:1083–1093.

Sablin, E. P., F. J. Kull, R. Cooke, R. D. Vale, and R. J. Fletterick. 1996. Crystal structure of the motor domain of the kinesin-related motor ncd. *Nature.* 380:555–559.

Schnurr, B., F. Gittes, F. C. MacKintosh, and C. F. Schmidt. 1997. Determining microscopic viscoelasticity in flexible and semiflexible polymer networks from thermal fluctuations. *Macromolecules.* 30:7781–7792.

Siegman, A. E. 1986. Lasers. University Science Books, Mill Valley, CA.

Simmons, R. M., J. T. Finer, S. Chu, and J. A. Spudich. 1996. Quantitative measurements of force and displacement using an optical trap. *Biophys. J.* 70:1813–1822.

Smith, F. H. 1955. Microscopic interferometry. *Research (London).* 8:385–395.

Smith, S. B., Y. J. Cui, and C. Bustamante. 1996. Overstretching B-DNA: the elastic response of individual double-stranded and single-stranded DNA molecules. *Science.* 271:795–799.

Stewart, R. J., J. P. Thaler, and L. S. Goldstein. 1993. Direction of microtubule movement is an intrinsic property of the motor domains of kinesin heavy chain and *Drosophila* ncd protein. *Proc. Natl. Acad. Sci. USA.* 90:5209–5213.

Svoboda, K., and S. M. Block. 1994a. Biological applications of optical forces. *Annu. Rev. Biophys. Biomol. Struct.* 23:247–285.

Svoboda, K., and S. M. Block. 1994b. Force and velocity measured for single kinesin molecules. *Cell.* 77:773–784.

Svoboda, K., C. F. Schmidt, B. J. Schnapp, and S. M. Block. 1993. Direct observation of kinesin stepping by optical trapping interferometry. *Nature.* 365:721–727.

Vale, R. D., T. Funatsu, D. W. Pierce, L. Romberg, Y. Harada, and T. Yanagida. 1996. Direct observation of single kinesin molecules moving along microtubules. *Nature.* 380:451–453.

Vale, R. D., T. S. Reese, and M. P. Sheetz. 1985. Identification of a novel force-generating protein, kinesin, involved in microtubule-based motility. *Cell.* 42:39–50.

VanBuren, P., S. S. Work, and D. M. Warshaw. 1994. Enhanced force generation by smooth muscle myosin in vitro. *Proc. Natl. Acad. Sci. USA.* 91:202–205.

Visscher, K., S. P. Gross, and S. M. Block. 1996. Construction of multiple-beam optical traps with nanometer-resolution position sensing. *IEEE J. Quantum Electron.* 2:1066–1076.

Walker, R. A., E. D. Salmon, and S. A. Endow. 1990. The *Drosophila* claret segregation protein is a minus-end directed motor molecule. *Nature.* 347:780–782.

Wang, Z., S. Khan, and M. P. Sheetz. 1995. Single cytoplasmic dynein molecule movements: characterization and comparison with kinesin. *Biophys. J.* 69:2011–2023.

Wang, M. D., H. Yin, R. Landick, J. Gelles, and S. M. Block. 1997. Stretching DNA with optical tweezers. *Biophys. J.* 72:1335–1346.

Williams, R. C., Jr., and J. C. Lee. 1982. Preparation of tubulin from brain. *Methods Enzymol.* 85(Pt B):376–385.

Wilson, T. 1988. Enhanced differential phase contrast imaging in scanning microscopy using a quadrant detector. *Optik.* 80:167–170.

Yin, H., M. D. Wang, K. Svoboda, R. Landick, S. M. Block, and J. Gelles. 1995. Transcription against an applied force. *Science.* 270:1653–1657.

Yoshida, A., and T. Asakura. 1974. Electromagnetic field near the focus of Gaussian beams. *Optik.* 41:281–292.

Force Generation in Single Conventional Actomyosin Complexes under High Dynamic Load

Yasuharu Takagi,*[†] Earl E. Homsher,[‡] Yale E. Goldman,*[§] and Henry Shuman*[§]

*Pennsylvania Muscle Institute, and [†]Department of Bioengineering, School of Engineering and Applied Science, University of Pennsylvania, Philadelphia, Pennsylvania 19104-6392; [‡]Department of Physiology, David Geffen School of Medicine at UCLA, Los Angeles, California 90095-1751; and [§]Department of Physiology, University of Pennsylvania School of Medicine, Philadelphia, Pennsylvania 19104-6085

ABSTRACT   The mechanical load borne by a molecular motor affects its force, sliding distance, and its rate of energy transduction. The control of ATPase activity by the mechanical load on a muscle tunes its efficiency to the immediate task, increasing ATP hydrolysis as the power output increases at forces less than isometric (the Fenn effect) and suppressing ATP hydrolysis when the force is greater than isometric. In this work, we used a novel 'isometric' optical clamp to study the mechanics of myosin II molecules to detect the reaction steps that depend on the dynamic properties of the load. An actin filament suspended between two beads and held in separate optical traps is brought close to a surface that is sparsely coated with motor proteins on pedestals of silica beads. A feedback system increases the effective stiffness of the actin by clamping the force on one of the beads and moving the other bead electrooptically. Forces measured during actomyosin interactions are increased at higher effective stiffness. The results indicate that single myosin molecules transduce energy nearly as efficiently as whole muscle and that the mechanical control of the ATP hydrolysis rate is in part exerted by reversal of the force-generating actomyosin transition under high load without net utilization of ATP.

INTRODUCTION

Optical trap studies of individual molecular motors have contributed to our understanding of the mechanism of muscle contraction and motility of unconventional myosins (1,2). Forces and displacements produced by single myosin molecules (3–8), the effects of mutations and truncations (9–14), and modulation by imposed force (3–5,12,15–18) have all been addressed. For studies of conventional myosin, the "three-bead assay" is particularly effective (3). In this method, an actin filament, suspended between two beads that are held by separate traps, is brought close to a surface that is sparsely coated with motor proteins deposited on pedestals of silica beads.

Previous investigations using the three-bead assay have recorded single myosin II displacements varying from 4 to 11 nm and unitary isometric forces varying from ~0.8 to 7 pN (3–5,10,12,19,20). Taking a typical average unitary force ($F \cong 3$ pN) and unitary displacement ($d \cong 10$ nm) from previous single molecule measurements and assuming that the mechanical energy stored in actomyosin is that of a Hookean (linear) spring, the maximum work (W) that can be performed during a single actomyosin interaction is given by $W = 1/2 \times F \times d = 1/2 \times 3$ pN $\times 10$ nm $= 15 \times 10^{-21}$ J. The measured thermodynamic efficiency of contracting whole muscles measured during contraction was as high as 50%–60% (21–23). Assuming that myosin molecules

hydrolyze one ATP molecule per mechanical cycle and that the free energy of ATP hydrolysis (ΔG_{ATP}) $= 100 \times 10^{-21}$ J, the maximum mechanical work performed per ATPase cycle in muscle is 50–60 $\times 10^{-21}$ J. Thus, the typical maximal work estimate from single molecule studies (15×10^{-21} J) is only 25%–30% of that observed in living muscles. Possible reasons for this discrepancy are that the myosin's mechanical elasticity is not linear or that the previously measured forces and displacements are lowered by an artifact such as attenuation by the actin-bead or the myosin-bead connection compliance (24,25).

The optical traps used in motility research are typically 2–6% as stiff as the actomyosin complex (stiffness (κ_{am}) $\cong$ 1 pN/nm of displacement). Thus in most optical traps, in this low stiffness/high compliance regime, the displacement caused by an actomyosin interaction is measured under small load. Finer et al. (3,19) used a feedback loop in a three-bead experiment to maintain (clamp) the position of one of the beads, in a fixed place to reduce the motion. The force measured in this experiment for whole conventional myosin molecules averaged 3.5 pN. The actin still moved significantly, however, when myosin attached because the linkage stiffness between the bead and the actin filament is comparable to the actomyosin stiffness. The reduction below the true isometric force that would be measured if the actin movements were prevented is unknown. The bead-actin compliance is highly nonlinear (24,25) and includes the effect of the bead rotating under the tangential force of the filament (24), but it can be reduced by applying high prestretch between the two beads suspending the filament. Linkage stiffness values up to 0.26 pN/nm using a pre-tension of 2 pN have been achieved (5). The link-

Submitted June 9, 2005, and accepted for publication November 4, 2005.

Address reprint requests to Henry Shuman, PhD, B400 Richards Bldg., 3700 Hamilton Walk, Philadelphia, PA 19104-6085. Tel.: 215-898-3408; Fax: 215-898-2653; E-mail: shuman@mail.med.upenn.edu.

Yasuharu Takagi's present address is Laboratory of Molecular Physiology, National Heart, Lung and Blood Institute, National Institutes of Health, Bldg. 10, Rm. 202, Bethesda, MD 20892-1762.

doi: 10.1529/biophysj.105.068429

age stiffness increases further at higher pre-tensions, but this is limited by the adhesion strength of noncovalent bonds and the effect of bead rotation under variable force. Thus high pre-tension does not eliminate the effects of end compliance.

In this work, a different approach is taken to reduce and vary the amount of actin motion occurring when myosin attaches. Based on the three-bead assay electrooptical feedback clamps the actin filament and one of the beads, termed the transducer bead, in position by changing the force on the other, motor, bead. The effective stiffness of the trap is increased, enabling testing of the effect of this stiffness on the developed force. The results indicate that force developed by single actomyosin interactions is considerably increased when the dynamic stiffness of actin presented to the myosin molecule is increased.

The rate of energy liberation during muscle contraction increases above that of an isometric contraction when the load is reduced and the filaments are allowed to slide (26). This observation was important in the development of the cross-bridge theory of contraction (27). The control of biochemical reactions by the external load indicates that a sarcomere does not release a fixed amount of free energy with each contraction. For several types of myosin, the rate of ADP release is dependent on mechanical strain in the expected direction (17,18,28,29). The properties expected for an actomyosin ATPase cycle with strain-dependent, rate-limiting ADP release are high ADP affinity, weak coupling between ADP binding and actomyosin affinity, high duty ratio (a large fraction of each actomyosin cycle attached to actin), and a conformational change associated with ADP release (30). Skeletal muscle myosin does not fit into this pattern because it has a low duty ratio, ($<5\%$) (31–34), strong coupling between ADP binding and actomyosin affinity, weak ADP affinity (35), and no conformational change associated with ADP release (36). Thus other steps in the actomyosin ATPase cycle may be the crucial strain-dependent steps in muscle.

By varying the gain of the feedback loop in the actin clamp system, the dynamic stiffness and loading rate of the myosin attachment can be varied. A high dynamic stiffness is expected to probe actomyosin interactions toward the beginning of the power stroke, and low dynamic stiffness should emphasize interactions toward the end of the power stroke. Surprisingly, these results show that higher dynamic stiffness results in shorter duration actomyosin attachment events. This observation suggests that mechanochemical modulation does indeed take place at the beginning of the power stroke. Some of these results have been presented in abstract form (37–42).

MATERIALS AND METHODS

Optical trap layout

The dual beam laser tweezers (Fig. 1) is based on an inverted microscope (Diaphot 300; Nikon, Tokyo, Japan) equipped with a high numerical aper-

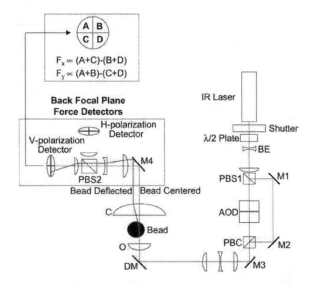

FIGURE 1 Optical scheme of the dual-beam optical trap. The optical tweezers setup incorporates a single FCBar Nd-YAG laser diode ($\lambda = 1064$ nm, Spectra-Physics Lasers) split into two traps with different polarizations. A two-dimensional AOD (Brimrose Corporation of America) enables rapid (~3 μs) electronic control of the horizontally polarized trap. Force measurements of the trapped beads are collected using the BFP force detectors (44,47). These detectors are quadrant photodiodes (Current Designs, Inc.) placed at a plane optically conjugate to the BFP of the condenser. The intensity distribution at the quadrant detectors indicates the deflection of the bead relative to the center of the trap and, therefore, proportional to the applied force. Only the vertical polarization (i.e., the motor trap) of the two beams is shown for simplicity on this diagram after PBC. Abbreviations are: M, mirror; DM, dichroic mirror; PBS, polarization beam splitter; PBC, polarization beam combiner; BE, beam expander; O, objective; C, condenser; $\lambda/2$ Plate, half-wave plate.

ture (NA) (1.3) objective lens (Nikon, Fluor 100$\times$ oil immersion). A single laser beam from an Nd:YAG laser (Spectra Physics, Mountain View, CA) is split into two orthogonally polarized beams to produce two optical traps (43), designated the "motor trap" and the "transducer trap". The position of the motor trap can be controlled rapidly (~3 μs) with an acoustooptic deflector (AOD) (Brimrose Corporation of America, Baltimore, MD), and the transducer trap position can be controlled manually.

After passing through the polystyrene beads used as handles to manipulate actin at the specimen plane, the scattered beams are collected with a 1.4-NA oil immersion condenser lens (Nikon). The back focal plane (BFP) of the oil condenser is imaged onto a pair of quadrant photodiode detectors (Current Designs, Philadelphia, PA; Light Spot Quadrant Detector System), one each for motor and transducer trap (43,44). Deflections of the optical trap light by the two beads are proportional to the forces on the beads rather than their positions (44–47). The output currents from the four quadrants of the photodiodes are converted to four voltages, which are combined to yield two signals proportional to forces in the x- and y-directions (48).

The sample chamber is mounted on a piezoelectric stage (Queensgate Instruments, Torquay, UK) for fine adjustment of x-, y-, and z-positions.

Preparation of proteins

Myosin

Full-length rabbit skeletal myosin was prepared as described by Margossian and Lowey (49), and modified by Harada et al. (50), stored before use in

50% glycerol and 0.6 M KCl at −20°C, and diluted in a buffered salt solution (0.5 M KCl, 5 mM MgCl₂, 0.1 mM EGTA, 10 mM DTT, 10 mM Hepes, pH 7.0) to 0.05 μg/ml for coating the experimental chamber surface.

G-actin

Actin from rabbit skeletal muscle was prepared essentially as described by Pardee and Spudich (51) and further purified by gel filtration (52) using a Superdex 200-pg column (Amersham Biosciences, Piscataway, NJ).

Biotinylated G-actin

Globular (G-) actin was biotinylated with ~50% efficiency on Cys-374 in 5 mM Tris-Cl, 0.2 mM CaCl₂, 0.2 mM NaN₃, 0.1 mM Tris(2-carboxyethyl)-phosphine (Sigma-Aldrich, St. Louis, MO) with a 10-fold excess of EZ-Link PEO-Iodoacetyl biotin ((+)-Biotinyl-iodoacetamidyl-3, 6-dioxaoctanediamine) (Pierce, Rockford, IL) at 4°C in the dark. After 45 min, the actin was polymerized by addition of 0.1 M KCl and 2 mM MgCl₂ and left overnight at 4°C. The biotinylated filamentous (F-) actin was then pelleted by centrifugation at 200,000 × g for 30 min, resuspended in depolymerizing buffer (2 mM Tris-Cl, pH 8.0 at 25°C, 0.2 mM ATP, 0.5 mM DTT, 0.1 mM CaCl₂, 1 mM NaN₃), dialyzed against depolymerizing buffer for 48 h, gel filtered as for unmodified actin, and stored as frozen aliquots at −80°C.

Preparation of biotinylated F-actin

A total of 20 μM biotinylated F-actin was prepared by mixing G-actin from rabbit skeletal muscle with biotinylated G-actin (1:10 biotin/unlabeled) (6) in polymerization buffer (0.5 M KCl, 10 mM MgCl₂, 10 mM EGTA, 100 mM Imidizole, pH 7.0), with an excess of phalloidin-tetramethyl-rhodamine B isothiocyanate (TRITC-Phalloidin) (Fluka Chemicals, Milwaukee, WI). The in vitro motility of actin filaments with 10% biotinylated actin on heavy meromyosin (HMM) was indistinguishable from that of unlabeled actin.

NeutrAvidin-biotinylated polystyrene beads

NeutrAvidin biotin binding protein was used as a cross-linker between the biotin labeled polystyrene beads and the biotin F-actin as previous protocols (6,53). Briefly, 9 μl of 1 μm (diameter) biotin beads (Sigma-Aldrich, 1% vol.) were incubated with 30 μl of 10 mg/ml NeutrAvidin (Pierce) for 1 h. The incubated beads were washed and sedimented (13,000 × g for 5 min) six times in 2 mg/ml BSA (Sigma-Aldrich) in low ionic strength buffer, Buffer A (20 mM KCl, 5 mM MgCl₂, 0.1 mM EGTA, 10 mM DTT, 10 mM Hepes, pH 7.0) to reduce nonspecific protein binding and to remove free NeutrAvidin from the supernatant. The NeutrAvidin-biotin beads were resuspended in 300 μl of Buffer A and aspirated using a 27G 1/2″ needle attached to a 1 ml syringe to disrupt aggregates.

Experimental chamber construction

Pedestal coverslips were made as described in Veigel et al. (25). Dehydrated 1.9 μm silica microspheres (Bangs Laboratories, Fishers, IN) were suspended in amyl acetate (Ernest F. Fullum, Latham, NY) and 0.1% "super clean" nitrocellulose solution (Ernest F. Fullum, 11180). The mixture was then "smeared" over a 25 × 40 mm microscope cover glass, using the side of a pipette tip, and air dried. An assay chamber was constructed using a microscope slide, pedestal coverslip, double-sided adhesive tape, and vacuum grease. The constructed chamber volume was ~20 μl.

The chamber was filled with the following sequence of solutions: 0.05 μg/ml of rabbit skeletal muscle myosin for 10 min, 2 mg/ml BSA in Buffer A for 10 min (×2), sheared actin (5 μM; sheared 10 times using a 27G 1/2″ needle attached to a 1 ml syringe) for 10 min, 1 mM ATP in Buffer A for 3 min (×2), and then Buffer A for 1 min (×4).

In vitro force assay ("three-bead assay") protocol

Tetramethyl-rhodamine phalloidin-biotinylated F-actin (2 nM) and covalently linked NeutrAvidin beads (Polysciences, Warrington, PA; final content = 0.0015% by volume) were suspended in Buffer A with 10 μM ATP, 2.5 mg/ml glucose, 0.1 mg/ml glucose oxidase (Sigma-Aldrich), 0.02 mg/ml catalase (Sigma-Aldrich), and 20 nM TRITC-phalloidin. The actin-bead mixture was incubated at room temperature for 30 min and then added to the chamber. Bead-actin-bead assemblies were found in the chamber and then trapped using two optical traps. The optical trap forces are calibrated as described below. The trapped actin filament could be pre-tensioned to ~5 pN for sustained periods (>45 min) with these linkages. The assembly was then lowered near to the surface of a silica bead and several positions at the top of a pedestal sampled for actomyosin interactions by translating the piezoelectric stage. Approximately one out of 10 scanned pedestals yielded actomyosin interactions. Each experimental chamber with the actin-bead mixture was used for up to 2 h.

Trap stiffness and photodiode sensitivity calibration

To convert the voltage signals acquired from the quadrant photodiode detectors into their force and displacement equivalents, calibrations of trap stiffness (κ_{trap}) and detector sensitivity (C) were performed for every bead-actin-bead assembly used to collect actomyosin interactions. After a bead-actin-bead complex was captured by the traps, the tension in the actin filament was reduced to zero, so that the Brownian motions of the two beads became uncorrelated. This calibration was performed at a height near to the top of a silica pedestal. The signals corresponding to the Brownian motion of each bead were collected for 5 s at 20 kHz sampling rate. Trap stiffness (κ_{trap}) and detector sensitivity (C) were calculated from the power spectrum of these fluctuations as previously reported (54). The trap stiffness and photodiode sensitivity calibration are described further in the Supplementary Material.

Isometric force clamp

To reduce the effect of end compliance and to apply different loading conditions at the single molecule level, we used the three-bead assay incorporating a novel feedback scheme, the isometric clamp (Fig. 2 a). Rather than clamp a single bead at a fixed position as has been done in the past (3–5), we used a different strategy in which the entire bead-actin-bead assembly was clamped at a fixed length. This increases the effective dynamic stiffness presented to the myosin molecule (38). In this case one of the beads, termed the transducer bead, detected force whereas the other bead, the motor bead, opposed the detected force to maintain the bead-actin-bead assembly at its initial position. The force signal measured from the transducer bead was fed through an analog integrating feedback circuit (designed and assembled in-house) to move the motor trap position using the AOD. The rate at which the motor trap moves depends on the integral gain of the feedback. When the feedback loop settles, the transducer bead and the actin filament are both returned to their preattachment positions. Altering the gain of the feedback loop, thus the settling time of the transducer bead, varies the initial loading rate experienced by the actomyosin interaction.

Isometric force clamp adjustment

After calibration of the bead forces, each bead-actin-bead complex was stretched to ~5 pN tension. A command square wave signal was applied to a summing junction of the transducer trap signal, and the integral feedback loop gain was adjusted to set the half-time, τ_f, for settling of the feedback error signal to 1 or 10 ms (Fig. S2 in the Supplementary Material). The different integral gain settings were used to collect actomyosin interactions under high ($\tau_f = 1$ ms) and moderate ($\tau_f = 10$ ms) dynamic stiffness.

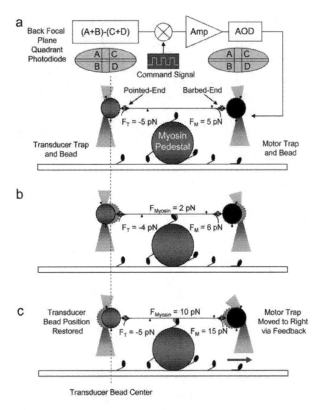

FIGURE 2 Step-by-step motion of the isometric clamp. The sequential mechanism of the isometric force clamp with and without an actomyosin attachment can be summarized as follows. Without an actomyosin attachment, as in panel *a*, the clamp is stable so that the average laser forces on the two beads are equal and opposite. Even without an actomyosin attachment, the clamp responds to Brownian motions by reducing the fluctuations of the transducer bead while increasing the fluctuations of the motor bead. When myosin attaches and undergoes a power stroke, as in panel *b*, the actin and both beads immediately move toward the pointed end, decreasing the tension between the myosin and the transducer bead and increasing it between the myosin and the motor bead. As an example, the force produced by the myosin is shown as 2 pN in panel *b*. The amplified integral of the error signal from the transducer detector moves the motor trap to the right, further increasing the tension on the motor bead such that the transducer bead is returned toward its preattachment position, as shown in panel *c*. When the feedback loop settles, the transducer bead is returned to its preattachment position returning the actin, and therefore the myosin too, to its prepower stroke position. The measurement of the motor bead force (15 pN) is therefore equal to the exact force exerted by the myosin (10 pN) plus the pre-tension (5 pN), which can be subtracted from the measurement since this offset applied to the actin filament is known from the calibration process. Returning the actin to its preattachment position restores the myosin to its isometric condition while effectively eliminating the end compliance at the transducer bead/actin interface. When the myosin detaches, the actin and both beads move rapidly to the right. The amplified negative error signal from the transducer causes the motor trap to move left, decreasing the force on the motor bead and returning it to the state shown in panel *a*. The darker circles show the current bead position, and the lighter circles with a dashed circumference show the previous bead positions.

Zero crossing data analysis

Interactions with the feedback operating were selected using zero crossing analysis, whereby interactions were scored as episodes when the motor force did not cross zero for longer than a minimum duration threshold, usually 1 ms. The peak amplitude for each episode was calculated after smoothing groups of five consecutive data points (50 μs sampling interval), and the event duration was defined as the time from the first (positive) zero crossing (Fig. 3, *time B*) to the time of most negative slope before its subsequent (negative-going) zero crossing (Fig. 3, *time C*).

Recording without feedback and covariance data analysis

Displacement produced by myosin was also measured under low loads in experiments performed without using feedback.

Actomyosin events, collected without feedback, were analyzed using a covariance threshold, similar to the analysis method used by Mehta et al. (55). The average covariance (*cov*) of the motor, F_m, and transducer, F_t, force signals was calculated for a time window centered on each data point of the force traces using the equation

$$\mathrm{cov}(F_t, F_m) = \langle (F_t - \langle F_t \rangle)(F_m - \langle F_m \rangle) \rangle = \langle F_t F_m \rangle - \langle F_t \rangle \langle F_m \rangle,$$
(1)

where $\langle \ \rangle$ is an average over the time window. The means of F_t, F_m, and $F_t F_m$ are given by

$$\langle F_t \rangle = \frac{1}{2n+1} \sum_{-n}^{+n} F_t; \quad \langle F_m \rangle = \frac{1}{2n+1} \sum_{-n}^{+n} F_m,$$
(2)

and

$$\langle F_t F_m \rangle = \frac{1}{2n+1} \sum_{-n}^{+n} (F_t F_m).$$
(3)

The covariance is large in the absence of a myosin linkage and decreases during attachment of myosin and actin.

The histogram of the covariance is bimodal with one peak centered about a high average covariance, representing the detached state, and another peak

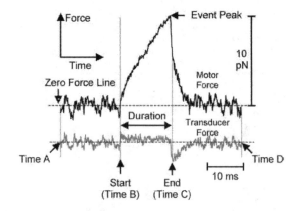

FIGURE 3 Expanded view of a single actomyosin interaction ($\tau_f = 1$ ms; 10 μM ATP). Force and duration of unitary actomyosin interactions are detected using a zero crossing analysis method. Positive force episodes are defined as the period when the force trace is above zero, and negative episodes as those that are below zero, for more than 1 ms. Each positive episode starts at a negative-to-positive zero crossing of the force on the motor bead (*B*). The peak of the episode is determined by fitting a quadratic curve to five consecutive data points. After the force peak the force rapidly declines. The time between the start (*B*) and the point at which the rate of force decline is fastest (*C*) is defined as the duration of an episode.

with a low average covariance, representing the attached state (Fig. S3 in the Supplementary Material). To determine durations and amplitudes of acto-myosin events, a covariance threshold to separate the noise from events was determined from the covariance histogram. Any instance where the variance decreased by more than two standard deviations away from the high variance peak of the detached state was scored as the start of an event. The end of the event was determined as the subsequent time point when the covariance increased back to the detached state peak.

RESULTS

Actomyosin interactions under varying dynamic load

Fig. 4, *a* and *b*, illustrates typical force and displacement measurements of unitary actomyosin interactions using the

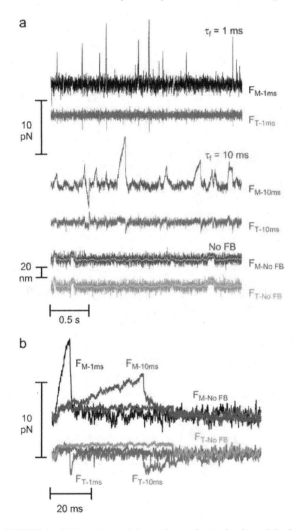

FIGURE 4   Unitary actomyosin interactions under varying dynamic loads. (*a*) 2.5 s of raw data and (*b*) raw data of typical single events. Forces on motor beads (*dark colors*) and the transducer beads (*lighter colors*) during actomyosin interactions collected with the feedback gain of the isometric clamp at $\tau_f = 1$ ms (*blue*), $\tau_f = 10$ ms (*red*), and with the feedback turned off (*green*). The forces on beads for individual interactions were extracted from panel *a*, shown on an expanded timescale. Data collected with 10 μM ATP.

isometric clamp at high and moderate dynamic stiffness and with feedback turned off, allowing actin displacement. The Brownian fluctuations in bead forces (or positions) were smaller during events than the fluctuations in the intervals between events or in the absence of myosin (not shown). The amplitudes of actomyosin force greatly increased and the durations of actomyosin interactions decreased as the rate of loading was increased (Fig. 4, *a* and *b*, no feedback: *green*, $\tau_f = 10$ ms: *red*, $\tau_f = 1$ ms: *blue*). Peak forces ranged up to 17 pN for high gain feedback and up to 10 pN for moderate gain feedback. These forces are markedly greater than those in earlier reports (3–5,20). Both positive and negative force events were observed for all three loads. However, the events were predominantly in one direction for each actin filament tested. Actin filaments with predominantly positive force events, assumed to be bound to the motor bead at their barbed ends, were analyzed further.

With the transducer bead feedback, both beads exhibited an initial step (Fig. 3, *time B*) before the integral feedback moved the motor trap and bead. Thereafter, the motor bead force rose, whereas the transducer bead force declined. At time *C* (Fig. 3) the forces returned to the pre-tension base-lines in two phases. The first phase corresponds to the rapid motion of the beads with a time constant defined by the trap stiffness and beads' viscous drag. The second phase corresponds to slower motion of beads as the feedback loop restored the transducer signal to zero. The initial steps and initial slopes of the bead forces varied widely as the myosin and actin bound over a range of relative axial positions. In most events the transducer bead signal failed to reach the zero force before time *C*. An interpretation of the bead forces and motions in terms of actin and myosin forces and motions are discussed later.

Without feedback (traps fixed in position), the motions of the two beads (Fig. 4, *a* and *b*, *green traces*) were the same as each other and their positions were constant during the actomyosin interaction apart from Brownian noise, as previously reported (3–5).

Synchronized dynamic load actomyosin interactions

To clarify the details of the mechanical events under high dynamic load ($\tau_f = 1$ ms), the forces and variances of the actomyosin interactions with durations >5 ms were ensemble averaged by synchronizing interactions at their beginnings (Fig. 5 *a*) and ends (Fig. 5 *b*) (29) ($n_{episodes} = 1870$). As with individual events, the averaged motor bead forces suddenly increased at the start of the event. The average motor force increased by 0.627 ± 0.004 pN and the transducer bead force fell by 0.421 ± 0.003 pN at the beginning of the events, equivalent to an average displacement of the actin filament by 7.46 ± 0.06 nm and 7.02 ± 0.06 nm respectively, comparable to the previously reported myosin power stroke (3–6,20,25). The force of the motor bead in the

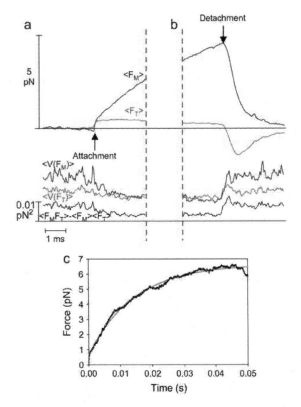

FIGURE 5 Synchronized events for $\tau_f = 1$ ms, (*a*) beginnings and (*b*) ends of events including all events longer than 5 ms. (*c*) Synchronized beginnings of events longer than 50 ms. The average forces and variances on the motor bead (*dark blue*) and transducer bead (*light blue*) from individual events that are longer than 5 ms synchronized at the start (*a*, first zero crossing) and end (*b*, most negative slope before its subsequent zero crossing) of an interaction ($\tau_f = 1$ ms and 10 μM ATP). The black trace is the covariance of the motor and transducer bead. Both variances and covariance decrease at the time of attachment and increase at the time of detachment, representing the change in stiffness accompanying actomyosin interactions. (*c*) Events longer than 50 ms were synchronized to determine the apparent rise time of events under high dynamic load.

averaged trace was still rising at 4.52 ± 0.003 pN at the end of the event (Fig. 5 *b*). Just before detachment, the transducer bead position was restored to 3.50 ± 0.03 nm, 50% of its initial deflection. The stiffness of the average single myosin molecule is thus 4.52 pN ÷ (7.02–3.50 nm) ≅ 1.3 ± 0.02 pN/nm. This stiffness includes the compliance between the molecule and the substrate but not the bead-actin series compliance. The averaged variances of the motor and transducer signals, and their covariance, all decreased at the start of the event, but not instantaneously. They increased more abruptly at the end of an event, suggesting that actomyosin detachment occurs quickly after force reaches the peak. Assuming that all the mechanical elements are linear over the full range of the interaction, the plateau force extrapolated from the average motor force trace was 7.02 nm × 4.52 pN ÷ (7.02–3.50 nm) = 9.0 ± 0.02 pN. This is lower than the largest individual events of 17 pN, but markedly greater than ear-

lier estimates of myosin II active force production (1–4 pN) (3–5,20).

The rise time of the events with feedback was determined by fitting a step and an exponential curve to the ensemble average of events whose durations were longer than a minimum threshold of 50 ms (Fig. 5 *c*; $n_{events} = 30$). The minimum event duration of 50 ms for this analysis was long enough that the isometric plateau forces were nearly reached, yet short enough to include sufficient events to give an average with relatively low noise. Ensemble averages for events in the range 30–60 ms gave similar rise times. The time required for the exponential to reach half its maximum, $t_{1/2} = 10$ ms, for the exponential portion of the increase in motor bead force is 10 times longer than the calibrated feedback response time ($\tau_f = 1$ ms), indicating that the response time of the feedback loop is appreciably slowed when myosin attaches to actin. The slowing depends on the stiffness of the optical traps, the actin-bead linkages, and the actomyosin. A model of the isometric clamp (described in the Appendix), based on a simple mechanical model of a single myosin molecule exerting a force on an actin filament (25), produces a comparable force trace when the myosin stiffness = 1.1 pN/nm and trap stiffness = 0.07 pN/nm only if the linkage stiffness ≥1 pN/nm. The isometric plateau force measured from events longer than 50 ms is ~6.5 pN (Fig. 5 *c*), smaller than the plateau isometric force extrapolated from shorter events (Fig. 5 *a*), suggesting that the process leading to termination of events is hastened by an increased load or smaller displacement.

Analysis of actomyosin interactions under varying external load

In the absence of feedback, interactions were scored by the decrease of the covariance of the transducer and motor bead force fluctuations (55). This method proved unreliable for brief (<20 ms) isometric interactions. Instead, isometric interactions were selected using the zero crossing analysis, whereby interactions were scored as episodes when the motor force did not cross zero for longer than a minimum duration threshold of 1 ms (Fig. 3) (38). The duration of 1 ms was chosen to be short enough to include many episodes due to Brownian noise.

Episodes were compiled into histograms of durations and peak forces (Fig. 6). Event termination rates were determined from the duration histograms. Histograms of durations of force episodes (Fig. 6 *a*) were fit with two exponential components. The rate constant for the faster of these exponentials was independent of dynamic load and was nearly the same as that for Brownian noise acquired in the absence of interactions or with the actin filament moved 100 nm away from the site of the actomyosin interaction (Fig. 6, *a* and *b*). The rate constant for the slower exponential component was strongly dependent on dynamic load (Fig. 6 *a*). The durations at intercepts of the fast and slow exponentials were ~7 ms

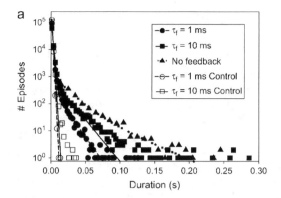

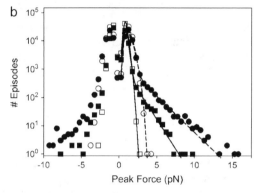

FIGURE 6 (*a*) Duration and (*b*) peak force histograms compiled using experimental data of unitary actomyosin interaction under varying dynamic loads. Histograms are compiled using the data analyzed by the zero crossing analysis. The histogram of actomyosin interactions (*solid symbols*) is plotted together with the histogram compiled using data collected away from pedestals (in the absence of myosin), which represent the Brownian fluctuations (*open symbols*). (*a*) The curve-fit for the duration histogram for actomyosin events was calculated by adding two exponential curves—the first representing the Brownian noise and the second the probable actomyosin events limited by a single rate-limiting step. (*b*) The peak force histogram for actomyosin events appeared to have a merged Gaussian and exponential profile. The initial part (i.e., lower forces) of the peak force histogram was curve fit by two Gaussian curves. A narrow Gaussian curve, which accounts for the 1-ms duration threshold for episodes, was subtracted from a broader Gaussian curve, which accounts for the Brownian noise, both centered about zero. The higher peak forces of the histogram were modeled with an exponential curve, assuming that the myosin detachment is limited by a single rate-limiting step.

and ~15 ms for the high and moderate feedback gains, respectively. Episodes with durations shorter then the values of the intercepts are noise, and episodes with longer durations are interpreted to be actomyosin interactions. The termination rates at 10 μM ATP for high gain feedback actomyosin interactions was 129.5 $\pm$ 4.6 s^{-1} ($n_{episodes} = 106,384$, $n_{pedestal} = 12$), for moderate gain feedback, 65.9 $\pm$ 1.9 s^{-1} ($n_{episodes} = 80,729$, $n_{pedestal} = 10$), and no feedback, 46.8 $\pm$ 1.6 s^{-1} ($n_{episodes} = 1402$, $n_{pedestal} = 9$) (mean $\pm$ SD), confirming the earlier observation that the rates are load dependent. A small fraction (<1%) of long duration events for both load settings did not fit on the double exponential curve and may represent an additional termination process.

The histograms of peak motor forces with the isometric clamp could also be fit with two probability distributions. The high force components were approximately fit by single exponentials that extended to 17 pN and 10 pN for the high and moderate force feedback, respectively (Fig. 6 *b*). The lower force components were well fit by a Gaussian peak centered on zero force with a narrow depression at zero force due to the exclusion of brief, <1 ms, episodes. This peak was identical to the distribution due to Brownian noise obtained in the absence of interactions. If the Gaussian curves represent Brownian noise, then episodes with peak forces greater than the force at the intercept of the Gaussian and exponential curves, 2.5 pN and 3.5 pN for the high and moderate force feedback respectively, are likely due to actomyosin interactions.

DISCUSSION

Actomyosin mechanics

The most striking feature of the results with the isometric clamp is that interactions at high feedback are significantly shorter and reach higher peak forces than interactions with moderate or no feedback. The change in interaction duration is almost certainly a consequence of the effect of force on the kinetics of the actomyosin ATPase and is discussed below. The actomyosin isometric force, 9.0 pN, is higher than previous measurements (3–5,20), probably a consequence of the reduction in end compliance by the feedback in the isometric clamp in addition to high actin pre-tension. Our single molecule isometric force measurement is close to that estimated from a recent high-resolution fiber mechanics study (10 pN) by Piazzesi et al. (56), with both results indicating that the isometric force of single myosin head is greater than previous estimates. The estimated myosin power stroke length, 7.0 nm, was comparable to that previously reported (1,2). The average actomyosin stiffness was 1.3 pN/nm, ~2–10 times larger than previously reported for double-headed myosins in optical trap measurements (20,25,57,58) but at the low end of the 1–5 pN/nm range determined from muscle fibers under isometric conditions (56,59,60). The single molecule and fiber stiffness may differ for at least two reasons. First myosin attached to a nitrocellulose surface may have a lower stiffness than when incorporated into a thick filament. A second possibility is that the stiffness may change during the power stroke. A myosin head in an isometric fiber remains near the beginning of its power stroke. However even at the highest dynamic load in these experiments, the myosin is pulled on average only half of the way back to the beginning of the power stroke. If the stiffness decreases by a factor of two halfway through the power stroke then the extrapolated isometric force would be higher.

The energy transduced into mechanical work per molecular interaction is in the range of $W = 1/2 \times F \times d = 31.5 \times 10^{-21}$ J (Force (F) = 9.0 pN; Displacement (d) = 7 nm). W is higher if the stiffness is higher near the beginning of the power stroke. Assuming actomyosin interactions performing work hydrolyze one ATP molecule, liberating 80–100 10^{-21} J (23) on average, these estimates of single molecule work capacity are closer to the 50–60% thermodynamic efficiency of intact muscle (61).

The isometric force produced by a skinned rabbit fiber at 20°C is ~200 kN/m^2 of cross sectional area (62). Vertebrate thick filaments contain ~300 myosin heads on each side of the M-line, and there are ~4.8 × 10^{14} filaments/m^2 in skinned rabbit psoas fibers. These values give the average force per head as ~1.4 pN in the rabbit fiber if all of the heads are attached within the half sarcomere and producing the same force throughout the cycle. A unitary force of 9.0 pN implies that ~16% of the heads in muscle are attached and contributing to the force during isometric contraction, at the lower end of previous estimates (63–65).

Actomyosin kinetics

The change in the duration of events observed in the raw data and the event termination rates determined from the duration histograms are direct reflections of the change in the kinetics of actomyosin at differing dynamic loads. The event durations decreased as the dynamic load was increased (Figs. 4 *a* and 6 *a*), contrary to the results for smooth muscle myosin in which the release of ADP from the actomyosin-ADP complex was delayed at moderate load (18). The termination rate without feedback yields the second order rate constant for ATP binding of ~4.7 × 10^6 M^{-1}s^{-1}, similar to the value found for acto-S1 ATPase in solution (4.0 × 10^6 M^{-1}s^{-1}) (66) in earlier single skeletal myosin molecule studies and skinned isometric fiber studies (3,5,6,10,20,25,67).

If each event requires binding of ATP at the end of the interaction to detach the myosin, the application of dynamic load would speed the binding of ATP. However, this interpretation contradicts the Fenn effect (26) and measurement of ATP hydrolysis in isometrically or eccentrically contracting muscles (21). An alternative interpretation is that under the influence of external force, myosin detachment from actin can be independent of ATP binding. Two possibilities are that after generation of force, myosin is forcibly detached from actin before the end of its enzymatic cycle and, second, that the actomyosin power stroke is reversed by the load and myosin then detaches at the beginning of its power stroke without performing net work.

Kinetic modeling

To analyze the duration and force histogram data further, histograms generated with several kinetic schemes were numerically modeled and compared to the data. The models were based on variations of a general scheme shown below. Two detached states (D_1 and D_4) and two attached force-producing states (A_2 and A_3) are considered. The first attached state (A_2) is a short-lived intermediate and the second (A_3) is a longer-lived state (at low [ATP]). The first detached state (D_1) represents myosin that has completed the biochemical cycle with bound ATP or ADP and inorganic phosphate, and the second (D_4) represents myosin that has been forcibly detached from actin and has ADP or no nucleotide bound. The two "lumped" force-producing states can be associated with the biochemical states of actomyosin and product release after ATP hydrolysis in several possible ways. Experiments performed to discriminate among these possibilities will be described in a forthcoming publication. Schemes with a single force-producing state did not fit the data as well as those with two force-producing states. The predominant mechanical cycle for unloaded events is D_1, A_2, A_3, then back to D_1. Since the event durations were dependent on dynamic load, one or more of the transition rates in the model depend on load (68). For example, the transition rate for forcible detachment from the long-lived state A_3 under load, F, to the detached state, D_4, is

$$k_{34}(F) = k_{34}(0)\exp\left[\frac{Fd_{34}}{kT}\right], \qquad (4)$$

where $k_{34}(0)$ is the detachment rate in the absence of load and d_{34} is the distance in the direction of the applied force between the minimum free energy of attached state A_3 and the maximum free energy of the transition state between A_3 and D_4 (69).

The differential equations which express the changes in the probabilities of finding a cross-bridge in one of the states were used to model the data in two ways (discussed in detail in the Appendix). In the first, a time sequence of randomly occurring actomyosin events was generated with a Monte Carlo simulation (Fig. 7, *a* and *b*), using the model in the Appendix. The advantage of the Monte Carlo method is that it generates simulated data with Brownian noise that could be analyzed using the zero crossing method and directly compared with the analysis of the experimental data. In the second method of simulation, the differential equations were solved numerically to determine the probability distribution of events (Fig. 8, *a* and *b*). The direct solutions of the probability densities could be generated far faster than with the Monte Carlo method so that model parameters were iteratively optimized to best fit the experimental data. The models incorporated values of parameters such as the trap stiffness and the feedback time responses, with and without myosin attached as determined in the Results section. The variance of the Brownian motion of the bead-actin-bead assembly determined from the data recorded in the absence of interactions was 8 nm. It was assumed that myosin could bind to five monomers within a target zone along the filament with relative binding probabilities as described in Steffen et al. (70). The duration and force histograms were computed for each of the monomer positions and for a range of the relative actomyosin strains before attachment and then averaged. The

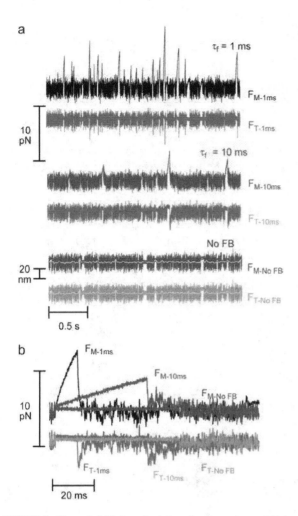

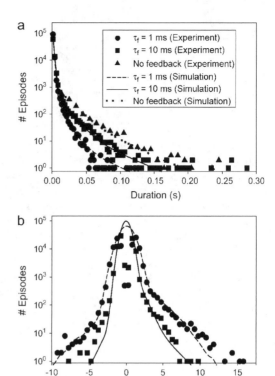

FIGURE 7   Theoretical Markov simulation of unitary actomyosin interactions under varying dynamic loads. (*a*) 2.5 s of simulated events and (*b*) typical single simulated events. A time sequence of randomly occurring actomyosin events under varying dynamic load was generated with a Monte Carlo simulation using the model in the Appendix and was plotted as in Fig. 4. Similarly, forces on motor beads (*dark colors*) and the transducer beads (*lighter colors*) during actomyosin interactions simulated with the feedback gain of the isometric clamp at $\tau_f = 1$ ms (*blue*), $\tau_f = 10$ ms (*red*), and with the feedback turned off (*green*). The forces on beads for individual interactions were extracted from panel *a* and shown on an expanded timescale.

FIGURE 8   Simulated (*a*) duration and (*b*) peak force distributions at two different feedback gains and no feedback using load-dependent k_{21} and k_{23}. The differential equations included in the Appendix were solved numerically to determine the probability distribution of events for a load-dependent model.

rate constants, transition distances, power stroke, and actomyosin stiffness were globally adjusted (i.e., kept the same for all three feedback settings) so that the sum of the reduced χ_r^2 terms for the individual force and duration traces for both feedback time constants and no feedback was minimized using a conjugant gradient algorithm in MathCad (MathSoft Engineering and Education, Cambridge, MA).

Models were also tested in which either none of the transition rates, single transition rates from among k_{21}, k_{23}, k_{32}, k_{31}, and k_{34}, or combinations of the transition rates were allowed to depend on load. The best global agreement to the peak force and duration histograms at both high and moderate feedback gain was found with both rates k_{21} and

k_{23} dependent on load (Fig. 8 and Table 1). The next best fits were obtained with model variants in which either the rate k_{34} or k_{21} depended on load. A model in which both rates k_{34} and k_{21} depend on load did not improve the fits over that with only k_{21} dependent on load. The fit parameters for different models and the sum of χ_r^2s for these models are listed in Table 1. These results suggest that reversal of the power stroke is the predominant load-dependent step in the muscle cross-bridge cycle at high forces and that the application of load slows down the transition into the later states of the actomyosin ATPase cycle. However since kinetic schemes that include forced detachment fit the data with reasonable fidelity, this mechanism may also play a role in determining the actomyosin interaction kinetics.

A kinetic model with no load-dependent rate constants predicts that the durations of the events at high ($\tau_f = 1$ ms) and moderate ($\tau_f = 10$ ms) dynamic loads are equal but the peak forces differ by 10-fold. The measured peak forces for the high and moderate loads do not differ that much (twofold difference). Permitting k_{21} to be load dependent improves the fitting because at high dynamic load (fast feedback), the predominant force-bearing state (short-lived A_2) is early in the working stroke, but at moderate load, the reaction tends to progress farther to the longer-lived state A_3. Dependence

TABLE 1 Best fit parameters for load-independent and -dependent models

Scheme	k_{21} s^{-1}	k_{23} s^{-1}	k_{31} s^{-1}	k_{34} s^{-1}	Power stroke, d (nm)	Stiffness,κ (pN/nm)	$\Sigma\chi^2$
Load-independent rates	145.2	10.79	40.51	0	5.17	0.60	30.47
k_{23} dependent on load	119.1	13.23 e$^{(-0.71F(pN))}$	33.51	0	6.67	0.62	23.91
k_{34} dependent on load	123.0	6.63	29.06	1.036e$^{(0.57F(pN))}$	5.67	0.68	21.80
k_{21} dependent on load	52.41e$^{(0.18F(pN))}$	0.26	10.73	0	8.31	0.74	22.03
k_{21} and k_{23} dependent on load	51.87e$^{(0.22F(pN))}$	4.27e$^{(-0.81F(pN))}$	23.64	0	9.17	0.75	17.53

of k_{21} on load reduces the peak force in state A_2, thus bringing the simulated forces at moderate and high feedback closer together, as observed. The load dependency of k_{21} can be interpreted as an increased rate of reversal of force production and detachment under high load without utilization of ATP. The load dependency of k_{23} can be interpreted as a mechanism to populate the earlier parts of the actomyosin ATPase cycle, rather than the latter, such that the reversal of force production could occur frequently. This reversal of work would be an effective method for muscle to increase its energy efficiency during isometric contraction and contribute to the Fenn effect.

Comparison to other studies

The force (load) and strain on individual actomyosin heads can be changed by several means: passively or actively changing trap stiffness (5,12); taking advantage of the natural variability of attachment due to Brownian motion before the interaction (17); moving the stage (or the traps) before or after an interaction (18); or as here, by positioning a trap under servo control (3,19,37). The advantage of the method here is that the load is automatically adjusted to restore the position of the actomyosin to the beginning of its power stroke, mimicking an isometric contraction in muscle. However, the load is not applied instantaneously. By altering the gain of the feedback, we can probe the active actomyosin interaction during different stages of the actomyosin cycle. At the higher dynamic load, the actomyosin spends $\sim$10 ms near the end of its power stroke before it is pulled back toward the beginning, whereas at moderate load the acto-

are more likely to occur at higher dynamic load. The isometric clamp thus allows probing of the strain- and load-dependent kinetics of the enzymatic cycle of motor proteins while at the same time increasing the effective stiffness of the trap.

If the myosin power stroke is rapid compared to the dynamics of the feedback loop, as is the case with skeletal myosin II used here, then most of the stroke takes place before the feedback loop responds and the optical traps pull the actin back toward its original position. This suggests that the traps are doing the work rather than the myosin. But the feedback never pulls the actin back farther than its original position and, assuming that an individual interaction is maintained during the restoration of position, then the final position of the actin and the structural state of the myosin will be the same as if the myosin generated its impulse on to an immovable actin. Thus the work done on compliances (either internal ones, e.g., subfragment 2, or external, e.g., attachment to the substrate) is the same as myosin would do in the isometric situation.

We conclude that the single myosin molecules transduce energy nearly as efficiently as whole muscle and that the control of ATPase activity by the mechanical load on a single myosin tunes its efficiency specific to its immediate task.

APPENDIX

Model for isometric clamp feedback loop

The equations describing the motion of a bead-actin-bead assembly held in two traps were modified from Veigel et al. (25) to include the movement of the motor trap under the control of an integral gain feedback loop.

$$\frac{\partial X_L}{\partial t} = \frac{-\kappa_{trap}X_L + \kappa_{con}\left[\left(\dfrac{\kappa_{con}X_L + \kappa_{am}d_{am} + \kappa_{con}X_R}{2\kappa_{con} + \kappa_{am}}\right) - X_L\right] + LF_L}{\gamma} \tag{A1}$$

$$\frac{\partial X_R}{\partial t} = \frac{-\kappa_{trap}(X_R - X_{RT}) + \kappa_{con}\left[\left(\dfrac{\kappa_{con}X_L + \kappa_{am}d_{am} + \kappa_{con}X_R}{2\kappa_{con} + \kappa_{am}}\right) - X_R\right] + LF_R}{\gamma} \tag{A2}$$

myosin spends $\sim$100 ms near the end of its power stroke; thus transitions that preferentially take place near the end of the power stroke are more likely to occur under moderate dynamic load than at higher feedback, and transitions that preferentially take place at the beginning of the power stroke

$$\frac{\partial X_{RT}}{\partial t} = -2\pi f_i X_L, \tag{A3}$$

where X_L is the position of the transducer (left) bead, X_R is the position of the motor (right) bead, X_{RT} is the position of the motor (right) trap, the position of the left trap is zero, κ_{trap} is the stiffness of each trap, κ_{con} is the actin to bead connection stiffness, κ_{am} is the actomyosin stiffness, d_{am} is the power stroke of the actomyosin, LF_L and LF_R are random Langevin forces that lead to the Brownian motion, γ is the viscous drag for each bead, and f_1 is the unity gain frequency for integral gain of the feedback loop. Initially, before attachment, the beads and traps positions are on average

$$X_L = X_R = X_{LT} = X_{RT} = 0. \quad (A4)$$

After myosin attachment and completion of the power stroke, the two beads move rapidly in the absence of feedback or before the feedback loop has time to act to new positions expressed as

$$X_L = X_R = \frac{\kappa_{con}\kappa_{am}d_{am}}{2\kappa_{trap}\kappa_{con} + \kappa_{trap}\kappa_{am} + \kappa_{con}\kappa_{am}}. \quad (A5)$$

The force imposed on the actomyosin by the static traps is

$$F_{am} = \frac{-2\kappa_{trap}\kappa_{con}\kappa_{am}d_{am}}{2\kappa_{trap}\kappa_{con} + \kappa_{trap}\kappa_{am} + \kappa_{con}\kappa_{am}}. \quad (A6)$$

After the feedback loop acts and settles to its steady-state position, the bead and right-hand (motor) trap positions are

$$X_L = 0 \quad (A7)$$

$$X_R = \frac{-\kappa_{am}d_{am}}{\kappa_{con}} \quad (A8)$$

$$X_{RT} = \frac{-\kappa_{am}d_{am}(\kappa_{trap} + \kappa_{con})}{\kappa_{trap}\kappa_{con}}. \quad (A9)$$

The forces on the right-hand (motor) bead and on actomyosin are

$$F_R = F_{am} = -\kappa_{am}d_{am}. \quad (A10)$$

The forces on the myosin and the beads settle after the power stroke with a single exponential time constant as long as all the mechanical elements are independent of load,

$$F(t) = F_0 + F_1(1 - e^{(-gt)}). \quad (A11)$$

The force borne by the actomyosin is approximately

$$F_{am} = \frac{2\kappa_{trap}\kappa_{con}\kappa_{am}d_{am}}{2\kappa_{trap}\kappa_{con} + \kappa_{trap}\kappa_{am} + \kappa_{con}\kappa_{am}} + \kappa_{am}d_{am}(1 - e^{(-gt)}), \quad (A12)$$

where g is the smallest eigenvalue of Eqs. A1–A3.

For $\kappa_{trap} = 0.07$ pN/nm, $\kappa_{con} \geq 1$ pN/nm, $\kappa_{am} = 1.1$ pN/nm, $d_{am} = 7$ nm, $\gamma = 1 \times 10^{-5}$ pN s/nm, and $f_1 = 1300/s$, the settling rate is $g = 69/s$ with bound actomyosin and $g = 690/s$ in the absence of an attachment in agreement with the rates measured in Fig. 6.

In this model, we assume that when myosin generates force under load it deforms an internal series elastic component with stiffness κ_{am}, i.e., by either stretching S2 or bending the lever arm. Other compliant elements such as those that may arise from the myosin-nitrocellulose linkage that are in series with the myosin head would reduce the measured stiffness. Our measurements of unitary actomyosin stiffness, step size, isometric force, and work are likely underestimates of their true values.

Kinetic model for duration of actomyosin events

The equations governing the time dependence of probabilities for finding the actomyosin in each of the states in Scheme 1 are

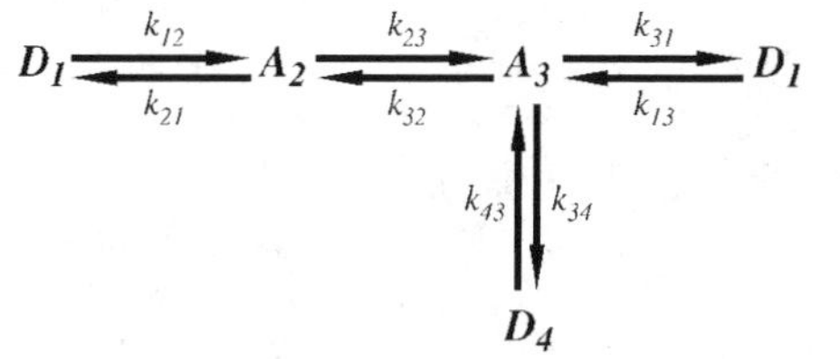

SCHEME 1 Minimal kinetic scheme for numerical model. The models were based on variations of a general scheme with detached states (D_1 and D_4) and two attached force-producing states (A_2 and A_3), whereby the first attached state, A_2, is a short-lived intermediate and the sequential state, A_3, is a longer-lived state, for example at low [ATP].

$$\frac{d}{dt}P_{D_1}(t) = -P_D(t)k_{12} + P_{A_2}(t)k_{21}(F) + P_{A_3}(t)k_{31}(F) \quad (A13)$$

$$\frac{d}{dt}P_{A_2}(t) = P_D(t)k_{12} - P_{A_2}(t)k_{21}(F) - P_{A_3}(t)k_{23}(F)$$
$$+ P_{A_3}(t)k_{32}(F) \quad (A14)$$

$$\frac{d}{dt}P_{A_3}(t) = P_{A_2}(t)k_{23}(F) - P_{A_3}(t)k_{31}(F) - P_{A_3}(t)k_{32}(F) \quad (A15)$$

$$\frac{d}{dt}P_{D_4}(t) = P_{A_3}(t)k_{34}(F) - P_{D_4}(t)k_{43}, \quad (A16)$$

where any or all of k_{21} k_{23}, k_{32}, k_{31}, and k_{34} might depend on the force, F, acting on the actomyosin. The explicit form for force dependence is

$$k_{xy}(F) = k_{xy}(0)\exp\left[\frac{Fd_{xy}}{kT}\right], \quad (A17)$$

where d_{xy} is a characteristic distance.

Mathematical approximation of the event distribution

To estimate the event distributions at differing mechanical parameters, feedback gains, and strain-dependent transition rates, the sets of equations governing the feedback loop and actomyosin must be solved simultaneously. Two methods were used to derive force and duration histograms from Eqs. A1–A3 and A13–A16.

Monte Carlo simulation

The first, a Monte Carlo method, calculates the discrete time sequence of forces and actomyosin states directly from a numerical solution of the stochastic differential equations. The time sequence of events was then analyzed with the zero crossing routine used for data analysis. The stochastic differential equations were transformed to an intrinsic form of difference equations (71) and solved numerically for a sequence of time points, t_i. The values of the Langevin forces, LF, acting on the beads were determined at each time point, t_i, from two independent Gaussian distributed pseudorandom number generators each with zero mean and variance that is determined from γ, the Stokes drag coefficient of the bead. The time points at which the actomyosin bound and changed states were determined by comparing the probabilities of transitions between states, $(t_{i+1} - t_i)k_{xy}(F(t_i))$ to a random number uniformly distributed between 0 and 1, generated at each time point. The stochastic model to simulate individual events was programmed using a LabVIEW 5.1 (National Instruments, Austin, TX) environment.

The solutions to differential equations A1–A3 and A13–A16 for $\kappa_{trap} = 0.06$ pN/nm, $\kappa_{con} = 0.4$ pN/nm, $\kappa_{am} = 0.75$ pN/nm, $d_{am} = 9.2$ nm,

$\gamma = 1 \times 10^{-5}$ pN s/nm, and three feedback gains $f_1 = 1300/s$, $130/s$, and $0/s$ are shown in Fig. 7 and compared with experimental data (Fig. 4).

Numerical probability distribution

In the second method the probability density for the durations of events is determined by directly solving the differential equations Eqs. A13–A16 modified to exclude transitions from unattached (D) to attached force-producing states (A) for the sum $P_{A_2}(t) + P_{A_3}(t)$ with initial probabilities$P_{D_1}(0) = P_{D_1}(0) = 0$, $P_{A_2}(0) = 1$, and $P_{A_3}(0) = 0$. This assumes that the beginning of an event corresponds to the binding of actin to myosin in A_2 and ends when myosin detaches from either A_2 or A_3. The probabilities incorporating duration and the load-dependent transition rates are given by

$$\frac{d}{dt}P_{A_2}(t) = -P_{A_2}(t)k_{21}e^{\delta_{21}F_{am}(t)} - P_{A_2}(t)k_{23}e^{\delta_{23}F_{am}(t)}$$
$$+ P_{A_3}(t)k_{32}e^{\delta_{32}F_{am}(t)} \qquad (A18)$$

$$\frac{d}{dt}P_{A_3}(t) = P_{A_2}(t)k_{23}e^{\delta_{23}F_{am}(t)} - P_{A_3}(t)k_{31}e^{\delta_{31}F_{am}(t)}$$
$$- P_{A_3}(t)k_{32}e^{\delta_{32}F_{am}(t)} - P_{A_3}(t)k_{34}e^{\delta_{34}F_{am}(t)}, \qquad (A19)$$

where the load borne by the actomyosin changes with time as described in Eqs. A12 and A20. The transition rates, k_{xy}, are the unloaded transition rates (units $= s^{-1}$). δ_{xy} (units $= pN^{-1}$) $= d_{xy}/kT$ and express the load- (or strain-) dependent change in transition rates. To make the model more realistic, it also incorporates the probabilities that myosin attaches randomly to the actin filament at target sites spaced at 5.5-nm repeat distance (70). The time-dependent force borne by the actomyosin is

$$F_{am}(t) = \frac{2\kappa_{trap}\kappa_{con}\kappa_{am}(d_{am} + B)}{2\kappa_{trap}\kappa_{con} + \kappa_{trap}\kappa_{am} + \kappa_{con}\kappa_{am}}$$
$$+ \kappa_{am}(d_{am} + B)(1 - e^{(-gt)}), \qquad (A20)$$

where B is a random variable that incorporates the distance along the filament between the myosin and its binding site. The probability distributions, $P_{A_2}(t, B)$ and $P_{A_3}(t, B)$, are solved with the differential equation solver *Stiffr* of Mathcad 2001 Professional (Mathsoft Engineering and Education) for each value of the relative actin and myosin position, B. The total probability for finding bound actomyosin after time t, $P(t)$, is computed from a weighted sum of $P_{A_2}(t, B) + P_{A_3}(t, B)$ over the distribution of B. The force histogram is obtained from the duration histogram since the force is dependent on duration.

SUPPLEMENTARY MATERIAL

An online supplement to this article can be found by visiting BJ Online at http://www.biophysj.org.

We thank the members of the Goldman and Shuman laboratories, as well as the P.M.I., especially Dr. E. M. Ostap for discussions, Dr. D. Safer for actin purification and labeling, and Dr. E. M. De La Cruz for suggestions on F-actin polymerization and handling.

This work was supported by grants from the National Institutes of Health.

REFERENCES

1. Knight, A. E., C. Veigel, C. Chambers, and J. E. Molloy. 2001. Analysis of single-molecule mechanical recordings: application to actomyosin interactions. *Prog. Biophys. Mol. Biol.* 77:45–72.

2. Tyska, M. J., and D. M. Warshaw. 2002. The myosin power stroke. *Cell Motil. Cytoskeleton.* 51:1–15.

3. Finer, J. T., R. M. Simmons, and J. A. Spudich. 1994. Single myosin molecule mechanics: piconewton forces and nanometre steps. *Nature.* 386:113–119.

4. Molloy, J. E., J. E. Burns, J. Kendrick-Jones, R. T. Tregear, and D. C. S. White. 1995. Movement and force produced by single myosin head. *Nature.* 378:209–212.

5. Guilford, W. H., D. E. Dupuis, G. Kennedy, J. Wu, J. B. Patlak, and D. M. Warshaw. 1997. Smooth muscle and skeletal muscle myosins produce similar forces and displacements in the laser trap. *Biophys. J.* 72:1006–1021.

6. Ishijima, A., H. Kojima, T. Funatsu, M. Tokunaga, H. Higuchi, H. Tanaka, and T. Yanagida. 1998. Simultaneous observation of individual ATPase and mechanical events by a single myosin molecule during interaction with actin. *Cell.* 92:161–171.

7. Tanaka, H., A. Ishijima, M. Honda, K. Saito, and T. Yanagida. 1998. Orientation dependence of displacements by a single one-headed myosin relative to the actin filament. *Biophys. J.* 75:1886–1894.

8. Baker, J. E., C. Brosseau, P. B. Joel, and D. M. Warshaw. 2002. The biochemical kinetics underlying actin movement generated by one and many skeletal muscle myosin molecules. *Biophys. J.* 82:2134–2147.

9. Lauzon, A.-M., M. J. Tyska, A. S. Rovner, Y. Freyzon, D. M. Warshaw, and K. M. Trybus. 1998. A 7-amino-acid insert in the heavy chain nucleotide binding loop alters the kinetics of smooth muscle myosin in the laser trap. *J. Muscle Res. Cell Motil.* 19:825–837.

10. Tyska, M. J., D. E. Dupuis, W. H. Guilford, J. B. Patlak, G. S. Waller, K. M. Trybus, D. M. Warshaw, and S. Lowey. 1999. Two heads of myosin are better than one for generating force and motion. *Proc. Natl. Acad. Sci. USA.* 96:4402–4407.

11. Molloy, J. E., J. Kendrick-Jones, C. Veigel, and R. T. Tregear. 2000. An unexpectedly large working stroke from chymotryptic fragments of myosin II. *FEBS Lett.* 480:293–297.

12. Warshaw, D. M., W. H. Guilford, Y. Freyzon, E. Krementsova, K. A. Palmiter, M. J. Tyska, J. E. Baker, and K. M. Trybus. 2000. The light chain binding domain of expressed smooth muscle heavy meromyosin acts as a mechanical lever. *J. Biol. Chem.* 275:37167–37172.

13. Purcell, T. J., C. Morris, J. A. Spudich, and H. L. Sweeney. 2002. Role of the lever arm in the processive stepping of myosin V. *Proc. Natl. Acad. Sci. USA.* 99:14159–14164.

14. Sakamoto, T., F. Wang, S. Schmitz, Y. Xu, Q. Xu, J. E. Molloy, C. Veigel, and J. R. Sellers. 2003. Neck length and processivity of myosin V. *J. Biol. Chem.* 278:29201–29207.

15. Mehta, A. D., R. S. Rock, M. Rief, J. A. Spudich, M. S. Mooseker, and R. E. Cheney. 1999. Myosin-V is a processive actin-based motor. *Nature.* 400:590–593.

16. Rief, M., R. S. Rock, A. D. Mehta, M. S. Mooseker, R. E. Cheney, and J. A. Spudich. 2000. Myosin-V stepping kinetics: a molecular model for processivity. *Proc. Natl. Acad. Sci. USA.* 97:9482–9486.

17. Veigel, C., F. Wang, M. L. Bartoo, J. R. Sellers, and J. E. Molloy. 2002. The gated gait of the processive molecular motor, myosin V. *Nat. Cell Biol.* 4:59–65.

18. Veigel, C., J. E. Molloy, S. Schmitz, and J. Kendrick-Jones. 2003. Load-dependent kinetics of force production by smooth muscle myosin measured with optical tweezers. *Nat. Cell Biol.* 5:980–986.

19. Finer, J. T., A. D. Mehta, and J. A. Spudich. 1995. Characterization of single actin-myosin interactions. *Biophys. J.* 68:291S–296S.

20. Molloy, J. E., J. E. Burns, J. C. Sparrow, R. T. Tregear, J. Kendrick-Jones, and D. C. S. White. 1995. Single-molecule mechanics of heavy meromyosin and S1 interacting with rabbit or Drosophila actins using optical tweezers. *Biophys. J.* 68:298S–303S.

21. Woledge, R. C., N. A. Curtin, and E. Homsher. 1985. Energetic aspects of muscle contraction. *Monogr. Physiol. Soc.* 41:1–357.

22. Cooke, R. 1997. Actomyosin interaction in striated muscle. *Physiol. Rev.* 77:671–697.

23. Howard, J. 2001. Mechanics of Motor Proteins and the Cytoskeleton. Sinauer Associates, Sunderland, MA.

24. Dupuis, D. E., W. H. Guilford, J. Wu, and D. M. Warshaw. 1997. Actin filament mechanics in the laser trap. *J. Muscle Res. Cell Motil.* 18:17–30.

25. Veigel, C., M. L. Bartoo, D. C. S. White, J. C. Sparrow, and J. E. Molloy. 1998. The stiffness of rabbit skeletal actomyosin cross-bridges determined with an optical tweezers transducer. *Biophys. J.* 75:1424–1438.

26. Fenn, W. O. 1923. A quantitative comparison between the energy liberated and the work performed by the isolated sartorius muscle of the frog. *J. Physiol. (Lond.).* 58:175–203.

27. Huxley, A. F. 1957. Muscle structure and theories of contraction. *Prog. Biophys. Biophys. Chem.* 7:255–318.

28. De La Cruz, E. M., A. L. Wells, S. S. Rosenfeld, E. M. Ostap, and H. L. Sweeney. 1999. The kinetic mechanism of myosin V. *Proc. Natl. Acad. Sci. USA.* 96:13726–13731.

29. Veigel, C., L. M. Coluccio, J. D. Jontes, J. C. Sparrow, R. A. Milligan, and J. E. Molloy. 1999. The motor protein myosin-I produces its working stroke in two steps. *Nature.* 398:530–533.

30. Nyitrai, M., and M. A. Geeves. 2004. Adenosine diphosphate and strain sensitivity in myosin motors. *Philos. Trans. R. Soc. Lond. B Biol. Sci.* 359:1867–1877.

31. Uyeda, T. Q. P., S. J. Kron, and J. A. Spudich. 1990. Myosin step size. Estimation from slow sliding movement of actin over low densities of heavy meromyosin. *J. Mol. Biol.* 214:699–710.

32. Harris, D. E., and D. M. Warshaw. 1993. Smooth and skeletal muscle myosin both exhibit low duty cycles at zero load *in vitro*. *J. Biol. Chem.* 268:14764–14768.

33. Hopkins, S. C., C. Sabido-David, J. E. T. Corrie, M. Irving, and Y. E. Goldman. 1998. Fluorescence polarization transients from rhodamine isomers on the myosin regulatory light chain in skeletal muscle fibers. *Biophys. J.* 74:3093–3110.

34. De La Cruz, E. M., and E. M. Ostap. 2004. Relating biochemistry and function in the myosin superfamily. *Curr. Opin. Cell Biol.* 16:61–67.

35. Dantzig, J. A., M. G. Hibberd, D. R. Trentham, and Y. E. Goldman. 1991. Cross-bridge kinetics in the presence of MgADP investigated by photolysis of caged ATP in rabbit psoas muscle fibres. *J. Physiol.* 432:639–680.

36. Gollub, J., C. R. Cremo, and R. Cooke. 1996. ADP release produces a rotation of the neck region of smooth myosin but not skeletal myosin. *Nat. Struct. Biol.* 3:796–802.

37. Takagi, Y., Y. E. Goldman, and H. Shuman. 2000. Single molecule force of myosin II measured with a novel optical trap system that eliminates linkage compliance. *Biophys. J.* 78:235a. (Abstr.).

38. Takagi, Y., E. E. Homsher, Y. E. Goldman, and H. Shuman. 2001. Analysis of isometric force measurements of myosin II using a modified mean-variance analysis. *Biophys. J.* 80:78a. (Abstr.).

39. Takagi, Y., E. E. Homsher, Y. E. Goldman, and H. Shuman. 2002. Probing the transduction mechanism of rabbit skeletal myosin II under isometric condition using an optical trap. *Biophys. J.* 82:373a. (Abstr.).

40. Takagi, Y., Y. E. Goldman, and H. Shuman. 2003. Mechanics of rabbit skeletal actomyosin in an isometric force clamp near rigor. *Biophys. J.* 84:247a. (Abstr.).

41. Takagi, Y., E. E. Homsher, Y. E. Goldman, and H. Shuman. 2004. ATP and phosphate dependence of single rabbit skeletal actomyosin interactions under differing loads. *Biophys. J.* 86:54a. (Abstr.).

42. Takagi, Y., Y. E. Goldman, and H. Shuman. 2005. A Markov process model to predict the mechanics and kinetics of single rabbit skeletal isometric actomyosin interactions. *Biophys. J.* 88:634a. (Abstr.).

43. Visscher, K., S. P. Gross, and S. M. Block. 1996. Construction of multiple-beam optical traps with nanometer-resolution position sensing. *IEEE J. Select. Topics Quant. Electr.* 2:1066–1076.

44. Smith, S. B., Y. Cui, and C. Bustamante. 1996. Overstretching B-DNA: the elastic response of individual double-stranded and single-stranded DNA molecules. *Science.* 271:795–799.

45. Block, S. M., L. S. Goldstein, and B. J. Schnapp. 1990. Bead movement by single kinesin molecules studied with optical tweezers. *Nature.* 348:348–352.

46. Gittes, F., and C. F. Schmidt. 1998. Signals and noise in micromechanical measurements. *Methods Cell Biol.* 55:129–156.

47. Allersma, M. W., F. Gittes, M. J. deCastro, R. J. Stewart, and C. F. Schmidt. 1998. Two-dimensional tracking of ncd motility by back focal plane interferometry. *Biophys. J.* 74:1074–1085.

48. Simmons, R. M., J. T. Finer, S. Chu, and J. A. Spudich. 1996. Quantitative measurements of forces and displacement using an optical trap. *Biophys. J.* 70:1813–1822.

49. Margossian, S. S., and S. Lowey. 1982. Preparation of myosin and its subfragments from rabbit skeletal muscle. *Methods Enzymol.* 85:55–71.

50. Harada, Y., K. Sakurada, T. Aoki, D. D. Thomas, and T. Yanagida. 1990. Mechanochemical coupling in actomyosin energy transduction studied by in vitro movement assay. *J. Mol. Biol.* 216:49–68.

51. Pardee, J. D., and J. A. Spudich. 1982. Purification of muscle actin. *Methods Cell Biol.* 24:271–289.

52. MacLean-Fletcher, S., and T. D. Pollard. 1980. Identification of a factor in conventional muscle actin preparations which inhibits actin filament self-association. *Biochem. Biophys. Res. Commun.* 96:18–27.

53. Rock, R. S., M. Rief, A. D. Mehta, and J. A. Spudich. 2000. In vitro assays of processive myosin motors. *Methods.* 22:373–381.

54. Svoboda, K., and S. M. Block. 1994. Biological applications of optical forces. *Annu. Rev. Biophys. Biomol. Struct.* 23:247–285.

55. Mehta, A. D., J. T. Finer, and J. A. Spudich. 1997. Detection of single-molecule interactions using correlated thermal diffusion. *Proc. Natl. Acad. Sci. USA.* 94:7927–7931.

56. Piazzesi, G., L. Lucii, and V. Lombardi. 2002. The size and the speed of the working stroke of muscle myosin and its dependence on the force. *J. Physiol.* 545:145–151.

57. Nishizaka, T., H. Miyata, H. Yoshikawa, S. Ishiwata, and K. Kinosita Jr. 1995. Unbinding force of a single motor molecule of muscle measured using optical tweezers. *Nature.* 377:251–254.

58. Nishizaka, T., R. Seo, H. Tadakuma, K. Kinosita Jr., and S. Ishiwata. 2000. Characterization of single actomyosin rigor bonds: load dependence of lifetime and mechanical properties. *Biophys. J.* 79:962–974.

59. Huxley, A. F., and S. Tideswell. 1996. Filament compliance and tension transients in muscle. *J. Muscle Res. Cell Motil.* 17:507–511.

60. Barclay, C. J. 1998. Estimation of cross-bridge stiffness from maximum thermodynamic efficiency. *J. Muscle Res. Cell Motil.* 19:855–864.

61. Kushmerick, M. J., and R. E. Davies. 1969. The chemical energetics of muscle contraction. II. The chemistry, efficiency and power of maximally working sartorius muscles. *Proc. R. Soc. Lond. B. Biol. Sci.* 174:315–353.

62. Hibberd, M. G., J. A. Dantzig, D. R. Trentham, and Y. E. Goldman. 1985. Phosphate release and force generation in skeletal muscle fibers. *Science.* 228:1317–1319.

63. Matsubara, I., and G. F. Elliott. 1972. X-ray diffraction studies on skinned single fibres of frog skeletal muscle. *J. Mol. Biol.* 72:657–669.

64. Cooke, R., M. S. Crowder, and D. D. Thomas. 1982. Orientation of spin labels attached to cross-bridges in contracting muscle fibres. *Nature.* 300:776–778.

65. Huxley, A. F. 2000. Cross-bridge action: present views, prospects, and unknowns. *J. Biomech.* 33:1189–1195.

66. White, H. D., and E. W. Taylor. 1976. Energetics and mechanism of actomyosin adenosine triphosphatase. *Biochemistry.* 15:5818–5826.

67. Goldman, Y. E. 1987. Kinetics of the actomyosin ATPase in muscle fibers. *Annu. Rev. Physiol.* 49:637–654.

68. Bell, G. I. 1978. Models for the specific adhesion of cells to cells. *Science.* 200:618–627.

69. Eisenberg, E., T. L. Hill, and Y. Chen. 1980. Cross-bridge model of muscle contraction. Quantitative analysis. *Biophys. J.* 29:195–227.

70. Steffen, W., D. Smith, R. Simmons, and J. Sleep. 2001. Mapping the actin filament with myosin. *Proc. Natl. Acad. Sci. USA.* 98:14949–14954.

71. Press, W. H., S. A. Teukolsky, W. T. Vetterling, and B. P. Flannery. 1992. Numerical Recipes in C: The Art of Scientific Computing, 2nd Ed. Cambridge University Press, New York.

4

Modeling Forces and Torques

The forces and torques exerted on a particle within optical tweezers arise from the change in the trapping beam's momentum as it is transmitted through, or scattered by, the particle. However, although the underlying physical principles may be simple, the implementation of a precise quantitative model is not. In essence, the complexity arises because the trapped particles are of a size comparable to the optical wavelength. Hence, the particles are too large to be treated as point objects in a spatially varying field and too small to simply model the transmission of the light in terms of ray optics.

Based on his own earlier work, Ashkin, in his seminal paper [1] (see **PAPER 1.5**), summarized how either the calculation of the dipole force applicable to atoms or a ray-optical model could be used to gain a qualitative insight into the force responsible for trapping within single-beam optical tweezers. The first attempts to accurately model the forces exerted by a single-beam trap acting on a micron-sized particle adopted a ray-optical approach [2]. To accurately account for extremely tight focusing, a Gaussian beam model was incorporated, within which the radius of curvature of the phase front was used to determine the ray direction as it entered the sphere. The refraction of the light rays upon both entry and exit gives the total momentum exchange between the light and particle. The total force can then be calculated by integrating over a set of light rays representing the trapping beam, exploiting any symmetries of the configuration to minimize the required computational load. Using such models, Ashkin showed that the axial trapping efficiency of optical tweezers could be enhanced by using an annular mode [3] (see **PAPER 3.2**); see also Ref. [4]. These annular modes eliminate the on-axis rays that, since they are not refracted, give no trapping force.

By allowing for partial absorption of the light as it was transmitted through the particle, it was also possible to calculate the torque that could be exerted on the particle when using a Laguerre–Gaussian mode, carrying orbital angular momentum [5]. Despite the seemingly large nature of the approximations, Wright et al. showed these ray-optical models perform quite well and, especially for larger particles, seem to be in both

qualitative and quantitative agreement with the forces actually measured [6], [7] (see **PAPER 4.1**).

Optical tweezers for larger particles can therefore be modeled using modified ray tracing, but what about the smaller particles [8]? Rohrbach and Stelzer showed that, in general, submicron-sized particles can be treated as points within a spatially varying field [9] (see **PAPER 4.2**). However, to obtain accurate results, one needs to understand precisely the structure of the tightly focused beams that are typical in most tweezers geometries [10,11]. The resulting optically induced force acting on the particle can be accurately calculated as the sum of the gradient and scattering forces, and this allows the trapping force to be calculated for various exotic or aberrated beams [12].

Within the past 10 years, modeling within optical tweezers has benefited from the availability of high-performance computing. Finite-difference, time-domain (FDTD) modeling can be applied to solve Maxwell's equations, giving values for the electromagnetic fields at each point of a fine mesh that covers the region of the focused beam, either with or without the trapped object in place [13]. The force of the particle can then be calculated either from the change in the momentum content of the fields before and after transmission, or by calculation of the Lorentz force at the particle surface, see Zakharian et al. [14] (see **PAPER 4.3**). However, neither approach to the calculation of forces is without difficulty in its execution. Barnett and Loudon have shown that there are differing forms of the stress tensor [15] (see **PAPER 4.4**), which can make the correct calculation of the radiation pressure a subtle exercise. These issues are closely related to the apparent controversy on the form of the optical momentum for light beams traveling through a medium—the Abraham-Minkowski dilemma [16–19]. In most, and possibly all, cases the two formulations appear to give identical results [15] but there are some experiments, mainly involving pulsed light sources, in which the two formulations may give rise to different predictions [20]. Early implementations of FDTD exploited the symmetry of the problem so that the trapping could be modeled on a two-dimensional (2D) rather than three-dimensional (3D) grid, thereby easing the computational load. Subsequent models have been run fully in 3D but even now are largely restricted to modeling steady-state situations.

Another computational tool that has found favor in the past few years is that of the T-matrix. In this approach both the incident and transmitted/scattered beams can be expressed as a superposition of appropriate modes. The transformation caused by the trapped particle can then be described in terms of a coupling, or transfer, matrix. The effect of any given particle on the incident light is described by its characteristic T-matrix; see Nieminen et al. [21] (see **PAPER 4.5**), and Simpson and Hanna [22] (see **PAPER 4.6**). Although potentially complicated to calculate, once done, the matrix can be used to rapidly calculate the transmitted light as either the beam type or its position is changed. As with the FDTD approach, the net force is calculated as the difference between the momentum flux of the incident and transmitted/scattered light fields. However, in that the particle is treated in terms of a single matrix, it is not possible to model the potential influence of short pulses where the entry and exit of the light are comparable to the pulse duration.

Another problem associated with both FDTD and T-matrix is that the calculations depend critically on the exact form of the beam as it encounters the object. Although this beam is perhaps well defined as it enters the microscope objective, the real effects of focusing with an aberrated and finite aperture, lens are hard to define. Within FDTD, the vast difference in size between initial beam and trapped particle makes it virtually impossible to model the whole setup on a single grid. Similarly, within a T-matrix, the standard modes do not necessarily provide an accurate description of the tightly focused beam. Most recently, groups have begun to use hybrid models incorporating ray-tracing or FDTD with T-matrix models [23].

In most situations, the dominant trapping force is that of the gradient force that acts to draw the dielectric particle to the position of highest intensity. For extended area beams with a uniform intensity, one needs to consider the additional forces that influence the particle's motion while they are confined to the areas of high intensity. One example of this situation is an optical vortex trap where the helical-phase beam produces a uniform intensity ring of light in the trapping plane. Dielectric particles are confined to the ring but experience an azimuthal force, causing them to orbit around the beam axis (e.g., O'Neil et al. [24], see **PAPER 6.5**). This force can be analyzed in various ways and is perhaps the most obvious manifestation of the orbital angular momentum of light [25]. More generally, it is indicative of a transverse phase gradient, meaning that the local wave vector and associated optical momentum flow both include transverse components. Isotropic scattering of light by the particle results in a transverse

recoil. This phase-gradient induced force has recently been modeled and observed explicitly [26,27].

Modeling the action of a light beam upon a single trapped particle is just the beginning. There are many other forces also at work within optical tweezers—both optical and hydrodynamic. One of these, optical binding, was first reported by Burns et al. in the late 1980s [28] (see **PAPER 5.7**). In essence, it embodies situations where the light scattered from one trapped object is sufficient to exert a significant force on neighboring particles. A configuration where optical binding is important is in optical traps formed by two counterpropagating beams. Such traps were readily implemented by Constable et al. who aligned two opposing single-mode optical fibers with their facets separated by a few tens of microns [29] (see **PAPER 8.1**). In these configurations, it has been observed that a number of particles can be simultaneously confined to the common axis of the beams, and that the motion of these particles is coupled. As shown by Tatarkova et al., the refraction of light by one particle exerts an appreciable force on the next particle in the chain [30] (see **PAPER 5.8**). In certain situations it is possible to create a bistable configuration where the distance between two confined particles exhibits a marked hysteresis with respect to the separation of the fiber facets. As shown by Metzger et al., the experimental results show excellent agreement with a theory based on a calculation of the Lorentz force associated with the field distribution [31] (see **PAPER 4.7**).

The same twin fiber configuration forms the basis of the optical stretcher developed by Guck et al. [32] (see **PAPER 2.2**). Cells passing through the counterpropagating beams are stretched by an amount corresponding to their stiffness, which itself may relate to their disease state. However, at a more subtle level, the detailed calculation of the optically induced stress exerted on a deformable dielectric is still an area of ongoing study [33,34].

Another important force acting between multiple particles trapped in optical tweezers arises from their hydrodynamic coupling, both to each other and between individual particles and the walls of the sample cell [35]. When superficially examined, the motion of each trapped particle appears to be that of an overdamped simple harmonic oscillator, which can be accurately described by a power spectrum; see Berg-Sørensen and Flyvbjerg [36] (see **PAPER 4.8**). The hydrodynamic coupling means that the motion of one particle exerts a force on neighboring particles, leading to differential damping of their eigenmodes. The first observations of hydrodynamic interactions related to coupling between two particles in optical tweezers was by Meiners and Quake, who measured the difference in decay rates of the symmetric stretch compared to the common motion eigenmodes [37] (see **PAPER 5.5**). These experiments used quadrant photodiodes to measure the position of the particles and hence were restricted, by complexity, to two particles. In recent years, groups have used video cameras to measure the positions of many particles simultaneously [38,39]. Perhaps surprisingly, the experimental observations match the predicted behavior to a few percent.

Finally, although the motion of trapped particles in most tweezers systems is overdamped by the surrounding fluid, it is now possible to trap objects in air [40]. As such, the systems are no longer overdamped, and in high-power optical traps it is possible to observe the onset of a resonant effect. As shown by Di Leonardo et al., in an air-damped tweezers, it is possible to parametrically excite the resonance, revealing again extremely close agreement of the experimental observations with the simple differential equations governing the system [41] (see **PAPER 4.9**).

Although the concept of optical tweezers and the origin of the force produced are easy to explain using either ray-optical or gradient-force models, the precise calculation of the forces operating in any trap is complicated. Nevertheless, following from simple models, it seems that the rapid advances in computational power mean that precise calculation and prediction of the performance of individual traps is now just about possible. Given this complexity, it is perhaps surprising that by assuming that each trap behaves as a simple harmonic potential, one is still able to predict the highly complex behavior of the interactions between multiple objects coupled via both hydrodynamic and optical forces.

Endnotes

1. A. Ashkin, J.M. Dziedzic, J.E. Bjorkman, and S. Chu, Observation of a single-beam gradient force optical traps, *Opt. Lett.* **11**, 288–290 (1986).
2. W.H. Wright, G.J. Sonek, Y. Tadir, and M.W. Berns, Laser trapping in cell biology, *IEEE J. Quant. Electron.* **26**, 2148–2157 (1990).
3. A. Ashkin, Forces of a single beam gradient laser trap on a dielectric sphere in the ray optics regime, *Biophys. J.* **61**, 569–582 (1992).
4. R. Gussgard, T. Lindmo, and I. Brevik, Calculation of the trapping force in a strongly focused laser beam, *J. Opt. Soc. Am. B* **9**, 1922–1930 (1992).
5. N.B. Simpson, L. Allen, and M.J. Padgett, Optical tweezers and optical spanners with Laguerre-Gaussian modes, *J. Mod. Opt.* **43**, 2485–2491 (1996).
6. W.H. Wright, G.J. Sonek, and M.W. Berns, Radiation trapping forces of microspheres with optical tweezers, *Appl. Phys. Lett.* **63**, 715–717 (1993).
7. W.H. Wright, G.J. Sonek, and M.W. Berns, Parametric study of the forces on microspheres held by optical tweezers, *Appl. Opt.* **33**, 1735–1748 (1994).
8. K.F. Ren, G. Grehan, and G. Gouesbet, Prediction of reverse radiation pressure by generalized Lorenz-Mie theory, *Appl. Opt.* **35**, 2702–2710 (1996).
9. A. Rohrbach and E.H.K. Stelzer, Optical trapping of dielectric particles in arbitrary fields, *J. Opt. Soc. A* **18**, 839–853 (2001).
10. J.P. Barton, D.R. Alexander, and S.A. Schaub, Theoretical determination of net-radiation force and torque for a spherical particle illuminated by a focused laser beam, *J. Appl. Phys.* **66**, 4594–4602 (1989).
11. J.P. Barton and D.R. Alexander, 5th order corrected electromagnetic field components for a fundamental Gaussian beam, *J. Appl. Phys.* **66**, 2800–2802 (1989).
12. A. Rohrbach and E.H.K. Stelzer, Trapping forces, force constants, and potential depths for dielectric spheres in the presence of spherical aberrations, *Appl. Opt.* **41**, 2494–2507 (2002).
13. R.C. Gauthier, Computation of the optical trapping force using an FDTD based technique, *Opt. Express* **13**, 3707–3718 (2005).
14. A.R. Zakharian, P. Polynkin, M. Mansuripur, and J.V. Moloney, Single-beam trapping of micro-beads in polarized light: Numerical simulations, *Opt. Express* **14**, 3660–3676 (2006).
15. S.M. Barnett and R. Loudon, On the electromagnetic force on a dielectric medium, *J. Phys. B* **39**, S671–S684 (2006).
16. I. Brevik, Experiments in phenomenological electrodynamics and the electromagnetic stress tensor, *Phys. Lett.* **52**, 133–201 (1979).
17. U. Leonhardt, Optics—Momentum in an uncertain light, *Nature* **444**, 823–824 (2006).
18. M. Mansuripur, Radiation pressure and the linear momentum of the electromagnetic field in magnetic media, *Opt. Express* **15**, 13502–13517 (2007).
19. R.N. C. Pfeifer, T.A. Nieminen, N.R. Heckenberg, and H. Rubinsztein-Dunlop, Colloquium: Momentum of an electromagnetic wave in dielectric media, *Rev. Mod. Phys.* **79**, 1197–1216 (2007).
20. M. Padgett, S.M. Barnett, and R. Loudon The angular momentum of light inside a dielectric, *J. Mod. Opt.* **50**, 1555–1562 (2003).
21. T.A. Nieminen, V.L.Y. Loke, A.B. Stilgoe, G. Knoner, A.M. Branczyk, N.R. Heckenberg, and H. Rubinsztein-Dunlop, Optical tweezers computational toolbox, *J. Opt. A—Pure Appl. Opt.* **9**, S196–S203 (2007).
22. S.H. Simpson and S. Hanna, Optical trapping of spheroidal particles in Gaussian beams, *J. Opt. Soc. Am. A* **24**, 430–443 (2007).
23. V.L.Y. Loke, T.A. Nieminen, S.J. Parkin, N.R. Heckenberg, and H. Rubinsztein-Dunlop, FDFD/T-matrix hybrid method, *J. Quant. Spect. Rad. Trans.* **106**, 274–284 (2007).
24. A.T. O'Neil, I. MacVicar, L. Allen, and M.J. Padgett, Intrinsic and extrinsic nature of the orbital angular momentum of a light beam, *Phys. Rev. Lett.* **88**, 053601 (2002).
25. L. Allen, M.J. Padgett, and M. Babiker, The orbital angular momentum of light, *Prog. Opt.* **39**, 291–372 (1999).
26. Y. Roichman, B. Sun, Y. Roichman, J. Amato-Grill, and D.G. Grier, Optical forces arising from phase gradients, *Phys. Rev. Lett.* **100**, 013602 (2008).

27. A. Jesacher, C. Maurer, A. Schwaighofer, S. Bernet, and M. Ritsch-Marte, Full phase and amplitude control of holographic optical tweezers with high efficiency, *Opt. Express* **16**, 4479–4486 (2008).

28. M.M. Burns, J.M. Fournier, and J.A. Golovchenko, Optical binding, *Phys. Rev. Lett.* **63**, 1233–1236 (1989).

29. A. Constable, J. Kim, J. Mervis, F. Zarinetchi, and M. Prentiss, Demonstration of a fibreoptic light force trap, *Opt. Lett.* **18**, 1867–1869 (1993).

30. S.A. Tatarkova, A.E. Carruthers, and K. Dholakia, One-dimensional optically bound arrays of microscopic particles, *Phys. Rev. Lett.* **89**, 283901 (2002).

31. N.K. Metzger, K. Dholakia, and E.M. Wright, Observation of bistability and hysteresis in optical binding of two dielectric spheres, *Phys. Rev. Lett.* **96**, 068102 (2006).

32. J. Guck, R. Ananthakrishnan, H. Mahmood, T.J. Moon, C.C. Cunningham, and J. Kas, The optical stretcher: A novel laser tool to micromanipulate cells, *Biophys. J.* **81**, 767–784 (2001).

33. A. Casner, J.P. Delville, and I. Brevik, Asymmetric optical radiation pressure effects on liquid interfaces under intense illumination, *J. Opt. Soc. Am. B* **20**, 2355–2362 (2003).

34. A. Casner and J.P. Delville, Laser-induced hydrodynamic instability of fluid interfaces, *Phys. Rev. Lett.* **90**, 144503 (2003).

35. A. Pralle, E.L. Florin, E.H.K. Stelzer, and J.K.H. Horber, Local viscosity probed by photonic force microscopy, *Appl. Phys. A* **66**, S71–S73 (1998).

36. K. Berg-Sørensen and H. Flyvbjerg, Power spectrum analysis for optical tweezers, *Rev. Sci. Instrum.* **75**, 594–612 (2004).

37. J.C. Meiners and S.R. Quake, Direct measurement of hydrodynamic cross correlations between two particles in an external potential, *Phys. Rev. Lett.* **82**, 2211–2214 (1999).

38. M. Polin, D.G. Grier, and S.R. Quake, Anomalous vibrational dispersion in holographically trapped colloidal arrays, *Phys. Rev. Lett.* **96** 088101 (2006).

39. R. Di Leonardo, S. Keen, J. Leach, C.D. Saunter, G.D. Love, G. Ruocco, and M.J. Padgett, Eigenmodes of a hydrodynamically coupled micron-size multiple-particle ring, *Phys. Rev. E* **76** 061402 (2007).

40. R. Omori, T. Kobayashi, and A. Suzuki, Observation of a single-beam gradient-force optical trap for dielectric particles in air, *Opt. Lett.* **22**, 816–818 (1997).

41. R. Di Leonardo, G. Ruocco, J. Leach, M.J. Padgett, A.J. Wright, J.M. Girkin, D.R. Burnham, and D. McGloin, Parametric resonance of optically trapped aerosols, *Phys. Rev. Lett.* **99** 010601 (2007).

Parametric study of the forces on microspheres held by optical tweezers

W. H. Wright, G. J. Sonek, and M. W. Berns

Optical-trapping forces exerted on polystyrene microspheres are predicted and measured as a function of sphere size, laser spot size, and laser beam polarization. Axial and transverse forces are in good and excellent agreement, respectively, with a ray-optics model when the sphere diameter is ≥ 10 μm. Results are compared with results from an electromagnetic model when the sphere size is ≤ 1 μm. Axial trapping performance is found to be optimum when the numerical aperture of the objective lens is as large as possible, and when the trapped sphere is located just below the chamber cover slip. Forces in the transverse direction are not as sensitive to parametric variations as are the axial forces. These results are important as a first-order approximation to the forces that can be applied either directly to biological objects or by means of microsphere handles attached to the biological specimen.

Introduction

Optical trapping is a novel technique that utilizes radiation pressure to control and manipulate microscopic particles, including biological cells and microorganisms. "Optical tweezers" is the common term used to describe the optical-force generation and confinement by a highly focused laser beam. The single-beam gradient-force optical trap, also known as optical tweezers, was first demonstrated by Ashkin et al.[1] It consists of a single laser beam that is focused by a microscope objective lens to a spot diameter $d \sim \lambda$ in the object plane of a microscope. When the laser beam is focused tightly, large intensity gradients are created in both the axial and transverse directions. In the presence of such a beam, a dielectric particle, such as a biological cell, is polarized by the electric field and drawn into the beam focus, where it becomes confined. The forces exerted by the trap are sufficient to move cells and intracellular objects without physical contact. Thus it should be possible to keep a biological sample sterile while optical manipulation is being performed. In addition, because an optical-trapping manipulator can be easily integrated into a standard microscope,

other diagnostic experiments on biological samples would be facilitated.

In its simplest implementation an optical trap functions as a manipulator for optically confined objects. When calibrated the optical trap becomes a microforce transducer, or tensometer, that can be used, for example, to measure the forces exerted on a trapped object as a result of various molecular motors within the cell or by the flagellum of swimming sperm. Recent reports also have established the manipulative capability of optical traps in important biological applications, including the trapping of cells, bacteria, and viruses,[2] cell sorting and classification,[3] cell fusion,[4] and intracellular surgery.[5] Optical traps have also been combined with other laser tools, for example, to perform microsurgery and trapping simultaneously on the same specimen, such as chromosomes.[6] Additional work has been done to establish optical traps as effective microforce transducers, including the measurement of the force produced by a single molecular motor in *Reticulomyxa*,[7] a single kinesin motor with an *in vitro* assay,[8] and the estimation of the force produced by a single kinesin motor in a sea urchin axoneme.[9] Knowledge of the applied trapping forces in terms of their magnitudes and directions is important for predicting and understanding of the interaction of the focused laser beam with various cells and organelles regardless of whether the trap is used as a manipulator or as a transducer.

In this paper a parametric study is presented that compares predicted and measured trapping forces with microspherical test particles for a combination of system parameters and describes the performance

The authors are with the Beckman Laser Institute and Medical Clinic, University of California, Irvine, Irvine, California 92715. W. H. Wright and G. J. Sonek are also with the Department of Electrical and Computer Engineering, University of California, Irvine.

Received 27 April 1993; revision received 27 September 1993.

0003-6935/94/091735-14$06.00/0.

characteristics of optical traps in terms of a nondimensional trapping-efficiency factor Q. The parameters investigated in the present study include beam convergence angle, laser spot size, sphere radius, and polarization state. While a number of investigators have calculated the forces exerted on large dielectric microspheres using a ray-optics approach,[10–13] to date we know of no parametric study done to measure the actual forces exerted on particles located in an optical trap. Experimental evidence is limited to measurement of the transverse trapping Q for 3-μm-diameter polystyrene microspheres[14] and axial trapping Q measurements on dielectric microspheres of 1- and 10-μm diameter.[15] Also a ray-optics model can be used only to calculate the expected forces on large ($d > 10\lambda$) particles. To circumvent this limitation, we present an electromagnetic-field model for the optical trap for calculating the axial and transverse forces exerted on the microsphere. These results when compared with a geometrical-optics model demonstrate that the electromagnetic model is more suitable for describing the laser–particle interaction when the particle diameter is $\leq 1\,\mu$m. Experimental measurements of trapping forces on microspheres are presented and correlated with the above analytical models in the regions where they apply. Although in practice calibration of optical traps is usually performed *in situ*, there is still a need to compare the forces exerted on particles by the optical tweezers with rigorous theoretical models. Discrepancies between the measured forces and the theoretical models may indicate the presence of other forces, including radiometric forces, in addition to the optical trapping forces.

The use of dielectric microspheres as a model system for biological specimens is justified for several reasons. First, it is extremely difficult to develop an analytical model, especially one that uses electromagnetic theory, to calculate the forces exerted on an inhomogeneous structure of arbitrary shape. Second, regardless of the type of particle analyzed, the general trends in the parametric analysis should be applicable to all particles. Finally, dielectric particles such as microspheres can be attached to biological specimens as a handle to facilitate trapping[8,16] and to move the focal point of the tweezers outside of the biological specimen to minimize laser-induced damage. By studying laser–microsphere interactions at the microscopic level and obtaining a quantitative understanding of the optical-trapping process, we should be able to design and develop optical traps for specific biomedical applications. The details of the analytical and experimental studies of the single-beam, gradient-force optical trap are presented in the following sections.

Calculation of the Trapping Efficiency

Calculation of the trapping efficiency Q can be performed with either a simple ray-optics (RO) model or a more complex electromagnetic (EM) force model. The RO model, because it follows the laws of geometrical optics, is applicable only to large diameter ($d > 10$

μm) spheres. In comparison the EM model accounts for the sphere size as well the nature of the trapping beam and is best suited for describing the laser–particle interactions for small ($d \leq 1\,\mu$m) particles. Both of these models are considered herein for predicting the optical forces exerted on particles in the large- and small-diameter size regimes.

The principle relationship between the trapping force and power is $F = Qn_1P/c$, where Q is a nondimensional efficiency parameter, n_1 is the refractive index of the surrounding medium, P is the laser power, and c is the speed of light in free space. For typical values of n_1, P, and Q for particles in an optical trap, F is of the order of pN. To increase the trapping force on the specimen, one can potentially increase Q, n_1, or P. The surrounding medium is usually some form of culture medium and generally cannot be altered without adversely affecting the cell. Trapping power can be increased, up to a limit, until absorption by the cell or the surrounding medium causes significant heating and subsequent thermal damage. The remaining term Q has the greatest influence on the performance of the trap and even determines if the cell is trapped. The efficiency parameter includes the effect that the laser beam has on the trap performance through parameters such as convergence angle, spot size, wavelength, polarization, and beam profile. It also takes into account the optical properties of the trapped object, including size, shape, and relative refractive index with respect to the surrounding medium. Understanding how the laser beam influences trapping is particularly useful because the convergence angle, spot size, wavelength, and polarization, for example, can be used as design parameters to specify an optimum trapping configuration.

Early modeling efforts that used a quasi-ray-optics approach were carried out by several researchers over a limited parameter space.[10–12] A more generalized strictly RO model that treated the case of tightly focused laser beams focused to a point with an arbitrary intensity distribution was developed by Ashkin.[13] This strict RO model is used to perform the RO calculations presented in this paper and in other recent work.[15] Figure 1 shows the geometry for this model. An incident ray passing through the microscope objective aperture is focused to a spot on the laser beam axis. The maximum convergence angle of the ray is determined by the objective numerical aperture. When the ray strikes the sphere, a fraction of the momentum carried by it is reflected (p_iR term) and the remainder is transmitted (p_iT term) into the sphere. No absorption is assumed for the sphere. The transmitted fraction then produces both an infinite number of internal reflections within the sphere, as well as an infinite number of emergent refracted rays that escape the sphere. The expressions for the momentum transferred to the sphere by the emergent rays in the direction parallel to the incident ray (p_s) and perpendicular to the incident ray

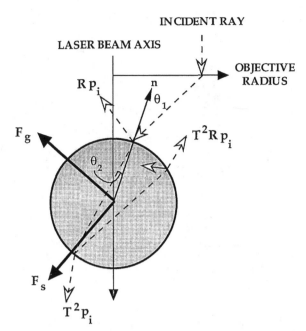

INCIDENT RAY

LASER BEAM AXIS

OBJECTIVE
RADIUS

$R\,p_i$

n

θ_1

F_g

θ_2

$T^2 R\,p_i$

F_s

$T^2 p_i$

Fig. 1. Ray-optics (RO) model for optical trapping (after Ashkin[13]). The scattering force F_s points in the direction of the incident ray, while the gradient force F_g is perpendicular to the direction of the incident ray. The intensities of the reflected and refracted rays are determined by the Fresnel transmission T and reflection R coefficients at a dielectric boundary, and by the power P of the incident ray. The angle of incidence is θ_1 and the angle of refraction is θ_2.

(p_g) can be written as[13,17,18]

$$p_s = \left\{ 1 + R\cos 2\theta_1 - \frac{T^2[\cos 2(\theta_1 - \theta_2) + R\cos 2\theta_1]}{1 + R^2 + 2R\cos\theta_2} \right\} p_i,$$

$$= Q_s p_i, \qquad (1)$$

$$p_g = \left\{ R\sin 2\theta_1 - \frac{T^2[\sin 2(\theta_1 - \theta_2) + R\sin 2\theta_1]}{1 + R^2 + 2R\cos\theta_2} \right\} p_i,$$

$$= Q_g p_i, \qquad (2)$$

where θ_1 is the angle of incidence, θ_2 is the angle of refraction, R and T are the Fresnel reflection and transmission coefficients, respectively, and p_i is the incident momentum. The scattering and gradient forces are then defined as[13]

$$F_s = n_1 Q_s P/c, \qquad (3)$$

$$F_g = n_1 Q_g P/c. \qquad (4)$$

From Eqs. (3) and (4) it is seen that the trapping efficiency Q is simply the fraction of the momentum transferred to the sphere by the emergent rays.

For the calculations presented herein, axial forces are obtained by vector-summing of the contributions of all rays with convergence angles ranging from zero to $\phi_{max} = \arcsin(\text{N.A.}/n_1)$, where ϕ_{max} is the maximum convergence angle over the entire aperture of the microscope objective. This integral summation is performed with the trapezoidal rule and convergence is obtained when successive approximations to the integral differ by less than 10^{-3}. A Gaussian intensity profile of the form $I(r) = I_0 \exp(-2r^2/\omega_0^2)$ is assumed at the microscope objective aperture, where ω_0 is the radius of the laser beam, with a beam waist-to-aperture radius ratio of 1.0. The laser beam is taken to be linearly polarized.

Transverse force calculations with the RO model are performed with the same method employed for the axial calculations above. However, in this case the calculation is complicated by the complexity in determining the direction of the incident ray as it strikes the sphere, the fraction of the ray in each of the two orthogonal polarizations, and the magnitude of the forces along the coordinate system that defines the particle–trap interaction, as explained in Ref. 13.

One of the limitations of the RO approach described above is that as λ approaches zero, the force exerted on the sphere is predicted to be independent of particle size.[19] On the other hand, when the sphere size approaches λ the geometrical-optics approximation is no longer valid and a complete electromagnetic description must be used. To date a number of authors have calculated the optical forces on spherical particles using an EM-field model.[20,21] However, none of these works have addressed the case for a focused laser spot size of the order of $\lambda/2n$, as representative of the optical trap described herein. To determine the trapping forces exerted on a sphere by a focused laser beam, the Lorenz–Mie theory[22] is extended to account for the use of arbitrary incident fields, as previously described.[23] The key steps in this process include (1) the infinite-series representation of the incident and scattered fields using the Ricatti–Bessel and spherical harmonic functions; (2) solution for the scattering coefficients by evaluation of the boundary conditions at the sphere surface; (3) determination of an exact expression for the radial component of the incident Gaussian beam; and (4) derivation of the expansion coefficients that describe the incident Gaussian beam. Once the scattering and expansion coefficients are known, the EM forces can be determined from the Maxwell stress tensor with expressions previously derived.[21] Because the laser beam used to hold the sphere is tightly focused, the paraxial approximation for the incident Gaussian laser beam is not valid. An accurate description of the EM fields requires the use of additional terms that describe both the transverse and axial field components present at the beam focus. It is important to realize, however, that the field expressions for the Gaussian beam are not exact solutions to the Maxwell equations regardless of the level of approximation used. Therefore there will be some error in the predicted forces, and it becomes significant when evaluating the case of a large particle held within a trapping beam.

The above EM-force calculations are performed at any arbitrary point for the laser beam in the coordinate system defined by the center of the microsphere.

Numerical integration of the surface integrals that define the expansion coefficients for the incident Gaussian beam were performed with the trapezoidal rule, in a manner similar to the RO calculations. The model employed fifth-order corrections for the radial component of the incident Gaussian beam that have been previously derived.[24] The procedure for calculating the trapping forces from the electromagnetic field expressions is detailed in Appendix A. In Fig. 2 the axial trapping efficiency Q is plotted as a function of axial position for several laser spot sizes that are in the size range used for optical trapping. This calculation assumed a 1-μm-diameter amorphous silica microsphere ($n = 1.45$) trapped in water, at a wavelength of 1.06 μm. The axial position is the location of the sphere center with respect to the laser beam waist. A negative trapping efficiency indicates a trapping, or restoring, force directed back to the laser beam focus. Clearly the best trapping is obtained when the spot size is as small as possible, as seen in the case of the 0.4-μm laser spot of Fig. 2, and maximum trapping Q occurs when the focal point is located outside the sphere. In Fig. 3 the axial trapping efficiency Q is plotted again as a function of axial position for sphere sizes ranging from 0.1 to 0.5 μm in radius. As previously observed,[15] the axial trapping efficiency is a function of sphere size, with the maximum trapping Q increasing in magnitude from 0.001 to 0.034 as the sphere radius increases from 0.1 to 0.5 μm. The dependence of axial trapping efficiency in the Mie regime on relative refractive index is shown in Fig. 4. The relative index is the ratio of the particle refractive index to the index of the surrounding medium (water, $n = 1.33$). The axial Q is a maximum for relative index values in the range of 1.1–1.2, corresponding to the relative index of most biological cells immersed in water, and decreases for very small (1.05) and very large (>1.3) relative indexes.

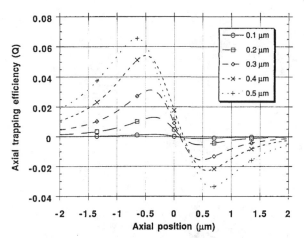

Fig. 3. Predicted axial trapping efficiency Q as a function of axial position for several microsphere radii. An amorphous silica microsphere suspended in water and trapped at a wavelength of 1.06 μm with a spot size of 0.4 μm was assumed.

Experimental Design

In this section the design of the experiments required to determine the performance characteristics of the optical trap are described. This includes the design of a laser microscope to implement optical trapping, a description of the test particles used to study trapping performance, and a description of the scanning-knife-edge measurement technique for determining laser spot size. Last, the method of measurement of axial and transverse trapping forces by the techniques of gravity and viscous drag-force measurements are presented.

Laser-Trapping Microscope

The design of the laser trapping microscope used in our experimental studies is summarized here and is depicted in Fig. 5. Further details regarding the general setup of an optical trap are described in Ref. 25. A Quantronix Model 116 Nd:YAG laser operating at a wavelength of 1.06 μm and having a TEM$_{00}$

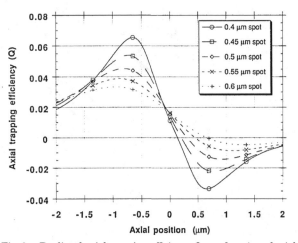

Fig. 2. Predicted axial trapping efficiency Q as a function of axial position for several laser spot sizes representative of optical traps. A 1-μm-diameter amorphous silica ($n = 1.45$) microsphere suspended in water and trapped at a wavelength of 1.06 μm was assumed.

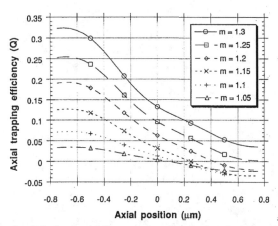

Fig. 4. Calculated axial force as a function of refractive index with the EM model. The sphere diameter was assumed to be 1 μm, and the wavelength was 1.06 μm.

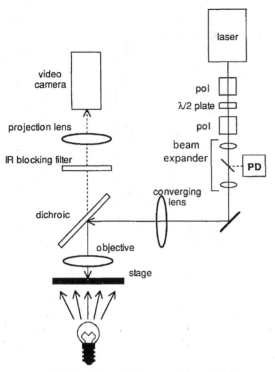

Fig. 5. Schematic diagram of the optical trap.

mode structure is the source for the trapping beam. The laser has an initial beam diameter of 0.7 mm and a far-field divergence angle of 1.1 mrad (half angle). A polarizer is placed immediately after the output aperture of the laser to create a linearly polarized beam. The intensity of the laser beam is varied with a half-wave plate followed by another polarizing prism. The beam size is then increased to a diameter of ~5 mm and collimated using a 6× beam expander. A beam splitter is used at this location to monitor the laser power during the experiment. After the beam is reflected off several mirrors, a 20-cm focal-length lens is then used to focus the beam to a spot diameter of ~22 µm at the image plane of the objective, located inside a Zeiss Universal microscope. The beam then diverges from the focal point ~160 mm to the entrance pupil of the objective, filling the entire aperture and creating the optical potential well that is required for trapping at the specimen plane. Between the focusing lens and the objective, a dichroic filter is used to separate the transmitted image from the trapping beam. All experiments are performed while observing the projected image on a video monitor, using a camera mounted on the trinocular port of the microscope. We determined the transmittance through the microscope objective using the method of Misawa *et al.*[26] A matched set of Zeiss Neofluar 100×/1.3 ph 3 objectives were first used to determine the transmission of a single objective. This objective, in turn, was used to determine the transmittance of the other objectives. The measured transmittances, and the corresponding objectives, are 0.60, Neofluar 100×/1.3 ph 3; 0.63, Neo-

fluar 100×/1.3; 0.43, Plan 100×/1.25; 0.33, Plan 100×/1.0; 0.16, Plan 100×/0.8.

Test Particles

All polystyrene microspheres were obtained from Duke Scientific (Palo Alto, Calif.), except for the 10-µm-diameter microspheres used for the transverse force measurements, which were obtained from Coulter Electronics (Hialeah, Fla.). Particle sizes and their tolerances for the Duke Scientific microspheres were 1.031 ± 0.02 µm, 5.007 ± 0.038 µm, 9.870 ± 0.057 µm, 15.00 ± 0.07 µm, and 20.49 ± 0.20 µm. The refractive index of the polystyrene microspheres was calculated with the dispersion equation for polystyrene to be 1.57 at a wavelength of 1.06 µm:

$$n = a + b/\lambda_0^2, \tag{5}$$

where $a = 1.5683$, $b = 10.087 \times 10^{-4}$, and wavelength is in centimeters.[27] The microspheres were assumed to be nonabsorbing. Amorphous silica microspheres obtained from Bangs Laboratories (Carmel, Ind.) and having a diameter of 1 ± 0.08 µm and a refractive index of 1.45 were also employed. Microspheres were suspended in distilled water for the axial-force measurements and in an 11% solution of NaCl for the transverse-force measurements. The addition of 0.2% Triton X-100 to the suspension media prevented the microspheres from sticking to the glass surfaces of the trapping chamber.

Laser Spot Size Measurement

The laser spot size produced in the above system is an important parameter in the characterization of an optical trap. To obtain a measurement of the spot size, we used a microscope objective with a high N.A. to focus radiation of $\lambda = 1.06$ µm to a spot of ~0.4 µm $(= \lambda/2n)$ in the object plane. The focused laser spot then was scanned by a knife edge attached to a translation stage driven by a piezoelectric translator.[28] The knife edge consisted of a chromium test target, deposited on glass, with edges that were smooth to less than 0.5 µm as observed by light microscopy. A differential micrometer was attached to the z-axis stage micrometer to provide the resolution necessary to position the knife edge in the laser beam waist.

If the laser beam is assumed to have a Gaussian intensity distribution, the laser power transmitted by a partially occluding knife edge is given by[29]

$$P(x) = \frac{P_0}{\omega} \sqrt{\frac{2}{\pi}} \int_x^\infty \exp\left(-\frac{2x^2}{\omega^2}\right) dx, \tag{6}$$

where P_0 is the unoccluded laser power, x is the location of the knife edge, and ω is the spot radius. After evaluating the integral with the complementary error function, the transmitted power can be expressed as

$$P(x) = (P_0/2)\mathrm{erfc}(2^{1/2}x/\omega). \tag{7}$$

To solve for the spot size, the knife-edge positions are measured when the transmitted power is 10% and 90% of P_0. Solving the above equation for these two power levels leads to

$$\omega = 0.7830(x_{10\%} - x_{90\%}). \qquad (8)$$

The distance between the $0.1P_0$ and $0.9P_0$ positions was measured by counting the number of fringes from an interferogram of the displaced translation stage.

Force Measurements

To measure the axial trapping efficiency, we loaded microspheres into a thin, rectangular glass chamber (Vitro Dynamics) having an internal opening of 20 μm for the 1-μm-diameter spheres, or 50 μm for all other sphere sizes, by capillary action. A microsphere was trapped and moved to a position just below the upper glass surface to minimize spherical aberration. The power of the trapping beam was then decreased until the microsphere was observed to fall out of the trap. The force on the microsphere was taken to be the difference between the gravitational and buoyant forces, given by $F_g - F_b = V_s(\rho_s g - \gamma)$, where V_s is the volume of the sphere, ρ_s is the density of the sphere, g is the gravitational constant, and γ is the specific weight of the surrounding fluid. The axial trapping efficiency is then calculated with the expression $Q = Fc/n_1 P$. For each set of parameters, 10 separate measurements were made and averaged.

Measurement of the transverse trapping forces were performed with a viscous-drag-force technique as the microsphere was moved through the fluid. The peak transverse force is measured simply by observing the speed at which the particle falls out of the optical trap in the presence of a drag force produced by the surrounding viscous medium. This is taken to be the maximum force exerted on the particle. In a Newtonian (constant viscosity) fluid with a Reynolds number below 1, the Stokes law states that the force on a sphere is $F = 6\pi r v \mu$, where r is the radius of the sphere and v is the velocity and μ the viscosity of the surrounding medium (0.01 dyn

s/cm^2 for water at 20 °C). However, a sphere in a viscous fluid moving close to a cover slip experiences an increase in drag because of the no-slip boundary condition at the coverslip. For a sphere, the Stokes law is then modified such that

$$F = \frac{6\pi r v \mu}{1 - \frac{9}{16}\left(\frac{r}{h}\right) + \frac{1}{8}\left(\frac{r}{h}\right)^3 - \frac{45}{256}\left(\frac{r}{h}\right)^4 - \frac{1}{16}\left(\frac{r}{h}\right)^5}, \qquad (9)$$

where h is the distance from the cover slip to the center of the sphere.[30] Microspheres were placed in a chamber fitted with a #1 thickness (~ 170 μm) cover slip at the top and suspended in an 11% NaCl solution so that they would come to rest just below the cover-slip surface. This was necessary to bring the microspheres within reach of the optical trap. During a force measurement, the particle under study was held in the center of the optical field with the trap. Motion was created by translation the stage in the transverse direction with a Zeiss scanning stage with a speed resolution of 25 μm/s. Again, for each data point 10 separate measurements were recorded and averaged.

Results

This section provides the results of our parametric study of the axial and transverse trapping efficiency. The parametric study of the trapping efficiency includes the comparison of measured Q values with predictions from both the RO and EM force models. For this study the parameters chosen were the beam convergence angle, laser spot size, microsphere size, and beam polarization, because they provide the greatest influence on the trapping efficiency Q.

Table 1 summarizes the measurement of the laser spot size formed by various high-N.A. oil-immersion microscope objectives at a wavelength of 1.06 μm. For the calculation of Q in the EM case, the experimental spot size ω_0 is the variable parameter of interest, while the beam convergence (cone) angle θ, determined from the lens numerical aperture (N.A.), is the variable parameter in the RO case. The convergence angle θ can be estimated from the expression $\theta \sim$

Table 1.   Measured Spot Size for Various High Numerical Aperture Microscope Objectives and the Corresponding Maximum Backward Axial Trapping Q

Objective	Spot Size[a] (ω_o)	Q for 1-μm-Spheres	EM Model Q	Cone Angle	Q for 20-μm-Spheres	RO Model Q
Zeiss Neofluar 100×/1.3	0.39	-0.006 ± 0.001	-0.034	60.0°	-0.10 ± 0.018	-0.14
Zeiss Neofluar 100×/1.3[b]	0.44	-0.004 ± 0.001	-0.024[c]	60.0°	-0.080 ± 0.016	-0.14[c]
Zeiss Plan 100×/1.25[d]	0.54	-0.007 ± 0.001	-0.0096	56.4°	-0.080 ± 0.014	-0.12
Zeiss Plan 100×/1.0[d]	0.53	-0.005 ± 0.001	-0.011	41.8°	-0.058 ± 0.011	-0.051
Zeiss Plan 100×/0.8[d]	0.61	-0.003 ± 0.001	-0.0041	32.2°	-0.034 ± 0.006	-0.023

[a]In micrometers, ± 0.03 μm.

[b]The effect of the phase ring was not accounted for in the model calculations.   Previous calculations[12] indicate that the difference in Q is $\sim 2\%$.

[c]Phase-contrast objective (ph 3).

[d]This objective had an internal aperture that determined the effective numerical aperture.   The aperture was continuously variable from 0.8 to 1.25.

arcsin (N.A./n), where n is the refractive index of the matching medium, or from the expression $\theta \sim \lambda/\pi n \omega_0$ in the paraxial case for values of $\theta \leq 30°$. In the RO case, the laser focus is taken as a geometric point. It can be seen that there is a general trend of decreasing spot size with increasing numerical aperture. The results for the 100×/1.3 Neofluar (Zeiss) indicate that the spot size approaches the minimum radius predicted by the expression $\lambda/2n$, and that there is essentially no difference between the 100×/1.3 Neofluar objectives with or without phase rings.

Axial Trapping Efficiency

Table 1 shows that the N.A., which determines the convergence angle in the RO model and the spot size in the EM model, has a dramatic effect on the maximum axial trapping Q. The maximum axial trapping Q, which represents the restoring force directed back towards the beam focus, was measured for a 20-μm-diameter polystyrene sphere and a 1-μm-diameter amorphous silica sphere, suspended in water, at a trapping wavelength of 1.06 μm. The convergence angle and spot size were determined by the choice of the microscope objective used. For the 20-μm-diameter sphere, the trapping efficiency increased in magnitude from 0.034 to 0.10 as the numerical aperture increased from 0.8 to 1.3. A similar change was observed for the 1-μm-diameter spheres, where Q increased in magnitude from 0.003 to 0.006. Also shown in Table 1 are the predicted Q's from the RO and EM models. The EM calculation was performed for the 1-μm silica microspheres, while the RO calculation assumed the use of polystyrene spheres. The greatest source of uncertainty in the measured Q was from the measurement of the power at the trapping site, estimated to be ~17%. In addition, the remaining uncertainty was due to the statistical nature of the measurement and was equal to the standard error of the mean.

It can be seen that the 20-μm-diameter spheres are a good approximation to the RO regime throughout the range of numerical apertures that were evaluated, and this indicates that the measurements are in good agreement with the theoretical predictions. In comparison, the 1-μm-diameter sphere measurements do not agree with the calculated RO Q's, as expected, and are not in good agreement with the EM model predictions. The predicted EM model Q's are higher by as much as a factor of 6 compared with our measurements. The maximum axial trapping Q predicted by the EM model decreases to -0.022 if the spot size is taken to be 0.45 μm instead of 0.4 μm for the Neofluar 100×/1.3 objective. The remaining discrepancy between the predicted and measured axial Q for the 1-μm microsphere may be due in part to spherical aberration in the 100× oil-immersion lens, because an ideal lens has been assumed in the calculations. Measurement of the axial Q's for the small particles is also hampered by the difficulty one encounters in determining the point at which they

actually fall out of the trap. However, this is not reflected in the uncertainty in measuring the 1-μm axial Q's, which are similar on a percentage basis with the uncertainty for the 20-μm-sphere measurements. Finally, the predicted axial Q by the EM model will have some error because of the approximations used to describe the Gaussian laser beam.

The maximum backwards axial trapping Q for polystyrene and amorphous silica microspheres ranging in size from 1 to 20 μm was measured and is shown in Table 2. Axial trapping efficiency is a strong function of sphere size, increasing in magnitude from 0.006 to 0.099 as the particle size increases from 1 to 20 μm. This effect has been predicted[13] and observed[15] previously; however, the latter study utilized a larger-diameter (20-μm) sphere to obtain better correlation with the RO model. In that case, the predicted value[15] for Q of -0.14 for a 20-μm sphere is roughly within experimental error for the measured Q of -0.099 ± 0.015 presented here. In the case of the 1-μm-diameter silica microsphere, our measured value of -0.006 ± 0.001 is significantly lower than the EM model prediction[15] of -0.034, for the reasons explained above. It was not possible to compare directly the measured axial Q's on both 1-μm-diameter polystyrene and silica microspheres, which have refractive indexes of 1.57 and 1.45, respectively. Because the polystyrene spheres have a density close to water, it was difficult to discern when the sphere fell out of the trap in the axial direction. The effects of relative refractive index on calculated axial Q have been shown in Fig. 4. The general trend illustrated by Table 2 is that the maximum backwards trapping efficiency increases with sphere size until the RO regime is reached, after which the value of Q becomes constant and is independent of particle size.

Axial trapping efficiency also is dependent on the distance the trapped particle is below the glass surface of the trapping chamber. In Table 3 the maximum backwards trapping Q is listed as a function of the axial depth for 1-μm-diameter silica and 10-μm-diameter polystyrene microspheres for a 1.3-N.A. objective. The trapping efficiency decreases steadily with depth in both the 1- and 10-μm particle cases. Eventually a distance below the cover slip is reached where the sphere can no longer be trapped. The change in Q as a function of depth within the trapping

Table 2. Maximum Backward Axial Trapping Q in Water as a Function of Sphere Diameter[a]

Sphere Type	Diameter (μm)	Q
Amorphous silica	1	-0.006 ± 0.001
Polystyrene	5	-0.070 ± 0.015
Polystyrene	10	-0.077 ± 0.014
Polystyrene	15	-0.091 ± 0.015
Polystyrene	20	-0.10 ± 0.015

[a]Trapping wavelength of 1.06 μm. Microscope objective was Zeiss Neofluor 100×/1.3 without phase ring.

Table 3. Maximum Backward Axial Trapping Q as a Function of the Numerical Aperture and the Distance Below the Cover Glass[a]

Particle	Numerical Aperture	Distance Below Glass (μm)	Q
1 μm, silica	1.3[b]	0.5	−0.006 ± 0.001
		4	−0.005 ± 0.001
		8	−0.004 ± 0.001
		12	−0.004 ± 0.001
10 μm, polystyrene	1.3[b]	5	−0.077 ± 0.014
		10	−0.059 ± 0.011
		15	−0.050 ± 0.010
		20	−0.047 ± 0.009
		25	−0.042 ± 0.008
		35	−0.034 ± 0.007
10 μm, polystyrene	1.25	5	−0.084 ± 0.015
		55	−0.028 ± 0.006
10 μm, polystyrene	0.8	5	−0.014 ± 0.003
		55	−0.013 ± 0.003

[a]Microspheres in water.
[b]No phase ring.

chamber is most likely a result of objective spherical aberration and cannot be predicted using either the RO or EM models in their current configuration. To confirm the role of spherical aberration in decreasing the trapping Q when working below the cover slip surface, we measured the trapping efficiency for 10-μm-diameter microspheres for two additional numerical apertures and depths (Table 3). With the

1.25-N.A. setting on a Plan 100× objective, the trapping efficiency decreased by 66%, from 0.084 to 0.028, as the sphere distance below the cover slip increased from 5 to 55 μm. However, when the internal aperture of the lens was changed to 0.8, the trapping efficiency decreased by only 7%, from 0.014 to 0.013, as the sphere distance below the cover slip increased from 5 to 55 μm. By decreasing the aperture of the microscope objective and blocking the rays with the largest convergence angle, the spherical aberration decreases while at the same time the depth of field increases.

Transverse Trapping Efficiency

Using the drag-force technique, we measured the maximum trapping force for polystyrene spheres with diameters ranging from 1–20 μm (Fig. 6) at several different power levels with the Neofluar 100×/1.3 objective. The slope of the best-fit line through the data points, which is proportional to the trapping efficiency Q, is given in Table 4. Equation (9) was used to correct the measured Q values shown in Figs. 6 and 7 and in Tables 4 and 5. In our experiments, the sphere was approximately 10 μm below the cover slip when it fell out of the trap, although this distance was not controlled well. Thus, from Eq. (9), the drag-correction factors for the sphere diameters examined in this study were calculated to be 1.39 for the 20-μm-diameter sphere, 1.23 for the 10-μm-diameter sphere, 1.13 for the 5-μm-

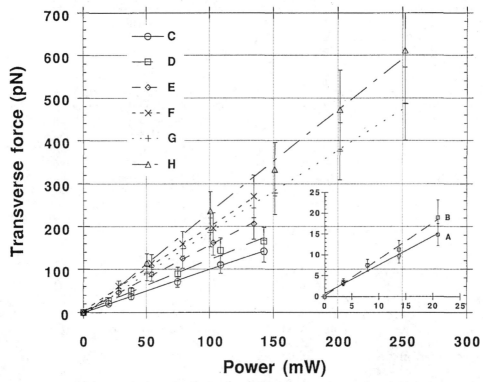

Fig. 6.   Transverse, or viscous, drag force as a function of power for polystyrene microspheres suspended in water of varying diameter and motion parallel ∥ and perpendicular ⊥ to the laser beam polarization.   Data for the 1-μm-diameter microsphere is shown in the inset. A, 1 μm (⊥); B, 1 μm (∥); C, 5 μm (⊥); D, 5 μm (∥); E, 10 μm (⊥); F, 10 μm (∥); G, 20 μm (⊥); H, 20 μm (∥).

Table 4. Maximum Transverse Trapping Q as a Function of Sphere Diameter[a]

Sphere Diameter and Polarization[b]	Measured Q	Predicted Q	Model
1 μm			
∥	0.15	0.16	EM
⊥	0.13	0.15	EM
5 μm			
∥	0.21	0.36	RO
⊥	0.19	0.29	RO
10 μm			
∥	0.37	0.36	RO
⊥	0.29	0.29	RO
20 μm			
∥	0.41	0.36	RO
⊥	0.32	0 29	RO

[a]Polystyrene microspheres in water; trapping wavelength of 1.06 μm and a spot size of ≈ 0.4 μm, based on the data in Fig. 3.

[b]The symbol ∥ denotes parallel and ⊥ denotes perpendicular to the polarization direction of the laser beam.

Table 5. Maximum Transverse Trapping Q as a Function of Maximum Convergence Angle[a]

Objective	Cone Angle	Measured Q	RO Model Predicted Q
Zeiss Neofluor 100×/1.3	60°	0.32	0.29
Zeiss Plan 100×/1.25[b]	56°	0.31	0.27
Zeiss Plan 100×/1.0[b]	41°	0.26	0.18
Zeiss Plan 100×/0.8[b]	32°	0.28	0.13

[a]20 μm polystyrene microspheres in water; trapping wavelength of 1.06 μm and sphere motion perpendicular to the laser beam polarization.

[b]This objective had an internal aperture that determined the effective numerical aperture. The aperture was continuously variable from 0.8 to 1.25.

diameter sphere, and 1.03 for the 1-μm-diameter sphere. As expected, the trapping efficiency increases as the sphere becomes larger. Figure 6 also shows the difference in forces resulting from spheres moving parallel to the E-field polarization axis compared with motion perpendicular to the E-field polarization. Moving the sphere in the direction perpendicular to the laser beam polarization results in a force of up to 20% lower when compared with that for motion parallel to the polarization direction. The difference in force resulting from the laser beam polarization is a consequence of the Fresnel equations that describe the transmission and reflection of light at a dielectric interface. Note that the Q values presented in Table 4 should be regarded as approximations, because increasing the power changes the axial trap location, hence the tranverse Q. So a single Q value cannot be derived for a given particle size over all laser powers.

Calculation of the maximum transverse Q must be

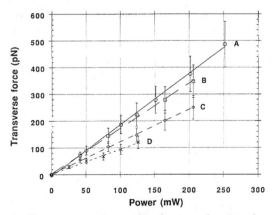

Fig. 7. Transverse, or viscous, drag force as a function of power and cone angle for 20-μm-diameter polystyrene microspheres suspended in water. Cone angles are A: 60°, B: 56°, C: 41°, and D: 32°.

done for the axial location where the force is entirely transverse to the direction of beam propagation if a meaningful comparison with experimental measurements is to be obtained, as previously noted.[13] Hence the predicted Q from the RO model was used for comparison with the measured Q for the 20-μm-diameter sphere (Table 4). Agreement between the predicted and experimental Q's is reasonable, given the uncertainties in the power measurements and in determination of the distance from the cover slip when the sphere fell out of the trap. The results validate the RO model for the largest diameter microspheres; they also indicate that the RO regime exists all the way down to at least a diameter of 10 μm, and perhaps as far as 5 μm. For the 1-μm-diameter microspheres, the EM model predicts a trapping Q of 0.16 for motion parallel to the beam polarization, somewhat larger than the measured value of 0.15. When the microsphere was translated perpendicular to the laser beam polarization, the EM model prediction of 0.15 exceeded the measured Q value of 0.13.

The transverse trapping force was examined also as a function of the laser beam spot size using the 20-μm-diameter microspheres, as shown in Fig. 7. The force was measured in the direction perpendicular to the beam polarization. It can be seen that the largest forces, hence the highest transverse trapping Q, 0.32 (Table 5), was obtained when the microsphere was trapped with the 100× Neofluar/1.3 objective, which produces a 60° cone angle. Q decreased as the numerical aperture decreased, reaching a value of 0.28 when the 100× Plan/0.8 objective was used for trapping. As mentioned above, the large error bars for the experimental data points taken at high laser powers (>150 mW), as shown in Figs. 6 and 7, have two origins, including the variations in trap laser power (∼17%) and the statistical variation in the measurements themselves. Table 5 provides a listing of Q obtained from the slope of the line that best fits the data sets in Fig. 7, along with the objective used and the expected Q predicted by the RO model. The data in Table 5 indicates that the measured Q for the 20-μm-diameter microspheres was consistently higher by *10–100%* than predicted by the RO model. Part of this discrepancy (∼20%) may be explained by the uncertainty in the force measurement; another variation of ~ 10% resulted from the uncertainty in the sphere–cover slip distance.

Discussion

To understand the behavior of particles within optical traps, we first examine the forces that act on a particle. These forces are primarily radiation, gravity, buoyancy, Stokes drag, and radiometric forces. Gravity and buoyancy are constant and are a function only of the particle and the surrounding medium. The Stokes drag force depends on the particle size, shape, and velocity through the surrounding fluid. In addition to the size and shape of the particle, the trapping and radiometric forces are sensitive to the parameters that describe the focused laser beam, most notably the beam convergence angle, spot size, beam profile, polarization, and the relative refractive index. Gravity, buoyancy, and Stokes drag can be used to measure the optical forces exerted on a trapped particle in both the axial and transverse directions. A comparison of the measured forces with the RO and EM model predictions can reveal the presence of other forces, such as radiometric forces, if there are any discrepancies beyond those explained by experimental error.

As the trapping forces for small particles are known to be a function of laser spot size, particular attention was directed to the measurement of the actual spot size formed by the trap in the object plane of the microscope. Our data indicate that the best objective in terms of the one that produces the smallest spot size is the Zeiss Neofluar 100×/1.3 numerical aperture. This is also confirmed by the data for the axial trapping efficiency, as shown in Table 1. We note that in the RO regime both the numerical aperture of the objective and the extent of the entrance aperture filling are critical parameters. For example, one can tailor the light distribution by overfilling the objective or by using a different input mode to increase the value of the axial Q, as noted in Ref. 13. A substantial increase in axial Q also can be realized with high-numerical-aperture water-immersion objectives that give convergence angles out to ~ 70°.[13] The present measurement of the spot size assumed that the laser beam profile had circular symmetry. In addition, because the depth of field of a $\lambda/2n$ spot size is less than λ, our measurement may overestimate the actual spot size. This can be seen, for example, in Table 1, in the measurement of the spot size produced by the Plan 100×/1.25 objective with the internal aperture completely open. Improvement in the axial positioning of the knife edge by use of another piezoelectric transducer instead of the differential micrometer used in this study would improve the accuracy of the spot size measurement.

The choice of the objective used for optical trapping is not limited to the size of the spot formed in the optical field. Table 3 indicates that the axial trapping efficiency is also a strong function of the location of the focused spot with respect to the sample chamber cover slip. This is not surprising, because micro-

scope objectives intended for biological research usually are designed to image objects attached or very close to the coverslip to minimize spherical aberration. This work emphasizes the importance of using well-corrected optics and microscope objectives and trapping objects close to the coverslip surface, if one is to minimize the deleterious effect of spherical aberration on the trapping efficiency. However, predicting the effect of the microscope objective aberrations on the trapping efficiency is difficult and was not attempted here. A related drawback to the use of single-beam gradient-force traps is that they are limited in the distance by which the particle can be translated vertically, given the limited working distance of high-numerical-aperture objectives. The maximum depth of the chamber for trapping on an upright microscope is that distance for which the microsphere can still be grasped by the trap while it is resting on the bottom of the sample chamber

Our experimental results for the axial trapping efficiency support the model predictions that Q increases as r increases, as well as the increase in Q as the numerical aperture increases. Of the two parameters, both the theoretical calculations and the experimental data indicate that Q is most sensitive to the numerical aperture when the particle is large. In fact, it is the numerical aperture that determines whether or not trapping is possible. These results reinforce the conclusion that strong three-dimensional trapping requires the largest possible convergence angle for the RO regime and the smallest possible spot size for the EM regime, which implies well-corrected, high-numerical-aperture objectives. The dependence of Q on r becomes significant as the particle size approaches the wavelength, as previously observed.[15] This can be seen in Table 2 where a 5-μm-diameter microsphere has an axial Q ~29% less than the value obtained for a 20-μm-diameter microsphere. However, the Q for a 1-μm-diameter microsphere is more than an order of magnitude less than that produced by the 20-μm microsphere.

Trapping in the transverse direction was found to be relatively insensitive to the size of the trapped microsphere when compared with the results for the axial trapping. The measured trapping Q was found to vary by a factor of only ~2.7 (from 0.15 to 0.40) as the sphere diameter was varied from 1 to 20 μm (Table 4). Also, the convergence angle appears to play a less significant role in changing the transverse trapping Q when compared with axial trapping, as can be seen from the data in Table 5. For the transverse case, the trapping Q decreased by only a factor of ~ 1.1 for a change in the cone angle from 60° to 32°. This result is initially surprising, as it was previously predicted that the transverse Q should increase when the convergence angle increases.[13] However, this prediction was made for a particle in the transverse plane located through the center of the particle. Our measurement of the transverse Q involves dragging the particle through a fluid under conditions where the net axial force is zero. Thus the axial location of the particle changes in response to the applied forces, as previously noted.[13] The predicted transverse Q's for the microspheres measured herein were determined with the condition that the net axial force is zero. Under these conditions, the calculated transverse Q decreases by approximately 50% as the cone angle decreases from 60° to 32°. The final parameter examined, laser beam polarization, was found both theoretically and experimentally to have a significant effect on the trapping efficiency, as high as ~20% for 10-μm-diameter microspheres. In any case, even this difference can be eliminated if the laser beam is converted to a circularly polarized beam before it enters the microscope.

Differences between measured and predicted RO Q values, as shown in Tables 4 and 5, can be reconciled given our ~20% uncertainty in the force measurement (primarily a result of the laser power measurement) and the uncertainty in the actual distance of the sphere from the cover slip when it fell out of the tweezers. Actual measurement of the axial distance from the microsphere to the cover slip at dropout is critical, as indicated by Eq. (9). As an example, for a 20-μm-diameter sphere, the correction factors to the drag force are 1.61 for a distance of 5 μm, 1.39 for a distance of 10 μm, and 1.28 for a distance of 15 μm from the cover slip to the vertex of the sphere. Alternatively, transverse forces could be measured for spheres well below the cover slip surface; however, our data (Table 3) indicate that other factors such as spherical aberration become significant and complicate analysis of the data.

An obvious question arises as to which theoretical model is best for predicting the trapping efficiency Q. Our recent work[15] suggests that the RO and EM models have two regions of applicability. For large particles ($d \geq 10$ μm), the RO model has sufficient accuracy and is the simplest to use. When the particle diameter is less than 10 μm, the RO model becomes less accurate, and the more complex EM model must be used. Unfortunately, errors in the EM field calculations become prohibitive even for fifth-order field corrections when the particle diameter is larger than ~1 μm for a laser spot size on the order of $\lambda/2n$.[15] Thus, there is still a range of 1–10 μm of particle diameters for which neither the RO nor EM model is sufficiently accurate to describe the laser–particle interaction. More work is needed to assess the accuracy of the EM model in the regime ≤1 μm. For very small particles (<30 nm), a simple dipole model[1] appears to fit the EM model axial force calculations very well,[15] based on the r^3 force dependence.

As stated in the introduction, the primary interest in determining the trapping efficiency is to predict the magnitude of the trapping forces on biological objects. We note, however, that real biological specimens can deviate significantly from sphericity, and this may result in Q's smaller than those that might be measured or predicted for purely spherical objects

having the same size and relative refractive index. For example, when optical tweezers are used to trap isolated mitochondria from *Reticulomyxa* in water, a maximum transverse trapping force of 5.8×10^{-9} dyn was measured with 1 mW of laser power.[7] The corresponding transverse trapping efficiency Q was 0.013. In this cell, because the mitochondria are of ~0.3-μm diameter, their trapping Q is approximately 4× less than an equivalently sized polystyrene microsphere, assuming that the transverse Q scales as a function of r in the range of 0.1–0.5 μm. The forces required to hold human sperm in a three-dimensional optical trap have been measured to be in the range of 10–60 pN.[31] However, the corresponding Q's cannot be determined because the trapping power was not specified. These two measurements illustrate the diverse requirements that must be satisfied in the design of an optical trapping instrument. Sperm trapping requires a three-dimensional trap that has a strong backwards axial Q to prevent the sperm from escaping in the weakest (i.e., axial) direction. We conclude, therefore, that the numerical aperture is a critical design parameter for sperm trapping and should be optimized for that case. However, trapping a dielectric bead or mitochondria attached to a microtubule does not require a backward axial Q because the particle is already constrained in that direction. Instead, it may be more advantageous to increase the spot size so that the transverse Q is increased, for example, when trapping in the plane passing through the center of the trapped particle.[13]

Absorption sets the maximum power threshold for trapping, and, thus the maximum force that can be exerted on the specimen. The absorbed laser energy can be determined with the EM model presented here if we solve for the internal fields and determine the source term for heating. Once the source term is known as a function of position, the temperature rise can be predicted. For a mammalian cell, a temperature increase of ~8° from 37 °C will result in death. The temperature changes inside a cell could be monitored by measuring the temperature-dependent shift in emission wavelength of a previously calibrated dye molecule. The dye molecule could be targeted to various structures on and within the cell by appropriate choices of antibodies. Absorption of the trapping beam by the particle may also produce deleterious effects through photochemical processes. It has been suggested[32] that cellular damage may be the result of the pumping of the transition from ground state oxygen to singlet oxygen, creating a highly reactive and destructive free radical. Experiments to confirm this hypothesis, either by direct measurement of the singlet oxygen phosphorescence at 1.26 μm or by observation of biochemical or morphological changes in a trapped cell or organelle are still required.

Conclusions

In summary the strict ray optics (RO) model[13] for optical tweezers has been shown to be highly accurate for the prediction of axial forces and reasonably accurate for prediction in the case of transverse forces in the size regime > 10 μm. The RO model has a reduced accuracy below 10 μm; for 1-μm-diameter microspheres, the measured Q was *comparable* for transverse trapping and was lower by nearly a factor of 5 for axial trapping. The EM model, as presented, has some problems in the region > 1 μm and more work is needed to assess its accuracy in the size regime below 1 μm. Our parametric study of optical tweezers confirms that the axial trapping efficiency is strongly dependent on the numerical aperture of the trapping laser beam. Obtaining the largest numerical aperture, which implies large convergence angles and the smallest spot size, is of paramount concern when a robust three-dimensional optical trap is required. The sphere size also plays an important role in optical trapping, especially if there is a need to manipulate objects far below the surface of the cover slip. Transverse trapping in general does not exhibit the same degree of sensitivity to the parameters examined herein. For example, this study found only a variation by a factor of ~ 2.2 for a change in the sphere diameter by $20\times$ and a factor of ~ 1.1 for a change in the objective cone angle from 32°–60°. The general trends illustrated by the microsphere calculations and measurements are certainly applicable to biological objects, although the actual magnitudes for Q are likely to be less, possibly by as much as an order of magnitude when compared with similarly sized microspheres. Measurement of the parametric dependence of trapping for biological specimens, such as cells and cellular organelles, is still required to determine the magnitude of Q's applicable to these objects.

Appendix A

Outline of the Electromagnetic-Model Algorithm

To determine the trapping forces which result when a highly focused Gaussian laser beam is incident on a spherical particle, we carry out the following procedure: First, the force on the sphere is described in terms of an infinite series involving expansion coefficients that describe the incident laser beam and the interaction of the beam with the sphere. Next, the calculation of the expansion coefficients is described, along with the scheme used to perform the surface integration. Finally, the algorithm to calculate the Ricatti–Bessel functions is presented. The computations were performed on a Compaq Deskpro 386/25 microcomputer containing a dedicated floating-point processing board (Microway Number Smasher-860, Kingston, Mass.).

To calculate the force on the sphere, the Maxwell stress tensor is integrated over a closed surface surrounding the sphere[21]

$$\langle \mathbf{F} \rangle = \frac{1}{4\pi} \int_0^{2\pi} \int_0^{\pi} \left\langle \left\{ \varepsilon^{(I)} E_r \mathbf{E} + B_r \mathbf{B} - \frac{1}{2} (\varepsilon^{(I)} E^2 + B^2) \hat{r} \right\} \right\rangle \\ \times r^2 \sin\theta d\theta d\phi \bigg|_{r>a}, \qquad (A1)$$

where $\varepsilon^{(I)}$ is the permittivity of the surrounding medium, r is the radius of the closed surface, $\mathbf{E}$ and $\mathbf{B}$ are the fields at the surface of the sphere, and E_r and B_r are the radial field components. For $r \gg a$, where a is the radius of the sphere, an analytical solution to the surface integral can be obtained by substituting the series expressions for the electromagnetic fields given by Eqs. (A1–A12) in Ref. 21 into Eq. (A1). For the axial force, the result can be written in terms of the coefficients that describe the incident and scattered fields as[21]

$$\langle \mathbf{F}_z \rangle = -\frac{a^2 E_0^2 \alpha^2}{8\pi} \sum_{l=1}^{\infty} \sum_{m=-l}^{l} Im \left[l(l+2) \right. \\ \times \sqrt{\frac{(l-m+1)(l+m+1)}{(2l+3)(2l+1)}} (2\varepsilon^{(I)} c_{l+1,m} c_{lm}{}^* \\ + \varepsilon^{(I)} c_{l+1,m} C_{lm}{}^* + \varepsilon^{(I)} C_{l+1,m} c_{l,m}{}^* + 2d_{l+1,m} d_{lm}{}^* \\ + d_{l+1,m} D_{lm}{}^* + D_{l+1,m} d_{l,m}{}^*) + \sqrt{\varepsilon^{(I)}} m (2c_{l,m} d_{lm}{}^* \\ \left. + c_{l,m} D_{lm}{}^* + C_{l,m} d_{l,m}{}^*) \right], \qquad (A2)$$

where $\alpha = 2\pi n a / \lambda$, n is the refractive index of the surrounding medium, λ is the wavelength, and $*$ indicates the complex conjugate. The constant E_0^2 is given by

$$E_0^2 = \frac{16P}{\sqrt{\varepsilon^{(I)}} c \omega_0^2 (1 + s^2 + 1.5 s^4)}, \qquad (A3)$$

where ω_0 is the spot size of the laser beam, $s = 1/k\omega_0$, and k is the wave number. The coefficients C_{lm} and D_{lm} describe the incident fields, and c_{lm} and d_{lm} are coefficients that describe the interaction of the incident field with the sphere. The expressions for $\mathbf{F}_x$ and $\mathbf{F}_y$ are also provided in Ref. 21.

The C_{lm} and D_{lm} coefficients are determined by integrating the product of the radial incident electric or magnetic field and the spherical harmonic function over a spherical surface with radius equal to the sphere $(r = a)$ to yield[23]

$$C_{lm} = \frac{a^2}{l(l+1)\psi_l(k^{(I)}a)} \int_0^{2\pi} \int_0^{\pi} \sin\theta E_r^{(i)}(a, \theta, \phi) \\ \times Y_{lm}{}^*(\theta, \phi) d\theta d\phi, \qquad (A4)$$

$$D_{lm} = \frac{a^2}{l(l+1)\psi_l(k^{(I)}a)} \int_0^{2\pi} \int_0^{\pi} \sin\theta B_r^{(i)}(a, \theta, \phi) \\ \times Y_{lm}{}^*(\theta, \phi) d\theta d\phi, \qquad (A5)$$

where ψ_l is a Ricatti–Bessel function and $k^{(I)}$ is the wave number in the medium. The spherical harmonic function $Y_{lm}{}^*(\theta, \phi)$ was calculated according to an algorithm presented in Ref. 33. Cartesian components of the incident electric ($E^{(i)}$) and magnetic ($B^{(i)}$) fields were previously derived for the fifth-order case by Barton *et al.*[23] and are given by Eqs. (25)–(30) in Ref. 24. The surface integration was performed with an extended trapezoidal rule:

$$\int_{x_l}^{x_N} f(x)\mathrm{d}x = h\left[\frac{1}{2}f_1 + f_2 + f_3 + \cdots + f_{N-1} + \frac{1}{2}f_N\right]$$
$$+ O\left[\frac{(b-a)^3 f''}{N^2}\right], \tag{A6}$$

according to a previously described algorithm.[33] The advantage of this rule as compared with Gaussian quadrature is that the integration can be refined by doubling N without losing the benefit of previous calculations, while convergence of the integration is monitored. Convergence was obtained when the change between successive evaluations was less than 1×10^{-6}. For slowly converging functions, the integration was terminated when $N = 1000$.

The coefficients c_{lm} and d_{lm} that describe the scattered fields were then found by applying the boundary conditions for the tangential components of the electric and magnetic fields at the sphere surface. The coeficients are written in terms of Ricatti–Bessel functions and the coefficients that describe the incident field:

$$c_{lm} = \frac{\psi_l'(k^{(II)}a)\psi_l(k^{(I)}a) - \bar{n}\psi_l(k^{(II)}a)\psi_l'(k^{(I)}a)}{\bar{n}\psi_l(k^{(II)}a)\xi_l^{(1)'}(k^{(I)}a) - \psi_l'(k^{(II)}a)\xi_l^{(1)}(k^{(I)}a)}C_{lm}, \tag{A7}$$

$$d_{lm} = \frac{\bar{n}\psi_l'(k^{(II)}a)\psi_l(k^{(I)}a) - \psi_l(k^{(II)}a)\psi_l'(k^{(I)}a)}{\psi_l(k^{(II)}a)\xi_l^{(1)'}(k^{(I)}a) - \bar{n}\psi_l'(k^{(II)}a)\xi_l^{(1)}(k^{(I)}a)}D_{lm}, \tag{A8}$$

where $k^{(II)}$ is the wave number inside the sphere, and $\bar{n}$ is equal to $(n_2/n_1)^{1/2}$. The Ricatti–Bessel functions $\xi_1(X) = Xh_l^{(1)}(X)$ and $\psi_l(X) = Xj_l(X)$, along with their derivatives (where j_l and $h_l^{(1)}$ are spherical Bessel functions of the first and third kind, respectively) were calculated with a previously defined algorithm[34] by the method of upward recurrence, because the recurrence relationship

$$\psi_{n+1}(X) = \frac{2n+1}{X}\psi_n(X) - \psi_{n-1}(X) \tag{A9}$$

is stable up to $l_{\max}$, which is given by

$$l_{\max} = X + 4X^{1/3} + 2, \tag{A10}$$

where the size parameter $X = 2\pi na/\lambda$. The same value of $l_{\max}$ also governed the number of terms summed in Eq. (A2).

We thank A. Ashkin for helpful discussions regarding the calculation of forces using the ray-optics model and S. Schaub for advice regarding the electromagnetic field calculations. We also are indebted to K. Svoboda for pointing out the correction to the viscous drag force when moving particles close to a boundary surface. This work was supported by grants from the Whitaker Foundation, National Science Foundation (BIR-9121325), National Institutes of Health (5P41 RR01192-12), U.S. Department of Energy (DE-FG03-91ER61227), U.S. Department of Defense Office of Naval Research ONR (N00014-91-C-0134), and by the Beckman Laser Institute endowment.

References

1. A. Ashkin, J. M. Dziedzic, J. E. Bjorkholm, and S. Chu, "Observation of a single-beam gradient force optical trap for dielectric particles," Opt. Lett. **11**, 288–290 (1986).
2. A. Ashkin and J. M. Dziedzic, "Optical trapping and manipulation of viruses and bacteria," Science **235**, 1517–1520 (1987); A. Ashkin, J. M. Dziedzic, and T. M. Yamane, "Optical trapping and manipulation of single cells using infrared laser beams," Nature (London) **330**, 769–771 (1987).
3. T. N. Buican, M. J. Smith, H. A. Crissman, G. C. Salzman, C. C. Stewart, and J. C. Martin, "Automated single-cell manipulation and sorting by light trapping," Appl. Opt. **26**, 5311–5316 (1987).
4. R. Steubing, S. Cheng, W. H. Wright, Y. Numajiri, and M. W. Berns, "Laser-induced cell fusion in combination with optical tweezers: the laser–cell fusion trap," Cytometry **12**, 505–510 (1991).
5. M. W. Berns, W. H. Wright, B. J. Tromberg, G. A. Profeta, J. J. Andrews, and R. J. Walter, "Use of a laser-induced optical force trap to study chromosome movement on the mitotic spindle," Proc. Natl. Acad. Sci. USA **86**, 4539–4543 (1989); A. Ashkin and J. M. Dziedzic, "Internal cell manipulation using infrared laser traps," Proc. Natl. Acad. Sci. USA **86**, 7914–7918 (1989).
6. H. Liang, W. H. Wright, S. Cheng, W. He, and M. W. Berns, "Micromanipulation of chromosomes in PTK2 cells using laser microsurgery (optical scalpel) in combination with laser-induced optical forces (optical tweezers)," Exp. Cell Res. **204**, 110–120 (1993).
7. A. Ashkin, K. Schultze, J. M. Dziedzic, U. Euteneuer, and M. Schliwa, "Force generation of organelle transport measured *in vivo* by an infrared laser trap," Nature (London) **348**, 346–348 (1990).
8. S. C. Kuo and M. P. Sheetz, "Force of single kinesin molecules measured with optical tweezers," Science **260**, 232–234 (1993).
9. S. Block, L. S. B. Goldstein, and B. J. Schnapp, "Using optical tweezers to investigate kinesin-based motility *in vitro*," J. Cell Biol. **109**, 81a (1989).
10. W. H. Wright, G. J. Sonek, Y. Tadir, and M. W. Berns, "Laser trapping in cell biology," IEEE J. Quantum Electron. **26**, 2148–2157 (1990).
11. T. C. Bakker Schut, G. Hesselink, B. G. de Grooth, and J. Greve, "Experimental and theoretical investigations on the validity of the geometric optics model for calculating the stability of optical traps," Cytometry **12**, 479–485 (1991).
12. K. Visscher and G. J. Brakenhoff, "Single-beam optical trapping integrated in a confocal microscope for biological applications," Cytometry **12**, 486–491 (1991).
13. A. Ashkin, "Forces of a single-beam gradient laser trap on a dielectric sphere in the ray optics regime," Biophys. J. **61**, 569–582 (1992).
14. S. Sato, M. Ohyumi, H. Shibata, H. Inaba, and Y. Ogawa, "Optical trapping of small particles using a 1.3-μm compact InGaAsP diode laser," Opt. Lett. **16**, 282–284 (1991).
15. W. H. Wright, G. J. Sonek, and M. W. Berns, "Radiation trapping forces on microspheres with optical tweezers," Appl. Phys. Lett. **63**, 715–717 (1993).

16. S. Chu, "Laser manipulation of atoms and particles," Science **253**, 861–866 (1991).

17. G. Roosen, "La levitation optique de spheres," Can. J. Phys. **57**, 1260–1279 (1979).

18. G. Roosen, "A theoretical and experimental study of the stable equilibrium positions of spheres levitated by two horizontal laser beams," Opt. Commun. **21**, 189–194 (1977).

19. H. C. van de Hulst, *Light Scattering by Small Particles* (Dover, New York, 1981), Sec. 4.5.

20. J. S. Kim and S. S. Lee, "Scattering of laser beams and the optical potential well for a homogeneous sphere," J. Opt. Soc. Am. **73**, 303–312 (1983).

21. J. P. Barton, D. R. Alexander, and S. A. Schaub, "Theoretical determination of net radiation force and torque for a spherical particle illuminated by a focused laser beam," J. Appl. Phys. **66**, 4594–4602 (1989).

22. M. Born and E. Wolf, *Principles of Optics* (Pergamon, Oxford, 1987), Sec. 13.5.1.

23. J. P. Barton, D. R. Alexander, and S. A. Schaub, "Internal and near-surface electromagnetic fields for a spherical particle irradiated by a focused laser beam," J. Appl. Phys. **64**, 1632–1639 (1988).

24. J. P. Barton and D. R. Alexander, "Fifth-order corrected electromagnetic field components for a fundamental Gaussian beam," J. Appl. Phys. **66**, 2800–2802 (1989).

25. R. S. Afzal and E. B. Treacy, "Optical tweezers using a diode laser," Rev. Sci. Instrum. **63**, 2157–2163 (1992).

26. H. Misawa, M. Koshioka, K. Sasaki, N. Kitamura, and H. Masuhara, "Three-dimensional optical trapping and laser ablation of a single polymer latex particle in water," J. Appl. Phys. **70**, 3829–3836 (1991).

27. J. B. Bateman, E. J. Weneck, and D. C. Eshler, "Determination of particle size and concentration from spectrophotometric transmission," J. Colloid Sci. **14**, 308–329 (1959).

28. A. H. Firester, M. E. Heller, and P. Sheng, "Knife-edge scanning measurements of subwavelength focused light beams," Appl. Opt. **16**, 1971–1974 (1977).

29. Y. Suzaki and A. Tachibana, "Measurement of the μm-sized radius of Gaussian laser beam using the scanning knife edge," Appl. Opt. **14**, 2809–2810 (1975).

30. J. Happel and H. Brenner, *Low Reynolds Number Hydrodynamics with Special Applications to Particulate Media* (Prentice-Hall, Englewood Cliffs, N.J., 1965), p. 327.

31. E. M. Bonder, J. M. Colon, J. M. Dziedzic, and A. Ashkin, "Force production by swimming sperm: analysis using optical tweezers," J. Cell Biol. **111**, 421a (1990).

32. S. M. Block, "Optical tweezers: a new tool for biophysics," in *Noninvasive Techniques in Cell Biology*, J. K. Foskett and S. Grinstein, eds. (Wiley-Liss, New York, 1990), pp. 375–402.

33. W. H. Press, B. P. Flannery, S. A. Teukolsky, and W. T. Vetterling, *Numerical Recipes: The Art of Scientific Computing* (Cambridge U. Press, New York, 1986).

34. C. F. Bohren and D. R. Huffman, *Absorption and Scattering of Light by Small Particles* (Wiley, New York, 1983), p. 478.

Optical trapping of dielectric particles in arbitrary fields

Alexander Rohrbach and Ernst H. K. Stelzer

European Molecular Biology Laboratory, Meyerhofstrasse 1, Postfach I0.2209, D-69117 Heidelberg, Germany

Received May 8, 2000; revised manuscript received October 17, 2000; accepted November 6, 2000

We present a new method to calculate trapping forces of dielectric particles with diameters $D \leq \lambda$ in arbitrary electromagnetic, time-invariant fields. The two components of the optical force, the gradient force and the scattering force, are determined separately. Both the arbitrary incident field and the scatterer are represented by plane-wave spectra. The scattering force is determined by means of the momentum transfer in either single- or double-scattering processes. Therefore the second-order Born series is evaluated and solved in the frequency domain by Ewald constructions. Numerical results of our two-force-component approach and an established calculation method are compared and show satisfying agreement. Our procedure is applied to investigate axial trapping by focused waves experiencing effects of aperture illumination and refractive-index mismatch. © 2001 Optical Society of America

OCIS codes: 290.0290, 180.0180, 260.0260.

1. INTRODUCTION

In 1970 Ashkin demonstrated that particles can be accelerated and trapped by radiation pressure.[1] Sixteen years later, he proved that three-dimensional trapping of a dielectric particle is possible with use of a single, highly focused laser beam.[2] Since then optical tweezers have become an indispensable tool for manipulating small particles without any mechanical contact. At a wavelength of ~ 1.06 μm, small probes inside a cell or biological objects such as cells or cell organelles can be held and moved by exertion of forces as low as several piconewtons without damaging the cell. Applications are, e.g., trapping of viruses and bacteria,[3] force measurements of molecular motors,[4–8] induced cell fusion,[9] studies of chromosome movement,[10] and tweezing and cutting in immunology and molecular genetics.[11] Small spheres held by optical tweezers have been used as probes to scan their two-dimensional[12] or three-dimensional environment in optical or photonic force microscopy,[13–15] where fluorescence emission and coherent scattering of the particle are exploited to determine the probe's position.[16]

The interaction of the particle with the impinging light is defined by a momentum transfer that is due to scattering, absorption, or emission of photons and by the fact that electric dipoles are drawn toward the highest amplitude of an electromagnetic field (dipole force). These forces were described theoretically already in 1909 by Debye,[17] who applied the theory of Mie[18] to describe the radiation pressure on spheres. Generally, these forces are summarized by the Maxwell stress tensor, which is a derivation of the conservation laws of electromagnetic energy and momentum.[19,20] However, a separation into the more popular terms scattering force and gradient force seems to be desirable. To the authors' knowledge a derivation of these two forces from basic electromagnetic equations has not been presented thus far.

First attempts to describe the forces on dielectric spheres in a focused laser beam theoretically were made by applying a ray-optics approach, which is valid for sphere diameters of $D \approx 10\lambda_0$ and larger (λ_0 is vacuum wavelength). This calculation using Snell's law and the Fresnel formulas does not take different sphere sizes into account since the phase front of the incident spherical wave is assumed not to change its shape in the focus.[21] Extending this approach by using an incident Gaussian beam, a similar ansatz delivers reasonable results[22,23] and fits well with experimental data. Gaussian optics satisfy the Helmholtz equation for small beam-divergence angles [i.e., when $\sin(x) \approx x$] and were used for weak focusing of a laser beam[24] to calculate the radiation pressure with Mie theory. Gaussian beam optics improved by higher-order corrections[25] were used in an electromagnetic-field model to calculate trapping forces of highly focused laser beams.[26,27] Here expansion coefficients were calculated for the radial components of the incident and the scattered fields (expressed by Ricatti–Bessel functions known from Mie theory), which are used to determine trapping forces by means of the Maxwell stress tensor. This model delivers good results as long as the sphere diameter is not larger than λ_0. However, it is restricted in applicability because realistic incident-field distributions cannot be simulated (e.g., the influence of the objective lens aperture, which cuts off the tails of a highly focused Gaussian beam or phase distortions such as spherical aberrations). These are reasons for the discrepancy between measured data and results simulated with the procedure outlined above.[28,29]

It should be noted, however, that in practice the majority of trapable particles such as macromolecules, some cell components, and small latex beads are smaller than 0.5 μm. Thus Harada and Asakura[30] investigated the range of validity for Rayleigh particles where $D \ll \lambda_0$ is required. Since the particles scatter isotropically, a simple dipole model for the scattering force can be used. For a strongly focused Gaussian beam, the optical forces are determined mainly by the gradient force, which draws the particle to the point of the highest beam intensity. When a Gaussian standing wave is used, even weak focusing can be used to achieve a stable trap.[31,32]

For larger diameters and higher refractive indices of the particle, the maximum phase shift inside the object increases and scattering at the particle becomes anisotropic. The influence of the phase shift is considered by the Rayleigh–Debye theory (also referred to as Rayleigh–Gans approximation or first Born approximation), which was treated in detail by Kerker.[33]

In this paper we extend the Rayleigh–Debye theory to include second-order scattering, which is equivalent to the well-known Born series expansion in quantum mechanics. A similar approach was pursued by Shifrin,[34] Aquista,[35] and Colak et al.,[36] where electromagnetic scattering of focused Gaussian beams at optical tenuous particles was investigated. The second-order term considers a stronger interaction between incident field and particles and thus allows the treatment of larger particles. We show furthermore that the scattering force and the gradient force can be derived from the Maxwell equations without requiring the detour to the Maxwell stress tensor. We show that adding the scattering force and the gradient force leads to nearly the same result as determining the force by means of the stress-tensor and Mie theory, although our two force components were calculated separately and plane-wave expansions (Fourier theory) instead of Ricatti–Bessel functions (Mie theory) were taken.

The main advantage of our approach is that optical forces of arbitrary wave fields on arbitrary-shaped dielectric particles can be calculated as long as the maximum phase shift $k_0(n_s - n_m)2a$ produced by the particle is smaller than $\pi/3$. This is still valid, for example, for a particle with radius $a = \lambda/(2n_m)$ and index $n_s = 1.57$ in water ($n_m = 1.33$). The approach includes objects with smooth edges or a filamentous structure. In addition, we can make use of scatter spectra to determine the three-dimensional position of the scatterer in the focus with nanometer resolution.[16,37]

This study is split into sections such that the general problem of force calculations, which is normally solved by means of the stress tensor, can now be divided into separate modules. The main modules are the descriptions of highly focused fields, scattering at dielectric particles, and the total optical force obtained by two components. The modules are described in such a way that each can be replaced by another calculation method without destroying the idea of our approach. All equations can also be solved analytically.

In Section 2 we explain how arbitrary electromagnetic fields can be calculated with the concept of Ewald spheres and pupil functions in the frequency domain. In Section 3 the scattering force and the gradient force are derived from the electromagnetic force acting on a dipole. The temporal behavior of the momentum vector is considered. In Section 4 we treat the polarizability of a dielectric particle. Section 5 explains the scattering and the momentum transfer at a dielectric particle. We solve the Born series to the second order in the frequency domain, still maintaining the concept of Ewald spheres (Ewald constructions). In Section 6 we show computed scattering efficiencies, intensity line scans of focal distributions, the

axial behavior of the trapping force, and trapping efficiencies for different sphere sizes and refractive indices. In Section 7 we briefly consider changes in phase and amplitude of the incident wave which affect the trapping efficiency. In Section 8 we present our conclusions.

2. GENERATION OF ARBITRARY FOCAL DISTRIBUTIONS

The vectorial wave equation can be derived by combining the two time-dependent Maxwell equations. In this study the MKS system is used. Separating the time-harmonic $\exp(i\omega t)$ leads to the Helmholtz equation:

$$\nabla^2 \mathbf{E}(\mathbf{r}) + n(\mathbf{r})^2 k_0^2 \mathbf{E}(\mathbf{r}) = 0. \tag{1}$$

Here $k_0 = 2\pi/\lambda_0$ is the vacuum wave number with vacuum wavelength λ_0 and $n(\mathbf{r})^2 = \mu_r \varepsilon_r \approx \varepsilon_r$ is the refractive-index distribution. The speed of light in the medium is $v = c/n = (\mu_0 \varepsilon_0 \mu_r \varepsilon_r)^{-1/2}$. The relative magnetic permeability is $\mu_r \approx 1$ for optical materials. The relative electric permittivity is generally complex $\varepsilon_r = \varepsilon' + i\varepsilon''$. In this study no absorption is assumed and ε_r is real ($\varepsilon'' = 0$).

The space-variant refractive index can be split into a constant term n_m of a homogeneous medium and a spatially varying term $\Delta n(\mathbf{r})$, hence

$$n(\mathbf{r})^2 = n_m^2 - \Delta n(\mathbf{r})^2. \tag{2}$$

Consider first a homogeneous medium where $\Delta n(\mathbf{r})^2 = 0$. Define furthermore both $\mathbf{k} = (k_x, k_y, k_z)$ as the wave vector and $k_n = nk_0$ as the wave number in a medium with arbitrary refractive index n. The Fourier transform of the electric-field vector is

$$\widetilde{\mathbf{E}}(k_x, k_y, k_z) = \iiint \mathbf{E}(x, y, z)$$
$$\times \exp[i(k_x x + k_y y + k_z z)]\mathrm{d}x\,\mathrm{d}y\,\mathrm{d}z. \tag{3}$$

Using the Fourier relationship $\nabla^2 \mathbf{E}(\mathbf{r}) \to -|\mathbf{k}|^2 \widetilde{\mathbf{E}}(\mathbf{k})$, we find the Fourier transform of the Helmholtz equation:

$$(|\mathbf{k}|^2 - k_n^2)\widetilde{\mathbf{E}}(\mathbf{k}) = 0 \;\Rightarrow\; |\mathbf{k}|^2 = k_x^2 + k_y^2 + k_z^2 = k_n^2. \tag{4}$$

This condition defines the Ewald sphere. It denotes that all solutions of the Helmholtz equation lie on a spherical surface with radius $k_n = n2\pi/\lambda_0$. Every point on the spherical surface $\widetilde{\mathbf{E}}(\mathbf{k}_p)\delta(|\mathbf{k}_p| - k_n)$ specifies the propagation direction of a plane wave with field strength $\widetilde{\mathbf{E}}(k_{xp}, k_{yp}, k_{zp})$.

Let θ be the angle to the optical axis that defines the propagation direction along the z or k_z direction. The resulting relations

$$k_\perp = (k_x^2 + k_y^2)^{1/2} = k_n \sin(\theta), \tag{5a}$$

$$k_z = (k_n^2 - k_x^2 - k_y^2)^{1/2} = k_n \cos(\theta), \tag{5b}$$

will be used throughout the paper. The case $k_\perp < k_n$ defines propagating waves, and the case $k_\perp > k_n$ defines evanescent waves.

The three-dimensional Fourier transform $\widetilde{\mathbf{E}}(k_x, k_y, k_z)$ of an electric field satisfying relation (4) can be expressed in terms of an angular spectrum representation $\widetilde{\mathbf{E}}_{\pm}(k_x, k_y)$, where the subscripts define the propagation direction along the positive or negative z axis:

$$\widetilde{\mathbf{E}}(k_x, k_y, k_z) = \widetilde{\mathbf{E}}_+(k_x, k_y)\delta[k_z - (k_n^2 - k_x^2 - k_y^2)^{1/2}]$$
$$+ \widetilde{\mathbf{E}}_-(k_x, k_y)\delta[k_z + (k_n^2 - k_x^2 - k_y^2)^{1/2}].$$
(6)

Thus the angular spectrum can be written as

$$\widetilde{\mathbf{E}}_+(k_x, k_y) = \int_0^\infty \widetilde{\mathbf{E}}(k_x, k_y, k_z)\mathrm{d}k_z,$$

$$\widetilde{\mathbf{E}}_-(k_x, k_y) = \int_{-\infty}^0 \widetilde{\mathbf{E}}(k_x, k_y, k_z)\mathrm{d}k_z.$$
(7)

According to McCutchen[38] and Goodman,[39] the field distribution in the focal region of an ideal lens can be obtained by taking the three-dimensional Fourier transform of a segment of the Ewald sphere:

$$\mathbf{E}_f(x, y, z)$$

$$= \frac{1}{(2\pi)^3}\int\int\int \mathbf{A}[k_x', k_y', (k_n'^2 - k_x'^2 - k_y'^2)^{1/2}]$$

$$\times \exp(-i\mathbf{k'r})\mathrm{d}k_x'\mathrm{d}k_y'\mathrm{d}k_z'.$$
(8)

The segment is described by the pupil function $\mathbf{A}(k_x', k_y', k_z')$, which is the generalized aperture. The prime denotes a scaling in k space with $s = 2\mathrm{NA}/\lambda$ such that $(k_x', k_y', k_z') = (sk_x, sk_y, sk_z)$. Hence the extent of $\mathbf{A}(\mathbf{k})$ on the sphere scales with the numerical aperture $\mathrm{NA} = n_m\sin(\theta_{max})$ of the lens, and the shape of $\mathbf{A}(\mathbf{k})$ describes the field distribution in the aperture plane of the lens. $\mathbf{A}(\mathbf{k})$ can be separated further into functions $E_0(\mathbf{k})$, $T(\mathbf{k})$, $\mathbf{P}(\mathbf{k})$, and $B(\mathbf{k})$ that respectively describe the field strength, the transmission, the polarization, and the apodization of the incident electric field at the aperture. Thus, we write for the pupil function with $k_\perp \leq k_0 \mathrm{NA}$,

$$\mathbf{A}(\mathbf{k}) = E_0(\mathbf{k})T(\mathbf{k})\mathbf{P}(\mathbf{k})B(\mathbf{k})$$

$$= \mathbf{A}_+(k_x, k_y)\delta[k_z - (k_n^2 - k_x^2 - k_y^2)^{1/2}].$$
(9)

The apodization function that obeys the sine condition is written as

$$B(\mathbf{k}) = (k_n/k_z)^{1/2} = \cos(\theta)^{-1/2}.$$
(10)

The polarization function $\mathbf{P}(\mathbf{k}) = [P_x(\mathbf{k}), P_y(\mathbf{k}), P_z(\mathbf{k})]$ describes the components of the electric field vector of a polarized beam that passes a deflecting optical element. According to Mansuripur,[40,41] and Sheppard and Larkin[42] the three components of $\mathbf{P}(\mathbf{k})$ for a linearly x-polarized beam can be expressed by

$$P_x(\mathbf{k}) = (k_x^2 k_z/k_n + k_y^2)/k_\perp^2,$$
(11a)

$$P_y(\mathbf{k}) = -k_x k_y[1 - (k_z/k_n)]/k_\perp^2,$$
(11b)

$$P_z(\mathbf{k}) = k_x/k_n.$$
(11c)

Figure 1 shows the electric field's x and z components, which are the Fourier transforms of the components $A_x(\mathbf{k})$ and $A_z(\mathbf{k})$ of the pupil function. An example: For

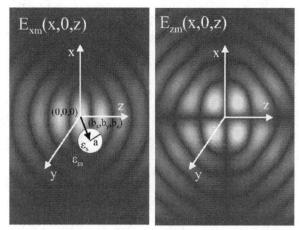

Fig. 1. Coordinate system of propagation and scattering: a dielectric spherical object with permittivity ε_s and radius a is located in a medium with permittivity ε_m. The object is shifted by the vector $\mathbf{b} = (b_x, b_y, b_z)$ from the origin $(0, 0, 0)$. The electric field's x component $\mathbf{E}_{xm}$ (left) and z component $\mathbf{E}_{zm}$ (right) of a highly focused wave, traveling along the z direction are shown in the background. Bright areas indicate a larger field amplitude.

a numerical aperture of $\mathrm{NA} = 0.9n_m$ and an x-polarized field, the energy of the z component is 18% of the total energy in the focal plane, and the y component's energy is $<1\%$. The smaller the NA, the weaker the strength of the z component (see also link [43] for further illustration).

The components of the polarization function for the magnetic fields are the same as for the electric field but rotated by $\pi/2$ about the propagation axis. In the following we confine ourselves to an incident x-polarized beam. The complex transmission function $T(\mathbf{k})$ describes any change of transmission at the aperture that is due to amplitude or phase filters.

The field $\mathbf{E}_f(x, y, z = f)$ in the focal plane can be easily computed by applying a Fourier transform of the pupil plane decribed by the function $\mathbf{A}_+(k_x, k_y)$. Starting with $\mathbf{E}_f(x, y, z = f)$, the field $\mathbf{E}(x, y, z = f + \Delta z)$ can be calculated by multiplying a phase function in the frequency domain:[39]

$$\widetilde{\mathbf{E}}(k_x, k_y, f + \Delta z) = \widetilde{\mathbf{E}}(k_x, k_y, f)\exp[-i\Delta z k_z(k_x, k_y)].$$
(12)

The phase function on the right-hand side acts as a propagator, and k_z is defined by Eq. (5b). Multiplying the phase function in the frequency domain corresponds to a convolution with a Huygens wavelet in the space domain (which is a spherical wave in the far field).[44]

3. CALCULATION OF OPTICAL FORCES

The electromagnetic force density $\mathbf{f}(\mathbf{r}, t)$ acting on a small polarizable particle or volume element with dipole moment density $\mathbf{P}(\mathbf{r}, t)$ is given by

$$\mathbf{f}(\mathbf{r}, t) = [\mathbf{P}(\mathbf{r}, t)\nabla]\mathbf{E}_m(\mathbf{r}, t) + \frac{\partial\mathbf{P}(\mathbf{r}, t)}{\partial t} \times \mathbf{B}_m(\mathbf{r}, t),$$
(13)

where $\mathbf{E}_m$ and $\mathbf{B}_m$ are the electric and the magnetic fields in a homogeneous medium.[45] This expression is derived from the total electromagnetic force on a charged particle $\mathbf{f} = q(\mathbf{E} + \mathbf{v} \times \mathbf{B})$.

The dipole moment density is the polarization of the particle (in units of As/m^2):

$$\mathbf{P}(\mathbf{r}, t) = \varepsilon_m \alpha \mathbf{E}_m(\mathbf{r}, t) \qquad (14)$$

is assumed to be proportional to the incident field $\mathbf{E}_m$ and the polarizability α. The permittivity in the medium is $\varepsilon_m = n_m^2 \varepsilon_0$. The tensor α will be treated in more detail later. Using the vector identity $(\mathbf{E}\nabla)\mathbf{E} = \nabla(\mathbf{E}^2/2) - \mathbf{E} \times (\nabla \times \mathbf{E})$ and the Maxwell equation $\nabla \times \mathbf{E} = -\partial/\partial t \mathbf{B}$, we can rewrite Eq. (13) as

$$\mathbf{f}(\mathbf{r}, t) = \alpha \varepsilon_m (\nabla |\mathbf{E}_m(\mathbf{r}, t)|^2/2)$$

$$+ \alpha \varepsilon_m \frac{\partial}{\partial t} [\mathbf{E}_m(\mathbf{r}, t) \times \mathbf{B}_m(\mathbf{r}, t)]. \qquad (15)$$

The vector product of the magnetic and the electric field is proportional to the Poynting vector

$$\mathbf{S}(\mathbf{r}, t) = n_m \varepsilon_0 c^2 [\mathbf{E}_m(\mathbf{r}, t) \times \mathbf{B}_m(\mathbf{r}, t)] \qquad (16)$$

that describes the energy transport of light in propagation direction. The time-average modulus of the Poynting vector is the intensity $I(\mathbf{r})$, which describes the number of photons per time and area in units of voltamperes per square meter:

$$I(r) = \langle |\mathbf{S}(\mathbf{r}, t)| \rangle_T = n_m \varepsilon_0 c^2/2 |\mathbf{E}_m(\mathbf{r}) \times \mathbf{B}_m(\mathbf{r})|$$

$$= n_m \varepsilon_0 c/2 |\mathbf{E}_m(\mathbf{r})|^2. \qquad (17)$$

Photons carry this energy as well as a momentum $\mathbf{m}$ per volume at velocity c/n_m. The momentum density (Abraham's form) given in units of nanoseconds per cubic meter is[45,46]

$$\mathbf{m}(\mathbf{r}, t) = \mathbf{S}(\mathbf{r}, t)/c^2 = \varepsilon_m/n_m [\mathbf{E}_m(\mathbf{r}, t) \times \mathbf{B}_m(\mathbf{r}, t)]. \qquad (18)$$

Having Eq. (15) at the back of the mind, we ask the following important question: What is the meaning of $\partial/\partial t \mathbf{m}(\mathbf{r}, t)$? Several authors[30,47] have interpreted $\partial \mathbf{m}/\partial t = c^{-2}\partial \mathbf{S}/\partial t$ as the temporal change of intensity. Hence $\partial \mathbf{m}/\partial t$ would be zero for a time-invariant illumination. However, this is true in free space but not in the presence of a scatterer. The change of momentum density per unit time is affected by the change of the k vectors when light is scattered. According to Gordon,[45] the change of momentum density can be regarded as the difference of two processes:

$$\Delta \mathbf{m}(\mathbf{r}) = [\langle \mathbf{m}(\mathbf{r}, t_{\text{after}}) \rangle_T - \langle \mathbf{m}(\mathbf{r}, t_{\text{before}}) \rangle_T]. \qquad (19)$$

The time indices indicate the state after and before the scattering process. $\Delta \mathbf{m}(\mathbf{r}, t) = \Delta \mathbf{m}(\mathbf{r})$ will be temporally constant if $\langle (\mathbf{E}_m \times \mathbf{B}_m) \rangle_T = (1/2)[\mathbf{E}_m(\mathbf{r}) \times \mathbf{B}_m(\mathbf{r})]$ does not change during Δt. The resulting force $\mathbf{f} = \Delta \mathbf{m}/\Delta t$ per volume element will not change during Δt, either. The particle will be accelerated as a reaction to the acting force. With increasing speed of the particle, the momentum transfer and the force on the particle will decrease. However, this effect is negligible for most applications.

Averaging the force density $\mathbf{f}(\mathbf{r}) = \langle \mathbf{f}(\mathbf{r}, t) \rangle_T$ over one temporal period $T = 1/\omega$ and using Eqs. (16) and (17), we can write Eq. (15) in the form

$$\mathbf{f}(\mathbf{r}) = \frac{\alpha n_m}{2c} \nabla I_m(\mathbf{r}) + \alpha n_m \frac{\Delta m(\mathbf{r})}{\Delta t} = \mathbf{f}_{\text{grad}}(\mathbf{r}) + \mathbf{f}_{\text{scat}}(\mathbf{r}). \qquad (20)$$

The two components of the total force per unit volume are commonly called the gradient force, resulting from the intensity gradient, and the scattering force, which describes the momentum transfer that is due to scattering. The strength of both forces is dependent on the polarizability α of the particle. Both forces stem from the total electromagnetic force on a dipole introduced in Eq. (13), whose two terms do not correspond to the gradient force and the scattering force, respectively.

4. POLARIZABILITY OF A DIELECTRIC PARTICLE

Consider now a small homogeneous nonabsorbing dielectric body with an extent much smaller than the wavelength. The volume of the particle is V, its permittivity is $\varepsilon_s = \varepsilon_0 n_s^2$, and that of the immersion medium is $\varepsilon_m = \varepsilon_0 n_m^2$ (see Fig. 1). The polarizability α is a scalar for isotropic dielectric bodies. For a sphere with radius a in a homogeneous field $\mathbf{E}_m$, the dipole moment $\mathbf{p}$ is parallel to $\mathbf{E}_m$ and can be calculated by means of the polarization $\mathbf{P} = (\varepsilon_s - \varepsilon_m)\mathbf{E}_{\text{int}}$. The internal field $\mathbf{E}_{\text{int}}$ of a spherical dielectric body is proportional to the incident field $\mathbf{E}_m$:[48,49]

$$\mathbf{E}_{\text{int}} = [(3\varepsilon_m)/(\varepsilon_s + 2\varepsilon_m)]\mathbf{E}_m. \qquad (21)$$

Integration of $\mathbf{P}$ over the spherical volume $V_s = (4/3)a^3\pi$ delivers the total dipole moment $\mathbf{p}$ of the small body

$$\mathbf{p} = \int \mathbf{P}dV = 4\pi a^3 \varepsilon_m [(\varepsilon_s - \varepsilon_m)/(\varepsilon_s + 2\varepsilon_m)]\mathbf{E}_m. \qquad (22)$$

For each volume element this relation describes the interaction of adjacent molecules leading to a dipole moment $\mathbf{p} = V_s \varepsilon_m \alpha \mathbf{E}_m$ induced by the incident field. Hence α is

$$\alpha = 3(\varepsilon_s - \varepsilon_m)/(\varepsilon_s + 2\varepsilon_m) = 3(m^2 - 1)/(m^2 + 2). \qquad (23)$$

This is the Clausius–Mossotti relation (also Lorentz–Lorenz formula) describing the influence of the field induced by adjacent dipoles in the low-frequency case in a dielectric medium with permittivities ε_s and ε_m. Here $m = n_s/n_m$ denotes the relative refractive index of the dielectric. It has been shown by Lalor and Wolf[50] and Born and Wolf[61] that interference of the incident field E_m with the total effective field inside the dielectric will cancel out the incident field (Ewald–Oseen extinction theorem) and results in the internal field, which is proportional to the polarizability described by the Lorentz–Lorenz formula.

Among others, Draine and Flatau[52] demonstrated that the Lorentz–Lorenz formula can be extended in its applicability such that it is also valid for larger dielectric bodies of arbitrary shape. This is possible with the discrete-dipole approximation, which replaces the dielectric continuum by a finite lattice of polarizable points. The method can be further improved with a coupled-multipole formulation, where higher-order coupling of electric and magnetic dipoles inside the particle is considered.[53]

In the following we will make use of the formulation in Eq. (23) for the polarizability of a dielectric body of arbitrary shape.

5. MOMENTUM TRANSFER DUE TO SCATTERING

The scattering-force density $\mathbf{f}_{scat}(\mathbf{r})$ per volume element was introduced to describe the momentum transfer from the incident field $\mathbf{E}_m(\mathbf{r})$ to the scattered field $\mathbf{E}_s(\mathbf{r})$, which is characterized by the shape and the refractive index n_s of the scatterer. A change of direction or length of an incident k vector—which is related to the momentum vector $\mathbf{k}_i \hbar$ ($\hbar = $ Planck constant$/2\pi$)—is the result of a single- or a multiple-scattering process. The momentum transfer $\mathbf{g} \cdot \hbar$ can be described as the difference between the incident vector and the scattered k vector:

$$\mathbf{g} = \mathbf{k}_s - \mathbf{k}_i, \qquad \mathbf{k}_s = \mathbf{k}_s^{(1)} + \mathbf{k}_s^{(2)} + \dots . \quad (24)$$

The scattered k vector can be a result of multiple-scattering events. Since every k vector represents a single plane wave, it is our aim to describe the total scattering of an arbitrary incident wave at an arbitrary-shaped particle by using plane-wave spectra. The momentum spectra for electromagnetic fields in homogeneous space lie on Ewald spheres.

By inserting Eq. (2) into Eq. (1), we find for the total electric field $\mathbf{E}(\mathbf{r})$ with the space-variant refractive-index step $\Delta n(\mathbf{r})^2 = \Delta n^2 s(\mathbf{r}) = (n_m^2 - n_s^2)s(\mathbf{r})$

$$(\nabla^2 + n_m^2 k_0^2)\mathbf{E}(\mathbf{r}) = \Delta n^2 s(\mathbf{r})k_0^2 \mathbf{E}(\mathbf{r})$$
$$= (n_m^2 - n_s^2)s(\mathbf{r})k_0^2 \mathbf{E}_{int}(\mathbf{r})$$
$$= \alpha s(\mathbf{r})n_m^2 k_0^2 \mathbf{E}(\mathbf{r}). \quad (25)$$

The shape function $s(\mathbf{r})$ defines the three-dimensional shape of the homogeneous dielectric scatterer. The inhomogeneity on the right-hand side is the polarization induced inside the volume of $s(\mathbf{r})$, thus defining the total dipole moment of the dielectric particle. Since the right-hand side of the first line is zero outside the region of the scatterer, the total field $\mathbf{E}(\mathbf{r})$ can be replaced by the internal field $\mathbf{E}_{int}(\mathbf{r})$ [see Eq. (21) for the homogeneous field]. Owing to this and according to Eq. (14), the strength of that internal polarization is proportional to αn_m^2. The inhomogeneous Helmholtz equation can be transformed into a Fredholm integral equation of the second kind:

$$\mathbf{E}(\mathbf{r}) = \mathbf{E}_m(\mathbf{r}) + \alpha k_n^2 \int s(\mathbf{r}')\mathbf{E}(\mathbf{r}')G(\mathbf{r}, \mathbf{r}')d^3\mathbf{r}'$$
$$= \mathbf{E}_m(\mathbf{r}) + \mathbf{E}_s(\mathbf{r}). \quad (26)$$

The total field $\mathbf{E}(\mathbf{r})$ consists of the unperturbated field $\mathbf{E}_m(\mathbf{r})$ (homogeneous solution) and the scattered field $\mathbf{E}_s(\mathbf{r})$ (inhomogeneous solution). The free-space Green's function in the MKS system $G(\mathbf{r}, \mathbf{r}') = \exp(ik_n|\mathbf{r} - \mathbf{r}'|)/(4\pi|\mathbf{r} - \mathbf{r}'|)$ is a shift-invariant spherical wave. An approach similar to that of Eq. (26) was used by Shifrin,[34] Aquista,[35] and Colak *et al.*[36] to calculate scattered electromagnetic fields. Since $\alpha/4\pi$ is a small quan-

tity (the factor $1/4\pi$ stems from the Green's function), we solve Eq. (26) by successive iterations. Constructing a power series for the total field $\mathbf{E}(\mathbf{r})$, we set

$$\mathbf{E}(\mathbf{r}) = \sum_{p=0}^{\infty} \alpha^p \mathbf{E}_s^{(p)}(\mathbf{r})$$
$$= \mathbf{E}_s^{(0)}(\mathbf{r}) + [\alpha \mathbf{E}_s^{(1)}(\mathbf{r}) + \alpha^2 \mathbf{E}_s^{(2)}(\mathbf{r}) + \dots]. \quad (27)$$

Here the order p denotes the number of scattering processes experienced by $\mathbf{E}(\mathbf{r})$. Comparing Eqs. (26) and (27) we find the 0th-order scattered field $\mathbf{E}_s^{(0)}$ as the incident field $\mathbf{E}_m$. The scattered field can be written explicitly as follows:

$$\mathbf{E}_s(\mathbf{r}) = \alpha k_n^2 \int s(\mathbf{r}')\mathbf{E}_s^{(0)}(\mathbf{r}')G(\mathbf{r}, \mathbf{r}')d^3r'$$
$$+ \alpha^2 k_n^2 \int s(\mathbf{r}')\mathbf{E}_s^{(1)}(\mathbf{r}')G(\mathbf{r}, \mathbf{r}')d^3r' + \dots . \quad (28)$$

Taking only the first-order scattered field $\mathbf{E}_s^{(1)}$ and neglecting orders $p > 1$ of the series is a valid approach for two cases: The first is the case of a weak scatterer, i.e., the index n_s is close to n_m and thus the polarizability is small ($\alpha \ll 1$). This is known as Rayleigh–Debye (also Rayleigh–Gans) scattering[54] as well as the Born approximation. The second case is that of a dielectric interface between two media with different refractive indices, leading to a single change of the momentum vector for arbitrary real indices ($n_s - n_m$).[55,56] In this case the first order of the Born series is exact.

To determine the scattered field iteratively, we perform a three-dimensional Fourier transform of Eq. (28). The result is

$$\widetilde{\mathbf{E}}_s(k_x, k_y, k_z)$$
$$= \frac{\alpha k_n^2}{(2\pi)^3} \widetilde{G}(\mathbf{k}) \int \widetilde{\mathbf{E}}_s^{(0)}(\mathbf{k}')\widetilde{s}(\mathbf{k} - \mathbf{k}')d^3\mathbf{k}'$$
$$+ \frac{\alpha^2 k_n^2}{(2\pi)^3} \widetilde{G}(\mathbf{k}) \int \widetilde{\mathbf{E}}_s^{(1)}(\mathbf{k}')\widetilde{s}(\mathbf{k} - \mathbf{k}')d^3\mathbf{k}' + \dots, \quad (29)$$

with the Green's-function Fourier transform $\widetilde{G}(\mathbf{k}) = 1/(|\mathbf{k}|^2 - k_n^2)$. Using the Weyl expansion,[57,58] we can rewrite $\widetilde{G}(\mathbf{k})$ in a manner that describes the Green's function acting in the positive or negative z direction (see Appendix A):

$$\widetilde{G}_{\pm}(k_x, k_y, k_z) = \frac{i2\pi^2 \delta[(k_n^2 - k_x^2 - k_y^2) \mp k_z]}{\pm(k_n^2 - k_x^2 - k_y^2)^{1/2}}. \quad (30)$$

The first-order scattered field can be expressed equivalently to Eq. (7) by a forward/backward (subscripts $+/-$, respectively) scattered angular spectrum:

$$\widetilde{\mathbf{E}}_{s\pm}^{(1)}(k_x, k_y)$$

$$= \frac{\pm i k_n^2}{4\pi (k_n^2 - k_x^2 - k_y^2)^{1/2}} \times \int\int \widetilde{\mathbf{E}}_+^{(0)}(k_x', k_y')$$

$$\cdot \widetilde{s}[k_x - k_x', k_y - k_y', \pm(k_n^2 - k_x^2 - k_y^2)^{1/2}$$

$$- (k_n^2 - k_x'^2 - k_{yz}'^2)^{1/2}] dk_x' dk_y' \qquad (31)$$

Here the spectrum of the first-order scattered field $\mathbf{E}_s^{(1)}$ at $z = 0$ is obtained through a convolution of the spectrum of the incident field [see Eq. (7)] and the Fourier transform of the shape function. This convolution is two-dimensional since $k_z = (k_n^2 - k_x^2 - k_y^2)^{1/2}$. The sifting property of the δ function was used two times after integration over both k_z and k_z'.

After inserting expression (31) into the second-order scattered field $\widetilde{E}_s^{(2)}(\mathbf{k})$ and making use of the δ function in $\widetilde{G}_\pm(k)$, we get (see Appendix A)

$$\widetilde{\mathbf{E}}_{s\pm}^{(2)}(k_x, k_y) = \frac{\mp k_n^4}{(4\pi)^2 (k_n^2 - k_x^2 - k_y^2)^{1/2}}$$

$$\times \int\int\int\int \frac{\widetilde{\mathbf{E}}_+^{(0)}(k_x'', k_{yz}'')}{(k_n^2 - k_x''^2 - k_y''^2)^{1/2}}$$

$$\cdot (\widetilde{s}(\mathbf{k}' - \mathbf{k}_+'') - \widetilde{s}(\mathbf{k}' - \mathbf{k}'')) dk_x'' dk_y''$$

$$\cdot \widetilde{s}(\mathbf{k} - \mathbf{k}_\pm') dk_x' dk_y', \qquad (32)$$

where

$$\mathbf{k}' - \mathbf{k}_\pm'' = [k_x' - k_x'', k_y' - k_y'', (k_n^2 - k_x'^2 - k_y'^2)^{1/2}$$

$$\mp (k_n^2 - k_x''^2 - k_y''^2)^{1/2}]$$

and $\mathbf{k} - \mathbf{k}_\pm'$, respectively.

If the scatterer is a homogeneous sphere of radius a with shape function $s(x, y, z) = \text{step}[a - (x^2 + y^2 + z^2)^{1/2}]$ [step($\mathbf{r}$) is the Heaviside step function], then the Fourier transform of $s(\mathbf{r})$ is

$$\widetilde{s}(k_x, k_y, k_z) = 3[\sin(ka) - (ka)\cos(ka)]/(ka)^3 V(a), \qquad (33)$$

where $k = |\mathbf{k}| = (k_x^2 + k_y^2 + k_z^2)^{1/2}$ and $V(a) = (4/3)\pi a^3$ is the volume of the sphere. A particle that is shifted from the origin by the vector $\mathbf{b} = (b_x, b_y, b_z = 0)$ has a laterally modulated spectrum:

$$\int\int\int s(\mathbf{r} - (b_x, b_y, 0)) \exp(i\mathbf{k}\mathbf{r}) d^3 r$$

$$= \widetilde{s}(\mathbf{k}) \exp[i(k_x b_x + k_y b_y)]. \qquad (34)$$

If the particle is shifted in the z direction, its spectrum will not be modulated in z, but the incident wave will be propagated in z by the distance b_z according to Eq. (12). This now means that the spectrum of the incident wave is modulated in z.

The shape of $\widetilde{s}$ (here it is a spherical Bessel function) in the (x, z) plane is indicated as the background of Fig. 2. $\widetilde{s}$ represents the probability density for all possible momentum transfers $\mathbf{g} = \mathbf{k}_s - \mathbf{k}_i$ defined by the three-dimensional refractive-index distribution of the scatterer. From this spectrum, physically reasonable entries are selected by the intersection with Ewald spheres.

The interpretation of Eqs. (31) and (32) is illustrated in Fig. 2. Equation (31) describes a convolution of two spherical surfaces. This means that the first-order scattered spectrum is obtained by shifting an Ewald sphere in the direction of the incident wave vector and storing all points of intersection of the δ surface and the object spectrum $\widetilde{s}(\mathbf{k})$. This is done for all incident plane-wave components weighted with $\widetilde{\mathbf{E}}_m(\mathbf{k}_i)$. The incident spectrum is limited by focusing, which means that the segment of the Ewald sphere has the half-angle $\sin(\theta) = \text{NA}/n_m$. An additional convolution is performed in Eq. (32), where each scattering direction with strength $\widetilde{E}(-\mathbf{k}_i)\widetilde{s}(\mathbf{k}_s^{(1)} - \mathbf{k}_i)$ is modified by another change of direction and strength, resulting in $\widetilde{E}(-\mathbf{k}_i)\widetilde{s}(\mathbf{k}_s^{(1)} - \mathbf{k}_i)\widetilde{s}(\mathbf{k}_s^{(2)} - \mathbf{k}_s^{(1)})$.

In general, when a linearly x-polarized beam is focused, the components of the electric field $E_{mx}(\mathbf{r})$, $E_{my}(\mathbf{r})$, and $E_{mz}(\mathbf{r})$ are all nonzero. However, even for the technically strongest possible focusing with $\sin(\theta) \approx 0.95$ [see Eq. (5a)], the contribution $\int\int |E_{my}(\mathbf{r})|^2 dx dy$ is <1% and can be neglected. The remaining x and z components of $\mathbf{E}_m$ (see Fig. 1) make a dipole oscillating inside the scatterer in both the x and the z direction. Hence a dipole oriented in the x direction emits radiations with angular strengths $[(k_n - k_x^2)^{1/2}/k_n] = \cos(\theta)$ and a dipole oriented in the z direction with strength $[(k_n - k_z^2)^{1/2}/k_n] = \sin(\theta)$ (θ is the angle to the z axis). The two parts $\widetilde{\mathbf{E}}_s^{\to x}(k_x, k_y)$ and $\widetilde{\mathbf{E}}_s^{\to z}(k_x, k_y)$ of the scattered field are generated by the incident components $[\widetilde{E}_{mx}, 0, 0]$ and $[0, 0, \widetilde{E}_{mz}]$. The two scattered fields again have two vec-

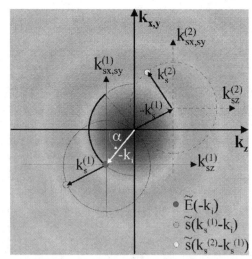

Fig. 2. First (second-) order momentum transfer from a dielectric sphere onto the incoming field (onto the first-order scattered field). The spectrum of plane waves $\widetilde{E}(-\mathbf{k}_i)$ of the incoming field is indicated by the bold circle segment with half-angle $\sin(\alpha) = \text{NA}/n_m$. The momentum spectrum of the sphere $\widetilde{s}(k_x, k_y, k_z)$ is shown as the background, where dark indicates a larger transfer probability. The first-order momentum transfer for a single plane wave obeys $\mathbf{g} = \mathbf{k}_s^{(1)} - \mathbf{k}_i$, with transfer strength $\widetilde{E}(-\mathbf{k}_i)\widetilde{s}(\mathbf{k}_s^{(1)} - \mathbf{k}_i)$; the second-order momentum transfer is given by $\mathbf{k}_s^{(2)} - \mathbf{k}_s^{(1)}$, with transfer strength $\widetilde{E}(-\mathbf{k}_i)\widetilde{s}(\mathbf{k}_s^{(1)} - \mathbf{k}_i)\widetilde{s}(\mathbf{k}_s^{(2)} - \mathbf{k}_s^{(1)})$. Owing to their constant length, all wave vectors $\mathbf{k}_s$ end on Ewald spheres, all having the same radius (coherent scattering). This is indicated by a solid circle for first-order scattering and by a dashed circle for second-order scattering.

tor components $\widetilde{\mathbf{E}}_s^{\to x} = [\cos(\theta_x)\widetilde{E}_s^{\to x}, 0, \sin(\theta_x)\widetilde{E}_s^{\to x}]$ and $\widetilde{\mathbf{E}}_s^{\to z} = [\cos(\theta_x)\widetilde{E}_s^{\to z}, 0, \sin(\theta_x)\widetilde{E}_s^{\to z}]$, which change according to the angle θ_x between k_x and k_z. They are added coherently as

$$\widetilde{\mathbf{E}}_s(k_x, k_y) = \widetilde{\mathbf{E}}_s^{\to x}(k_x, k_y)D_x(k_x, k_y)$$
$$+ \widetilde{\mathbf{E}}_s^{\to z}(k_x, k_y)D_z(k_x, k_y), \quad (35)$$

with polarization functions of a dipole $D_x = (k_n - k_x^2)^{1/2}/k_n$ and $D_z = (k_n - k_z^2)^{1/2}/k_n$. The parts $\widetilde{\mathbf{E}}_s^{\to x}$ and $\widetilde{\mathbf{E}}_s^{\to z}$ are given by a series as introduced in Eq. (29). For a weakly focused beam, i.e., where the paraxial approximation is valid [$\sin(\theta) < 0.5$], even the component $\widetilde{\mathbf{E}}_s^{\to z}$ can be neglected.

The scattered intensity $\widetilde{I}_s(k_x, k_y)$ in the direction $\mathbf{k} = [k_x, k_y, (k_n^2 - k_x^2 - k_y^2)^{1/2}]$ is obtained by addition of all scattering terms as follows:

$$\widetilde{I}_s = \varepsilon_0 c |\widetilde{\mathbf{E}}_s^{\to x} D_x + \widetilde{\mathbf{E}}_s^{\to z} D_z|^2$$
$$= \varepsilon_0 c (|\widetilde{E}_{sx}^{\to x} D_x + \widetilde{E}_{sx}^{\to z} D_z|^2 + |\widetilde{E}_{sz}^{\to x} D_x + \widetilde{E}_{sz}^{\to z} D_z|^2)$$
$$= \varepsilon_0 c |(\alpha \widetilde{\mathbf{E}}_s^{(1)} + \alpha^2 \widetilde{\mathbf{E}}_s^{(2)} + \ldots)^{\to x} D_x$$
$$+ (\alpha \widetilde{\mathbf{E}}_s^{(1)} + \alpha^2 \widetilde{\mathbf{E}}_s^{(2)} + \ldots)^{\to z} D_z|^2. \quad (36)$$

The intensity at the focus of the incident beam is $I_m = \varepsilon_0 c |\mathbf{E}_m(0,0,0)|^2$ and $\widetilde{I}_s^+(k_x, k_y) = c\varepsilon_0 |\widetilde{\mathbf{E}}_s^+(k_x, k_y)|^2$. Integrating the normalized modulus square of the scattered field $\widetilde{I}_s(k_x, k_y)/I_m$ over the whole 4π space or, equivalently, both $\widetilde{I}_s^+(k_x, k_y, k_z > 0)/I_m$ and $\widetilde{I}_s^-(k_x, k_y, k_z < 0)/I_m$ over their respective 2π half-spaces yields the scattering cross section C_{sca}, which describes the area in which energy is scattered. It is given by

$$C_{sca} = \frac{1}{I_m k_n^4} \int_{-k_n}^{k_n} \int_{-k_n}^{k_n} [\widetilde{I}_s^+(k_x, k_y)$$
$$+ \widetilde{I}_s^-(k_x, k_y)] \frac{k_n}{(k_n^2 - k_x^2 - k_y^2)^{1/2}} dk_x dk_y$$
$$= \frac{1}{I_m k_n^2} \int_0^{2\pi} \int_0^{\pi/2} I_s^+(\theta, \varphi)\sin(\theta) d\theta d\varphi + \frac{1}{I_m k_n^2}$$
$$\times \int_0^{2\pi} \int_{\pi/2}^{\pi} I_s^-(\theta, \varphi)\sin(\theta) d\theta d\varphi. \quad (37)$$

In the first line, integration is performed in the Cartesian Fourier space, where the integration limits are $\pm k_n = \pm k_0 n_m$ and the Jacobi determinant is $k_n/k_z = k_n/(k_n^2 - k_x^2 - k_y^2)^{1/2} = 1/\cos(\theta)$. This corresponds to an integration over all entries on the Ewald sphere with radius k_n. Entries in the region $|\mathbf{k}| > k_n$ are zero and do not contribute to the integration in the first line of Eq. (37). Integration can also be performed in polar coordinates as described in the second line of Eq. (37).

The fraction of incident energy that is scattered by a particle is described by the scattering efficiency Q_{sca}. It is defined as the scattering cross section C_{sca} divided by the geometrical cross section C_{geo} (which is $a^2 \pi$ for a sphere):

$$Q_{sca} = C_{sca}/C_{geo}. \quad (38)$$

Please note that Q_{sca} is not the efficiency of the scattering force F_{scat}, which would be $Q_{scat}(\mathbf{r}) = F_{scat}(\mathbf{r}) \cdot n_m \cdot P/c$. The momentum $\hbar\langle k_{sz}\rangle$ carried by the scattered wave in the z direction is the mean value of all k_z components weighted by the values of $\widetilde{I}_s(k_x, k_y)$ on the Ewald sphere. After normalizing by C_{sca}, one finds

$$\langle k_{sz}\rangle = \langle k_{tz}\rangle - \langle k_{rz}\rangle$$
$$= \frac{1}{C_{sca}I_m k_n^4} \int_{-k_n}^{k_n} \int_{-k_n}^{k_n} k_z(k_x, k_y)[\widetilde{I}_s^+(k_x, k_y)$$
$$+ \widetilde{I}_s^-(k_x, k_y)] \frac{k_n}{(k_n^2 - k_x^2 - k_y^2)^{1/2}} dk_x dk_y. \quad (39)$$

Since the scattered wave can be split into a (t)ransmitted and a (r)eflected part, the momentum also can be split into $\langle k_{tz}\rangle$ and $\langle k_{rz}\rangle$. The term $\langle k_{sz}\rangle/k_n = \langle (k_n^2 - k_{sx}^2 - k_{sy}^2)^{1/2}\rangle/k_n = \langle\cos(\theta_s)\rangle$ is called the asymmetry parameter. θ_s denotes the scattering angle.

The total momentum transfer in the z direction is the momentum difference between the incident and the scattered wave. This is proportional to

$$C_{pr,z} = C_{sca}\langle g_z\rangle \frac{1}{k_n} = C_{sca}[\langle k_{iz}\rangle - (\langle k_{tz}\rangle - \langle k_{rz}\rangle)] \frac{1}{k_n}, \quad (40)$$

which is equivalent to the more common formulation of radiation pressure $C_{pr,z} = \langle\cos(\theta_i)\rangle C_{ext} - \langle\cos(\theta_s)\rangle C_{sca}$, which utilizes the cross sections for pressure, extinction, and scattering. The momentum transfer in the lateral directions $\langle g_x\rangle$ and $\langle g_y\rangle$ can be determined in the same manner as in Eqs. (39) and (40). The momentum transfer for each photon, in units of $1.054 \times 10^{-34} N/s$, in an arbitrary direction is then described as

$$\mathbf{g}\hbar = (\langle g_x\rangle \mathbf{e}_x + \langle g_y\rangle \mathbf{e}_y + \langle g_z\rangle \mathbf{e}_z)\hbar, \quad (41)$$

where $\mathbf{e}_x$, $\mathbf{e}_y$ and $\mathbf{e}_z$ are unit direction vectors. For an incident-field distribution and a scatterer located on axis that are both symmetric in x or y, we find $\langle g_x\rangle = 0$ or $\langle g_y\rangle = 0$, respectively.

Let the power $P_{sca} = I_m C_{sca}([P] = Nm/s)$ describe the total energy that is scattered per unit time. The energy of one photon is $ck_0\hbar$, so the rate of photons scattering is then $P_{sca}/(ck_0\hbar)$. Hence the total momentum transfer per unit time describes the total force acting on a particle that is due to scattering:

$$\mathbf{F}_{scat} = \frac{P_{sca}\mathbf{g}\hbar}{ck_0\hbar} = \frac{n_m}{c} P_{sca}\mathbf{g}/k_n. \quad (42)$$

The term $P_{sca}n_m/c$ describes the force acting in all directions, which results from the fraction α of the scattered field [see Eq. (26)]. The dimensionless term $(\mathbf{g}/k_n)$ describes the component of the force that acts in the direction of $\mathbf{g}$. Its magnitude includes the asymmetry parameter introduced above such that $|\mathbf{g}/k_n| = \langle\cos(\theta_i)\rangle - \langle\cos(\theta_s)\rangle$. If we now compare Eq. (42) with the second term on the right-hand side of Eq. (20), we identify the momentum transfer $P_{\text{sca}} \cdot \mathbf{g}$ with the change of the momentum vector $\mathbf{M} := \int \mathbf{m} dV = \int \mathbf{S}/c^2 dV$:

$$\mathbf{F}_{\text{scat}} = \frac{n_m}{c} P_{\text{sca}}(\mathbf{g}/k_n) = n_m \alpha \frac{\Delta M}{\Delta t}. \qquad (43)$$

Here the right-hand side implies that only the fraction α of photons has changed its direction out of the total number of photons that is proportional to both I_m and $|\mathbf{M}|$. The left-hand side expresses exactly the same with different physical parameters.

Hence the total electromagnetic force $\mathbf{F}(\mathbf{r})$ can be obtained by integrating over all volume elements dV inside the scatterer:

$$\mathbf{F}(\mathbf{r}) = \iiint\limits_V \mathbf{f}(\mathbf{r}')dV'$$

$$= \mathbf{F}_{\text{grad}}(\mathbf{r}) + \mathbf{F}_{\text{scat}}(\mathbf{r})$$

$$= \iiint\limits_V \frac{\alpha n_m}{2c} \nabla I_m(\mathbf{r}')dV' + \frac{n_m}{c} I_m C_{\text{sca}}(\mathbf{g}/k_n).$$

$$(44)$$

The gradient force $\mathbf{F}_{\text{grad}} = (F_{\text{grad},x}, F_{\text{grad},y}, F_{\text{grad},z})$ can easily be determined by computing the gradient in x, y, and z of the intensity $I_m(x, y, z)$ for each volume element and summing up all gradient values. This operation is performed separately for the intensities generated by the x and the z components of the electric field. Alternatively, $\mathbf{F}_{\text{grad}}$ can be obtained by applying the divergence theorem

$$\mathbf{F}_{\text{grad}}(\mathbf{r}) = \frac{\alpha n_m}{2c} \iiint\limits_V \nabla I_m(\mathbf{r}')dV'$$

$$= \frac{\alpha n_m}{2c} \oiint\limits_{\partial V} I_m(\mathbf{r}')\mathbf{n}_A dA, \qquad (45)$$

where $\mathbf{n}_A$ represents the normal on the surface element dA.

6. CALCULATION RESULTS

A. Comparison of Scattering Efficiencies

The accuracy of the scattering efficiency Q_{sca} computed with use of the second-order Born approximation was tested for two different refractive indices and for sphere radii from 100 to 800 nm. The results for Q_{sca} were compared with those obtained by Mie theory.[59] The incident plane wave has a vacuum wavelength of $\lambda_0 = 1.064\,\mu\text{m}$, and the immersion medium is water ($n_m = 1.33$). Hence the largest sphere diameter is two wavelengths λ_n. Figure 3 shows that the second-order Born approximation delivers good results up to $a = 450\,\text{nm}$ when the index of the scatterer is $n_s = 1.57$ and up to $a = 700\,\text{nm}$ when the index is $n_s = 1.46$. Please note that at $\lambda_0 = 1\,\mu\text{m}$, $n_s = 1.57$ is the index of polystyrene or latex and $n_s = 1.46$ is the index of fused silica.

As a next step we compared the optical force $F_{\text{scat}}(z = 0)$ due to momentum transfer from a weakly focused Gaussian wave computed by use of our approach with results for $F_{\text{scat}}(z)$ from Harada and Asakura[30] and Zemánek et al.[31] Good agreement was found (results not shown).

B. Comparison of Focal Distributions from Highly Focused Beams

To validate our two-component-force model for highly focused waves, we compared numerical results with those of an established theory that is able to treat the interaction of an incident electromagnetic wave with a dielectric scatterer. Such a theory is the generalized Lorenz–Mie theory (GLMT), which was developed by Barton et al.[26] to compute optical forces on spheres in focused Gaussian beams. The theory uses fifth-order corrected electromagnetic-field components for a Gaussian beam, which satisfy with sufficient accuracy both the Helmholtz equation and the boundary conditions at a spherical surface provided that the sphere radius is smaller than λ.

However, the diameter of a lens focusing a Gaussian beam is finite and thus cuts off the tails of the Gaussian beam. In the focus a pure Gaussian profile is not achieved and focal line scans reveal undulations in both axial and lateral directions. The undulations become stronger as the beam waist in the focus becomes narrower (e.g., by beam expansion at the aperture plane).

Focal intensity distributions were calculated by using the theory introduced in Section 1 of this paper. We computed Eq. (12) for $\Delta z = -5\,\mu\text{m}... + 5\,\mu\text{m}$ by using a 256^2 fast Fourier transform with a spatial sampling of 16 pixels per wavelength (per $800\,\text{nm} = \lambda_0/1.33$). The result was compared with a rigorous electromagnetic calculation by Török[60] for a NA = 1.2 lens obeying the sine condition. An exact agreement between the two calcula-

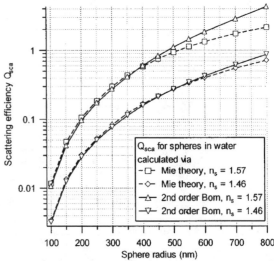

Fig. 3. Comparison of scattering efficiencies $Q_{\text{sca}} = C_{\text{sca}}/(a^2\pi)$ for different sphere radii a and refractive indices n_s computed with Mie theory and with the second-order Born approximation. The scattering efficiencies correspond to the scattering of a plane wave in water ($\lambda_0 = 1.064\,\mu\text{m}$, $n_m = 1.33$) by nonabsorbing dielectric spheres.

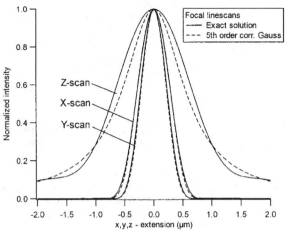

Fig. 4. Comparison of the intensity distribution of a highly focused, x-polarized beam calculated with fifth-order corrected Gaussian beam optics and with our exact method. The immersion medium is water ($n_m = 1.33$), the wavelength is $\lambda_0 = 1.064\,\mu$m. The line scans in x, y, and z correspond to an ideal focus of an NA = 1.2 lens. The circular lens aperture with radius R was illuminated by a Gaussian beam with waist $w = 0.8R$.

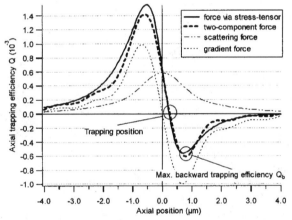

Fig. 5. Comparison of results derived by different calculation methods for axial forces as a function of axial position. A homogeneous dielectric particle ($a = 100$ nm, $n_s = 1.57$) is moved along the z axis through the focus of an x-polarized Gaussian beam with $w_{0x} = 0.59\,\mu$m. The wavelength is $\lambda_0 = 1.064\,\mu$m in vacuum, and the refractive index in the medium is $n_m = 1.33$. The solid curve shows the axial normalized trapping force Q (i.e., trapping efficiency) determined by Maxwells stress tensor and Mie theory; the dashed curve shows the result of our computation with the two-component procedure: It is the sum of the scattering (dotted–dashed curve) and the gradient (dotted curve) forces. Positive values of Q denote forces pushing the sphere in the propagation direction of the light; a negative Q denotes a force that pulls in the opposite direction.

tion results was established. Computation of spherical aberrated focii also yielded the same result for the two calculation methods.

Our computation method for beams of arbitrary shape—which we assume to be exact—is compared with the fifth-order Gaussian optics approach. Figure 4 shows line scans in x,y, and z through an x-polarized highly focused beam. First, the broadening in the x direction that is due to the z component of the electric field in the focus

can be established (see Fig. 1, where the modulus square of both components $\mathbf{E}_x$ and $\mathbf{E}_z$ must be added). Second, the agreement between the two calculation methods is quite good in the lateral directions but not in the axial direction. Although both axial line scans reveal the same width at the $1/e^2$ intensity, the intensity gradients differ significantly. In the plotted region of $\pm 2\,\mu$m, the relative difference between the two z gradients varies by as much as $\pm 60\%$. This has consequences for the gradient forces acting on particles.

C. Comparison with Force Calculations by Means of the Stress Tensor

To perform a useful comparison of forces obtained by GLMT and by the two-component model, we chose a Gaussian beam with a beam waist w_0 that was only 55% of the aperture radius (i.e., only a small part of the Gaussian tails are cut off). The focusing of the x-polarized beam with a NA = 1.2 lens generated a beam waist in the focus of $w_{0x} = 0.598\,\mu$m and $w_{0y} = 0.526\,\mu$m. The wavelength is $\lambda_0 = 1.064\,\mu$m in vacuum and $\lambda_n = 0.800\,\mu$m in water ($n_m = 1.33$). λ_0 and $n_m = 1.33$ are the parameters kept constant in all of the following investigations. The fifth-order corrected Gaussian beam as well as the force calculation through solution of the stress tensor were performed by Jonás and Zemánek (see Ref. 31).

To describe the trapping force $\mathbf{F}(\mathbf{r})$ independent of the total incident optical power P, we define the trapping efficiency $\mathbf{Q} = (Q_x, Q_y, Q_z)$ as follows:

$$\mathbf{Q}(\mathbf{r}) = \mathbf{F}(\mathbf{r})\frac{c}{Pn_m} = \mathbf{Q}_{\text{grad}}(\mathbf{r}) + \mathbf{Q}_{\text{scat}}(\mathbf{r}). \qquad (46)$$

The experimentally most interesting and critical direction in which trapping forces act is the axial z direction. Thus we compared the axial trapping efficiency $\mathbf{Q}(z) = \mathbf{e}_z Q(z)$ around the geometrical focus where $z = 0$.

Figure 5 shows the trapping efficiency $Q(z)$ of a spherical particle with radius $a = 100$ nm and index $n_s = 1.57$ that is moved through the focus along the z axis. Differences between the two thick curves arise from the fact that the dashed curve has slight undulations that stem from an incident-light distribution that is not purely Gaussian. This curve is the sum of the scattering and the gradient forces $F_{\text{scat}}(z)$ and $F_{\text{grad}}(z)$, which are also shown in Fig. 5. The scattering force is always positive and pushes the particle in the propagation direction of the incident light. The gradient force is negative behind the geometrical focus and pulls the sphere back toward the focus. A force equilibrium is reached behind the focus when $F_{\text{scat}}(z) + F_{\text{grad}}(z) = 0$; this position is called the trapping position. The total minimum of $Q(z) = F(z)c/(n_mP)$ is termed the maximum backward trapping efficiency Q_b.

Our model was further tested for different sphere radii and different refractive indices at a very narrow focus of the incident beam of $w_{0x} = 0.52\,\mu$m ($w_{0y} = 0.42\,\mu$m). This corresponds to an incident Gaussian beam in the aperture of which the waist is 80% of the aperture radius and focusing with a NA = 1.2 (intensity line scans through the focus are shown in Fig. 4). The range of the sphere radii is $a = 25\,$nm to $a = 500\,$nm, and the refractive indices are $n_s = 1.46$ and $n_s = 1.57$. The maximum phase shift $\Delta\phi$ experienced by the wave penetrating the scatterer is

$$\Delta\phi = 2a/\lambda_0(n_s - n_m)2\pi. \qquad (47)$$

The second-order scattering force was calculated only for $\Delta\theta > 0.01 \cdot 2\pi$; for $\Delta\phi$ smaller than $0.01 \cdot 2\pi$ a first-order calculation led to the same result and was therefore used to speed computations. The value $\Delta\phi = 0.01 \cdot 2\pi$ was found empirically and obeys approximately the condition for Rayleigh–Debye scattering $\Delta\phi \ll 1$.[54]

The absolute minimum of $Q(z)$, which is referred to as the maximum backward trapping efficiency $Q_b(z)$, characterizes the behavior of the force along z. This is the strongest (repulsive) force a particle can experience in the direction opposite to the light pressure. Figure 6 describes the change of $Q_b(z)$ as a function of particle radius for calculation by means of both the stress tensor and the two-component approach. The calculation results agree at $n_s = 1.46$ for all sphere sizes to within 10–20% but differ significantly for $n_s = 1.57$ and sphere radii larger than 0.3 μm.

7. APPLICATIONS

We now apply the algorithm to calculate trapping forces for typical problems that play an important role in trapping experiments.[15,29,61] Effects such as underillumination and overillumination of a lens aperture, refractive-index mismatch at an interface, and the asymmetric shape of a two-dimensional trapping potential are briefly outlined.

An objective lens (NA = 1.2, water immersion) fulfilling the sine condition $B(\mathbf{k}) = \cos(\theta)^{-1/2}$ [see Eq. (10) and Ref. 62] generates a focus with full width at half-maximum (FWHM) along x of FWHM$_x = 1.28\,\mu$m and

along y of FWHM$_y = 1.04\,\mu$m when the beam waist w_0 of the incident Gaussian beam is 80% of the aperture radius R ($\lambda_0 = 1.064\,\mu$m, x-polarized light). Overillumination of the aperture by a Gaussian beam with waist $w_0 = 2R$ leads to an even narrower focus with FWHM$_x = 1.18\,\mu$m and FWHM$_y = 0.88\,\mu$m. This results in a steeper intensity gradient $\nabla I(\mathbf{r})$ in all directions and therefore in a stronger gradient force $\mathbf{Q}_{\text{grad}}(\mathbf{r})$. The change in the total trapping efficiency relative to that of an incident beam with $w_0 = 0.8R$ is plotted for different sphere sizes in Fig. 7. For a refractive index $n_s = 1.46$ the more homogeneous illumination of the aperture leads to an increase of $Q_b(z)$ by 20–40%. For $n_s = 1.57$ the effect of the aperture overillumination depends more strongly on the sphere size: The increase is 20–40% for smaller spheres and as much as 60–80% for spheres with $a \geqslant 200\,$nm.

An arrangement leading to a deterioration of the trapping efficiency $Q(z)$ is the use of an oil-immersion lens for trapping in water or aqueous solution. The refractive-index mismatch between oil ($n_i = 1.52$) and water ($n_m = 1.33$) causes spherical aberrations (SA's) of the focus. The effect of SA gets worse for larger distances d between the focus and the mismatched interface. The spherical wave front incident on the interface experiences a phase disturbance $\Delta\psi(k_x, k_y, d)$, which can be described as follows:[60]

$$\Delta\psi = d\,2\pi/\lambda(n_m \cos\theta_m - n_i \cos\theta_i). \qquad (48)$$

This change of the phase $\Delta\psi$ is the change of the k_z components of the k vectors at the interface that is due to refraction: $\Delta k_z = k_{zm} - k_{zi}$. According to Snell's law, the lateral components $k_\perp$ ($\perp = x, y$) remain constant $k_{\perp i} = k_{\perp m}$. By adding the phase disturbance $\Delta\psi$ to the spherical phase of the incident wave and matching the angle-dependent transmission according to the Fresnel formulas, we computed the resulting focal distributions at

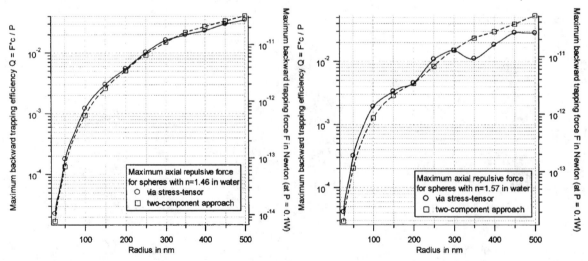

Fig. 6.   Maximum backward trapping efficiency and trapping force of a sphere as a function of the radius.   Spheres with indices $n = 1.46$ (fused silica) and $n = 1.57$ (latex) are trapped by a strongly focused x-polarized beam.   The Gaussian beam waist in x is $w_{0x} = 0.52\,\mu$m at a wavelength of $\lambda_n = 1.064\,\mu$m/1.33 = 0.800 μm in water.   The solid curves represent the result of a Mie theory computation; the dashed curves are results obtained with the two-component procedure.   Corresponding forces are in the piconewton range at a laser power of $P = 100\,$mW.

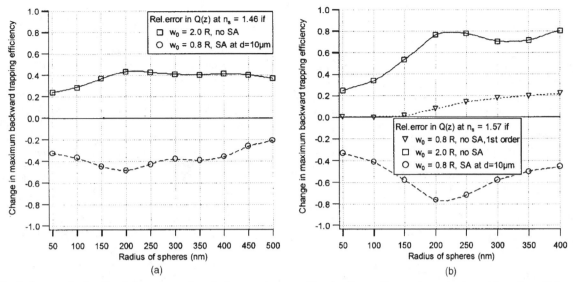

Fig. 7. Influence of refractive-index mismatch and of an aperture overillumination on the maximum backward trapping efficiency as a function of sphere radius. The difference is given relative to the case shown in Fig. 6 (a Gaussian illumination with a beam waist $w_0 = 0.8R$, no refractive-index mismatch). The curves are plotted as follows: For no SA, but a Gaussian illumination with beam waist $w_0 = 2.0R$ (solid curves) (R = aperture radius), for SA at $d = 10\,\mu m$, Gaussian illumination with beam waist $w_0 = 0.8R$ (dashed curves), and for beam waist $w_0 = 0.8R$, no SA, first-order Born approximation for the scattering force (dotted curve). Left, spheres with index $n_s = 1.46$; right, spheres with index $n_s = 1.57$.

$d = 10\,\mu m$. As a consequence of the disturbed wave, the intensity in the focal region is altered and the trapping efficiencies of the gradient and scattering forces change. The result is shown in Fig. 7 for different sphere sizes and two refractive indices. A deterioration of $Q(z)$ by 20–50% for the lower index and up to 80% for the higher refractive index is ascertainable mainly as a result of the smaller axial intensity gradient.

The effect of a weaker scattering force, which was determined only by the first-order Born approximation, is also shown in Fig. 7 on the right. Hence the influence of the gradient force is stronger, and this results in larger values for Q_b.

Optical trapping of a particle with a single highly focused beam is possible only by generating a force equilibrium in all directions. In other words, the particle must rest at the bottom of a potential well (provided that no other forces act on the particle). The depth of the well is proportional to the incident intensity, and the steepness and the shape of the potential walls are determined by the intensity gradient $\nabla I(\mathbf{r})$. The light pressure (the scattering) affects the potential in the direction of light propagation and decreases the height of the potential well in that direction.

The trapping potential $W(\mathbf{r})$ of a particle at position $\mathbf{r}$ is given by the integral over the scalar product of force and path. However, only $\mathbf{F}_{grad}(\mathbf{r})$ is a conservative force (the integral is path independent), whereas integration of $\mathbf{F}_{scat}(\mathbf{r})$ depends on the path. Since the most interesting behavior of trapping forces and potentials are those acting along the z direction, we calculated the two-dimensional axial trapping potential by integration along z:

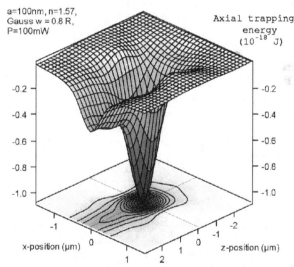

Fig. 8. Two-dimensional axial trapping potential for a spherical particle ($a = 100$ nm, $n_s = 1.57$) in a focus from an ideal lens with NA = 1.2. The asymmetry in the z direction can be seen clearly. To escape from the trap, a trapped particle must have an energy equal to the difference between its thermal energy and the energy minimum in the $z = +3\,\mu m$ plane. The strength of the potential is scaled in atto-Joule (aJ = 10^{-18} J) for an incident-light power of 100 mW.

$$W_z(x, 0, z) - W_z(x, 0, z_0)$$
$$= -\int_{z_0}^{z} [F_{grad,z}(x, 0, z') + F_{scat,z}(x, 0, z')]\mathrm{d}z'. \tag{49}$$

The reference position $\mathbf{r}_0 = (x, y, z_0)$ is chosen such that $W(\mathbf{r}_0) \to 0$; i.e., at a sufficiently large axial distance to the focus the optical force $\mathbf{F}(\mathbf{r}_0)$ on the particle is negligible. Figure 8 shows the computed resulting energy landscape for a spherical particle moving in the z direction through the focal region. The NA of the lens was 1.2, the radius of the particle was $a = 100\,\text{nm}$, and the refractive index was $n_s = 1.57$. The laser energy entering the immersion medium was 100 mW. The two-dimensional representation of the trapping potential gives an impression of the shape, height, and symmetry of the trap. To escape from the trap, a trapped particle must have an energy equal to the difference between its thermal energy and the energy minimum in the $z = +3$ μm plane.

8. SUMMARY AND CONCLUSION

We have shown in this paper how arbitrary electromagnetic fields can be expressed in the frequency domain by means of Ewald spheres. Following the approach of McCutchen,[38] the field in the focus of an ideal lens can be obtained by the Fourier transform of the pupil function. The latter can be generated by projecting the aperture function, the polarization function, and the apodization function onto an Ewald sphere. Three Ewald spheres are needed to represent the angular momentum spectrum for three polarization components of a focused wave.

We have derived a two-component expression for the total electromagnetic force acting on a dielectric particle. In contrast to other authors, we interpret the component called the scattering force as the temporal change of the momentum vector in the presence of a scatterer. The change in momentum is given by the length of the mean k vectors before and after scattering averaged over the incident and the scattered spectrum, respectively.

The scatter spectrum is obtained by first- and second-order Born scattering of the incident wave at the dielectric particle. The first- and second-order scatter spectra can be found iteratively by convolving Ewald spheres. The other component of the total force, called the gradient force, is computed by integrating (numerically) the local intensity gradient of each volume element inside the scatterer.

We showed that the second-order Born approximation yields results for the scattering efficiency Q_{sca}, which are close to those obtained by Mie theory for sphere sizes with a central phase shift of $\Delta\phi \leq 1$. For spheres with $\Delta\phi > 1$, higher orders of the Born series should improve accuracy. Whereas the second order overrates the total scattering for larger spheres (see Fig. 3), a third order should decrease the scattering efficiency—a possibility due to the coherent addition of fields obtained by each scattering order.

Intensity distributions of highly focused beams compare well: Line scans through the focus in the lateral direction reveal good agreement between the method presented in this paper and fifth-order corrected Gaussian optics. The width of the focus from a NA = 1.2 lens was $w_{0x} = 0.598\,\mu$m and $w_{0y} = 0.526\,\mu$m. However, the difference between these calculation methods for the gradient in the z direction varied by up to $\pm 60\%$ in the trapping region within 2 μm of the focus. This affected the gradient forces $F_{\text{grad}}(z)$ in nearly the same manner.

The total optical force computed with our method was compared with the electromagnetic forces obtained by solving the stress tensor with the generalized Lorenz–Mie theory (GLMT). The test was performed for sphere diameters from 50 nm to 1.0 μm and refractive indices of 1.46 and 1.57 at a wavelength of $\lambda_0 = 1.064\,\mu$m. The agreement between the two theories is satisfying.

For spheres with index $n_s = 1.46$ this difference of 10–20% is constant over the investigated size range of $a = 25$–500 nm. For the larger index and spheres with $a > 300$ nm, the maximum backward trapping efficiencies may differ by up to 100% ($a \approx 350$ nm). There are two possible explanations for this discrepancy: First, our method to determine the gradient force does not consider the change of the field inside the sphere beyond the first-order Born approximation and thus leads to incorrect intensity gradients for larger spheres ($a > 300$ nm, $n_s = 1.57$). Second, for larger sphere radii and higher refractive indices the second-order Born approximation yields good results for the scattering efficiency Q_{sca} but is inexact for specific angular scattering described by $I_s(\theta)$, especially for angles $\theta > 45°$.

The Born scattering series is not able to treat resonances, where a large number of interaction processes is necessary to accumulate much energy inside the scatterer. The second-order Born approximation is a sufficiently exact method to calculate scattering efficiencies Q_{sca} and radiation pressure efficiencies $Q_{\text{pr}} = Q_{\text{sca}}\langle\cos\theta\rangle$ for plane and focused waves when the phase shift by the particle is $\Delta\phi \leq 0.9$. On the other hand, the method of Barton et al.[26] delivers only approximate values for the maximum backward trapping efficiency, owing to an incorrect intensity gradient in z that results from the fifth-order Gaussian optics model.

To indicate the wide applicability of our two-component procedure, we investigated the influence of refractive-index mismatch at an interface and varying aperture illumination on the trapping efficiency. The altered focal distributions were computed, and trapping efficiencies were determined for different sphere sizes and refractive indices. We calculate the effect of underillumination and overillumination, which strongly influences the trapping efficiency (at constant power transmitted by the lens). When (experimentally determined) trapping efficiencies are compared, the degree of overillumination of the aperture should be stated, since the exact focal width is often difficult to measure. Moreover, in the presence of spherical aberrations, the trapping efficiency will decrease significantly in comparison with an unaberrated trapping focus. Finally, a two-dimensional trapping potential generated for a sphere with $a = 100$ nm in an ideal focus was shown.

To our knowledge, calculations treating effects such as these have not been presented before and cannot be modeled with the method of Barton et al.[26]

Although many papers have been published dealing with calculation of trapping forces and scattering of electromagnetic waves at all kind of particles, we have found none that could take typical experimental trapping conditions into account for dielectric particles with $D \leq \lambda$. Even though further efforts might be useful to improve the accuracy and computation speed of the procedure, the two-component approach presented in this study offers the opportunity to analyze a variety of applications, especially when Fourier spectra and Fourier relations can be brought into play. The following conclusions are of particular interest:

1. The splitting of the trapping force into two components delivers a better understanding of trapping efficiencies and offers the possibility of engineering diverse trapping potentials.

2. Plane-wave propagation and scattering by means of Ewald spheres allows the treatment of many kinds of incident waves and scatterers. Ewald constructions are illustrative and easy to interpret, especially with respect to momentum transfer.

3. The approach easily provides the far-field interference between the scattered and the unscattered wave, thus offering the possibility of regaining the exact object position.[16]

4. Computation is relatively easy for particles located off axis.

5. The two-component approach is extendable without major effort to absorbing or small metallic particles as well as to inhomogeneous scatterers.

APPENDIX A

1. The Green's Function in the Frequency Domain

To make use of the sifting property of the Dirac δ function, we derive an alternative expression for the Fourier transform of the free-space Green's function. The result is presented in Eq. (30). Using this form and integrating in the k_z direction, we can reduce the dimension of the field representation to a two-dimensional angular spectrum representation. The six-dimensional second-order scattering integral can be thus reduced to a four-dimensional integral.

The Green's function for the Helmholtz equation in the MKS system is defined through the relation

$$(\nabla^2 + k_n^2)G(\mathbf{r} - \mathbf{r}') = -\delta(\mathbf{r} - \mathbf{r}'). \qquad \text{(A1)}$$

After a Fourier transform,

$$(-\mathbf{k} \cdot \mathbf{k} + k_n^2)\widetilde{G}(\mathbf{k}) = -1, \qquad \text{(A2)}$$

we find directly that

$$\widetilde{G}(k_x, k_y, k_z) = \frac{1}{|\mathbf{k}|^2 - k_n^2} = \frac{1}{k_x^2 + k_y^2 + k_z^2 - k_n^2}. \qquad \text{(A3)}$$

Thus the Green's function with $R := |\mathbf{r} - \mathbf{r}'|$ can be expressed either by its inverse Fourier transform,

$$\frac{\exp(ik_n R)}{4\pi R} = \frac{1}{(2\pi)^3}\int\int\int \frac{1}{k_x^2 + k_y^2 + k_z^2 - k_n^2}$$
$$\times \exp[-i(k_x x + k_y y + k_z z)]\mathrm{d}k_x \mathrm{d}k_y \mathrm{d}k_z. \qquad \text{(A4)}$$

or by the Weyl expansion,

$$\frac{\exp(ik_n R)}{4\pi R} = \frac{i}{8\pi^2}\int\int \frac{\exp[-iz|(k_n^2 - k_x^2 - k_y^2)^{1/2}|]}{|(k_n^2 - k_x^2 - k_y^2)^{1/2}|}$$
$$\times \exp[-i(k_x x + k_y y)]\mathrm{d}k_x \mathrm{d}k_y. \qquad \text{(A5)}$$

Setting Eq. (A4) equal to Eq. (A5) and performing the Fourier transform in x and y yields

$$\frac{1}{(2\pi)}\int \frac{1}{k_x^2 + k_y^2 + k_z^2 - k_n^2}\exp[-ik_z z]\mathrm{d}k_z$$
$$= \frac{i}{2}\frac{\exp[-iz|(k_n^2 - k_x^2 - k_y^2)^{1/2}|]}{|(k_n^2 - k_x^2 - k_y^2)^{1/2}|}. \qquad \text{(A6)}$$

Finally, we perform the Fourier transform in z and find for $\widetilde{G}_\pm(k_x, k_y, k_z)$

$$\frac{1}{k_x^2 + k_y^2 + k_z^2 - k_n^2}$$
$$= \frac{i}{2}2\pi\int \frac{\exp[-iz|(k_n^2 - k_x^2 - k_y^2)^{1/2}|]}{|(k_n^2 - k_x^2 - k_y^2)^{1/2}|}\exp[ik_z z]\mathrm{d}z. \qquad \text{(A7)}$$

Using the representation of the Dirac δ function $2\pi\delta(k_1 - k_2) = \int\exp[iz(k_1 - k_2)]\mathrm{d}z$ leads to $\widetilde{G}_\pm(k_x, k_y, k_z)$ as given in Eq. (30).

2. Lateral Scatter Spectra for an Incident Plane Wave

We derive the angular scatter spectra in both the first and the second order for a single incident plane wave to understand expressions (31) and (32). The Fourier transform of a plane wave traveling in the positive z direction can be written as

$$\widetilde{\mathbf{E}}_s^{(0)}(k_x, k_y, k_z) = \mathbf{E}_0\delta(k_x, k_y, k_z - k_n). \qquad \text{(A8)}$$

Insert Eq. (A8) into Eq. (29) to get the first-order scattered field in the frequency domain,

$$\widetilde{\mathbf{E}}_s^{(1)}(k_x, k_y, k_z) = \frac{k_n^2}{(2\pi)^3}\widetilde{G}(\mathbf{k})\mathbf{E}_0\widetilde{s}(k_x, k_y, k_z - k_n), \qquad \text{(A9)}$$

and for the lateral scatter spectrum

$$\widetilde{\mathbf{E}}_{s\pm}^{(1)}(k_x, k_y) = \frac{\pm ik_n^2}{4\pi\sqrt{k_n^2 - k_\perp^2}}\mathbf{E}_0\widetilde{s}(k_x, k_y,$$
$$\pm \sqrt{k_n^2 - k_\perp^2} - k_n). \qquad \text{(A10)}$$

For the second-order scattered field we obtain

$$\widetilde{\mathbf{E}}_s^{(2)}(k_x, k_y, k_z) = \left[\frac{k_n^2}{(2\pi)^3}\right]^2\widetilde{G}(\mathbf{k})\int \mathbf{E}_0\widetilde{s}(k_x', k_y', k_z' - k_n)$$
$$\times \widetilde{G}(\mathbf{k}')\widetilde{s}(\mathbf{k} - \mathbf{k}')\mathrm{d}^3\mathbf{k}'. \qquad \text{(A11)}$$

Inserting the modified version of the Green's function $\widetilde{G}_\pm(k_x, k_y, k_z)$ from Eq. (A7) yields

$$\widetilde{\mathbf{E}}_s^{(2)}(k_x, k_y, k_z)$$

$$= \frac{-k_n^4 \delta(|\sqrt{k_n^2 - k_\perp^2}| - k_z)}{(4\pi)^2 |\sqrt{k_n^2 - k_\perp^2}|}$$

$$\times \iiint \frac{\mathbf{E}_0 \delta(\sqrt{k_n^2 - k_\perp'^2} - k_z')}{\sqrt{k_n^2 - k_\perp'^2}}$$

$$\cdot [\tilde{s}(k_x', k_y', k_z' - k_n) - \tilde{s}(k_x', k_y', -k_z' - k_n)]$$

$$\cdot \tilde{s}(k_x - k_x', k_y - k_y', k_z - k_z') dk' dk_y' dk_z'. \quad (A12)$$

Using two times the sifting property of the δ function delivers for the lateral spectrum

$$\widetilde{\mathbf{E}}_{s\pm}^{(2)}(k_x, k_y) = \frac{\mp k_n^4}{(4\pi)^2 \sqrt{k_n^2 - k_\perp^2}} \iint \frac{\mathbf{E}_0}{\sqrt{k_n^2 - k_\perp'^2}}$$

$$\cdot [\tilde{s}(k_x', k_y', \sqrt{k_n^2 - k_\perp'^2} - k_n)$$

$$- \tilde{s}(k_x', k_y', -\sqrt{k_n^2 - k_\perp'^2} - k_n)]$$

$$\cdot \tilde{s}(k_x - k_x', k_y - k_y', \pm\sqrt{k_n^2 - k_\perp^2}$$

$$\mp \sqrt{k_n^2 - k_\perp'^2}) dk_x' dk_y'. \quad (A13)$$

ACKKNOWLEDGMENTS

The authors thank Jim Swoger, Christian Tischer, Wolfgang Singer, Hanns Harney, Pavel Zemánek, and Alexandr Jonás for fruitful discussions and comments. Further thanks go to Pavel Zemánek and Alexandr Jonás for calculating the trapping forces via the stress tensor and to Peter Török for delivering results from rigorous diffraction calculations of fields in the focal region.

The authors may be reached at the address on the title page; by phone, 49-6221-387123; by fax, 49-6221-387306; or by e-mail, alexander.rohrbach@EMBL-Heidelberg.de.

REFERENCES

1. A. Ashkin, "Acceleration and trapping of particles by radiation pressure," Phys. Rev. Lett. **24**, 156–159 (1970).
2. A. Ashkin, J. M. Dziedzic, J. E. Bjorkholm, and S. Chu, "Observation of a single-beam gradient force optical trap for dielectric particles," Opt. Lett. **11**, 288–290 (1986).
3. A. Ashkin and J. M. Dziedzic, "Optical trapping and manipulation of viruses and bacteria," Science **235**, 1517–1520 (1987).
4. S. M. Block, D. F. Blair, and H. C. Berg, "Compliance of bacterial flagella measured with optical tweezers," Nature (London) **338**, 514–518 (1989).
5. S. M. Block, L. S. Goldstein, and B. J. Schnapp, "Bead movement by single kinesin molecules studied with optical tweezers," Nature (London) **348**, 348–352 (1990).
6. A. Ashkin, K. Schutze, J. M. Dziedzic, U. Euteneuer, and M. Schliwa, "Force generation of organelle transport measured *in vivo* by an infrared laser trap," Nature (London) **348**, 346–348 (1990).
7. S. C. Kuo and M. P. Sheetz, "Force of single kinesin molecules measured with optical tweezers," Science **260**, 232–234 (1993).
8. K. Svoboda, C. F. Schmidt, B. J. Schnapp, and S. M. Block, "Direct observation of kinesin stepping by optical trapping interferometry," Nature **365**, 721–727 (1993).
9. R. W. Steubing, S. Cheng, W. H. Wright, Y. Numajiri, and M. W. Berns, "Laser induced cell fusion in combination with optical tweezers: the laser cell fusion trap," Cytometry **12**, 505–510 (1991).
10. M. W. Berns, W. H. Wright, B. J. Tromberg, G. A. Profeta, J. J. Andrews, and R. J. Walter, "Use of a laser-induced optical force trap to study chromosome movement on the mitotic spindle," Proc. Natl. Acad. Sci. U.S.A. **86**, 7914–7918 (1989).
11. S. Seeger, S. Monajembashi, K. J. Hutter, G. Futterman, J. Wolfrum, and K. O. Greulich, "Application of laser optical tweezers in immunology and molecular genetics," Cytometry **12**, 497–504 (1991).
12. L. P. Ghislain and W. W. Webb, "Scanning-force microscope based on an optical trap," Opt. Lett. **18**, 1678–1680 (1993).
13. A. Pralle, E.-L. Florin, E. H. K. Stelzer, and J. K. H. Hörber, "Local viscosity probed by photonic force microscopy," Appl. Phys. A **66**, P71–P73 (1998).
14. E.-L. Florin, A. Pralle, J. K. H. Hörber, and E. H. K. Stelzer, "Photonic force microscope (PFM) based on optical tweezers and two-photon excitation for biological applications," J. Struct. Biol. **119**, 202–211 (1997).
15. E.-L. Florin, A. Pralle, E. H. K. Stelzer, and J. K. H. Hörber, "Photonic force microscope calibration by thermal noise analysis," Appl. Phys. A: Solids Surf. **66**, S75–S78 (1998).
16. A. Pralle, M. Prummer, E.-L. Florin, E. H. K. Stelzer, and J. K. H. Hörber, "Three-dimensional position tracking for optical tweezers by forward scattered light," Microsc. Res. Tech. **44**, 378–386 (1999).
17. P. Debye, "Der Lichtdruck auf Kugeln von beliebige Material," Ann. Phys. **30**, 57–136 (1909).
18. G. Mie, "Beitraege zur Optik trueber Medien speziell Kolloidaler Metalloesungen," Ann. Phys. **25**, 377–445 (1908).
19. J. D. Jackson, *Classical Electrodynamics*, 2nd ed. (Wiley, New York, 1975), p. 236.
20. P. Mulser, "Radiation pressure on microscopic bodies," J. Opt. Soc. Am. B **2**, 1814–1829 (1985).
21. A. Ashkin, "Forces of a single-beam gradient laser trap on a dielectric sphere in the ray-optics regime," Biophys. J. **61**, 569–582 (1992).
22. R. Gussgard, T. Lindmo, and I. Brevik, "Calculation of the trapping force in a strongly focused laser beam," J. Opt. Soc. Am. B **9**, 1922–1930 (1992).
23. T. Wohland, A. Rosin, and E. H. K. Stelzer, "Theoretical determination of the influence of the polarization on forces exerted by optical tweezers," Optik **102**, 181–190 (1996).
24. J. S. Kim and S. S. Lee, "Scattering of laser beams and the optical potential well for a homogeneous sphere," J. Opt. Soc. Am. **73**, 303–312 (1983).
25. J. P. Barton and D. R. Alexander, "Fifth-order corrected electromagnetic field components for a fundamental Gaussian beam," J. Appl. Phys. **66**, 2800–2802 (1989).
26. J. P. Barton, D. R. Alexander, and S. A. Schaub, "Theoretical determination of net radiation force and torque for a spherical particle illuminated by a focused laser beam," J. Appl. Phys. **66**, 4594–4602 (1989).
27. F. Ren, G. Gréhan, and G. Gouesbet, "Radiation pressure forces exerted on a particle located arbitrarily in a Gaussian beam by using the generalized Lorentz–Mie theory, and associated resonance effects," Opt. Commun. **108**, 343–354 (1994).
28. W. H. Wright, G. J. Sonek, and M. W. Berns, "Radiation trapping forces on microspheres with optical tweezers," Appl. Phys. Lett. **63**, 715–717 (1993).
29. W. H. Wright, G. J. Sonek and M. W. Berns, "Parametric study of the forces on microspheres held by optical tweezers," Appl. Opt. **33**, 1735–1748 (1994).
30. Y. Harada and T. Asakura, "Radiation forces on a dielectric sphere in the Rayleigh scattering regime," Opt. Commun. **124**, 529–541 (1996).
31. P. Zemánek, A. Jonás, L. Srámek, and M. Liska, "Optical trapping of Rayleigh particles using a Gaussian standing wave," Opt. Commun. **151**, 273–285 (1998).
32. P. Zemánek, A. Jonás, L. Srámek, and M. Liska, "Optical-trapping of nanoparticles and microparticles by a Gaussian standing wave," Opt. Lett. **24**, 1448–1450 (1999).

33. M. Kerker, *The Scattering of Light*, 1st ed. (Academic, New York, 1969).

34. K. S. Shifrin, *Scattering of Light in a Turbid Medium*, 1st ed. (Nauka, Moscow, 1951) N. T. t. T. F.-. (1968).

35. C. Acquista, "Light Scattering by tenuous particles: a generalization of the Rayleigh–Gans–Rocard approach," Appl. Opt. **15**, 2932–2936 (1976).

36. S. Colak, C. Yeh and L. W. Casperson, "Scattering of focused beams by tenuous particles," Appl. Opt. **18**, 294–302 (1979).

37. M. W. Allersma, F. Gittes, M. J. deCastro, R. J. Stewart, and C. F. Schmidt, "Two-dimensional tracking of ncd motility by back focal plane interferometry," Biophys. J. **74**, 1074–1085 (1998).

38. C. W. McCutchen, "Generalized aperture and the three-dimensional diffraction image," J. Opt. Soc. Am. **54**, 240–244 (1964).

39. J. W. Goodman, *Introduction to Fourier Optics*, 1st ed. (McGraw-Hill, San Francisco, Calif., 1968), pp. 48–56.

40. M. Mansuripur, "Distribution of light at and near the focus of high-numerical-aperture objectives," J. Opt. Soc. Am. A **3**, 2086–2093 (1986).

41. M. Mansuripur, "Certain computational aspects of vector diffraction problems," J. Opt. Soc. Am. A **6**, 786–805 (1989).

42. C. J. R. Sheppard and K. G. Larkin, "Vectorial pupil functions and vectorial transfer functions," Optik **107**, 79–87 (1997).

43. A. Rohrbach, www.embl-heidelberg.de/~rohrbach.

44. J. E. Harvey, "Fourier treatment of near-field scalar diffraction theory," Am. J. Phys. **47**, 974–980 (1979).

45. J. P. Gordon, "Radiation forces and momenta in dielectric media," Phys. Rev. **8**, 14–21 (1973).

46. J. D. Jackson, *Classical Electrodynamics*, 2nd ed. (Wiley, New York, 1975), p. 239.

47. K. Visscher and G. J. Brakenhoff, "Theoretical study of optically induced forces on spherical particles in a single beam trap I: Rayleigh scatterers," Optik **89**, 174–180 (1992).

48. J. D. Jackson, *Classical Electrodynamics*, 2nd ed. (Wiley, New York, 1975), p. 151.

49. C. Bohren and D. R. Huffman, *Absorption and Scattering of Light by Small Particles* (Wiley Science Paperback, New York, 1998), p. 137.

50. E. Lalor and E. Wolf, "Exact solution of the equation of molecular optics for refraction and reflection of an electromagnetic wave on a semi-infinite dielectric," J. Opt. Soc. Am. **62**, 1165–1174 (1972).

51. M. Born and E. Wolf, *Principles of Optics*, 7th ed. (Cambridge U. Press, New York, 1999).

52. B. T. Draine and P. J. Flatau, "Discrete-dipole approximation for scattering calculations," J. Opt. Soc. Am. A **11**, 1491–1498 (1994).

53. T. Lemaire, "Coupled-multipole formulation for the treatment of electromagnetic scattering by a small dielectric particle of arbitrary shape," J. Opt. Soc. Am. A **14**, 470–474 (1997).

54. M. Kerker, "Rayleigh–Debye scattering," in *The Scattering of Light*, E. M. Loebl, ed., 1st ed. (Academic, New York, 1969), p. 414.

55. W. Singer and K.-H. Brenner, "Transition of a scalar field at a refracting surface in the generalized Kirchhoff diffraction theory," J. Opt. Soc. Am. A **12**, 1913–1919 (1995).

56. A. Rohrbach and W. Singer, "Scattering of a scalar field at dielectric surfaces by Born series expansion," J. Opt. Soc. Am. A **15**, 2651–2659 (1998).

57. H. Weyl, "Ausbreitung elektromagnetischer Wellen ueber einem ebenen Leiter," Ann. Phys. **60**, 481–500 (1919).

58. L. Mandel and E. Wolf, *Optical Coherence and Quantum Optics* (Cambridge U. Press, New York, 1995).

59. Mie scattering, www.omlc.ogi.edu/calc/mie_calc.html.

60. P. Török, P. Varga, Z. Laczik, and G. R. Booker, "Electromagnetic diffraction of light focused through planar interface between materials of mismatched refractive indices: an integral representation," J. Opt. Soc. Am. A **12**, 325–332 (1995).

61. H. Felgner, O. Müller, and M. Schliwa, "Calibration of light forces in optical tweezers," Appl. Opt. **34**, 977–982 (1995).

62. M. Gu, P. C. Ke, and X. S. Gan, "Trapping force by a high numerical-aperture microscope objective obeying the sine condition," Rev. Sci. Instrum. **68**, 3666–3668 (1997).

Single-beam trapping of micro-beads in polarized light: Numerical simulations

A.R. Zakharian, P. Polynkin, M. Mansuripur, and J.V. Moloney

College of Optical Sciences, University of Arizona, Tucson, Arizona 85721

armis@u.arizona.edu

Abstract: Using numerical solutions of Maxwell's equations in conjunction with the Lorentz law of force, we compute the electromagnetic force distribution in and around a dielectric micro-sphere trapped by a focused laser beam. Dependence of the optical trap's stiffness on the polarization state of the incident beam is analyzed for particles suspended in air or immersed in water, under conditions similar to those realized in practical optical tweezers. A comparison of the simulation results with available experimental data reveals the merit of one physical model relative to two competing models; the three models arise from different interpretations of the same physical picture.

OCIS codes: (260.2110) Electromagnetic theory; (140.7010) Optical trapping; (000.4430) Numerical computation

References and links

1. M. Mansuripur, "Radiation Pressure and the linear momentum of the electromagnetic field," Opt. Express **12,** 5375–5401 (2004), http://www.opticsexpress.org/abstract.cfm?id=81636.
2. M. Mansuripur, A.R. Zakharian and J.V. Moloney, "Radiation Pressure on a dielectric wedge," Opt. Express **13,** 2064–2074 (2005), http://www.opticsexpress.org/abstract.cfm?id=83011.
3. M. Mansuripur, "Radiation Pressure and the linear momentum of light in dispersive dielectric media," Opt. Express **13,** 2245–2250 (2005), http://www.opticsexpress.org/abstract.cfm?id=83032.
4. M. Mansuripur, "Angular momentum of circularly polarized light in dielectric media," Opt. Express **13,** 5315–5324 (2005), http://www.opticsexpress.org/abstract.cfm?id=84895.
5. S. M. Barnett and R. Loudon, "On the electromagnetic force on a dielectric medium," submitted to J. Phys. B: At. Mol. Phys. (January 2006).
6. A.R. Zakharian, M. Mansuripur and J.V. Moloney, "Radiation Pressure and the distribution of the electromagnetic force in dielectric media," Opt. Express **13,** 2321–2336 (2005), http://www.opticsexpress.org/abstract.cfm?id=83272.
7. R. Gauthier, "Computation of the optical trapping force using an FDTD based technique," Opt. Express **13,** 3707–3718 (2005), http://www.opticsexpress.org/abstract.cfm?id=83817.
8. A. Rohrbach and E.H.K. Stelzer, "Three-dimensional position detection of optically trapped dielectric particles," J. Appl. Phys. **91,** 5474–5488 (2002).
9. A. Rohrbach, "Stiffness of Optical Traps: Quantitative agreement between experiment and electromagnetic theory," Phys. Rev. Lett. **95,** 168102 (2005).
10. W.H. Wright, G.J. Sonek, and M.W. Berns, "Radiation trapping forces on microspheres with optical tweezers," Appl. Phys. Lett. **63,** 715–717 (1993).
11. W.H. Wright, G.J. Sonek, and M.W. Berns, "Parametric study of the forces on microspheres held by optical tweezers," Appl. Opt. **33,** 1735–1748 (1994).
12. A. Rohrbach and E. H. K. Stelzer, "Trapping forces, force constants, and potential depths for dielectric spheres in the presence of spherical aberrations," Appl. Opt. **41,** 2494–2507 (2002).
13. D. Ganic, X. Gan and M. Gu, "Exact radiation trapping force calculation based on vectorial diffraction theory," Opt. Express **12,** 2670–2675 (2004), http://www.opticsexpress.org/abstract.cfm?id=80240.
14. P.W. Barber and S.C. Hill, *Light Scattering by Particles: Computational Methods* (World Scientific Publishing Co. 1990).

1. Introduction

Computation of the force of radiation on a given object, through evaluation of the electromagnetic field distribution according to Maxwell's equations, followed by a direct application of the Lorentz law of force, has been described in Ref. [1]-[4]. In particular, for an isotropic, piecewise-homogeneous dielectric medium, the total force was shown to result from the force of the magnetic field acting on the induced bound current density, $\mathbf{J}_b \times \mathbf{B} = [(1 - 1/\varepsilon)\nabla \times \mathbf{H}] \times \mathbf{B}$, and from the force exerted by the electric component of the light field on the induced bound charge density at the interfaces between media of differing relative permittivity ε. The contribution of the E-field component of the Lorentz force is thus specified (Ref. [1]) by the force density $\rho_b \mathbf{E} = (\varepsilon_0 \nabla \cdot \mathbf{E})\mathbf{E}$. (Note: As far as the total force and torque of radiation on a solid object are concerned, Barnett and Loudon Ref. [5] have recently shown that an alternative formulation of the Lorentz law - one that is often used in the radiation pressure literature - leads to exactly the same results. We discuss the relation between these two formulations in the Appendix, and extend the proof of their equivalence to the case of solid objects immersed in a liquid, which is a prime concern of the present paper.)

In a previous publication Ref. [6] we described the numerical implementation of the aforementioned approach to the computation of the Lorentz force, based on the Finite-Difference Time-Domain (FDTD) solution of Maxwell's equations. (An alternative application of the FDTD method to problems of radiation pressure may be found in Ref. [7].) Our application of the FDTD method to the computation of the force exerted by a focused laser beam on a spherical dielectric particle immersed in a liquid showed that the forces experienced by the liquid layer immediately at the particle's surface could impact the overall force experienced by the particle. A detailed discussion of the (conceptual) separation of the bound charges on the surface of the particle from the bound charges induced in the surrounding liquid at the solid-liquid interface is given in Ref. [2]. The analysis indicated that the contribution to the $\rho_b \mathbf{E}$ part of the Lorentz force by the component of the E-field perpendicular to the interface can be computed in different ways, depending on the assumptions made concerning the nature of the electromagnetic and hydrodynamic interactions between the solid particle and its liquid environment.

In the present study we apply the FDTD method to analyze the polarization dependence of the interaction between a focused laser beam and a small spherical particle trapped either in the air or in a liquid host medium (water). In the latter case, results from different methods of computing the contributions to the net force by bound charges at the solid-liquid interface will be compared, with the goal of quantifying the effects that might be possible to differentiate in experiments Ref. [8]-[9]. The polarization dependence of optical tweezers has been studied in the past, and the dependence of trap stiffness on polarization direction is well documented Ref. [10]-[13]. The goal of the present paper is not a re-evaluation of the existing models, but rather a demonstration of the applicability of our own new model Ref. [1]-[4] to the problem of trap stiffness anisotropy. Also, upon comparing our numerical results with experimental data, we gain further insight into the nature of the Lorentz force acting on solid objects in liquid environments, and identify a preferred interpretation of the proposed physical model. In a nutshell, there exists an ambiguity as to the nature of the effective force at the surface of the sphere, with at least three different models in contention. Carefully examining the trapping force on a dielectric sphere immersed in water provides enough information to establish one of the three competing models and rule out the other two.

The paper is organized as follows. Section 2 discusses example cases used to validate our numerical computations. In sections 3 and 4 we compare the computed stiffness of the trap for two orthogonal linear polarizations of a beam illuminating a micro-sphere suspended in air and in water, respectively. Summary and conclusions are presented in section 5.

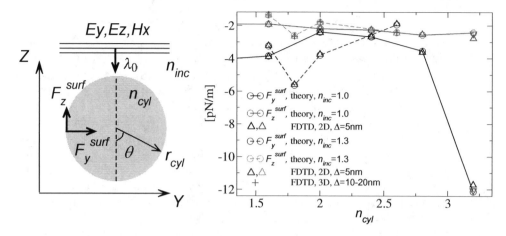

Fig. 1. Surface force integrated over the left-half of the cylinder as function of the cylinder's refractive index n_{cyl}. The incident plane-wave has vacuum wavelength $\lambda_0 = 0.65\mu m$, and the incidence medium's refractive index is $n_{inc} = 1.0$ (solid lines) or $n_{inc} = 1.3$ (dashed lines). The exact results of the Lorentz-Mie theory (circles) are compared with those of our FDTD-based method (triangles and crosses). In the case of cylinder immersed in water, the charges induced at the solid-liquid interface were lumped together, then subjected to the average **E**-field at the interface; in other words, no attempts were made in these calculations to distinguish the force of the light's **E**-field on the solid surface from that on the adjacent liquid. For each FDTD simulation the grid resolution Δ is indicated.

2. Comparison to exact solutions

To validate our numerical procedures, we first consider the problem of computing the surface forces exerted by a plane-wave incident on a dielectric cylinder, with propagation direction being perpendicular to the cylinder axis. We consider in the YZ-plane of incidence a p-polarized plane-wave (E_y, E_z, H_x), for which the electric-field component normal to the cylinder surface, $\mathbf{E}_\perp$, is discontinuous; see Fig. 1. The free-space wavelength of the incident light is $\lambda_0 = 0.65\mu m$ and the radius of the cylindrical rod is $r_{cyl} = 1.0\mu m$. The surface forces computed in our FDTD simulations can be compared with those obtained from the Lorentz-Mie theory of light scattering, Ref. [14]. In computing the Lorentz force of the **E**-field on the interfacial (bound) charges induced on the cylinder surface we used the average (Ref. [2]) of the **E**-field normal to the surface (the tangential component of the **E**-field is continuous across the interface); the surface-charge density is assumed to be proportional to the discontinuity of the $\mathbf{E}_\perp$ field at the interface. While in the Lorentz-Mie theory $\mathbf{E}_\perp$ at the surface is readily available, on the FDTD grid the cylinder surface is approximated as a discrete staircase, and the surface force density components (F_y^{surf}, F_z^{surf}) are evaluated along the coordinate axes, Ref. [6]. Figure 1 shows, for two different cases, the surface force integrated over one-half of the cylinder surface ($180° \leq \theta \leq 360°$), as function of the refractive index n_{cyl} of the cylinder. The agreement between exact theory and numerical simulation is remarkable. Our numerical discretization error is of the order of 1-2% for n_{cyl} ranging from 1.5 to 3.4, consistent with the 30-60 points per wavelength discretization and first-order convergence due to staircase approximation of the geometry. For the high index-contrast case of $n_{inc}/n_{cyl} = 1/3.4$ the error (Fig. 1, $\triangle$) was found to be dominated by the inadequate convergence to a time-harmonic state; the error was reduced (Fig. 1, $\triangledown$) when we increased the integration time.

In the second test problem we used the exact solutions for radiation pressure distribution in a solid dielectric prism illuminated by a Gaussian beam of light at Brewster's incidence, Ref. [2].

Table 1. Definitions of the surface charge and surface force densities for methods I,II and III.

Method	I	II	III
σ_2	$\varepsilon_0(E_{1\perp} - E_{2\perp})$	$\varepsilon_0(E_{g\perp} - E_{2\perp})$	$\varepsilon_0(E_{g\perp} - E_{2\perp})$
$F_{\parallel}^{surf}$	$\sigma_2 E_{\parallel}$	$\sigma_2 E_{\parallel}$	$\sigma_2 E_{\parallel}$
$F_{\perp}^{surf}$	$\sigma_2(E_{1\perp} + E_{2\perp})/2$	$\sigma_2(E_{1\perp} + E_{2\perp})/2$	$\sigma_2(E_{g\perp} + E_{2\perp})/2$

When the prism is immersed in water, the interfacial charges induced on the solid and liquid sides of the interface must be distinguished from each other. The interaction of these charged layers with the local **E**-field determines their contribution to the net force acting on the prism. Following the notation introduced in Ref. [2] we enumerate in Table 1 the three approaches to the computation of the surface force density acting on the solid side of the solid-liquid interface. Here $(F_{\parallel}^{surf}, F_{\perp}^{surf})$ are the surface force density components parallel and perpendicular to the solid's surface; σ_2 is the surface charge density belonging to the solid side; $E_{\parallel}, E_{1\perp}, E_{2\perp}$ are, respectively, the **E**-field components parallel to the interface, normal to the interface on the liquid side, and normal to the interface on the solid side; and $\varepsilon_0 E_{g\perp} = \varepsilon_0\varepsilon_1 E_{1\perp} = \varepsilon_0\varepsilon_2 E_{2\perp}$ is the electric displacement field $\mathbf{D}_\perp$ normal to the interface. Method *III* isolates the force acting on the solid side of the interface by introducing a small (artificial) gap between the solid and the liquid, then using the gap field $E_{g\perp}$ (derived from the continuity of the normal component of the displacement field **D**) to evaluate both the charge density σ_2 and the average **E**-field that acts on the solid; this corresponds to Eq.(21) in Ref. [2]. While the conceptual introduction of a small gap at the solid-liquid interface is essential for the calculation of σ_2, it gives rise to a mutual attractive force between the two charged layers thus pushed apart. This force, which is included in the total force experienced by the solid object as calculated by Method *III*, is probably inactive in practice and should be excluded. In Method *II* the contribution of this attractive force between the two charged layers (induced on the solid and liquid sides of the interface) is removed by using $(E_{1\perp} + E_{2\perp})/2$ as the effective perpendicular field acting on the surface charge density σ_2 on the solid side of the interface; see Eq.(24) in Ref. [2]. In Method

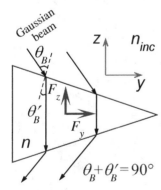

Fig. 2. Dielectric prism of refractive index n immersed in a host medium of refractive index n_{inc}.

Net force [pN/m]	$n_{inc} = 1$	$n_{inc} = 1.33$		
		I	II	III
F_y ($\Delta = 20nm$)	16.63	4.83×10^{-3}	22.38	30.95
F_y ($\Delta = 10nm$)	16.64	——	22.42	31.08
F_y (*exact*)	16.66	4.05×10^{-3}	22.45	31.09
F_z ($\Delta = 20nm$)	-0.047	3.13×10^{-2}	-0.02	-0.16
F_z ($\Delta = 10nm$)	-0.026	——	-0.016	-0.001
F_z (*exact*)	0.0	0.0	0.0	0.0

Table 2. Radiation force on the prism of Fig. 2, computed with the exact method and with the FDTD simulations. A p-polarized (H_x, E_y, E_z) Gaussian beam illuminates the prism at Brewster's angle θ_B. The net radiation force exerted on the prism is denoted by (F_y, F_z). The value of the mesh parameter Δ used in each simulation is indicated.

I the normal components of the **E**-field on the liquid and solid sides of the interface are used to compute the normal component of the average **E**-field as well as the induced charge density at the interface. In other words, the charges on the solid and liquid sides of the interface are lumped together, then subjected to the forces of the tangential **E**-field, which is continuous at the interface, and the perpendicular **E**-field, which is discontinuous at the interface (and, therefore, in need of averaging). Method *I*, which corresponds to Eq.(27) in Ref. [2], is also the method used in the case of a cylindrical rod immersed in water discussed earlier in conjunction with Fig. 1.

We use a Gaussian beam ($\lambda_0 = 0.65\mu m$) having a full-width of $7.8\mu m$ at the $1/e$ point of the field amplitude, incident on the prism at Brewster's angle $\theta_B = \arctan(n/n_{inc})$; see Fig. 2. In the case of incidence from the free-space, $n_{inc} = 1.0$, the prism has refractive index $n = 1.5$, while for $n_{inc} = 1.33$ (water) the prism index is $n = 1.995$. The incident field amplitude's peak value is $E_0 = 10^3/n_{inc}$ V/m. Table 2 compares the total radiation force (consisting of the surface force density integrated over both surfaces, and the volume force density integrated over the prism volume) exerted on the prism, obtained with the exact method and with the FDTD simulations. As shown in Fig. 2, the total force has a component F_y directed along the bisector of the prism's apex, and a second component F_z perpendicular to this bisector in the plane of incidence. (The exact value of F_z is zero in all cases.) For a prism immersed in water ($n_{inc} = 1.33$) the results obtained with the aforementioned methods *I*, *II*, and *III* of force computation are tabulated. The numerical results are generally in good agreement with the exact solutions; the largest disagreements occur in the case of Method *I* used in conjunction with the immersed prism, where (due to the specific set of parameters chosen for the simulation) the exact value of F_y happens to be so small (less than $10^{-2} pN/m$) that it falls below the numerical accuracy of our simulations. In the computations with method *III* the separation of the charges on the liquid and solid sides of the interface was simulated using an actual air-gap (width = 2Δ). The numerical and exact solutions in general are in good agreement with less than 1% error in the net force magnitude when $\Delta \leq 20nm$.

3. Single-beam optical trap in the air

We present computed results of the electromagnetic force exerted on a dielectric micro-sphere at and near the focus of a laser beam in the free space ($n_{inc} = 1.0$). The incident beam, obtained by focusing a $\lambda_0 = 532nm$ (or $1064nm$) plane-wave through a $0.9NA$, $5mm$ focal-length objective lens, propagates in the negative z-direction, as shown in Fig. 3. The plane-wave entering the objective's pupil is linearly polarized along either the x-axis or the y-axis. The total power of the incident beam is $P = \int S_z dx dy = 1.0W$. Figure 4 shows the Poynting vector distribution (**S**-field) in the XZ-plane for a micro-sphere of refractive index $n = 1.5$ and diameter $d = 460nm$, offset by (x,y,z)-offset=$(250,0,50)nm$ from the focal point of the y-polarized incident beam. Positive (negative) values of the z-offset represent the particle being displaced into the converging (diverging) half-space of the focused beam. Upon scattering from the micro-sphere, the large positive momentum acquired by the incident light along the x-axis, seen in Fig. 4, results in a net F_x force directed towards the beam axis. Figure 5 shows the computed (F_x, F_z) force components experienced by the micro-sphere as functions of the sphere center's offset from the focal point of the lens (y-offset = 0). The top (bottom) row shows the case of an x-polarized (y-polarized) incident beam. Due to symmetry, y-offset is set to zero, and offsets along the positive half of the x-axis are the only ones considered. The y component of the force was found to be zero (within the numerical accuracy of the simulations) for these linearly-polarized beams. Figure 5 indicates that the particle is trapped laterally, but not vertically along the z-axis, since $F_z < 0$ for the entire range of offsets shown. (Note: A 460nm-diameter glass bead weighs approximately $1.0fN$. If the focused laser beam shines on the bead from below, and

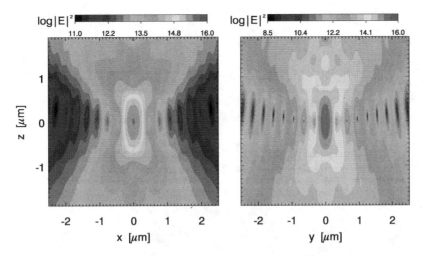

Fig. 3. Computed distributions of the electric field intensity $|\mathbf{E}|^2$ (units: $[V^2/m^2]$) in the xz- and yz-planes near the focus of a 0.9NA, $f = 5.0mm$, diffraction-limited objective. The $\lambda_0 = 532nm$ plane-wave illuminating the entrance pupil of the lens is linearly polarized along the x-axis.

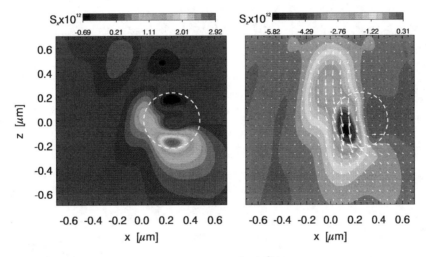

Fig. 4. Distribution of the Poynting vector $\mathbf{S}$ (units: $[W/m^2]$) in and around a glass micro-sphere ($n = 1.5$, $d = 460nm$). The focused beam, obtained by sending a linearly-polarized plane-wave (polarization along y) through a 0.9NA objective, propagates along the negative z-axis. Sphere center offset from the focal point: $(250, 0, 50)nm$. The (S_x, S_z) vector-field is superimposed on the color-coded S_z plot on the right-hand side.

if the laser power level is adjusted to a few microwatts, then the radiation pressure will work against the force of gravity to hold the bead in a stable trap along the z-axis.) The computed anisotropy of this trap in the lateral direction is $s_l = 1 - (\kappa_x/\kappa_y) = -0.15$, where κ_x and κ_y are the trap stiffness coefficients along the x-axis given by $\partial F_x/\partial x$ for x- and y-polarized beams, respectively, Ref. [12]. The aforementioned s_l was computed at the x-offset value of $50nm$, where z-offset $\approx 0\mu m$ is chosen to yield the maximum of F_x in the vicinity of the center of the small rectangles depicted in Fig. 5, where F_x is fairly insensitive to small variations of z. The computed stiffness anisotropy is plotted in Fig. 6 versus the particle diameter d for spherical beads having $n = 1.5$, trapped under a $\lambda_0 = 1064nm$ focused beam through a $0.9NA$ objective.

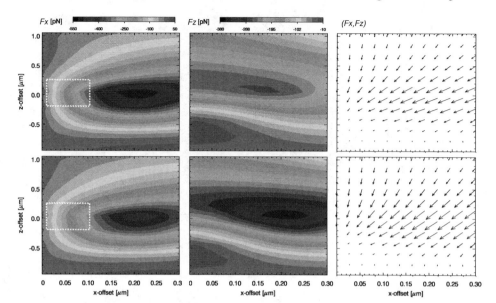

Fig. 5. Plots of the net force components (F_x, F_z) experienced by a glass micro-sphere $(d = 460nm, n = 1.5)$ versus the offset from the focal point in the xz-plane. The incidence medium is air, $\lambda_0 = 532nm$, the objective lens NA is 0.9, and the incident beam's power is $P = 1.0W$. Top row: x-polarization, bottom row: y-polarization. The stiffness coefficients κ_x, κ_y are computed at the center of the small rectangles shown on the left-hand-side of the F_x plots.

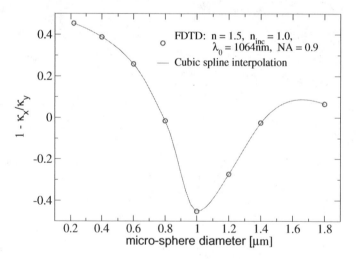

Fig. 6. Computed trap stiffness anisotropy $s_l = 1 - (\kappa_x/\kappa_y)$ versus particle diameter d, for micro-spheres of refractive index $n = 1.5$ trapped in the air with a $\lambda_0 = 1064nm$ laser beam focused through a $0.9NA$ objective lens. The stiffness is computed at x-offset $= 50nm$, z-offset $\approx 0\mu m$, where, for the chosen value of x-offset, the lateral trapping force F_x is at a maximum.

For the offset ranges and particle diameters considered, the radiation force along the z-axis, F_z, was found to be negative (i.e., inverted traps are necessary to achieve stable trapping). The lateral trap anisotropy s_l, evaluated at x-offset = $50nm$ (with z-offset adjusted to yield maximum lateral trapping force F_x), is seen in Fig. 6 to be positive for small particle diameters, negative when the particle size is comparable to the wavelength of the trap beam, and weakly positive for $d > 1.4\mu m$.

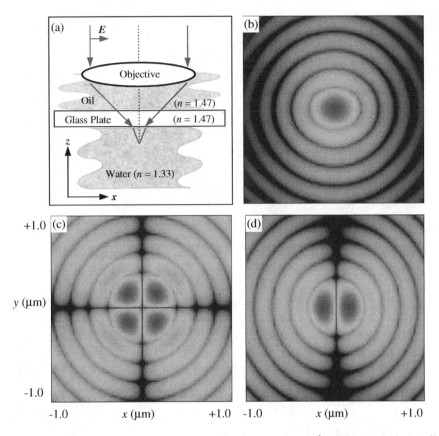

Fig. 7. (a) A linearly-polarized Gaussian beam having wavelength λ_0=532nm and 1/e (amplitude) radius r_0= 4.0mm is focused through a 1.4NA oil immersion objective lens. The lens has focal length f=3.0mm and aperture radius (at the entrance pupil) R_a=2.85mm; the refractive index of the immersion oil is 1.47. A glass plate of the same index as the oil separates the oil from the water (n_{water}=1.33), where the focused spot is used to trap various dielectric beads. The marginal rays are lost by total internal reflection at the glass-water interface; there is also some degree of apodization due to Fresnel reflection at this interface, but the phase aberrations induced by the transition from oil/glass to water are ignored in our calculations. The neglect of the spherical aberrations thus induced is justifiable, so long as the trap is not too far from the oil/water interface, Ref. [12]. (b)-(d) Logarithmic plots of the intensity distribution for x-, y-, and z-components of the E-field at the focal plane. The relative peak intensities are $|E_x|^2$:$|E_y|^2$: $|E_z|^2 \approx$ 1000 : 9 : 200. The total intensity distribution at the focal plane, $I(x,y) = |E_x|^2 + |E_y|^2 + |E_z|^2$, (not shown) is elongated in the x-direction; the full-width at half-maximum intensity (FWHM) of the focused spot is 300nm along x and 196nm along y. The transmission efficiency of the oil/glass-to-water interface is 92.6%; that is, the overall loss of optical power to total-internal and Fresnel reflections at this interface is 7.4%.

4.   Single-beam optical trap in water

In computing the optical trap properties in a liquid host medium (refractive index $n_{inc} = 1.33$), a collimated laser beam having $\lambda_0 = 532nm$ (or $1064nm$) was focused through an $NA \approx 1.4$ immersion objective designed for diffraction-limited focusing within an immersion liquid of refractive index $n_{oil} = 1.47$; see Figs. 7, 8. In our first set of simulations corresponding to $\lambda_0 = 532nm$, the spherical particle has $d = 460nm$, $n = 1.5$, and the total power of the incident beam is $P = 1.0W$. For Methods *I*, *II*, and *III* of computing the surface-force discussed in section 2, Figs. 9-11 show the net force components (F_x, F_z) exerted on the micro-sphere illuminated by *x*- or *y*-polarized light. For most of the offset range considered in Fig. 9, the F_x force component computed with method *I* for *x*-polarized light is opposite in direction to the F_x computed for *y*-polarized light, indicating the impossibility of lateral trapping with *x*-polarization. Such a marked difference in the behavior of F_x for different polarization states exhibited by method *I*, contradicts the experimental observations which show trapping (with similar strengths) for both polarization states. We conclude, therefore, that Method *I* is unphysical and must be abandoned. [The root of the problem with Method *I* can be traced to the lumping together of the solid and liquid charges at the interface, which weakens the negative contribution of F_x^{surf}, thus allowing the positive contribution by the magnetic part of the Lorentz force (acting on the micro-sphere volume) to push the particle away from the focal point. In contrast, for *y*-polarized light, the contribution of the magnetic Lorentz force to F_x is negative, thus enabling lateral trapping.]

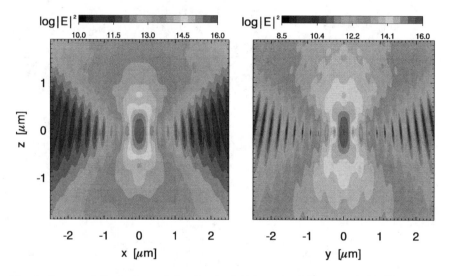

Fig. 8. Computed distributions of the electric field intensity $|\mathbf{E}|^2$ (units: $[V^2/m^2]$) in the *xz*- and *yz*-planes near the focus of a $\lambda_0 = 532nm$ beam in water. The focused spot is obtained by sending an *x*-polarized plane-wave through an oil-immersion $\approx 1.4NA$ objective. The oil and water are separated by a thin glass slide, index-matched to the immersion oil, Fig. 7.

The force distributions computed with Method *II* (Fig. 10) and Method *III* (Fig. 11) show qualitatively similar behavior for the two polarization states, and indicate trapping along both *x*- and *z*-directions. F_x is strongest at lateral offset values on the order of one micro-sphere radius. While both methods *II* and *III* result in trapping of the micro-bead, the ratio of the maximum lateral restoring forces F_x (sampled inside the dashed rectangles in Figs. 10 and 11) for *x*- and *y*-polarized beams is found to be 0.92 for method *II* and 1.2 for method *III*. Thus, although Figs. 10 and 11 exhibit qualitatively similar behavior, their quantitative estimates of the restoring forces are sufficiently different to enable one to distinguish Method *II* from Method *III* based

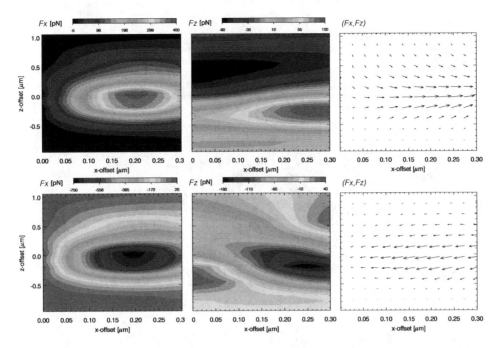

Fig. 9. Plots of the net force components (F_x, F_z), computed with Method *I*, for a glass micro-bead ($d = 460nm$, $n = 1.5$) versus the offset from the focal point. The host medium is water ($n_{inc} = 1.33$), $\lambda_0 = 532nm$, the objective lens *NA* is ≈ 1.4, and the incident beam's power is $P = 1.0W$. Top row: *x*-polarization, bottom row: *y*-polarization. The non-trapping behavior of *x*-polarized beam, which is contrary to experimental observations, indicates the invalidity of Method *I* used in these calculations.

on experimentally determined values of stiffness anisotropy.

Figure 12 shows the dependence of the trap stiffness anisotropy $s_l = 1 - (\kappa_x/\kappa_y)$ on bead diameter d, computed with Method *II* for polystyrene micro-beads ($n = 1.57$) trapped in water ($n_{inc} = 1.33$) under a $\lambda_0 = 1064nm$ laser beam focused through an oil-immersion $\approx 1.4NA$ objective lens (with a glass slide used to separate oil from water, as shown in Fig. 7). For $d < 850nm$, the lateral trap stiffness is found to be smaller for the particle offsets along the polarization direction, that is, $s_l > 0$. For larger particles ($850 < d < 1400nm$) the trap stiffness in the *x*-direction becomes greater for *x*-polarization than that for *y*-polarization, i.e., $s_l < 0$. The trap stiffness anisotropy reaches a minimum before returning to positive values for $d > 1400nm$.

Superimposed on Fig. 12 are two sets of experimental data. The green triangles correspond to measurements carried out with a system similar to that depicted in Fig. 7, operating at $\lambda_0 = 1064nm$. In practice, this system suffered from chromatic (and possibly spherical) aberrations; defects that are not taken into account in our computer simulations. The agreement with theoretical calculations is good for the $d = 1260nm$ particle, but not so good for $d = 1510nm$ and $d = 1900nm$ particles. The second set of experimental data (solid blue circles in Fig. 12) is from Table II of Rohrbach Ref. [9]. These were obtained with a 1.2*NA* water immersion objective, corrected for all aberrations and, therefore, presumably operating in the diffraction limit. All other experimental parameters were the same as those used in our simulations. (Considering the loss of marginal rays at the oil/water interface, apodization due to Fresnel reflection losses, and the relatively small beam diameter at the entrance pupil of the simulated lens, our focused spot should be fairly close to that used in Rohrbach's experiments.)

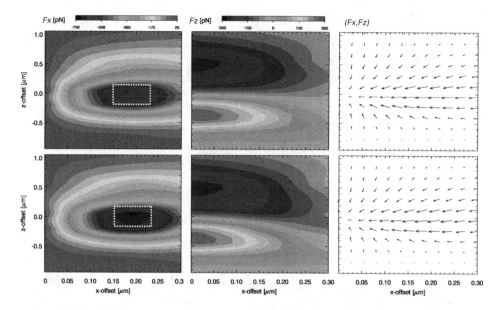

Fig. 10. Plots of the net force components (F_x, F_z), computed with Method *II*, for a glass micro-bead ($d = 460nm$, $n = 1.5$) versus the offset from the focal point. The host medium is water ($n_{inc} = 1.33$), $\lambda_0 = 532nm$, the objective lens *NA* is ≈ 1.4, and the incident beam's power is $P = 1.0W$. Top row: *x*-polarization, bottom row: *y*-polarization. The lateral trapping force F_x at the center of the small rectangle in the case of *x*-polarization is weaker than the corresponding F_x in the case of *y*-polarization (ratio = 0.92).

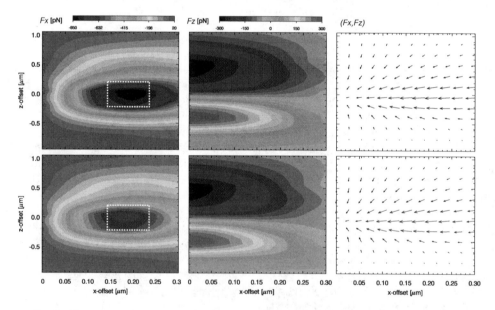

Fig. 11. Plots of the net force components (F_x, F_z), computed with Method *III*, for a glass micro-bead ($d = 460nm$, $n = 1.5$) versus the offset from the focal point. The host medium is water ($n_{inc} = 1.33$), $\lambda_0 = 532nm$, the objective lens *NA* is ≈ 1.4, and the incident beam's power is $P = 1.0W$. Top row: *x*-polarization, bottom row: *y*-polarization. The lateral trapping force F_x at the center of the small rectangle in the case of *x*-polarization is stronger than the corresponding F_x in the case of *y*-polarization (ratio = 1.2).

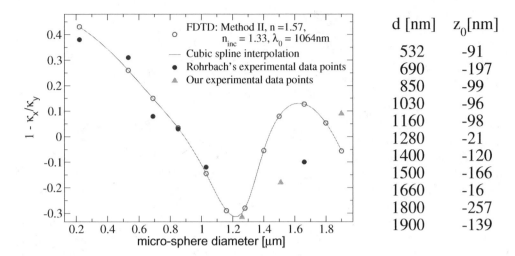

d [nm]	z_0 [nm]
532	-91
690	-197
850	-99
1030	-96
1160	-98
1280	-21
1400	-120
1500	-166
1660	-16
1800	-257
1900	-139

Fig. 12. Trap stiffness anisotropy $s_l = 1 - (\kappa_x/\kappa_y)$ versus particle diameter d, computed with method *II* for polystyrene micro-beads ($n = 1.57$) trapped in water ($n_{inc} = 1.33$) under a $\lambda_0 = 1064nm$ laser beam focused through an oil-immersion $\approx 1.4NA$ objective lens. Numerically, κ_x and κ_y were evaluated with a forward finite-difference approximation at x-offset $= 50 - 100nm$, offset discretization $\Delta x = 50nm$, and z-offset $= z_0$ where $F_z(x = 0, z_0) = 0$. The computed value of z_0 for each particle is listed on the right-hand side. The solid triangles (green) represent experimental data obtained with a system similar to that depicted in Fig. 7, operating at $\lambda_0 = 1064nm$, and suffering from chromatic (and possibly spherical) aberrations. The solid circles (blue) represent experimental data from Table II of Rohrbach Ref. [9]. Although Rohrbach uses a 1.2*NA* water immersion objective in his experiments, the comparison is warranted here because our simulated 1.4*NA* focused beam, upon transmission from oil to water, loses its marginal rays and acquires characteristics that are not too far from those of a 1.2NA diffraction-limited focused spot. (Our experimental methodology is similar to that of Rohrbach, the only difference being that we use the oil-immersion objective depicted in Fig. 7(a), whereas Rohrbach uses a water-immersion objective.)

The agreement between theory and experiment is remarkable in this case, except for the $d = 1660nm$ particle. (We repeated the simulation for the $d = 1660$nm particle using the exact NA used in Rohrbach's experiments; the resulting value of s_l, however, did not change very much.) While it is possible that the larger beads used in the aforementioned experiments have been described inaccurately (i.e., n and d deviating from the specifications), and while it is true that the aberrations of our own measurement system need to be understood and properly modeled, we cannot rule out the possibility that the radiation forces acting on the surrounding liquid (and ignored in the simulations) could have impacted the results of these measurements.

We mention in passing that, in the "Rayleigh particle" limit ($d \ll \lambda_0/n_{inc}$), the Lorentz force density $(\varepsilon_0 \nabla \cdot \mathbf{E})\mathbf{E} + \mathbf{J}_b \times \mathbf{B}$ reduces, in the dipole approximation, to $(\mathbf{p} \cdot \nabla)\mathbf{E} + \partial\mathbf{p}/\partial t \times \mathbf{B}$, assuming linear dependence of the polarization $\mathbf{p}$ on the local $\mathbf{E}$-field, namely, $\mathbf{p}(\mathbf{r},t) = \alpha_0\mathbf{E}(\mathbf{r},t)$, and neglecting the small terms due to the inhomogeneity of the magnetic field. For the smaller particles, the experimental data was found in Ref. [9] to be in good agreement with the computed results based on the Rayleigh-Gans approximation.

5. Concluding remarks

We have computed the radiation force of $\lambda_0 = 532nm$ and $\lambda_0 = 1064nm$ linearly polarized laser beams focused through air ($NA = 0.9$) and through water ($NA \approx 1.4$) on various spherical micro-particles. For example, the restoring force acting on a small glass bead ($d = 460nm$, $n = 1.5$, $\lambda_0 = 532nm$) trapped in the air was found to be roughly 10-20% stronger when the polarization was aligned with, rather than perpendicular to, the particle offset direction. This and related predictions (described in Section 3) now await experimental verification to confirm the validity of the assumptions underlying our theoretical model.

The radiation force on a small glass bead ($d = 460nm$, $n = 1.5$, $\lambda_0 = 532nm$) immersed in water was evaluated using three different models for the partitioning of the radiation force between the solid particle and its liquid environment. Method *I*, which lumps together the induced bound charges on the solid and liquid sides of the interface, was ruled out on account of its unsubstantiated prediction of a strong polarization-dependence of the trap behavior. While methods *II* and *III* both predict the trapping of the bead under *x*- and *y*-polarized beams, the ratio of the maximum lateral restoring force for polarization directions parallel and perpendicular to the particle offset direction was found to be 0.92 for method *II* and 1.2 for method *III*. The dependence of the trap stiffness anisotropy on the physical model underlying the computation is thus seen to be large enough to enable one to accept or reject specific models. In the case of trapping polystyrene beads ($n = 1.57$) in water under a $\lambda_0 = 1064nm$ focused beam, comparison with available experimental data strongly suggests that Method *II* is the correct method of computing the force of radiation on objects immersed in a liquid environment. (Method *III* yielded stiffness anisotropies of 0.07, -0.004, -0.1, -0.25, -0.34, -0.06, -0.007 for particle diameters d = 532nm, 690nm, 850nm, 1030nm, 1280nm, 1500nm, 1660nm, respectively.) For small particles (d less than or equal to λ_0), we found excellent agreement between Method *II* and experiment; see Fig. 12. The disagreement between theory and experiment for larger particles may indicate the need for more accurate measurements in this regime. Alternatively, the deviation of the observed stiffness anisotropy of larger particles from model calculations may be a hint that the radiation forces acting on the surrounding liquid and/or the attendant hydrodynamic effects should not be ignored in such calculations.

Appendix. Equivalence of total force (and torque) for two formulations of the Lorentz law

The Lorentz law of force, $\mathbf{F} = q(\mathbf{E} + \mathbf{V} \times \mathbf{B})$, may be written in two different ways for a medium in which the macroscopic version of Maxwell's equations are satisfied. The two formulations are:

$$\mathbf{F}_1(\mathbf{r}) = -(\nabla \cdot \mathbf{P})\mathbf{E} + (\partial \mathbf{P}/\partial t) \times \mathbf{B} \tag{A1}$$

$$\mathbf{F}_2(\mathbf{r}) = (\mathbf{P} \cdot \nabla)\mathbf{E} + (\partial \mathbf{P}/\partial t) \times \mathbf{B} \tag{A2}$$

In an isotropic, homogeneous medium of dielectric constant ε where the electric displacement field $\mathbf{D}(\mathbf{r}) = \varepsilon_0 \mathbf{E}(\mathbf{r}) + \mathbf{P}(\mathbf{r}) = \varepsilon_0 \varepsilon \mathbf{E}(\mathbf{r})$, one may write $\mathbf{P}(\mathbf{r}) = \varepsilon_0(\varepsilon - 1)\mathbf{E}(\mathbf{r})$. In the absence of free charges and free currents in such a medium, Maxwell's first equation, $\nabla \cdot \mathbf{D}(\mathbf{r}) = 0$, implies that the volume density of bound charges within the medium is zero, that is, $\rho_b = -\nabla \cdot \mathbf{P}(\mathbf{r}) = 0$. This leaves for the Lorentz force density in the bulk of the medium, according to Eq.(A1), only the magnetic contribution, namely, $\mathbf{F}_1(\mathbf{r}) = (\partial \mathbf{P}/\partial t) \times \mathbf{B}$. Equation (A2), however, leads to an entirely different result. Using Maxwell's equations in conjunction with the Lorentz law as expressed in Eq.(A2) yields:

$$\mathbf{F}_2(\mathbf{r}) = \frac{1}{4}\varepsilon_0(\varepsilon - 1)\nabla(|E_x|^2 + |E_y|^2 + |E_z|^2). \tag{A3}$$

In other words, the Lorentz force density in the bulk of a homogeneous medium according to the second formulation is proportional to the gradient of the E-field intensity, irrespective of the state of polarization of the optical field.

At the boundaries of isotropic media, the two formulations again lead to substantially different force densities. The contribution of the B-field to the surface force is negligible as, for non-magnetic media ($\mu = 1$), the B-field is continuous at the boundary; also, the discontinuity of $\partial \mathbf{P}/\partial t$, if any, is finite. In contrast, the perpendicular component of the E-field at the boundary, $E_\perp$, has a sharp discontinuity, which results in a Dirac δ-function behavior for the E-field gradient $\nabla E_\perp(\mathbf{r})$. When evaluating the integral of $(\mathbf{P} \cdot \nabla)\mathbf{E}$ across the boundary between the free space and a medium of dielectric constant ε, one finds a force density (per unit interfacial area) $\mathbf{F}_2(\mathbf{r}) = \frac{1}{2} P_\perp (\mathbf{E}_{a\perp} - \mathbf{E}_{b\perp})$; here $P_\perp$ is the normal component of the medium's polarization density at the surface, $\mathbf{E}_{a\perp}$ is the perpendicular E-field in the free space region just outside the medium, and $\mathbf{E}_{b\perp}$ is the perpendicular E-field within the medium just beneath the surface, Ref. [5]. Note that $P_\perp$, being identical to the bound surface charge density σ_b, may be derived from the discontinuity in the magnitude of $E_\perp$ at the surface, namely, $P_\perp = \sigma_b = \varepsilon_0(E_{a\perp} - E_{b\perp})$.

In the first formulation based on Eq.(A1), the surface charge density σ_b derived from $-(\nabla \cdot \mathbf{P})$ is equal to $P_\perp$, as mentioned before. This σ_b, however, must be multiplied by the effective E-field at the boundary to yield the surface force density (per unit area) $\mathbf{F}_1(\mathbf{r}) = \sigma_b \mathbf{E}$. The tangential component $\mathbf{E}_\parallel$ of the E-field is continuous at the boundary and, therefore, defined unambiguously. The perpendicular component, however, can be shown to be exactly equal to the average $\mathbf{E}_\perp$ at the boundary, namely, $\mathbf{E}_\perp = \frac{1}{2}(\mathbf{E}_{a\perp} + \mathbf{E}_{b\perp})$.

We give two examples from electrostatics to demonstrate the difference between the two formulations of the Lorentz law as concerns the E-field contribution to the force within isotropic media. In the first example, shown on the left-hand side of Fig. A1, the medium is heterogeneous with a dielectric constant $\varepsilon(x)$. Let the D-field be constant and oriented along the x-axis, that is, $\mathbf{D}(x) = D_0 \hat{\mathbf{x}} = [\varepsilon_0 E_x(x) + P_x(x)]\hat{\mathbf{x}} = \varepsilon_0 \varepsilon(x) E_x(x) \hat{\mathbf{x}}$. (Clearly $\nabla \cdot \mathbf{D} = 0$ and $\nabla \times \mathbf{E} = 0$, as required by Maxwell's electrostatic equations.) From the first formulation, Eq.(A1), we find:

$$F_{1x}^{surface}(x=0) = -\frac{1}{2}P_x(0)[E_0 + E_x(0)] = -(D_0/2\varepsilon_0)[1 + 1/\varepsilon(0)]P_x(0); \qquad x = 0 \quad \text{(A4)}$$

$$F_{1x}^{bulk}(x) = -E_x(x)dP_x(x)/dx = (1/2\varepsilon_0)d[D_0 - P_x(x)]^2/dx; \qquad 0 < x < L \quad \text{(A5)}$$

$$F_{1x}^{surface}(x=L) = (D_0/2\varepsilon_0)[1 + 1/\varepsilon(L)]P_x(L); \qquad x = L \quad \text{(A6)}$$

The second formulation, Eq.(A2), yields:

$$F_{2x}^{surface}(x=0) = \frac{1}{2}P_x(0)[E_x(0) - E_0] = -(D_0/2\varepsilon_0)[1 - 1/\varepsilon(0)]P_x(0); \qquad x = 0 \quad \text{(A7)}$$

$$F_{2x}^{bulk}(x) = P_x(x)dE_x(x)/dx = -(1/2\varepsilon_0)dP_x^2(x)/dx; \qquad 0 < x < L \quad \text{(A8)}$$

$$F_{2x}^{surface}(x=L) = (D_0/2\varepsilon_0)[1 - 1/\varepsilon(L)]P_x(L); \qquad x = L \quad \text{(A9)}$$

Clearly the two formulations predict different distributions for the force density, both at the surface and in the bulk. However, the total force - obtained by adding the surface contributions to the integral of the force density within the bulk- turns out to be exactly the same in the two formulations ($\mathbf{F}^{total} = 0$).

In the second example, shown on the right-hand-side of Fig. A1, the medium is homogeneous (dielectric constant$=\varepsilon$), and the D-field profile is $\mathbf{D}(\mathbf{r}) = (D_0 r_0/r)\hat{\theta}$. In the first formulation, based on Eq.(A1), the bulk force is zero, and the surface force density is

$$\mathbf{F}_1^{surface}(r, \theta = \theta_{0,1}) = \pm(1/2\varepsilon_0)(1 - 1/\varepsilon^2)(D_0 r_0/r)^2 \hat{\theta}; \qquad \theta = \theta_0, \theta_1 \quad \text{(A10)}$$

In the second formulation, Eq.(A2) yields:

$$\mathbf{F}_2^{surface}(r, \theta = \theta_{0,1}) = \pm(1/2\varepsilon_0)(1 - 1/\varepsilon)^2(D_0 r_0/r)^2 \hat{\theta}; \qquad \theta = \theta_0, \theta_1 \quad \text{(A11)}$$

$$\mathbf{F}_2^{bulk}(r, \theta) = -(1/\varepsilon_0\varepsilon)(1 - 1/\varepsilon)(D_0^2 r_0^2/r^3)\hat{\mathbf{r}}; \qquad \theta_0 < \theta < \theta_1 \quad \text{(A12)}$$

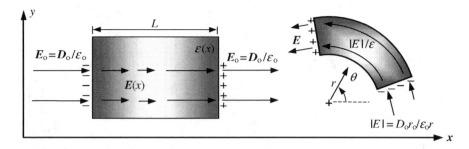

Fig. A1. Examples of electrostatic field distributions in two dielectric media. On the left-hand side, the medium's dielectric constant $\varepsilon(x)$ varies along the x-axis while the electric displacement $\mathbf{D} = D_0\hat{\mathbf{x}}$ remains constant. On the right-hand side, ε is constant within the crescent-shaped medium $(r_0 < r < r_1, \theta_0 < \theta < \theta_1)$, while the azimuthally oriented $\mathbf{D}(\mathbf{r}) = (D_0 r_0/r)\hat{\theta}$ decreases with the inverse of the radius r.

Once again, the force distributions in the two formulations are quite different, but the total force is exactly the same, that is,

$$\mathbf{F}^{total} = -(1/\varepsilon_0)(1 - 1/\varepsilon^2)(D_0 r_0/r)^2 \sin[(\theta_1 - \theta_0)/2]\hat{\mathbf{r}}. \tag{A13}$$

Next we prove the generality of the above results by showing that the total force obtained from Eqs. (A1) and (A2) is the same under all circumstances. The proof of equivalence between the two formulations with regard to total force (and also total torque) is due to Barnett and Loudon Ref. [5], who also clarified the role of surface forces in the second formulation. In what follows we outline this proof of equivalence along the same lines as originally suggested by Barnett and Loudon Ref. [5], then extend their results to objects that are immersed in a liquid (or surrounded by an isotropic, homogeneous medium of a differing index of refraction).

Consider the difference between $\mathbf{F}_1(\mathbf{r})$ and $\mathbf{F}_2(\mathbf{r})$, written as

$$\Delta\mathbf{F}(\mathbf{r}) = \mathbf{F}_2(\mathbf{r}) - \mathbf{F}_1(\mathbf{r}) = (\mathbf{P}\cdot\nabla)\mathbf{E} + (\nabla\cdot\mathbf{P})\mathbf{E} = \partial(P_x\mathbf{E})/\partial x + \partial(P_y\mathbf{E})/\partial y + \partial(P_z\mathbf{E})/\partial z. \tag{A14}$$

When $\int\Delta\mathbf{F}(\mathbf{r})dxdydz$ over the volume of an object is being evaluated, individual terms on the right-hand-side of Eq.(A14), being complete differentials, produce expressions such as $(P_x\mathbf{E})|_{x\_max} - (P_x\mathbf{E})|_{x\_min}$ after a single integration over the proper variable. These expressions then reduce to zero because x_{min}, x_{max} are either in the free-space region outside the object, where $\mathbf{P} = 0$, or they are outside the beam's boundary, where $\mathbf{E} = \mathbf{P} = 0$. Either way, the volume integral of $\Delta\mathbf{F}(\mathbf{r})$ evaluates to zero, yielding $\int\mathbf{F}_1(\mathbf{r})dxdydz = \int\mathbf{F}_2(\mathbf{r})dxdydz$.

If the object of interest happens to be immersed in (or surrounded by) a medium other than the free space, where, for example, the condition $(P_x\mathbf{E})|_{x\_max} = 0$ cannot be ascertained, one must proceed to assume the existence of a narrow gap between the object and its surroundings, as shown in Fig. A2. Under such circumstances, the radiation force exerted on the boundary of the object should be computed using $\mathbf{E}_g(\mathbf{r})$, the gap field, rather than $\mathbf{E}_a(\mathbf{r})$, the field in the immersion medium just outside the object. However, the introduction of the gap creates oppositely charged layers (across the gap), and the force of attraction between these proximate charges accounts for the difference between $\mathbf{F}^{surface}(\mathbf{r})$ computed using $\mathbf{E}_g(\mathbf{r})$ and $\mathbf{E}_b(\mathbf{r})$ on the one hand, and that computed using $\mathbf{E}_a(\mathbf{r})$ and $\mathbf{E}_b(\mathbf{r})$ on the other hand. Once the effect of this attractive force on the object is discounted, the remaining force may be computed by ignoring the gap field and assuming, for the first formulation, that the effective E-field is $\frac{1}{2}(\mathbf{E}_a + \mathbf{E}_b)$, while, for the second formulation, the E-field discontinuity is $\mathbf{E}_a - \mathbf{E}_b$. [Note: In the first formulation, using $\frac{1}{2}(\mathbf{E}_a + \mathbf{E}_b)$ for the effective field leads to Method II discussed in Section 2, whereas using $\frac{1}{2}(\mathbf{E}_g + \mathbf{E}_b)$ leads to Method III.]

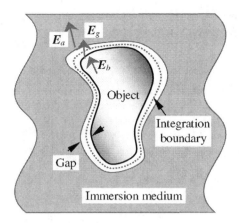

Fig. A2. When the object is surrounded by a medium other than the free space, a narrow gap may be imagined to exist between the object and its surroundings, so that, within the gap, the polarization density $\mathbf{P}(\mathbf{r})$ may be set to zero. The integration boundary is then placed within the gap to ensure that the integrated force densities corresponding to Eqs. (A1) and (A2) lead to the same total force.

Finally, we show that the total torque $\mathbf{T}^{total}$ exerted on a given object is independent of whether $\mathbf{F}_1(\mathbf{r})$ or $\mathbf{F}_2(\mathbf{r})$ is used to compute the torque. In the following derivation, the fact that $\mathbf{P}(\mathbf{r}) = 0$ when $\mathbf{r}$ is outside the proper boundaries of the object is used in both steps. In the first step, terms containing the integrals of complete differentials are omitted. (As before, such integrals reduce to expressions that vanish outside the object's boundaries.) In the second step, integration by parts is used to simplify the integrands.

$$
\begin{aligned}
\Delta\mathbf{T}^{total} &= \iiint \mathbf{r} \times \Delta\mathbf{F}(\mathbf{r})\,dxdydz \\
&= \iiint \mathbf{r} \times [\partial(P_x\mathbf{E})/\partial x + \partial(P_y\mathbf{E})/\partial y + \partial(P_z\mathbf{E})/\partial z]\,dxdydz \\
&= \iiint \{[y\partial(P_yE_z)/\partial y - z\partial(P_zE_y)/\partial z]\hat{\mathbf{x}} + [z\partial(P_zE_x)/\partial z - x\partial(P_xE_z)/\partial x]\hat{\mathbf{y}} \\
&\quad + [x\partial(P_xE_y)/\partial x - y\partial(P_yE_x)/\partial y]\hat{\mathbf{z}}\}\,dxdydz \\
&= -\iiint [(P_yE_z - P_zE_y)\hat{\mathbf{x}} + (P_zE_x - P_xE_z)\hat{\mathbf{y}} + (P_xE_y - P_yE_x)\hat{\mathbf{z}}]\,dxdydz \\
&= \iiint (\mathbf{E} \times \mathbf{P})\,dxdydz.
\end{aligned}
\tag{A15}
$$

In an isotropic medium $\mathbf{E}$ and $\mathbf{P}$ will be parallel to each other and, therefore, their cross-product appearing on the right-hand-side of Eq.(A15) will vanish. Consequently $\Delta\mathbf{T}^{total} = 0$, completing the proof that both formulations of the Lorentz law yield the same overall torque on the object.

It is remarkable that the two formulations in Eqs. (A1) and (A2) yield the same total force (and torque) exerted on a given object under quite general circumstances. In fact, until recently, the present authors had relied solely on the first formulation, assuming (falsely) that the second formulation is incapable of producing consistent answers for certain well-defined problems Ref. [1]-[4]. This situation has now changed with the equivalence proof provided by Barnett and Loudon Ref. [5]. It appears, therefore, that the classical electromagnetic theory (with its reliance on the macroscopic properties of matter, such as polarization density $\mathbf{P}$, displacement field $\mathbf{D}$, dielectric constant ε, etc.) cannot decide which formula, if any, provides the correct force distribution throughout the object (even as both formulas yield the correct values for total force and torque). A resolution of this interesting problem may have to await the verdict of future experiments.

Acknowledgments

We are grateful to Rodney Loudon and Stephen Barnett for illuminating discussions and, in particular, for sharing the preprint of their latest manuscript, Ref. [5], with us. Thanks are also due to Alexander Rohrbach for his extensive comments on an early draft of this paper, and to Koen Visscher for granting us access to his laboratory. This work has been supported by the Air Force Office of Scientific Research (AFOSR) contracts FA9550-04-1-0213, FA9550-04-1-0355, and F49620-03-1-0194.

On the electromagnetic force on a dielectric medium

Stephen M Barnett[1] and Rodney Loudon[2]

[1] SUPA, Department of Physics, University of Strathclyde, Glasgow G4 0NG, UK
[2] Electronic Systems Engineering, University of Essex, Colchester CO4 3SQ, UK

E-mail: steve@phys.strath.ac.uk and loudr@essex.ac.uk

Received 27 January 2006, in final form 2 March 2006
Published 24 July 2006
Online at stacks.iop.org/JPhysB/39/S671

Abstract
The study of the mechanical effects of light on a dielectric medium has led to
two quite distinct forms for the force density: one based on the microscopic
distribution of charges and the other on the distribution of atomic dipoles. Both
approaches are based directly on the Lorentz force, but it has been suggested
that they lead, in a number of cases, to significantly different predictions. In
this paper we address this paradoxical situation and show that in the majority
of problems the force densities lead to identical results. Where the theories
do differ we attempt to determine which of the two descriptions is the more
reliable.

This paper is respectfully dedicated to the memory of Edwin Power

1. Introduction: the two force densities

The problem of calculating the force induced on a dielectric material by an electromagnetic
field has a surprisingly long and complicated history. It is intimately tied up with such thorny
issues as the correct form of the momentum of the electromagnetic field inside a medium
[1]. Recent approaches have focused directly on the Lorentz force as the means by which to
calculate the force on the medium. The problem, however, is that there are two competing
forms of the Lorentz force density: one, based on treating the medium as formed from
individual charges, has the form

$$\mathbf{f}^c = -(\nabla \cdot \mathbf{P})\mathbf{E} + \dot{\mathbf{P}} \times \mathbf{B}, \tag{1}$$

while the other, based on treating the medium as being formed from individual dipoles, has
the form

$$\mathbf{f}^d = (\mathbf{P} \cdot \nabla)\mathbf{E} + \dot{\mathbf{P}} \times \mathbf{B}. \tag{2}$$

The first of these follows directly from the application of the Lorentz force law to the bound
charges comprising the dielectric and has been applied to study the forces and torques on
dielectric media in a wide variety of situations [2, 3]. The second follows from the force on a

medium comprising point dipoles [4]; equivalent forms have also been given by Penfield and Haus [5] and by Landau, Lifshitz and Pitaevskii [6]. We have used it to study in some detail the forces acting on dielectric media [7–9]. It has been suggested [10] that the different force densities make distinct predictions concerning the interaction of light with the dielectric and it is important, therefore, to explore thoroughly the fundamental bases for these expressions and to investigate in detail the differences between them. This is the aim of the present paper. We shall work within the framework of classical electromagnetism, but will discuss briefly the implications for quantum phenomena at the end of the paper.

A good and safe place to start is with the microscopic Maxwell equations which, in SI units, take the familiar form:

$$\nabla \cdot \mathbf{E} = \frac{\varrho}{\varepsilon_0} \qquad \nabla \cdot \mathbf{B} = 0 \qquad \nabla \times \mathbf{E} = -\dot{\mathbf{B}} \qquad \nabla \times \mathbf{B} = \frac{1}{c^2}\dot{\mathbf{E}} + \mu_0 \mathbf{J}. \tag{3}$$

Here $\mathbf{E}$ and $\mathbf{B}$ are the full microscopic electric and magnetic fields, and ϱ and $\mathbf{J}$ are the true electric charge and current densities with contributions for every charged particle present, including all of those in the atoms forming any host medium. The description is completed by the form of the force exerted on the charges due to any electric and magnetic fields. We are interested in the total or nett force exerted on all of the charges in a given volume and, for this reason, it is simplest to work with a force density, or force per unit volume, in the form

$$\mathbf{f}^L = \varrho \mathbf{E} + \mathbf{J} \times \mathbf{B}. \tag{4}$$

One of our tasks will be to work from this force to obtain an appropriate form for the force on a dielectric. It is interesting to note that H A Lorentz himself wrote this force density [11], albeit in a more old-fashioned notation.

We can understand the origin of the two rival force densities (1) and (2) by looking at the dielectric medium on the microscopic scale. We begin by introducing the displacement and polarization fields, $\mathbf{D}$ and $\mathbf{P}$, which are related by

$$\mathbf{D} = \varepsilon_0 \mathbf{E} + \mathbf{P}. \tag{5}$$

We note that the dielectric medium is electrically neutral and that all of the charges are bound and so set $\nabla \cdot \mathbf{D} = 0$. This, together with the first Maxwell equation, leads us to

$$\nabla \cdot \mathbf{P} = -\varrho. \tag{6}$$

Taking the divergence of the final Maxwell equation then leads us to

$$\nabla \cdot \dot{\mathbf{P}} = \nabla \cdot \mathbf{J}. \tag{7}$$

The relation (6) constrains only the longitudinal part of $\mathbf{P}$ and hence we can, without loss of generality, use (7) to define $\mathbf{P}$ so that it obeys

$$\dot{\mathbf{P}} = \mathbf{J}. \tag{8}$$

Substituting these relationships into the Lorentz force density gives us, directly, the charge-based force density $\mathbf{f}^c$ given in (1).

An alternative approach treats not the individual charges, but the individual atomic electric dipoles. The force on a single point dipole $\mathbf{d}$ is

$$\mathbf{F} = (\mathbf{d} \cdot \nabla)\mathbf{E} + \dot{\mathbf{d}} \times \mathbf{B}, \tag{9}$$

so that the force density is

$$\mathbf{f}^{\text{spd}}(\mathbf{r}) = ((\mathbf{d} \cdot \nabla)\mathbf{E}(\mathbf{r}) + \dot{\mathbf{d}} \times \mathbf{B}(\mathbf{r}))\delta(\mathbf{r} - \mathbf{R}), \tag{10}$$

where $\mathbf{R}$ is the position of the dipole. We note that the polarization $\mathbf{P}$ is also the dipole density

$$\mathbf{P}(\mathbf{r}) = \sum_i \mathbf{d}_i \delta(\mathbf{r} - \mathbf{R}_i), \tag{11}$$

where the sum runs over all the dipoles. It follows immediately that the total force density is $\mathbf{f}^d$ given in (2). It seems that both of the candidate force densities are sensible and that any difference between them is likely to be subtle. This does indeed turn out to be the case. We will find that in most situations the two force densities lead to the same predictions. As a clue to where they do not, we note that the polarization $\mathbf{P}$, as it has been introduced here, is the full microscopic polarization and not the more familiar macroscopic polarization, related to the electric field by an effective susceptibility. We will conclude that it is in the replacement of this microscopic polarization by a macroscopic polarization that the source of the difficulties lies. This conclusion has been reached before, of course, and Sommerfeld even went as far as stating that 'ponderable bodies with their continuous material constants ε, μ are simply convenient abstractions and are not physical realities' [12]. The macroscopic dielectric constant ε can be used but care needs to be taken.

2. Is there really a problem?

It is quite possible, of course, that there is no problem. This would be the case if, despite appearances, all possible predictions derived on the basis of the force densities (1) and (2) were identical. For most problems this does indeed turn out to be the case and we need to search very carefully to find any situations in which there are different predictions.

2.1. The microscopic scale

In the introduction we derived the two candidate force densities on the basis of rival, but good, theoretical models based on either the microscopic distribution of charges or on the microscopic distribution of point dipoles. For this reason we would expect the two force densities to give the same predictions at least at the microscopic level in which we consider a single point dipole at position $\mathbf{R}$. The polarization field associated with this dipole is

$$\mathbf{P}(\mathbf{r}) = \mathbf{d}\delta(\mathbf{r} - \mathbf{R}), \tag{12}$$

and so the two force densities are

$$\mathbf{f}^c = -(\mathbf{d} \cdot \nabla\delta(\mathbf{r} - \mathbf{R}))\mathbf{E} + \dot{\mathbf{d}} \times \mathbf{B}\delta(\mathbf{r} - \mathbf{R})$$
$$\mathbf{f}^d = ((\mathbf{d} \cdot \nabla)\mathbf{E} + \dot{\mathbf{d}} \times \mathbf{B})\delta(\mathbf{r} - \mathbf{R}). \tag{13}$$

Despite the differences of form, the total *force* exerted on the dipole is the same for the two force densities. To see this we integrate the two force densities over a volume containing the dipole:

$$\mathbf{F}^d = \int \mathbf{f}^d \, \mathrm{d}V = (\mathbf{d} \cdot \nabla)\mathbf{E} + \dot{\mathbf{d}} \times \mathbf{B}$$

$$\mathbf{F}^c = \int \mathbf{f}^c \, \mathrm{d}V = -\int \mathbf{E}(\mathbf{d} \cdot \nabla\delta(\mathbf{r} - \mathbf{R})) \, \mathrm{d}V + \dot{\mathbf{d}} \times \mathbf{B}$$
$$= (\mathbf{d} \cdot \nabla)\mathbf{E} + \dot{\mathbf{d}} \times \mathbf{B}$$
$$= \mathbf{F}^d, \tag{14}$$

where we have used integration by parts in order to treat the derivatives of the delta function. It is clear that, at least at the microscopic level, the two force densities give rise to the same force and should, therefore, be regarded as equivalent.

2.2. The macroscopic scale

At the opposite extreme to the single point dipole is the total force exerted on a macroscopic dielectric medium. We consider a dielectric surrounded by vacuum and calculate the total force associated with each of the two densities by integrating over the volume of the dielectric. In doing so, we need to be careful about defining the integration volume. We select a volume V, bounded by the surface S_j, which contains the dielectric and a thin volume of the surrounding vacuum, so that dielectric–vacuum interface is fully contained within the integration volume. There is, of course, no force exerted in the vacuum region and the polarization there is identically zero. It is convenient to work with Cartesian components of the force, labelled by an index i, which takes the three values x, y and z. We will also employ the familiar convention in which a summation is implied over the three values, corresponding to the Cartesian coordinates, of any index appearing twice [13]. The difference between the two total forces is then

$$F_i^d - F_i^c = \int_V (P_j \nabla_j + (\nabla_j P_j)) E_i \, dV$$

$$= \int_V \nabla_j (P_j E_i) \, dV. \tag{15}$$

We should note that the fields, **E** and **P**, are the microscopic fields; we shall introduce the familiar averaged macroscopic fields in the following subsection. Gauss's theorem allows us to write this difference of forces as an integral over the surface of the integration volume:

$$F_i^d - F_i^c = \int P_j E_i \, dS_j$$

$$= 0, \tag{16}$$

as the polarization is zero on the integration surface.

As with the point dipole, the difference in the total force on the dielectric given by the two force densities is *zero*. This is reasonable as $\mathbf{F}^d$ is based on the force exerted on the (centre of mass of) each dipole, while $\mathbf{F}^c$ is based on the force acting on each charge. The nett effect in each case is the same total force. It has been suggested that the total forces calculated on the basis of $\mathbf{f}^c$ or $\mathbf{f}^d$ are sometimes different [10]. This is at odds, however, with the simple proof presented here that they should be the same. We shall explain this contradiction in section 2.4.

2.3. The mesoscopic scale

The results of the preceding two sections suggest that there might not be a significant difference between the two force densities. This happy situation does not survive, however, when we examine the force acting on a small volume v of dielectric, enclosed by the surface s_j. The volume is supposed to be very large compared to the spacing of the individual dipoles, but small compared with the scales on which any externally applied electromagnetic field varies. The use of such volumes is familiar from elementary classes on electromagnetism where it is used to derive macroscopic fields by averaging over the volume [14].

The difference between the forces acting on our small volume is simply

$$F_i^d - F_i^c = \int P_j E_i \, ds_j. \tag{17}$$

If we keep P_j as the microscopic polarization, then we can make the integral zero by carefully threading our surface between the dipoles. This is impractical for real calculations, however, and if we wish to employ a macroscopic polarization ($\bar{\mathbf{P}}$), as we must, then the difference persists and needs to be resolved. Might it be that this difference can be removed by introducing a distribution of charges onto the surface of our volume? The answer would appear to be no. To see this we suppose that our volume is within a larger region in which the material susceptibility is a constant. It is then clear that $\bar{\mathbf{P}}$ will be continuous, $\nabla \cdot \bar{\mathbf{P}}$ will be zero, and that there will be no buildup of charge on the surface of our volume v. It follows that there will be no surface force that we can easily add to restore the agreement between the charge and dipole forms of the force.

We note that direct application of Maxwell's equations to (2) leads to the equivalent form [4]

$$f_i^d = P_j \nabla_i E_j + \frac{\partial}{\partial t} (\mathbf{P} \times \mathbf{B})_i. \tag{18}$$

This form is useful in electrostatic problems, where the time derivative vanishes, and on introducing the macroscopic fields and writing $\bar{\mathbf{P}} = \varepsilon_0 (\varepsilon - 1) \bar{\mathbf{E}}$ we reach the simple form

$$\mathbf{f}^d = \tfrac{1}{2} \varepsilon_0 (\varepsilon - 1) \nabla \bar{E}^2. \tag{19}$$

For a beam of (quasi) monochromatic radiation, it is convenient to introduce the complex macroscopic electric field $\bar{\mathcal{E}}$, the real part of which is the electric field. The corresponding macroscopic polarization is then $\bar{\mathcal{P}} = \varepsilon_0 (\varepsilon - 1) \bar{\mathcal{E}}$ and leads to distinct forms for the force density:

$$\begin{aligned} f_i^c &= \tfrac{1}{2} \varepsilon_0 (\varepsilon - 1) \operatorname{Re}(\bar{\mathcal{E}}_j^* \nabla_i \bar{\mathcal{E}}_j - \bar{\mathcal{E}}_j^* \nabla_j \bar{\mathcal{E}}_i) \\ f_i^d &= \tfrac{1}{4} \varepsilon_0 (\varepsilon - 1) \nabla_i |\bar{\mathcal{E}}|^2. \end{aligned} \tag{20}$$

Consider a linearly polarized beam propagating in the z-direction in which the y-component of the electric field is zero. If the beam is not too tightly focused then there will be a component of the electric field pointing in the x-direction and a very much smaller component along the z-direction, associated with the transverse confinement of the beam, which can be neglected for the purposes of this exercise. Both of the force densities (20) predict a force tending to constrict the dielectric within the beam. The difference, however, is that $f_x^c = 0$, while $f_x^d = \tfrac{1}{4} \varepsilon_0 (\varepsilon - 1) \partial |\bar{\mathcal{E}}|^2 / \partial x$ [2].

2.4. Dielectric interfaces

It is clear from the foregoing analysis that the introduction of the macroscopic electric and polarization fields introduces a significant difference between the two force densities and that this is manifest at the mesoscopic scale. It is important to check that the use of macroscopic fields does not invalidate the conclusion, derived in section 2.2, that the nett forces are the same. This leads us to investigate the forces acting at dielectric interfaces.

The problem is most simply addressed by reference to a classic and exactly solvable problem in electrostatics: the attraction by a point charge on a large piece of the dielectric material. We consider a semi-infinite dielectric medium, with the dielectric constant ε, filling the half-space $z < 0$. We shall work in cylindrical polar coordinates (ρ, ϕ, z) and consider a point charge q placed at $(0, 0, d)$ (see figure 1). The macroscopic electric field is readily calculated using the method images and leads to [15]

$$\bar{\mathbf{E}}^{\text{out}} = \frac{1}{4\pi\varepsilon_0} \left[\hat{\boldsymbol{\rho}} \left(\frac{q\rho}{R_1^3} + \frac{q'\rho}{R_2^3} \right) + \hat{\mathbf{z}} \left(\frac{q(z-d)}{R_1^3} + \frac{q'(z+d)}{R_2^3} \right) \right] \tag{21}$$

outside the dielectric ($z > 0$) and

$$\bar{\mathbf{E}}^{\text{in}} = \frac{q''}{4\pi\varepsilon_0\varepsilon} \left(\hat{\boldsymbol{\rho}} \frac{\rho}{R_1^3} + \hat{\mathbf{z}} \frac{(z-d)}{R_2^3} \right) \tag{22}$$

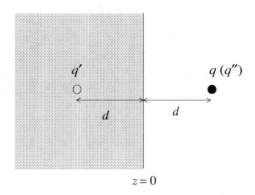

Figure 1. Configuration for calculating the force exerted by a charge q on a dielectric medium. The electric field can be calculated by the method of images: the field outside the dielectric is that due to the charge q and its image q' and the field inside the dielectric is equivalent to that associated with a charge q'' at the position of q.

inside the dielectric ($z < 0$). Here $\hat{\rho}$ and $\hat{\mathbf{z}}$ are unit vectors in the radial and z-directions and the two distances are $R_1 = [\rho^2 + (z - d)^2]^{1/2}$ and $R_2 = [\rho^2 + (z + d)^2]^{1/2}$. The charges q' and q'' are related to q by

$$q' = -\left(\frac{\varepsilon - 1}{\varepsilon + 1}\right) q \qquad q'' = \left(\frac{2\varepsilon}{\varepsilon + 1}\right) q. \tag{23}$$

Our task is to calculate the force exerted on the dielectric and no less than five possible methods come to mind. These are (i) to extend the image concept by calculating the force exerted by the charge q on its image q', (ii) to calculate the force exerted by q on the surface charge density induced by q, (iii) to use the force density $-(\nabla \cdot \bar{\mathbf{P}})\bar{\mathbf{E}}$, (iv) to use the force density $(\bar{\mathbf{P}} \cdot \nabla)\bar{\mathbf{E}}$ and (v) to use the force density $\frac{1}{2}\varepsilon_0(\varepsilon - 1)\nabla \bar{E}^2$. We shall examine each of these in turn.

2.4.1. Image method. The method of images replaces the dielectric by a point charge q' positioned at $(0, 0 - d)$ (see figure 1). The force exerted on the dielectric should be equivalent to that exerted by the charge q on q'. This gives

$$\mathbf{F} = \frac{q'q}{4\pi\varepsilon_0(2d)^2}(-\hat{\mathbf{z}}) = \left(\frac{\varepsilon - 1}{\varepsilon + 1}\right)\frac{q^2}{16\pi\varepsilon_0 d^2}\hat{\mathbf{z}}. \tag{24}$$

2.4.2. Force exerted on the surface charges. The point charge induces a charge distribution on the dielectric surface. This polarization–charge density is [15]

$$\sigma_{\text{pol}} = -\frac{q}{2\pi}\left(\frac{\varepsilon - 1}{\varepsilon + 1}\right)\frac{d}{(\rho^2 + d^2)^{3/2}}. \tag{25}$$

The force exerted by the point charge q on the surface charges is

$$\mathbf{F} = \int_0^{2\pi} d\phi \int_0^\infty \rho \, d\rho \, \sigma_{\text{pol}} \frac{qd}{4\pi\varepsilon_0(\rho^2 + d^2)^{3/2}}(-\hat{\mathbf{z}}), \tag{26}$$

where we have used the cylindrical symmetry to see that the force points along the z-axis. Substituting for σ_{pol} gives

$$\mathbf{F} = \frac{q^2 d^2}{4\pi\varepsilon_0}\left(\frac{\varepsilon - 1}{\varepsilon + 1}\right)\int_0^\infty \rho \, d\rho \, \frac{1}{(\rho^2 + d^2)^3}\hat{\mathbf{z}} = \left(\frac{\varepsilon - 1}{\varepsilon + 1}\right)\frac{q^2}{16\pi\varepsilon_0 d^2}\hat{\mathbf{z}}, \tag{27}$$

which agrees with (24).

2.4.3. Force on the charges, $-(\nabla \cdot \bar{\mathbf{P}})\bar{\mathbf{E}}$.   The term $\nabla \cdot \bar{\mathbf{P}}$ is intimately related to the charge density (6). For our problem, it is simply related to σ_{pol}:

$$-\nabla \cdot \bar{\mathbf{P}} = \sigma_{\text{pol}}\delta(z). \tag{28}$$

Hence the force density is

$$\mathbf{f}^c = \sigma_{\text{pol}}\delta(z)\bar{\mathbf{E}} \tag{29}$$

and the total force on the dielectric is

$$\mathbf{F} = \int \mathbf{f}^c \, dV = \int_0^{2\pi} d\phi \int_0^\infty \rho \, d\rho \, \sigma_{\text{pol}}\bar{\mathbf{E}}(z=0). \tag{30}$$

The field $\bar{\mathbf{E}}$ is the macroscopic field obtained by a local volume average. It is clear, therefore, that $\bar{\mathbf{E}}(\rho, \phi, 0)$ should be equal to its local average, that is

$$\bar{\mathbf{E}}(\rho, \phi, 0) = \tfrac{1}{2}[\bar{\mathbf{E}}(\rho, \phi, 0^+) + \bar{\mathbf{E}}(\rho, \phi, 0^-)]. \tag{31}$$

Again we can invoke the cylindrical symmetry of the problem to write

$$\begin{aligned}
\mathbf{F} &= \int_0^{2\pi}\int_0^\infty \rho \, d\rho \, \sigma_{\text{pol}}\frac{1}{2}[\bar{E}_z(\rho, \phi, 0^+) + \bar{E}_z(\rho, \phi, 0^-)]\hat{\mathbf{z}} \\
&= 2\pi \int_0^\infty \rho \, d\rho \, \sigma_{\text{pol}}\frac{1}{8\pi\varepsilon_0}\frac{(-q + q' - q'')d}{(\rho^2 + d^2)^{3/2}}\hat{\mathbf{z}} \\
&= \left(\frac{\varepsilon - 1}{\varepsilon + 1}\right)\frac{q^2}{16\pi\varepsilon_0 d^2}\hat{\mathbf{z}},
\end{aligned} \tag{32}$$

which is in exact agreement with (24) and (27).

2.4.4. Force on the dipoles, $(\bar{\mathbf{P}} \cdot \nabla)\bar{\mathbf{E}}$.   In contrast with the preceding calculations, the force density $(\bar{\mathbf{P}} \cdot \nabla)\bar{\mathbf{E}}$ leads to two contributions to the total force: one due to the bulk dielectric and a second surface contribution due to the discontinuity of $\bar{E}_z$ at the dielectric interface.

We begin by examining the bulk contribution. Inside the dielectric we can write $\bar{\mathbf{P}} = \varepsilon_0(\varepsilon - 1)\bar{\mathbf{E}}$ and hence integrating $(\bar{\mathbf{P}} \cdot \nabla)\bar{\mathbf{E}}$ over the dielectric volume is equivalent to integrating $\frac{1}{2}\varepsilon_0(\varepsilon - 1)\nabla\bar{E}^2$. Hence the bulk contribution to the force is

$$\begin{aligned}
\mathbf{F}^{\text{bulk}} &= \int_0^{2\pi} d\phi \int_0^\infty \rho \, d\rho \int_{-\infty}^0 dz \frac{1}{2}\varepsilon_0(\varepsilon - 1)\frac{\partial}{\partial z}\bar{E}^2\hat{\mathbf{z}} \\
&= \int_0^{2\pi} d\phi \int_0^\infty \rho \, d\rho \frac{1}{2}\varepsilon_0(\varepsilon - 1)\bar{E}^2(z=0^-)\hat{\mathbf{z}} \\
&= \frac{q^2(\varepsilon - 1)}{8\pi\varepsilon_0 d^2(\varepsilon + 1)^2}\hat{\mathbf{z}}.
\end{aligned} \tag{33}$$

The surface contribution arises from the derivative of $\bar{\mathbf{E}}$ across the interface leading to the force density

$$\bar{P}_z\delta(z)(\bar{E}_z(z=0^+) - \bar{E}_z(0^-))\hat{\mathbf{z}} \tag{34}$$

corresponding to the surface force

$$\begin{aligned}
\mathbf{F}^{\text{surface}} &= \int_0^{2\pi} d\phi \int_0^\infty \rho \, d\rho \int_{-\infty}^\infty dz \, \bar{P}_z\delta(z)(\bar{E}_z(z=0^+) - \bar{E}_z(0^-))\hat{\mathbf{z}} \\
&= 2\pi \int_0^\infty \rho \, d\rho \, \bar{P}_z(\bar{E}_z(z=0^+) - \bar{E}_z(0^-))\hat{\mathbf{z}}.
\end{aligned} \tag{35}$$

The field $\bar{\mathbf{P}}$ is the macroscopic field obtained by a local volume average (in much the same way as for $\bar{\mathbf{E}}$ above). It is clear, therefore that $\bar{P}_z(\rho, \phi, 0)$ should be equal to

$$\bar{P}_z(\rho, \phi, 0) = \tfrac{1}{2}[\bar{P}_z(\rho, \phi, 0^+) + \bar{P}_z(\rho, \phi, 0^-)]$$
$$= \tfrac{1}{2}\bar{P}_z(\rho, \phi, 0^-). \tag{36}$$

Substituting this into (35) gives

$$\mathbf{F}^{\text{surface}} = \pi \int_0^\infty \rho \, d\rho \frac{(\varepsilon - 1)q''(-d)}{32\pi^2 \varepsilon_0 \varepsilon (\rho^2 + d^2)^3} \left(-qd + q'd + \frac{q''d}{\varepsilon} \right) \hat{\mathbf{z}}$$
$$= \frac{q^2(\varepsilon - 1)^2}{16\pi \varepsilon_0 (\varepsilon + 1)^2 d^2} \hat{\mathbf{z}}. \tag{37}$$

Adding the bulk and surface contributions gives the total force:

$$\mathbf{F} = \mathbf{F}^{\text{bulk}} + \mathbf{F}^{\text{surface}}$$
$$= \left(\frac{\varepsilon - 1}{\varepsilon + 1} \right) \frac{q^2}{16\pi \varepsilon_0 d^2} \hat{\mathbf{z}}, \tag{38}$$

which agrees exactly with (24), (27) and (32).

We have arrived at precisely the same result in four very different ways, each suggestive of a different physical interpretation. These are (i) the force on an effective point charge, (ii) the force on a surface density of bound charges due to the Coulomb interaction with our point charge (iii) the force due to the (average) electric field at the boundary acting on a surface density of charges and (iv) the Coulomb force on the elementary dipoles forming the dielectric. The last of these comprises both surface and bulk contributions. Distinguishing between $\mathbf{f}^c$ and $\mathbf{f}^d$ on the basis of the total force is, therefore, impossible.

2.4.5. Force based on permittivity, $\tfrac{1}{2}\varepsilon_0(\varepsilon - 1)\nabla \bar{E}^2$. It is tempting, even if only for computational convenience, to work with the form (19) of the force density. There is, however, a cautionary note that should be mentioned and this concerns its application at boundaries. To illustrate the point we proceed to calculate, one final time, the force exerted by our charge q on its neighbouring dielectric.

As in section 2.4.4, we find that the force comprises a bulk and a surface contribution. The bulk contribution is clearly the same as obtained using $(\bar{\mathbf{P}} \cdot \nabla)\bar{\mathbf{E}}$:

$$\mathbf{F}^{\text{bulk}} = \frac{q^2(\varepsilon - 1)}{8\pi \varepsilon_0 d^2(\varepsilon + 1)^2} \hat{\mathbf{z}}. \tag{39}$$

A surface contribution arises from the derivative of the discontinuous $\bar{E}^2$ at the interface. It follows that the surface contribution to the force density is

$$\tfrac{1}{2}\varepsilon_0(\varepsilon - 1)\delta(z)(\bar{E}^2(\rho, \phi, 0^+) - \bar{E}^2(\rho, \phi, 0^-))\hat{\mathbf{z}} \tag{40}$$

corresponding to the surface force

$$\mathbf{F}^{\text{surface}} = \int dV \frac{1}{2}\varepsilon_0(\varepsilon(\mathbf{r}) - 1)\delta(z)(\bar{E}^2(\rho, \phi, 0^+) - \bar{E}^2(\rho, \phi, 0^-))\hat{\mathbf{z}}$$
$$= 2\pi \int_0^\infty \rho \, d\rho \frac{1}{2}\varepsilon_0(\varepsilon(z=0) - 1)(\bar{E}^2(z=0^+) - \bar{E}^2(z=0^-))\hat{\mathbf{z}}. \tag{41}$$

At this stage we are stuck! The reason is that we have no physical principle by which to determine the appropriate form of $\varepsilon(z = 0)$. *If* we simply let

$$\varepsilon(z = 0) = \tfrac{1}{2}[\varepsilon(z = 0^+) + \varepsilon(z = 0^-)] = \tfrac{1}{2}(\varepsilon + 1) \tag{42}$$

then we get into difficulties. This substitution leads to a surface force

$$\mathbf{F}'^{\text{surface}} = 2\pi \int_0^\infty \rho \, d\rho \frac{1}{4} \frac{(\varepsilon - 1)}{16\pi^2 \varepsilon_0 (\rho^2 + d^2)^3} \frac{q''^2 (\varepsilon^2 - 1)}{\varepsilon^2} \hat{\mathbf{z}}$$
$$= \frac{(\varepsilon - 1)^2 q^2}{32\pi \varepsilon_0 (\varepsilon + 1) d^2} \hat{\mathbf{z}}, \tag{43}$$

and addition to the bulk contribution gives the total force

$$\mathbf{F}' = \left(\frac{\varepsilon - 1}{\varepsilon + 1} \right) \frac{q^2}{16\pi \varepsilon_0 d^2} \left(\frac{\varepsilon^2 + 3}{2(\varepsilon + 1)} \right) \hat{\mathbf{z}}. \tag{44}$$

This differs from the expressions calculated above and it is certainly incorrect.

The problem lies *not* with the force density $\frac{1}{2}\varepsilon_0(\varepsilon - 1)\nabla \bar{E}^2$ and certainly not with the form $(\bar{\mathbf{P}} \cdot \nabla)\bar{\mathbf{E}}$. Rather it is with the unjustified, if appealing, replacement (42). The forms of $\bar{\mathbf{E}}$ and $\bar{\mathbf{P}}$ follow from their definition as local volume averages, but there is no corresponding physical motivation for (42). We can demonstrate the error rather directly by considering the force density in the form (18):

$$f_i^d = \bar{P}_j \nabla_i \bar{E}_j. \tag{45}$$

For our problem of a charge acting on a dielectric, this leads to the surface force density

$$\bar{P}_j(\rho, \phi, z)(\bar{E}_j(\rho, \phi, 0^+) - \bar{E}_j(\rho, \phi, 0^-))\delta(z)\hat{\mathbf{z}}$$
$$= \frac{1}{2}\bar{P}_z(\rho, \phi, 0^-)(\bar{E}_z(\rho, \phi, 0^+) - \bar{E}_z(\rho, \phi, 0^-))\delta(z)\hat{\mathbf{z}}$$
$$= \frac{1}{2}\varepsilon_0(\varepsilon - 1)\bar{E}_z(\rho, \phi, 0^-)(\bar{E}_z(\rho, \phi, 0^+) - \bar{E}_z(\rho, \phi, 0^-))\delta(z)\hat{\mathbf{z}}, \tag{46}$$

where we have used the fact that only the z-component of $\bar{\mathbf{E}}$ is discontinuous at the interface. Using the form $\frac{1}{2}\varepsilon_0(\varepsilon - 1)\nabla \bar{E}^2$ together with the *incorrect* procedure of using the average value of ε, however, leads to the force density

$$\frac{1}{2}\varepsilon_0(\varepsilon(\rho, \phi, z) - 1)\left(\bar{E}_z^2(\rho, \phi, 0^+) - \bar{E}_z^2(\rho, \phi, 0^-) \right)\delta(z)\hat{\mathbf{z}}$$
$$= \frac{1}{2}\varepsilon_0(\varepsilon - 1)(\bar{E}_z(\rho, \phi, 0^+) + \bar{E}_z(\rho, \phi, 0^-))$$
$$\times (\bar{E}_z(\rho, \phi, 0^+) - \bar{E}_z(\rho, \phi, 0^-))\delta(z)\hat{\mathbf{z}}, \tag{47}$$

which contains an additional and incorrect contribution proportional to the z-component of the electric field *outside* the dielectric.

It has been stated that use of the dipole form of the force density produces inconsistencies when applied to problems such as the torque exerted on a dielectric slab and the force exerted on a dielectric wedge [10]. It seems to us that this conclusion arises from the application of the incorrect averaging procedure (42). We have demonstrated by explicit calculation that the correct boundary conditions together with the force density (2) produce forces that are both consistent and in full agreement with those calculated using (1). We will present these calculations elsewhere.

3. A resolution?

We have seen that use of either of our candidate force densities, (1) and (2) produces precisely the same total force, but that this can be formed from apparently different contributions. Only at the level of the force density itself does any significant difference remain. In particular, the force density $\mathbf{f}^c$ seems to predict polarization-dependent forces in a homogeneous medium while $\mathbf{f}^d$ does not. It may be argued that force density is not truly an observable and that only forces on determined volumes have physical significance. Alternatively, it is possible that a satisfactory resolution can only be obtained by appealing to experiment. In this section, however, we attempt to advise on which force density is likely to be more reliable.

It has been pointed out that 'one must avoid the use of ad hoc formulas and, instead, embrace the universal form of the Lorentz force $\varrho(\mathbf{r}, t)\mathbf{E}(\mathbf{r}, t)$, where ϱ is the local charge density' [2]. This seems an eminently sensible suggestion and we will use this fundamental approach to evaluate the force on a volume v of dielectric as introduced in section 2.3.

It is clear that the difference between the two candidate forces arises solely from the electric parts of the respective force densities. For this reason it suffices to consider only the electric part of the Lorentz force density:

$$\mathbf{f}_E^L = \varrho\mathbf{E}. \tag{48}$$

We stress that this force includes all charges comprising the medium and the full microscopic electric field. We can calculate the electric force acting on the matter within our volume v by integration:

$$\mathbf{F}_E^L = \int_v \varrho(\mathbf{r})\mathbf{E}(\mathbf{r})\,\mathrm{d}V. \tag{49}$$

The electric field $\mathbf{E}$ can be divided into two parts, one arising from sources within the volume v and the second arising from sources outside. The former, which includes the fields responsible for the local integrity of the medium, can be neglected by appealing to Newton's third law of motion: they cannot affect the centre of mass of the volume and so do not contribute to the nett force. In using (49) we need only include the electric field due to sources *outside* of the volume v. These fields will have a much slower variation than those due to sources inside the volume and, provided that the volume v is not too big, we can replace $\mathbf{E}$ by the local macroscopic electric field, $\bar{\mathbf{E}}$, due to the externally applied field and the averaged affect of the material surrounding our volume. The slow variation of this field means that we can replace $\mathbf{E}(\mathbf{r})$ by the Taylor expansion of $\bar{\mathbf{E}}(\mathbf{r})$. Let $\mathbf{r}_0$ be a point within the volume v. We can then write the electric field as

$$\mathbf{E}(\mathbf{r}) \approx \bar{\mathbf{E}}(\mathbf{r}_0) + ((\mathbf{r} - \mathbf{r}_0) \cdot \nabla)\,\bar{\mathbf{E}}(\mathbf{r}_0) + \cdots, \tag{50}$$

where $\nabla\bar{\mathbf{E}}(\mathbf{r}_0)$ denotes the derivatives of $\bar{\mathbf{E}}(\mathbf{r})$ evaluated at $\mathbf{r}_0$. Note that no such averaging or introduction of a Taylor expansion is possible for the local charge density ϱ as this quantity varies wildly on Ångstrom lengthscales. The electric field also varies on these scales but, as noted above, the rapidly varying part of the electric field does not contribute to the force.

If we substitute the expanded electric field (50) into the electric Lorentz force acting on our volume of medium (49) then we obtain

$$\mathbf{F}_E^L = \int_v \varrho(\mathbf{r})\bar{\mathbf{E}}(\mathbf{r}_0)\,\mathrm{d}V + \int_v \varrho(\mathbf{r})\,((\mathbf{r} - \mathbf{r}_0) \cdot \nabla)\,\bar{\mathbf{E}}(\mathbf{r}_0)\,\mathrm{d}V$$
$$= \left(\int_v \varrho(\mathbf{r})\,\mathrm{d}V\right)\bar{\mathbf{E}}(\mathbf{r}_0) + \left(\int_v \varrho(\mathbf{r})(\mathbf{r} - \mathbf{r}_0)\,\mathrm{d}V\right) \cdot \nabla_0\bar{\mathbf{E}}(\mathbf{r}_0), \tag{51}$$

where ∇_0 denotes the 'del' operator for the coordinate $\mathbf{r}_0$. The first term is zero as the atomic dipoles comprising the medium are all neutrally charged and hence a volume containing just these will be electrically neutral so that[3]

$$\int_v \varrho(\mathbf{r})\,\mathrm{d}V = 0. \tag{52}$$

---

[3] Indeed were this not the case then the force on the volume would have to include the Coulomb interaction with the necessarily residually charged surrounding dielectric.

The integral in the second term does not vanish, but rather gives the macroscopic polarization as it is usually introduced

$$\bar{\mathbf{P}}(\mathbf{r}_0) = \frac{1}{v} \int_v \varrho(\mathbf{r})(\mathbf{r} - \mathbf{r}_0) \, dV. \tag{53}$$

It is this quantity, and not the microscopic polarization $\mathbf{P}$, that we can write in terms of a dielectric constant:

$$\bar{\mathbf{P}}(\mathbf{r}) = \varepsilon_0(\varepsilon(\mathbf{r}) - 1)\bar{\mathbf{E}}(\mathbf{r}). \tag{54}$$

The electric force acting on our small volume of dielectric is therefore

$$\mathbf{F}_E^L(\mathbf{r}_0) = v(\bar{\mathbf{P}}(\mathbf{r}_0) \cdot \nabla_0)\bar{\mathbf{E}}(\mathbf{r}_0). \tag{55}$$

Dividing this by the volume v then gives the electric force density at a general point $\mathbf{r}$ as

$$\mathbf{f}_E^L(\mathbf{r}) = (\bar{\mathbf{P}}(\mathbf{r}) \cdot \nabla)\bar{\mathbf{E}}(\mathbf{r}), \tag{56}$$

which is the same as the expression based on microscopic dipoles (2). It seems that starting with the fundamental Lorentz force density leads to the force density $\mathbf{f}^d$ and not to $\mathbf{f}^c$, as was suggested in [2]. This suggests that if we wish to work with macroscopic electric and polarization fields, related by a material susceptibility, then we probably should use $\mathbf{f}^d$ and not $\mathbf{f}^c$. Using $\mathbf{f}^d$, moreover, obviates the need to introduce any surface charges at dielectric interfaces.

The problem with $\mathbf{f}^c$ appears when we introduce the macroscopic polarization. Averaging $\mathbf{P}$ over a small volume to get $\bar{\mathbf{P}}$ is not a problem, but averaging the term $\nabla \cdot \mathbf{P} = -\varrho$ does *not* give $\nabla \cdot \bar{\mathbf{P}}$. The force density $\mathbf{f}^d$, by contrast, does not contain any spatial derivatives of the microscopic polarization and so the averaging procedure can be applied safely. An analysis based on the universal form of the Lorentz force suggests, somewhat surprisingly, that the force density is $\mathbf{f}^d$ rather than $\mathbf{f}^c$.

4. Electromagnetic torque

An electromagnetic field can also exert a torque on a dielectric and this has been used in demonstrations of both the orbital and spin angular momenta of light [16]. It has also been proposed that this mechanism might shed some light on the form of optical angular momentum within a dielectric [17].

We can use the preceding discussion to elucidate the form of the electromagnetic torque acting on a dielectric medium. We start by considering the single dipole introduced in section 2.1. The force density $\mathbf{f}^d$ gives the nett force on the *centre of mass* of our point dipole. The associated torque density, about the origin of coordinates, acting on this centre of mass is $\mathbf{r} \times \mathbf{f}^d$ and to this we need to add the internal torque $\mathbf{P} \times \mathbf{E}$, which acts to orient the dipole:

$$\begin{aligned}
\mathbf{t}^d &= \mathbf{r} \times \mathbf{f}^d + \mathbf{P} \times \mathbf{E} \\
&= \mathbf{r} \times ((\mathbf{d} \cdot \nabla)\mathbf{E} + \dot{\mathbf{d}} \times \mathbf{B})\delta(\mathbf{r} - \mathbf{R}) + \mathbf{d} \times \mathbf{E}\delta(\mathbf{r} - \mathbf{R}). \tag{57}
\end{aligned}$$

Integrating this over a volume containing the dipole gives the total torque

$$\mathbf{T}^d = \mathbf{R} \times ((\mathbf{d} \cdot \nabla)\mathbf{E} + \dot{\mathbf{d}} \times \mathbf{B}) + \mathbf{d} \times \mathbf{E}. \tag{58}$$

We might expect that the force, $\mathbf{f}^c$, being based on the microscopic distribution of charges should also give the correct torque. In this case, however, there is no separation of a centre of mass and so the full torque density is

$$\mathbf{t}^c = \mathbf{r} \times [-\mathbf{E}(\mathbf{d} \cdot \nabla \delta(\mathbf{r} - \mathbf{R})) + \dot{\mathbf{d}} \times \mathbf{B}\delta(\mathbf{r} - \mathbf{R})]. \tag{59}$$

The total torque is again obtained by integrating over a volume containing the dipole:

$$
\begin{aligned}
\mathbf{T}^c &= -\int \mathbf{r} \times \mathbf{E}(\mathbf{d} \cdot \nabla \delta(\mathbf{r} - \mathbf{R})) \, dV + \int \mathbf{r} \times (\dot{\mathbf{d}} \times \mathbf{B}) \delta(\mathbf{r} - \mathbf{R}) \, dV \\
&= \int \delta(\mathbf{r} - \mathbf{R}) \mathbf{d} \cdot \nabla(\mathbf{r} \times \mathbf{E}) \, dV + \mathbf{R} \times (\dot{\mathbf{d}} \times \mathbf{B}) \\
&= \mathbf{T}^d.
\end{aligned}
\tag{60}
$$

At the microscopic level, both force densities give the correct torque. The only distinction is that using $\mathbf{f}^d$ requires us to add the contribution associated with orientation of the dipole.

At the other extreme we would like to calculate the nett torque on a dielectric medium by integrating the torque density (57) or (59) expressed in terms of the macroscopic fields:

$$
\begin{aligned}
\mathbf{t}^d &= \mathbf{r} \times ((\bar{\mathbf{P}} \cdot \nabla)\bar{\mathbf{E}} + \dot{\bar{\mathbf{P}}} \times \bar{\mathbf{B}}) + \bar{\mathbf{P}} \times \bar{\mathbf{E}} \\
\mathbf{t}^c &= \mathbf{r} \times ((-\nabla \cdot \bar{\mathbf{P}})\bar{\mathbf{E}} + \dot{\bar{\mathbf{P}}} \times \bar{\mathbf{B}}).
\end{aligned}
\tag{61}
$$

This leads to the torque

$$
\begin{aligned}
\mathbf{T}^d &= \int dV [\mathbf{r} \times ((\bar{\mathbf{P}} \cdot \nabla)\bar{\mathbf{E}} + \dot{\bar{\mathbf{P}}} \times \bar{\mathbf{B}}) + \bar{\mathbf{P}} \times \bar{\mathbf{E}}] \\
\mathbf{T}^c &= \int dV [\mathbf{r} \times ((-\nabla \cdot \bar{\mathbf{P}})\bar{\mathbf{E}} + \dot{\bar{\mathbf{P}}} \times \bar{\mathbf{B}})] \\
&= -\int (\mathbf{r} \times \bar{\mathbf{E}})\bar{\mathbf{P}} \cdot d\mathbf{S} + \int dV ((\bar{\mathbf{P}} \cdot \nabla)(\mathbf{r} \times \bar{\mathbf{E}}) + \mathbf{r} \times (\dot{\bar{\mathbf{P}}} \times \bar{\mathbf{B}})) \\
&= \mathbf{T}^d
\end{aligned}
\tag{62}
$$

if we choose the volume to include all of the dielectric and its surfaces so that $\bar{\mathbf{P}} = 0$ on the surface of the integration volume. As with the total force, we find that the total torque is the same for either torque density provided we apply the correct boundary conditions at any interfaces.

5. Conclusion

We have considered the form of the electromagnetic force on a dielectric medium. Two rival forms, (1) and (2), have been suggested for the force density and it is highly desirable to determine the extent to which they are different and lead to distinct predictions. Contrary to recent claims we have shown that both force densities lead to the same total forces and torques, although the component parts of the forces have rather different forms. An experiment designed to measure a total force or torque cannot distinguish between the two rival force densities. The origin of the inconsistencies noted previously [10] appears to be the incorrect replacement of the dielectric constant at an interface by the average of its values on either side. We have shown, by reference to a classic problem in electrostatics, that this does indeed lead to an apparently erroneous distinction. We will report further examples from optics elsewhere.

Much of the interest in the force on a dielectric stems from an experiment by Ashkin and Dziedzic in which the propagation of laser pulses was reported to form bulges on the surface of water [18]. The estimated radiation pressure force at the *peak* laser power of 4 kW used in the experiments is 4×10^{-6}N. They interpreted the effect as caused by a longitudinal force associated with the change in photon momentum as the light passes from air to water. Gordon proposed an alternative explanation in which the bulge is caused by inward radial forces associated with the Gaussian intensity profile of the laser beam within the dielectric [4]. This is essentially the tube-of-toothpaste effect, where a transverse squeeze produces a longitudinal flow [8]. Gordon's analysis showed that the longitudinal force on the surface is negligible and that this transverse force produces an average effective outward force on the surface of magnitude

$$F = \frac{2}{c} \left(\frac{\eta - 1}{\eta + 1} \right) Q \approx 10^{-9} \text{N}, \tag{63}$$

where Q is the average beam power (1 W) and $\eta = \sqrt{\varepsilon} = 1.33$ is the refractive index. Note that, although the physical origin of the radial force is quite distinct, its effect on the surface is the same as that of the longitudinal force used by Ashkin and Dziedzic in their numerical estimate quoted above.

Mansuripur has recently proposed a further explanation of the experiment [10]. He assumes a focused laser beam of elliptical intensity profile incident normally on the liquid surface. The force arises from an interaction of the normal component of the optical electric field with a surface charge density from the normal field discontinuity and it has a longitudinal character. Its magnitude depends on the optical polarization direction relative to the elliptical axes but its maximum value is approximately 4.4×10^{-12} N for the parameter values as used above. This is nearly three orders of magnitude smaller than the Gordon result and it does not provide a quantitative explanation of the observed bulges. Gordon's result is based on (2) and Mansuripur's on (1). Our work suggests that the nett forces should be the same in both cases and that one of the two mechanisms suggested is therefore incorrect. We will return to this point elsewhere.

Finally, we note that our analysis of the forces acting on a dielectric has assumed a linear isotropic medium described by the real dielectric constant ε. It is straightforward to quantize the electromagnetic field in the presence of such a medium and so we could use the quantized forms of the force densities to explore such effects as Casimir forces and the Casimir–Polder interaction [19, 20]. These forces are usually obtained by considering the modification of the vacuum field by the new boundary conditions. The quantized force densities, however, should provide a more mechanical method of calculation.

Acknowledgments

We are grateful to Masud Mansuripur for helpful discussions and an enlightening correspondence in which the issues addressed in this paper were made clear. We also thank Roberta Zambrini and Sarah Croke for helpful suggestions. This work was supported by the UK Engineering and Physical Sciences Research Council.

References

[1] Brevik I 1979 *Phys. Rep.* **52** 133
[2] Mansuripur M 2004 *Opt. Express* **12** 5375
[3] Mansuripur M, Zakharian A R and Moloney J V 2005 *Opt. Express* **13** 2064
[4] Gordon J P 1973 *Phys. Rev. A* **8** 14
[5] Penfield P and Haus H A 1967 *Electrodynamics of Moving Media* (Cambridge, MA: MIT Press)
[6] Landau L D, Lifshitz E M and Pitaevskii 1984 *Electrodynamics of Continuous Media* (Oxford: Heinemann)
[7] Loudon R 2003 *Phys. Rev. A* **68** 013806
[8] Loudon R 2004 *Fortschr. Phys.* **52** 1134
[9] Loudon R, Barnett S M and Baxter C 2005 *Phys. Rev. A* **71** 063802
[10] Mansuripur M 2005 *Proc. SPIE* **5930** 154
[11] Lorentz H A 1909 *The Theory of Electrons* (Leipzig: Teubner)
[12] Sommerfeld A 1952 *Electrodynamics* (New York: Academic)

[13] Stephenson G and Radmore P M 1990 *Mathematical Methods for Engineering and Science Students* (Cambridge: Cambridge University Press)
[14] Grant I S and Phillips W R 1975 *Electromagnetism* (New York: Wiley)
[15] Jackson J D 1999 *Classical Electrodynamics* 3rd edn (New York: Wiley)
[16] Allen L, Barnett S M and Padgett M J 2003 *Optical Angular Momentum* (Bristol: Institute of Physics Publishing)
[17] Padgett M, Barnett S M and Loudon R 2003 *J. Mod. Opt.* **50** 1555
[18] Ashkin A and Dziedzic J M 1973 *Phys. Rev. Lett.* **30** 139
[19] Power E A 1964 *Introductory Quantum Electrodynamics* (London: Longmans)
[20] Milonni P W 1994 *The Quantum Vacuum* (San Diego, CA: Academic)

Optical tweezers computational toolbox

Timo A Nieminen, Vincent L Y Loke, Alexander B Stilgoe, Gregor Knöner, Agata M Brańczyk, Norman R Heckenberg and Halina Rubinsztein-Dunlop

Centre for Biophotonics and Laser Science, School of Physical Sciences, The University of Queensland, Brisbane QLD 4072, Australia

Received 5 March 2007, accepted for publication 13 April 2007
Published 24 July 2007
Online at stacks.iop.org/JOptA/9/S196

Abstract
We describe a toolbox, implemented in Matlab, for the computational modelling of optical tweezers. The toolbox is designed for the calculation of optical forces and torques, and can be used for both spherical and nonspherical particles, in both Gaussian and other beams. The toolbox might also be useful for light scattering using either Lorenz–Mie theory or the T-matrix method.

Keywords: optical tweezers, laser trapping, light scattering, Mie theory, T-matrix method

(Some figures in this article are in colour only in the electronic version)

1. Introduction

Computational modelling provides an important bridge between theory and experiment—apart from the simplest cases, computational methods must be used to obtain quantitative results from theory for comparison with experimental results. This is very much the case for optical trapping, where the size range of typical particles trapped and manipulated in optical tweezers occupies the gap between the geometric optics and Rayleigh scattering regimes, necessitating the application of electromagnetic theory. Although, in principle, the simplest cases—the trapping and manipulation of homogeneous and isotropic microspheres—have an analytical solution—generalized Lorenz–Mie theory, significant computational effort is still required to obtain quantitative results. Unfortunately, the mathematical complexity of Lorenz–Mie theory presents a significant barrier to entry for the novice, and is likely to be a major contributor to the lagging of rigorous computational modelling of optical tweezers compared to experiment.

If we further consider the calculation of optical forces and torques on non-spherical particles—for example, if we wish to consider optical torques on and rotational alignment of non-spherical microparticles, the mathematical difficulty is considerably greater. Interestingly, one of the most efficient methods for calculating optical forces and torques on non-spherical particles in optical traps is closely allied to Lorenz–Mie theory—the T-matrix method (Waterman 1971,

Mishchenko 1991, Nieminen *et al* 2003a). However, while the Mie scattering coefficients have a relatively simple analytical form, albeit involving special functions, the T-matrix requires considerable numerical effort for its calculation. It is not surprising that the comprehensive bibliographic database on computational light scattering using the T-matrix method by Mishchenko *et al* (2004) lists only four papers applying the method to optical tweezers (Bayoudh *et al* 2003, Bishop *et al* 2003, Nieminen *et al* 2001a, 2001b). Since the compilation of this bibliography, other papers have appeared in which this is done (Nieminen *et al* 2004, Simpson and Hanna 2007, Singer *et al* 2006), but they are few in number.

Since the potential benefits of precise and accurate computational modelling of optical trapping is clear, both for spherical and non-spherical particles, we believe that the release of a freely available computational toolbox will be valuable to the optical trapping community.

We describe such a toolbox, implemented in Matlab. We outline the theory underlying the computational methods, the mathematics and the algorithms, the toolbox itself, and typical usage, and present some example results. The toolbox can be obtained at http://www.physics.uq.edu.au/people/nieminen/software.html at the time of publication. Since such software projects tend to evolve over time, and we certainly intend that this one will do so, potential users are advised to check the accompanying documentation. Along these lines, we describe our plans for future development. Of course, we welcome input, feedback, and contributions from the optical trapping community.

2. Fundamentals

2.1. Optical tweezers as a scattering problem and the T-matrix method

The optical forces and torques that allow trapping and manipulation of microparticles in beams of light result from the transfer of momentum and angular momentum from the electromagnetic field to the particle—the particle alters the momentum or angular momentum flux of the beam through scattering. Thus, the problem of calculating optical forces and torques is essentially a problem of computational light scattering. In some ways, it is a simple problem: the incident field is monochromatic, there is usually only a single trapped particle, which is finite in extent, and speeds are so much smaller than the speed of light that we can for most purposes neglect Doppler shifts and assume we have a steady-state monochromatic single-scattering problem.

Although typical particles inconveniently are of sizes lying within the gap between the regimes of applicability of small-particle approximations (Rayleigh scattering) and large-particle approximations (geometric optics), the particles of choice are often homogeneous isotropic spheres, for which an analytical solution to the scattering problem is available—the Lorenz–Mie solution (Lorenz 1890, Mie 1908). While the application of Lorenz–Mie theory requires significant computational effort, the methods are well known.

The greatest difficulty encountered results from the incident beam being a tightly focused beam. The original Lorenz–Mie theory was developed for scattering of plane waves, and its extension to non-plane illumination is usually called generalized Lorenz–Mie theory (GLMT) (Gouesbet and Grehan 1982), which has seen significant use for modelling the optical trapping of homogeneous isotropic spheres (Ren *et al* 1996, Wohland *et al* 1996, Maia Neto and Nussenzweig 2000, Mazolli *et al* 2003, Lock 2004a, 2004b, Knöner *et al* 2006, Neves *et al* 2006). The same name is sometimes used for the extension of Lorenz–Mie theory to non-spherical, but still separable, geometries such as spheroids (Han and Wu 2001, Han *et al* 2003).

The source of the difficulty lies in the usual paraxial representations of laser beams being solutions of the scalar paraxial wave equation rather than solutions of the vector Helmholtz equation. Our method of choice is to use a least-squares fit to produce a Helmholtz beam with a far field matching that expected from the incident beam being focused by the objective (Nieminen *et al* 2003b).

At this point, we can write the incident field in terms of a discrete basis set of functions $\psi_n^{(\text{inc})}$, where n is the mode index labelling the functions, each of which is a solution of the Helmholtz equation,

$$U_{\text{inc}} = \sum_n^\infty a_n \psi_n^{(\text{inc})}, \tag{1}$$

where a_n are the expansion coefficients for the incident wave. In practice, the sum must be truncated at some finite n_{max}, which places restrictions on the convergence behaviour of useful basis sets. A similar expansion is possible for the scattered wave, and we can write

$$U_{\text{scat}} = \sum_k^\infty p_k \psi_k^{(\text{scat})}, \tag{2}$$

where p_k are the expansion coefficients for the scattered wave.

As long as the response of the scatterer—the trapped particle in this case—is linear, the relation between the incident and scattered fields must be linear, and can be written as a simple matrix equation

$$p_k = \sum_n^\infty T_{kn} a_n \tag{3}$$

or, in more concise notation,

$$\mathbf{P} = \mathbf{TA} \tag{4}$$

where T_{kn} are the elements of the T-matrix. This is the foundation of both GLMT and the T-matrix method. In GLMT, the T-matrix $\mathbf{T}$ is diagonal, whereas for non-spherical particles, it is not.

When the scatterer is finite and compact, the most useful set of basis functions is vector spherical wavefunctions (VSWFs) (Waterman 1971, Mishchenko 1991, Nieminen *et al* 2003a, 2003b). Since the VSWFs are a discrete basis, this method lends itself well to representation of the fields on a digital computer, especially since their convergence is well behaved and known (Brock 2001).

The T-matrix depends only on the properties of the particle—its composition, size, shape, and orientation—and the wavelength, and is otherwise independent of the incident field. This means that for any particular particle the T-matrix only needs to be calculated once, and can then be used for repeated calculations of optical force and torque. This is the key point that makes this a highly attractive method for modelling optical trapping and micromanipulation, since we are typically interested in the optical force and torque as a function of position within the trap, even if we are merely trying to find the equilibrium position and orientation within the trap. Thus, calculations must be performed for varying incident illumination, which can be done very easily with the T-matrix method. This provides a significant advantage over many other methods of calculating scattering, where the entire calculation needs to be repeated. This is perhaps the reason that while optical forces and torques have been successfully modelled using methods such as the finite-difference time-domain method (FDTD), the finite-element method (FEM), or other methods (White 2000a, 2000b, Hoekstra *et al* 2001, Collett *et al* 2003, Gauthier 2005, Chaumet *et al* 2005, Sun *et al* 2006, Wong and Ratner 2006), the practical application of such work has been limited.

Since, as noted above, the optical forces and torques result from differences between the incoming and outgoing fluxes of electromagnetic momentum and angular momentum, calculation of these fluxes is required. This can be done by integration of the Maxwell stress tensor, and its moment for the torque, over a surface surrounding the particle. Fortunately, in the T-matrix method, the bulk of this integral can be performed analytically, exploiting the orthogonality properties of the VSWFs. In this way, the calculation can be reduced to sums of products of the expansion coefficients of the fields.

2.2. Controversies in electromagnetic theory

At this point, two controversies in macroscopic classical electromagnetic theory intrude. The first of these is the Abraham–Minkowski controversy, concerning the momentum of an electromagnetic wave in a material medium (Minkowski 1908, Abraham 1909, 1910, Jackson 1999, Pfeifer *et al* 2006, Leonhardt 2006). Abraham's approach can be summarized as calling P/nc the electromagnetic momentum flux, where P is the power, n the refractive index, and c the speed of light in free space. The quantum equivalent is the momentum of a photon in a material medium being $\hbar k/n^2 = \hbar k_0/n$. Minkowski, on the other hand, gives nP/c as the electromagnetic momentum flux, or $\hbar k = n\hbar k_0$ per photon.

While the difference between these two might seem disconcerting when one is intending to calculate the optical force on a particle in a dielectric medium, it is necessary to realize that we are not seeking the *electromagnetic* force but the *total* force due to the incident beam. At least as far as optical tweezers are concerned, where we have time-harmonic steady-state illumination in a medium that can be simply characterized by a refractive index, and are satisfied with the time-averaged force, the total force is the same regardless of our choice of the Abraham or Minkowski expression for the momentum (Jones 1978).

In the Minkowski picture, in our circumstances, other than the refractive index of the medium not being equal to one, the stationary medium is treated identically to free space. Thus, the total momentum is equal to the Minkowski momentum nP/c.

In the Abraham picture, on the other hand, we find a more physical view of the medium, and the electromagnetic momentum of the wave and the momentum carried by the wave of induced dielectric polarization are considered separately. The sum of these momenta is equal to, again, the Minkowski momentum nP/c.

Thus, for practical purposes, we take nP/c to be the total momentum flux.

The second controversy is the angular momentum density of circularly polarized electromagnetic waves (Humblet 1943, Stewart 2005, Pfeifer *et al* 2006). On the one hand, we can begin with the assumption that the angular momentum density is the moment of the momentum density, $\mathbf{r} \times (\mathbf{E} \times \mathbf{H})/c$, which results in a circularly polarized plane wave carrying zero angular momentum in the direction of propagation. On the other hand, we can begin with the Lagrangian for an electromagnetic radiation field, and obtain the canonical stress tensor and an angular momentum tensor that can be divided into spin and orbital components (Jauch and Rohrlich 1976). For a circularly polarized plane wave, the component of the angular momentum flux in the direction of propagation would be I/ω, where I is the irradiance and ω the angular frequency, in disagreement with the first result.

However, it is not the angular momentum density as such that we are interested in, but the total angular momentum flux through a spherical surface surrounding the particle, which is equal to the torque exerted on the trapped particle. For the electromagnetic fields used in optical tweezers, this integrated flux is the same for both choices of angular momentum density (Humblet 1943, Crichton and Marston 2000, Zambrini and Barnett 2005). Since the formula given in section 5 for the torque is for the total angular momentum flux, it is unaffected by this controversy.

Crichton and Marston (2000) also show that for monochromatic radiation the division into spin and orbital angular momenta is gauge invariant, and observable, with it being possible to obtain the spin from measurement of the Stokes parameters. Since the torque due to spin is of practical interest (Nieminen *et al* 2001a, Bishop *et al* 2003, 2004), it is worthwhile to calculate this as well as the total torque.

Whether or not the angular momentum flux is altered by the presence of a dielectric medium has also been considered (Padgett *et al* 2003, Mansuripur 2005), essentially an angular momentum version of the Abraham–Minkowski controversy. As far as optical tweezers are concerned, it is immaterial whether or not the angular momentum is carried electromagnetically or by the medium, as long as we know the total rate of transfer of angular momentum to the particle in the trap. Where the total momentum flux is given by the Minkowski momentum, the total angular momentum flux is unchanged by entry into a dielectric medium (Brevik 1979, Padgett *et al* 2003). Thus, we can take the total angular momentum flux of a circularly polarized finite beam to be P/ω, whether it is in free space or a dielectric medium.

2.3. Further reading

The mathematics of optical tweezers is largely that of light scattering. The literature on scattering is extensive, often highly specialized, and frequently difficult to digest. Fortunately for the newcomer, Bohren (1995) gives an engaging and accessible introduction. The classic treatise by van de Hulst (1957) is another good starting point, especially for its lucid coverage of Lorenz–Mie theory.

The T-matrix method is often used synonymously to mean the extended boundary condition method (EBCM), introduced by Waterman (1971) as a method to calculate the T-matrix. However, the EBCM is just that—a method which one can use to calculate the T-matrix—and the T-matrix method is more general, as pointed out by Kahnert (2003) and Nieminen *et al* (2003a). The T-matrix method is described by Mishchenko *et al* (2002) and Tsang (2001), and the interested reader can consult the bibliography by Mishchenko *et al* (2004) for other works.

We recommend the review by Kahnert (2003) for readers interested in a more general coverage of numerical methods in light scattering, including FDTD and FEM. The shorter review by Wriedt (1998) may also be of interest.

The reader may well come to the conclusion that the T-matrix method is more mathematically esoteric than most other methods. This is, unfortunately, the price we pay for its strengths, which include, in addition to the previously mentioned efficiency with regard to repeated calculations for the same particle and the ease of calculating the force and torque from the fields, it being naturally suited to scattering in open domains.

Apart from the works noted earlier, in which Lorenz–Mie theory, the T-matrix method, or some other rigorous electromagnetic method is used to calculate optical forces in optical tweezers, we can also mention related works. Firstly, the T-matrix method has been used for the calculation of forces in optical binding (Simpson and Hanna 2006, Grzegorczyk *et al* 2006).

Secondly, approximate methods such as geometric optics or Rayleigh scattering can also prove useful, for large and small particles respectively. The classic paper by Ashkin (1992) is an excellent introduction to the modelling of optical tweezers using geometric optics, and Harada and Asakura (1996) discuss forces in the Rayleigh regime and the limits of applicability of the Rayleigh approximation in optical tweezers.

Although the practical impact of the Abraham–Minkowski and angular momentum controversies on optical tweezers is negligible, the converse does not appear to be true—the controversies continue to be debated, more actively than ever before, and this appears to be driven at least in part by the widespread application of electromagnetic forces and torques in optical tweezers.

In our opinion, the Abraham–Minkowski controversy is essentially one of semantics—what portion of the total momentum is to be labelled 'electromagnetic', and what portion is to be labelled 'material'; Penfield and Haus (1967), de Groot and Suttorp (1972), Gordon (1973), and others have shown that the division of the total energy–momentum tensor into electromagnetic and material parts is effectively arbitrary (Pfeifer *et al* 2006). That the transport of energy necessitates the transfer of momentum, with a momentum flux of P/v, where v is the speed of energy transport, was shown as long ago as 1873 by Umov, with the result that the Minkowski momentum gives the total momentum flux. Unfortunately, Umov's work does not seem to be readily available in English (Umov 1950).

In the context of optical tweezers, Gordon's (1973) work is usefully enlightening. We can consider a finite beam passing through a medium composed of molecules. At the edges of the beam, where $\nabla |\mathbf{E}|^2$ is non-zero, an optical gradient force will act on the molecules. However, in a steady-state case, the net force must be zero. Since the only forces acting on the molecules at the edge of the beam are the electromagnetic force and forces due to the surrounding molecules—pressure, there must be an increase in pressure within the beam. If a particle is trapped by the beam, the force exerted on the particle will be the sum of the electromagnetic force and this pressure force, the total being equal to the change in the Minkowski momentum.

Nonetheless, the controversy continues to be actively debated (Mansuripur 2004, Barnett and Loudon 2006, Leonhardt 2006, Mansuripur 2007, Loudon and Barnett 2006). Clearly, the last word is yet to be said on this issue.

The second controversy, concerning angular momentum, appears to arise from two sources. Firstly, it is natural to assume that the angular momentum density is $\mathbf{r} \times (\mathbf{E} \times \mathbf{H})/c$. Secondly, the conservation law for angular momentum obtained via Lagrangian field theory is for the total angular momentum, and identification of the integrand in the integral as the angular momentum density is uncertain.

As far as the total angular momentum flux is concerned, both pictures give exactly the same result for a bounded beam (Humblet 1943, Soper 1976). Jackson (1999) gives a problem (7.28) allowing the student to explicitly demonstrate this. The division of the angular momentum density of an electromagnetic field into spin and orbital parts is, in general, not gauge invariant, and it is common to transform the integral of the angular momentum density into a gauge-invariant form, yielding the integral of $\mathbf{r} \times (\mathbf{E} \times \mathbf{H})/c$. Jauch and Rohrlich (1976) carefully point out that this transformation requires the dropping of surface terms at infinity. The reverse of this procedure, obtaining the spin and orbital term starting from $\mathbf{r} \times (\mathbf{E} \times \mathbf{H})/c$, involving the same surface terms, had already been shown by Humblet (1943). These surface terms disappear for finite beams and only in the case of an unbounded wave—such as an infinite plane wave—does a possible discrepancy exist.

The gauge invariance of the separation of the angular momentum into spin and orbital angular momentum has also attracted attention. However, this does not affect the torque exerted on a particle, and, in any case, the separation appears to be gauge invariant for a monochromatic wave (Crichton and Marston 2000, Barnett 2002).

3. Incident field

The natural choice of coordinate system for optical tweezers is spherical coordinates centred on the trapped particle. Thus, the incoming and outgoing fields can be expanded in terms of incoming and outgoing vector spherical wavefunctions (VSWFs):

$$\mathbf{E}_{\text{in}} = \sum_{n=1}^{\infty} \sum_{m=-n}^{n} a_{nm}\mathbf{M}_{nm}^{(2)}(k\mathbf{r}) + b_{nm}\mathbf{N}_{nm}^{(2)}(k\mathbf{r}), \qquad (5)$$

$$\mathbf{E}_{\text{out}} = \sum_{n=1}^{\infty} \sum_{m=-n}^{n} p_{nm}\mathbf{M}_{nm}^{(1)}(k\mathbf{r}) + q_{nm}\mathbf{N}_{nm}^{(1)}(k\mathbf{r}) \qquad (6)$$

where the VSWFs are

$$\mathbf{M}_{nm}^{(1,2)}(k\mathbf{r}) = N_n h_n^{(1,2)}(kr)\mathbf{C}_{nm}(\theta,\phi)$$

$$\mathbf{N}_{nm}^{(1,2)}(k\mathbf{r}) = \frac{h_n^{(1,2)}(kr)}{krN_n}\mathbf{P}_{nm}(\theta,\phi)$$

$$+ N_n \left(h_{n-1}^{(1,2)}(kr) - \frac{nh_n^{(1,2)}(kr)}{kr} \right)\mathbf{B}_{nm}(\theta,\phi) \qquad (7)$$

where $h_n^{(1,2)}(kr)$ are spherical Hankel functions of the first and second kinds, $N_n = [n(n+1)]^{-1/2}$ are normalization constants, $\mathbf{B}_{nm}(\theta,\phi) = \mathbf{r}\nabla Y_n^m(\theta,\phi)$, $\mathbf{C}_{nm}(\theta,\phi) = \nabla \times (\mathbf{r}Y_n^m(\theta,\phi))$, and $\mathbf{P}_{nm}(\theta,\phi) = \hat{\mathbf{r}}Y_n^m(\theta,\phi)$ are the vector spherical harmonics (Waterman 1971, Mishchenko 1991, Nieminen *et al* 2003a, 2003b), and $Y_n^m(\theta,\phi)$ are normalized scalar spherical harmonics. The usual polar spherical coordinates are used, where θ is the co-latitude, measured from the $+z$ axis, and ϕ is the azimuth, measured from the $+x$ axis towards the $+y$ axis.

$\mathbf{M}_{nm}^{(1)}$ and $\mathbf{N}_{nm}^{(1)}$ are outward-propagating TE and TM multipole fields, while $\mathbf{M}_{nm}^{(2)}$ and $\mathbf{N}_{nm}^{(2)}$ are the corresponding inward-propagating multipole fields. Since these wavefunctions are purely incoming and purely outgoing, each has a singularity at the origin. Since fields that are free of singularities are of interest, it is useful to define the singularity-free regular vector spherical wavefunctions:

$$\mathbf{RgM}_{nm}(k\mathbf{r}) = \tfrac{1}{2}[\mathbf{M}_{nm}^{(1)}(k\mathbf{r}) + \mathbf{M}_{nm}^{(2)}(k\mathbf{r})], \qquad (8)$$

$$\mathbf{RgN}_{nm}(k\mathrm{r}) = \tfrac{1}{2}[\mathbf{N}_{nm}^{(1)}(k\mathrm{r}) + \mathbf{N}_{nm}^{(2)}(k\mathrm{r})]. \qquad (9)$$

Although it is usual to expand the incident field in terms of the regular VSWFs, and the scattered field in terms of outgoing VSWFs, this results in both the incident and scattered waves carrying momentum and angular momentum away from the system. Since we are primarily interested in the transport of momentum and angular momentum by the fields (and energy, too, if the particle is absorbing), we separate the total field into purely incoming and outgoing portions in order to calculate these. The user of the code can choose whether the incident–scattered or incoming–outgoing representation is used otherwise.

We use an over-determined point-matching scheme to find the expansion coefficients a_{nm} and b_{nm} describing the incident beam (Nieminen *et al* 2003b), providing stable and robust numerical performance and convergence.

Finally, one needs to be able to calculate the force and torque for the same particle in the same trapping beam, but at different positions or orientations. The transformations of the VSWFs under rotation of the coordinate system or translation of the origin of the coordinate system are known (Brock 2001, Videen 2000, Gumerov and Duraiswami 2003, Choi *et al* 1999). It is sufficient to find the VSWF expansion of the incident beam for a single origin and orientation, and then use translations and rotations to find the new VSWF expansions about other points (Nieminen *et al* 2003b, Doicu and Wriedt 1997, Moine and Stout 2005). Since the transformation matrices for rotation and translations along the z-axis are sparse, while the transformation matrices for arbitrary translations are full, the most efficient way to carry out an arbitrary translation is by a combination of rotation and axial translation. The transformation matrices for both rotations and axial translations can be efficiently computed using recursive methods (Videen 2000, Gumerov and Duraiswami 2003, Choi *et al* 1999).

3.1. Implementation

Firstly, it is necessary to provide routines to calculate the special functions involved. These include the following.

(i) `sbesselj.m`, `sbesselh.m`, `sbesselh1.m`, and `sbesselh2.m` for the calculation of spherical Bessel and Hankel functions. These make use of the Matlab functions for cylindrical Bessel functions.
(ii) `sparm.m` for scalar spherical harmonics and their angular partial derivatives.
(iii) `vsh.m` for vector spherical harmonics.
(iv) `vswf.m` for vector spherical wavefunctions.

Secondly, routines must be provided to find the expansion coefficients, or beam shape coefficients, a_{nm} and b_{nm} for the trapping beam. These are the following.

(i) `bsc_pointmatch_farfield.m` and `bsc_pointmatch_focalplane.m`, described in (Nieminen *et al* 2003b), which can calculate the expansion coefficients for Gaussian beams, Laguerre–Gauss modes, and bi-Gaussian beams. Since these routines are much faster for rotationally symmetric beams, such as Laguerre–Gauss beams, a routine, `lgmodes.m`, that can provide the Laguerre–Gauss decomposition of an arbitrary paraxial beam is also provided.

(ii) `bsc_plane.m`, for the expansion coefficients of a plane wave. This is not especially useful for optical trapping, but makes the toolbox more versatile, improving its usability for more general light scattering calculations.

Thirdly, the transformation matrices for the expansion coefficients under rotations and translations must be calculated. Routines include the following.

(i) `wigner_rotation_matrix.m`, implementing the algorithm given by Choi *et al* (1999).
(ii) `translate_z.m`, implementing the algorithm given by Videen (2000).

4. *T*-matrix

For spherical particles, the usual Mie coefficients can be rapidly calculated. For non-spherical particles, a more intensive numerical effort is required. We use a least-squares overdetermined point-matching method (Nieminen *et al* 2003a). For axisymmetric particles, the method is relatively fast. However, as is common for many methods of calculating the T-matrix, particles cannot have extreme aspect ratios, and must be simple in shape. Typical particle shapes that we have used are spheroids and cylinders, and aspect ratios of up to four give good results. Although it can take a long time to calculate the T-matrix for general non-axisymmetric particles, it is possible to make use of symmetries such as mirror symmetry and discrete rotational symmetry to greatly speed up the calculation (Kahnert 2005, Nieminen *et al* 2006). We include a symmetry-optimized T-matrix routine for cubes.

Expanding the range of particles for which we can calculate the T-matrix is one of our current active research efforts, and we plan to include routines for anisotropic and inhomogeneous particles, and particles with highly complex geometries.

Once the T-matrix is calculated, the scattered field coefficients are simply found by a matrix–vector multiplication of the T-matrix and a vector of the incident field coefficients.

4.1. Implementation

Our T-matrix routines include the following.

(i) `tmatrix_mie.m`, calculating the Mie coefficients for homogeneous isotropic spheres.
(ii) `tmatrix_pm.m`, our general point-matching T-matrix routine.
(iii) `tmatrix_pm_cube.m`, the symmetry-optimized cube code.

5. Optical force and torque

As noted earlier, the integrals of the momentum and angular momentum fluxes reduce to sums of products of the expansion coefficients. It is sufficient to give the formulae for the z-components of the fields, as given, for example, by Crichton and Marston (2000). We use the same formulae to calculate the x and y components of the optical force and torque, using $90°$ rotations of the coordinate system (Choi *et al* 1999). It is also possible to directly calculate the x and y components using similar, but more complicated, formulae (Farsund and Felderhof 1996).

The axial trapping efficiency Q is

$$Q = \frac{2}{P} \sum_{n=1}^{\infty} \sum_{m=-n}^{n} \frac{m}{n(n+1)} \mathrm{Re}(a_{nm}^{\star} b_{nm} - p_{nm}^{\star} q_{nm})$$

$$- \frac{1}{n+1} \left[\frac{n(n+2)(n-m+1)(n+m+1)}{(2n+1)(2n+3)} \right]^{\frac{1}{2}}$$

$$\times \mathrm{Re}(a_{nm} a_{n+1,m}^{\star} + b_{nm} b_{n+1,m}^{\star} - p_{nm} p_{n+1,m}^{\star}$$

$$- q_{nm} q_{n+1,m}^{\star}) \qquad (10)$$

in units of $n\hbar k$ per photon, where n is the refractive index of the medium in which the trapped particles are suspended. This can be converted to SI units by multiplying by nP/c, where P is the beam power and c is the speed of light in free space.

The *torque efficiency*, or normalized torque, about the z-axis acting on a scatterer is

$$\tau_z = \sum_{n=1}^{\infty} \sum_{m=-n}^{n} m(|a_{nm}|^2 + |b_{nm}|^2 - |p_{nm}|^2 - |q_{nm}|^2)/P \quad (11)$$

in units of $\hbar$ per photon, where

$$P = \sum_{n=1}^{\infty} \sum_{m=-n}^{n} |a_{nm}|^2 + |b_{nm}|^2 \qquad (12)$$

is proportional to the incident power (omitting a unit conversion factor which will depend on whether SI, Gaussian, or other units are used). This torque includes contributions from both spin and orbital components, which can both be calculated by similar formulae (Crichton and Marston 2000). Again, one can convert these values to SI units by multiplying by P/ω, where ω is the optical frequency.

5.1. Implementation

One routine, `forcetorque.m`, is provided for the calculation of the force, torque and spin transfer. The orbital angular momentum transfer is the difference between the torque and the spin transfer. The incoming and outgoing power (the difference being the absorbed power) can be readily calculated directly from the expansion coefficients, as can be seen from (12).

6. Miscellaneous routines

A number of other routines that do not fall into the above categories are included. These include the following.

(i) Examples of use.

(ii) Routines for conversion of coordinates and vectors from Cartesian to spherical and spherical to Cartesian.

(iii) Routines to automate common tasks, such as finding the equilibrium position of a trapped particle, spring constants, and force maps.

(iv) Functions required by other routines.

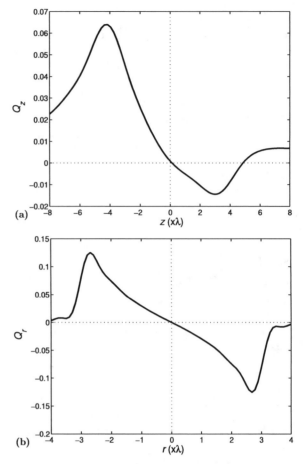

Figure 1. Gaussian trap (`example_gaussian.m`). Force on a sphere in a Gaussian beam trap. The half-angle of convergence of the $1/e^2$ edge of the beam is 50°, corresponding to a numerical aperture of 1.02. The particle has a relative refractive index equal to $n = 1.59$ in water, and has a radius of 2.5 λ, corresponding to a diameter of 4.0 μm if trapped at 1064 nm in water. (a) The axial trapping efficiency as a function of axial displacement and (b) the transverse trapping efficiency as a function of transverse displacement from the equilibrium point.

7. Typical use of the toolbox

Typically, for a given trap and particle, a T-matrix routine (usually `tmatrix_mie.m`) will be run once. Next, the expansion coefficients for the beam are found. Depending on the interests of the user, a function automating some common task, such as finding the equilibrium position within the trap, might be used, or the user might directly use the rotation and translation routines to enable calculation of the force or torque at desired positions within the trap.

The speed of calculation depends on the size of the beam, the size of the particle, and the distance of the particle from the focal point of the beam. Even for a wide beam and a large distance, the force and torque at a particular position can typically be calculated in much less than one second.

More complex tasks are possible, such as finding the optical force as a function of some property of the particle, which can, for example, be used to determine the refractive index of a microsphere (Knöner *et al* 2006).

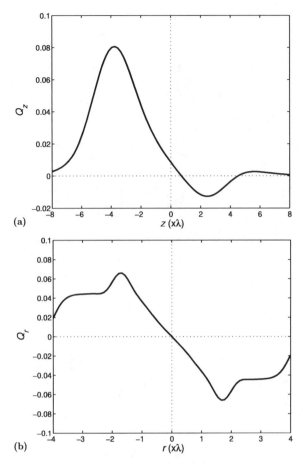

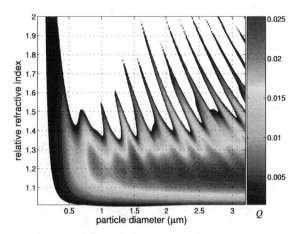

Figure 3. Trapping landscape (`example_landscape.m`). The maximum axial restoring force for displacement in the direction of beam propagation is shown, in terms of the trapping efficiency as a function of relative refractive index and microsphere diameter. The trapping beam is at 1064 nm and is focused by an NA = 1.2 objective. This type of calculation is quite slow, as the trapping force as a function of axial displacement must be found for a grid of combinations of relative refractive index and sphere diameter. On the left-hand side, we can see that the trapping force rapidly becomes very small as the particle size becomes small—the gradient force is proportional to the volume of the particle for small particles. In the upper portion, we can see that whether or not the particle can be trapped strongly depends on the size—for particular sizes, reflection is minimized, and even high index particles can be trapped.

Figure 2. Laguerre–Gauss trap (`example_lg.m`). Force on a sphere in a Laguerre–Gauss beam trap. The half-angle of convergence of the $1/e^2$ outer edge of the beam is 50°, as in figure 1. The sphere is identical to that in figure 1. (a) The axial trapping efficiency as a function of axial displacement and (b) the transverse trapping efficiency as a function of transverse displacement from the equilibrium point. Compared with the Gaussian beam trap, the radial force begins to drop off at smaller radial displacements, due to the far side of the ring-shaped beam no longer interacting with the particle.

Figures 1–4 demonstrate some of the capabilities of the toolbox. Figure 1 shows a simple application—the determination of the force as a function of axial displacement from the equilibrium position in a Gaussian beam trap. Figure 2 shows a similar result, but for a particle trapped in a Laguerre–Gauss LG_{03} beam. Figure 3 shows a more complex application, with repeated calculations (each similar to the one shown in figure 1(a)) being used to determine the effect of the combination of relative refractive index and particle size on trapping. Finally, figure 4 shows the trapping of a non-spherical particle, a cube.

Agreement with precision experimental measurements suggests that errors of less than 1% are expected.

8. Future development

We are actively engaged in work to extend the range of particles for which we can model trapping. This currently includes birefringent particles and particles of arbitrary

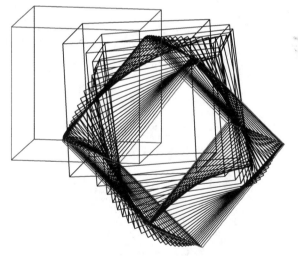

Figure 4. Optical trapping of a cube (`example_cube.m`). A sequence showing the optical trapping of a cube. The cube has faces of $2\lambda/n_{\text{medium}}$ across, and has a refractive index of $n = 1.59$, and is trapped in water. Since the force and torque depend on the orientation as well as position, a simple way to find the equilibrium position and orientation is to 'release' the cube and calculate the change in position and orientation for appropriate time steps. The cube can be assumed to always be moving at terminal velocity and terminal angular velocity (Nieminen *et al* 2001a). The cube begins face up, centred on the focal plane of the beam, and to one side. The cube is pulled into the trap and assumes a corner-up orientation. The symmetry optimizations allow the calculation of the T-matrix in 20 min; otherwise, 30 h would be required. Once the T-matrix is found, successive calculations of the force and torque require far less time, on the order of a second or so.

geometry. Routines to calculate the *T*-matrices for such particles will be included in the main code when available.

Other areas in which we aim to further improve the toolbox are robust handling of incorrect or suspect input, more automation of tasks, and GUI tools.

We also expect feedback from the optical trapping and micromanipulation community to help us add useful routines and features.

References

Abraham M 1909 *Rend. Circolo Mat. Palermo* **28** 1–28
Abraham M 1910 *Rend. Circolo Mat. Palermo* **30** 33–46
Ashkin A 1992 *Biophys. J.* **61** 569–82
Barnett S M 2002 *J. Opt. B: Quantum Semiclass. Opt.* **4** S7–16
Barnett S M and Loudon R 2006 *J. Phys. B: At. Mol. Opt. Phys.* **39** S671–84
Bayoudh S, Nieminen T A, Heckenberg N R and Rubinsztein-Dunlop H 2003 *J. Mod. Opt.* **50** 1581–90
Bishop A I, Nieminen T A, Heckenberg N R and Rubinsztein-Dunlop H 2003 *Phys. Rev.* A **68** 033802
Bishop A I, Nieminen T A, Heckenberg N R and Rubinsztein-Dunlop H 2004 *Phys. Rev. Lett.* **92** 198104
Bohren C F 1995 *Handbook of Optics* 2nd edn, vol 1, ed M Bass, E W Van Stryland, D R Williams and W L Wolfe (New York: McGraw-Hill) chapter 6, pp 6.1–21
Brevik I 1979 *Phys. Rep.* **52** 133–201
Brock B C 2001 Using vector spherical harmonics to compute antenna mutual impedance from measured or computed fields *Sandia Report SAND2000-2217-Revised* Sandia National Laboratories, Albuquerque, NM
Chaumet P C, Rahmani A, Sentenac A and Bryant G W 2005 *Phys. Rev.* E **72** 046708
Choi C H, Ivanic J, Gordon M S and Ruedenberg K 1999 *J. Chem. Phys.* **111** 8825–31
Collett W L, Ventrice C A and Mahajan S M 2003 *Appl. Phys. Lett.* **82** 2730–2
Crichton J H and Marston P L 2000 *Electron. J. Diff. Eqns Conf.* **04** 37–50
de Groot S R and Suttorp L G 1972 *Foundations of Electrodynamics* (Amsterdam: North-Holland)
Doicu A and Wriedt T 1997 *Appl. Opt.* **36** 2971–8
Farsund Ø and Felderhof B U 1996 *Physica* A **227** 108–30
Gauthier R C 2005 *Opt. Express* **13** 3707–18
Gordon J P 1973 *Phys. Rev.* A **8** 14–21
Gouesbet G and Grehan G 1982 *J. Opt. (Paris)* **13** 97–103
Grzegorczyk T M, Kemp B A and Kong J A 2006 *J. Opt. Soc. Am.* A **23** 2324–30
Gumerov N A and Duraiswami R 2003 *SIAM J. Scientific Comput.* **25** 1344–81
Han Y, Gréhan G and Gouesbet G 2003 *Appl. Opt.* **42** 6621–9
Han Y and Wu Z 2001 *Appl. Opt.* **40** 2501–9
Harada Y and Asakura T 1996 *Opt. Commun.* **124** 529–41
Hoekstra A G, Frijlink M, Waters L B F M and Sloot P M A 2001 *J. Opt. Soc. Am.* A **18** 1944–53
Humblet J 1943 *Physica* **10** 585–603
Jackson J D 1999 *Classical Electrodynamics* 3rd edn (New York: Wiley)
Jauch J M and Rohrlich F 1976 *The Theory of Photons and Electrons* 2nd edn (New York: Springer)
Jones R V 1978 *Proc. R. Soc. Lond.* A **360** 365–71
Kahnert F M 2003 *J. Quant. Spectrosc. Radiat. Transfer* **79/80** 775–824
Kahnert M 2005 *J. Opt. Soc. Am.* A **22** 1187–99
Knöner G, Parkin S, Nieminen T A, Heckenberg N R and Rubinsztein-Dunlop H 2006 *Phys. Rev. Lett.* **97** 157402
Leonhardt U 2006 *Nature* **444** 823–4

Lock J A 2004a *Appl. Opt.* **43** 2532–44
Lock J A 2004b *Appl. Opt.* **43** 2545–54
Lorenz L 1890 *Vidensk. Selsk. Skr.* **6** 2–62
Loudon R and Barnett S M 2006 *Opt. Express* **14** 11855–69
Maia Neto P A and Nussenzweig H M 2000 *Europhys. Lett.* **50** 702–8
Mansuripur M 2004 *Opt. Express* **12** 5375–401
Mansuripur M 2005 *Opt. Express* **13** 5315–24
Mansuripur M 2007 *Opt. Express* **15** 2677–82
Mazolli A, Maia Neto P A and Nussenzveig H M 2003 *Proc. R. Soc. Lond.* A **459** 3021–41
Mie G 1908 *Ann. Phys., Lpz.* **25** 377–445
Minkowski H 1908 *Nachr. Gess. Wiss. Gött. Math. Phys. Klasse* **1908** 53–111
Mishchenko M I 1991 *J. Opt. Soc. Am.* A **8** 871–82
Mishchenko M I, Travis L D and Lacis A A 2002 *Scattering, Absorption, and Emission of Light by Small Particles* (Cambridge: Cambridge University Press)
Mishchenko M I, Videen G, Babenko V A, Khlebtsov N G and Wriedt T 2004 *J. Quant. Spectrosc. Radiat. Transfer* **88** 357–406
Moine O and Stout B 2005 *J. Opt. Soc. Am.* B **22** 1620–31
Neves A A R, Fontes A, Pozzo L d, de Thomaz A A, Chillce E, Rodriguez E, Barbosa L C and Cesar C L 2006 *Opt. Express* **14** 13101–6
Nieminen T A, Heckenberg N R and Rubinsztein-Dunlop H 2001a *J. Mod. Opt.* **48** 405–13
Nieminen T A, Heckenberg N R and Rubinsztein-Dunlop H 2004 *Proc. SPIE* **5514** 514–23
Nieminen T A, Loke V L Y, Brańczyk A M, Heckenberg N R and Rubinsztein-Dunlop H 2006 *PIERS Online* **2** 442–6
Nieminen T A, Rubinsztein-Dunlop H and Heckenberg N R 2001b *J. Quant. Spectrosc. Radiat. Transfer* **70** 627–37
Nieminen T A, Rubinsztein-Dunlop H and Heckenberg N R 2003a *J. Quant. Spectrosc. Radiat. Transfer* **79/80** 1019–29
Nieminen T A, Rubinsztein-Dunlop H and Heckenberg N R 2003b *J. Quant. Spectrosc. Radiat. Transfer* **79/80** 1005–17
Nieminen T A, Rubinsztein-Dunlop H, Heckenberg N R and Bishop A I 2001a *Comput. Phys. Commun.* **142** 468–71
Padgett M, Barnett S M and Loudon R 2003 *J. Mod. Opt.* **50** 1555–62
Penfield P Jr and Haus H A 1967 *Electrodynamics of Moving Media* (Cambridge, MA: MIT)
Pfeifer R N C, Nieminen T A, Heckenberg N R and Rubinsztein-Dunlop H 2006 *Proc. SPIE* **6326** 63260H
Ren K F, Gréhan G and Gouesbet G 1996 *Appl. Opt.* **35** 2702–10
Simpson S H and Hanna S 2006 *J. Opt. Soc. Am.* A **23** 1419–31
Simpson S H and Hanna S 2007 *J. Opt. Soc. Am.* A **24** 430–43
Singer W, Nieminen T A, Gibson U J, Heckenberg N R and Rubinsztein-Dunlop H 2006 *Phys. Rev.* E **73** 021911
Soper D E 1976 *Classical Field Theory* (New York: Wiley)
Stewart A M 2005 *Eur. J. Phys.* **26** 635–41
Sun W, Pan S and Jiang Y 2006 *J. Mod. Opt.* **53** 2691–700
Tsang L 2001 *Scattering of Electromagnetic Waves* 3 volumes (New York: Wiley)
Umov N A 1950 *Izbrannye Sochineniya (Selected Works)* (Moscow: Gostexizdat) (in Russian)
van de Hulst H C 1957 *Light scattering by small particles* (New York: Wiley)
Videen G 2000 Light Scattering from Microstructures number 534 *Lecture Notes in Physics* ed F Moreno and F González (Berlin: Springer) chapter 5, pp 81–96
Waterman P C 1971 *Phys. Rev.* D **3** 825–39
White D A 2000a *J. Comput. Phys.* **159** 13–37
White D A 2000b *Comput. Phys. Commun.* **128** 558–64
Wohland T, Rosin A and Stelzer E H K 1996 *Optik* **102** 181–90
Wong V and Ratner M A 2006 *J. Opt. Soc. Am.* B **23** 1801–14
Wriedt T 1998 *Part. Part. Syst. Charact.* **15** 67–74
Zambrini R and Barnett S M 2005 *J. Mod. Opt.* **52** 1045–52

Optical trapping of spheroidal particles in Gaussian beams

Stephen H. Simpson and Simon Hanna

H. H. Wills Physics Laboratory, University of Bristol, Tyndall Avenue, Bristol, BS8 1TL, UK

Received April 10, 2006; revised July 5, 2006; accepted August 1, 2006;
posted August 25, 2006 (Doc. ID 69820); published January 10, 2007

The T matrix method is used to compute equilibrium positions and orientations for spheroidal particles trapped in Gaussian light beams. It is observed that there is a qualitative difference between the behavior of prolate and oblate ellipsoids in linearly polarized Gaussian beams; the former generally orient with the symmetry axis parallel to the beam except at very small particle sizes, while the latter orient with the symmetry axis perpendicular to the beam. In the presence of a circularly polarized beam, it is demonstrated that oblate ellipsoids will experience a torque about the beam axis. However, for a limited range of particle sizes, where the particle dimensions are comparable with the beam waist, the particles are predicted to rotate in a sense *counter* to the sense of rotation of the circular polarization. This unusual prediction is discussed in some detail.
© 2007 Optical Society of America

OCIS codes: 140.7010, 290.5850.

1. INTRODUCTION

Since their introduction in the 1980s,[1] optical tweezers have found extensive applications from fundamental studies of the properties of light to the manipulation of biological structures.[2] With the introduction of high-quality spatial light modulators, holographic optical tweezers have become a reality,[3] and systems are now being developed capable of generating and controlling large arrays of optical traps. In a previous paper,[4] we described a method, based on T matrix theory, for calculating the forces on and between spherical particles in multiple optical traps. In this paper, we explore the forces and torques operating during the optical trapping of spheroidal particles.

A spheroid is defined as an ellipsoid with uniaxial symmetry or, alternatively, as a sphere that has been uniformly deformed parallel to one axis. One example of this that has appeared in the optical trapping literature is the chloroplast, which may be modeled as an oblate spheroid.[5] The symmetry axis of the spheroid is referred to variously as the unique, primary, or principal axis. In fact, a general ellipsoid has two other principal axes, which in the case of the spheroid, will be equal in length. We will refer to these as the secondary axes. The aspect ratio of the spheroid δ is given by

$$\delta = a/b, \tag{1}$$

where a and b are the lengths of the primary and secondary axes, respectively. We will consider the trapping of both prolate spheroids, for which $\delta > 1$, and oblate spheroids, where $\delta < 1$.

Optical trapping of spherical particles is well understood and has been extensively documented both in terms of experimental observation and in terms of theoretical analysis.[6–8] Qualitative discussion of the phenomenon frequently involves expressions such as gradient force, scattering force, and radiation pressure, each of which serves to describe a different aspect of the changes in electromagnetic momentum that take place as a result of the scattering of light from a particle. (N.B. we use the term "radiation pressure" to denote the force associated with the net flow of electromagnetic energy down the beam and not the total pressure, as used by some authors). An analogous vocabulary does not exist for the description of torque-related phenomena. This is unfortunate, since it is the induced torque that is responsible for the equilibrium orientation of a spheroidal particle in a laser beam.

A helpful principle to bear in mind is one that forms the basis of the theory of optical waveguides; it is often energetically favorable for the overlap between the refractive index distribution and the light intensity to be maximized, hence the propensity of light to be confined to the guide. This idea gives rise to the concept of gradient force and in the case of geometrically anisotropic particles suggests that an elongated particle will tend to align itself in an elongated intensity distribution (such as that furnished by a Gaussian beam), and that optical torques will be exerted in order to achieve this arrangement.

A second principle that has a part to play in the trapping of spheroidal particles is related to the fact that, in the quasi-static limit, i.e., for particles much smaller than a wavelength, the scattering properties of an ellipsoid can be expressed in terms of a polarizability tensor.[9] The potential energy of an anisotropic point polarizability in an electromagnetic field is minimized by aligning the eigenvector corresponding to the highest eigenvalue of the polarizability tensor parallel to the polarization vector of the electric field.[10] While the quasi-static limit does not generally pertain in the present case, the behavior might be expected to be mimicked to some extent.

A number of experimental studies of anisotropic systems have been performed. For example, Bishop *et al.*[11] and Bonin *et al.*[12] have observed the reorientation of the long axis of a microcylinder toward the polarization direction of the incident beam for particles lying close to a horizontal interface. Similar orientational properties have been observed for flat, cross-shaped particles by Galajda and Ormos.[13] Analogous effects have also been observed for particles whose anisotropy stems from their refractive index as opposed to their shape. For example, experiments performed by La Porta and Wang[14] show that near-spherical quartz particles tend to align their optic axes with the polarization vector of the electric field by virtue of their small, positive birefringence.

There are a number of different approaches that could be adopted in gaining theoretical insight into the phenomena outlined above. This is reflected in the variety of computational methods that have been employed in the literature. For length scales greatly in excess of one wavelength, optical interactions can be approximated by a ray-based model, as demonstrated by Kim and Kim.[15] Another approach involves the calculation of radiation pressures by considering photon flux equations.[16,17] Both of these methods have their own advantages and disadvantages and rely on the appropriate conditions pertaining. For example, the ray-tracing approach assumes that interfaces are approximately flat on the wavelength scale, which limits the size of scattering bodies that can be considered. Inaccurate answers may also be obtained for small particles using the photon flux model, since multiple reflections are not generally included.

In the present paper, the T matrix method[18,19] is employed; the particles considered will be both prolate and oblate spheroids of an optically isotropic material, i.e., silica glass. The T matrix method involves an expansion of the incident and scattered fields in the trap as a series of orthogonal vector spherical harmonics. The expansion coefficients of the scattered field are given by a linear transformation of the incident field coefficients, the transformation being determined by the boundary condition on the surface of the scatterer. In theory, the method provides exact solutions to Maxwell's equations, the only approximation being the truncation of the series expansions for the various fields. The method has been employed previously by Bishop *et al.*[11] to calculate the optical torques on cylindrical particles. In a previous paper, we used T matrices to calculate the forces arising between pairs of spherical particles in optical traps.[4] Here, we use a similar method to calculate equilibrium trapping positions and orientations for a range of spheroids with different aspect ratios and examine the stability of their equilibrium positions with respect to translational and orientational perturbations. We also use hydrodynamic theory, through the use of friction tensors, to convert the calculated forces and torques to linear and angular velocities. However, the effect of Brownian motion will be reserved for a future publication.

2. DETAILS OF THE CALCULATION METHOD

Extensive accounts of T matrix theory may be found elsewhere.[18,19] Some of the details of the implementation used in producing the following results were given previously.[4] A brief summary is included here.

As mentioned above, the scattering process is incorporated into T matrix theory by way of a linear transformation; the matrix representing this transformation has elements determined by the boundary conditions on the surface of the scatterer as expressed by the extended boundary condition method. Translation and rotation of fields may also be achieved by linear transformations, and the relevant matrix elements are determined, respectively, by the translation-addition theorem and the rotation theorem for vector spherical wave functions (VSWFs), respectively. The details of both theorems may be found in Refs. 18 and 19. The final part of the calculation involves the evaluation of the forces and torques on a particle. Formally, this involves a surface integral of a vector-valued function of the Maxwell stress tensor.[10,20] The procedure for performing the calculations follows.

A. Vector Spherical Wave Function Expansion

The incident and scattered fields are written in terms of VSWFs in the following way:

$$\mathbf{E}^{\text{inc}}(\mathbf{r}) = \sum_{n=1}^{\infty} \sum_{m=-n}^{n} [a_{mn}\text{Rg}\mathbf{M}_{mn}(k\mathbf{r}) + b_{mn}\text{Rg}\mathbf{N}_{mn}(k\mathbf{r})],$$

(2a)

$$\mathbf{E}^{\text{sca}}(\mathbf{r}) = \sum_{n=1}^{\infty} \sum_{m=-n}^{n} [p_{mn}\mathbf{M}_{mn}(k\mathbf{r}) + q_{mn}\mathbf{N}_{mn}(k\mathbf{r})]. \quad (2b)$$

a_{mn}, b_{mn}, p_{mn}, and q_{mn} are a set of complex coefficients, which are to be determined. $\text{Rg}\mathbf{M}_{mn}$, $\text{Rg}\mathbf{N}_{mn}$, $\mathbf{M}_{mn}$, and $\mathbf{N}_{mn}$ are a complete set of VSWFs given previously,[4] k is the modulus of the wave vector in the background medium, and $\mathbf{r}$ is the relevant position vector. The summations in Eqs. (2a) and (2b) are truncated at a predetermined value of n (i.e., n max), which will depend on a convergence criterion (see below).

The first requirement in the scattering calculation is to obtain coefficients for the VSWF expansion of the incident field, which is taken to be a Gaussian beam. For reasons of accuracy, it is advisable to use one of the higher-order Davis beam approximations. In the following study, the fifth-order beam is used, coefficients for which may be found in the literature[20–22] and are reproduced in Appendix A for completeness.

B. *T* Matrix Description

Under the extended boundary condition method (EBCM), the T matrix for general particles is given by a set of integrals of vector cross products of vector spherical harmonics taken over the surface of the scatterer.[18] When the particle is spherical, this integral can be evaluated analytically resulting in a diagonal matrix, whose elements are given by the familiar Mie coefficients. For nonspherically symmetric particles, the integration must be performed numerically, although when there is rotational symmetry, the azimuthal part of the surface integral may be performed analytically leaving a one-dimensional integral that can be evaluated by quadrature.

Integrals for obtaining elements of the T matrices for spheroidal particles are given in Ref. 18. It should be noted that the integrands involved are oscillatory, so that a sufficient number of points must be included in the quadrature to resolve the variation. The precise number of points necessary will depend on n max, the value of n for which the field expansions are truncated. A useful test of the computer code written to carry out the integration for spheroidal particles is to check that the analytical values for spherical particles are reproduced when the principal axes are equal. In the present case, it is found that, for n max < 20, the quadrature achieves good agreement if approximately 10,000 or more points are used. For a code written in a compiled language, such as Fortran, the entire matrix may be evaluated in under a minute on a desktop computer, for n max $= 20$.

C. Beam Transformations

The translation and rotation theorems[18] may be used to move the incident beam relative to the particle, so that the required position and orientation of the particle are achieved relative to the trap. When applying these theorems numerically, it should be noted that large translations will affect convergence.[4] In other words, the larger the distance over which the beam is translated, the greater the value of n max that will be required to achieve a desired level of accuracy (see below).

D. Scattered Field Calculation

The VSWF expansion of the scattered field was given above [Eq. (2b)]. The coefficients p_{mn} and q_{mn} are obtained by matrix multiplication of the T matrix by the transformed incident field coefficients, i.e.,

$$\tilde{p} = T \cdot \tilde{a}, \tag{3}$$

where $\tilde{a}$ is a vector consisting of an ordered set of the coefficients, a_{mn} and b_{mn}, from Eq. (2a) after suitable rotations and translations have been applied, and $\tilde{p}$ is an equivalent vector comprising the p_{mn} and q_{mn} coefficients. This step is most efficiently carried out by using an optimized matrix multiplication routine such as found in the LAPACK linear algebra package.[23]

E. Forces and Torques

The net force on a particle is found by integrating the product of the Maxwell stress tensor, $\underline{\underline{T}}_M$, and the unit normal over a surface enclosing the particle[10,24]:

$$\mathbf{F} = \int_S \langle \underline{\underline{T}}_M(\mathbf{r}) \rangle \cdot d\mathbf{S}, \tag{4}$$

where

$$\langle \underline{\underline{T}}_M(\mathbf{r}) \rangle = \frac{1}{2} \mathfrak{R} \left\{ \varepsilon \mathbf{E}(\mathbf{r}) \mathbf{E}^*(\mathbf{r}) + \mu \mathbf{H}(\mathbf{r}) \mathbf{H}^*(\mathbf{r}) \right.$$
$$\left. - \frac{1}{2} [\varepsilon |\mathbf{E}(\mathbf{r})|^2 + \mu |\mathbf{H}(\mathbf{r})|^2] \underline{\underline{I}} \right\}. \tag{5}$$

$\mathbf{E}^*$ and $\mathbf{H}^*$ represent the complex conjugates of the electric and magnetic fields, and $\mathbf{E}\mathbf{E}^*$ is a dyadic product. The double underline indicates a second-rank tensor, angle brackets denote a time average for harmonic fields, and $\underline{\underline{I}}$ is the unit tensor. A similar technique is used for calculating torques:

$$\mathbf{\Gamma} = - \int_S r d\mathbf{S} \cdot [\langle \underline{\underline{T}}_M(\mathbf{r}) \rangle \times \hat{\mathbf{r}}]. \tag{6}$$

All of the fields involved in the above equations are total fields, and the material parameters are those of the medium in which the particle is immersed.

For single particles, the above integrals can be calculated in the far field, where they amount to a sum of products of the VSWF expansion coefficients, a_{mn}, b_{mn}, p_{mn}, and q_{mn}. Such equations have been developed for spheres by several authors.[8,20,25–27] In each case, the equations have structural similarities. Most conspicuously, if the coefficients in the scattered field expansion are zero, i.e., if there is no particle present, or if the refractive index of the particle is identical to that of the background medium, the calculated forces and torques will also be zero as should be the case. This essential property is replicated by the numerical integration used throughout this paper.

The equations most easily adapted to our purposes are those due to Barton *et al.*[8] The basis functions used by Barton *et al.* are normalized differently from those used here; accounting for the differences in normalization leads to the following equations connecting the two sets of expansion coefficients (the dashed coefficients are from Ref. 8):

$$a'_{mn} = \frac{-i a_{mn} n_m}{k^2 A^2 [n(n+1)]^{1/2}}, \quad p'_{mn} = \frac{-i p_{mn} n_m}{k^2 A^2 [n(n+1)]^{1/2}},$$

$$b'_{mn} = \frac{b_{mn}}{k^2 A^2 [n(n+1)]^{1/2}}, \quad q'_{mn} = \frac{q_{mn}}{k^2 A^2 [n(n+1)]^{1/2}}, \tag{7}$$

where A is a size parameter for the sphere in the derivation of Barton *et al.*, and n_m is the refractive index of the medium. Substitution of the above relationships into Eqs. (10)–(12) of Ref. 8 gives a set of expressions for the torque, which are generally applicable to *any* shape of the particle:

$$\Gamma_x = \frac{-E_0^2 \epsilon_m}{8\pi k^3} \sum_{n=1}^{\infty} \sum_{m=-n}^{n} \mathfrak{R} \left\{ \sqrt{(n-m)(n+m+1)} \right.$$
$$\times \left[q_{mn} q^*_{m+1,n} + p_{mn} p^*_{m+1,n} \right.$$
$$\left. + \frac{1}{2} (q_{mn} b^*_{m+1,n} + q_{m+1,n} b^*_{mn} + p_{mn} a^*_{m+1,n} + p_{m+1,n} a^*_{mn}) \right] \right\}, \tag{8a}$$

$$\Gamma_y = \frac{-E_0^2 \epsilon_m}{8\pi k^3} \sum_{n=1}^{\infty} \sum_{m=-n}^{n} \mathfrak{I} \left\{ \sqrt{(n-m)(n+m+1)} \right.$$
$$\times \left[q_{mn} q^*_{m+1,n} + p_{mn} p^*_{m+1,n} \right.$$
$$\left. + \frac{1}{2} (q_{mn} b^*_{m+1,n} - q_{m+1,n} b^*_{mn} + p_{mn} a^*_{m+1,n} - p_{m+1,n} a^*_{mn}) \right] \right\}, \tag{8b}$$

$$\Gamma_z = \frac{-E_0^2 \epsilon_m}{8\pi k^3} \sum_{n=1}^{\infty} \sum_{m=-n}^{n} m[|q_{mn}|^2 + |p_{mn}|^2 + \Re(p_{mn}a_{mn}^*$$

$$+ q_{mn}b_{mn}^*)]. \tag{8c}$$

E_0 is the magnitude of the electric field at the beam focal point, and ϵ_m is the permittivity of the ambient medium; the asterisks denote complex conjugates.

In the work presented in Section 3, the integrals were performed both numerically and analytically. For the torque calculations, precise correspondence was found between the results of numerical integration and Eqs. (8a)–(8c). For the force calculations, it was found similarly that numerical integration of Eq. (4) produced agreement with the expressions derived by Moine and Stout[20] and also with force expressions (not shown) following from the work of Barton *et al.*[8]

F. Convergence

Before continuing, it is worth emphasizing a further point concerning convergence. As mentioned above, the translation of incident fields has a deleterious effect on convergence, such that a higher value of n max is generally required. Fortunately, no such effect is noticeable when the incident fields are rotated. However, increasing the aspect ratio of the particles causes further problems with the convergence. The underlying causes of this have been analyzed by a number of authors.[28–30] In essence, the problems stem from the use of spherical harmonic expansions in scattering problems involving nonspherical symmetry, which inevitably requires the use of high-order terms.

Figure 1 gives some quantitative insight into the issues concerning convergence encountered in this work. Values of the fractional change in calculated force with increasing n max are given for a variety of silica glass ellipsoids ($n_m = 1.45$) displaced by 0.6 μm below the focal point of the beam. In all calculations, the background medium is

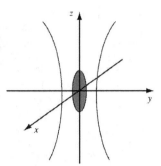

Fig. 2. Gaussian beam geometry. Light is incident from above traveling in the negative z direction and polarized parallel to the x axis. A prolate spheroid with a major axis parallel to z is shown.

taken to be water ($n_m = 1.33$), the frequency of the incident radiation is 3.75×10^{14} Hz, which corresponds to a wavelength of 600 nm in water, and the radius of the beam waist is 0.3 μm. Data are presented for spheres with a radius of $\rho = 1\,\mu$m (open circles) and 2 μm (triangles), prolate spheroids with an aspect ratio of $\delta = 2$, and an equivalent radius of $\rho_e = 1\,\mu$m with the symmetry axis aligned parallel to the x axis (diamonds) and the z axis (squares) and larger ellipsoids with $\delta = 2$ and $\rho_e = 2\,\mu$m aligned parallel to the z axis (solid circles). The geometry is illustrated in Fig. 2. The equivalent radius is defined in such a way that a spheroid with $\rho_e = 1\,\mu$m has the same volume as a sphere with $\rho = 1\,\mu$m.

As previously,[4] we take the criterion for convergence as

$$\left| \frac{|F^i| - |F^{i-1}|}{|F^{i-1}|} \right| \leq 0.001, \tag{9}$$

where $|F^i|$ is the modulus of the force calculated for n max$=i$. For the spherical particles, it can be seen that convergence is achieved for n max≥ 13. The rate of convergence is poorer for anisotropic particles, especially when they present their smallest profile to the incoming beam. For example, force calculations on prolate spheroids with $\rho_e = 1\,\mu$m oriented parallel to the z or beam axis converge for n max≥ 18; whereas similar particles with an equivalent radius of 2 μm fail to converge over the range of n max studied. Taking these considerations into account, attention will be focused, in this paper, on spheroids with aspect ratios in the range of $0.5 \leq \delta \leq 2.0$ and with a maximum volume equivalent to that of a sphere of a 1 μm radius. Under these conditions, convergence is reliable and all calculations can be performed with n max $= 18$.

3. RESULTS

An equilibrium configuration of an arbitrary particle can be defined as any position and orientation for which all the forces and torques on the particle vanish and can therefore be specified by three positional and three orientational coordinates. When the forces are those associated with an external field of some sort, sufficient conditions for equilibrium can be stated in terms of the symmetry of the system, i.e., of the fields and particles. For example, if the system contains a mirror plane, then the force perpendicular to this plane must be zero. Similarly, the torque

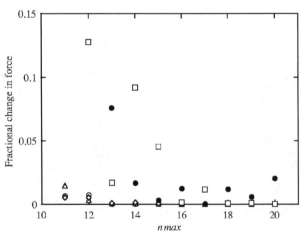

Fig. 1. Convergence characteristics for force calculations of single silica particles in Gaussian beams: spherical particles with $\rho = 1\,\mu$m (open circles) and 2 μm (triangles); prolate spheroids with $\delta = 2$ and $\rho_e = 1\,\mu$m and the long axis parallel to the x axis (diamonds) or the z axis (squares); prolate spheroids with $\delta = 2$ and $\rho_e = 2\,\mu$m and the long axis parallel to the z axis (solid circles). In each case, the particles are displaced below the focus by 0.6 μm with the beam incident from above.

about any axis contained within a mirror plane must also be zero. Since a laser beam does not have a plane of symmetry perpendicular to its axis, positional equilibrium in this direction does not result from symmetry requirements but can be seen, for example, as arising from a balancing of radiation pressure and gradient forces. Hence, in general, symmetry conditions are sufficient, but not necessary, for determining equilibrium configurations.

If the beam is linearly polarized, then, in terms of its interaction with a dielectric particle, it effectively contains two mirror planes through its axis, one containing the polarization vector of the electric field, the second orthogonal to the first. If the particle is similarly symmetric, then there must be equilibrium positions that lie on the beam axis. With these considerations in mind, it can be seen that a spheroidal particle will have at least three equilibrium configurations, each one corresponding to one of the permissible orientations of the principal axes with respect to the beam and polarization directions and each having an equilibrium position at a different point along the beam axis.

For these configurations to trap the particle, the equilibrium must be stable with respect to both translation and rotation. In the following sections, equilibrium configurations for spheroidal particles are calculated, and their stability is analyzed by examining the effect of small displacements and rotations on the forces and torques felt by each particle. Figure 2 shows the coordinate system that is being used. The incident beam is directed down the z axis and has a focal point at the origin; it is polarized parallel to the x axis. The translational stability of the equilibrium position to small perturbations parallel to z is indicated by the sign of the translational stiffness parameter given by

$$K^t_{z\beta} = \frac{\partial Q_{F,z}}{\partial z},$$ (10)

where $Q_{F,z}$ is the z component of the trapping efficiency for a particle aligned along axis β defined for the *force*, F_z, by

$$Q_{F,z} = \frac{F_z c}{n_m P}.$$ (11)

c is the speed of light *in vacuo*, n_m is the refractive index of the background medium, and P is the power in the beam. A negative value of the stiffness parameter implies a force directed back, toward the equilibrium position.

The rotational stability of a uniaxial particle is examined by reference to two similar parameters, each quantifying the change in torque experienced by the particle as it undergoes infinitesimal rotations from the equilibrium orientation about each of the axes perpendicular to its symmetry axis. For example, the rotations relevant to the ellipsoid drawn in Fig. 2 are about the x and y axes. The rotational stiffness parameters are given by

$$K^r_{\alpha\beta} = \frac{\partial Q_{T,\alpha}}{\partial \theta_\alpha},$$ (12)

where the symmetry axis of the spheroid is aligned with axis β, and rotations are about axis α. α and β refer to the coordinate axes, x, y, and z. $Q_{T,\alpha}$ is a trapping efficiency defined for the *torque* T_α about axis α; θ_α is the angle through which the particle is rotated about that axis. We define the torque trapping efficiency in a similar way to the *force* trapping efficiency:

$$Q_{T,\alpha} = \frac{T_\alpha c}{n_m P \lambda},$$ (13)

where λ is the wavelength of the incident beam in the surrounding medium. As with the force trapping efficiency, this expression is dimensionless, and also has the advantage that, unlike some definitions in the literature, it gives zero in the absence of a particle or if the particle has the same refractive index as the medium. The derivatives in Eq. (12) are taken such that a positive θ_α corresponds to a clockwise rotation about the named axis. As in the case of translations, rotational stability is indicated by negative values of the stiffness parameters.

The stiffness parameters discussed above quantify the changes in force, or torque, associated with small displacements about the equilibrium configurations. However, it is sometimes more informative to consider changes in directly observable quantities. For particles of the sizes considered here, immersed in water, typical values for Reynolds numbers are in the vicinity of 10^{-4}. In this regime, terminal velocities (angular or linear) are reached virtually instantaneously. The forces are then related to velocities and torques to rotational velocities by the translational and rotational friction tensors, respectively. Specifically,

$$\mathbf{F} = \underline{\underline{\mathbf{f}}}^t \cdot \mathbf{v},$$ (14a)

$$\mathbf{T} = \underline{\underline{\mathbf{f}}}^r \cdot \boldsymbol{\omega},$$ (14b)

where $\mathbf{F}$ is the force; $\mathbf{T}$ is the torque; $\mathbf{f}^r$ and $\mathbf{f}^t$ refer to the rotational and translational friction tensors, respectively; and $\mathbf{v}$ and $\boldsymbol{\omega}$ are the linear and angular velocities. Since attention is being focused here on uniaxial particles, these tensors have one eigenvalue expressing the frictional forces felt when the spheroids are rotated about their symmetry axes and a further two identical eigenvalues relating to rotations about the perpendicular axes. Since symmetry prevents optical forces from imposing torques about the unique axis of the particle, only the perpendicular eigenvalues are of interest.

Values for these elements of the rotational friction tensor may be obtained by evaluating[31]

$$f^r_{\text{perp}} = \frac{16\pi\eta}{3}\left[\frac{a^2 + b^2}{\alpha_a a^2 + \alpha_b b^2}\right],$$ (15)

in which

$$\alpha_a = \int_0^\infty \frac{d\lambda}{(a^2 + \lambda)^{3/2}(b^2 + \lambda)},$$ (16)

$$\alpha_b = \int_0^\infty \frac{d\lambda}{(a^2 + \lambda)(b^2 + \lambda)^{3/2}}. \quad (17)$$

a and b are the primary and secondary radii of the spheroid (as defined above), and η is the viscosity of the surrounding fluid.

As might be anticipated, both from physical intuition and from the appearance of the values of the radii in the denominators of α_a and α_b, the value of the perpendicular element of the friction tensor increases rapidly with equivalent radius; i.e., the larger the particle, the greater the resistance to motion. Representative values of the rotational friction coefficient, f^r_{perp}, are given in Fig. 3 for spheroids of varying aspect ratios. It can be seen to vary over 4 orders of magnitude as the equivalent radius changes by a factor of 10. Since the angular velocity relating to an applied torque is given by dividing by f^r_{perp}, it can be seen that there will be a corresponding decrease in angular velocity with an equivalent radius.

In combination with Eqs. (11) and (13), Eqs. (14a) and (14b) provide a sensitive method of quantifying the trap stiffnesses in terms of angular and translational velocities. Thus, by analogy with Eq. (12), we may define an angular velocity gradient as

$$G_{\alpha\beta} = \frac{\partial \omega_\alpha}{\partial \theta_\alpha}. \quad (18)$$

This gives a more sensitive indication of the susceptibility of a particle to motion under an applied torque and of how rapidly a displaced particle will recover its equilibrium configuration than quantities defined in terms of the torque alone.

A. Prolate Spheroids

Figure 4 shows maps of the trapping positions of prolate spheroids oriented in each of the three orthogonal, torque-free directions. The positions are measured with respect to an origin at the focus of the beam. The vertical axis represents the aspect ratio of the ellipsoids δ, while the horizontal axis is the equivalent radius ρ_e, i.e., the radius of a sphere of equivalent volume to the ellipsoid. Graphs showing the trapping positions of spheres, and spheroids with $\delta = 2$, have been extracted and plotted in Fig. 5. When the symmetry axes of the spheroids are perpendicu-

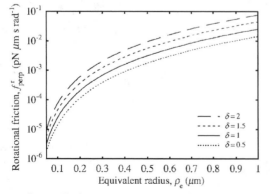

Fig. 3. Perpendicular component of the rotational friction tensor f^r_{perp} as a function of equivalent radius for spheroids with a range of aspect ratios.

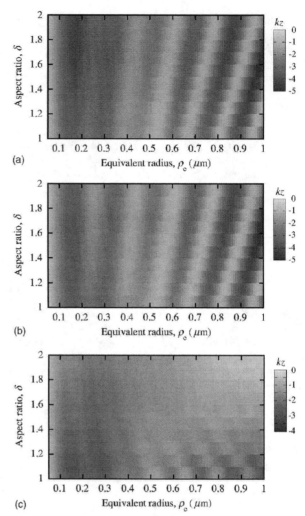

Fig. 4. (Color online) Maps showing the equilibrium trapping position, in dimensionless units kz, for prolate spheroidal silica particles with aspect ratios $\delta \leq 2$ and equivalent radii $\rho_e \leq 1\ \mu m$ with the symmetry axis of the particle aligned parallel to (a) x, (b) y, and (c) z axes.

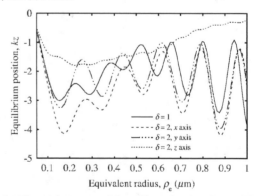

Fig. 5. Line plots extracted from Fig. 4 showing the equilibrium trapping position as a function of equivalent radius ρ_e for spheres ($\delta = 1$) and prolate spheroids with $\delta = 2$ oriented with symmetry axis parallel to each of the coordinate axes.

lar to the beam axis, the variation of trapping position with the equivalent radius is oscillatory, as might be expected from knowledge of the Mie theory and similar to the trapping behavior of a sphere. The period of the oscillations increases with the increasing aspect ratio. In fact, the period increases approximately in proportion with the cube root of the aspect ratio. In other words, the periods may be brought into line approximately by plotting the data against a minor radius instead of the equivalent radius. The minor differences that may be observed between Figs. 4(a) and 4(b) arise because the beam is polarized in the *x*–*z* plane.

The equilibrium positions for spheroids, whose symmetry axis is parallel to the beam, are radically different from those of the orthogonal orientations. This distinction has a simple explanation in terms of the concepts discussed in Section 1. The shape of the focal region of the Gaussian beam is elongated along the *z* axis (see Fig. 2). Vertically aligned prolate spheroids are able to overlap this intensity distribution more effectively than horizontal ones. In addition, the vertically aligned particles present a smaller cross section to the incoming beam thus diminishing the effect of downward radiation pressure. The reduction in downward radiation pressure combined with the increased energetic advantage of overlapping the focus more fully means that the vertically aligned particles tend to sit higher in the beam than do the horizontal particles. Evidently, this effect is enhanced as the volume of the particle is increased.

Figure 6 shows the translational stiffness, $K^t_{z\beta}$, for prolate spheroids with $\delta = 2$ in each of the orthogonal orientations and for spheres. The stiffness is evaluated for small displacements from the equilibrium position. In each case, it is negative, indicating that the trap is stable with respect to translational perturbation and, beyond a certain point related to the diameter of the beam, tends to diminish in magnitude with increasing particle size. There are no significant changes in similar plots for spheroids with lower aspect ratios.

The measures of rotational stiffness are more interesting. Figure 7 shows the rotational stiffness parameters, $K^r_{\alpha\beta}$, relevant to each of the three distinct equilibrium orientations for prolate spheroids with $\delta = 2$, the components being indicated in the figure. For the moment, results for small spheroids are ignored—they are not discernible in the main figure. Figure 7(a) shows that the configuration with the particle aligned with the symmetry axis parallel to the polarization vector is generally stable with respect

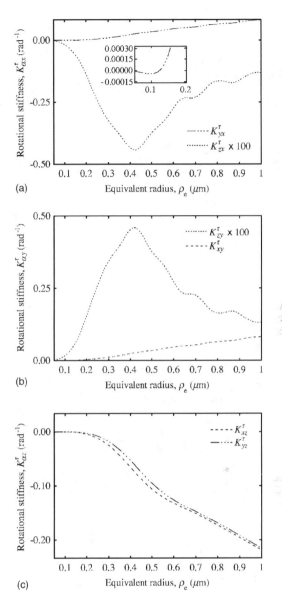

Fig. 7. Rotational stiffness parameters, $K^r_{\alpha\beta}$, for prolate silica spheroids with $\delta = 2$ aligned with the symmetry axis parallel to (a) *x*, (b) *y*, and (c) *z* axes. For each alignment, parameters are shown for rotations about each of the other two axes.

to rotations about the beam axis but apparently unstable with respect to rotation about the *y* axis. In other words, equilibrium points for this orientation are saddle points: small fluctuations are likely to result in a reorientation toward the vertical, but in the *x*–*y* plane, there is a preference for alignment with the polarization vector. In Fig. 7(b), the configuration with the symmetry axis perpendicular to both the beam axis and the polarization vector is shown to be unstable with respect to both relevant rotations, while alignment of the symmetry axis with the beam axis leads to stability in all respects [Fig. 7(c)]. All of these findings are in accordance with physical intuition. There are energetic reasons why the particle will try to maximize its overlap with the intensity distribution, and if it is constrained from doing this, it will reorient itself parallel to the polarization vector of the electric field.

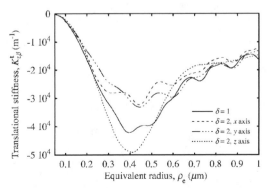

Fig. 6. Translational stiffness, $K^t_{z\beta}$, for spheres and prolate spheroids of silica with $\delta = 2$ oriented parallel to each of the coordinate axes.

At smaller volumes, however, the situation is slightly different. This is apparent in the inset to Fig. 7(a); in which there appears to be a slight restoring force toward the x–y plane for rotations about the y axis. The effect is seen more clearly in Fig. 8, in which the angular velocity gradients, $G_{\alpha\beta}$, have been plotted for the same rotations as shown in Fig. 7. For small spheroids ($\rho_e < 0.122\ \mu$m), Fig. 8(a) shows that the configuration with the symmetry axis aligned parallel to the polarization vector is stable. Conversely, Fig. 8(c), showing the results for the vertically oriented spheroids indicates that this configuration is unstable for small ellipsoids against rotations about the y axis. Combining these observations demonstrates that, for sufficiently small spheroids, the preferred orientation is parallel to the polarization vector. This behavior can be

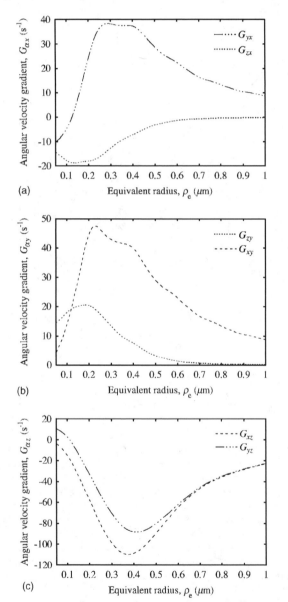

Fig. 8. Rotational velocity gradients, $G_{\alpha\beta}$, for prolate silica spheroids with $\delta=2$ aligned with the symmetry axis parallel to (a) x, (b) y, and (c) z axes. As with Fig. 7, parameters are shown for each alignment for rotations about each of the remaining axes.

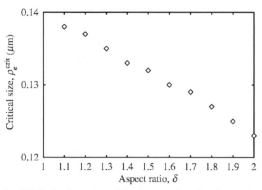

Fig. 9. Critical value of particle size, as given by the equivalent radius, ρ_e^{crit}, at which the transition occurs between stable orientation of the prolate spheroid parallel to the polarization vector (small particles) and stable orientation parallel to the beam (large particles).

qualitatively explained by realizing that, for small spheroids, the variation in intensity across the length of the particle becomes negligible, and the impact of orientation on the overlap with the intensity profile is reduced. Consequently, the advantages to alignment with the polarization vector become relatively more significant, and below a certain threshold, this configuration will be selected. Figure 9 shows how this transition point varies with the aspect ratio. As might be predicted, the lower the aspect ratio, the broader the range of equivalent radii over which the horizontal orientation is stable, but this is associated with a weakening of the restoring torques.

B. Oblate Spheroids

The trapping behavior of oblate spheroids is qualitatively different from that of the prolate variety. However, the same principles can be seen to be active. Accordingly, the particles seek to maximize their overlap with the intensity profile of the beam while simultaneously aligning themselves in some way with the polarization vector of the electric field. Unlike prolate spheroids, oblate spheroids can satisfy both of these requirements at the same time. In a static field, an oblate dielectric spheroid has a uniaxial polarizability tensor, whose eigenvalues are greater for vectors perpendicular to the symmetry axis than they are for vectors parallel to it. In other words, the symmetry axis is the *short* axis, and perpendicular axes may be described as *long* axes. It is possible, therefore, to align a direction of higher polarizability with the electric field while maximizing the overlap with the intensity profile by orienting the symmetry axis perpendicular to both the electric polarization vector and the beam axis. It is found that this is the stable configuration for oblate spheroids of any volume within the range considered here. This trapping orientation has been reported for spinach chloroplasts (*Spinacia oleracea*), which crudely resemble oblate spheroids.[5]

Figure 10 shows the equilibrium trapping positions for spheroids aligned in the manner described above. The variation of trapping position with equivalent radius is again oscillatory, but this time, increasing the anisotropy, i.e., decreasing the aspect ratio, appears to reduce the amplitude of the regular oscillations. Reducing the aspect ratio also reduces the period of the oscillations continuing the trend established for prolate spheroids. The stability of these configurations is demonstrated in Fig. 11 where the translational stiffness $K_{z\beta}^t$ [Fig. 11(a)], rotational stiffness $K_{\alpha y}^r$ [Fig. 11(b)], and angular velocity gradients $G_{\alpha y}$ [Fig. 11(c)] are plotted for spheroids with aspect ratios of 0.5. As might be expected, the angular stiffness of these traps is significantly greater for rotations that take the symmetry axis toward the beam axis, thus decreasing the overlap with the intensity profile, than it is for rotations that move the symmetry axis toward the polarization direction. It would appear that the effect of anisotropy in the intensity profile on the orientation is substantially greater than the effect of the polarization direction. It is worth noting that the orthogonal orientation with a symmetry axis parallel to the polarization vector (not shown here) is unstable to perturbations about the beam axis.

The feature that makes the trapping of oblate spheroids distinct from prolate ones is that the oblate spheroids align with their symmetry axes perpendicular to the beam axis, while prolate spheroids generally adopt the parallel alignment (with the exception of the smallest particles noted above). This leads to the existence of restoring torques for rotations of the oblate spheroids about the beam axis. Their presence can be inferred by the rotational stiffness data plotted in Fig. 11(b), but is demonstrated explicitly in Fig. 12(a) in which the variation of torque is plotted as a function of the angular deviation from the polarization vector for a spheroid with ρ_e

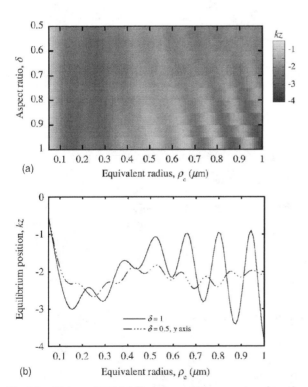

Fig. 10. (Color online) (a) Trapping position, expressed as the dimensionless parameter kz, of oblate silica spheroids oriented with the symmetry axis parallel to the y axis and secondary (major) axes parallel to the polarization vector (x axis) and beam axis as a function of the aspect ratio and equivalent radius. (b) Line plots extracted from (a) for spheres and oblate spheroids with $\delta = 0.5$.

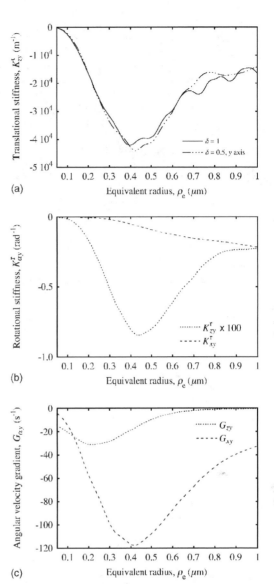

Fig. 11. (a) Translational stiffness parameter parallel to the beam axis, K_{zy}^t, for spheres and oblate spheroids of silica with $\delta = 0.5$ oriented with the symmetry axis parallel to the y axis and secondary axes parallel to the polarization vector (x axis) and the beam axis as a function of equivalent radius. (b) Rotational stiffness parameters, $K_{\alpha y}^r$, for the same oblate spheroid as in (a) for rotations about the x and z axes. (c) Similar plots of the angular velocity gradients $G_{\alpha y}$.

$= 0.5$ μm and $\delta = 0.5$. As can be seen, the variation of torque with an angle is sinusoidal; in fact, the torque is proportional to $\sin 2\phi$ where ϕ is the angle between the beam polarization vector and the long axis of the spheroid, which is exactly the result expected for a point polarizability in such a field. That the torque is negative implies that any particle not aligned with its long axis parallel to the polarization direction will rotate backward until it is. The only exception will be particles sitting at 90° to the polarization vector, which will not move until Brownian motion nudges them out of unstable equilibrium. Changes in the setting angle are accompanied by changes in the component of the force parallel to the beam axis [Fig. 12(b)]. This is a consequence of the fact that the equilibrium positions of oblate spheroids in the trapping beam are dependent on the orientation of the particle just as they were for the prolate spheroids.

The variation of torque with angle leads to a propensity for the particle to rotate with the plane of polarization. The equation of motion for such a particle in the low Reynolds number regime exposed to a continuously varying plane of polarization is given by[12]

$$f^r_{\text{perp}} \dot{\theta} = T_z \sin[2(\Omega t - \theta)], \tag{19}$$

where Ω is the angular velocity of the plane of polarization. For sufficiently large values of T_z or sufficiently small values of Ω, this equation has steady-state solutions:

$$\theta = \Omega t - \phi, \tag{20}$$

where ϕ now represents a constant offset between the polarization vector and the long axis of the particle in the x–y plane. It follows from Eqs. (19) and (20) that ϕ is given by

$$\sin \phi = \frac{f^r_{\text{perp}} \Omega}{T_z}, \tag{21}$$

and that there will be a maximum value of Ω given by

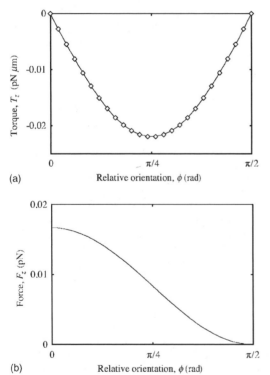

(a)

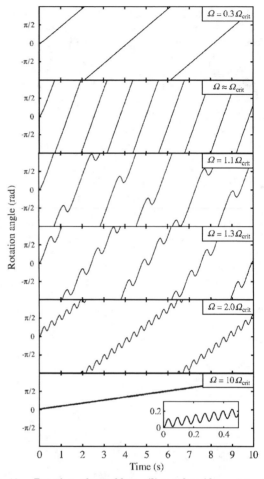

Fig. 13. Rotation of an oblate silica spheroid, $\rho_e = 0.5\,\mu$m, $\delta = 0.5$, induced by a rotating polarization vector with frequency Ω. The spheroid is oriented with its symmetry axis in the x–y plane.

$$\Omega_{\text{crit}} = \frac{T_z}{f^r_{\text{perp}}} \tag{22}$$

corresponding to $\phi = 90°$ and beyond which the steady-state solution breaks down.

Figure 13 shows solutions of Eq. (19) obtained by numerical integration for an oblate particle ($\delta = 0.5$) for several values of Ω. In this case, $\Omega_{\text{crit}} \simeq 5.2$ rad/s, and as the frequency of rotation of the polarization plane is increased, the rate of rotation of the particle increases until this value is reached. Beyond this point, the motion acquires an oscillatory component superposed on top of the steady rotation, a consequence of the particle intermittently failing to keep up with the changing polarization. Increasing Ω further results in a gradual reduction in both the amplitude of the oscillation and the rate of rotation of the particle. In the high-frequency limit, any response of the particle to the changing polarization will likely be masked by Brownian fluctuations.

C. Circularly Polarized Light

In the situation described above, the transfer of angular momentum to the particle occurs as a consequence of the motion of the particle itself. If the particle were held sta-

(b)

Fig. 12. (a) Torque, T_z, induced by a 1 mW beam on an oblate silica spheroid with $\rho_e = 0.5\,\mu$m, and $\delta = 0.5$. The particle lies with its symmetry axis in the x–y plane and an angle ϕ between the long axis of the spheroid and the polarization vector. The solid curve shows the calculated values; the diamonds indicate a sine curve, suitably scaled, for comparison. The height of the particle in the beam corresponds to the equilibrium position when the particle is held with the symmetry axis parallel to the polarization vector, i.e., $\phi = \pi/2$. (b) z component of force on the particle as ϕ varies.

tionary, it would feel a sinusoidally varying torque that averaged to zero over an integer number of periods and that, indeed, is what is observed in the high-frequency limit of Eq. (19). It is interesting, therefore, to consider the behavior of the particle in a circularly polarized beam. This situation is, however, subtly different from the rotating polarization described above. As has been discussed elsewhere, a circularly polarized *plane wave* carries no angular momentum.[32] It will, however, induce torques on a variety of particles, from the quartz wave plates of Beth[33] to absorbing particles[25,34] or particles having anisotropic refractive indices.[35] Circularly polarized beams have also been observed to impart angular momentum to trapped particles having either geometric or optical anisotropy. This behavior has been the subject of some controversy, the resolution of which has been attributed to the boundary conditions on the material or, in the case of beams, to edge effects.[25] It will be discussed in Section 4.

It is a straightforward matter to incorporate circular polarization in the *T* matrix method, in which case, forces and torques may be evaluated as before. Here, we examine, specifically, the case of oblate silica spheroids, oriented as above, with the symmetry axis in the *x–y* plane. Figure 14 shows the trapping positions for such spheroids with $\delta=0.5$ in circularly polarized light as compared with the linear polarization parallel and perpendicular to the symmetry axis of the spheroid. In the presence of the circularly polarized beam, it may be seen that the trapping positions are approximately midway between the equilibrium positions for the two linear polarizations of beam as might have been expected.

The torques experienced by these particles have also been calculated, and the rotation frequencies have been determined for 1 mW beams using Eq. (14b). The results are shown in Fig. 15. According to these calculations, the torque experienced by these particles is approximately one tenth of that corresponding to the maximum values associated with linearly polarized beams. The values obtained for spheres are negligible as they should be. However, the most intriguing aspect of the results shown in Fig. 15 is that, while for very small and large particles, the rotation is in the same sense as the angular momen-

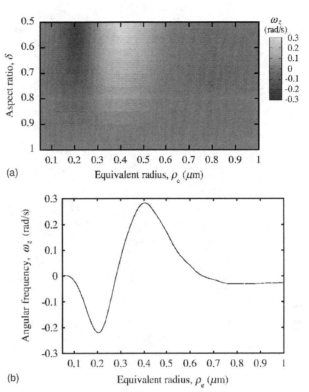

(a)

(b)

Fig. 15. (Color online) Frequency of rotation of oblate silica particles trapped in a 1 mW circularly polarized Gaussian beam. The particles are oriented with their symmetry axes in the *x–y* plane. (a) Map showing the frequency as a function of aspect ratio δ and equivalent radius ρ_e. (b) Single plot taken from (a) $\delta=0.5$.

tum of the beam (negative in this case), there is a region of the plot, for intermediate particle sizes, in which the sense of rotation is *opposite* to the sense of angular momentum of the beam completely counter to what might normally be expected.

The unusual nature of the results shown in Fig. 15 made it important to attempt some independent verification of the predictions. This was achieved in two ways: first by evaluating the torques analytically using the expressions in Eqs. (8a)–(8c) and second, by using finite-difference time-domain (FDTD) calculations. In the case of the analytical calculations, the agreement was essentially exact provided that *n* max was sufficiently large. Very minor deviations were observed using the FDTD approach, which could be attributed to the discrete nature of the lattice used. The FDTD results will be reported elsewhere.[36]

4. DISCUSSION

The calculations presented above have demonstrated that the trapping of spheroids can be qualitatively thought of in terms of a propensity for the particles to overlap the intensity distribution of the incident field and, where possible, align themselves with the plane of polarization, each in the presence of a prevailing downward radiation pressure. Small prolate spheroids therefore tend to preferentially align themselves with the polarization direction, while larger particles align themselves with the

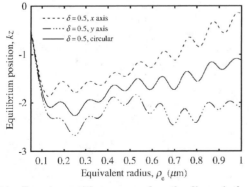

Fig. 14. Trapping position expressed as the dimensionless parameter *kz* of an oblate silica spheroid oriented with its symmetry axis in the *x–y* plane. The plot compares the effect of a circularly polarized beam with the linear polarization aligned parallel (*x* axis) and perpendicular (*y* axis) to the symmetry axis of the particle.

beam axis. The transition between these equilibrium orientations has been seen to be dependent on the volume and the aspect ratio of the spheroids. Oblate spheroids behave differently, having only one equilibrium orientation that satisfies both of the previously mentioned conditions. In this configuration, oblate particles are sensitive to the plane of polarization and will seek to align themselves to it, enabling them to be rotated as desired. Unexpected behavior is observed in the presence of circularly polarized beams in which oblate spheroids of certain sizes acquire angular momenta in the opposite direction to that which would be expected from a consideration of the angular momentum of the beam.

Many of these phenomena, with the exception of the last mentioned, have been observed experimentally in systems that approximate to the ones considered here. As mentioned in Section 1, the flat particles studied by Galajda and Ormos[13] and the chloroplasts of Bayoudh et al.[5] have been observed to behave similarly to the oblate ellipsoids studied here. The vertical equilibrium orientation of elongated particles has been observed by several people (see, for example, Refs. 11 and 12); the horizontal alignment of very small particles accords with physical intuition although direct observation may be complicated by the size of the particles in question.

The unusual rotational behavior of oblate spheroids in circularly polarized beams remains more elusive and deserves some discussion. An instinctive reaction to this behavior would be that it infringes the requirement that angular momentum be conserved, and that the calculations must be flawed in some way; the beam has angular momentum in one sense but imparts it in the opposite sense. A more nuanced view can be arrived at by considering the underlying mechanisms in more depth. Barnett[37] has developed a mathematical framework for understanding the transport of linear and angular electromagnetic momentum. Within this framework, linear and angular momentum can be thought of as examples of conserved quantities like energy. They therefore have associated with them densities from which they derive by way of volume integrals, i.e., following Barnett:

$$\mathbf{j} = \mathbf{D} \times \mathbf{B}, \tag{23a}$$

$$\mathbf{l} = \mathbf{r} \times \mathbf{D} \times \mathbf{B}, \tag{23b}$$

where $\mathbf{j}$ is the electromagnetic momentum density, and $\mathbf{l}$ is its angular counterpart. These densities obey continuity relationships that express the conservation of the quantity in question:

$$\frac{\partial j_i}{\partial t} + \frac{\partial T_{M,ij}}{\partial x_j} = 0, \tag{24a}$$

$$\frac{\partial l_i}{\partial t} - \frac{\partial M_{ij}}{\partial x_j} = 0. \tag{24b}$$

$\underline{T}_M$ is the Maxwell stress tensor from Eq. (4), while $\underline{\mathbf{M}}$ is the tensor given by $\underline{\mathbf{M}} = \underline{T}_M \times \mathbf{r}$ appearing in Eq. (5). The Einstein summation convention has been used. The first term appearing in both Eqs. (24a) and (24b) is the time rate of change of the pertaining momentum density, while

the second term gives the gradient of the density flux. Integrating Eqs. (24a) and (24b) over regions of space gives, for the first terms, the time rate of change of the total linear or angular momentum within a prescribed volume. The spatial derivatives in the second terms can be replaced by surface integrals by invoking the divergence theorem, giving the terms appearing on the right-hand side of Eqs. (4) and (5). They can now be interpreted as density fluxes across the prescribed surface.

In light of this, the meaning of Eqs. (4) and (5) becomes clearer. The surface integrals on the right-hand side give the flux of the appropriate momentum density across a surface bounding the scattering object. By virtue of the continuity equations, which, in turn, are expressions of underlying conservation laws, this flux is equivalent to the rate of change of the total linear or angular momentum inside the surface that occurs as a result of the scattering process. Since total angular (or linear) momentum is conserved, the rates of change of electromagnetic angular (or linear) momentum are equated to the mechanical torques (or forces) experienced by the particle.

These considerations suggest that the interactions between beams and particles may be more complicated than initially thought. Light beams *do not* possess a fixed quantity of momentum that can be partially transferred to scattering bodies in such a way as to increase the momentum of the particle at the cost of the beam. On the contrary, the total momentum of an unperturbed beam is proportional to its length. In the frequency domain, therefore, it is infinite, and making statements about its conservation is not meaningful. What determines the torque experienced by the particle is the net angular momentum *flux* arising as a consequence of scattering, and this need not be in accordance with intuition.

This last point can be appreciated by considering a more generally accepted example. When an elongated particle (prolate spheroid, cylinder, etc.) is held horizontally in a linearly polarized Gaussian beam, it will orient itself with its long axis parallel to the electric field polarization. If it is rotated about the beam axis away from the equilibrium orientation and held in that position, it will feel a restoring torque that is the result of an angular momentum flux across a surface bounding the particle. It is therefore known that a beam that does not carry angular momentum can, in the presence of a scattering particle, give rise to a linearly increasing angular momentum density within a surface bounding the particle. It would seem, therefore, not so very unusual that a beam carrying angular momentum in one sense should be capable of imparting a relatively small torque on a particle of the opposite sense.

The situation is further complicated by the controversy that has, for some time, surrounded the existence, or otherwise, of angular momentum in circularly polarized light (see Ref. 32 for a review of the debate). It is commonly accepted that although circularly polarized *plane* waves do not carry angular momentum, finite sections of them do, and that the transport of angular momentum in circularly polarized light occurs as a consequence of boundary effects. The fact that the counterrotation of oblate spheroids observed above, occurs in a size range in which the size of the beam waist falls between two radii of the spheroid is, therefore, also likely to be significant. In this respect, we would expect the behavior in circularly polarized beams to differ from that in beams with phase singularities, i.e., optical vortices, in which transfer of angular momentum will occur irrespective of the existence of boundaries.

The above paragraphs are intended to suggest that, although the counterrotation of certain sizes of oblate spheroids in circularly polarized beams predicted here may appear to be against intuition, it is not significantly more unusual than other, more generally accepted, phenomena. However, the precise nature of the mechanisms giving rise to these results remains unclear and requires further analysis.

Another, potentially fruitful, avenue of future investigation involves the coupling of rotational and translational motion suggested by Fig. 12. In this case, the rotation of an oblate spheroid about the beam axis results in significant changes in the vertical component of force on the particle. In fact, although not demonstrated here, this will be equally true for the rotation of any principal axis of any spheroid from its equilibrium orientation. Similarly, translating the center of a particle away from its equilibrium position will induce a torque as well as a restoring force. This has implications for the dynamics of trapped anisotropic particles and will have consequences both for the spectrum of Brownian motion that will be observed and for the behavior of such particles in moving traps. An understanding of this coupling will be of great importance when using dynamic holographic techniques to assemble anisotropic nanoparticles into complex structures.

5. CONCLUSIONS

We have presented a study of the forces and torques on spheroidal particles in Gaussian optical traps calculated using T matrix theory. In linearly polarized beams, the general principle that operates is that of maximizing the overlap between the particle and the most intense parts of the beam. Thus, prolate spheroids generally align with the symmetry axis parallel to the beam; whereas oblate spheroids always assume a perpendicular orientation. A second principle operating is that of aligning the long axis of the particle with the polarization direction. For oblate particles, both conditions may be satisfied simultaneously. However, for the prolate case, alignment with the polarization vector is only possible for small particles, i.e., those whose long axis is small compared with the beam waist. In circularly polarized light, oblate particles experience a torque about the beam axis leading to continuing rotation. However, the sense of the rotation varies depending on the relative dimensions of the particle and the beam waist.

APPENDIX A: GAUSSIAN BEAM REPRESENTATION

The coefficients in the VSWF expansion of the fifth-order Davis beam approximation to the incident Gaussian beam are given by

$$a_{1n} = a_{-1n} = b_{1n} = -b_{1n} = -(-i)^{n+1}[4\pi(2n+1)]^{1/2}g_{5,n},$$

$$\tag{A1}$$

where

$$g_{5,n} = g_{3,n} + \exp[-s^2(n-1)(n+2)](n-1)^2(n+2)^2s^8$$

$$\times[10 - 5(n-1)(n+2)s^2 + 0.5(n-1)^2(n+2)^2s^4], \quad \text{(A2)}$$

$$g_{3,n} = g_{1,n} + \exp[-s^2(n-1)(n+2)](n-1)(n+2)s^4$$

$$\times[3 - (n-1)(n+2)s^2], \tag{A3}$$

$$g_{1,n} = \exp[-s^2(n-1)(n+2)]. \tag{A4}$$

ACKNOWLEDGMENTS

The authors thank J. P. Barton and Massimo Antognozzi for helpful comments. They are grateful to the United Kingdom Research Councils (RCUK) for financial support. This work was performed as part of the Basic Technology project: A Dynamic Holographic Assembler.

S. Hanna's e-mail address is s.hanna@bristol.ac.uk.

REFERENCES

1. A. Ashkin, J. M. Dziedzic, J. E. Bjorkholm, and S. Chu, "Observation of a single-beam gradient force optical trap for dielectric particles," Opt. Lett. **11**, 288–290 (1986).
2. J. E. Molloy and M. J. Padgett, "Lights, action: optical tweezers," Contemp. Phys. **43**, 241–258 (2002).
3. D. G. Grier, "A revolution in optical manipulation," Nature **424**, 810–816 (2003).
4. S. H. Simpson and S. Hanna, "Numerical calculation of interparticle forces arising in association with holographic assembly," J. Opt. Soc. Am. A **23**, 1419–1431 (2006).
5. S. Bayoudh, T. A. Nieminen, N. R. Heckenberg, and H. Rubinsztein-Dunlop, "Orientation of biological cells using plane-polarized Gaussian beam optical tweezers," J. Mod. Opt. **50**, 1581–1590 (2003).
6. A. Ashkin, "Acceleration and trapping of particles by radiation pressure," Phys. Rev. Lett. **24**, 156–159 (1970).
7. K. Ren, G. Grehan, and G. Gouesbet, "Radiation pressure forces exerted on a particle arbitrarily located in a Gaussian beam by using the generalized Lorenz–Mie theory, and associated resonance effects," Opt. Commun. **108**, 343–354 (1994).
8. J. P. Barton, D. R. Alexander, and S. A. Schaub, "Theoretical determination of net radiation force and torque for a spherical particle illuminated by a focused laser beam," J. Appl. Phys. **66**, 4594–4602 (1989).
9. C. J. F. Böttcher, *Theory of Electric Polarization*, Vol. 1 of Dielectrics in Static Fields, 2nd ed. (Elsevier, 1973).
10. J. D. Jackson, *Classical Electrodynamics*, 2nd ed. (Wiley, 1975).
11. A. I. Bishop, T. A. Nieminen, N. R. Heckenberg, and H. Rubinsztein-Dunlop, "Optical application and measurement of torque on microparticles of isotropic nonabsorbing material," Phys. Rev. A **68**, 033802 (2003).
12. K. D. Bonin, B. Kourmanov, and T. G. Walker, "Light torque nanocontrol, nanomotors and nanorockers," Opt. Express **10**, 984–989 (2002).
13. P. Galajda and P. Ormos, "Orientation of flat particles in optical tweezers by linearly polarized light," Opt. Express **11**, 446–451 (2003).
14. A. La Porta and M. D. Wang, "Optical torque wrench: angular trapping, rotation, and torque detection of quartz microparticles," Phys. Rev. Lett. **92**, 190801 (2004).

15. J.-S. Kim and S.-W. Kim, "Dynamic motion analysis of optically trapped nonspherical particles with off-axis position and arbitrary orientation," Appl. Opt. **39**, 4327–4332 (2000).

16. R. C. Gauthier, "Theoretical investigation of the optical trapping force and torque on cylindrical micro-objects," J. Opt. Soc. Am. B **14**, 3323–3333 (1997).

17. R. C. Gauthier, "Optical levitation and trapping of a micro-optic inclined end-surface cylindrical spinner," Appl. Opt. **40**, 1961–1973 (2001).

18. M. I. Mishchenko, L. D. Travis, and A. A. Lacis, *Scattering, Absorption and Emission of Light by Small Particles* (Cambridge U. Press, 2002).

19. L. Tsang, J. A. Kong, and R. T. Shin, *Theory of Microwave Remote Sensing* (Wiley, 1985).

20. O. Moine and B. Stout, "Optical force calculations in arbitrary beams by use of the vector addition theorem," J. Opt. Soc. Am. B **22**, 1620–1631 (2005).

21. G. Gouesbet, J. A. Lock, and G. Grehan, "Partial wave representations of laser beams for use in light scattering calculations," Appl. Opt. **34**, 2133–2143 (1995).

22. G. Gouesbet, "Partial-wave expansions and properties of axisymmetric light beams," Appl. Opt. **35**, 1543–1555 (1996).

23. E. Anderson, Z. Bai, C. Bischof, S. Blackford, J. Demmel, J. Dongarra, J. Du Croz, A. Greenbaum, S. Hammarling, A. McKenney, and D. Sorensen, *LAPACK Users' Guide*, 3rd ed. (Society for Industrial and Applied Mathematics, 1999).

24. F. Melia, *Electrodynamics* (The University of Chicago Press, 2001).

25. P. L. Marston and J. H. Crichton, "Radiation torque on a sphere caused by a circularly-polarized electromagnetic wave," Phys. Rev. A **30**, 2508–2516 (1984).

26. S. Chang and S. S. Lee, "Optical torque exerted on a homogeneous sphere levitated in the circularly polarized fundamental-mode of a laser beam," J. Opt. Soc. Am. B **2**, 1853–1860 (1985).

27. G. Gouesbet, B. Maheu, and G. Grehan, "Light scattering from a sphere arbitrarily located in a Gaussian beam, using a Bromwich formulation," J. Opt. Soc. Am. A **5**, 1427–1443 (1988).

28. P. Barber, "Resonance electromagnetic absorption by nonspherical dielectric objects," IEEE Trans. Microwave Theory Tech. **25**, 373–381 (1977).

29. A. Mugnai and W. Wiscombe, "Scattering of radiation by moderately nonspherical particles," J. Atmos. Sci. **37**, 1291–1307 (1980).

30. V. Varadan, A. Lakhtakia, and V. Varadan, "Scattering by 3-dimensional anisotropic scatterers," IEEE Trans. Antennas Propag. **37**, 800–802 (1989).

31. T. G. M. van der Ven, *Colloidal Hydrodynamics* (Academic, 1989).

32. A. M. Stewart, "Angular momentum of the electromagnetic field: the plane wave paradox resolved," Eur. J. Phys. **26**, 635–641 (2005).

33. R. A. Beth, "Mechanical detection and measurement of the angular momentum of light," Phys. Rev. **50**, 115–125 (1936).

34. M. E. J. Friese, J. Enger, H. Rubinsztein-Dunlop, and N. R. Heckenberg, "Optical angular-momentum transfer to trapped absorbing particles," Phys. Rev. A **54**, 1593–1596 (1996).

35. M. E. J. Friese, T. A. Nieminen, N. R. Heckenberg, and H. Rubinsztein-Dunlop, "Optical alignment and spinning of laser-trapped microscopic particles," Nature **394**, 348–350 (1998).

36. D. Benito, S. H. Simpson, and S. Hanna are preparing a paper to be called "FDTD calculations of forces and torques on ellipsoidal particles."

37. S. M. Barnett, "Optical angular-momentum flux," J. Opt. B: Quantum Semiclassical Opt. **4**, S7–S16 (2002).

Observation of Bistability and Hysteresis in Optical Binding of Two Dielectric Spheres

N. K. Metzger,[1,*] K. Dholakia,[1,2] and E. M. Wright[1,2]

[1]*SUPA, School of Physics and Astronomy, University of St. Andrews, North Haugh, St. Andrews KY16 9SS, United Kingdom*
[2]*College of Optical Sciences and Department of Physics, University of Arizona, Tucson, Arizona 85721, USA*
(Received 13 July 2005; published 16 February 2006)

Using a dual-beam fiber optic trap, we have experimentally observed bistability and hysteresis in the equilibrium separations of a pair of optically bound dielectric spheres in one dimension. These observations are in agreement with our coupled system model in which the dielectric spheres modify the field propagation, and the field self-consistently determines the optical forces on the spheres. Our results reveal hitherto unsuspected complexity in the coupled light-sphere system.

DOI: 10.1103/PhysRevLett.96.068102   PACS numbers: 87.80.Cc

1. Introduction.—Light-matter interactions may be used to dictate the organization and manipulation of colloidal and biological matter at the microscopic level. An inhomogeneous optical field permits dielectric spheres of higher refractive index than their surrounding medium to be trapped in three dimensions in the field maxima [1,2] primarily through the dipole interaction. This allows physicists, chemists, and biologists to explore a range of fundamental phenomena. This includes many seminal studies involving just one or two trapped objects, namely, thermally activated escape from a metastable state for a single particle exposed to a double well potential [3], studies of optical angular momentum, stochastic resonance [4], and various studies of colloidal behavior in external potentials. Furthermore, hydrodynamic interactions between two particles have been shown to display interesting correlation behavior in their positions [5].

An emergent but poorly understood phenomena is that the interaction of light and matter may lead to self-organization of particles into arrays and create analogues to atomic systems [6,7]. This is termed "optically bound matter." Such self-organized systems are likely to have impact across the biological and colloidal sciences and, indeed, may possibly permit large scale self-assembly in up to three dimensions. In particular, the two-particle interactions between an object and its nearest neighbor create a self-consistent and homogeneous solution that allows an optical geometry to, in principle, create a colloidal array ranging for a few to several hundred objects.

Golovchenko and co-workers [6,7] investigated systems where the interaction of coherently induced dipole moments of the spheres were said to interact to bind matter. In addition, optical organization through interactions of optical scattering in the beam propagation direction has been recently observed and allows interactions between microparticles separated by distances and order of magnitude larger than their individual diameters [8–10]. This latter form of optical binding offers the promise for new studies of hydrodynamic interactions and, indeed, of large scale organization of matter but is presently not well understood. A detailed study of the two-particle optically bound system

would offer a new and important step to truly addressing this issue, and we believe it to be a suitable starting point for larger, more complex forms of binding.

In this Letter we present a detailed investigation of a one-dimensional optically bound system of an isolated pair of colloidal microspheres held in a dual-beam, optical-fiber trap. Careful investigation of the equilibria positions in this system reveals a hitherto unsuspected complexity: Namely, we observe bistability in the sphere separations dependent on the refractive-index difference between spheres and host medium, and hysteresis in the trap equilibrium separations as the fiber separation is varied. These observations match well with numerical solutions based on the coupled equations for the light propagation and the forces acting on the spheres, where an expression for the optical forces is employed that is derived from the Lorentz force formula.

Bistability and bifurcation are ubiquitous in several physical and biological systems and are closely linked with the concept of feedback. In the optical domain bistability is usually linked with the notion of nonlinearity but one can observe classical bistability with no explicit nonlinearity, for example, the radiation pressure from the intracavity field on a moving mirror [11]. Competition between parameters such as dispersion and nonlinearity in a wide variety of physical systems can ultimately lead to the coexistence of several stable solutions that may each be energetically favorable. Our observations for a two-particle optically bound system constitute the first realization of bifurcation and bistability that is inherently linked with the coupled nature of the problem and the direct interplay between radiation pressure and the light redistribution by each constituent microsphere with accompanying positive feedback. This results in a novel bistable optically trapped system showing bifurcation of the equilibrium separations and hysteresis of the interparticle separation when we adiabatically vary the fiber separation.

2. Coupled systems model.—Our model comprises two laser fields of wavelength λ counterpropagating (CP) along the z axis that interact with a pair of transparent dielectric spheres of refractive index n_s, diameter d, with centers at longitudinal positions z_j, $j = 1, 2$, and that are immersed in a host medium of refractive index n_h. The CP fields originate from two fibers (see Fig. 1) aligned along the z axis and with ends located at $z = -D_f/2$ for the forward field and $z = D_f/2$ for the backward field, with D_f being the fiber spacing, and the output fields being modeled as identical collimated Gaussians of spot size w_0 and power P_0. By virtue of the symmetry of the applied laser fields, we seek a solution for the configuration of the two spheres, which is also symmetric around $z = 0$, with $z_2 = -z_1 = D/2 > 0$, D being the sphere separation. The spatial evolution of the CP fields for a given configuration of the two spheres is modeled using the paraxial wave theory described in Ref. [9]. The paraxial theory, which fully incorporates the focusing effect of the spheres on the fields, is valid for sphere diameters and Gaussian spot sizes larger than an optical wavelength, meaning that the present theory is specialized to Mie scatterers.

In the next step the CP fields for a given configuration of the two spheres are in turn used to calculate the optical force F_z acting along the z axis on each sphere using the Lorentz force based approach of Ref. [12], assuming that the two CP beams are mutually incoherent, thereby neglecting any interference between them. A detailed account of this force calculation will be given in a forthcoming publication. For each CP field the calculated optical force acts along the direction of that field. Optical binding arises from the fact that the force acting on a given sphere is composed of two components along the z axis, one from the laser field whose beam waist is closest to the given sphere and a second oppositely directed force arising from the CP laser field that is partly refocused onto the given sphere by the other sphere. Balancing of these two forces by the refocusing of the spheres provides an intuitive explanation of optical binding [8].

We have numerically solved the coupled equations for the CP fields and the optical forces acting on the spheres to find the equilibrium sphere spacings where the net force F_z acting on each sphere is zero. An equilibrium spacing is stable when $dF_z/dD < 0$ at the zero crossing. In a previous paper we employed a simplified approach to calculating the equilibrium sphere positions [9] in which the force was assumed proportional to the laser intensity via a scattering coefficient. Although this approximate model gave qualitative agreement with previous experiments, it fails to account for the observed bistability in optical binding. Finally, we note that in our model we assume that the spheres remain well confined in the plane transverse to the z axis by virtue of the optical forces provided by the transverse structure of the CP fields, so that the sphere motion is confined to the z axis. With this caveat the equilibrium spacings do not depend on the laser power, which agrees with our experimental observations and previous studies [10].

The dashed lines in Figs. 2(a) and 3 show the numerically predicted sphere equilibrium separations. For a laser

wavelength of 1070 nm the fiber mode spot size was chosen as $w_0 = 3.4$ μm, and $d = 3$ μm diameter spheres were used with refractive index $n_s = 1.41$. In the experiment the host index was varied (see below), giving a controllable index mismatch $\Delta n = n_s - n_h$ between the spheres and host medium. The dotted line in Fig. 2(a) shows that for a fixed fiber separation $D_f = 90$ μm, when $\Delta n < 0.076$ only one equilibrium is present, whereas for $\Delta n > 0.076$ three solutions appear, the middle solution being found to be unstable. Thus, the coupled light-matter system exhibits regions of bistability, namely, two stable solutions for a given set of parameters. Physically, bistability in the optical binding is possible in the coupled system due to feedback: Changing the sphere separation alters the electromagnetic field distribution via the focusing properties of the spheres, which in turn alters the forces on the spheres. (Optically bound matter is thereby nonlinear in a manner analogous to Einstein's theory of gravity in which "matter tells space how to curve and space tells matter how to move.") Because of this feedback the forces on the spheres, viewed as a function of sphere separation, can become highly nonlinear, and give rise to bistability. We note that the bistability predicted here for two spheres is wholly distinct from the "bistable trapping" found by Lyons and Sonek for a single sphere [13]. That refers to the existence of two stable trapping positions for a *single* object in a trap rather than any interplay between two or more adjacent trapped objects. We note too that Singer and colleagues [10] observed a bistability between an optically bound state and a standard linear trapped array (with touching particles). The dashed line in Fig. 3 shows the predicted equilibrium separations as a function of fiber separation D_f for fixed index mismatch $\Delta n = 0.0924$ in the bistable region, with the negative sloped middle branch being found unstable. Here we see an upper switch point at $D_f = 120$ μm beyond which there is only one stable solution, and a lower switch point at $D_f = 65$ μm. Based on this, we expect that if the fiber separation is slowly cycled from below the lower switch point to above the upper switch point and this process reversed, the sphere separation should trace out a hysteresis loop as the sphere separation follows the local stable equilibrium.

3. Experiment.—Light at 1070 nm from a ytterbium fiber laser was split into two equal beams and coupled into two single mode fibers (measured mode field diameter at 1070 nm of 6.8 μm) to form a dual-beam fiber trap. Prior to launch within the fibers, two variable neutral density filters ensured equal CP beam intensities in the trap region with the power from each fiber variable in the range from 100 to 200 mW. Both fiber ends were mounted face to face on a glass cover slip. One fiber ($F1$) was kept stationary and the opposing ($F2$) was adjustable through a motorized micropositioning stage to vary the fiber separation (see Fig. 1). The emitted light fields were overlapped transverse to the beam propagation axis in the (x-y) plane

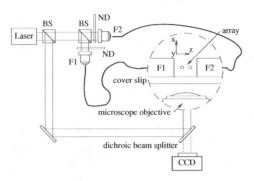

FIG. 1. Experimental setup: Laser and first beam splitter (BS) for optical tweezers coupled through a dichroic beam splitter in a 60× microscope objective. Second beam splitter (BS) for fiber coupling (F1 and F2) with neutral density filters (ND). The magnified inset shows the side view of both fibers mounted on the cover slip, F1 stationary and F2 adjustable via a micro-positioner. The array is formed between the two fiber faces and observed from underneath the setup through the microscope objective and the dichroic beam splitter with a CCD camera.

and could be separated along the z axis. By carefully choosing an optical path difference well above the laser coherence length (<1 mm), the beams were rendered mutually incoherent so that standing wave effects were eliminated [10].

Silica microspheres of $d = 3$ μm (with an estimated refractive index, related to information provided by Bangs laboratories and [14], of $n = 1.41$) in a host solution were added on the cover slip between the two fibers. The refractive-index difference Δn of the host solution with respect to the sphere was successively varied in our experiment by using D_2O or a deionized water and sucrose solution, which was measured with a refractometer [10]. A key parameter in the formation of optically bound arrays is the refractive-index difference Δn between the sphere and the host medium.

A 60× microscope objective and CCD camera were used to capture images onto a computer. Using a LABVIEW program, we tracked the positions of the particles with a relative accuracy of better than ± 0.5 μm and the fiber separations with an absolute accuracy of better than ± 3 μm and extracted the experimental data presented. Additionally, optical tweezers were incorporated through the observation microscope in order to load the fiber trap and initialize the sphere positions, and were then turned off.

In the first experiments the stable equilibrium of two 3 μm silica spheres were determined experimentally for a fixed fiber separation of 90 ± 3 μm and varying refractive-index mismatch Δn. Figure 2(b) shows an example observation of the two fibers F1 and F2 at the same spacing with two distinct sphere spacings D1 and D2. The overall mean values of the sphere separation taken over on average 12 data sets of about 300 measurements each with typically 3 different sphere pairs are shown as crosses in Fig. 2(a), and the associated spread in the measurements are indicated by vertical bars. We see that there is good

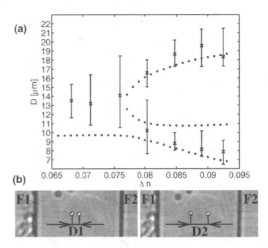

FIG. 2. (a) Plot of numerical data (dots) and experimental results (crosses with error bars) at a power of 100 mW. The crosses indicate the overall mean separation D, error bars represent the distribution of the mean values with their standard deviations (typically 12 data sets with about 3 different sphere combinations) of different measurements (typically 300) for a fiber separation of 90 ± 3 μm. For a refractive-index difference Δn from 0.095 down to 0.08 two stable positions can be observed. Between $\Delta n \sim 0.008$–0.07 the bistability of the systems ceases, resulting in a larger fluctuation in the data and in only one stable position for Δn smaller than 0.07. (b) Two stable positions with a separation $D1(\Delta n) > D2(\Delta n)$ within a dual-beam fiber trap (F1 and F2; the picture shows part of the cladding of each fiber).

overall agreement between the experimental data and the stable equilibria obtained from the theory, and, in particular, we see that for $\Delta n > 0.075$ bistability is evident in the experiment. Furthermore, as expected, the fluctuations are largest closest to the critical point where the new solutions appear. It is clearly seen that the deviation between theory and experiment is largest for smaller index mismatches Δn. This is understood by realizing that as the index mismatch decreases, the net optical forces acting on the spheres get smaller, so the equilibria are created by cancellation of ever smaller forces due to the CP fields. In this situation the numerical equilibria become more and more sensitive to the precise material parameters, whereas for larger index mismatches the equilibria are more robust against the precise parameters. Nevertheless, there is very good correlation between the numerics and experiment and excellent evidence for bistability. We note that the detailed structure of the bistability depends on the parameter values. For example, for a fiber separation of 100 μm an upper branch persists and the new solutions have equilibrium spacings less than the upper branch. This inverted case compared to Fig. 2(a) has also been seen and shows also bistability in good agreement with the coupled systems model.

In a second experiment we investigated the hysteresis predicted in the numerical results (dots) in Fig. 3. For this experiment the separation between the fiber ends was

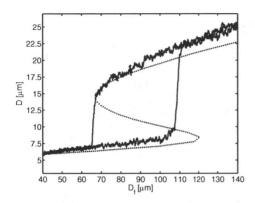

FIG. 3. Parametric plot of the equilibrium sphere separation D with varying fiber separation D_f showing clear hysteresis. The dots are the results of the numerical simulations, and the solid line is the experiment.

varied from 140 to 40 μm and back to 140 μm. If the fiber spacing is changed too quickly, the hysteresis loop washes out and eventually vanishes. For a relatively slow velocity, 1 μm/ sec , the spheres were found to adiabatically follow the changing fiber separation in the system.

The solid line in Fig. 3 shows the sphere separation plotted parametrically as a function of fiber separation as the fiber separation is slowly cycled, and there is excellent overall agreement between the numerics and experiment. The agreement is best for smaller fiber separations but dwindles for larger separations. The explanation for this is that for large fiber separations the optical forces acting on the spheres are getting ever smaller, meaning that vibrations and imperfections in the system, and not included in the model, can play a bigger role and deviations between theory and experiment are not unexpected. To put this in context, the Rayleigh range for the light fields emitted by the fibers is around 30 μm so that for a fiber separation of 100 μm the fields are considerably reduced compared to the input. Thus, at the upper switching point we see that the system switches early, which can be attributed to enhanced sensitivity to external perturbations and noise around the switching point. In particular, the data are clearly noisier on the upper branch for fiber separations between $D_f = 110$–140 μm in comparison to the lower branch for fiber separations between $D_f = 40$–60 μm. We note that the detailed numerical hysteresis loop is sensitive to the parameters used, in particular, the upper switching point can change by many microns with a small change in refractive index. This is also reflected in our experimental observations, where the hysteresis loop for different pairs of spheres is sensitive to the nominal size and refractive-index differences within one batch. Nonetheless, this experiment clearly demonstrates that hysteresis can occur in optically bound systems.

Our model predicts that the sphere equilibria should be independent of power, and we do observe this in our experiments over a range of 60 to 200 mW. At the lowest powers the system is far noisier, as mechanical vibrations, evaporative flow in our open sample cell, and other external perturbations are more able to induce a premature transition between the stable branches, or even (at large fiber separations) a total loss of particles from the trap. For this reason, the experiment shown in Fig. 3 was performed at the relatively high power of 200 mW, allowing us access to data at very large fiber separations.

4. Conclusion.—In this Letter we have reported observations and simulations of bistability in an optically bound array. In particular, using a dual fiber trap, we mapped out the region of bistable solutions as a function of the refractive-index difference between the spheres and the host solution. By slowly cycling the fiber separation, we showed that the system exhibits hysteresis as the array adiabatically follows the local equilibrium separation. Our simulations based on the spheres coupled to the light field were shown to yield good agreement with the experimental results.

The authors acknowledge many discussions with Professor Masud Mansuripur (College of Optical Sciences at the University of Arizona). This work was supported by the European Commission Sixth Framework Programme—NEST ADVENTURE Activity—through ATOM-3D (Contract No. 508952) and the European Science Foundation EUROCORES Programme SONS (project NOMSAN) by funds for the U.K. EPSRC and the EC Sixth Framework Programme.

*Corresponding author.
Electronic address: nkm2@st-and.ac.uk

[1] A. Ashkin, Phys. Rev. Lett. **24**, 156 (1970).
[2] A. Ashkin *et al.*, Opt. Lett. **11**, 288 (1986).
[3] L. I. McCann, M. Dykman, and B. Golding, Nature (London) **402**, 785 (1999).
[4] A. Simon and A. Libchaber, Phys. Rev. Lett. **68**, 3375 (1992).
[5] J. C. Meiners and S. R. Quake, Phys. Rev. Lett. **82**, 2211 (1999).
[6] M. M. Burns, J.-M. Fournier, and J. A. Golovchenko, Phys. Rev. Lett. **63**, 1233 (1989).
[7] M. M. Burns, J.-M. Fournier, and J. A. Golovchenko, Science **249**, 749 (1990).
[8] S. A. Tatarkova, A. E. Carruthers, and K. Dholakia, Phys. Rev. Lett. **89**, 283901 (2002).
[9] D. McGloin *et al.*, Phys. Rev. E **69**, 021403 (2004).
[10] W. Singer *et al.*, J. Opt. Soc. Am. B **20**, 1568 (2003).
[11] A. Dorsel, J. D. McCullen, P. Meystre, E. Vignes, and H. Walther, Phys. Rev. Lett. **51**, 1550 (1983).
[12] A. R. Zakharian, M. Mansuripur, and J. V. Moloney, Opt. Express **13**, 2321 (2005).
[13] E. R. Lyons and G. J. Sonek, Appl. Phys. Lett. **66**, 1584 (1995).
[14] M. Ibisate *et al.*, Langmuir **18**, 1942 (2002).

Power spectrum analysis for optical tweezers

Kirstine Berg-Sørensen[a]
The Niels Bohr Institute, Blegdamsvej 17, DK-2100 Copenhagen Ø, Denmark

Henrik Flyvbjerg[b]
Plant Research Department, Risø National Laboratory, DK-4000 Roskilde, Denmark

(Received 7 April 2003; accepted 1 December 2003)

The force exerted by an optical trap on a dielectric bead in a fluid is often found by fitting a Lorentzian to the power spectrum of Brownian motion of the bead in the trap. We present explicit functions of the experimental power spectrum that give the values of the parameters fitted, including error bars and correlations, for the best such χ^2 fit in a given frequency range. We use these functions to determine the information content of various parts of the power spectrum, and find, at odds with lore, much information at relatively high frequencies. Applying the method to real data, we obtain perfect fits and calibrate tweezers with less than 1% error when the trapping force is not too strong. Relatively strong traps have power spectra that cannot be fitted properly with *any* Lorentzian, we find. This underscores the need for better understanding of the power spectrum than the Lorentzian provides. This is achieved using old and new theory for Brownian motion in an incompressible fluid, and new results for a popular photodetection system. The trap and photodetection system are then calibrated simultaneously in a manner that makes optical tweezers a tool of precision for force spectroscopy, local viscometry, and probably other applications. © *2004 American Institute of Physics.* [DOI: 10.1063/1.1645654]

I. INTRODUCTION

Optical tweezers are used in many contexts in biological physics,[1] e.g., in single molecule studies of molecular motors[2,3] and other proteins and polymers,[4–7] and in surgery at the cellular level,[8] to name a few. In some of these contexts, the tweezers are only used to grab and hold something. In other contexts, they are used to exert a prescribed force, or to measure force: pico-Newton forces are measured or exerted,[4,7,9–11] local viscosity is measured,[12–14] or properties of a system are deduced from the Brownian motion of its parts.[15–17] While relative relationships for, e.g., force versus displacement can be calculated theoretically in a decent approximation,[18] the absolute value of the force cannot. It depends too much on experimental circumstances. So calibration is necessary. Good calibration is also a test that the tweezers, detection system, and data acquisition work as they are supposed to. As tweezer technology evolves and applications multiply, the need for good calibration methods will undoubtedly grow.

There are a number of ways to calibrate optical traps. (We use *trap* and tweezers as synonyms.) They are discussed in several excellent texts.[19–25] The most reliable procedure interprets the power spectrum of Brownian motion of a bead in the trap. This is conventionally done with the Einstein–Ornstein–Uhlenbeck theory of Brownian motion, which predicts a Lorentzian spectrum. The stochastic distribution of experimental spectral values about this Lorentzian is also known theoretically (Sec. III). This fact, and the simplicity of the Lorentzian form, permit us to give explicit analytical

results for the values of fitted parameters as functions of the experimental power spectrum (Sec. IV). Thus, when the power spectrum is legitimately modeled with a Lorentzian, these results fit it with ease, insight, and error bars on the parameter values found. The trap's strength is consequently known with precision that may be limited mainly by the calibration of the position detection system (Sec. V).

Very similar results might be obtained also when *aliasing* is taken into account (Sec. VI), but are not that interesting because the Lorentzian is a low-frequency approximation (Sec. VII) that can exploit only a fraction of the information content of the power spectrum (Sec. VIII). Even worse is the fact that even at intermediate trapping strengths some photodetection systems have no low-frequency window at all in which a Lorentzian can be fitted properly.

Thus a correct theory for the power spectrum is needed when the Lorentzian fails, and in general improves precision. We show that the power spectrum can be fully understood by combining known and new results on Brownian motion in incompressible fluids (Secs. IX–XII) with new results for typical photodetection systems (Sec. XIII). Once the effects of aliasing and antialiasing filters are accounted for (Sec. XIV), we have a procedure for how to calibrate tweezers (Secs. XV and XVI) which, e.g., adds decades of interpretable spectrum to local viscometry measurements. It makes details of the power spectrum of complex systems interpretable.[17] And it may open up the way for new applications of tweezers by making them a tool of precision.

Thus the thrust of this article is theory for how to analyze experimental data, with examples of how this is done. Readers who might wish to apply this analysis to their own data may do so easily with our MATLAB program that fits and

[a]Electronic mail: berg@alf.nbi.dk
[b]Electronic mail: henrik.flyvbjerg@risoe.dk

Reprinted with permission from K. Berg-Sørensen and H. Flyvbjerg, "Power spectrum analysis for optical tweezers," Rev. Sci. Instrum. 75, 594–612 (2004). Copyright 2004, American Institute of Physics.

plots at the click of a button. Reference 26 documents this program.

In Appendix A is a collection of notations and characteristic values of quantities in this article. Appendices B–D describe manipulations and tests of data which ensure that the data quality matches the precision that the theory can extract from good data. These procedures are an integral part of the practical application of our theoretical results. Appendix E explores how maximum likelihood estimation changes with the data compression we employ. Appendix F contains a long calculation. Appendices G and H expand on two technical points.

II. MATERIALS AND METHODS

The experimental data analyzed here were obtained with the optical tweezer setup described in Ref. 27. The laser light had wavelength $\lambda = 1064$ nm and positions were detected with a silicon *PIN* photodiode, S5981 from Hamamatsu, which is a popular choice because of its large active area, 10×10 mm. The trapped microsphere was from a batch of uniform microspheres from Bang Laboratories, Inc., catalog code No. PS04N, Bangs lot No. 1013, Inventory No. L920902A, with density of 1.050 g/ml and diameter of 1.05 ± 0.01 μm.[28]

The equipment was tested for electronic and mechanical noise in two "null tests" described in Appendix B. These tests set a bound on electronic noise, on the laser beam's pointing instability, and on much, but not all, mechanical noise. Crosstalk between *x*- and *y*-channel data was diagnosed and removed by a linear transformation described in Appendix C. A model-independent data analysis that tests our assumption of a harmonic trapping potential was done and is described in Appendix D. This test can indicate, but not prove, that the potential actually is harmonic, so its role is to warn us against analyzing data that seem to not satisfy this essential assumption. The data analyzed here pass this and another test [Fig. 1(b)] to perfection.

III. SIMPLE THEORY RECAPITULATED

The Einstein–Ornstein–Uhlenbeck theory of Brownian motion[29] describes the motion of the bead in a harmonic trapping potential with the following Langevin equation:

$$m\ddot{x}(t) + \gamma_0 \dot{x}(t) + \kappa x(t) = (2k_B T \gamma_0)^{1/2} \eta(t), \tag{1}$$

given here in one dimension for simplicity. Here $x(t)$ is the trajectory of the Brownian particle, m is its mass, γ_0 its friction coefficient, $-\kappa x(t)$ the harmonic force from the trap, and $(2k_B T \gamma_0)^{1/2} \eta(t)$ a random Gaussian process that represents Brownian forces at absolute temperature T; for all t and t':

$$\langle \eta(t) \rangle = 0; \quad \langle \eta(t) \eta(t') \rangle = \delta(t - t'). \tag{2}$$

Stokes's law for a spherical particle gives

$$\gamma_0 = 6\pi \rho \nu R, \tag{3}$$

where $\rho\nu$ is the fluid's shear viscosity, ρ the fluid's density, ν its kinematic viscosity, and R the sphere's radius.

The characteristic time for loss of kinetic energy through friction, $t_{\text{inert}} \equiv m/\gamma_0$, is 1000 times shorter than our experimental time resolution at 16 kHz sampling rate. We consequently follow Einstein and drop the inertial term in Eq. (1), so it then reads

$$\dot{x}(t) + 2\pi f_c x(t) = (2D)^{1/2} \eta(t), \tag{4}$$

where the *corner frequency*,

$$f_c \equiv \kappa/(2\pi\gamma_0), \tag{5}$$

has been introduced, and Einstein's equation,

$$D = k_B T/\gamma_0 \tag{6}$$

relating the diffusion constant, Boltzmann energy, and friction coefficient has been used.

After recording $x(t)$ for time T_{msr}, we Fourier transform $x(t)$ and $\eta(t)$,

$$\tilde{x}_k = \int_{-T_{\text{msr}}/2}^{T_{\text{msr}}/2} dt\, e^{i2\pi f_k t} x(t), \quad f_k \equiv k/T_{\text{msr}}, \quad k \text{ integer.} \tag{7}$$

Equation (4) gives the path as a function of noise,

$$\tilde{x}_k = \frac{(2D)^{1/2} \tilde{\eta}_k}{2\pi(f_c - if_k)}.$$

(When $\dot{x}(t)$ is Fourier transformed, partial integration gives a contribution from the ends of the interval of integration which we ignore. This leakage term (Ref. 30, Sec. 12.7) is truly negligible in our case because the power spectral density in Eq. (10) is a smooth function without spikes or other abrupt changes in value.)

From Eq. (2) it follows that

$$\langle \tilde{\eta}_k \rangle = 0; \quad \langle \tilde{\eta}_k^* \tilde{\eta}_\ell \rangle = T_{\text{msr}} \delta_{k,\ell}; \quad \langle |\tilde{\eta}_k|^4 \rangle = 2T_{\text{msr}}^2. \tag{8}$$

Since $\eta(t)$ is an uncorrelated Gaussian process, $(\text{Re }\tilde{\eta}_k)_{k=0,1,...}$ and $(\text{Im }\tilde{\eta}_k)_{k=1,2,...}$ are uncorrelated random variables with Gaussian distribution. Consequently, $(|\tilde{\eta}_k|^2)_{k=1,2,...}$ are uncorrelated non-negative random variables with *exponential* distribution. Hence so are experimental values for the power spectrum,

$$P_k^{(\text{ex})} \equiv |\tilde{x}_k|^2 / T_{\text{msr}} = \frac{D/(2\pi^2 T_{\text{msr}}) |\tilde{\eta}_k|^2}{f_c^2 + f_k^2} \tag{9}$$

for $k > 0$. Their expected value is a Lorentzian,

$$P_k \equiv \langle P_k^{(\text{ex})} \rangle = \frac{D/(2\pi^2)}{f_c^2 + f_k^2}, \tag{10}$$

and because $P_k^{(\text{ex})}$ is exponentially distributed,

$$\sigma[P_k^{(\text{ex})}] = \langle (P_k^{(\text{ex})} - P_k)^2 \rangle^{1/2} = P_k. \tag{11}$$

IV. LEAST-SQUARES FITTING OF LORENTZIAN

Experimentally, we sample $x(t)$ with frequency f_{sample} for time T_{msr}. From the resulting time series $x_j \equiv x(t_j)$, $j = 1,...,N$, we form the *discrete* Fourier transform,

$$\hat{x}_k \equiv \Delta t \sum_{j=1}^{N} e^{i2\pi f_k t_j} x_j = \Delta t \sum_{j=1}^{N} e^{i2\pi jk/N} x_j, \tag{12}$$

$k = -N/2 + 1,...,N/2$, where $\Delta t \equiv 1/f_{\text{sample}}$, $t_j \equiv j\Delta t$, and $N\Delta t = T_{\text{msr}}$. This discrete Fourier transform is a good approximation to the continuous one, Eq. (7), for frequencies $|f_k| \ll f_{\text{sample}}$. Consequently, the experimental power spectrum

$$P_k^{(\mathrm{ex})} \equiv |\hat{x}_k|^2/T_{\mathrm{msr}}$$

obeys the same statistics as $|\tilde{x}_k|^2/T_{\mathrm{msr}}$; see Sec. VI for details.

Least-squares fitting in its simplest form presupposes that each data point is "drawn" from a Gaussian distribution and that different data points are statistically independent. The second condition is satisfied by P_k in Eq. (9), but the first is not, since P_k is exponentially distributed. The solution is *data compression*, which results in a smaller data set with less noise, *and*, by way of the central limit theorem, in normally distributed data.

Data compression by *windowing* is common and has its advantages.[30] When the number n_w of windows used is large, the values of the compressed power spectrum are statistically independent and Gaussian distributed, and can be used in the formulas below. Windowing always compresses to equidistant points on the frequency axis. "Blocking" is an alternative method without this constraint, and hence is useful for data display with the logarithmic frequency axis. It replaces a "block" of n_b consecutive data points $(f, P^{(\mathrm{ex})}(f))$ with a single new "data point" $(\bar{f}, \bar{P}^{(\mathrm{ex})}(f))$, with coordinates that simply are block averages.[30] When n_b is so large that we can ignore terms of nonleading power in n_b, $\bar{P}^{(\mathrm{ex})}(\bar{f})$ is Gaussian distributed with $\langle \bar{P}^{(\mathrm{ex})}(\bar{f}) \rangle = P(\bar{f})$ and $\sigma(\bar{P}^{(\mathrm{ex})}(\bar{f})) = P(\bar{f})/\sqrt{n_b}$.

In the following, it is understood that data have been blocked (or windowed, or both), but we leave out the overbar to keep the notation simple. We fit by minimizing

$$\chi^2 = \sum_k \left(\frac{P_k^{(\mathrm{ex})} - P_k}{P_k/\sqrt{n_b}} \right)^2 = n_b n_w \sum_k \left(\frac{P_k^{(\mathrm{ex})}}{P_k} - 1 \right)^2 ;$$

see Appendix E for background. This χ^2 can be minimized analytically: The theoretical spectrum can be written $P_k = (a + bf_k^2)^{-1}$ with a and b positive parameters to be fitted, so χ^2 is a quadratic function of a and b. Minimization gives

$$f_c = (a/b)^{1/2} = \left(\frac{S_{0,1}S_{2,2} - S_{1,1}S_{1,2}}{S_{1,1}S_{0,2} - S_{0,1}S_{1,2}} \right)^{1/2}, \tag{13}$$

$$\frac{DT_{\mathrm{msr}}}{2\pi^2} = 1/b = \frac{S_{0,2}S_{2,2} - S_{1,2}^2}{S_{1,1}S_{0,2} - S_{0,1}S_{1,2}},$$

$$\frac{\chi^2_{\min}}{n_b} = S_{0,0} - \frac{S_{0,1}^2 S_{2,2} + S_{1,1}^2 S_{0,2} - 2 S_{0,1} S_{1,1} S_{1,2}}{S_{0,2} S_{2,2} - S_{1,2}^2}, \tag{14}$$

where we have introduced the sums

$$S_{p,q} \equiv \sum_k f_k^{2p} P_k^{(\mathrm{ex})q}.$$

V. FROM MILLIVOLTS TO NANOMETERS: CALIBRATING LENGTH SCALES

Note that we fit both D and f_c to the power spectrum of the x coordinate, and, in an independent fit, to the power spectrum of the y coordinate. The position detection system's output has a somewhat arbitrary amplitude that depends linearly on laser power and the three independent amplifier settings for the voltages measured, $V_{\mathrm{I}} - V_{\mathrm{II}} - V_{\mathrm{III}} + V_{\mathrm{IV}}$, V_{I}

$+ V_{\mathrm{II}} - V_{\mathrm{III}} - V_{\mathrm{IV}}$, and V_z defined in Appendix C. This results in three arbitrary measures of length, one for each direction, x, y, and z. We determine two of the three corresponding conversion factors to units of physical length by equating the fitted values for D, which is determined in arbitrary units, to the value in physical units known from Einstein's relation, Eq. (6).

The fitted values of D can be determined with high precision, as demonstrated below. Its value in physical units, however, is not known with similar precision in some biophysical experiments. While the temperature T can be known very well, the value of γ_0 is a source of error in Einstein's relation. Microspheres are commercially available with radius R known to within 1% and similar precision of the spherical shape, so Stokes law, Eq. (3), applies. But the value of the dynamic viscosity $\rho\nu$ of the fluid, in which the experiments take place, may not be known with the same precision. Additives such as glucose, BSA, and casein change the value of $\rho\nu$ dramatically. Five percent of glucose, e.g., changes the viscosity of water by a factor of 1.12.[32] Similarly, 5% of NaCl changes the viscosity of water by a factor of 1.19. Such concentrations of additives occur, e.g., in studies of single kinesin molecules[33] and in studies of single myosin molecules.[34] In such cases it is better to determine the conversion factors from arbitrary units to physical units of length by independent measurements, as was done in, e.g., Refs. 23 and 35. The fitted values of D may then serve as either a check of consistency for the method, or as independent determination of the value of γ_0, hence of $\rho\nu$, which may also be calculated if one knows the concentrations of additives and how they affect the viscosity.

VI. ALIASED LORENTZIAN

We will now understand the effect of finite sampling time better than we did in Sec. IV: In an experiment we sample $x(t)$ at discrete times $t_j = j\Delta t$, $\Delta t \equiv 1/f_{\mathrm{sample}}$. We consequently solve Eq. (4) in the time interval $t_j \leq t \leq t_{j+1}$ for given noise to find the *effective* Einstein–Ornstein–Uhlenbeck theory for discretely sampled data. We find

$$x_{j+1} = c x_j + \Delta x \, \eta_j, \tag{15}$$

with

$$\langle \eta_j \rangle = 0; \quad \langle \eta_i \eta_j \rangle = \delta_{i,j} \quad \text{for all } i, j. \tag{16}$$

Here we have introduced

$$\eta_j \equiv \left(\frac{4\pi f_c}{1 - c^2} \right)^{1/2} \int_{t_j}^{t_{j+1}} dt \, e^{-2\pi f_c(t_{j+1} - t)} \eta(t), \tag{17}$$

$$c \equiv \exp(-\pi f_c/f_{\mathrm{Nyq}}); \quad f_{\mathrm{Nyq}} \equiv f_{\mathrm{sample}}/2, \tag{18}$$

and

$$\Delta x \equiv \left(\frac{(1 - c^2)D}{2\pi f_c} \right)^{1/2}. \tag{19}$$

Application of the discrete Fourier transform, Eq. (12), to x and η in Eq. (15) transforms Eq. (15) to

$$e^{i2\pi k/N} \hat{x}_k = c \hat{x}_k + \Delta x \, \hat{\eta}_k, \tag{20}$$

while the Fourier transformed version of Eq. (16) is

$$\langle \hat{\eta}_k \rangle = 0; \quad \langle \hat{\eta}_k^* \hat{\eta}_\ell \rangle = T \Delta t \, \delta_{k,\ell} \qquad (21)$$

for all $k, \ell \in \{N/2+1,...,N/2\}$. So now our experimental estimate for the power spectrum is

$$P_k^{(\mathrm{ex})} = |\hat{x}_k|^2/T = \frac{(\Delta x)^2 |\hat{\eta}_k|^2/T}{1+c^2-2c\cos(2\pi k/N)}. \qquad (22)$$

The expected value for this spectrum is

$$P_k \equiv \langle P_k^{(\mathrm{ex})} \rangle = \langle |\hat{x}_k|^2/T \rangle = \frac{(\Delta x)^2 \Delta t}{1+c^2-2c\cos(2\pi k/N)}, \qquad (23)$$

and its root-mean-square deviation is

$$\sigma(P_k^{(\mathrm{ex})}) = P_k, \qquad (24)$$

identical in form to Eq. (11) because $|\hat{\eta}_k|^2$ like $|\tilde{\eta}_k|^2$ is exponentially distributed.

Equation (23) gives the function that replaces the Lorentzian in the case of finite sampling frequency, and it should fit the experimental spectrum for all frequencies of $0 < f_k \leq f_{\mathrm{Nyq}}$ if the simple theory discussed here is correct. Least-squares fitting of Eq. (23) to experimental data can be done analytically, once and for all, and it results in expressions very similar to those in Sec. IV.

We note that for $f_c \ll f_{\mathrm{Nyq}}$ and $|f_k| \ll f_{\mathrm{Nyq}}$, Eq. (23) to leading order in f_c/f_{Nyq} and f_k/f_{Nyq} becomes the Lorentzian in Eq. (10). So the approximation done in Sec. IV when we fitted the Lorentzian to the experimental spectrum has now been understood within the same simple theory for Brownian motion by accounting for the finite sampling frequency. The effect of the latter is maximal at $f_{\mathrm{Nyq}} = f_{N/2}$ where, for same f_c and D, $P_{N/2} = 2.47 P(f_{N/2})$, i.e., at f_{Nyq} the finite sampling rate increases the power spectrum by 247% over its Lorentzian value.

With this understanding, it seems more correct to replace the Lorentzian altogether with Eq. (23), the so-called *aliased Lorentzian*; see Appendix H for more about aliasing. More so because so-called antialiasing filters do *not* change the aliased Lorentzian back into a Lorentzian; see Fig. 11. The outcome of this replacement can be determined without actually doing it from a simple phenomenological plot of the experimental power spectrum. If an aliased Lorentzian fits the experimental power spectrum $P_k^{(\mathrm{ex})}$, a plot of $P_k^{(\mathrm{ex})-1}$ vs $\cos(2\pi k/N)$ falls onto a straight line, according to Eq. (23). Figure 1 shows our experimental power spectrum plotted in this manner. Clearly, the data points do not fall onto a straight line. They do increasingly for smaller beads and sampling rates, especially for a different photodetection system and/or shorter laser wavelength, as explained below.

VII. LIMITS ON LORENTZIANS

Over which range of frequencies f_k should the sums $S_{p,q}$ be done? Stochastic errors are minimized by maximizing this range, but systematic errors limit the range: At low frequencies the experimental power spectrum typically is contaminated by low-frequency noise external to the experiment; see Appendix B. At high frequencies there are three concerns: (i) The Lorentzian is a good approximation only for $f^2 \ll f_{\mathrm{Nyq}}^2$, where $f_{\mathrm{Nyq}} \equiv f_{\mathrm{sample}}/2$, as we have seen. As we have also

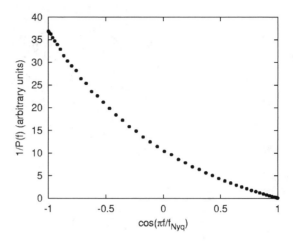

FIG. 1. Inverse experimental power spectrum, $P_k^{(\mathrm{ex})-1}$ plotted vs $\cos(\pi f/f_{\mathrm{Nyq}})$. Plotted this way, an aliased Lorentzian form would fall onto a straight line, see Eq. (23). In case one uses an oversampling delta–sigma data converter, aliasing and electronic filtering do not occur below the output frequency. So for that case one should plot $P_k^{(\mathrm{ex})-1}$ vs f^2. Then a pure Lorentzian form falls onto a straight line, and effects of frequency-dependent hydrodynamical friction and unintended filtering (see Secs. IX–XIII) show up as curvature.

seen, one can trade this approximation for an exact result, an aliased Lorentzian, but the latter does not describe the data. Other effects are in play. (ii) Some position detection systems, including ours, unintentionally cause significant low-pass filtering.[36] (iii) The Einstein–Ornstein–Uhlenbeck theory of Brownian motion is only a low-frequency approximation when used for liquids; the hydrodynamically correct spectrum is not Lorentzian.

Thus one should fit with a Lorentzian in an interval $[f_{\min}, f_{\max}]$ that avoids these systematic errors at high and low frequencies while minimizing stochastic errors of the fitted parameters. To this end, we give the stochastic errors' dependence on $[f_{\min}, f_{\max}]$.

VIII. INFORMATION CONTENT OF THE SPECTRUM

Given Eqs. (13) and (14), propagation of errors gives (see Appendix F)

$$\frac{\sigma(f_c)}{f_c} = \frac{s_{f_c}(x_{\min}, x_{\max})}{\sqrt{\pi f_c T_{\mathrm{msr}}}}, \qquad (25)$$

$$\frac{\sigma(D)}{D} = \left(\frac{1+\pi/2}{\pi f_c T_{\mathrm{msr}}} \right)^{1/2} s_D(x_{\min}, x_{\max}), \qquad (26)$$

and the covariance $\langle (f_c - \langle f_c \rangle)(D - \langle D \rangle) \rangle \equiv \langle f_c D \rangle_c$ is

$$\frac{\langle f_c D \rangle_c}{\sigma(f_c)\sigma(D)} = \left(\frac{v(x_{\min}, x_{\max})}{u(x_{\min}, x_{\max})} \right)^{1/2}.$$

Here, $x_{\min} \equiv f_{\min}/f_c$, $x_{\max} \equiv f_{\max}/f_c$, and we have introduced the dimensionless functions,

$$s_{f_c}(x_1,x_2) \equiv \left(\frac{\pi}{u(x_1,x_2)-v(x_1,x_2)} \right)^{1/2}, \qquad (27)$$

$$s_D(x_1,x_2) \equiv \left(\frac{u(x_1,x_2)}{(1+\pi/2)(x_2-x_1)} \right)^{1/2} s_{f_c}(x_1,x_2),$$

$$u(x_1,x_2) \equiv \frac{2x_2}{1+x_2^2} - \frac{2x_1}{1+x_1^2} + 2\arctan\left(\frac{x_2-x_1}{1+x_1x_2} \right),$$

$$v(x_1,x_2) \equiv \frac{4}{x_2-x_1}\arctan^2\left(\frac{x_2-x_1}{1+x_1x_2} \right). \qquad (28)$$

The function s_{f_c} is normalized such that $s_{f_c}(0,\infty)=1$. Thus $s_{f_c}(x_{min},x_{max})\geq 1$, because maximum precision is achieved only by fitting to the whole spectrum. Less will do in practice, and do well, as Figs. 2 and 3 illustrate.

Figure 2 has $f_c=357$ Hz, $f_{min}=110$ Hz, and $f_{max}=1$ kHz, hence $s_{f_c}(x_{min},x_{max})=s_{f_c}(0.31,2.80)=2.4$. For comparison, $s_{f_c}(0,2.80)=1.8$. So, given our value for f_{max}, our nonvanishing value for f_{min} costs us a 30% increase in the error bar for the value we find for f_c. On the other hand, $s_{f_c}(0.31,\infty)=1.26$. So, given our value for f_{min}, by increasing f_{max} we could reduce the stochastic error for f_c by almost a factor of 2, if systematic errors did not prevent this. This is despite $f_{max}\simeq 3f_c$. To harvest this extra information, one needs a better understanding of the power spectrum at these frequencies than the Lorentzian provides.

Systematic errors may leave no frequency range at all in which one can properly fit a Lorentzian. A data set with almost twice larger corner frequency illustrates this, although it is sampled three times faster. When a Lorentzian is fitted to this power spectrum, f_c decreases as f_{max} is increased, and support for the fit vanishes although $f_{max}^2\ll f_{Nyq}^2$ is satisfied; see Table I. In this case, proper calibration is impossible without better understanding of the power spectrum than the Lorentzian provides. Below, this understanding is provided, and calibration is achieved using the very same data set.

IX. FRICTION FELT BY A MICROSPHERE MOVING IN AN INCOMPRESSIBLE FLUID

When a rigid body moves through a dense fluid like water, the friction between the body and fluid depends on the body's past motion, since that determines the fluid's present motion. For a sphere performing linear harmonic motion $x(t)$ with cyclic frequency $\omega=2\pi f$ in an incompressible fluid and at vanishing Reynold's number, the (Navier–) Stokes equations were solved analytically and give a "frictional" force,[37,38]

$$F_{fric} = -\gamma_0\left(1+\frac{R}{\delta}\right)\dot{x} - \left(3\pi\rho R^2\delta + \frac{2}{3}\pi\rho R^3\right)\ddot{x}, \qquad (29)$$

where only the term containing $\dot{x}$ dissipates energy; the term containing $\ddot{x}$ is inertial force from entrained fluid. The notation is the same as above: γ_0 is the friction coefficient of Stokes' law for linear motion with constant velocity, Eq. (3), $\rho=1.0$ g/cm is the density of water at room temperature, $\nu=1.0$ μm^2/s is the kinematic friction coefficient of water, and $2R=1.05$ μm is the diameter of the sphere we used. Thus $\gamma_0=9\times10^{-6}$ g/s. The *penetration depth* δ character-

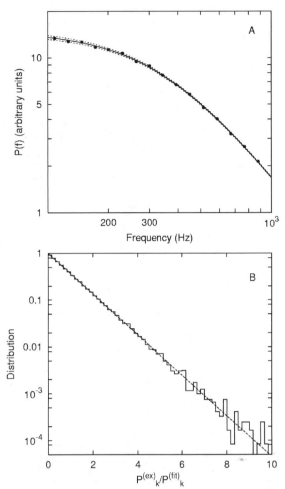

FIG. 2. (a) Lorentzian fitted to a power spectrum in the interval [110 Hz, 1 kHz] yielding $f_c=357\pm3$ Hz and $D=585\pm4$ (arb units)^2/s. (The position detection system's arbitrary units of length are calibrated in Sec. V.) The power spectrum in this interval, already an average of $n_w=5$ spectra, was blocked by a factor of $n_b=517$ to $N'=29$ points evenly distributed on the linear axis, then fitted using Eqs. (13) and (25). Dashed lines indicate ±one standard deviation of the theoretical curve. Statistical support (Refs. 30 and 31) for the fit shown here is 60%. The experimental spectrum has $f_{Nyq}=8$ kHz. (b) Histogram of $P_k^{(ex)}/P_k$ for f_k in the frequency range of the fit, $P_k^{(ex)}$ the unblocked experimental power-spectral values at f_k, and P_k its expected value, the fitted theory's value at f_k. According to theory, Eqs. (6) and (7), this ratio is exponentially distributed. Dashed line: $y=\exp(-x)$. Perfect agreement between theory and data is seen over all four decades of probability shown.

izes the exponential decrease of the fluid's velocity field as a function of the distance from the oscillating sphere. It is frequency dependent,

$$\delta(f) = (\nu/\pi f)^{1/2} = R(f_\nu/f)^{1/2}, \qquad (30)$$

and large compared to R for the frequencies we consider, $f_\nu \equiv \nu/(\pi R^2)=1.3$ MHz. This and other notations are given in Appendix A.

X. BEYOND EINSTEIN: BROWNIAN MOTION IN AN INCOMPRESSIBLE FLUID

Since Fourier decomposition describes any trajectory as a sum of linear oscillatory motions, the friction in Eq. (29)

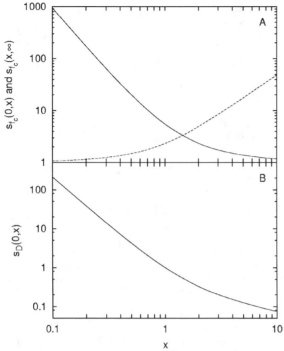

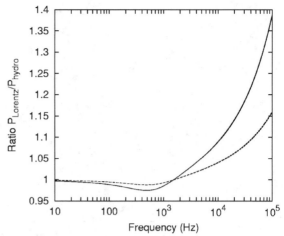

FIG. 4. Ratio between the Lorentzian power spectrum and the hydrodynamically correct power spectrum $P_{\text{hydro}}(f)$ in Eq. (32) evaluated with same parameter value $f_c = 666$ Hz. Solid line: Case of $2R = 1.05$ μm like that in this article. Dashed line: Case of $2R = 500$ nm. Simple Einstein–Ornstein–Uhlenbeck theory is a better approximation for smaller objects and at lower frequencies. A change in f_c to ≈ 370 Hz does not change these curves at higher frequencies, but shifts the location of their minima to values just below the new value for f_c.

FIG. 3. (a) Ratio $s_{fc}(x_{\min}, x_{\max})$ between $\sigma(f_c)$ and the theoretical minimum for $\sigma(f_c)$, the former from a fit of a Lorentzian to an experimental power spectrum in the interval $[f_{\min}, f_{\max}]$, the latter from fitting to the same spectrum in $[0, \infty]$, assuming it is known there. Solid line: $s_{fc}(0, x_{\max})$ vs $x = x_{\max} = f_{\max}/f_c$. Dashed line: $s_{fc}(x_{\min}, \infty)$ vs $x = x_{\min} = f_{\min}/f_c$. Here $x_{\max} = \infty$ only means that $f_{\max}$ is so much larger than f_c that the experimental spectrum's information content regarding f_c is essentially exhausted. Thus one can simultaneously have $f_{\max}^2 \ll f_{\text{Nyq}}^2$, ensuring aliasing is negligible for $f \lesssim f_{\max}$. Note that $s_{fc}(1, \infty) < s_{fc}(0,2)$, i.e., there is more information about the f_c value in interval (f_c, ∞) than in interval $(0, 2f_c)$. (b) Graph of $s_D(0,x)$ showing $\sigma(D)/D$ vanishes as $x_{\max} \to \infty$, where $s_D(0,x) \sim x_{\max}^{-(1/2)}$ according to Eq. (28). Both (a) and (b) illustrate the great amount of information located in the high-frequency part of the spectrum.

also appears in the frequency representation of the *generalized Langevin equation* describing the Brownian motion of a harmonically trapped sphere in an incompressible fluid,[39]

$$[m(-i2\pi f)^2 + \gamma_{\text{Stokes}}(f)(-i2\pi f) + \kappa]\tilde{x}(f)$$

$$= [2k_B T \operatorname{Re} \gamma_{\text{Stokes}}(f)]^{1/2} \tilde{\eta}(f),$$

$$\gamma_{\text{Stokes}}(f) = \frac{F_{\text{fric}}}{i2\pi f \tilde{x}(f)} = \gamma_0 \left(1 + (1-i)\frac{R}{\delta} - i\frac{2R^2}{9\delta^2}\right), \quad (31)$$

which becomes Einstein–Ornstein–Uhlenbeck theory, Eq. (1), in the limit of $f \to 0$. Here, as above, m is the mass of the

TABLE I. Parameter values of the Lorentzian fit (not shown) as a function of $f_{\max}$. The experimental power spectrum fitted to was obtained at a larger laser intensity than the spectrum shown in Fig. 2 and $f_{\text{sample}} = 25$ kHz. Data points in the experimental power spectrum were blocked to 750 equidistant points in the range of 110 Hz–25 kHz before fitting the Lorentzian to the points in the interval $[f_{\min}, f_{\max}]$ with $f_{\min} = 110$ Hz and $f_{\max}$ listed.

$f_{\max}$ (kHz)	f_c (Hz)	D (arb units)^2/s	Support (%)
1	641 ± 10	429 ± 9	37
2	630 ± 6	420 ± 4	9
3	610 ± 5	405 ± 2	0

sphere, κ the spring constant of the harmonic trapping force, $k_B T$ the Boltzmann energy, Re denotes the "real part of," and $\tilde{\eta}(f)$ is the Fourier transform of an uncorrelated random process $\eta(t)$, normalized like in Eq. (2) in order to show explicitly the frequency dependence of the Brownian noise that makes up the right-hand side of Eq. (31).

Experimentally, we monitor $x(t)$ for a long, but finite, time T_{msr}. Fourier transformation on this time interval, Eq. (7), gives the experimental power spectrum,

$$P_k^{(\text{ex})} = \frac{|\tilde{x}_k|^2}{T_{\text{msr}}} = \frac{2k_B T \operatorname{Re} \gamma_{\text{Stokes}}(f_k) |\tilde{\eta}_k|^2 / T_{\text{msr}}}{|\kappa - i2\pi f_k \gamma_{\text{Stokes}}(f_k) - m(2\pi f_k)^2|^2},$$

where $(|\tilde{\eta}_k|^2)_{k=1,2,\dots}$, are uncorrelated non-negative random variables with exponential distribution that satisfy Eq. (8). Thus $(P_k^{(\text{ex})})_{k=1,2,\dots}$, are uncorrelated non-negative exponentially distributed random variables, each of which consequently has RMSD equal to its expected value. This property is unchanged by the filtering and aliasing applied below.

The expected value of $P_k^{(\text{ex})}$ is

$$P_{\text{hydro}}(f) = \frac{D/(2\pi^2)[1 + (f/f_v)^{1/2}]}{(f_c - f^{3/2}/f_v^{1/2} - f^2/f_m)^2 + (f + f^{3/2}/f_v^{1/2})^2}, \quad (32)$$

where $f_m \equiv \gamma_0/(2\pi m^*) \simeq 3f_v/2 = 1.9$ MHz since $m^* \equiv m + 2\pi \rho R^3/3 \simeq 3m/2$ for the polystyrene bead we use. (This simple relation between f_m and f_v might tempt one to eliminate one of these frequencies in favor of the other. They parameterize different physics however, f_v parameterizes the flow pattern established around a sphere undergoing linear harmonic oscillations in an incompressible fluid. This pattern is unrelated to the mass of the sphere. It need not have any, for that matter. f_m parameterizes the time it takes for friction to dissipate the kinetic energy of the sphere and the fluid it entrains. It depends on the mass of the sphere. By keeping both parameters in formulas, the physical origin of various terms remains clear.). This power spectrum contains the same two fitting parameters, f_c and D, as the Lorentzian of the Einstein–Ornstein–Uhlenbeck theory, but differs significantly from it, except at low frequencies, as shown in Fig. 4. The radius R of the bead now also occurs in f_ν and f_m, and not only through f_c, but it is not a parameter we must fit, because it is known to 1% uncertainty, and occurs only in terms that are so small that this small uncertainty of R has negligible effect on $P_{\text{hydro}}(f)$.

XI. FAXÉN'S FORMULA GENERALIZED TO LINEAR HARMONIC MOTION

The frictional force in Eq. (29) was derived by Stokes under the assumption that the oscillating sphere is infinitely deep inside the fluid volume. For optimal designs, the lens that focuses the light into an optical trap typically has a short focal length. So a microsphere caught in such a trap is typically near a microscope coverslip. Consequently, the hydrodynamical interaction between the microsphere and the essentially infinite surface of the coverslip must be accounted for. Faxén has done this for a sphere moving parallel to an infinite plane with *constant* velocity in an incompressible fluid bounded by the plane and asymptotically at rest at conditions of vanishing Reynold's number. Solving perturbatively in R/ℓ, where ℓ is the distance from the sphere's center to the plane, Faxén found,[40,41]

$$\gamma_{\text{Faxén}}(R/\ell) = \frac{\gamma_0}{1 - (9R/16\ell) + (R^3/8\ell^3) - (45R^4/256\ell^4) - (R^5/16\ell^5) + \cdots}.$$

There is no second-order term in the denominator, so this formula remains good to within 1% for $\ell > 3R$ if one ignores all but the first-order term. This first-order result was first obtained by Lorentz.[42] If his first-order calculation is repeated for a sphere undergoing linear oscillating motion parallel to a plane, one finds a friction formula that has Stokes' and Lorentz's as limiting cases[43]

$$\gamma(f, R/\ell) = \gamma_{\text{Stokes}}(f)\left(1 + \frac{9}{16}\frac{R}{\ell} \times \left[1 - \frac{1-i}{3}\frac{R}{\delta}\right.\right.$$
$$\left.\left. + \frac{2i}{9}\left(\frac{R}{\delta}\right)^2 - \frac{4}{3}(1 - e^{-(1-i)(2\ell-R)/\delta})\right]\right).$$

$$(33)$$

The effect of the infinite plane is to increase friction, but less so at larger frequencies where δ is smaller.

We measured ℓ by first focusing the microscope on the coverslip surface. Having established $\ell = 0$, the distance ℓ to the bead was determined with software provided for the microscope (Leica DM IRBE) by Leica. The software computes the distance moved by the oil-immersion microscope objective. This distance multiplied with the ratio of the refractive index of water to that of glass, 1.33/1.5, gives distance ℓ. The software gives ℓ within precision of 0.1 μm, to which must be added the independent error for determination of $\ell = 0$, which is also 0.1 μm, we found, from repeated determinations.

We had $R/\ell \approx 1/12$ when it was largest. So the bead's hydrodynamic interaction with the coverslip has an effect of 4% or less, large enough that we must account for it. Fortunately, this introduces no new fitting parameters, because we know the value of ℓ. Since ℓ occurs only in a term of at most 4% relative importance, any error in ℓ value affects the final result with a 0.04 times smaller error, in our case by one per mil, at most.

XII. PHYSICAL POWER SPECTRUM

By replacing $\gamma_{\text{Stokes}}(f)$ with $\gamma(f, R/\ell)$ in Eqs. (29) and (31), one obtains a power spectrum that accounts for all relevant physics of the bead in the trap, with the expected value

$$P_{\text{hydro}}(f; R/\ell) = \frac{D/(2\pi^2)\,\text{Re}\,\gamma/\gamma_0}{(f_c + f\,\text{Im}\,\gamma/\gamma_0 - f^2/f_m)^2 + (f\,\text{Re}\,\gamma/\gamma_0)^2},$$

$$(34)$$

where

$$\text{Re}\,\gamma/\gamma_0 = 1 + \sqrt{f/f_\nu} - \frac{3R}{16\ell}$$
$$+ \frac{3R}{4\ell}\exp\left(-\frac{2\ell}{R}\sqrt{f/f_\nu}\right)\cos\left(\frac{2\ell}{R}\sqrt{f/f_\nu}\right)$$

and

$$\text{Im}\,\gamma/\gamma_0 = -\sqrt{f/f_\nu} + \frac{3R}{4\ell}$$
$$\times \exp\left(-\frac{2\ell}{R}\sqrt{f/f_\nu}\right)\sin\left(\frac{2\ell}{R}\sqrt{f/f_\nu}\right).$$

We refer to this as the *physical* power spectrum. It differs from the *recorded* power spectrum because the data acquisition system contains filters, some intended, some not, and because the data acquisition system samples the resulting *filtered* spectrum only at discrete times to produce the spectrum recorded. Below, we discuss the effects of filtering and finite sampling frequency.

XIII. POSITION DETECTION SYSTEM IS A LOW-PASS FILTER

Silicon is transparent to infrared light. For this reason, position detection systems like ours have finite response times of the order of tens of microseconds. The delayed part of the signal decreases over time as a simple exponential, so our diode's characteristics is a sum of two terms:[36] (i) a small constant $\alpha^{(\text{diode})2}$, corresponding to the fraction $\alpha^{(\text{diode})}$ of response that is instantaneous, and (ii) a Lorentzian, corresponding to the delayed response,

$$\frac{P(f)}{P_0(f)} = \alpha^{(\text{diode})2} + \frac{1 - \alpha^{(\text{diode})2}}{1 + (f/f_{3\,\text{dB}}^{(\text{diode})})^2}. \qquad (35)$$

The parameters $\alpha^{(\text{diode})}$ and $f_{3\,\text{dB}}^{(\text{diode})}$ depend on (i) the laser's wavelength and intensity, (ii) the photodiode's thickness, material properties, and reverse bias, and (iii) line up of the laser beam and photodiode.[36,44,45] Consequently, the optimal way in which to determine the values of $f_{3\,\text{dB}}^{(\text{diode})}$ and $\alpha^{(\text{diode})}$ relevant for a recorded spectrum is to include them with f_c and D as third and fourth fitting parameters in a fit to the spectrum. The logic of this procedure is sound, even though we calibrate the position detection system with the same data from which we want to calibrate the trap; see Appendix G for details.

For the position detection system analyzed here, Eq. (35) describes the filtering effect of this system out to approximately 30 kHz with the precision achieved here. We used zero reverse bias across the photodiode. With reverse bias, Eq. (35) is good out to larger frequencies. In general, the filtering effect of the position detection system at larger frequencies is described by a more complicated expression than Eq. (35), an expression that accounts also for faster decreasing solutions to the diffusion equation for charge carriers in the photodiode.[44]

XIV. ALIASING AND ANTIALIASING FILTERS

Sampling of the power spectrum with finite sampling rate causes *aliasing*. Data acquisition electronics typically have built-in antialiasing filters. Delta–sigma data conversion systems use oversampling and "noise-shaping" filters to eliminate aliasing altogether, and the effect of their built-in filters is only seen near their readout frequency. Section XIV can be skipped by those using such data acquisition systems. For others, aliasing, antialiasing filters, and how they relate are discussed in some detail in Appendix H. An important point made there is that antialiasing filters do not prevent all aliasing, they do not prevent the aliasing accounted for here with Eq. (37).

Our data acquisition electronics have two built-in antialiasing filters, both first-order filters with roll-off frequencies, $f_{3\,\text{dB}}$, that we set as high as possible, 80 and 50 kHz, respectively. This is not the normal recommended setting, but it gives optimal conditions for observation of the physics of the problem and of unintentional filtering by the position detection system.

A first-order filter reduces the power of its input, $P_0(f)$, by a factor of

$$\frac{P(f)}{P_0(f)} = \frac{1}{1 + (f/f_{3\,\text{dB}})^2}. \qquad (36)$$

The effect of filters, intended or not, is accounted for by multiplying the physically correct power-spectral expected value in Eq. (34) with the characteristic function for each filter, i.e., with Eqs. (35) and (36)—the latter twice, once for

each of our filters. That done, aliasing is accounted for by summing the result, $P^{(\text{filtered})}$, over aliased frequencies,

$$P^{(\text{aliased})}(f) = \sum_{n=-\infty}^{\infty} P^{(\text{filtered})}(f + nf_{\text{sample}}), \qquad (37)$$

where in practice a finite number of terms exhausts the sum. The result, $P^{(\text{aliased})}(f)$, is our theory for the expected value of the experimental power spectrum recorded.

XV. HOW TO CALIBRATE TWEEZERS

The procedure described here in Sec. XV is implemented in freely available MATLAB software documented in Ref. 26.

The experimental spectrum to which we fitted in Fig. 5 is the result of *blocking*[30] a spectrum that is the average of five spectra, which were calculated from five time series recorded in five time windows, each with duration of $2^{18}\Delta t \simeq 16$ s ($\Delta t = 1/f_{\text{sample}} = 1/16\,000$ s). Similarly, Fig. 6 is based on four such time windows, each with duration of $2^{18}\Delta t \simeq 5$ s ($\Delta t = 1/f_{\text{sample}} = 1/50\,000$ s). In both cases consecutive windows were separated by several minutes. A study of power spectra and fitted parameters obtained from individual windows showed that neither drifted between windows, although the center of the trap did. Thus it is legitimate to average over five/four spectra as we did. The spectrum thus obtained was blocked on the linear frequency axis to $N' = 150$ data points, with each block containing approximately 870 points. Before any blocking was done, crosstalk between channels was eliminated in the manner described in Appendix C.

Our theory for the expected value of the power spectrum was fitted to the recorded experimental power spectrum using our theory for the scatter of the latter about the former: It scatters with standard deviation proportional to its expected value, i.e., Eqs. (E4) and (E5) apply with n_b replaced by $n_b n_w$, $n_w = 5$ (4), as described in Appendix E. Thus we fit by minimizing

$$\chi^2 = \sum_{k=1}^{N'} \left(\frac{P_k^{(\text{ex})} - P_k}{P_k/\sqrt{n_w n_b}}\right)^2 = n_w n_b \sum_{k=1}^{N'} \left(\frac{P_k^{(\text{ex})}}{P_k} - 1\right)^2, \qquad (38)$$

where the sum is over blocked data points and $N' \equiv N/(n_w n_b)$. The form of χ^2 does not suit standard least-squares fitting routines. However, exact rewriting yields a form that does,

$$\chi^2 = \sum_{k=1}^{N'} \left(\frac{P_k^{-1} - P_k^{(\text{ex})-1}}{\sigma_k}\right)^2 \qquad (39)$$

where $\sigma_k = P_k^{(\text{ex})-1}/(n_w n_b)^{1/2}$.

The solid line in Fig. 5(a) [Fig. 6(a)] shows $P_{\text{hydro}}(f; R/l)$, multiplied by the characteristic functions of the diode and electronic filters, aliased with $f_{\text{sample}} = 16$ kHz [50 kHz], then fitted to the experimental spectrum using f_c, D, $f_{3\,\text{dB}}^{(\text{diode})}$, and $\alpha^{(\text{diode})}$ as fitting parameters. The parameter values obtained from these least-squares fits are listed in Tables II and III. We see that both the strength of the trap, in the form of f_c, and conversion of the position detection system's arbitrary units to nanometers, are determined to within 1% precision or better. The latter conversion was found by

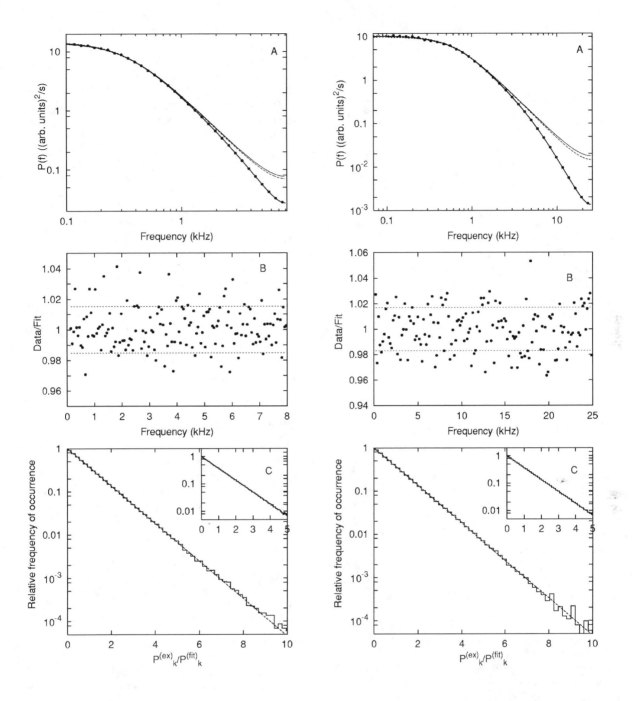

FIG. 5. Same data as in Fig. 2, here fitted with $f_{max} = f_{Nyq}$. (a) The thick solid line is the theoretical spectrum in Eq. (34), filtered and aliased, then fitted to the experimental spectrum in the interval [110 Hz, 8 kHz]. Statistical support for the fit is 96% (Refs. 30 and 31). The data points shown were obtained by blocking the experimental spectrum in intervals of equal length on the *logarithmic* axis, and hence are not the same as those shown in (b). Two dashed lines practically on top of the solid line delineate a vertical window of ± 1 standard deviation of Gaussian scatter of data. Thin solid and dashed lines that overshoot the data are aliased Lorentzian and aliased $P_{hydro}(f; R/\ell)$, respectively, unfiltered, and with same values for D and f_c as those shown by the thick solid line and given in Table II for the x coordinate. They illustrate the importance of filters and the frequency dependence of hydrodynamical friction. (b), (c) See the caption for Fig. 6.

FIG. 6. Same plots as in Fig. 5, but for data obtained with a stronger optical trap and $f_{Nyq} = 25$ kHz; see the values in Table III. The same power spectral data were used to obtain results given in Table I. Statistical support for the fit is 49% (Refs. 30 and 31). (a) See the caption for Fig. 5. (b) Values of the data fitted to, divided by fitted theory in order to visualize their scatter about a value 1. The two dashed lines delineate the vertical window of ± 1 standard deviation of Gaussian scatter. Thus 68% of the data points should fall between the two dashed lines if the data indeed are Gaussian distributed. They do. Further blocking will reduce the scatter to less than 1%. (c) Histogram of $N = 10^6$ [$N = 1.3 \times 10^6$ in Fig. 5(c)] experimental power-spectral values $P_k^{(ex)}$ measured in units of their expected values P_k, the latter being the fit shown in (a). Inset: Same data binned into a histogram with finer resolution, showing 99% of the data.

TABLE II. Values of fitted parameters for Fig. 5, based on blocking to 150 data points and $f_{min}=110$ Hz, $f_{max}=f_{Nyq}=8$ kHz. $D=0.41$ nm$^2/\mu$s was used to find the nanometer equivalent for diode output for R_x and R_y. The covariance between f_c and D was -0.95.

Parameter	x coordinate	y coordinate
f_c (Hz)	374 ± 2	383 ± 2
$f_{3\,dB}^{(diode)}$ (kHz)	6.73 ± 0.17	6.39 ± 0.14
D (arb units)^2/s	610 ± 9	584 ± 8
Arb unit equiv (nm)	26.0 ± 0.2	26.3 ± 0.2
$1-\alpha^{(diode)2}$	0.92 ± 0.02	0.91 ± 0.01
Support (%)	96	81

equating the value for D obtained in the fit with the value known from Einstein's relation, Eq. (6). Thus, the force $\kappa\Delta x$ exerted at distance Δx from the center of the trap is known to within only 1%–2% error due to calibration of the trap and diode. This error is typically negligible compared to the uncertainty of the position. Thus our calibration scheme essentially eliminates calibration errors from force measurements.

The x coordinate in Table II should be compared with the results in Fig. 2. They were obtained from the same data set. The two values for the corner frequency differ by four standard deviations, with the Lorentzian fit yielding the lower value, because it absorbs the effect of unintentional filtering in this manner, with a small systematic error as result. This point is borne out in Table I and should be compared with Table III's data for the x coordinate obtained from the same data set.

The two values found for $f_{3\,dB}^{(diode)}$ for the x and y coordinates, respectively, are indistinguishable in Table II as well as in Table III. This is what one would expect for a diode with four identical quadrants. It is a coincidence that the values differ only in the fourth digit, not shown in Table III. The two values found for f_c, on the other hand, differ by three to four standard deviations. They differ by 2% (4%), corresponding to an elliptical cross section of the beam with 1% (2%) difference between the lengths of the major and minor axes, or ellipticity of $[1-(374\pm2)/(383\pm2)]^{1/2}=0.15\pm0.03$, $\{[1-(666\pm5)/(637\pm5)]^{1/2}=0.21\pm0.04\}$. This ellipticity does not differ significantly from the 10% maximum ellipticity of the laser beam promised by the manufacturer. Also, trap ellipticity and laser ellipticity are not necessarily the same thing. Polarized laser beams, even if perfectly nonelliptical, tend to get focused onto elliptical, diffraction limited spots in the image plane. This might be the source of differences between the trap stiffness ellipticity

TABLE III. Values of fitted parameters for Fig. 6, based on fit to 150 blocked data points and $f_{min}=110$ Hz, $f_{max}=f_{Nyq}=25$ kHz. The covariance between f_c and D was -0.95.

Parameter	x coordinate	y coordinate
f_c (Hz)	666 ± 5	637 ± 5
$f_{3\,dB}^{(diode)}$ (kHz)	7.27 ± 0.04	7.27 ± 0.05
D (arb units)^2/s	447 ± 9	467 ± 9
Arb unit equiv (nm)	30.3 ± 0.3	29.6 ± 0.3
$1-\alpha^{(diode)2}$	0.928 ± 0.001	0.924 ± 0.001
Support (%)	49	52

one measures and the laser beam ellipticity specified by the manufacturer.

According to theory, the blocked experimental data points are statistically independent and normally distributed with known standard deviations. The "residual plot" in Fig. 5(b) [6(b)] shows the scatter about their expected value, fitted theory, in units of this expected value. Figures 5(b) and 6(b) show that the theoretical power spectrum presented here fits the experimental one perfectly.

Figures 5(c) and 6(c) provide a more radical illustration that the theoretical power spectrum used here really describes the expected value of the experimental spectrum: The "raw" experimental spectral values, i.e., unaveraged and unblocked values, were divided with the fitted theoretical value and binned into histograms that show that the raw experimental spectral values really *are* exponentially distributed about their expected value, as stated by theory. This is a powerful illustration of the correctness of the theory, as well as of the experiment: The histogram shows an exponential distribution over four decades obtained from experimental values that range over no less than seven decades: the four decades they scatter about their expected value, plus the three decades that this expected value varies with the frequency.

XVI. DISCUSSION

A. When precision is no concern

In many biological experiments, e.g., 10%–20% calibration error is of no concern because other sources of error are dominant. So the trapping force can be estimated with sufficient—albeit unknown—precision with the roughest calibration based on a Lorentzian spectrum. Freely available MATLAB software[26] will do this as was shown in Sec. IV.

B. When precision is a concern

When precision is a concern an optical trap can advantageously be calibrated as was demonstrated above. Our recommendations—which we have implemented in freely available MATLAB software[26]—thus are the following.

(a) When plotting the experimental power spectrum, compress data by blocking to show fewer data points with smaller scatter about their expected value. After all, we know *a priori* that the expected value is a smoothly varying function of the frequency. So the data can advantageously reflect this.

(b) Plot P_{xy} (defined in Appendix C) and use its minimalization as a criterion for good alignment of the diode with the laser beam.

(c) If P_{xy} cannot be made to vanish, find a linear coordinate transformation to a frame of reference in which it does vanish, and work in this frame of reference.

(d) The frequency dependence of the friction coefficient and of the Brownian noise should be taken into account. Not only is that correct theory, but using it costs nothing: No new fitting parameters are introduced with it.

(e) Data acquisition electronics contain antialiasing filters.

Their effect is known, or easily measured with a signal generator, so it is costs nothing to account for it when it affects the power spectrum recorded. If the filters are set to have minimal effect, that also minimizes the effect of imprecise knowledge about these filters' parameter values.

(f) One should be aware that one's position detection system may have frequency-dependent sensitivity, hence may act as an unintended low-pass filter. Ours, a Si *PIN* diode used with a 1064 nm laser, does, and it is by far the most important filter in our setup. However, since we know the form of its characteristic, we can calibrate its parameters from the very power spectrum we wanted to fit. This is the optimal way in which to determine these parameters, because their values depend on experimental circumstances. If one calibrates with $f_{max}^2 \ll f_{3\,dB}^{(diode)2}$, one should use the approximation in Eq. (G1). If one uses another kind of quadrant photodiode and/or laser wavelength, this filter effect may be different or absent.[44,45]

(g) Aliasing due to finite sampling frequency always occurs, unless ones data acquisition system uses oversampling. Aliasing is easily accounted for, however, and doing so costs nothing if the theory one aliases is also correct at frequencies $f > f_{Nyq}$ that contribute through aliasing to the spectrum below f_{Nyq}. No new fitting parameters are introduced, only f_{Nyq}, which is known to high precision.

(h) Leakage,[30] on the other hand, is truly negligible because the power spectrum of a trapped bead is a smooth and slowly varying function. So there is no need to introduce window functions that reduce leakage. Consequently, overlapping data windows that compensate for loss of information caused by window functions are also not needed. However, if built into one's data acquisition software, they can be used as they were intended: for quick, on-line data compression. The correlations they introduce in the resulting power spectrum are negligible if a very large number of windows is used.

(i) The scatter of experimental power spectral values about their expected values is known theoretically. So it costs nothing to use correct error bars, and doing so yields correct stochastic error bars on fitted parameters such as f_c and D, and, last but not least, use of correct error bars makes it possible to obtain statistical support for fits.

(j) Plots like Figs. 1, 2(b), 5(b) and 5(c), 6(b) and 6(c), and 7–10 are well worth doing. They provide simple, strong, virtually model-independent checks that will catch many kinds of errors in one's experiment and initial data analysis, if present.

(k) If the fitted value of D is used to calibrate the length scale of displacements measured, one should be aware that the viscosity of the fluid in which one measures may differ significantly from that of pure water at the same temperature. If this is so, reliable calculation or

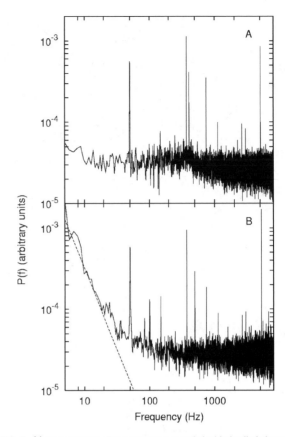

FIG. 7. (a) Dark spectrum: Power spectrum recorded with the diode in total darkness, a measure of the electronic noise level. The spike at 50 Hz is caused by the power supply. All values are a factor of $10^2 - 10^4$ below our calibration spectra. (b) Light spectrum: Power spectrum recorded with the trap's laser light impinging directly onto the photodiode with no microsphere in the trap. The dashed line at low frequencies has a slope of -2.

measurement of its viscosity is needed, or a direct measurement of length scales by moving a fixed bead with a piezo stage.

(l) Finally, one should beware that the procedure described here calibrates the *center* of the trap. This makes it valid anywhere near the center where the trapping potential is harmonic. That includes everywhere Brownian motion took our bead during calibration, it seems from Fig. 10. How to calibrate off center, at a given displacement along the beam axis, is a separate project of practical interest, but was not addressed here.

ACKNOWLEDGMENTS

This work was prompted by Simon Tolić-Nørrelykke, who wanted to calibrate tweezers in a statistically correct manner. The authors are grateful to Steve Block, Ernst-Ludwig Florin, Joe Howard, and Christoph Schmidt for sharing with them their expertise on tweezers and calibration. The authors also appreciate help with experimental techniques from Jakob Kisbye Dreyer.

APPENDIX A: NOTATION

For convenience, our notation and characteristic values of parameters and variables are given in Table IV.

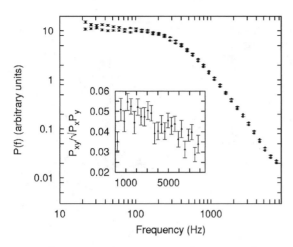

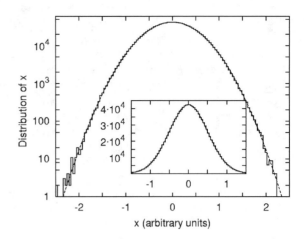

FIG. 8. Experimental power spectra for x and y coordinates (lower and upper data points, respectively). Points plotted here are averages over "blocks" of points from the original power spectrum; see Appendix E. The error bars were calculated from data within blocks. Since the block intervals were chosen to be of equal size on the logarithmic frequency axis used here, the number of data points in a block grows exponentially with the frequency. Consequently, the error bars decrease exponentially with an increase in frequency, and range from small at low frequencies to nondiscernible at intermediate and large frequencies. Inset: Experimental values of the dimensionless cross correlation function $P_{xy}/(P_x P_y)^{1/2}$ introduced in Sec. C 1, as a function of the frequency. (Same power spectral data as in Figs. 2 and 5.)

FIG. 10. Linear–log plot of a histogram of the positions that occur in a time series for a bead in a trap. Data are the same as those used in Figs. 2, 8, and 9. Superimposed is the Gaussian distribution with the same second moment as the data. Inset: Linear–linear plot of the same histogram and Gaussian. The 4.5 decades of Gaussian behavior seen here demonstrates that the trap's potential is harmonic up to $10k_B T$ at least. (Same data as was used in Figs. 2 and 5.)

APPENDIX B: NULL TESTS FOR NOISE

We did two simple null tests of our equipment before we recorded power spectra for the bead in the trap. We recorded a "dark spectrum," the power spectrum generated when the diode is kept in total darkness; see Fig. 7(a). This is a measurement of the equipment's electronic noise. We see the spectrum is flat, except for a spike at 50 Hz from the power supply, and at a few higher frequencies, 400 Hz in particular. All values are a factor of $10^2–10^4$ below that of our calibration spectra, hence noise may contribute from 1% to 10% to the spectra, since amplitudes, not spectra, add up. However, the spikes are so narrow and few in addition to being small

that they do not matter statistically in our calibration spectra, and also do not show above their noise. The spike at ~7 kHz was an exception, but it was so narrow that it easily was filtered out manually and had negligible consequence for the statistics of the calibration spectrum.

We also recorded a "light spectrum," see Fig. 7(b). Compared with the dark spectrum, this light spectrum shows significant low-frequency noise, plus a peak at 100 Hz, probably caused by stray light. Apart from stray light, the difference is caused by the limited pointing stability of the unscattered laser beam and the optics it passes through relative to the photodiode. The low-frequency noise seems to fall off as f^{-2}, which is what one would observe if the direction of the laser were doing a slow random walk about its average direction driven by white noise.

Although only the microsphere is missing, this light spectrum is not a direct measurement of *all* noise in the system apart from the Brownian noise of the sphere. Mechanical vibrations, e.g., are transmitted to the fluid volume, but not to the light spectrum because the fluid is transparent. They *are*, however, transmitted by the fluid to the sphere's spectrum when the sphere is present. So this is noise that occurs in the experiment, but not in the light spectrum. Strictly speaking, the light spectrum therefore only provides a lower bound on "all noise but the sphere's Brownian." It may nevertheless be a good approximation to all noise, although proof of this is missing.

By choosing $f_{min} = 110$ Hz in calibration fits, we leave low-frequency noise entirely out of the calibration. The light spectrum's extra features relative to those in Fig. 7(a) above this f_{min} value all fall a factor of 10^{-4} below the calibration spectrum's power at the same frequencies, hence add less than 1% to the calibration spectrum.

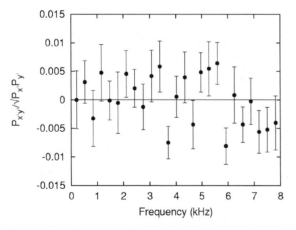

FIG. 9. $P_{x'y'}(f)/[P_x(f)P_y(f)]^{1/2}$ as a function of f at a minimum with respect to (b,c) of $\Sigma_f P_{x'y'}(f)^2/[P_x(f)P_y(f)]$. Note that the quantity plotted, and hence $P_{x'y'}(f)$, are both zero within errors. More precisely, statistical support for the hypothesis that it vanishes is 10%. (Same data as was used in Figs. 2 and 5.)

TABLE IV. Notation used and characteristic values of quantities encountered for data set shown in Figs. 2, 5, and 6. Note: The thermal velocity and the half width of the trap both refer to a single coordinate of motion, i.e., to motion in one dimension.

Quantity	Notation	Equal to	Value (Figs. 2 and 5)	Value (Fig. 6)
Sampling frequency	f_{sample}		16 kHz	50 kHz
Nyquist frequency	f_{Nyq}	$f_{sample}/2$	8 kHz	25 kHz
Corner frequency	f_c	$\kappa/(2\pi\gamma)$	~370 Hz	~670 Hz
Frequency where $\delta = R$	f_ν	$\nu/(\pi R^2)$	1.3 MHz	
Frequency of inertial relaxation	f_m	$\gamma/(2\pi m^*) \approx \gamma/(3\pi m)$	1.9 MHz	
Minimum fitted frequency	f_{min}		110 Hz	
Maximum fitted frequency	f_{max}		8 kHz	25 kHz
Diode frequency	$f_{3\,dB}^{(diode)}$		6.8 kHz	7.3 kHz
Total No. of data points	N	$n_w \times 2^{18}$	1.3×10^6	1.0×10^6
Time between measurements	Δt	f_{sample}^{-1}	62.5 μs	20.0 μs
Total duration of measurements	T_{msr}	$N\Delta t$	82 s	21 s
No. of data windows	n_w		5	4
Duration of one data window			16.4 s	5.2 s
Time between windows			1 min	
No. of points of block	n_b		517, 861	869
No. of blocked points	N'		29, 150	150
Diameter of bead	$2R$		1.05 μm	
Density of bead and water	ρ		1.0 g/cm^3	
Mass of bead	m	$4\pi R^3 \rho/3$	6.1×10^{-13} g	
Hydrodynamical mass	m^*	$m + 2\pi R^3 \rho/3$	9.1×10^{-13} g	
Thermal energy	$k_B T$		4.1 pN nm	
Thermal velocity	$\langle v^2 \rangle^{1/2}$	$k_B T/m$	3 mm/s	
Kinematic viscosity	ν		1.0 μm$^2/\mu$s	
Reynolds number	N_{Re}	$R\langle v^2 \rangle^{1/2}/\nu$	1.4×10^{-3}	
Drag/friction coefficient	γ	$6\pi\rho\nu R$	9×10^{-6} g/s	
Trap stiffness	κ	$2\pi f_c \gamma$	0.021 pN/nm	0.038 pN/nm
Relaxation time in trap	t_{trap}	$\gamma/\kappa = (2\pi f_c)^{-1}$	0.5 ms	0.2 ms
Diffusion coefficient	D	$k_B T/\gamma$	0.41 nm$^2/\mu$s	
Inertial time scale	t_{inert}	m/γ	56 ns	
Half width of trap	$\langle x^2 \rangle^{1/2}$	$(k_B T/\kappa)^{1/2}$	14 nm	10 nm
Penetration depth	δ		Frequency dependent	

APPENDIX C: CROSSTALK BETWEEN CHANNELS

1. How to decorrelate channels

The photodiode that we used to measure the bead's position consists of four quadrants, each of which outputs voltage proportional to the amount of light impinging upon it. We number the quadrants I, II, III, and IV like the quadrants of a two-dimensional (2D) coordinate system, and denote their output voltages V_I, V_{II}, V_{III}, and V_{IV}. Then changes in the voltage and the ratios,

$$V_z \equiv V_I + V_{II} + V_{III} + V_{IV}, \tag{C1}$$

$$R_x \equiv (V_I - V_{II} - V_{III} + V_{IV})/V_z, \tag{C2}$$

$$R_y \equiv (V_I + V_{II} - V_{III} - V_{IV})/V_z, \tag{C3}$$

are, to good first approximation,[35,46–48] proportional to changes in the bead's position (z,x,y), with z the coordinate along the laser beam's axis.

Figure 8 shows that this approximation is not adequate when precision is desired. The experimentally recorded "coordinates" $R_x(t)$ and $R_y(t)$ are not independent. Their power spectra, $P_x(f) \equiv |\hat{R}_x(f)|^2$ and $P_y(f) \equiv |\hat{R}_y(f)|^2$, where the caret denotes discrete Fourier transformation, Eq. (12), are plotted in Fig. 8, together with $P_{xy}(f) \equiv \mathrm{Re}[\hat{R}_x(f)\hat{R}_y^*(f)]$. When x and y are uncorrelated degrees of freedom, so are R_x

and R_y, hence so are $\hat{R}_x$ and $\hat{R}_y$. Consequently, P_{xy} should vanish compared to P_x and P_y. We find $P_{xy}(f)/[P_x(f)P_y(f)]^{1/2} \approx 3\%-5\%$, however, as shown in the inset in Fig. 8.

Two explanations for crosstalk given below suggest that (R_x, R_y) is a linear function of (x,y), and vice versa, and that this function does not depend on time. Assuming this, we look for a linear transformation of (R_x, R_y) to a pair of coordinates (x', y') for which $P_{x'y'} = 0$. If we can find such a transformation, it does not matter what motivated the search for it: the transformed coordinates are the correct Cartesian coordinates in which to analyze the bead's motion and calibrate the trap. To find this transformation, we must find two real constants, b and c, such that the time series,

$$(x'(t), y'(t)) \equiv (R_x(t) + bR_y(t), R_y(t) + cR_x(t)), \tag{C4}$$

has the property that $P_{x'y'}(f) = 0$ for all f.

Clearly, constants b and c are greatly overdetermined. Nevertheless, we were able to find them for all time series that we have analyzed, and have also found that the solution is almost degenerate. This is because $P_x(f)$, $P_y(f)$, and $P_{xy}(f)$ are nearly proportional to each other. It probably also helped that before we recorded any of the data used here, we aligned the diode with the laser beam using as the alignment criterion $\langle R_x \rangle = \langle R_y \rangle = 0$ *and* P_{xy} minimal (all three computed and plotted on line).

In general, the transformation just defined gives

$$P_{x'} = P_x + 2bP_{xy} + b^2 P_y, \tag{C5}$$

$$P_{y'} = P_y + 2cP_{xy} + c^2 P_x, \tag{C6}$$

$$P_{x'y'} = (1 + bc)P_{xy} + cP_x + bP_y. \tag{C7}$$

We found $(b,c) = (0.47, -0.56)$ by minimizing $\Sigma_f P_{x'y'}(f)^2/[P_x(f)P_y(f)]$ with respect to b and c. As desired, we found that at the minimum $P_{x'y'}(f)/[P_x(f)P_y(f)]^{1/2} = 0$ was satisfied for all f to within experimental error of this quantity; see Fig. 9.

Having determined b and c in this manner, the resulting power spectra, $P_{x'}$ and $P_{y'}$, for the uncoupled coordinates (x', y') are the spectra that we analyzed in the manner described in the body of this article. So we drop the prime from the notation hereafter, but it should always be understood implicitly.

2. Possible origins of crosstalk

If the parabolic trapping potential V is perfectly rotationally symmetric about the beam axis (chosen as the z axis), $V(x', y', z) = v(x'^2 + y'^2, z)$, the bead's equation of motion decouples no matter which pair of Cartesian coordinates (x', y') we use, as long as they are orthogonal to the beam axis. If the parabolic trap is *not* rotationally symmetric, but elliptic about the beam axis, decoupling is achieved in coordinates (x', y') that coincide with the major and minor axes of the ellipse.

Figure 8 shows that P_x and P_y are approximately proportional to each other: The data set for one function is shifted vertically relative to the data set for the other function by an amount approximately independent of frequency f. This means that the two channels have nearly the same corner frequency. So the trap is nearly rotationally symmetric.

With an asymmetric trap excluded, the simplest explanation for the constant ratio between P_x and P_y is a difference in sensitivity of the photodiode with respect to the two directions. This would come about if the four quadrants of the diode are not identically sensitive, and this would also explain the nonvanishing values for P_{xy}, including its nearly constant ratio to P_x and P_y: R_x becomes linearly correlated with R_y if we introduce independent sensitivities s_i for each of the quadrants, $V_i = s_i L_i$, $i =$ I, II, III, and IV, where L_i is the amount of light impinging on the ith quadrant. If, e.g., a spot of light moves in the x direction, $L_I + L_{II}$ remains constant, as does $L_{III} + L_{IV}$ and V_z. The ratio R_y changes value when $s_I \neq s_{II}$ and/or $s_{III} \neq s_{IV}$, however. Note that such asymmetry needs not be a property of the diode itself. All four quadrants could be identical, but have a nonlinear relationship between input light intensity and output voltage. In that case, less-than-perfect centering of the laser beam on the diode will cause different amounts of light to fall on different quadrants, and hence make them respond with different sensitivity to the small changes in light that correspond to movement of the bead.

Another explanation, which does not exclude the first, could be small asymmetry in the spot of light scattered by the bead onto the photodiode. That would cause different amounts of light to shift between quadrants for identical shifts of the bead in the x and y directions. If, furthermore, some of that asymmetrically scattered light falls beyond the edge of the quadrant diode, then a shift of the bead in the x direction will change R_y, and hence register as a correlated change in y.

APPENDIX D: MODEL-INDEPENDENT DATA ANALYSIS

Histograms of x and y positions of the decorrelated time series were consistent with a harmonic trapping potential up to $10 k_B T$, at least, as shown in Fig. 10 for the x channel.

The parabola through the data in the lin–log plot shows that the data are modeled well with $V(x) = \frac{1}{2}\kappa x^2$ in the range of x values visited. It is very satisfying that a model-independent data analysis can point so precisely to a specific model. If we determined the value of κ in this manner, however, we may find too low a value. This is because the true distribution $\propto \exp[-V(x)/k_B T]$ was smeared to a wider one by low-frequency vibrations which are external to the experiment in the sense that they do not originate in the bead's thermal motion. We may also find too large a value for κ because low-pass frequency filters artificially narrow the distribution of positions recorded. Furthermore, since fits like that in Fig. 10 do not calibrate the position detection system, the units for x and y remain arbitrary until an independent calibration of the position detection system is carried out, e.g., by finding f_c. Force measurements would consequently contain errors that originate in that calibration as well, were we to do one.

APPENDIX E: MAXIMUM-LIKELIHOOD FITTING AND DATA COMPRESSION

Suppose we data compress a power spectrum by blocking, and only then fit to it. How should that be done, and what is the approximation introduced by this?

In an unprocessed (uncompressed) power spectrum, the power spectral values are *exponentially* distributed. Least-squares fitting presupposes that the data are Gaussian distributed. With sufficient compression, a spectrum whose values were exponentially distributed will turn into a spectrum with much less scatter and Gaussian distributed values by virtue of the central limit theorem. So after compression, least-squares fitting can be applied. What is the approximation involved, we should ask, and what is the precision to expect, we *must* ask, in view of the precision we achieve in calibrating to compressed spectra, and hence want to claim for the calibration method described.

In order to answer these questions, we first observe that maximum-likelihood fitting to exponentially distributed data is equivalent to minimizing

$$\mathcal{F} \equiv -\log p = \sum_k (P_k^{(ex)}/P_k + \log P_k). \tag{E1}$$

Now consider the contribution from one block of data to $\mathcal{F}$ in Eq. (E1) before blocking has been done,

$$\Delta\mathcal{F}(\text{block}) = \sum_{f \in \text{block}} \left(\frac{P^{(\text{ex})}(f)}{P(f)} + \log P(f) \right). \qquad (E2)$$

In this equation we expand $P(f)$ at $f = \bar{f}$ to second order in $f - \bar{f}$, and find

$$\Delta\mathcal{F}(\text{block}) = n_b \left[\frac{\bar{P}^{(\text{ex})}(\bar{f})}{P(\bar{f})} + \log P(\bar{f}) \right.$$
$$\left. - \frac{1}{24} \left(\frac{n_b \Delta f P'(\bar{f})}{P(\bar{f})} \right)^2 \right], \qquad (E3)$$

where the last term on the right-hand side was obtained by replacing the sum $\Sigma_{f \in \text{block}}$ with an integral over f between $\bar{f} \pm \frac{1}{2} n_b \Delta f$.

Note that $n_b \Delta f P'(\bar{f})/P(\bar{f})$ is the relative change in $P(f)$ across a block. Since $\Delta f = 1/T = 0.06$ Hz while $P'(f)/P(f) = \mathcal{O}(f_c^{-1})$, it is possible to choose n_b large, e.g., $n_b = 500$, and still have the last term in Eq. (E3) negligible, so that we are left with the same form as Eq. (E1) for maximum-likelihood estimation of the theory's parameters from given, now blocked, experimental data.

Calculation entirely like the one leading from Eqs. (E2)–(E3) shows that the expected value and RMSD for $\bar{P}^{(\text{ex})}(\bar{f})$ are

$$\langle \bar{P}^{(\text{ex})}(\bar{f}) \rangle = P(\bar{f}), \qquad (E4)$$

$$\sigma[\bar{P}^{(\text{ex})}(\bar{f})] = \sigma[P^{(\text{ex})}(\bar{f})]/\sqrt{n_b} = P(\bar{f})/\sqrt{n_b}, \qquad (E5)$$

to the orders in $n_b \Delta f$ given. In this last identity, Eq. (E5), if the power spectrum blocked was already windowed, its values were not exponentially distributed, and n_b should be replaced with $n_w n_b$ in the case of nonoverlapping rectangular windows, and with $9 n_w n_b / 11$ in the case of Hanning windows (Ref. 30, p. 428).

For large n_b, or, if $n_w \neq 1$, for large $n_w n_b$, the central limit theorem tells us that $\bar{P}^{(\text{ex})}(\bar{f})$ is Gaussian distributed with the expected value and root-mean-square deviations given in Eqs. (E4) and (E5). Fitting to data that are known to be Gaussian distributed is usually done with the method of least squares. So it is natural to ask how the method of least squares relates to the maximum-likelihood estimation discussed above. According to Eq. (E3) with its last term neglected, maximum-likelihood estimation based on blocked data amounts to minimization of

$$\bar{\mathcal{F}} \equiv n_b \sum_{\bar{f}} \left(\frac{\bar{P}^{(\text{ex})}(\bar{f})}{P(\bar{f})} + \log P(\bar{f}) \right). \qquad (E6)$$

A brief calculation shows that this is equivalent to minimization of

$$\bar{\mathcal{F}}_2 \equiv \frac{1}{2} \chi^2 + \sum_{\bar{f}} \log P(\bar{f}), \qquad (E7)$$

where

$$\chi^2 \equiv n_b \sum_{\bar{f}} \left(\frac{\bar{P}^{(\text{ex})}(\bar{f}) - P(\bar{f})}{P(\bar{f})} \right)^2. \qquad (E8)$$

$\bar{\mathcal{F}}_2$ is precisely the expression one must minimize with respect to fitting parameters in the function P when these parameters are maximum likelihood estimated from a set of experimental data $[\bar{f}, \bar{P}^{(\text{ex})}(\bar{f})]_{\bar{f},...}$, that are normally distributed with the theoretical expected value and root-mean-square deviation given in Eqs. (E4) and (E5).

Thus we see that maximum-likelihood estimation of P simplifies to χ^2 minimization only when n_b is so large that one can ignore the last term in Eq. (E7) compared to the first. This last term occurs because our theory gives both the expected value for the data *and* the data's root-mean-square deviation for this expected value. Thus the parameters of our theory occur also in the root-mean-square deviation, the logarithm of which is the second term in $\bar{\mathcal{F}}_2$ above. In textbook derivations of ordinary least-squares fitting, this term is independent of the fitted theory's parameters, e.g., because experimentally measured error bars are used, hence only the first term, χ^2, is minimized.

Since $n_b n_w$ ranged from 2500 to 4350 in our data analysis, we could neglect the second term in $\bar{\mathcal{F}}_2$ relative to the first term, and fit by minimizing only χ^2. Factors other than this approximation limited our precision. However, with data and equipment other than those discussed here, we have encountered situations where χ^2 fitting clearly was not adequate, and we had to minimize the full expression for $\bar{\mathcal{F}}_2$.[44]

APPENDIX F: CALCULATION OF $\sigma(f_c)$

Equations (25) and (27) give f_c as a function of the experimental power spectrum $(P_k^{(\text{ex})})_{k=1,...,N'}$. Since $P_k^{(\text{ex})}$ is a random variable, so is the value we find for f_c. We determine $\sigma(f_c)$ by the usual method of linear propagation of errors. The calculation is long, but is simplified by a convenient choice of notation.

$$\sigma^2(f_c^2) = \sum_{k=1}^{N'} \left(\frac{\partial f_c^2}{\partial P_k^{(\text{ex})}} \right)^2_{P^{(\text{ex})}=P} \sigma^2(P_k^{(\text{ex})}), \qquad (F1)$$

$$\frac{\partial f_c^2}{\partial P_k^{(\text{ex})}} = \sum_{p,q} \frac{\partial f_c^2}{\partial S_{p,q}} \frac{\partial S_{p,q}}{\partial P_k^{(\text{ex})}}, \qquad (F2)$$

$$\frac{\partial S_{p,q}}{\partial P_k^{(\text{ex})}} = q f_k^{2p} P_k^{(\text{ex})q-1}, \qquad (F3)$$

$$f_c^2 = \mathcal{N}/\mathcal{D}, \qquad (F4)$$

where we have introduced an explicit notation for the numerator and the denominator,

$$\mathcal{N} \equiv S_{0,1} S_{2,2} - S_{1,1} S_{1,2}, \qquad (F5)$$

$$\mathcal{D} \equiv S_{0,2} S_{1,1} - S_{0,1} S_{1,2}. \qquad (F6)$$

Then

$$\frac{\partial f_c^2}{\partial S_{p,q}} = \mathcal{D}^{-1}\left(\frac{\partial \mathcal{N}}{\partial S_{p,q}} - f_c^2 \frac{\partial \mathcal{D}}{\partial S_{p,q}}\right), \tag{F7}$$

and thus

$$\frac{\partial f_c^2}{\partial P_k^{(ex)}} = \mathcal{D}^{-1}(\check{S}_{2,2} - \check{S}_{1,2}(f_c^2 + f_k^2) - 2\check{S}_{1,1}(f_c^2 + f_k^2)P_k^{(ex)}$$

$$+ 2\check{S}_{0,1}(f_c^2 + f_k^2)^2 P_k^{(ex)}), \tag{F8}$$

where we have introduced the notation,

$$\check{S}_{p,q} \equiv \sum_{k=1}^{N'} (f_c^2 + f_k^2)^p P_k^{(ex)q}, \tag{F9}$$

and note that $\check{S}_{0,q} = S_{0,q}$, and $S_{0,0} = N'$. In this notation,

$$\mathcal{D} = \check{S}_{0,2}\check{S}_{1,1} - \check{S}_{0,1}\check{S}_{1,2}. \tag{F10}$$

Evaluated at

$$P_k^{(ex)} = P_k = b^{-1}(f_c^2 + f_k^2)^{-1}, \tag{F11}$$

$$\check{S}_{p,q} = b^{-q}\check{S}_{p-q,0}, \tag{F12}$$

and consequently,

$$\left(\frac{\partial f_c^2}{\partial P_k^{(ex)}}\right)_{P^{(ex)} = P} = b^{-2}\mathcal{D}^{-1}(\check{S}_{-1,0}(f_c^2 + f_k^2) - \check{S}_{0,0}), \tag{F13}$$

with

$$\mathcal{D}(P^{(ex)} = P) = b^{-3}(\check{S}_{-2,0}\check{S}_{0,0} - \check{S}_{-1,0}^2). \tag{F14}$$

Using

$$\sigma(P_k^{(ex)}) = \eta_b^{-1/2} P_k = n_b^{-(1/2)} b^{-1}(f_c^2 + f_k^2)^{-1} \tag{F15}$$

(n_b should be replaced with $n_b n_w$ if $n_w \neq 1$), we find

$$\sigma^2(f_c^2) = n_b^{-1} b^{-6} \mathcal{D}^{-2} \check{S}_{0,0}(\check{S}_{-2,0}\check{S}_{0,0} - \check{S}_{-1,0}^2),$$

$$= n_b^{-1} b^{-3}\mathcal{D}^{-1}\check{S}_{0,0}, \tag{F16}$$

$$= \frac{\check{S}_{0,0}}{n_b(\check{S}_{-2,0}\check{S}_{0,0} - \check{S}_{-1,0}^2)}.$$

Finally, we replace the sums with the integrals they approximate,

$$\check{S}_{p,0} = \sum_{k=1}^{N/n_b} (f_c^2 + f_k^2)^p = \frac{T}{n_b}\int_{f_{min}}^{f_{max}} df(f_c^2 + f^2)^p \tag{F17}$$

and have, with $x_{min} \equiv f_{min}/f_c$ and $x_{max} \equiv f_{max}/f_c$,

$$\check{S}_{0,0} = = \frac{Tf_c}{n_b}(x_{max} - x_{min}), \tag{F18}$$

$$\check{S}_{-1,0} = \frac{T}{n_b f_c}\arctan\left(\frac{x_{max} - x_{min}}{1 + x_{max}x_{min}}\right), \tag{F19}$$

$$\check{S}_{-2,0} = \frac{T}{2n_b f_c^3}\left[\frac{x_{max}}{1 + x_{max}^2} - \frac{x_{min}}{1 + x_{min}^2}\right. $$

$$\left. + \arctan\left(\frac{x_{max} - x_{min}}{1 + x_{max}x_{min}}\right)\right], \tag{F20}$$

from which Eq. (25) follows, when using

$$\frac{\sigma(f_c)}{f_c} = \left(\frac{\sigma^2(f_c^2)}{4f_c^4}\right)^{1/2}. \tag{F21}$$

APPENDIX G: FITTING $f_{3\,dB}^{(diode)}$ AND $\alpha^{(diode)}$

The fit shown in Fig. 5 gives $f_c = 0.37$ kHz and $f_{3\,dB}^{(diode)} = 6.8$ kHz, i.e., the latter is 19 times larger than the former. So the values of these two frequencies are sensitive to different parts of the power spectrum. This does not mean that the covariances between f_c, on the one hand, and $f_{3\,dB}^{(diode)}$ and $\alpha^{(diode)}$ on the other, are negligible. Both f_c and $f_{3\,dB}^{(diode)}$, for example, depend on a large range of frequencies, and consequently have significant covariance for realistic values of both. This is seen when fitting using a program that gives correlations, and can also be shown analytically for a Lorentzian fit with $f_c \ll f_{max} \lesssim f_{3\,dB}^{(diode)} \ll f_{Nyq}$ in a calculation analogous to the one done in Appendix F; see Sec. G 1 below. In units of $\sigma(f_c)\sigma(f_{3\,dB}^{(diode,eff)})$, one finds that the covariance of f_c and $f_{3\,dB}^{(diode,eff)}$ is $-2(f_c/f_{3\,dB}^{(diode,eff)})^{1/2}$, which takes values of -0.46 and -0.47 for the values of f_c and $f_{3\,dB}^{(diode,eff)}$ that we found above (Fig. 5, Table II) for the x and the y coordinates, respectively. In view of the approximation involved, this is in good agreement with the values of -0.55 and -0.54 found by the fitting program for the covariance between f_c and $f_{3\,dB}^{(diode)}$ in units of $\sigma(f_c)\sigma(f_{3\,dB}^{(diode)})$. The agreement is even better for the data set whose x-coordinate data are shown in Fig. 6. The fit shown there and the equivalent one for the y coordinate give -0.56, respectively, -0.55, for the correlation between f_c and $f_{3\,dB}^{(diode)}$ in units of $\sigma(f_c)\sigma(f_{3\,dB}^{(diode)})$. This compares very well with the analytical result $-2(f_c/f_{3\,dB}^{(diode,eff)})^{1/2} = -0.59$, respectively, -0.58, for the covariance between f_c and $f_{3\,dB}^{(diode,eff)}$ in units of $\sigma(f_c)\sigma(f_{3\,dB}^{(diode,eff)})$.

Since the covariance in units of $\sigma(f_c)\sigma(f_{3\,dB}^{(diode,eff)})$ is $-2(f_c/f_{3\,dB}^{(diode,eff)})^{1/2}$ in our analytical case based on Lorentzians, we see that one needs an unrealistically small ratio for $f_c/f_{3\,dB}^{(diode)}$ in order to have negligible covariance between these two parameters. We also see that because of their substantial covariance one cannot determine one correctly without determining the other with similar precision. Because of our rich data and well-fitting theory, we determine both f_c and $f_{3\,dB}^{(diode)}$ with the high precision listed in Tables II and III. This precision refers to the 68% probability interval for the parameter in question, with the values for the other three parameters floating; i.e., it is the most conservative, largest interval. So our procedure is quite sound despite the nonvanishing covariance of f_c with $f_{3\,dB}^{(diode)}$.

1. Low-frequency approximation for the diode characteristic function

From Tables II and III we see that less than 10% of the power in the spectrum, $\alpha^{(\text{diode})2}$, is unaffected by the finite response time of the diode. In this case, and for frequencies f for which $(f/f_{3\,\text{dB}}^{(\text{diode})})^2 \ll 1$, we have

$$\frac{P(f)}{P_0(f)} = \frac{1 + \alpha^{(\text{diode})2}(f/f_{3\,\text{dB}}^{(\text{diode})})^2}{1 + (f/f_{3\,\text{dB}}^{(\text{diode})})^2},$$

$$\approx \frac{1}{1 + (f/f_{3\,\text{dB}}^{(\text{diode,eff})})^2}, \tag{G1}$$

where we have introduced

$$f_{3\,\text{dB}}^{(\text{diode,eff})} \equiv (1 - \alpha^{(\text{diode})2})^{-(1/2)} f_{3\,\text{dB}}^{(\text{diode})}. \tag{G2}$$

The last expression in Eq. (G1) is a simple Lorentzian. Equation (G2) shows that in this case $\alpha^{(\text{diode})}$ combines with $f_{3\,\text{dB}}^{(\text{diode})}$ into a single parameter, $f_{3\,\text{dB}}^{(\text{diode,eff})}$, an effective 3 dB frequency of a first-order filter that describes the diode's characteristics.

One can use this approximation in a calculation analogous to the one done in Appendix F to obtain the analytical result used above: the covariance of f_c and $f_{3\,\text{dB}}^{(\text{diode,eff})}$, in units of $\sigma(f_c)\sigma(f_{3\,\text{dB}}^{(\text{diode,eff})})$, is $-2(f_c/f_{3\,\text{dB}}^{(\text{diode,eff})})^{1/2}$.

APPENDIX H: ALIASING AND ANTIALIASING

1. What aliasing is

When a signal is sampled at discrete times, t_j, with frequency f_{sample}, the sampling process cannot distinguish frequency components of the signal which differ from each other by integer multiples of the sampling frequency. They all add up to a single amplitude. This is seen as follows: Our experiment records a time series $(x_j)_{j=1,\dots,N}$ by sampling the continuous signal $x(t)$ with frequency f_{sample} for time T_{msr}. With $\tilde{x}_k$ the continuous Fourier transformed in Eq. (7), the continuous signal can be written in terms of its inverse Fourier transform,

$$x(t) = \frac{1}{T_{\text{msr}}} \sum_{n=-\infty}^{\infty} e^{-i2\pi t k/N} \tilde{x}_k. \tag{H1}$$

Inserting this in the *discrete* Fourier transform of our recorded time series, Eq. (12), and using

$$\frac{1}{N} \sum_{j=1}^{N} e^{i2\pi j(k-\ell)/N} = \sum_{n=-\infty}^{\infty} \delta_{k-\ell,nN}, \tag{H2}$$

where the right-hand side is equal to 1 for $k = \ell$ modulo N, and 0 otherwise, we find

$$\hat{x}_k = \sum_{n=-\infty}^{\infty} \tilde{x}_{k+nN}. \tag{H3}$$

Here the real and imaginary parts of $\tilde{x}_{k+nN}$ are uncorrelated random Gaussian variables with zero mean and common variance in both the Einstein–Ornstein–Uhlenbeck theory and the hydrodynamically correct theory, and for filtered versions of both theories. Hence, so are the real and imaginary parts of $\hat{x}_k$. Consequently,

$$P_k^{(\text{ex,aliased})} \equiv |\hat{x}_k|^2/T_{\text{msr}} \tag{H4}$$

is exponentially distributed on the real, non-negative numbers, with mean

$$P_k^{(\text{aliased})} = \langle |\hat{x}_k|^2 \rangle / T_{\text{msr}},$$

$$= \sum_{n=-\infty}^{\infty} \langle |\tilde{x}_{k+nN}|^2 \rangle / T_{\text{msr}},$$

$$= \sum_{n=-\infty}^{\infty} P_{k+nN}. \tag{H5}$$

$P^{(\text{aliased})}$ is obviously a periodic function of k with period N, i.e., of f_k with period f_{sample}. So it is sufficient to know its value in the interval $[-f_{\text{Nyq}}, f_{\text{Nyq}}]$, $f_{\text{Nyq}} = f_{\text{sample}}/2$.

On the other hand, Eq. (H5) shows that neither f_{Nyq} nor f_{sample} represents a sharp frequency cutoff. The value of $P_k^{(\text{aliased})}$ depends through $P_{k+nN}^{(\text{unaliased})}$, $n = \pm 1, \pm 2, \dots$, on frequency components $\tilde{x}_k$ of the signal $x(t)$ outside $[-f_{\text{Nyq}}, f_{\text{Nyq}}]$.

2. Example: Aliased Lorentzian

As a specific example, we consider the Lorentzian, for which

$$P_k^{(\text{aliased})} = \sum_{n=-\infty}^{\infty} \frac{D/(2\pi^2)}{f_c^2 + (f_k + nf_{\text{sample}})^2}$$

$$= \frac{(\Delta x)^2 \Delta t}{1 + c^2 - 2c\cos(2\pi k/N)}. \tag{H6}$$

In the case of $f_c \ll f < f_{\text{Nyq}}$, $P(f) \propto 1/f^2$, hence

$$P_{N/2}^{(\text{aliased})} \propto \sum_{n=-\infty}^{\infty} \frac{1}{(f_{\text{Nyq}} + 2nf_{\text{Nyq}})^2},$$

$$= \frac{2}{f_{\text{Nyq}}^2}\left(1 + \frac{1}{9} + \frac{1}{25} + \frac{1}{49} + \frac{1}{81} + \cdots\right)$$

$$\approx \frac{2.47}{f_{\text{Nyq}}^2}, \tag{H7}$$

i.e., aliasing adds almost 150% to the power spectrum near f_{Nyq}. This means that frequencies several times f_{Nyq} contribute significantly to the power spectrum near f_{Nyq}, no matter what value f_{Nyq} has, and one must consequently consider whether the model yielding the Lorentzian really is also valid at these higher frequencies, even if the model is known to be valid below f_{Nyq}.

3. What antialiasing is

Data acquisition electronics have built-in *antialiasing filters*. These filters prevent aliasing of electronic noise from much higher frequencies. However, if not all 3 dB frequencies of these filters are much larger than our highest frequency of interest, the power spectrum we *want* to measure is distorted by antialiasing. Since $f_{3\,\text{dB}} = f_{\text{Nyq}}$ is a popular choice, this distortion commonly occurs and is significant; see Fig. 11.

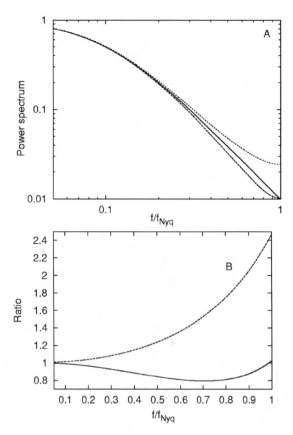

FIG. 11. (a) Solid line: Simple Lorentzian power spectrum with f_c = 374 Hz. Dotted line above the Lorentzian: Same simple Lorentzian, but aliased as if the signal were sampled at finite rate $f_{sample}=2f_{Nyq}$, hence a periodic function with period f_{sample}. Dashed line below the Lorentzian: Same simple Lorentzian, but filtered with a first-order filter with f_{3dB} $=f_{Nyq}$, then aliased as if the signal were sampled at finite rate f_{sample} $=2f_{Nyq}$. Note that the antialiasing filter with $f_{3dB}=f_{Nyq}$ does not prevent aliasing as such, it only suppresses large frequencies, so less power is aliased. (b) Dashed line: Ratio between the aliased Lorentzian and the simple Lorentzian, i.e., the ratio between power spectra of signals recorded with a finite sampling rate and recorded continuously. Solid line: Ratio between the filtered, aliased Lorentzian, and simple Lorentzian, i.e., the ratio between power spectra of the signal recorded with a finite sampling rate after filtering and recorded continuously with no filtering. Note that the antialiased spectrum falls up to 20% below the unaliased spectrum: Anti-aliasing filters are for electronic noise suppression, and are no substitute for a correct description of the aliasing caused by the finite sampling time.

It should be avoided when possible by choosing high values for f_{3dB}. Electronic noise *must* be filtered away, but this should be done with minimal effect on the information-containing part of the power spectrum. Although any anti-aliasing filter is easily accounted for in principle when its characteristic function is known, in practice parameter values of the characteristic function are not known with precision. However, if the filter settings are chosen to minimize filtering, the uncertainty that this lack of precision causes, appears on subdominant digits that describe an already small effect. So this lack of precision may not matter.

4. What antialiasing is not

Antialiasing filters are no substitute for a correct descrip-tion of the aliasing caused by finite sampling time, as Fig. 11

illustrates: If the theoretical power spectrum for a system is a simple Lorentzian, but the signal from this system is sampled at a finite rate, an aliased Lorentzian results. This is then what one should fit to the experimental spectrum. If one filters the signal before sampling it, say, with $f_{3dB}=f_{Nyq}$, that should also be accounted for. However, one should *not* assume that the filter prevents the aliasing caused by finite sampling time, and then fit a simple Lorentzian to the spec-trum of the filtered, sampled signal, to frequencies up to f_{Nyq}. Not if 20% error matters, because the two spectra dif-fer by that much for $0.6f_{Nyq}<f<0.8f_{Nyq}$. If one fits only to frequencies below $f_{max}=0.6f_{Nyq}$, a simple Lorentzian is about as bad an approximation to the filtered time series as to the unfiltered one. So in this range also is antialiasing no substitute for a theory that accounts for filters and sampling rates.

[1] A. Ashkin, Proc. Natl. Acad. Sci. U.S.A. **94**, 4853 (1997).
[2] S. M. Block, Nature (London) **386**, 217 (1997).
[3] A. D. Mehta, M. Rief, J. A. Spudich, D. A. Smith, and R. M. Simmons, Science **283**, 1689 (1999).
[4] S. B. Smith, Y. Cui, and C. Bustamante, Science **271**, 795 (1996).
[5] R. G. Larson, T. T. Perkins, D. E. Smith, and S. Chu, Phys. Rev. E **55**, 1794 (1997).
[6] M. S. Z. Kellermayer, S. B. Smith, H. L. Granzier, and C. Bustamante, Science **276**, 1112 (1997).
[7] M. D. Wang, H. Yin, R. Landick, J. Gelles, and S. M. Block, Biophys. J. **72**, 1335 (1997).
[8] M. W. Berns, Sci. Am. **April**, 52 (1998).
[9] M. D. Wang, M. J. Schnitzer, H. Yin, R. Landick, J. Gelles, and S. M. Block, Science **282**, 902 (1998).
[10] C. Veigel, M. L. Bartoo, D. C. S. White, J. C. Sparrow, and J. E. Molloy, Biophys. J. **75**, 1424 (1998).
[11] K. Visscher, M. J. Schnitzer, and S. M. Block, Nature (London) **400**, 184 (1999).
[12] B. Schnurr, F. Gittes, F. C. MacKintosh, and C. F. Schmidt, Macromol-ecules **30**, 7781 (1997).
[13] F. Gittes, B. Schnurr, P. D. Olmsted, F. C. MacKintosh, and C. F. Schmidt, Phys. Rev. Lett. **79**, 3286 (1997).
[14] I. M. Tolić-Nørrelykke, E.-L. Munteanu, G. Thon, L. Oddershede, and K. Berg-Sørensen, (unpublished).
[15] A. Pralle, E.-L. Florin, E. H. K. Stelzer, and J. K. H. Hörber, Appl. Phys. A: Mater. Sci. Process. **A66**, S71 (1998).
[16] L. Oddershede, H. Flyvbjerg, and K. Berg-Sørensen, J. Phys.: Condens. Matter **15**, S1737 (2003).
[17] K. Berg-Sørensen, L. Oddershede, and H. Flyvbjerg, Proc. SPIE (to be published).
[18] T. Tlusty, A. Meller, and R. Bar-Ziv, Phys. Rev. Lett. **81**, 1738 (1998).
[19] K. Svoboda and S. M. Block, Annu. Rev. Biophys. Biomol. Struct. **23**, 247 (1994).
[20] R. M. Simmons, J. T. Finer, S. Chu, and J. A. Spudich, Biophys. J. **70**, 1813 (1996).
[21] K. Visscher and S. M. Block, Methods Enzymol. **298**, 460 (1998).
[22] F. Gittes and C. F. Schmidt, Methods Cell Biol. **55**, 129 (1998).
[23] E.-L. Florin, A. Pralle, E. H. K. Stelzer, and J. K. H. Hörber, Appl. Phys. A: Mater. Sci. Process. **A66**, S75 (1998).
[24] W. Singer, S. Bernet, N. Hecker, and M. Ritsch-Marte, J. Mod. Opt. **47**, 2921 (2000).
[25] M. Capitanio, G. Romano, R. Ballerini, M. Giuntini, F. C. Pavone, D. Dunlap, and L. Finzi, Rev. Sci. Instrum. **73**, 1687 (2002).
[26] I. M. Tolić-Nørrelykke, K. Berg-Sørensen, and H. Flyvbjerg, Comput. Phys. Commun (to be published).
[27] L. Oddershede, S. Grego, S. F. Nørrelykke, and K. Berg-Sørensen, Probe Microsc. **2**, 129 (2001).
[28] A. Jones, Customer Service, Bangs Laboratories, Inc., 9025 Technology Drive, Fishers, IN 46038-2886; private communication (2002).
[29] R. Kubo, M. Toda, and N. Hashitsume, *Statistical Physics* (Springer, Heidelberg, 1985), Vol. 2.
[30] W. H. Press, B. P. Flannery, S. A. Teukolsky, and W. T. Vetterling, *Nu-*

merical Recipes. The Art of Scientific Computing (Cambridge University Press, Cambridge, 1986), Sec. 12.7.

[31] N. C. Barford, *Experimental Measurements: Precision, Error and Truth*, 2nd ed. (Wiley, New York, 1986).

[32] *Handbook of Chemistry and Physics*, 76th ed. edited by D. R. Lide (Chemical Rubber, Boca Raton, FL, 1995).

[33] M. J. Schnitzer and S. M. Block, Nature (London) **388**, 386 (1997).

[34] C. Veigel, L. M. Coluccio, J. D. Jontes, J. C. Sparrow, R. A. Milligan, and J. E. Molloy, Nature (London) **398**, 530 (1999).

[35] L. P. Ghislain, N. A. Switz, and W. W. Webb, Rev. Sci. Instrum. **65**, 2762 (1994).

[36] K. Berg-Sørensen, L. Oddershede, E.-L. Florin, and H. Flyvbjerg, J. Appl. Phys. **93**, 3167 (2003).

[37] G. G. Stokes, Trans. Cambridge Philos. Soc. **IX**, 8 (1851).

[38] L. D. Landau and E. M. Lifshitz, *Fluid Mechanics*, 2nd ed. (Butterworth and Heinemann, Oxford, 1987), problem 5.

[39] D. Bedeaux and P. Mazur, Physica (Amsterdam) **76**, 247 (1974).

[40] H. Faxén, Ark. Mat., Astron. Fys. **17**, 1 (1923).

[41] J. Happel and H. Brenner, *Low Reynolds Number Hydrodynamics* (Nijhoff The Hague, 1983), p. 327.

[42] H. A. Lorentz, *Abhandlungen über Theoretische Physik* (Druck und von B.-G. Teubner, Leipzig, 1906), Vol. 1.

[43] H. Flyvbjerg (unpublished).

[44] K. Berg-Sørensen, E. Peterman, C. Schmidt, and H. Flyvbjerg, (unpublished).

[45] E. J. G. Peterman, M. van Dijk, L. C. Kapitein, and C. F. Schmidt, Rev. Sci. Instrum. **74**, 3246 (2003).

[46] K. Visscher, S. P. Gross, and S. M. Block, IEEE J. Sel. Top. Quantum Electron. **2**, 1066 (1996).

[47] F. Gittes and C. F. Schmidt, Opt. Lett. **23**, 7 (1998).

[48] A. Pralle, M. Prummer, E.-L. Florin, E. H. K. Stelzer, and J. K. H. Hörber, Microsc. Res. Tech. **44**, 378 (1999).

Parametric Resonance of Optically Trapped Aerosols

R. Di Leonardo,[1,*] G. Ruocco,[1] J. Leach,[2] M. J. Padgett,[2] A. J. Wright,[3] J. M. Girkin,[3]
D. R. Burnham,[4,5] and D. McGloin[4,5]

[1]*INFM-CRS SOFT c/o Universitá di Roma "La Sapienza," I-00185, Roma, Italy*
[2]*SUPA, Department of Physics and Astronomy, University of Glasgow, Glasgow, Scotland, United Kingdom*
[3]*Institute of Photonics, SUPA, University of Strathclyde, Glasgow, Scotland, United Kingdom*
[4]*SUPA, School of Physics and Astronomy, University of St. Andrews, St. Andrews, Scotland, United Kingdom*
[5]*Electronic Engineering and Physics Division, University of Dundee, Dundee, Scotland, United Kingdom*
(Received 23 February 2007; published 5 July 2007; corrected 6 July 2007)

The Brownian dynamics of an optically trapped water droplet are investigated across the transition from over- to underdamped oscillations. The spectrum of position fluctuations evolves from a Lorentzian shape typical of overdamped systems (beads in liquid solvents) to a damped harmonic oscillator spectrum showing a resonance peak. In this later underdamped regime, we excite parametric resonance by periodically modulating the trapping power at twice the resonant frequency. The power spectra of position fluctuations are in excellent agreement with the obtained analytical solutions of a parametrically modulated Langevin equation.

DOI: 10.1103/PhysRevLett.99.010601   PACS numbers: 05.40.−a, 46.40.Ff, 82.70.Rr, 87.80.Cc

Parametric resonance provides an efficient and straightforward way to pump energy into an underdamped harmonic oscillator [1]. If the resonance frequency of an oscillator is dependent upon a number of parameters, modulating any of these at twice the natural oscillation frequency parametrically excites the resonance. Such behavior leads to surprising phenomena in the macroscopic world (pumping a swing, stability of vessels, surface waves in vibrated fluids) [2,3]. On the microscopic scale, where stochastic forces become important, one refers to Brownian parametric oscillators [4]. The parametric driving of Brownian systems has been shown to be at the origin of some peculiar behaviors such as the squeezing of thermal noise in Paul traps [5]. Parametrically excited oscillations have also been reported in a single-crystal silicon microelectromechanical system [6]. What makes parametric resonance useful is that it is easier to modulate a system parameter rather than to apply an oscillating driving force. Moreover, for finite but low damping rates, one never reaches a stationary state with the damping forces dissipating the input power, and, consequently, the amplitudes of oscillations diverge. Optically trapped microparticles constitute a beautiful example of a Brownian damped harmonic oscillator, and they are becoming an increasingly common tool for the investigation of different fields of basic and applied science [7]. Pumping mechanical energy into optically trapped particles could open the way to many applications. In optical tweezers, even though it is easy to periodically modulate the laser power, parametric excitation is usually ineffective because of the heavy damping action of the surrounding fluid.

It has been reported that modulating the laser power at the parametric resonant frequency in an overdamped system increased the amplitude of fluctuations [8,9]. However, these results have been difficult to reproduce and are in contrast to the predictions of the Langevin equation [10–12].

In this Letter, we demonstrate how parametric resonance can be excited in optically trapped water droplets suspended in air. We measure power spectra of position fluctuations and find an excellent agreement with the theoretical expectations based on Langevin dynamics. Besides providing a resolution of a still-controversial issue, our results are the first study of the dynamics of trapped particles in a gas-damped tweezers system, a configuration that is finding wider applications in the study of aerosol droplets and associated atmospheric chemistry [13].

The dynamics of an optically trapped droplet is described by the Langevin equation [14]:

$$\ddot{x}(t) + \Omega_0^2 x(t) + \Gamma_0 \dot{x}(t) = \xi(t), \qquad (1)$$

where $\Omega_0^2 = k/m$ is the angular frequency of the oscillator depending on trap stiffness k and particle mass m. $\Gamma_0 = 6\pi\eta a/C_c m$ is the viscous damping due to the medium viscosity η and depending on particle radius a and mass m. To correct Stokes' law for finite Knudsen number effects, we introduced the empirical slip correction factor C_c, with a 5.5%–1.6% reduction in drag for 3–10 μm diameter droplets, respectively [15].

The stochastic force ξ, due to thermal agitation of solvent molecules, is generally assumed to have a white noise power spectrum:

$$S_\xi(\omega) = \frac{\Gamma_0 K_B T}{\pi m}. \qquad (2)$$

By Fourier transforming (1) and using (2), we can easily obtain the power spectrum of position fluctuations [16]:

$$S_x(\omega) = \frac{K_B T}{k} \frac{1}{\pi} \frac{\Omega_0^2 \Gamma_0}{(\omega^2 - \Omega_0^2)^2 + \Gamma_0^2 \omega^2}. \qquad (3)$$

The ratio Γ_0/Ω_0 depends only slightly on particle size and, in solvents with waterlike viscosities, is always greater than 1 (the system is overdamped) up to power levels of some tens of watts. For typical trapping powers of order 10 mW in water, Γ_0/Ω_0 is typically >10. As a result, only those frequencies smaller than Γ_0, and hence much smaller than Ω_0, have a significant amplitude in the power spectrum. Under these conditions, we can therefore neglect ω^2 with respect to Ω_0^2 in the first term of the denominator in (3) and obtain the usual Lorentzian power spectrum characterized by ω^{-2} tails [16]. Such an overdamped condition precludes the possibility of exciting significant oscillations either directly or parametrically. To probe oscillations in the liquid-damped regime, we would need to be able to increase typical trap power by 4 orders of magnitude. A more feasible route is to reduce viscosity by 2 orders of magnitude, which can be readily obtained by trapping particles in air whose viscosity is approximately 1/55 of water ($\eta = 1.8 \times 10^{-5}$ Pa s) [15].

Our optical tweezers are based around an inverted microscope with a high numerical aperture oil immersion microscope objective (1.3 NA, 100 ×). The continuous wave laser is a Nd:YAG, frequency doubled to give 0–2 W of 532 nm light. To couple the beam into the air medium, a single cover slip is rested over the objective on a thin oil layer. In a method similar to Refs. [17,18], a water aerosol is produced using a nebulizer (3.4–6.0 μm) and injected into a sample chamber 30 mm in diameter and 10 mm deep, sealed by the cover slip. This isolates the droplets from convective air currents and create a near-saturated atmosphere within which the droplet size is stable. To obtain a saturated atmosphere, we decrease the vapor pressure of the droplets by adding salt to the water. In such conditions, the droplet quickly reaches an equilibrium size between condensation and evaporation and has a size stability within 2% over the trapping time [19]. After a few seconds, a droplet in the range of 4–7 μm is trapped at the beam focus; see Fig. 1. For our laser powers, this gives a trap resonance frequency in the vicinity of 2 kHz, and we can maintain the trap for 40 min. For particles trapped in fluid, a laser power of 10 mW is typical; however, to maximize the stiffness of our traps (and hence Ω_0), we use powers of 100 s mW. Despite this, we calculate a temperature increase of less than 1 K due to laser heating [17]. Though this does not significantly enhance evaporation, temperature gradients across the droplet, due to non-uniform heating, could initiate thermal Marangoni effects. However, being concerned with the center of mass dynamics occurring in the kilohertz region, none of these relatively slow phenomena disturbs the high frequency dynamics. A quadrant photodetector, placed in the back focal plane of the condenser lens, receives the light transmitted through the droplet. By measuring the imbalance of the light collected by the quadrants, the lateral displacement of the droplet is deduced with a bandwidth of several kilohertz and a precision of better than 5 nm [20]. The

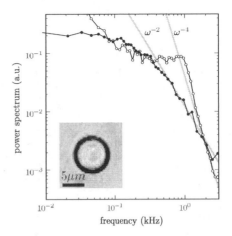

FIG. 1. The measured power spectra of trapped aerosol particle at two different powers. At lower power (black circles), it is overdamped, and the mean squared amplitude of the high frequency motion decays as ω^{-2}. At higher powers (white circles), the aerosol is underdamped, and the mean squared amplitude decays as ω^{-4}. Gray lines show the calculated slopes -2 and -4. The inset shows an optical image of a trapped aerosol particle.

stability of our system allows a series of power spectra to be obtained from the same droplet while scanning one parameter. The three reported experimental protocols (Figs. 1–3) will refer to three independently trapped droplets.

The power spectra of the measured displacement, for two different trap powers, are shown in Fig. 1. It is clear how the particle dynamics changes from an overdamped Lorentzian spectrum with a high frequency roll-off proportional to ω^{-2} to an underdamped regime with a faster roll-off ω^{-4} and the appearance of a resonance peak near 1 kHz (more clearly seen in Fig. 2, which is plotted on a linear scale). The emergence of such a peak arises from the

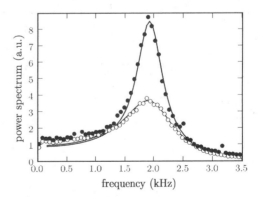

FIG. 2. The measure power spectrum of a trapped water droplet for no modulation of the laser power (white circles) and modulation at 3.9 kHz ($\Omega_1 \simeq 2\Omega_0$) (black circles). The peak is higher and narrower on the resonant condition, thus indicating parametric excitation. The solid line below the black circles is the predicted spectrum from (11).

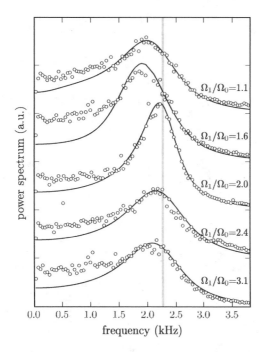

FIG. 3. Evolution of position power spectra on varying the modulation frequency Ω_1. Parametric excitation of oscillations is evident at the parametric resonance condition $\Omega_1/\Omega_0 = 2$. The solid lines are the theoretical predictions from (11). The vertical gray line indicates $\Omega_0/2\pi = 2.3$ kHz.

fact that the inertial terms in (1) are no longer negligible. As a consequence, an average trajectory starting away from the equilibrium position crosses the equilibrium position with a finite velocity. In this situation, the parametric resonance is excited by modulating the strength of the trapping potential. Ideally, the potential is made shallower when the particle traverses the equilibrium position and steeper again when the particle is far from the equilibrium position. This is maximally efficient when we modulate the potential at twice the natural oscillation frequency Ω_0. To consider this model in quantitative terms, we can rewrite (1) in the presence of a parametrically modulated external potential:

$$\ddot{x}(t) + \Omega_0^2[1 + gf(t)]x(t) + \Gamma_0\dot{x}(t) = \xi(t), \quad (4)$$

$$f(t + \mathcal{T}) = f(t), \qquad -1 < f(t) < 1, \quad (5)$$

where $0 < g < 1$ measures the strength of modulation. By Fourier transforming (4), we obtain

$$(-\omega^2 + \Omega_0^2 + i\omega\Gamma_0)\hat{x}(\omega)$$

$$+ \Omega_0^2 g \sum_{k=-\infty}^{\infty} a_k\hat{x}(\omega + k\Omega_1) = \hat{\xi}(\omega), \quad (6)$$

where a_k is the coefficient of the $k2\pi/\mathcal{T} = k\Omega_1$ frequency component of the Fourier series expansion of $f(t)$. It is clear from Eq. (6) how parametric modulation introduces a coupling between all of those frequencies

differing by an integer number of Ω_1. We now introduce the vectors $X_n(\omega) = \hat{x}(\omega + n\Omega_1)$ and $R_n(\omega) = \hat{\xi}(\omega + n\Omega_1)$ and write the recursive relations:

$$[-(\omega + n\Omega_1)^2 + \Omega_0^2 + i(\omega + n\Omega_1)\Gamma_0]X_n(\omega)$$

$$+ \Omega_0^2 g \sum_{k=-\infty}^{\infty} a_k X_{n+k}(\omega) = R_n. \quad (7)$$

To obtain the power spectrum $S_x(\omega)$, for each frequency ω we compute $X_0(\omega)$. This is coupled to all other components in the array X_n. However, the strength of the coupling will decay for large $|n|$, so that we can limit ourselves to a finite number of components and write the matrix equation for the array $\mathbf{X}(\omega) = [X_{-N}(\omega), \dots, X_N(\omega)]$:

$$\mathbf{G}^{-1}(\omega)\mathbf{X}(\omega) = \mathbf{R}(\omega), \quad (8)$$

with

$$G_{nk}^{-1}(\omega) = [-(\omega + n\Omega_1)^2 + \Omega_0^2 + i(\omega + n\Omega_1)\Gamma_0]\delta_{nk}$$

$$+ \Omega_0^2 g a_{k-n}. \quad (9)$$

By matrix inversion, we obtain the power spectrum as

$$\langle X_0^*(\omega)X_0(\omega')\rangle = \sum_{k,n=-N}^{N} G_{0k}^*(\omega)G_{0l}(\omega')\langle R_k^*(\omega)R_n(\omega')\rangle$$

$$= \frac{\Gamma_0 K_B T}{\pi m} \sum_{n=-N}^{N} |G_{0n}(\omega)|^2\delta(\omega - \omega'), \quad (10)$$

and from the definition of power spectrum:

$$S_x(\omega) = \frac{\Gamma_0 K_B T}{\pi m} \sum_{k=-N}^{N} |G_{0k}(\omega)|^2. \quad (11)$$

If Ω_0 and Γ_0 are known, we can use (11) to predict the power spectrum of a droplet in a modulated trap. The white circles in Fig. 2 show the measured power spectrum for a water droplet trapped with constant laser power. The presence of the peak suggests that we are in an underdamped regime. By fitting these data to Eq. (3), we can deduce the resonant frequency $\Omega_0/2\pi = 2.0$ kHz and the damping term $\Gamma_0 = 6.8$ kHz for our experimental conditions. The fitted value of Γ_0 corresponds to the Stokes drag on an aerosol droplet of radius 3.4 μm. We then apply the square-wave modulation of the trapping power adjusted to give the same average power as before, $\Omega_1 \simeq 2\Omega_0$ and $g = 0.4$. The black circles in Fig. 2 show the corresponding power spectrum of the motion. As expected, the resonance is excited, matching closely the expected behavior obtained by applying the measured parameters Ω_0, Γ_0, Ω_1, g to Eq. (11), supporting our interpretation of a parametric excitation of the resonance. Using these parameters with (11), we can make further predictions about the dynamics that can be verified by our observations. One comparison is the predicted and observed form of the power spectra as a function of the modulation frequency, both above and

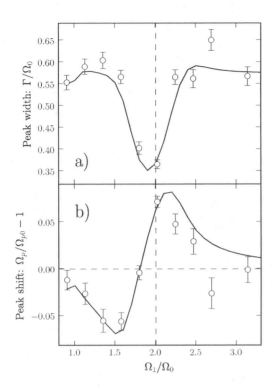

FIG. 4. (a) Peak full width at half maximum as a function of modulation frequency. The solid lines are the theoretical predictions from (11). (b) Peak position shift as a function of modulation frequency. Ω_{p0} is the peak position in the absence of modulation.

nut or Laguerre-Gaussian beams having zero on-axis intensity and improved axial trapping [21].

We have reported the first observation of a parametrically excited resonance within a Brownian oscillator. The demonstration of this effect within optical tweezers was made possible by relying on the low viscosity of air to lightly damp the motion of a trapped aerosol droplet. The detailed observed dynamics matches closely the power spectra predicted from a parametrically modulated Langevin equation.

We thank S. Ciuchi for many helpful discussions. D. M. thanks the Royal Society for support.

*roberto.dileonardo@phys.uniroma1.it

[1] C. Van den Broeck and I. Bena, in *Stochastic Processes in Physics, Chemistry and Biology* (Springer-Verlag, Berlin, 2000).
[2] L. Ruby, Am. J. Phys. **64**, 39 (1996).
[3] J. Bechhoefer and B. Johnson, Am. J. Phys. **64**, 1482 (1996).
[4] C. Zerbe, P. Jung, and P. Hänggi, Phys. Rev. E **49**, 3626 (1994).
[5] A. F. Izmailov, S. Arnold, S. Holler, and A. S. Myerson, Phys. Rev. E **52**, 1325 (1995).
[6] L. Turner *et al.* Nature (London) **396**, 149 (1998).
[7] D. G. Grier, Nature (London) **424**, 810 (2003).
[8] J. Joykutty, V. Mathur, V. Venkataraman, and V. Natarajan, Phys. Rev. Lett. **95**, 193902 (2005).
[9] V. Venkataraman, V. Natarajan, and N. Kumar, Phys. Rev. Lett. **98**, 189803 (2007).
[10] L. Pedersen and H. Flyvbjerg, Phys. Rev. Lett. **98**, 189801 (2007).
[11] Y. Deng, N. R. Forde, and J. Bechhoefer, Phys. Rev. Lett. **98**, 189802 (2007).
[12] Y. Deng, J. Bechhoefer, and N. R. Forde, J. Opt. A Pure Appl. Opt. (to be published).
[13] J. Buajarern, L. Mitchem, A. D. Ward, N. Nahler, D. McGloin, and J. P. Reid J. Chem. Phys. **125**, 114506 (2006).
[14] S. Chandrasekhar, Rev. Mod. Phys. **15**, 1 (1943).
[15] J. H. Seinfeld and S. N. Pandis, *Atmospheric Chemistry and Physics: Air Pollution to Climate Change* (Wiley, New York, 1997).
[16] M. C. Wang and G. E. Uhlenbeck, Rev. Mod. Phys. **17**, 323 (1945).
[17] R. J. Hopkins, L. Mitchem, A. D. Ward, and J. P. Reid, Phys. Chem. Chem. Phys. **6**, 4924 (2004).
[18] D. R. Burnham and D. McGloin, Opt. Express **14**, 4175 (2006).
[19] L. Mitchem *et al.*, J. Phys. Chem. A **110**, 8116 (2006).
[20] F. Gittes and C. Schmidt, Opt. Lett. **23**, 7 (1998).
[21] A. T. O'Neil and M. J. Padgett, Opt. Commun. **193**, 45 (2001).

below the parametric resonance condition; see Fig. 3. Again, there is an excellent agreement between the observed and predicted particle motions. In particular, the parametric excitation of oscillations manifests as a narrowing of the peak (or a reduced apparent damping Γ, defined as the full width at half maximum), occurring when modulating at twice Ω_0. Close to parametric resonance, a shift in peak position Ω_p is also apparent. Both of these signatures confirm our interpretation of the system as being a Brownian parametric oscillator; see Fig. 4.

We recognize that our observations relate to the study of the lateral motion of the trapped droplets. In keeping with other work, we note from examination of the video images that the axial movement of the trapped droplet occurs on much longer time scales, corresponding to frequencies in the subkilohertz region. This slow axial dynamics is responsible for the low frequency component of the spectra in Fig. 3. This reflects the comparatively weak axial trapping, possibly arising from aberrations associated with nonoptimized objectives. It may be possible to use dough-

5

Studies in Colloid Science

A colloidal suspension is simply particulate matter dispersed in a fluid. The role that colloids play in our everyday lives is huge. Typical examples of colloids include blood, milk, paint, and aerosols. The forces between particles that make up colloidal suspensions are very important in determining their behavior, and optical tweezers are a convenient method to probe such forces. Here we consider the work carried out on nonbiological particles.

The original optical tweezers paper by Ashkin et al. [1] (see **PAPER 1.5**) was very accurate in its prediction of optical tweezers saying that:

> They also open a new size regime to optical trapping encompassing macromolecules, colloids, small aerosols, and possibly biological particles. The results are of relevance to proposals for the trapping and cooling of atoms by resonance radiation pressure.

However, the early years of the technique largely focused on the biological applications of the techniques in preference to anything else. The exceptions were some studies in the theoretical modeling of optical tweezers [2], [3] (see **PAPER 4.1**), an issue that is still relevant in the literature (e.g., [4] **PAPER 4.5**; see also Chapter 4).

One of the first papers to move away from biology (although ultimately with biological applications in mind) was that by Svoboda and Block [5] (see **PAPER 5.1**), which dealt with gold nanoparticles (Ashkin et al. [1], see **PAPER 1.5**, had shown that this was possible using dielectric particles). This paper showed definitively that metallic particles could be trapped by optical tweezers (which is not intuitively obvious if one considers scattering forces). The thinking was that because of the large scattering forces, metallic particles could not be trapped in a single-beam gradient trap. However, the large polarizability of metals means that there are large enough gradient forces to compensate for the scattering. Svoboda and Block showed that gold nanoparticles (36 nm in diameter) trapped more strongly, by a factor of ~ 7, than latex spheres of a similar size.

The ability to trap metallic particles (including nanorods [6]) has a number of applications and is of particular topical interest with the advent of plasmonics and nanophotonics [7,8]. Note that one must be careful at what wavelength the particles are trapped, as too much absorption causes significant heating, as shown quantitatively in Ref. [9]. The use of metal particles is useful in areas such as surface-enhanced Raman spectroscopy (SERS) [10] and when creating handles that can be easily functionalized. It is also possible to trap related structures such as nanowires [11].

Interfaces can be difficult to probe with optical tweezers, primarily as the surface tension is usually far greater than the forces produced by a trap. It is, however, possible to manipulate a monolayer at a surface [12] and, through combination with other techniques, make measurements of thermodynamic, electrical, and viscoelastic properties. Surface tension can also be used as an aid to trapping [13–15] by confining the particles in a single plane. One of the most interesting examples of this type of work is that of Aveyard et al. [16] (see **PAPER 5.2**), in which the interactions between charged particles lying at an oil–water interface are moved closer together using the tweezers until one of the particles falls out of the trap. By making note of the distances and powers involved in this measurement, the long-range repulsive force between particles is found to decay as d^{-4} where d is a measure of the particle separation.

The multiplexing of optical tweezers using scanning or holographic techniques allows the experimenter to create arrays of particles. This gives rise to various possibilities with colloidal particles: examining the interaction between particles, creating arrays for template studies, and the organization of large arrays into complex structures. At the simplest level, one can create structures by sticking beads to each other [17] but one of the great hopes of optical tweezers was that they could be used to fabricate complex three-dimensional colloidal structures. A major goal, or at least a suggested goal, is to create photonic crystal structures fabricated point by point with the easy addition of defects or different types of materials. This idea remains more honored in the breach rather than the observance, as the ability of tweezers to create large-scale structures is poor given their limited field of view. Also, for three-dimensional structures, the ability to build up layer upon layer is rather slow and laborious and cannot be achieved using interference patterns robustly beyond one or two layers;

Mio and Marr [18] suggested that a template for larger structures can be created by making use of an optically trapped array of particles. In this approach, a scanning optical tweezers is used to trap an array of particles in a photopolymerizable solution. The particles can thus be set in a fixed pattern to be used either as a mask in lithographic processes [19,20] or to act as a template for three-dimensional structures—that is, a template nucleus.

A demonstration using two-dimensional arrays to seed three-dimensional growth was shown by Hoogenboom et al. [21]. The authors use a scanning technique to pick up and move a large number of both silica particles and latex particles with a metallic core (thus allowing substrates of mixed particles to be created). The particles are positioned close to a surface using the tweezers and then adhere to the surface due to electrostatic attraction. The surface can then be dried to create a well-ordered, patterned substrate with a resolution lower than the particle size.

Holographic optical patterns offer flexible optical patterns to enable the study of particle interactions and effects such as ratcheting, colloidal transport [22–25], and optical sorting [26] (see **PAPER 2.1**). Korda et al. [27] (see **PAPER 5.3**) examined colloidal transport through a two-dimensional optical lattice (a pinscape). They showed that by appropriate orientation of the optical potential to the direction of the colloidal flow, and appropriate flow rates, particles can become "locked in" to the optical potential as opposed to the direction dictated by the force driving the flow. This idea underpins the notion of fractionation and optical sorting of colloidal particles, in which the interaction of the flow forces with the optical forces dictates the size and refractive-index sorting abilities of the lattice. Linking the trap sites also aids the process, and newer techniques using time-sharing patterns may offer more flexibility [28].

A natural extension of the pinscape work has been to examine lattices that move. These give rise to ratchetlike behavior in which transport through the lattice is carried out in the absence of flow and is in one direction. This type of moving lattice enables particle "pumping" [29] and flux reversal [30], and allows studies to be carried out exploring statistical physics using microscopic particles.

Particle interactions can be studied by making use of the properties of optical tweezers: either their force-sensing capabilities or simply their ability to hold particles in a controllable manner. In this latter category is the work of Crocker [31] (see **PAPER 5.4**), in which the hydrodynamic corrections to the Brownian motion of two spheres is calculated by making use of the ability of tweezers to localize the beads in the same plane and away from other beads and walls, interactions that would mask the measured effect. Crocker and Grier used a similar technique [32] to measure the pair interaction potential of colloidal particle pairs. In this work, a

dual-beam optical tweezers is used to trap two charge-stabilized colloidal particles (the preparation of particles in such experiments is critical). The tweezers are blinked on and off and the particles are allowed to freely diffuse. The subsequent particle motions are recorded on video and then the trajectories are analyzed. The interaction potential between the particles was found to follow the Derjaguin, Verway, Landau, and Overbeek (DVLO) theory of colloidal interaction. Although we note that optical tweezers play only a minor part in this work, they do have a critical role in the precise positioning and monitoring of particles. This is also true also in another key paper [33] that looks at the control of particles though changes in system entropy.

Another form of the dual-beam tweezers measurement of particle interactions was carried out by Meiners and Quake [34] (see **PAPER 5.5**), in which they directly measured the cross-correlations, due to hydrodynamic coupling between two particles in neighboring traps. So in contrast to the Crocker [31] and Crocker and Grier [32] work in which the tweezers are used to position the beads and then let them go, Meiners and Quake account for the trapping potential and make use of a quadrant-detector position measurement to examine the particle position. They show that low-Reynolds-number fluidic systems have a "memory," in that the particles' cross-correlation function shows an anticorrelation.

These key papers have ultimately led to numerous studies based on the techniques outlined, including looking at shorter timescale hydrodynamic interactions [35] and interactions in more complicated geometries, such as rings of particles [36], optical binding [37], and anomalous effects [38]. Interactions can also be used to study magnetic interactions between particles. In **PAPER 5.6**, Furst and Gast [39] examine the micromechanical properties of dipolar chains of particles, where the dipoles are magnetically induced. Viscous drag methods are used to measure trapping forces and spring constants. Known tensions can then be applied to the particle chains and information such as tensile strength and rupturing stress can be measured.

The ability to measure particle fluctuations is not limited to linear motion, and particle rotation must also be considered. Martin et al. [40] examine translation-rotation coupling occurring due to hydrodynamic interactions between spheres. Meiners and Quake [34] and Bartlett et al. [41] showed that the translation–translation coupling had a $1/R$ dependence, where R is the distance between the spheres. The rotation–translation coupling is a shorter range effect, with a $1/R^2$ dependence. To carry out the measurement, the birefringent particles are held and positioned with optical tweezers. The particle positions are measured using a video microscopy technique making use of a center-finding algorithm. To measure the rotational variations, a cross-polarizer-analyzer setup is used and the particles are trapped using linearly polarized light at a fixed angle to the polarizer orientation. As the particle rotates, the intensity of its image changes and this is a measure of the particle rotation angle. These measurements lead to autocorrelation and cross-correlation functions between the various linear and rotational parameters and give a complete description of the hydrodynamic interactions, which may be of use when dealing with more complex systems such as swimming biological samples, that is, cells with beating cilia or in active systems such as microfluidic chips.

Volpe and Petrov [42] carried out a similar type of measurement but instead of looking at particle interactions they examined the Brownian motion of a rotating bead with the torque applied to the bead by a beam carrying orbital angular momentum, thus combining many facets of optical tweezers work to date. By measuring the auto- and cross-correlation functions of the particle, an absolute measurement of the torque can be made. In this case, values lower than those previously reported were measured, 4×10^{-21} nm.

Tweezers also play a role in more general soft-matter physics, in particular in areas such as microrheology [43]. **PAPER 6.8** examines mechanical properties of both polymer and colloidal systems [44]. Processes such as clouding [45], molecular accumulation [46], and liquid crystal analysis [47–49] have all benefited from the ability of tweezers to hold microscopic particles and measure interaction forces.

One of the more unusual subdisciplines of optical tweezers to develop in recent years is that of optical binding. This is an interaction between particles and the optical field that allows particles to be arranged in defined, regularly spaced patterns. This effect was first observed by Burns et al. [50] (see **PAPER 5.7**) and is caused by the interaction between gradient-force trapping of the particles coupled with the light scattered from the particles and the interaction of this light on neighboring particles. The paper illustrated an effect that is now referred to as transverse optical binding, in which the particles lie in a plane orthogonal to the direction of the incident beam. In the experiment, a line trap is created by focusing a Gaussian beam through a set of cylindrical lenses. Two particles are trapped in the line and they have a fixed separation distance that is correlated to the particle motion—that is, if one particle moves, the other will follow suit. This work was expanded upon in Ref. [51].

Although the first observations of binding came at the end of the 1980s, the idea never really sparked the imagination of the trapping community. This changed with the observation of longitudinal binding, in which the bound array of particles lies along the beam propagation direction, by Tatarkova et al. [52] (see **PAPER 5.8**). In this experiment, particles were trapped in a dual counterpropagating beam trap and chains up to about nine particles in length were observed. Interactions due to particle charge and interference effects were ruled out and the chains were found to be caused due to scattering effects. Similar results were observed around the same time by Singer et al. [53].

Since then, optical binding has become an active research area largely with a pure scientific goal at heart: understanding how light and matter interact. There may be applications, with concepts such as colloidal arrays [54] (often formed using on a surface using evanescent fields [55]), enhanced spectroscopic effects [10], and photonic crystals being discussed in the literature [56,57]. Other notable work includes the observation of optical bistability in bound arrays [58] (see **PAPER 4.7**), the measurement of forces between bound particles [37], and the suggestion that binding has been observed in air [59–61]. The theoretical framework for binding is also under development and is an active area of research [56,57,62–68].

The power of optical tweezers in the area of colloidal interactions is very elegantly shown by the ability to measure Casimir forces. This work carried out by the Bechinger group [69] makes use of an optically trapped colloidal bead that is brought very close to a surface. Total internal reflection microscopy (TIRM) is then used to measure forces analogous to the quantum electrodynamical Casimir force. These forces arise when two surfaces come close to each other in the presence of a binary (two-phase) liquid that is close to its critical point, where the liquids will demix. The forces are then thermodynamic in nature and occur due to concentration fluctuations that are confined between the two surfaces when they are close enough together. This measurement shows both the power and utility of optical tweezers and how they can be combined with other techniques to provide sensitive positioning and measurement devices. Furthermore, it illustrates that tweezers have a significant role to play in examining colloidal particles and their interactions in the coming years.

Endnotes

1. A. Ashkin, J.M. Dziedzic, J.E. Bjorkholm, and S. Chu, Observation of a single-beam gradient force optical trap for dielectric particles, *Opt. Lett.* **11**, 288–290 (1986).
2. W.H. Wright, G.J. Sonek, and M.W. Berns, Radiation trapping forces on microspheres with optical tweezers, *Appl. Phys. Lett.* **63**, 715–717 (1993).
3. W.H. Wright, G.J. Sonek, and M.W. Berns, Parametric study of the forces on microspheres held by optical tweezers, *Appl. Opt.* **33**, 1735–1748 (1994).
4. T.A. Nieminen, V.L.Y. Loke, A.B. Stilgoe, G. Knoner, A.M. Branczyk, N.R. Heckenberg, and H. Rubinsztein-Dunlop, Optical tweezers computational toolbox, *J. Opt. A—Pure Appl. Opt.* **9**, S196–S203 (2007).

5. K. Svoboda and S.M. Block, Optical trapping of metallic Rayleigh particles, *Opt. Lett.* **19**, 930–932 (1994).

6. M. Pelton, M.Z. Liu, H.Y. Kim, G. Smith, P. Guyot-Sionnest, and N.E. Scherer, Optical trapping and alignment of single gold nanorods by using plasmon resonances, *Opt. Lett.* **31**, 2075–2077 (2006).

7. A.S. Zelenina, R. Quidant, and M. Nieto-Vesperinas, Enhanced optical forces between coupled resonant metal nanoparticles, *Opt. Lett.* **32**, 1156–1158 (2007).

8. G. Volpe, R. Quidant, G. Badenes, and D. Petrov, Surface plasmon radiation forces, *Phys. Rev. Lett.* **96**, 238101 (2006).

9. Y. Seol, A.E. Carpenter, and T.T. Perkins, Gold nanoparticles: Enhanced optical trapping and sensitivity coupled with significant heating, *Opt. Lett.* **31**, 2429–2431 (2006).

10. F. Svedberg, Z.P. Li, H.X. Xu, and M. Kall, Creating hot nanoparticle pairs for surface-enhanced Raman spectroscopy through optical manipulation, *Nano Lett.* **6**, 2639–2641 (2006).

11. P.J. Pauzauskie, A. Radenovic, E. Trepagnier, H. Shroff, P.D. Yang, and J. Liphardt, Optical trapping and integration of semiconductor nanowire assemblies in water, *Nat. Mater.* **5**, 97–101 (2006).

12. S. Wurlitzer, C. Lautz, M. Liley, C. Duschl, and T.M. Fischer, Micromanipulation of Langmuir-monolayers with optical tweezers, *J. Phys. Chem. B.* **105**, 182–187 (2001).

13. R. Dasgupta, S. Ahlawat, and P.K. Gupta, Trapping of micron-sized objects at a liquid-air interface, *J. Opt. A—Pure Appl. Opt.* **9**, S189–S195 (2007).

14. A. Jesacher, S. Furhapter, C. Maurer, S. Bernet, and M. Ritsch-Marte, Holographic optical tweezers for object manipulations at an air-liquid surface, *Opt. Express* **14**, 6342–6352 (2006).

15. A. Jesacher, S. Furhapter, C. Maurer, S. Bernet, and M. Ritsch-Marte, Reverse orbiting of microparticles in optical vortices, *Opt. Lett.* **31**, 2824–2826 (2006).

16. R. Aveyard, B.P. Binks, J.H. Clint, P.D.I. Fletcher, T.S. Horozov, B. Neumann, V.N. Paunov, J. Annesley, S.W. Botchway, D. Nees, A.W. Parker, A.D. Ward, and A.N. Burgess, Measurement of long-range repulsive forces between charged particles at an oil-water interface, *Phys. Rev. Lett.* **88**, 246102 (2002).

17. S.H. Xu, Y.M. Li, L.R. Lou, H.T. Chen, and Z.W. Sun, Steady patterns of microparticles formed by optical tweezers, *Jpn. J. Appl. Phys. Part 1* **41**, 166–168 (2002).

18. C. Mio and D.W.M. Marr, Tailored surfaces using optically manipulated colloidal particles, *Langmuir* **15**, 8565–8568 (1999).

19. D.L.J. Vossen, D. Fific, J. Penninkhof, T. van Dillen, A. Polman, and A. van Blaaderen, Combined optical tweezers/ion beam technique to tune colloidal masks for nanolithography, *Nano Lett.* **5**, 1175–1179 (2005).

20. A. van Blaaderen, J.P. Hoogenboom, D.L.J. Vossen, A. Yethiraj, A. van der Horst, K. Visscher, and M. Dogterom, Colloidal epitaxy: Playing with the boundary conditions of colloidal crystallization, *Faraday Discuss.* **123**, 107–119 (2003).

21. J.P. Hoogenboom, D.L.J. Vossen, C. Faivre-Moskalenko, M. Dogterom, and A. van Blaaderen, Patterning surfaces with colloidal particles using optical tweezers, *Appl. Phys. Lett.* **80**, 4828–4830 (2002).

22. C.J.O. Reichhardt and C. Reichhardt, Ratchet cellular automata for colloids in dynamic traps, *Europhys. Lett.* **74**, 792–798 (2006).

23. A. Libal, C. Reichhardt, and C.J.O. Reichhardt, Realizing colloidal artificial ice on arrays of optical traps, *Phys. Rev. Lett.* **97**, 228302 (2006).

24. A. Libal, C. Reichhardt, B. Janko, and C.J.O. Reichhardt, Dynamics, rectification, and fractionation for colloids on flashing substrates, *Phys. Rev. Lett.* **96**, 188301 (2006).

25. Y. Roichman, V. Wong, and D.G. Grier, Colloidal transport through optical tweezer arrays, *Phys. Rev. E* **75**, 011407 (2007).

26. M.P. MacDonald, G.C. Spalding, and K. Dholakia, Microfluidic sorting in an optical lattice, *Nature* **426** 421–424 (2003).

27. P.T. Korda, M.B. Taylor, and D.G. Grier, Kinetically locked-in colloidal transport in an array of optical tweezers, *Phys. Rev. Lett.* **89**, 128301 (2002).

28. G. Milne, D. Rhodes, M. MacDonald, and K. Dholakia, Fractionation of polydisperse colloid with acousto-optically generated potential energy landscapes, *Opt. Lett.* **32**, 1144–1146 (2007).

29. B.A. Koss and D.G. Grier, Optical peristalsis, *Appl. Phys. Lett.* **82**, 3985–3987 (2003).

30. S.H. Lee and D.G. Grier, Flux reversal in a two-state symmetric optical thermal ratchet, *Phys. Rev. E* **71**, 060102 (2005).

31. J.C. Crocker, Measurement of the hydrodynamic corrections to the Brownian motion of two colloidal spheres, *J. Chem. Phys.* **106**, 2837–2840 (1997).

32. J.C. Crocker and D.G. Grier, Microscopic measurement of the pair interaction potential of charge-stabilized colloid, *Phys. Rev. Lett.* **73**, 352–355 (1994).

33. A.D. Dinsmore, A.G. Yodh, and D.J. Pine, Entropic control of particle motion using passive surface microstructures, *Nature* **383**, 239–242 (1996).

34. J.C. Meiners and S.R. Quake, Direct measurement of hydrodynamic cross correlations between two particles in an external potential, *Phys. Rev. Lett.* **82**, 2211–2214 (1999).

35. S. Henderson, S. Mitchell, and P. Bartlett, Propagation of hydrodynamic interactions in colloidal suspensions, *Phys. Rev. Lett.* **88**, 088302 (2002).

36. R. Di Leonardo, S. Keen, J. Leach, C.D. Saunter, G.D. Love, G. Ruocco, and M.J. Padgett, Eigenmodes of a hydrodynamically coupled micron-size multiple-particle ring, *Phys. Rev. E* **76**, 061402 (2007).

37. N.K. Metzger, R.F. Marchington, M. Mazilu, R.L. Smith, K. Dholakia, and E.M. Wright, Measurement of the restoring forces acting on two optically bound particles from normal mode correlations, *Phys. Rev. Lett.* **98**, 068102 (2007).

38. M. Polin, D.G. Grier, and S.R. Quake, Anomalous vibrational dispersion in holographically trapped colloidal arrays, *Phys. Rev. Lett.* **96**, 088101 (2006).

39. E.M. Furst and A.P. Gast, Micromechanics of dipolar chains using optical tweezers, *Phys. Rev. Lett.* **82**, 4130–4133 (1999).

40. S. Martin, M. Reichert, H. Stark, and T. Gisler, Direct observation of hydrodynamic rotation-translation coupling between two colloidal spheres, *Phys. Rev. Lett.* **97**, 248301 (2006).

41. P. Bartlett, S.I. Henderson, and S.J. Mitchell, Measurement of the hydrodynamic forces between two polymer-coated spheres, *Philos. Trans. R. Soc. A—Math. Phys. Eng. Sci.* **359**, 883–893 (2001).

42. G. Volpe and D. Petrov, Torque detection using Brownian fluctuations, *Phys. Rev. Lett.* **97**, 210603 (2006).

43. A.I. Bishop, T.A. Nieminen, N.R. Heckenberg, and H. Rubinsztein-Dunlop, Optical microrheology using rotating laser-trapped particles, *Phys. Rev. Lett.* **92**, 198104 (2004).

44. E.M. Furst, Applications of laser tweezers in complex fluid rheology, *Curr. Opin. Colloid Interface Sci.* **10**, 79–86 (2005).

45. P. Luchette, N. Abiy, and H.B. Mao, Microanalysis of clouding process at the single droplet level, *Sens. Actuator B—Chem.* **128**, 154–160 (2007).

46. W. Singer, T.A. Nieminen, N.R. Heckenberg, and H. Rubinsztein-Dunlop, Collecting single molecules with conventional optical tweezers, *Phys. Rev. E* **75**, 011916 (2007).

47. I.I. Smalyukh, A.N. Kuzmin, A.V. Kachynski, P.N. Prasad, and O.D. Lavrentovich, Optical trapping of colloidal particles and measurement of the defect line tension and colloidal forces in a thermotropic nematic liquid crystal, *Appl. Phys. Lett.* **86**, 021913 (2005).

48. M. Yada, J. Yamamoto, and H. Yokoyama, Direct observation of anisotropic interparticle forces in nematic colloids with optical tweezers, *Phys. Rev. Lett.* **92**, 185501 (2004).

49. S.A. Tatarkova, D.R. Burnham, A.K. Kirby, G.D. Love, and E.M. Terentjev, Colloidal interactions and transport in nematic liquid crystals, *Phys. Rev. Lett.* **98**, 157801 (2007).

50. M.M. Burns, J.M. Fournier, and J.A. Golovchenko, Optical binding, *Phys. Rev. Lett.* **63**, 1233–1236 (1989).

51. M.M. Burns, J.M. Fournier, and J.A. Golovchenko, Optical matter: Crystallization and binding in intense optical fields, *Science* **249**, 749–754 (1990).

52. S.A. Tatarkova, A.E. Carruthers, and K. Dholakia, One-dimensional optically bound arrays of microscopic particles, *Phys. Rev. Lett.* **89**, 283901 (2002).

53. W. Singer, M. Frick, S. Bernet, and M. Ritsch-Marte, Self-organized array of regularly spaced microbeads in a fiber-optical trap, *J. Opt. Soc. Am. B—Opt. Phys.* **20**, 1568–1574 (2003).

54. C.D. Mellor and C.D. Bain, Array formation in evanescent waves, *Chemphyschem.* **7**(2), 329–332 (2006).

55. C.D. Mellor, T.A. Fennerty, and C.D. Bain, Polarization effects in optically bound particle arrays, *Opt. Express* **14**, 10079–10088 (2006).

56. D. Maystre and P. Vincent, Making photonic crystals using trapping and binding optical forces on particles, *J. Opt. A—Pure Appl. Opt.* **8**, 1059–1066 (2006).

57. J. Ng, Z.F. Lin, C.T. Chan, and P. Sheng, Photonic clusters formed by dielectric microspheres: Numerical simulations, *Phys. Rev. B* **72**, 085130 (2005).

58. N.K. Metzger, K. Dholakia, and E.M. Wright, Observation of bistability and hysteresis in optical binding of two dielectric spheres, *Phys. Rev. Lett.* **96**, 068102 (2006).

59. M. Guillon, O. Moine, and B. Stout, Erratum: Longitudinal optical binding of high optical contrast microdroplets in air, *Phys. Rev. Lett.* **99**, 079901 (2007).

60. M. Guillon, O. Moine, and B. Stout, Longitudinal optical binding of high optical contrast microdroplets in air, *Phys. Rev. Lett.* **96**, 143902 (2006).

61. M.D. Summers, J.P. Reid, and D. McGloin, Optical guiding of aerosol droplets, *Opt. Express* **14**, 6373–6380 (2006).

62. V. Karasek and P. Zemanek, Analytical description of longitudinal optical binding of two spherical nanoparticles, *J. Opt. A—Pure Appl. Opt.* **9**, S215-S220 (2007).

63. J. Ng and C.T. Chan, Localized vibrational modes in optically bound structures, *Opt. Lett.* **31**, 2583–2585 (2006).

64. N.K. Metzger, E.M. Wright, and K. Dholakia, Theory and simulation of the bistable behaviour of optically bound particles in the Mie size regime, *New J. Phys.* **8** (2006).

65. V. Karasek, K. Dholakia, and P. Zemanek, Analysis of optical binding in one dimension, *Appl. Phys. B—Lasers Opt.* **84**, 149–156 (2006).

66. T.M. Grzegorczyk, B.A. Kemp, and J.A. Kong, Stable optical trapping based on optical binding forces, *Phys. Rev. Lett.* **96**, 113903 (2006).

67. T.M. Grzegorczyk, B.A. Kemp, and J.A. Kong, Passive guiding and sorting of small particles with optical binding forces, *Opt. Lett.* **31**, 3378–3380 (2006).

68. D. McGloin, A.E. Carruthers, K. Dholakia, and E.M. Wright, Optically bound microscopic particles in one dimension, *Phys. Rev. E* **69**, 021403 (2004).

69. C. Hertlein, L. Helden, A. Gambassi, S. Dietrich, and C. Bechinger, Direct measurement of critical Casimir forces, *Nature* **451**, 172–175 (2008).

Optical trapping of metallic Rayleigh particles

Karel Svoboda and Steven M. Block*

Rowland Institute for Science, Cambridge, Massachusetts 02142

Received February 15, 1994

Metallic objects reflect light and have generally been considered poor candidates for optical traps, particularly with optical tweezers, which rely on a gradient force to provide trapping. We demonstrate that stable trapping can occur with optical tweezers when they are used with small metallic Rayleigh particles. In this size regime, the scattering pictures for metals and dielectrics are similar, and the larger polarizability of metals implies that trapping forces are greater. The latter fact makes the use of metal particles attractive for certain biological applications. Comparison of trapping forces for latex and gold spheres demonstrates that the gradient force is the major determinant of trapping strength and that competing effects, such as scattering or radiometric forces, are relatively minor.

The single-beam gradient force optical trap, or optical tweezers,[1,2] is a powerful tool capable of remotely manipulating dielectric objects, with applications to biology,[3] chemistry,[4] and physics.[5] One fruitful approach has been to attach dielectric microspheres to biological material. Latex (polystyrene) or silica beads are highly refractile and hence are strongly trapped by optical tweezers.[3] As such, they function as handles. Dielectric beads that range in size from 25 nm to tens of micrometers[2] can be trapped.

Radiation pressure forces can be decomposed into several components.[3] Scattering and absorption forces are due to momentum transfer of scattered and absorbed photons, respectively. These forces are proportional to the light intensity and point along the direction of the incident beam, tending to destabilize the trap. Optical tweezers owe their trapping instead to the gradient force, which is proportional to the gradient of light intensity and points toward the focus. For stable trapping, the gradient force must overcome the scattering and absorption forces pointing down the beam axis. Although much effort has gone into the computation of optical forces,[6–8] quantitative comparisons with experiment have been largely unsatisfactory, except for large particles.[8,9] This may be due to the fact that small spherical aberrations, axial misalignments, etc. can significantly degrade trapping. Destabilization of trapping by radiometric forces, which are due to asymmetric heating,[10] has also been proposed.[3,8,11]

Because of the relatively large scattering and absorption forces, it has not generally been believed that metallic particles could be trapped with optical tweezers,[12,13] although the manipulation of gold handles in the plane of cell membranes has been reported.[14] Indeed, optical arrangements that rely on repulsive antitrapping effects have been proposed to cope with metals.[13,15] Here we show that gold particles can be trapped by optical tweezers in three dimensions in the Rayleigh regime ($\lambda >> a$, where a is the particle radius). We compare trapping forces of 36-nm-diameter gold and 38-nm-diameter latex particles and show that the ratio of trapping forces equal the ratio of polarizabilities, with gold particles trapping more strongly by a factor of ~7.

The trapping apparatus[16] incorporates interferometric position detection[17] of trapped particles (Fig. 1). Polarized laser light from a diode-pumped Nd:YLF laser (TFR, Spectra-Physics; $\lambda = 1.047$ μm, 2.5 W, TEM$_{00}$ mode) is coupled into an optical fiber and introduced into a modified inverted microscope (Axiovert 35, Zeiss) below the objective Wollaston prism. We computed the power at the specimen, P, by measuring the external power and multiplying it by an attenuation factor, determined separately.[3] The prism produces two beams with orthogonal polarization. These are focused by an objective (Plan Neofluar 100×, 1.3 N.A.) to two overlapping, diffraction-limited spots, whose centers are separated by ~250 nm. The two beams act as a single optical trap. A trapped object located asymmetrically between the two spots introduces a relative phase shift between the two beams. When the beams interfere in the condenser back focal plane, elliptically polarized light results. The degree of this ellipticity, given by the output voltage V_{out}, provides a sensitive measure of displacement along the Wollaston shear direction. We determined trapping of latex (Duke Scientific) or gold (Ted Pella, Inc.) particles by measuring the mean-square (MS) voltage output of the interferometer. When particles diffused into the trap, the MS voltage increased in a stepwise fashion (Fig. 1). The step height was independent of the number of particles already in the trap, indicating that trapped objects behaved as independent Brownian particles in a potential well. Trapped gold particles were simultaneously observed with video-enhanced differential-interference-contrast (DIC) microscopy [Fig. 2(a)].

Single 36-nm-diameter gold particles were trapped stably in three dimensions (Fig. 2). Smaller gold particles (nominally 28 nm) could also be trapped, but these consisted of mainly triangular disks, probably single crystals, and were not useful for quantitative analysis (data not shown). Below $P \sim 100$ mW, 36-nm particles were trapped for only ~5 s before escaping the trap. Higher powers were required for 38-nm latex particles. The dynamics of a trapped particle correspond to Brownian motion in a harmonic

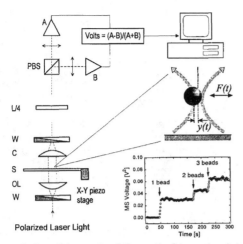

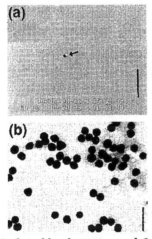

Fig. 1. Left: Schematic of the optical trapping interferometer. W's, Wollaston prisms; OL, objective lens; S, specimen plane; C, condenser; L/4, quarter-wave plate; PBS, polarizing beam-splitting cube; A, B, photodetectors/amplifiers. Center right: Schematic of the trapping geometry. The position of the bead with respect to the beam center, $y(t)$, is proportional to the normalized difference voltage; $F(t)$ is the force due to the moving fluid. Bottom right: MS difference over sum voltage [$\propto \langle y^2(t) \rangle$] of the position detector as a function of time. The trace shows stepwise increments due to the successive arrival of three gold particles. The points between steps are an averaging artifact and the variability in step height is due to particle heterogeneity.

Fig. 2. (a) Single gold sphere, trapped 2 μm from the coverglass, viewed with video-enhanced DIC microscopy (scale bar, 5 μm). (b) Transmission electron micrograph of gold spheres (scale bar, 100 nm). The particle diameter was 36.2 ± 2 nm (mean ± SD). Few aggregates were observed.

potential.[3] The power spectrum of position fluctuations is Lorentzian, with a corner frequency $f_0 = \kappa(2\pi\beta)^{-1}$, where κ is the effective spring constant of the trap, $\beta = 6\pi\mu a$ is the drag coefficient of the particle, and μ is the viscosity of water. By fitting the power spectrum, we could measure the value of κ along the shear direction (Fig. 3). We found $\kappa/P = (5.0 \pm 0.3) \times 10^{-6}$ pN nm^{-1} mW^{-1} [mean ± standard error of the mean (SEM); $N = 10$]. To measure the maximal (escape) force, F_{esc}, we moved the stage in a

sinusoidal fashion (amplitude $x_0 = 1$ μm) along the shear direction while keeping the trapping beam stationary. The escape force is $F_{esc} = 2\pi x_0 \beta f_{max}$, where f_{max} is the frequency at which the particle escapes the trap. The ratio of the trapping force for gold to latex particles is 6.58 ± 1.18 (mean ± SEM; $N = 15$) (Table 1). The error in the force measurements is accounted for by particle-size heterogeneity (diameter standard deviation; ±11% latex, ±6% gold).

Absolute trapping forces depend on details of the beam structure and are difficult to compute, but relative forces for different materials in the Rayleigh regime can easily be compared. For dielectric Rayleigh particles, the amplitude of the illuminating light is constant over the extent of the particle, and the electrostatic approximation can be used to compute particle polarizabilities.[18] The polarizability is

$$\alpha = 3V \frac{\hat{\varepsilon} - \varepsilon_m}{\hat{\varepsilon} + 2\varepsilon_m}, \quad (1)$$

where $\hat{\varepsilon}(\lambda) = \varepsilon_1(\lambda) + i\varepsilon_2(\lambda)$ and $\varepsilon_m = n_m^2$ are the dielectric constants of the particle and the surrounding medium, respectively, and $V = 4\pi a^3/3$ is the sphere volume. Equation (1) is valid when the skin depth $\delta = \lambda/2\pi k$ obeys $\delta \gg a$. Here k is the imaginary part of the index of refraction, $k = (-\varepsilon_1/2 + \sqrt{\varepsilon_1^2 + \varepsilon_2^2}/2)^{1/2}$. For gold, the dielectric constant at $\lambda = 1.047$ μm is $\hat{\varepsilon} = -54 + 5.9i$,[19] and $\delta = 23$ nm. Since $\delta \approx a = 18.1$ nm, we correct for the attenuation of the field by replacing V with $V' = 4\pi \int_0^a r^2 \exp[(r - a)/\delta] dr$ in Eq. (1). The scattering force is given by $F_{scat} = n_m \langle S \rangle C_{scat}/c$, where $C_{scat} = k_m^4 |\alpha|^2/4\pi$ is the scattering cross section, $\langle S \rangle$ is the time-averaged Poynting vector, and $k_m = 2\pi n_m/\lambda$ is the wave number in the surrounding medium. The absorption force is $F_{abs} = n_m \langle S \rangle C_{abs}/c$, where $C_{abs} = k_m \mathrm{Im}(\alpha)$ is the absorption cross section. The gradient force is $F_{grad} = (|\alpha|/2)\nabla\langle E^2 \rangle$. The phase shift between applied field and induced dipole is $\tan^{-1}[\mathrm{Im}(\alpha)/\mathrm{Re}(\alpha)] \approx 0.01$ and can be neglected in computing the time average.

Estimates of the various optical forces lead to insights into why metallic Rayleigh particles can be trapped. We assume the intensity profile to be $I(r) = I_0 \exp(-r^2/2r_0^2)$, where r_0 is the beam radius. The maximal gradient force then is $F_g^{(m)} = I_0 |\alpha| \sqrt{e}/(\varepsilon_m c r_0)$, the maximal scattering

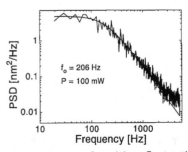

Fig. 3. Power spectrum of position fluctuations for a gold particle trapped at 100-mW power, 2 μm from the surface. The spectrum was normalized such that $\langle y^2(t) \rangle = 2\pi \int_0^\infty \mathrm{PSD}(f) df$, where PSD is the power spectral density.

Table 1. Computed Polarizabilities of Gold and Latex Particles

	Particles		
Quantity	36.2-nm Gold	38.0-nm Latex	Ratio
F_{max}/P	$5.80 \pm 0.36 \times 10^{-4}$ pN/mW	$0.88 \pm 0.11 \times 10^{-4}$ pN/mW	6.58 ± 1.18
$\alpha = 3V'(\hat{\varepsilon} - \varepsilon_m)/(\hat{\varepsilon} + 2\varepsilon_m)$	7.33×10^{-17} cm^3	1.03×10^{-17} cm^3	7.12

force is $F_{\mathrm{scat}}^{(m)} = 8\pi^3 I_0 \alpha^2 \varepsilon_m /(3c\lambda^4)$, and the absorption force is $F_{\mathrm{abs}}^{(m)} = 2\pi \varepsilon_m I_0 \operatorname{Im}(\alpha)/\lambda c$. Note that $F_g^{(m)}/F_{\mathrm{scat}}^{(m)} \propto F_{\mathrm{abs}}^{(m)}/F_{\mathrm{scat}}^{(m)} \propto a^{-3}$, that is, the gradient and absorption forces gain rapidly in significance as the particle size decreases. Using $r_0 = 0.5$ μm, $a = 20$ nm, and $\lambda = 1.05$ μm, we find that $F_g^{(m)} \approx 50 F_{\mathrm{scat}}^{(m)}$ and $F_{\mathrm{abs}}^{(m)} \approx F_{\mathrm{scat}}^{(m)}$. We estimate that $F_g^{(m)}/P \approx 4.8 \times 10^{-4}$ pN mW^{-1}, similar to the measured escape force (Table 1). This value is also in agreement with computations that use a fifth-order approximation of a high-N.A. Gaussian beam profile (data not shown).[9] Thus conventional theory predicts stable trapping of 40-nm and smaller gold particles. For larger conducting particles, two effects conspire to destabilize trapping. First, $F_{\mathrm{scat}}^{(m)}$ increases in significance compared with $F_g^{(m)}$. Second, because of the smallness of the skin depth, $F_g^{(m)}$ will grow with a weaker dependence than a^3, eventually settling on a^2.

Beyond a favorable force balance, trap energies must be significant compared with thermal energies to achieve stable trapping.[2] In an overdamped system, the escape time from a potential well of depth $U = I_0 |\alpha|/cn_m$ is given by $\tau_{\mathrm{esc}} = \tau_0 \exp(U/kT)$,[20] where $\tau_0 = \beta/\kappa$. For $P = 100$ mW and a Gaussian beam profile, $U/kT \approx 8$. Together with the measured trap stiffness, this predicts $\tau_{\mathrm{esc}} \approx 2$ s, close to the observed value (≈ 5 s).

The maximal gradient force is proportional to the polarizability of the Rayleigh particle. The computed polarizabilities for gold and latex particles are summarized in Table 1. Their ratio is 7.12, equal to the ratio of trapping forces (6.58 ± 1.18). The absorption coefficient, hence the radiometric force contribution, is 10^6-fold larger for gold than for latex particles at $\lambda = 1.047$ μm.[19,21] Because the difference in polarizabilities alone accounts for the difference in trapping force, we conclude that radiometric forces are insignificant in this size regime.

The fact that small gold particles can be trapped more strongly than dielectric particles is noteworthy. In many applications it may be advantageous to use the smallest possible handles, consistent with trapping the particles with appreciable force. An additional advantage of gold particles as handles is that they can be readily modified to provide diverse surface chemistries, including antibodies and organic coupling reagents. Possibly, nanofabricated gold tips may find application in the promising technique of optical trapping force microscopy,[22] where small, pointed probes at high trapping stiffnesses would be ideal. Gold particles can be trapped in high-index-of-refraction solvents. Finally, the measurement of trapping forces may be useful to measure the dielectric constants of colloidal particles.

This research was supported by the Rowland Institute for Science. We thank Joel Parks for enlightening discussions, Paul Burnett for help with electron microscopy, Mary Ann Nilsson and Jay Scarpetti for photography, and Bill Wright for force calculations with a fifth-order approximation to a high-N.A. Gaussian mode structure.

*Present address, Department of Molecular Biology, Princeton University, Princeton, New Jersey 08544.

Karel Svoboda is also with the Committee on Biophysics, Harvard University, Cambridge, Massachusetts 02138.

References

1. A. Ashkin, Phys. Rev. Lett. **40**, 729 (1978).
2. A. Ashkin, J. M. Dziedzic, J. E. Bjorkholm, and S. Chu, Opt. Lett. **11**, 288 (1986).
3. K. Svoboda and S. M. Block, Ann. Rev. Biophys. Biomol. Struct. **23**, 247 (1994).
4. H. Misawa, N. Kitamura, and H. Masuhara, J. Am. Chem. Soc. **113**, 7859 (1991).
5. A. Simon and A. Libchaber, Phys. Rev. Lett. **68**, 3375 (1992).
6. J. P. Barton, D. R. Alexander, and S. A. Schaub, J. Appl. Phys. **66**, 4594 (1989).
7. A. Ashkin, Biophys. J. **61**, 569 (1992).
8. W. H. Wright, G. J. Sonek, and M. W. Berns, Appl. Phys. Lett. **63**, 715 (1993).
9. W. H. Wright, G. J. Sonek, and M. W. Berns, Appl. Opt. **33**, 1735 (1994).
10. G. Hettner, Z. Phys. **37**, 179 (1926).
11. A. Ashkin and J. M. Dziedzic, Appl. Phys. Lett. **28**, 333 (1976).
12. A. Ashkin, Science **210**, 1081 (1980).
13. K. Sasaki, M. Koshioka, H. Misawa, N. Kitamura, and H. Masuhara, Appl. Phys. Lett. **60**, 807 (1992).
14. M. Edidin, S. C. Kuo, and M. P. Sheetz, Science **254**, 1379 (1991).
15. G. Roosen and C. Imbert, Opt. Commun. **26**, 432 (1978).
16. K. Svoboda, C. F. Schmidt, B. J. Schnapp, and S. M. Block, Nature (London) **365**, 721 (1993).
17. W. Denk and W. W. Webb, Appl. Opt. **29**, 2382 (1990).
18. C. F. Bohren and D. R. Huffman, *Absorption and Scattering of Light by Small Particles* (Wiley, New York, 1983).
19. H. J. Hagemann, W. Gudat, and C. Kunz, "Optical constants from the far infrared to the x-ray region: Mg, Al, Cu, Ag, Au, Bi, C, and Al$_2$O$_3$," DESY Rep. SR-74/7 (Hamburg, Germany, 1974), Table 5, p. 1.
20. N. G. van Kampen, *Stochastic Processes in Physics and Chemistry* (North-Holland, Amsterdam, 1992), p. 349.
21. R. H. Boundy and R. F. Boyer, eds., *Styrene, Its Polymers, Copolymers and Derivatives* (Reinhold, New York, 1952).
22. I. P. Ghislain and W. W. Webb, Opt. Lett. **18**, 1678 (1993).

Measurement of Long-Range Repulsive Forces between Charged Particles at an Oil-Water Interface

R. Aveyard, B. P. Binks, J. H. Clint, P. D. I. Fletcher,* T. S. Horozov, B. Neumann, and V. N. Paunov

Surfactant & Colloid Group, Department of Chemistry, University of Hull, Hull HU6 7RX, United Kingdom

J. Annesley, S. W. Botchway, D. Nees, A. W. Parker, and A. D. Ward

CLRC, Rutherford Appleton Laboratory, Chilton, Didcot, Oxon OX11 0QX, United Kingdom

A. N. Burgess

ICI PLC, P.O. Box 90, Wilton Centre, Middlesborough, Cleveland TS90 8JE, United Kingdom

(Received 11 October 2001; published 3 June 2002)

Using a laser tweezers method, we have determined the long-range repulsive force as a function of separation between two charged, spherical polystyrene particles ($2.7\,\mu$m diameter) present at a nonpolar oil-water interface. At large separations (6 to 12 μm between particle centers) the force is found to decay with distance to the power -4 and is insensitive to the ionic strength of the aqueous phase. The results are consistent with a model in which the repulsion arises primarily from the presence of a very small residual electric charge at the particle-oil interface. This charge corresponds to a fractional dissociation of the total ionizable (sulfate) groups present at the particle-oil surface of approximately 3×10^{-4}.

DOI: 10.1103/PhysRevLett.88.246102   PACS numbers: 68.05.Gh, 41.20.Cv, 42.50.Vk, 82.70.Dd

We have used a laser tweezers method to investigate the very long-range repulsive forces which have recently been observed to operate between particles present in monolayers at the nonpolar oil-water interface [1]. Such forces have general relevance to the properties of emulsions stabilized by adsorbed solid particles. Interparticle repulsion leads to high order in relatively dilute monolayers; an image of a monolayer of spherical polystyrene latex particles at the interface between water and a mixture of decane and undecane is shown in Fig. 1. The particles, which carry ionizable sulfate groups at their surfaces, have a diameter of 2.7 μm.

It has been shown [2,3] that the electrostatic repulsion between charged particles at an interface between water and a medium of low relative permittivity (air or oil) is enhanced over that between similar particles in bulk water. The origin of the effect is due to the asymmetry of the electrical double layer at that part of the particle surface in the aqueous phase, which gives rise to an effective dipole normal to the liquid surface. The dipole-dipole repulsion through the phase of low relative permittivity is longer range than that through the aqueous (electrolyte) phase, the latter being subject to counterion screening. Theoretical treatments of the phenomenon have subsequently been given by Hurd [4], by Earnshaw [5], and by Goulding and Hansen [6].

It has been demonstrated [1], however, that the "long range" repulsion arising from asymmetric double layers of adsorbed particles is insufficient to give rise to the kind of structuring seen in Fig. 1. The ultralong range repulsion between polystyrene particles has been attributed to a small amount of surface electric charge present at the oil/particle interface. We believe that this charge originates from the surface sulfate groups, a very small fraction

of which is stabilized by water trapped on the rough particle surface. The interparticle repulsion between two particles is thought to arise from unscreened charge-charge interactions acting through the oil phase together with interactions involving the charge on one particle and the image charge (in the water phase) of the other particle. Importantly, the repulsion between particles at the oil/water interface is insensitive to the presence of even high concentrations of electrolyte (1 mol dm^{-3}) in the aqueous phase. This observation is contrary to the predictions of the interparticle forces based on dipole-dipole interactions which

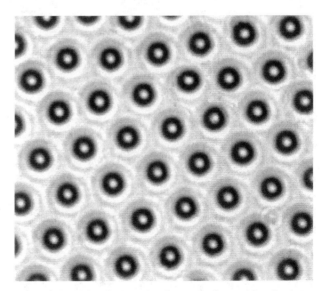

FIG. 1. Optical micrograph of a monolayer of polystyrene spheres, diameter 2.7 μm, at the interface between water and a mixture of decane and undecane in the trough (see Fig. 2); the average distance between particle centers is 5.8 μm. In the force experiments much more dilute films have been used.

are expected to decrease strongly when the ionic strength of the aqueous phase is increased [4]. Most recently Sun and Stirner have carried out molecular dynamics simulations of charged colloidal monolayers, and included finite size effects in their analysis [7]. They conclude that the long-range repulsion is mainly controlled by charge-charge interactions.

Direct measurements of the interaction potentials between two charged colloidal particles in the bulk of an aqueous suspension have been made as a function of particle separation by Crocker and Grier using an optical tweezers method [8]. We report in this Letter the first measurements using optical tweezers of the forces between a pair of charged polystyrene spheres (diameter 2.7 μm), trapped at the interface between an alkane mixture and water, as a function of the particle separation.

The polystyrene particle monolayers (see Fig. 1) were obtained by spreading the particles from a dispersion in a mixture of (mainly) isopropanol and water (300:1 by volume, respectively), at the interface between water and a mixture of decane (70.5 vol%) and undecane (29.5 vol%) in a small polytetrafluorethylene (PTFE) trough (see Fig. 2). The particles were monodisperse (2.7 μm diameter) and had a surface charge density, arising from the terminal sulfate groups, of 8.9 μC cm^{-2} (value supplied by the particle manufacturer Interfacial Dynamics Corp., USA). Force measurements were made for aqueous phases of either pure water (ionic strength $\approx 3 \times 10^{-6}$ M) or water containing 1 mM NaCl. The composition of the oil phase was chosen in order to match its viscosity to that of water. Monolayers used in the force measurements were much more dilute than in the system depicted in Fig. 1.

A complete description of the laser tweezers apparatus is given in Ref. [9]. The trough containing the spread monolayer was mounted on the stage of a Leica DM IRB inverted microscope. A variable power continuous wave Nd:YAG laser, operating at wavelength 1064 nm, entered

the trough from below and was focused by the microscope objective (40× magnification, numerical aperture 0.5, working distance 2 mm). The trap was formed at the waist of the beam focus located in the plane of the oil/water interface. The trough containing the monolayer is illustrated in Fig. 2. In the present configuration the single laser was employed to create two traps using fast response acousto-optic deflection (AOD). Each trap had a "dwell time" of 1 ms at each trap position with a transit time of 6 μs. The separation between the two traps was adjustable using computer control of the AOD. The microscope stage was equipped with piezoelectric translators which enabled computer controlled lateral movement of the stage, and hence of the monolayer in the trough. The particles in the monolayer were observed using a CCD camera, attached to the microscope, with fast frame image acquisition for subsequent analysis. All measurements were made at 21 °C.

Using a set laser power, P, two particles in the monolayer were trapped initially at a large separation. The separation was decreased to determine the separation at which one of the particles was released from a trap; this gives the interparticle force that is equal to the trapping force at the laser power P. This procedure was repeated for a range of values of P. The measured P values were converted to forces using the calibration method now to be described, and hence the force-distance relationship for two particles was obtained. In order to obtain the relationship between force f and laser power P, two particles were trapped at large separation (19 μm) such that their mutual interaction was negligible. A lateral oscillatory movement was then applied to the stage in a direction normal to the line between the trapped particle centers in order to create a velocity dependent viscous drag force on the particles. The stage movement had a smoothed triangular waveform, with a total displacement amplitude of 72 μm. The maximum stage velocity, v_{max}, was varied by adjustment of the frequency of the triangular waveform. The maximum (critical) value of v_{max}, v_{max}^*, just sufficient to free the particles from the traps, was determined for each value of P. The drag force is related to v_{max}^* by

$$f \approx 6\pi\eta R v_{max}^* , \qquad (1)$$

where η is the viscosity of the liquid (matched for the oil and water phases) and R is the particle radius. Equation (1) is only approximate since the presence of the oil-water interface perturbs the flow streamlines. However, from symmetry arguments it is expected to be valid in the case where the contact angle, θ, of the particle with the oil-water interface is 90°. The observed contact angle measured through the water phase (when the oil phase is octane) for similar particles is 75 $\pm$ 5° [1]. Although the problem of the drag coefficient of a particle in a liquid-liquid interface has not been solved yet, the case of particles at a liquid-gas interface has been considered theoretically [10]. Based on the results of this work, it is estimated that the use of the approximate Eq. (1)

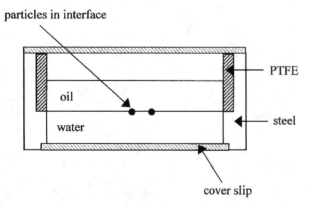

FIG. 2. Side view of the circular trough, internal diameter 25 mm. Pinning of the oil-water interface at the steel-PTFE junction enables the formation of a flat interface.

leads to an underestimation of the absolute magnitude of the drag force with a systematic error of less than 10%, similar to the experimental uncertainties in the measured forces. The results of the calibration are illustrated in Fig. 3, where plots of v_{max}^* against P (filled points) and the corresponding f against P (open points) are shown.

The measured interparticle force, f_{inter}, is depicted in Fig. 4 as a function of separation, L, between the centers of particles for aqueous phases containing either zero added electrolyte (ionic strength $\approx 3 \times 10^{-6}$ M) or 1 mM NaCl. It can be seen that the force scales as L^{-4} and is insensitive to the ionic strength of the aqueous phase. The experimental results are compared with two alternative theories. According to Stillinger [2], Pieranski [3], and Hurd [4], the long-range repulsion corresponds to an effective dipole-dipole interaction resulting from the asymmetry of the particle ionic atmosphere in water due to the presence of the particle/oil interface. Using Hurd's theory, taking into account the difference in relative dielectric constant of air and oil and the finite particle size, the force is

$$f_{inter} \approx \frac{6\varepsilon_{oil}q_{water}^2}{4\pi\varepsilon_0\varepsilon_{water}^2\kappa^2 L^4}, \qquad (2)$$

where ε_0 is the permittivity of free space, ε_{oil} and ε_{water} are the relative dielectric constants of oil and water, respectively, q_{water} is the charge of the water-immersed section of the particle, and κ^{-1} is the Debye screening length. When $\kappa R \ll 1$, $q_{water} = 2\pi R^2(1 + \cos\theta)\sigma\alpha_{water}$, where σ is the particle surface charge density corresponding to full dissociation and α_{water} is the degree of dissociation of the sulfate groups at the particle-water surface. However, in our experiment, $\kappa R \gg 1$; i.e., only the particle surface charges within a distance κ^{-1} from the three-phase con-

tact line effectively contribute to the long-range particle-particle interaction. The rest of the particle surface charge in water is screened similarly as in the bulk of the electrolyte and gives only a short-range (Debye-Hückel type) contribution to the interaction force. Thus, the effective value of the particle/water surface charge to be accounted for in Eq. (2) is $q_{water} = 2\pi R \sin\theta \kappa^{-1}\sigma\alpha_{water}$. In this case

$$f_{inter} \approx \frac{24\pi^2 R^2 \sin^2\theta\,\varepsilon_{oil}\sigma^2\alpha_{water}^2}{4\pi\varepsilon_0\varepsilon_{water}^2\kappa^4 L^4}. \qquad (3)$$

Equation (3) predicts that the expected force contribution from this effect should scale with κ^{-4}, corresponding to an inverse square scaling with the ionic strength of the aqueous phase. This scaling is not observed in the results of Fig. 4 where it can be seen that the force is the same for pure water (ionic strength approximately 3×10^{-6} M) and 1 mM NaCl. The dashed line of Fig. 4 shows the force calculated using Eq. (3) with the following known values for the parameters: $R = 1.35\ \mu$m, $\theta = 75°$, $\sigma = 8.9\ \mu$C cm^{-2}, $\varepsilon_{oil} = 2$, and ionic strength $= 3 \times 10^{-6}$ M (pure water). The unknown value of α_{water} was taken to be 0.25, typical of the values for ionic micelles of similar surface charge density. Hence, even for pure water, the calculated force is small compared to the experimental values. At higher ionic strengths, the calculated force is insignificant.

The solid curve of Fig. 4 was calculated according to the model presented by Aveyard *et al.* [1] in which the dominant repulsive force is postulated to occur as a result

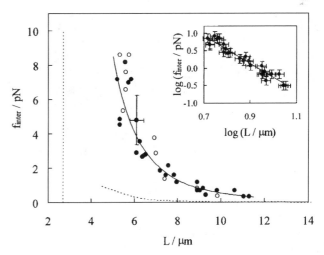

FIG. 4. Interparticle repulsive force versus center-to-center particle separation for aqueous phases containing zero added electrolyte (filled circles) and 1 mM NaCl (open circles). The vertical dashed line corresponds to contact of the particle surfaces. The curved dashed line is calculated using Eq. (3) with κ corresponding to pure water and $\alpha_{water} = 0.25$. The solid curve is calculated using Eq. (5) with the parameters given in the text. The inset shows a double logarithmic plot for the system with zero added electrolyte, for which the best-fit slope is -4.0 ± 0.5 for a confidence level of 95%.

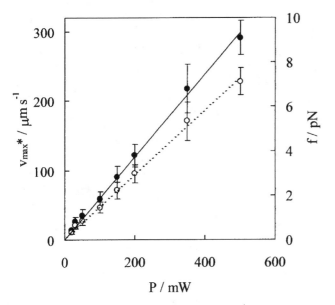

FIG. 3. Laser trap calibration plot showing v_{max}^* versus laser power P (left-hand ordinate, filled symbols) and the corresponding values of trap force f (right-hand ordinate, open symbols).

of the presence of a small amount of net charge at the particle-oil surface. This charge is represented by a point charge, q_{oil}, located at a distance ζ above the oil-water interface and is given by the product of the particle surface area immersed in the oil, the surface charge density σ and the fractional degree of dissociation, α_{oil}, of the ionizable groups at the oil-particle interface. Thus

$$q_{oil} = 2\pi R^2 \sigma (1 - \cos\theta)\alpha_{oil}. \tag{4}$$

The interaction force between particles is obtained by differentiation of Eq. A23 in Ref. [1],

$$f_{inter} \approx \frac{q_{oil}^2}{4\pi\varepsilon_{oil}\varepsilon_0}\left\{\frac{1}{L^2} - \frac{L}{(4\zeta + L^2)^{3/2}}\right\}. \tag{5}$$

The value of ζ is given by [1]

$$\zeta = R(3 + \cos\theta)/2. \tag{6}$$

The first term in Eq. (5) corresponds to the Coulombic interaction between two identical point charges, of magnitude q_{oil}, across the oil phase. The second term corresponds to the interaction between the second particle, charge q_{oil}, and the image charge of the first particle. At large separations such that $(\zeta/L)^2 \ll 1$, the interaction has the asymptotic form expected for a charge-dipole interaction

$$f_{inter} \approx \frac{6q_{oil}^2\zeta^2}{4\pi\varepsilon_{oil}\varepsilon_0 L^4}. \tag{7}$$

In agreement with Eqs. (5) and (7), the measured force is independent of the ionic strength of the aqueous phase and decays as L^{-4}, as shown in the inset log-log plot in Fig. 4, the slope of which is -4.0 ± 0.5 for a confidence level of 95%. The solid curve of Fig. 4 was obtained by using the known values for the parameters quoted above. The unknown value of the degree of dissociation α_{oil} was floated to obtain the best-fit value, which was found to be 3.3×10^{-4}. Within the precision of the data, the model accounts reasonably well for the experimentally measured interparticle forces, both for pure water and for 1 mM NaCl. The value of α obtained here (3.3×10^{-4}) is significantly smaller than the values of 0.01 [1] estimated by fitting the model described above to surface pressure data for similar particles at the octane-water interface. We be-

lieve that the observed difference results from the effect on surface pressure (measured at a lower range of particle separations than the force measurements described here) of surface active oligomers leached from the latex particles by the spreading solvent (isopropanol).

Additional experiments (results not shown) were made with similar latex particles of radius 1.95 μm and $\sigma = 7.5$ μC cm^{-2} with pure water. The forces, again found to scale as L^{-4}, were approximately twofold higher than those for the 1.35 μm particles shown in Fig. 4. The observed force increase is less than that predicted from Eq. (5) if it assumed that both θ and α_{oil} are equal for the two particle types. The forces measured for the larger particles are consistent with Eq. (5) if either θ is taken to be 55° (compared with 75° measured for the small particles) or α_{oil} is taken to be 2×10^{-4} (compared with 3.3×10^{-4} estimated for the small particles).

The main conclusions from this study are that charged polymer particles adsorbed at the oil-water interface exhibit a long-range repulsive force which scales as separation distance to the power -4 and is insensitive to the electrolyte content of the aqueous phase. The results are consistent with a model in which the repulsion arises primarily from the presence of a small residual electric charge at the particle-oil interface.

We thank ICI, Wilton, the Central Laser Facility, CLRC, and the Engineering and Physical Sciences Research Council for financial support.

---

*Corresponding author.
  Email address: P.D.Fletcher@hull.ac.uk

[1] R. Aveyard, J. H. Clint, D. Nees, and V. N. Paunov, Langmuir **16**, 1969 (2000).

[2] F. H. Stillinger, J. Chem. Phys. **35**, 1584 (1961).

[3] P. Pieranski, Phys. Rev. Lett. **45**, 569 (1980).

[4] A. J. Hurd, J. Phys. A **18**, L1055 (1985).

[5] J. C. Earnshaw, J. Phys. D **19**, 1863 (1986).

[6] D. Goulding and J-P. Hansen, Mol. Phys. **95**, 649 (1998).

[7] J. Sun and T. Stirner, Langmuir **17**, 3103 (2001).

[8] J. C. Crocker and D. G. Grier, Phys. Rev. Lett. **77**, 1897 (1996).

[9] D. Nees, S. W. Botchway, M. Towrie, A. D. Ward, A. W. Parker, and A. Burgess, Central Laser Facility, Rutherford Appleton Laboratory, Annual Report No. 209, 1999/2000.

[10] K. Danov, R. Aust, F. Durst, and U. Lange, J. Colloid Interface Sci. **175**, 36 (1995).

Kinetically Locked-In Colloidal Transport in an Array of Optical Tweezers

Pamela T. Korda, Michael B. Taylor,* and David G. Grier

Department of Physics, James Franck Institute and Institute for Biophysical Dynamics, The University of Chicago, Chicago, Illinois 60637

(Received 8 January 2002; published 3 September 2002)

We describe measurements of colloidal transport through arrays of micrometer-scale potential wells created with holographic optical tweezers. Varying the orientation of the trap array relative to the external driving force results in a hierarchy of lock-in transitions analogous to symmetry-selecting processes in a wide variety of systems. Focusing on colloid as a model system provides the first opportunity to observe the microscopic mechanisms of kinetic lock-in transitions and reveals a new class of statistically locked-in states. This particular realization also has immediate applications for continuously fractionating particles, biological cells, and macromolecules.

DOI: 10.1103/PhysRevLett.89.128301   PACS numbers: 82.70.Dd, 05.45.–a, 42.40.–i

Depending on the balance of forces, a particle driven across a corrugated potential energy landscape either flows with the driving force or else becomes locked-in to a symmetry-preferred route through the landscape. The emergence of kinetically locked-in states whose transport properties are invariant over a range of control parameters characterizes many systems and is referred to variously as phase-locking, mode-locking, and stochastic resonance. Examples arise in the electromigration of atoms on crystal surfaces [1], in flux creep through type-II superconductors [2,3], in flux tunneling through Josephson junction arrays [4], and in electron transport through charge density waves and two-dimensional electron gases [5]. Related problems abound in the theory of chemical kinetics and glass formation.

Despite their ubiquity, kinetically locked-in states and transitions among them have been observed directly only in numerical simulations. Their presence in experiments has been inferred indirectly from their influence on collective large-scale properties such as the magnetoresistance and Hall conductance of superconductors and two-dimensional electron gases. Consequently, most theoretical studies have addressed the collective transport properties of strongly coupled systems whose internal interactions modify the influence of the modulated potential and the external driving force. How kinetic lock-in affects single-particle transport has received far less attention.

This Letter describes observations of a hierarchy of kinetically locked-in states in the microscopic trajectories of individual colloidal particles flowing classically through large arrays of optical tweezers. Unlike previous studies on other systems which have found that locked-in states correspond to deterministically commensurate trajectories through the potential energy landscape, our observations also reveal a new class of statistically locked-in states. The locked-in states' ability to systematically and selectively deflect particles' trajectories suggests that optical trap arrays will be useful for continuously fractionating materials in suspension.

Previous studies [6] have created optical potential energy landscapes with static interference patterns and studied their influence on the equilibrium phase behavior of strongly interacting colloidal monolayers. The present study extends this approach to explore colloidal kinetics in adjustable arrays of discrete potential wells.

Our system, shown schematically in Fig. 1, consists of colloidal silica spheres $2a = 1.5$ μm in diameter (Bangs Labs Lot 4258) dispersed in a 20 μm-thick layer of deionized water sandwiched between horizontal glass surfaces. These spheres are considerably denser than water and readily sediment into a monolayer about 1 μm above the lower wall [7]. The edges of the sample volume are sealed to form a flow cell, with access provided by two glass tubes bonded to holes passing through the upper glass wall. These tubes also serve as reservoirs for colloid, water, and clean mixed-bed ion exchange

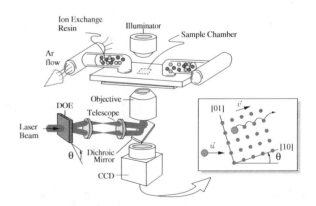

FIG. 1 (color online). Schematic diagram of the experimental system. Laser beams formed by a diffractive optical element (DOE) are transferred by a telescope to the input pupil of a high-NA objective lens which focuses each into an optical trap. The same lens is used in a conventional light microscopy system (illuminator, condenser, objective, video eyepiece, and CCD camera) to form images of spheres moving past the traps, as shown schematically in the inset. A spatial filter blocking the undiffracted laser light is omitted for clarity.

resin. Their ends are connected to continuous streams of humidified Ar gas which minimize the infiltration of airborne contaminants and enable us to drive the colloid back and forth through the channel. Blocking one of the gas streams causes a pressure imbalance which forces the dispersion through the sample chamber and past the $75 \times 58 \ \mu m^2$ field of view of a $100\times$ numerical aperture (NA) 1.4 oil-immersion objective mounted on an Olympus IMT-2 microscope base. Steady flows of up to $u = 100 \ \mu m/s$ can be sustained in this way for about 10 min.

We use precision digital video microscopy [8] to track the individual spheres' in-plane motion with a resolution of 10 nm at 1/60 s intervals. The resulting trajectory data allow us to monitor the spheres' progress through potential energy landscapes that we create with light.

Our optical potential landscapes are created with the holographic optical tweezer technique [9,10] in which a single beam of light is formed into arbitrary configurations of optical traps by a computer-designed diffractive beam splitter. Each beam created by the diffractive optical element (DOE) is focused by the objective lens into a diffraction-limited spot which acts as an optical tweezer [11] capable of stably trapping one of the silica spheres against gravity and random thermal forces. For the present experiments, we created a planar 10×10 array of optical traps on $2.4 \ \mu m$ centers using light from a frequency-doubled Nd : YVO$_4$ laser operating at 532 nm. The traps are focused into the plane of the monolayer to avoid displacing spheres vertically as they flow past. Each trap is powered by about $150 \ \mu W$, and their intensities vary by $\pm 20\%$ from the mean, as determined by imaging photometry. Rotating the DOE through an angle θ, as shown in Fig. 1, rotates the array of traps relative to the flow, $\vec{u}$, by the same amount, without otherwise affecting the traps' properties [10].

If the Stokes drag due to the flowing fluid greatly exceeds the optical tweezers' maximum trapping force, then colloidal particles flow past the array with their trajectories unperturbed. Conversely, if the trapping force dominates, then particles fall irreversibly into the first traps they encounter. Our observations are made under intermediate conditions for which trapping and viscous drag are nearly matched and particles hop readily from trap to trap. Our silica spheres enter the hopping state for flow speeds u in the range $40 \ \mu m/s < u < 80 \ \mu m/s$. The monolayer's areal density is low enough that typically only one or two spheres are in the array at any time. Their separations are large enough that hydrodynamic coupling between spheres should be negligible [12].

Figure 2 shows the superimposed trajectories of 300 particles flowing through a $9 \times 9 \ \mu m^2$ section of the field of view which includes one corner of the optical tweezer array. The flow drives spheres directly from left to right across the field of view, with small lateral deviations resulting from Brownian fluctuations. Those spheres passing within about $1.5 \ \mu m$ of the optical traps are drawn into the array's [10] rows and follow them to their ends. In

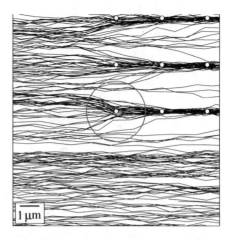

FIG. 2 (color online). Spheres flowing through an array of optical traps follow channels along the [10] direction of the trap lattice. White spots denote tweezer positions, and the $3 \ \mu m$-diameter circle indicates an individual tweezer's region of influence. Traces show the paths of 300 separate spheres tracked in 1/30 s intervals over a 3 min period.

this case, the [10] rows are aligned with the bulk flow, and the traps' principal influence is to herd particles into well-defined channels and to suppress their transverse fluctuations. The appearance of such commensurate trajectories through the array defines a channeling state, named by analogy to ion channeling through conventional crystals.

While plotted trajectories help to visualize individual particles' interactions with the optical traps, a complementary view of the array's overall influence is offered by the relative probability $P(\vec{r}) \, d^2r$ of finding a particle within d^2r of $\vec{r}$ at some time after it enters the field of view. Figure 3(a) shows data compiled from the trajectories of 18 601 spheres obtained under the same conditions as Fig. 2. They reveal that spheres are nearly 7 times more likely to be found in the rows of traps than at any point in the bulk flow outside of the array. The correspondingly low probability for finding spheres between the rows and the comparatively subtle modulation along the rows reveals that the time required for a sphere to hop from trap to trap along a row is so much shorter than the time needed for a transverse jump that the spheres essentially never leave the [10] rows. Once the spheres have hopped through the ranks of traps, they return to the bulk flow, their trajectories eventually blurring into each other through diffusion.

Figure 3(b) shows data from the same sample but with the traps oriented at $\theta = 9°$ with respect to the flow. Even though the flow is no longer aligned with the lattice, the spheres still closely follow the array's [10] rows. As a result, the channeling trajectories are systematically deflected away from the flow's direction and leave a distinct shadow downstream of the array. This insensitivity to orientation distinguishes the [10]-commensurate state as being kinetically locked-in and confirms the conjecture [2,5] that kinetic lock-in with systematic deflection can

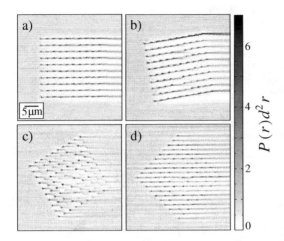

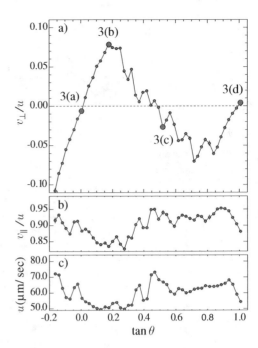

FIG. 3. Relative probability that a sphere will pass through a point in the field of view, when the [10] direction of the trap lattice is oriented at (a) $\theta = 0°$, (b) 9°, (c) 28°, and (d) 45°. In all figures, the external flow is from left to right.

FIG. 4 (color online). (a) Variation of the mean normalized particle speed perpendicular to the externally applied force as a function of array orientation. Emphasized data points correspond to conditions in Fig. 3. (b) Mean normalized longitudinal speed. (c) Mean flow speed in the bulk for each orientation studied.

occur as a single-body process rather than requiring the elasticity of an interacting monolayer.

When the trap array is rotated even further to $\theta = 28°$, as in Fig. 3(c), the particles no longer channel along the [10] direction. Although the spheres still spend more time in individual traps than in the bulk, they no longer follow clearly defined paths from one trap to another.

Rotating to $\theta = 45°$, as in Fig. 3(d), reveals another channeling state with particles following the array's diagonal [1$\bar{1}$] rows. Rotating away from 45° demonstrates this channeling state also to be locked-in. In principle, additional locked-in channeling states should appear at other angles corresponding to commensurate paths through the array [13]. In a system with square symmetry, commensurate orientations occur for rational values of $\tan\theta$.

To quantify the degree to which the array deflects spheres' trajectories, we compare the velocity $\vec{v}$ a particle attains while moving inside the array to its velocity $\vec{u}$ in the bulk flow. In particular, Fig. 4(a) shows the mean normalized transverse component of the in-array velocity $v_\perp(\theta)/u = [\vec{v}(\theta) \times \vec{u}]/(uv)$ which is roughly analogous to the Hall coefficient in electron transport. The monotonically positive slope of $v_\perp(\theta)/u$ in the range $|\tan\theta| < 0.2$ characterizes the domain over which the [10] state is locked-in, with increasing rotation yielding systematically increasing deflection.

After the deflection reaches its maximum at $\tan\theta = 0.2$, it decreases nonmonotonically to zero at the commensurate orientation $\tan\theta = 1/2$. Rotating the array beyond this point results in *retrograde* deflection. In contrast, no change of sign is predicted for the Hall coefficient of a periodically modulated two-dimensional electron gas with increasing magnetic field, although this may reflect the choice of sixfold rather than fourfold symmetry for the potential landscape in the available simulations [5]. If indeed such sign reversal can be ob-

tained through simple patterning of an electronic system, the effect would be unprecedented and could have widespread applications in magnetic data retrieval.

Beyond $\tan\theta = 0.85$, trajectories become locked-in to the commensurate channeling state along the [1$\bar{1}$] direction. The deflection returns to zero in this state when the [1$\bar{1}$] rows align with the external force at $\tan\theta = 1$. Quantitatively indistinguishable results for $v_\perp(\theta)/u$ were obtained for particles moving with different speeds in the range 50 μm/s $< u <$ 75 μm/s.

While $v_\perp/u$ is independent of u over the entire range of hopping conditions, such is not the case for the other component, $v_\parallel/u$. As can be seen from Figs. 4(b) and 4(c), the normalized longitudinal velocity is strongly correlated with u. Although structure in $v_\parallel(\theta)$ may reflect aspects of particles' hopping mechanisms, much as the magnetoresistance does for electron transport, it is masked in the available data by variations in u.

The [10] and [1$\bar{1}$] locked-in states are characterized by the positive slope they induce in $v_\perp(\theta)/u$. That other, smaller features also correspond to locked-in states becomes apparent in another representation of the data. Simulations [2] have demonstrated that kinetically locked-in states appear as plateaus in the θ dependence of the ratio $v_{[01]}/v_{[10]}$ of in-array speeds along the perpendicular [01] and [10] directions. Comparison with predictions for the circle map and related dynamic systems further suggests that the widest plateaus should be

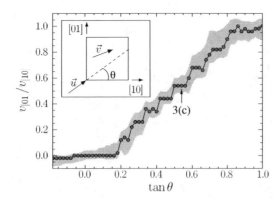

FIG. 5 (color online). Orientation dependence of particles' speeds along and normal to the [10] direction reveals a series of plateaus corresponding to kinetically locked-in states. Plotted points are the most probable, or mode, values from the distribution obtained from all trajectories at each angle. Shaded regions indicate the 99% confidence interval about these values.

centered at the simplest rational values of $\tan\theta$ and that the overall hierarchy of locked-in states should take the form of a Devil's staircase with increasing rotation [2,13]. The corresponding representation of our data appears in Fig. 5. As expected, the data display a series of kinetically locked-in states, with plateaus in $v_{[01]}/v_{[10]}$ corresponding to regions of positive slope in $v_\perp(\theta)/u$. The large plateaus around $\tan\theta = 0$ and $\tan\theta = 1$ correspond to the [10] and [1$\bar{1}$] locked-in states. However, the higher-order plateaus between $\tan\theta = 0.2$ and $\tan\theta = 0.8$ are not centered on simple rational values of $\tan\theta$. Instead, the commensurate orientations at $\tan\theta = 1/3$, $1/2$, and $2/3$ correspond to *transitions* between plateaus. Furthermore, the associated plateaus include nonchanneling states such as Fig. 3(c) which nonetheless are locked-in.

Nonchanneling transport in the plateaus of Fig. 5 suggests a previously unrecognized class of statistically locked-in states that are distinct from deterministically channeling states. Their absence from measurements on perfect atomic crystals and idealized molecular dynamics simulations suggests that they may result from quenched disorder in our optically defined potential energy landscape. The pattern of plateaus reflects symmetries in the potential energy landscape and so would not be affected by the individual potential wells' shapes [13]. Their statistical nature suggests a possible role for random thermal forcing. How disorder gives rise to the distribution of steps observed in Fig. 5 poses an outstanding challenge.

Beyond providing an experimental context in which to study the microscopic mechanisms of kinetic lock-in, the techniques introduced for this study also constitute a practical method for continuously fractionating mesoscopic materials. Particles in a heterogeneous suspension that interact more strongly with optical traps will be pushed to one side by an appropriately tuned array of traps. Particles that interact less strongly will pass through the same array undeflected. The deflected and undeflected fractions then can be collected in separate microfluidic channels and passed on to additional stages of optical traps for further stages of fractionation.

We are grateful to Franco Nori, Charles Reichhardt, Woowon Kang, and Sidney Nagel for enlightening conversations. This work was supported by the National Science Foundation through Grant No. DMR 9730189 and by the MRSEC program of the NSF through Grant No. DMR-9880595. Additional support was provided by a grant from Arryx, Inc. The diffractive optical element used in this work was fabricated by Gabriel Spalding, Steven Sheets, and Matthew Dearing to a design computed by Eric R. Dufresne [10].

*Department of Physics, Illinois Wesleyan University, Bloomington, IL 61702.

[1] O. Pierre-Louis and M. I. Haftel, Phys. Rev. Lett. **87**, 048701 (2001).

[2] C. Reichhardt and F. Nori, Phys. Rev. Lett. **82**, 414 (1999).

[3] M. Baert, V.V. Metlushko, R. Jonckheere, V.V. Moshchalkov, and Y. Bruynseraede, Phys. Rev. Lett. **74**, 3269 (1995); K. Harada, O. Kamimura, H. Kasai, T. Matsuda, A. Tonomura, and V.V. Moshchalkov, Science **274**, 1167 (1996); J. I. Martín, M. Vélez, J. Nogués, and I. K. Schuller, Phys. Rev. Lett. **79**, 1929 (1997); D. J. Morgan and J. B. Ketterson, Phys. Rev. Lett. **80**, 3614 (1998).

[4] K. D. Fisher, D. Stroud, and L. Janin, Phys. Rev. B **60**, 15 371 (1999); V. I. Marconi and D. Domínguez, Phys. Rev. B **63**, 174509 (2001).

[5] J. Wiersig and K.-H. Ahn, Phys. Rev. Lett. **87**, 026803 (2001).

[6] B. J. Ackerson and A. H. Chowdhury, Faraday Discuss. Chem. Soc. **83**, 1 (1987); M. M. Burns, J.-M. Fournier, and J. A. Golovchenko, Science **249**, 749 (1990); K. Loudiyi and B. J. Ackerson, Physica (Amsterdam) **184A**, 1 (1992); Q.-H. Wei, C. Bechinger, D. Rudhardt, and P. Leiderer, Phys. Rev. Lett. **81**, 2606 (1998).

[7] S. H. Behrens and D. G. Grier, Phys. Rev. E **64**, 050401 (2001); J. Chem. Phys. **115**, 6716 (2001).

[8] J. C. Crocker and D. G. Grier, J. Colloid Interface Sci. **179**, 298 (1996).

[9] E. R. Dufresne and D. G. Grier, Rev. Sci. Instrum. **69**, 1974 (1998).

[10] E. R. Dufresne, G. C. Spalding, M. T. Dearing, S. A. Sheets, and D. G. Grier, Rev. Sci. Instrum. **72**, 1810 (2001).

[11] A. Ashkin, J. M. Dziedzic, J. E. Bjorkholm, and S. Chu, Opt. Lett. **11**, 288 (1986).

[12] E. R. Dufresne, T. M. Squires, M. P. Brenner, and D. G. Grier, Phys. Rev. Lett. **85**, 3317 (2000).

[13] E. Ott, *Chaos in Dynamical Systems* (Cambridge University Press, New York, 1993).

Measurement of the hydrodynamic corrections to the Brownian motion of two colloidal spheres

John C. Crocker

James Franck Institute and Department of Physics, University of Chicago, Chicago, Illinois 60637

(Received 25 July 1996; accepted 14 November 1996)

The hydrodynamic coupling between two isolated 0.97 μm diameter polystyrene spheres is measured by reconstructing their Brownian motion using digital video microscopy. Blinking optical tweezers are used to facilitate data collection by positioning the spheres in the microscope's focal plane and in close proximity to one another. The observed separation dependence of the spheres' relative and center of mass diffusion coefficients agree with that predicted by low Reynolds number hydrodynamics and the Stokes-Einstein relation. © *1997 American Institute of Physics.* [S0021-9606(97)51307-9]

I. INTRODUCTION

The hydrodynamic interactions between colloidal particles are of tremendous importance to understanding such common colloid phenomena as sedimentation and aggregation. This makes colloidal hydrodynamics relevant to diverse industrial processes, from fluidized beds to blast furnaces, to paper making. Most of the previous experimental work consists of either measurements of the resistance coefficients of macroscopic objects at low Reynolds number,[1-3] the motion of a single microscopic particle near a planar wall,[4-6] or the non-Brownian dynamics of two microscopic spheres colliding in a shear flow.[7] The work presented here concerns the hydrodynamic coupling between two 0.97 μm polystyrene spheres suspended in water, as measured with digital video microscopy. Optical tweezers[8,9] were used to hold the two spheres near each other and in the same focal plane far from the walls of the sample container. The hydrodynamic coupling was measured by its effect, via the Stokes-Einstein relation, on the Brownian diffusivity of the pair of spheres. While the Stokes-Einstein relation has been well established to describe the diffusion of individual particles, there is some theoretical uncertainty regarding its validity in multi-particle systems.[10] This work is uniquely suited to verify the validity of the Stokes-Einstein relation for the two sphere system.

The hydrodynamic forces acting between two spheres at low Reynolds number have been calculated by a number of authors.[11-14] The goal of these calculations is to estimate the resistance to motion felt by a sphere of radius a in a fluid of viscosity η in the presence of a second sphere at a distance r. For a single isolated sphere, motion at a velocity v produces a drag force given by the familiar Stokes formula: $\vec{F} = -(6\pi\eta a)\vec{v}$. The ratio of the velocity to the drag force is termed, in general, the mobility coefficient. When a second sphere is added, the spherical symmetry of the Stokes case is reduced to axial symmetry and four mobility coefficients are needed to quantify the hydrodynamic drag forces. The spheres' motion can be decomposed into two parts: the relative motion of the two spheres with respect to each other and the common motion of the two spheres' center of mass. Each of these two motions are in turn described by a pair of mo-

bility coefficients: one for motion along the line of centers and one for motion perpendicular to it.

These four mobilities can be related to four corresponding diffusion coefficients by a generalized Stokes-Einstein relation. For a single sphere, the Stokes-Einstein relation predicts that the Brownian diffusivity is simply $k_B T$ multiplied by the mobility

$$D_0 = \frac{k_B T}{6\pi\eta a}. \tag{1}$$

For two spheres, Batchelor[11] gives asymptotic expressions for the four diffusivities as functions of the dimensionless center-center separation, $\rho = r/a$. If the two spheres have the same radius, no external applied torques, and are free to rotate in response to each other's flow field, the expressions for the diffusion relative to the center of mass are

$$D_{\text{RM}}^{\parallel} = \frac{D_0}{2}\left[1 - \frac{3}{2\rho} + \frac{1}{\rho^3} - \frac{15}{4\rho^4} + O(\rho^{-6})\right], \tag{2}$$

$$D_{\text{RM}}^{\perp} = \frac{D_0}{2}\left[1 - \frac{3}{4\rho} - \frac{1}{2\rho^3} + O(\rho^{-6})\right], \tag{3}$$

where $D_{\text{RM}}^{\parallel}$ is the diffusivity along the center-center line, and $D_{\text{RM}}^{\perp}$ is perpendicular to it.

The corresponding expressions for the diffusion of the center of mass itself are:

$$D_{\text{CM}}^{\parallel} = \frac{D_0}{2}\left[1 + \frac{3}{2\rho} - \frac{1}{\rho^3} - \frac{15}{4\rho^4} + O(\rho^{-6})\right], \tag{4}$$

$$D_{\text{CM}}^{\perp} = \frac{D_0}{2}\left[1 + \frac{3}{4\rho} + \frac{1}{2\rho^3} + O(\rho^{-6})\right], \tag{5}$$

where $D_{\text{CM}}^{\parallel}$ is the diffusivity along the center-center line and $D_{\text{CM}}^{\perp}$ is perpendicular to it.

Examination of the lowest order terms indicates that the relative diffusivities are suppressed while the center of mass diffusivities are enhanced. The suppression of the relative diffusivities is caused by the resistance of the fluid between the spheres to being sheared or squeezed out of the gap.

Conversely, the diffusion of the center of mass is enhanced by the fluid entrained by one sphere pulling along the other.

The lowest order terms in Eqs. (2)–(5) would suffice to accurately describe the diffusivities when the spheres are separated by more than a few radii. Comparison with other theoretical forms which describe the diffusivities at very small separations[10] suggest that Eqs. (2)–(5) accurately describe the diffusivities over the experimental range of separations ($\rho > 2.5$).

II. EXPERIMENTAL METHOD

The experiment was conducted using commercially available polystyrene sulfate spheres with a radius of 0.966 ± 0.01 μm (Duke Scientific, Cat. No. 5095A). Such spheres typically have titratable surface charges of more than 10^5 electron equivalents and are thus surrounded by a diffuse cloud of counterions whose thickness is determined by the electrolyte's Debye-Hückel screening length. This counterion cloud can increase the hydrodynamic radius of the spheres, resulting in a diffusivity significantly smaller than suggested by Eq. (1) and the nominal radius. Furthermore, overlap of counterion clouds produces a long-range electrostatic repulsion between spheres. To eliminate the confusion that such effects might cause in this experiment, the colloid was suspended in a 0.1 mM solution of HCl. In such an electrolyte, the electrostatic and van der Waals interactions between spheres should be negligible so long as they are more than 300 nm from contact and the single-sphere diffusivities should be within 1% of that predicted by Eq. (1).

Our sample cell was constructed by sealing the edges of a No. 1 cover-slip to the surface of a glass microscope slide with uv curing epoxy. Holes drilled through the slide connect the roughly $25 \times 10 \times 0.05$ mm sample volume to two reservoirs of dilute suspension. Prior to assembly, all the glass surfaces were stringently cleaned with an acid-peroxide wash, and rinsed well with deionized water. After filling, this sample cell was placed on an Olympus IMT-2 inverted video microscope using a $100\times$, numerical aperture 1.3 oil immersion objective. A $5\times$ projection eyepiece at the camera yields a total system magnification of 85 nm per CCD pixel. At such magnification, the colloidal spheres create bright images which are broadened by diffraction to about 15 pixels in diameter. These video images were recorded on an NEC PC-VCR model S-VHS video deck during the course of the experiment.

To measure the hydrodynamic coupling between spheres, we want to study the Brownian motion of a pair of spheres which are simultaneously near each other and far from all the other spheres and the walls of the sample cell. Unfortunately, in thermally distributed dilute suspensions such isolated pairs are both rare and short-lived, making the collection of statistically large data-sets difficult and time consuming. Further complicating matters is the fact that an optical microscope only provides a two-dimensional projection of the three-dimensional distribution of spheres. While video microscopy techniques are available for estimating the spheres' position in third dimension[15] they fail if the sphere

images overlap or if either sphere is more than about 1 μm out of the focal plane.

We avoided both of these difficulties by manipulating the spheres with a pair of optical gradient force traps, called optical tweezers. We can form such a trap by focusing a 30 mW, 780 nm wavelength laser beam to a diffraction-limited spot with our high numerical aperture objective lens. Near that spot the optical gradient forces acting on a sphere's light-induced electric dipole can be greater than both the thermal and radiation pressure forces, trapping it firmly in three dimensions.[8] We used two such optical traps to position a pair of colloid spheres near each other and in the same focal plane 12 μm from the lower glass wall and 40 μm from the upper wall.

We blinked the laser on and off at 6 Hz with an electromechanical shutter synchronized with the 30 Hz frame rate of our video camera. Thus we arranged for the tweezers to be turned off for three out of every five consecutive video frames. During the 110 ms interval when the tweezers are off, the spheres diffuse freely. When the tweezers are turned on, the spheres return to their trapped positions. By examining only those frames where the traps are turned off we can rapidly acquire large amounts of free diffusion data while the blinking traps hold the spheres near the desired configuration.

Furthermore, we can neglect the effect of the spheres' out of plane diffusion provided that it is small compared with their center-to-center separation. While the traps are turned off, the spheres diffuse out of the plane with an average displacement of about 350 nm. To reduce the effects of such out of plane motion, we discarded any data not acquired during the first 33 ms of a "blink" if the spheres were separated by less than 3 μm. This reduces the average out of plane motion for the remaining data to less than 200 nm.

During the course of the experiment, the separation between the traps was slowly increased, so that data were acquired at a range of separations. We were able to move the traps without changing their focus or introducing vignetting by varying the angle at which the beams enters the objective's back aperture.[9] A convenient method for such steering is to use a Keplerian telescope to create a plane conjugate to the objective lens' back aperture outside of the microscope body. Rotating the laser beam with a gimbal mounted mirror in that plane will move the trap about the field of view without vignetting or changing its focus.

After 100 min of video data were collected, the videotape was digitized with a Data Translation DT-3851A video digitizer installed in a 486-class personal computer. Video frames with the traps turned on were readily distinguishable by a increase in the background brightness due to scattered laser light. By controlling the video deck with the same computer, all the video frames could be digitized and the roughly 10^5 frames with the traps off could be written to disk automatically. Since the tweezers held the spheres in roughly the same position in the field of view, only a small portion of each video frame (170×80 pixels), needed to be stored and processed.

Once the images were digitized, we determined the

sphere locations by calculating the centroid of each sphere's brightness distribution in the image. This process can determine the two-dimensional coordinates of the spheres to better than 10 nm, and is described in detail elsewhere.[16] Since NTSC video is interlaced, the even and odd rows of each frame are exposed at different times, 1/60th of a second apart. Thus by determining the positions of the spheres in each set of rows (video fields) separately, we can measure their positions with a time resolution of 60 Hz.

III. RESULTS AND DISCUSSION

Some care must be taken when calculating the spheres' self-diffusivity D_0, due to the effects of the walls. For a sphere far from two parallel walls, the modified self-diffusivity is given by[17]

$$D_0' = D_0 \left[1 - \frac{9a}{16x_1} - \frac{9a}{16x_2} \right], \tag{6}$$

where x_1, x_2 are the distances from the sphere to the two walls. For our sample, which was temperature regulated at 23.0 ± 0.2 °C, Eq. (1) gives $D_0 = 0.484$ μm^2/s. Applying Eq. (6) yields a wall-corrected value which is 3% smaller: $D_0' = 0.470$ μm^2/s.

Since the long-range flow field around a pair of co-moving spheres has the same form as that for a single sphere, the same wall correction should apply. For this reason, we will use the wall-corrected self-diffusivity D_0' when calculating the center of mass diffusivities $D_{CM}^{\parallel}, D_{CM}^{\perp}$. Conversely, the flow fields far from a pair of oppositely moving spheres should partially cancel, resulting in much smaller coupling to distant walls. Thus we will use the uncorrected self-diffusivity D_0 when calculating the relative diffusivities $D_{RM}^{\parallel}, D_{RM}^{\perp}$.

To measure the diffusivities, we considered the motion of the spheres during the interval between pairs of adjacent video fields. Each blink of the optical traps lasted for six consecutive fields, providing five such pairs and thus five independent samples of the Brownian dynamics. In general, displacements caused by Brownian motion will be Gaussian distributed, with a half-width determined by the diffusivity. Specifically, the probability distribution for finding a one-dimensional displacement Δ is

$$P(\Delta) = \frac{1}{2\pi\sigma^2} e^{-\frac{\Delta^2}{2\sigma^2}}, \quad \text{where} \quad \sigma = \sqrt{2D\tau} \tag{7}$$

where D is the diffusivity, τ is the time interval between position measurements and σ is the distribution's root-mean-squared (rms) deviation.

To test the Batchelor[11] formulae, Eqs. (2)–(5), we decomposed the two sphere displacements during the $\tau = 16.7$ ms interval between fields into four separate components. The motion of the spheres' center of mass between the fields was decomposed into components parallel and perpendicular to center-center line in the initial field. Similarly, the spheres' motion relative to their center of mass provided the

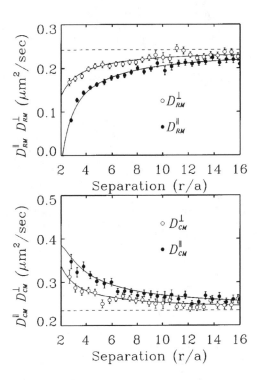

FIG. 1. The measured relative (top) and center of mass (bottom) diffusion coefficients for a pair of colloidal spheres of diameter $2a = 0.966$ μm as a function of dimensionless separation ρ. The solid curves indicate the theoretical prediction given by Eqs. (2)–(5). The dashed line indicates the asymptotic diffusivity $D_0/2$ (top) and $D_0'/2$ (bottom).

other two components. Computing these projections on all the video field pairs in the data-set yielded 1.2×10^5 sets of four displacements.

The four sets of displacements were partitioned according to the initial separation of spheres in the corresponding video fields. The diffusivities were then calculated via Eq. (7) from the rms value of the displacements within each partition. The results are shown in Fig. 1, and show excellent agreement with the theory curves given by Eqs. (2)–(5).

While such joint diffusivities of the pair of spheres are clearly affected by their proximity, the effect on the self-diffusion coefficients is much more subtle, as shown in Fig. 2. Here, rather than decomposing the motion of the two spheres into a differential and a common motion, as in Fig. 1, we consider the motion of each sphere separately. The displacements were, as before, divided into a component along the center-center line, and another one perpendicular to it. These components were partitioned as before, and the corresponding diffusivities calculated from their rms values via Eq. (7).

We can determine the theoretical prediction for such self-diffusion along the center-center line by summing Eq. (2) and Eq. (4). We find that the lower order corrections cancel, leaving

$$D^{\parallel} = D_0 \left[1 - \frac{15}{4\rho^4} + O(\rho^{-6}) \right] \tag{8}$$

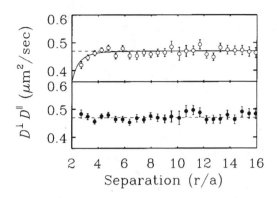

FIG. 2. The measured self-diffusion coefficients for each of a pair of colloidal diameter $2a = 0.966$ μm as a function of their dimensionless separation ρ. The top plot shows the diffusion coefficients for motion perpendicular to the center-center line; the bottom one for parallel motion. The solid curve indicates the theoretical prediction given by Eq. (8). The dashed line indicates the wall-corrected, asymptotic self-diffusivity D_0'.

which is consistent with the subtle downturn seen in Fig. 2. Repeating this calculation for the perpendicular self-diffusion by summing Eq. (3) and Eq. (5), yields

$$D^\perp = D_0[1 + O(\rho^{-6})] \qquad (9)$$

showing cancellation of all the corrections up to the sixth order considered by Batchelor. This nicely accounts for our failure to observe any separation dependence of the self-diffusion coefficient $D^\perp$.

The cancellation of the separation dependence of the self-diffusion coefficients can be easily understood. If we consider moving only one of a pair of spheres through a viscous fluid, the other sphere is free to move in response much the same way as the parcel of fluid it displaces. At low Reynolds number inertial effects due to any difference of mass are neglected, so the only difference between the second sphere and the fluid parcel is their resistance to deformation, which is presumably responsible for the leading

ρ^{-4} term in Eq. (8). The more complete cancellation for the perpendicular motion is due to the fact that in that case the motion of the second sphere, or fluid parcel, is mostly rotational rather than elongational.

By carefully examining the Brownian diffusivity of two isolated colloidal spheres, the hydrodynamic forces coupling them to one another were measured, and are in excellent agreement with theoretical expectations. These forces do not manifest themselves by a significant reduction in the magnitude of the self-diffusivity of either sphere alone. These forces do lead, however, to strong long-range correlations of the two spheres' Brownian motion.

ACKNOWLEDGMENTS

The author is pleased to acknowledge enlightening conversations with Amy Larsen and David Grier. This work was supported by the Grainger Foundation and the National Science Foundation under Grant Number DMR-9320378.

[1] Z. Adamczyk and T. G. M. van de Ven, J. Colloid Interface Sci. **96**, 204 (1983).

[2] G. F. Eveson, E. W. Hall, and S. G. Ward, Brit. J. Appl. Phys. **10**, 43 (1959).

[3] J. Happel and R. Pfeffer, A. I. Ch. E. J. **6**, 129 (1960).

[4] D. C. Prieve, S. G. Bike, and N. A. Frej, Faraday Discuss. **90**, 209 (1990).

[5] L. P. Faucheux and A. J. Libchaber, Phys. Rev. E **49**, 5158 (1994).

[6] Y. Grasselli and G. Bossis, J. Colloid Interface Sci. **170**, 269 (1995).

[7] K. Takamura, H. L. Goldsmith, and S. G. Mason, J. Colloid Interface Sci. **82**, 175 (1981).

[8] A. Ashkin, J. M. Dziedzic, J. E. Bjorkholm, and S. Chu, Opt. Lett. **11**, 288 (1986).

[9] R. S. Afzal and E. B. Treacy, Rev. Sci. Instrum. **63**, 2157 (1992).

[10] Theo G. M. van de Ven, *Colloidal Hydrodynamics* (San Diego, Academic, 1989).

[11] G. K. Batchelor, J. Fluid Mech. **74**, 1 (1976).

[12] M. Stimson and G. B. Jeffery, Proc. Roy. Soc. **111**, 110 (1926).

[13] H. Faxen, Z. Agnew. Math. Mech. **7**, 79 (1927).

[14] J. Happel and H. Brenner, *Low Reynolds Number Hydrodynamics* (Kluwer Academic, Norwell, 1991).

[15] J. C. Crocker and D. G. Grier, Phys. Rev. Lett. **73**, 352 (1994).

[16] J. C. Crocker and D. G. Grier, J. Colloid Interface Sci. **179**, 298 (1996).

[17] W. B. Russel, D. A. Saville, and W. R. Schowalter, *Colloidal Dispersions* (Cambridge University Press, Cambridge, 1989).

Direct Measurement of Hydrodynamic Cross Correlations between Two Particles in an External Potential

Jens-Christian Meiners and Stephen R. Quake

Department of Applied Physics, California Institute of Technology, Pasadena, California 91125
(Received 22 October 1998)

We report a direct measurement of the hydrodynamic interaction between two colloidal particles. Two micron-sized latex beads were held at varying distances in optical tweezers while their Brownian displacements were measured. In spite of the fact that fluid systems at low Reynolds number are generally considered to have no "memory," the cross-correlation function of the bead positions shows a pronounced, time-delayed anticorrelation. We show that the anticorrelations can be understood in terms of the standard Oseen tensor hydrodynamic coupling. [S0031-9007(99)08607-X]

PACS numbers: 82.70.Dd, 83.10.Pp, 87.80.Cc

Hydrodynamic interactions play a crucial role in many physically interesting systems, including colloidal suspensions, polymers in solution, and the microscopic dynamics of proteins. Colloids have a collective diffusion constant that is affected by the distribution of neighboring particles [1], while hydrodynamics interactions are a crucial ingredient in the theory of polymer dynamics [2]. Solvent hydrodynamic effects also have a strong influence on the microscopic dynamics and collective excitations of protein molecules [3,4]. It has been shown that secondary structural elements of a protein can move as collective groups [5]. Thus protein molecules have been treated as deformable Brownian particles, which are subject to friction and random forces from the surrounding solvent. Using such a model, Kitao *et al.* [6] found that, in particular, the dynamics of the low-frequency eigenmodes depend crucially on hydrodynamic effects. Furthermore, such hydrodynamic interactions are thought to play a key role in "steering" ligand-protein binding [7]. Experimentally, it can often be difficult to isolate the effects due to hydrodynamics since measurements are made on bulk systems with indirect methods.

Here we describe an experiment in which we directly studied hydrodynamic interactions between individual colloidal particles. Two microscopic latex beads were held a fixed distance apart in separate optical tweezers. The position fluctuations of the beads were measured, from which we calculated correlation and cross-correlation functions. Previous studies have used similar arrangements to study electrostatic forces between particles [8,9] and to measure the mutual diffusion constants of two particles [10]. We use it as a simple model system with which to study in detail the effects of hydrodynamic interactions between two particles. The tweezers function as harmonic potential wells and can thus approximate a variety of possible local forces. For example, one can imagine this system idealizing the dynamic motion of two subunits on a large protein complex.

The most striking feature of the experimental data is the presence of a pronounced time-delayed dip in the cross correlations. While it is counterintuitive that the motion of the spheres is anticorrelated, the time delay is also surprising in light of the fact that in fluid systems at low Reynolds number, dynamics are determined only by the instantaneous forces; there is no "memory" [11]. Furthermore, the hydrodynamic interaction does not introduce a propagation delay; it is represented by the Oseen tensor, which is derived directly from the Navier-Stokes equation and assumes instantaneous propagation of forces through the fluid. However, we show here that a stationary, time-independent external potential can impose time-delayed correlations between particles in solution and that one particle does "remember" where the other one was a short time before. The time delay is determined by the natural relaxation time of the harmonic well.

The notion of memory in these systems can be made precise in the formal context of control theory and linear systems. The concept of "observability" is a mathematical measure of whether or not a system has memory, i.e., whether its complete internal state at some point in the past can be determined from a measurement of its input and output variables [12]. Consider the case of two independent particles in potential wells in which one can measure the position of only one particle. Even if one knows the Brownian forces, the position of the second particle can never be calculated and the system is unobservable. However, introducing hydrodynamic coupling renders the system formally observable, and it is possible to calculate the position of the second particle solely from measurements of the first particle and knowledge of the Brownian forces. The past history of the second particle is encoded in the position of the first.

Experimentally, we studied an aqueous solution of fluorescent carboxyl-modified polystyrene latex spheres with a diameter of 1.0 ± 0.025 μm at a volume fraction of $\phi = 10^{-7}$. At such a low concentration additional spheres are typically several hundred μm away from the trapped beads and thus do not interfere with the measurement either by hydrodynamically coupling to the beads or by diffusing into the traps. For some experimental runs, 1 M of NaCl was added to the solution to discriminate between hydrodynamic interactions and possible effects of surface charges. However, no difference in the data obtained from either solution was discernible. This is consistent with the fact that although van der Waals and Coulombic forces are significant at this experiment's force sensitivity, they do not vary appreciably over the distance the beads move in their traps and thus do not contribute to the cross correlations. The solutions were hermetically sealed in a sample cell with a depth of approximately 100 μm and a width of 18 mm. The optical potential was applied by means of a dual-beam optical tweezers apparatus. Two orthogonally polarized beams from an Nd:YAG laser at $\lambda = 1064$ nm with an intensity of 80 mW each were focused with an immersion-oil microscope objective (Olympus PlanApo 60 $\times$ 1.4) into the sample, with the focal plane lying approximately at a depth of 20 μm inside the sample cell. Each of the laser beams holds one of the microspheres in its focus, providing the harmonic potential wells for our experiment. The lateral separation between focal spots and thus the mean separation E between the particles along the x axis was varied between 2 and 15 μm. The position of the beads was measured by imaging the light scattered from the spheres onto quadrant photodiodes. For this purpose, a microscope objective (20 $\times$ 0.4) is placed on the other side of the sample cell. A polarizing beam splitter separates the light from the two traps before it is focused onto the quadrant photodiodes. A sketch of the apparatus is shown in Fig. 1. To reduce the polarization cross talk and interference phenomena between the two traps, the two trapping beams are chopped alternately at a frequency of 100 kHz. Synchronous data acquisition yields positional data for each of the particles that were contaminated by less than a few parts per thousand from cross talk between the traps. Typically, 10^7 data points representing the position of the particles in their traps were acquired at a rate of 50 kHz for each measurement, allowing us to measure forces as low as 10 fN. Subsequent data processing consisted of subtraction of a base line stemming from the dark current of the photodiode and normalization by the photodiode sum intensity to account for laser power fluctuations. Eventually, the autocorrelation functions for each of the particles as well as their cross correlation was calculated. From the latter, an offset resulting from long-term drifts of the experimental apparatus was subtracted. Representative correlation functions are shown in Fig. 2.

The optical traps were calibrated by measuring the autocorrelation function of a bead in one trap with the other trap empty. One expects to find an exponential relaxation whose time constant τ_x is the friction coefficient of the bead divided by the lateral spring constant k of the trap. The friction coefficient is known to within a few percent, and thus the trap strength can be determined. The spring constants of the traps were balanced to within a few percent. In the experimen-

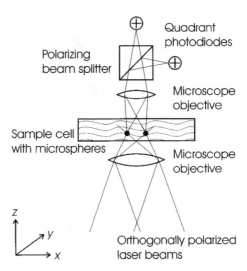

FIG. 1. Schematic diagram of the optical tweezer apparatus. Two orthogonally polarized laser beams are focused into the sample cell, where each of them holds a microsphere. The light scattered from the microspheres is collected with a second microscope objective, separated by a polarizing beam splitter, and focused onto a position sensitive quadrant photodiode. Data points are acquired with 20 μsec time resolution and an ultimate position resolution of ~1 nm. The force sensitivity is ~200 fN/$\sqrt{\text{Hz}}$.

tally obtained autocorrelation functions we also see a second exponential with a different time constant, both with and without a second bead present. This second time constant is typically an order of magnitude longer than τ_x, and the corresponding amplitude is about 20% of the principal exponential. We attribute this second time constant to the motion of the bead along the weaker z axis of the trap, which couples to a small

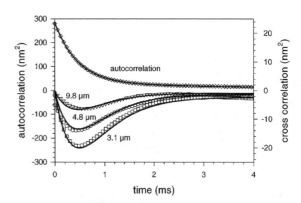

FIG. 2. Longitudinal correlation functions of the position of the two beads. The upper curve shows the autocorrelation function of a single bead in its trap, together with a double exponential fit. The lower curves show the cross-correlation functions of two beads held at separations of 9.8, 4.8, and 3.1 μm, respectively, together with the theoretically predicted curves, as detailed in the text. Only every third of the experimentally obtained data points is shown.

degree into the detector signal. This notion was verified by changing the depth of the plane of focus inside the sample cell, thus changing the trap strength in the z direction. The time constant of the second exponential changed accordingly. Since the two time constants differ vastly, the contributions from the motion of the bead along either axis are readily distinguishable. Indeed, a double exponential decay fits the experimentally obtained autocorrelation functions perfectly, yielding the time constants for motion in the x and z axes and a calibration factor for the amplitudes. A similar effect has been observed with localized dynamic light scattering [13].

For a theoretical framework to understand the auto- and cross-correlation functions, we utilize the Langevin equation [2] for the stochastic motion of particles in a fluid and external potential. The equations of motion for the particles are

$$\frac{d\mathbf{R}_n}{dt} = \sum_{m=1}^{2} \mathbf{H}_{nm}(\mathbf{R}_n - \mathbf{R}_m)[-k\mathbf{R}_m + \mathbf{f}_m(t)]. \quad (1)$$

Brownian forces are represented by randomly fluctuating functions $\mathbf{f}_m(t)$ which satisfy the following correlations:

$$\langle \mathbf{f}_m(t) \rangle = 0; \qquad \langle \mathbf{f}_n(t)\mathbf{f}_m(t') \rangle = 2\mathbf{H}_{nm}^{-1}k_B T \delta(t - t'). \quad (2)$$

The hydrodynamic interactions of the particles with the surrounding fluid are described by their mobility matrix $\mathbf{H}_{nm}$, which is also known as the Oseen tensor:

$$\mathbf{H}_{nn}(\mathbf{R}) = \frac{\mathbf{I}}{\zeta}; \qquad \mathbf{H}_{nm}(\mathbf{R}) = \frac{1}{8\pi\eta R}(\mathbf{I} + \hat{\mathbf{R}}\hat{\mathbf{R}}). \quad (3)$$

$\zeta = 6\pi\eta a$ is the friction coefficient of a sphere of radius a in a solvent with viscosity η, $\mathbf{I}$ denotes the 3×3 unity matrix, and $\hat{\mathbf{R}}$ is the unity vector parallel to $\mathbf{R}$. Higher-order corrections to the matrix elements in Eq. (3) are small, scaling as $(a/E)^4$ for the diagonal elements and $(a/E)^3$ for the off-diagonal elements [14]. Under our experimental conditions, these corrections are always smaller than 1% or 3.5%, respectively. Since the coupling in Eq. (3) is nonlinear, there is no general closed-form solution to Eq. (1). However, since individual beads move only with a rms amplitude of 16 nm in the trap, $\mathbf{R}_2 - \mathbf{R}_1 \approx E\hat{\mathbf{x}}$, and thus to a good approximation $\mathbf{H}_{nm}$ is constant and Eq. (1) is linear.

It is then a straightforward calculation to find the normal coordinates in which Eq. (1) decouples, and then the correlation functions for the vector components R_i ($i = x, y, z$) can be directly calculated:

$$\langle R_{1,i}(t)R_{1,j}(0) \rangle = \langle R_{2,i}(t)R_{2,j}(0) \rangle$$

$$= \delta_{ij}\frac{k_B T}{2k_i}(e^{-t(1+\varepsilon_i)/\tau_i} + e^{-t(1-\varepsilon_i)/\tau_i}), \quad (4)$$

$$\langle R_{1,i}(t)R_{2,j}(0) \rangle = \langle R_{2,i}(t)R_{1,j}(0) \rangle$$

$$= \delta_{ij}\frac{k_B T}{2k_i}(e^{-t(1+\varepsilon_i)/\tau_i} - e^{-t(1-\varepsilon_i)/\tau_i}), \quad (5)$$

where the fundamental relaxation time $\tau_i = \zeta/k_i$ is determined by the trap strength k_i and the friction of the bead ζ. The dimensionless parameter ε_i describes the ratio between the mobility of the beads and the strength of the hydrodynamic coupling between them, which amounts to $\varepsilon_x = 3a/2E$ for motion in the longitudinal axis of the beads and $\varepsilon_y = \varepsilon_z = 3a/4E$ for the transverse axis. In our experiment, a typical value of τ_x was 0.45 ms, which corresponds to a trap stiffness of 18.5 pN/μm.

Armed with the analytical expressions for the correlations between the spheres we can now interpret the experimental results. First, we note that the autocorrelation functions in Eq. (4) consist of two exponentials with equal amplitude and time constants that are very close to the fundamental relaxation time of the traps τ_x, compared to a single exponential decay with twice the amplitude and a relaxation time τ_x for a single trapped bead in absence of any hydrodynamic interactions. In fact, the change in the autocorrelation functions due to the presence of the second bead turns out to be so small that it is not noticeable in the experimentally obtained autocorrelation functions. However, the split in the time constants dominates the cross-correlation function [Eq. (5)]. Physically, it reflects the asymmetry of the hydrodynamic interaction: Since one sphere tends to drag the other in its wake, correlated fluctuations relax faster than anticorrelated fluctuations, in which the fluid between the spheres must be displaced.

Since τ_x is known from the trap calibration, the cross-correlation function [Eq. (5)] can be predicted exactly with no free parameters. A small correction accounting for the coupling in the z direction can also be computed. This correction is calculated with Eq. (5) using the time constant τ_z and the amplitude of the secondary exponential from the autocorrelation functions. It mostly affects the shape of the tails of the curves at longer times, while it remains below 3% near the minimum. The result of this procedure is shown for three representative curves together with the actual experimental data in Fig. 2. In the transverse direction (data not shown) the cross-correlation functions are in quantitative agreement with Eq. (5), verifying the directional dependence of ε.

The cross-correlation curves exhibit a time-delayed anticorrelation with a pronounced minimum at $t_{min} = (\tau_i/2\varepsilon_i)\ln[(1 + \varepsilon_i)/(1 - \varepsilon_i)] \approx \tau_i$. The depth of the minimum

$$\langle R_{1,i}(\tau_i)R_{2,i}(0) \rangle \approx -\frac{1}{e}\frac{k_B T}{k_i}\sinh(\varepsilon_i) \approx -\frac{1}{e}\frac{k_B T}{k_i}\varepsilon_i \quad (6)$$

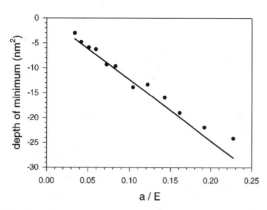

FIG. 3. Depth of the minimum in the longitudinal cross-correlation functions as a function of the ratio between the radius of the spheres a and their average separation E. The prediction from the linear approximation in Eq. (6) is shown as the straight line for comparison. No corrections for motion along the z axis have been included.

is in good approximation proportional to the strength of the hydrodynamic coupling ε, which in turn is inversely proportional to the separation between the beads. The shapes of the cross-correlation curves are almost self-similar; if we normalize them by the depth of their minimum, they are nearly indistinguishable. The error stemming from the linearizations in Eq. (6) is smaller than 2% for all data points.

Measuring directly the depth of the minimum in the longitudinal cross correlation as a function of a/E for a large number of different separations between the beads confirms the expected linear relationship from Eq. (6), as depicted in Fig. 3. Since the correction for motion along the z axis is negligible on such a short time scale, no such corrections have been included here. Corrections due to the slight mismatch of spring constants are of second order and are neglected. The last data point in Fig. 3 is in a statistically significant disagreement with the theoretical prediction, which corrections to the various linearizations in our model cannot account for. However, for that point the gap between the spheres is as small as the wavelength of the light used for the traps. Thus the deviation is

likely a systematic error due to double scattering from the spheres or optical near-field effects.

In conclusion, we have directly measured the effects of hydrodynamic coupling on the dynamics of two particles held in potential wells and shown that the observed time-delayed anticorrelation between the particles can be understood in the framework of Langevin dynamics. The hydrodynamically coupled spheres also serve as a general model system and might help in understanding microscopic biological dynamics. It is conceivable that proteins, organelles, or even cells use hydrodynamic correlations to synchronize signaling or other collective behavior.

We thank John Doyle, John Brady, Noel Corngold, and Michael Brenner for helpful discussions. This work was supported by NSF CAREER Grant No. PHY-9722417.

[1] T. G. M. van de Ven, *Colloidal Hydrodynamics* (Academic, London, 1989).

[2] M. Doi and S. F. Edwards, *The Theory of Polymer Dynamics* (Oxford Press, Oxford, 1994).

[3] J. G. de la Torre and V. A. Bloomfield, Q. Rev. Biophys. **14**, 81 (1981).

[4] S. Hayward and N. Go, Annu. Rev. Phys. Chem. **46**, 223 (1995).

[5] T. Ichiye and M. Karplus, Proteins Struct. Funct. Genet. **11**, 205 (1991).

[6] A. Kitao, F. Hirata, and N. Go, J. Chem. Phys. **158**, 447 (1991).

[7] D. Brune and S. Kim, Proc. Natl. Acad. Sci. U.S.A. **91**, 2930 (1994).

[8] J. C. Crocker and D. G. Grier, Phys. Rev. Lett. **73**, 352 (1994).

[9] T. Sugimoto, T. Takahashi, H. Itoh, S. Sato, and A. Muramatsu, Langmuir **13**, 5528 (1997).

[10] J. C. Crocker, J. Chem. Phys. **106**, 2837 (1997).

[11] E. M. Purcell, Am. J. Phys. **45**, 3 (1977).

[12] B. C. Kuo, *Automatic Control Systems* (Prentice-Hall, Englewood Cliffs, NJ, 1987).

[13] R. Bar-Ziv, A. Meller, T. Tlusty, E. Moses, J. Stavans, and S. A. Safran, Phys. Rev. Lett. **78**, 154 (1997).

[14] G. K. Batchelor, J. Fluid Mech. **74**, 1 (1976).

Micromechanics of Dipolar Chains Using Optical Tweezers

Eric M. Furst and Alice P. Gast*

Department of Chemical Engineering, Stanford University, Stanford, California 94305-5025
(Received 7 December 1998)

Here we present our initial study of the micromechanical properties of dipolar chains and columns in a magnetorheological (MR) suspension. Using dual-trap optical tweezers, we are able to directly measure the deformation of the dipolar chains parallel and perpendicular to the applied magnetic field. We observe the field dependence of the mechanical properties such as resistance to deformation, chain reorganization, and rupturing of the chains. These forms of energy dissipation are important for understanding and tuning the yield stress and rheological behavior of an MR suspension. [S0031-9007(99)09212-1]

PACS numbers: 82.70.Dd

Optical trapping techniques have rapidly grown into significant tools for studying the microscopic structure, mechanics, and interactions in biological, colloidal, and macromolecular systems. The first single-beam optical gradient force trap, or "laser tweezer," was realized by Ashkin and co-workers and later used to manipulate colloidal particles and living cells [1–3]. Since then, the ability to probe forces on the order of piconewtons down to subnanometer length scales using laser tweezers has led to direct measurements of the mechanics of motor proteins [4], the forces of transcription [5], thermal-scale forces in colloidal systems [6], and the elastic properties of polymers [7].

Here we demonstrate a new application of optical trapping techniques to study the mechanics and interactions of tunable, induced-dipolar colloidal suspensions through direct manipulation at the microscopic level. The magnetorheological suspensions we use consist of superparamagnetic particles dispersed in a nonmagnetic fluid. The particles acquire dipole moments in an external magnetic field $\mathbf{H}$, $\mathbf{m} = \frac{4}{3}\pi a^3 \mu_0 \chi \mathbf{H}$, where a and χ are, respectively, the particle radius and susceptibility. At field strengths sufficient to overcome thermal energy kT, the particles align in the field direction to form long chains. We characterize the strength of the interaction with $\lambda = U_{max}/kT$, the maximum attraction of two particles in contact and aligned in the field relative to kT, where

$$U = \frac{1}{4\pi\mu_0} \frac{\mathbf{m}_1 \cdot \mathbf{m}_2 - 3(\hat{\mathbf{r}} \cdot \mathbf{m}_1)(\hat{\mathbf{r}} \cdot \mathbf{m}_2)}{r^3} \quad (1)$$

is the point dipolar potential, $\mathbf{r}$ is the vector between particle centers, and $\hat{\mathbf{r}}$ is the unit vector $\mathbf{r}/r$.

In concentrated suspensions, the magnetic field-induced aggregation of the dispersed particles is responsible for a rapid rheological transition which exhibits a finite yield stress, analogous to the electrorheological (ER) effect [8]. The yield stress is attributed to the deformation and subsequent rupturing of particle chains, and motivates our interest in studying the mechanics of these processes [9,10]. This unique, tunable rheological response is

ideal for interfacing mechanical systems to electronic controls [11].

Our laser tweezers consist of two independently controlled traps, which enable us to manipulate magnetic chains directly through bending and stretching deformations. A single beam from the 488 nm line of a 5 W Ar^+ laser (Lexel Corporation) is expanded and passes through a polarizing cube splitter. The beams recombine to illuminate the back aperture of a 100× NA 1.2 oil immersion microscope objective, creating two regions of highly focused light in our sample. The sample-objective assembly is located at the center of Helmholtz coils. The intensity of each trap beam at the back aperture of the objective is approximately 10 mW, and the angle of incidence for each beam is controlled by a motorized gimbal mirror and 1:1 telescope. During our experiments, one trap is held stationary, while the other is translated either parallel or perpendicular to the magnetic field. Images are captured on a CCD camera (8 bit, 640 × 480 pixels), and stored with a S-VHS video recorder (JVC BR-S622U) for further image processing and analysis (NIH IMAGE).

We study magnetic chains of polystyrene (PS) microspheres embedded with monodomain (11 nm) iron oxide particles that cause the beads to exhibit a superparamagnetic response (Bangs Laboratories [12]). Using transmission electron microscopy (TEM), we find that the nominal size of the particles is 0.85 μm; however a small population of polydisperse beads with an average diameter of 0.4 μm is also present. Strong absorption and forward scattering forces from the iron oxide prevent us from directly trapping the beads forming the magnetic chains. Instead, we create "tethers" from streptavidin-coated nonmagnetic PS microspheres ($2a = 3.5$ μm) carrying a few biotinylated magnetic beads. Biotin and streptavidin interact with a high affinity, $K_a = 10^{15} M^{-1}$ [14], that is well suited here as a strong, inert adhesive. The field-induced chains incorporate the tethers somewhere along the chain, which are then captured by the laser traps and used to extend and bend the chains, as illustrated in Fig. 1. In our experiments, the magnetic and nonmagnetic beads

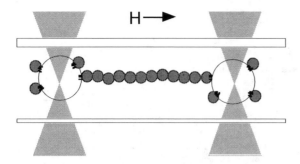

FIG. 1. Two laser beams are focused to form independent laser tweezers in a colloidal suspension between a coverslip and microscope slide. The tweezers hold streptavidin-coated PS tether particles attached to biotinylated magnetic particles chaining in the field. By translating the traps, we perform micromechanical measurements of dipolar chains.

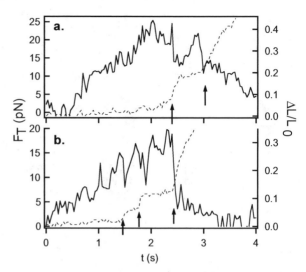

FIG. 2. Linear strains (dashed lines, right axis) and tensions (solid lines, left axis) of dipolar chains as the tether particles are pulled apart. Arrows mark chain reorganization and rupture events. (a) $\lambda = 480$ and (b) $\lambda = 740$.

are dispersed between a coverslip and microscope slide at volume fractions $\phi_{\mathrm{mag}} = 10^{-4}$ and $\phi_{\mathrm{PS}} = 10^{-6}$, respectively, and stabilized using 2.351 g/L sodium dodecyl sulfate (SDS).

We characterize the trapping force F_t on the tether particles for each tweezer by holding the trap stationary while translating the sample stage. The displacement of the particle in the trap, due to the viscous drag force imparted on the sphere by the surrounding fluid, is measured, correcting the drag coefficient to account for the hydrodynamic effect of the wall [15]. The center of the tether particles for both the calibration and the tension measurements of dipolar chains is found by locating the brightest point with an intensity-weighting algorithm, which is accurate to within one-tenth of a pixel (8 nm) [16]. We monitor the image processing algorithm for accuracy. In the calibration measurements, the drag force applied when the particle escapes is the maximum axial trapping force of the laser tweezer. The maximum measured trapping forces are $F_{t,1}^{\mathrm{max}} = 56 \pm 3$ pN and $F_{t,2}^{\mathrm{max}} = 53 \pm 2$ pN. Using the maximum displacement of the bead from the trap center, we find corresponding force constants for the traps of $k_1 = 4.5 \times 10^{-5}$ N/m and $k_2 = 4 \times 10^{-5}$ N/m.

We apply a tension F_T to the magnetic chains in the direction of the field and measure the tensile strain $\Delta L/L_0$ for the dipole strengths $\lambda = 480$, 740, and 1000. In all of the experiments discussed in this paper, the tether beads were incorporated into the magnetic chain at a sufficient distance from the ends to minimize finite-length effects of our measurements. We observe an increase of $\Delta L/L_0$ as the dipole strength decreases. For $\lambda = 480$ and $\lambda = 740$, reorganization of the chain can occur during which the tension increases and then releases by including an additional particle into the chain. The experiments described here have been reproduced at each field strength with consistent results.

Figures 2a and 2b show the tension exerted on a magnetic chain and the corresponding $\Delta L/L_0$ for $\lambda = 480$ and

$\lambda = 740$. The dashed line in Fig. 2a shows $\Delta L/L_0$ for $\lambda = 480$ increasing slowly at first, then faster after reaching a tension of ~ 25 pN. After this, a large increase in $\Delta L/L_0$ is observed, corresponding to the reorganization of the chain ($\Delta L = 0.89 \pm 0.08\ \mu$m) that relaxes the tension to ~ 13 pN, as shown. The observed chain reorganization is possibly due to the inclusion of a "satellite" particle near or attached to the chain. After this, the tension again increases until the chain ruptures at 22 pN.

In Fig. 2b, $\Delta L/L_0$ for $\lambda = 740$ increases slowly to 0.02, then steps rapidly to 0.05. An additional step is seen a short time later, from 0.065 to 0.11. Again, the steps correlate to reductions in the measured tension exerted on the chain after rapid increases. After the second rapid increase, the strain nearly plateaus, increasing only to 0.125, and finally breaking at a tension of 20 pN. It appears from the strain that satellite particles approximately 0.32 and 0.56 μm in diameter are included in this case. These figures are consistent with our TEM characterization of the particle sizes; however, at this time, we cannot optically resolve whether single or multiple particles are incorporated into the chain during reorganizations. Other mechanisms exist as well that could explain the increases, such as the discontinuous inclusion of a single large particle.

At the highest field strength, $\lambda = 1000$, $\Delta L/L_0$ increases monotonically to 0.031, at which time the tether particle escapes. The maximum tension applied to the chain is thus the trap strength, $F_T^{\mathrm{max}} = 53 \pm 2$ pN.

Qualitatively, the rupturing tension increases with field strength from $\lambda = 480$ and 740 to the $\lambda = 1000$ case. The fact that the tension did not increase for $\lambda = 480$ and 740 reflects a defect in the chain for the latter case possibly due to the inclusion of smaller particles during the reorganization events. The initial strain decreases with increasing interaction strength and exhibits reorganizations at $\lambda = 480$ and 740. Lateral chain fluctuations could be responsible for the dependence on the tensile strain with interaction strength, since stronger interactions suppress lateral thermal motion, and give the tendency toward straighter, more rigid chains [17,18]. The phenomenon of chain reorganization under applied tension has not been studied in detail until now. As a source of energy dissipation, it may have important implications for the yield stress of magnetorheological (MR) suspensions.

We calculate the interaction force between 0.85 μm electrostatically stabilized dipolar particles for comparison with our measurements of the breaking tension. Our treatment of the point-dipole interaction potential includes local field corrections from mutual induction between particles. The particle dipole moment is due to the applied and local fields, $\mathbf{H} = \mathbf{H}_0 + \mathbf{H}_{\text{loc}}$, where $\mathbf{H}_{\text{loc}} = [3\hat{\mathbf{r}}(\mathbf{m} \cdot \hat{\mathbf{r}}) - \mathbf{m}]/4\pi\mu_0 r^3$ is the dipolar field of a neighboring particle. The moment is found in a self-consistent manner, and we include long-range interactions along the chain [10,19]. For PS particles in SDS, we find the electrostatic double layer interaction using the Derjaguin approximation [20]. The surface potential is estimated as $\psi_0 = 40$ mV [21] and the Debye length is 0.29 nm^{-1}. The calculated attractive force rapidly achieves a maximum, then decreases. The maximum corresponds to the rupturing tension, since the chain would be mechanically unstable if pulled at higher forces. The calculated rupturing tensions are 24, 37, and 52 pN for $\lambda = 480$, 740, and 1000, respectively. In the $\lambda = 480$ and 740 cases, the calculations are in good agreement with experimental breaking tensions. The increased tension at $\lambda = 1000$ is consistent with the ability of the chain to resist breaking past the maximum trapping force of the tweezer. As a comparison, the maximum attractive force between particles neglecting the induced field along the chain yields maximum forces of 5.9, 9.2, and 13 pN for $\lambda = 480$, 740, and 1000, far below the measured values. This highlights the need to account for induction and long-range interactions.

In contrast to the relative deformability of single-particle dipolar chains, columns created from laterally coalesced chains show greater resistance to applied stresses. Figure 3 illustrates the increase in the tensile strain of a short dipolar column as the dipole strength is decreased. In Fig. 3, a column approximately 7 μm long is held with the two traps and subjected to an increasing tension. At $\lambda = 740$, 480, and 360, the translating tether particle releases from the trap before the column ruptures. While the column at $\lambda = 740$ and 480 shows no reorganization during the deformation, at 360 the chain rapidly stretches breaks in multiple places immediately before pulling the tether from the translating trap, resulting in an irreversible deformation characteristic of materials stretched past their yield point. This interesting result suggests that multiple reorganizations provide a mechanism for the chain to acquire a mechanically stronger configuration, analo-

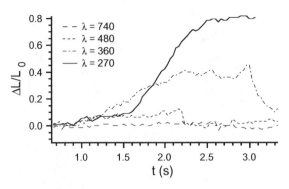

FIG. 3. $\Delta L/L_0$ for columnar chain aggregates. The resistance to deformation decreases as the dipole strength decreases.

gous to shear-induced hardening in polymeric materials. In contrast, at $\lambda = 270$ the chain offers an initial resistance and slow deformation, then we observe a succession of ruptures as multiple chain segments slide past one another. A short time after beginning this "slipping" motion, the column breaks completely, maintaining mechanical resistance at field strengths far below those observed for single chains. The amount of additional strength that lateral aggregation provides in the MR system is appreciable considering that the calculated maximum attractive force between two dipolar particles, including induction at $\lambda = 360$ and 270, is 17 and 13 pN, respectively.

Since magnetorheological systems are primarily useful for their ability to support shear stresses perpendicular to the applied magnetic field, a full understanding of the micromechanical properties and failure of dipolar chains deformed orthogonally to a field should provide insight into the macroscopic yield stress. As shown in Fig. 4, our experiments provide us with the unique ability to measure the deformation and rupture of chains by pulling one of the tether particles perpendicular to the applied field. We measure the rupturing tension and shear strain and compare them to calculated values of the interparticle force versus applied shear strain using methods described above, assuming that the chains undergo an affine deformation as the tether particle is pulled [9]. The maximum attractive force identifies the rupturing stress and strain of the chain.

We observe chains bending at field strengths $\lambda = 270$, 480, 740, and 1000, and, in every case, the chain breaks. We measure the strain between the two tether contact points immediately before the chain ruptures. The observed rupturing strain does not depend on field strength, and averages 0.16 ± 0.03. This is in excellent agreement with the calculated rupturing strain of 0.165, which is also field independent. These rupture strains are consistent with rupture strains in ER systems assuming affine deformation [9]; however, our values are less than calculations based on rigid chains [10]. It is interesting to note that, before rupturing, the chain often develops a discontinuity where the angle is much steeper than the rest

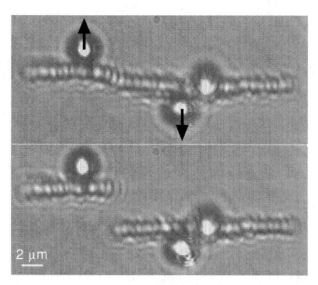

FIG. 4. Dipolar chain deformed orthogonally to **H** (left to right in images, $\lambda = 1000$) immediately before and after rupturing. The arrows indicate the tether particles held by the laser tweezers. The top tether is translating upwards, while the bottom tether is experiencing a restoring force from the laser trap.

of the bent chain, as shown in Fig. 4. Eventually the chain fails at this defect. The steepest angle in the chain before rupturing is independent of field strength, and varies from approximately 27° to 32°.

The stationary tether provides an accurate measurement of the force applied to the chain during the bending process. The maximum force applied to the dipolar chain before a rupture occurs increases with the dipole strength. We measure rupturing forces of 6.4 ± 3, 21 ± 3, 32 ± 3, and 45 ± 3 for $\lambda = 270$, 480, 740, and 1000, respectively, in agreement with the corresponding calculated rupturing tensions of 13, 23, 36, and 50 pN. The errors are estimated based on the uncertainty of the trap force constant. At $\lambda = 270$, the measured rupturing force is about half the calculated value of 13 pN. At lower dipole strengths, the increased lateral fluctuations in the chain [17] play a larger role in the rupturing process, since Brownian motion can move adjacent particles into a state of mechanical instability.

Manipulation of colloidal materials with optical traps has developed into a powerful *quantitative* technique for investigating fundamental interactions, structures, and mechanics. The micromechanics of dipolar chains investigated here have implications for accurate calculations of the macroscopic yield stress in magnetorheological suspensions. Reorganization events may play a particularly important role for configurational rearrangements that cause a shear-hardening effect in multiple-chain columns. Also, microscopic models of the rheological behavior that do not account for the increased chain strength due to induction

and chaining effects as well as lateral aggregation may significantly underestimate the yield stress. In the near future, we plan to continue our micromechanical experiments by examining the properties of novel *permanently linked* dipolar chains [22], which should have unique microscopic and rheological properties.

The authors are grateful to D. Grier, S. Chu, H. Babcock, and D. Smith for their optical trapping expertise, and thank M. Fermigier for fruitful discussions. Support by NASA (Grant No. NAG3-1887-1) is gratefully acknowledged.

*Author to whom all correspondence should be addressed.
[1] A. Ashkin, Science **210**, 1081 (1980).
[2] A. Ashkin, J. M. Dziedzic, J. E. Bjorkholm, and S. Chu, Opt. Lett. **11**, 288 (1986).
[3] A. Ashkin, J. M. Dziedzic, and T. Yamane, Nature (London) **330**, 769 (1987).
[4] J. T. Finer, R. M. Simmons, and J. A. Spudich, Nature (London) **368**, 113 (1994).
[5] H. Yin *et al.*, Science **270**, 1653 (1995).
[6] A. Larsen and D. G. Grier, Nature (London) **385**, 230 (1997).
[7] T. T. Perkins, D. E. Smith, R. G. Larson, and S. Chu, Science **268**, 83 (1995).
[8] Z. Shulman, R. Gorodkin, E. Korobko, and V. Gleb, J. Non-Newton. Fluid Mech. **8**, 29 (1981).
[9] D. Klingenberg and C. Zukoski, Langmuir **6**, 15 (1990).
[10] J. E. Martin and R. A. Anderson, J. Chem. Phys. **104**, 4814 (1996).
[11] O. Ashour, C. Rogers, and W. Kordonsky, J. Intell. Mater. Syst. Struct. **7**, 123 (1996).
[12] For a spherical particle, the susceptibility is $\chi = \chi_i/(1 + \frac{1}{3}\chi_i)$, where the intrinsic susceptibility of the iron oxide monodomains follows the classical Langevin form $\chi_i = \alpha[\coth(\beta H_0) - 1/\beta H_0]/H_0$. From Ref. [13], we find $\alpha = 9.37 \times 10^4$ A/m and $\beta = 9.43 \times 10^{-5}$ m/A, using an iron oxide volume fraction of 0.21 and an average iron oxide grain size of 11 nm.
[13] R. E. Rosensweig, *Ferrohydrodynamics* (Cambridge University Press, New York, 1985).
[14] N. M. Green, Adv. Protein Chem. **29**, 85 (1975).
[15] J. Happel and H. Brenner, *Low Reynolds Number Hydrodynamics* (Prentice-Hall, Englewood Cliffs, NJ, 1965).
[16] J. C. Crocker and D. G. Grier, J. Colloid Interface Sci. **179**, 298 (1996).
[17] E. M. Furst and A. P. Gast, Phys. Rev. E **58**, 3372 (1998).
[18] A. Silva, B. Bond, F. Plouraboué, and D. Wirtz, Phys. Rev. E **54**, 5502 (1996).
[19] H. Zhang and M. Widom, Phys. Rev. E **51**, 2099 (1995).
[20] B. Derjaguin, Trans. Faraday Soc. **36**, 203 (1940).
[21] W. Brown and J. Zhao, Macromolecules **26**, 2711 (1993).
[22] E. M. Furst, C. Suzuki, M. Fermigier, and A. P. Gast, Langmuir **14**, 7334 (1998).

Optical Binding

Michael M. Burns,[1] Jean-Marc Fournier,[1] and Jene A. Golovchenko[1,2]

[1]*Rowland Institute for Science, Cambridge, Massachusetts 02142*
[2]*Harvard University, Cambridge, Massachusetts 02138*

(Received 5 June 1989)

Significant forces between dielectric objects can be induced by intense optical fields. We discuss the origin of these forces which are very long range and oscillate in sign at the optical wavelength. A consequence is that light waves can serve to bind matter in new organized forms. We experimentally demonstrate the simplest case by observing a series of bound states between two 1.43-μm-diam plastic spheres in water, and discuss the extension to more complex cases.

PACS numbers: 41.10.Fs, 42.20.Ji, 78.90.+t

Ashkin[1-4] and others[5] have shown that gradients of time-averaged optical fields can produce forces on microscopic dielectric objects, and by fashioning proper optical gradients have demonstrated trapping. Here we call attention to the apparently little-known fact that intense optical fields can induce significant forces *between* microscopic aggregates of dielectric matter. These forces can result in new ordered states of matter with attendant possibilities for the manipulation and study of systems ranging from small "optical molecules" to extended condensed-matter systems.

The basic ideas can be illustrated in the idealized system of two interacting oscillators separated by a distance **r**, each consisting of a light particle of mass m and charge e, harmonically bound with resonant frequency ω_0 to a heavy mass of opposite charge. An external optical field, incident on these oscillators, is plane polarized with wave vector $|\mathbf{k}| = \omega/c$. For simplicity we consider the electric field vector such that $\mathbf{E} \times \mathbf{r}$ is perpendicular to **k**.

In a first approximation the internal motions of each oscillator respond to the Lorentz forces from the external fields alone, which in turn give rise to scattered fields. These scattered fields can be described as originating from a point dipole of time-dependent moment **p** located at the position of the heavy mass of each oscillator. The scattered fields from each oscillator acting on its already excited neighbor gives rise to mutual forces between the oscillators.

There are two types of forces that dominate the interaction between the oscillators. The first arises from the induced dipole moment of one oscillator acted on by the gradient of the scattered electric field from the other oscillator. This force is already of a much longer range than those of the standard van der Waals type, since the induced moment (due to the external field) stays constant as the oscillator separation is increased. Perhaps more interesting, however, is the second, magnetic, force that arises from the interaction between the induced currents in the two oscillators. Here the interaction simply results from a Lorentz-force term involving the cross product of the time derivative of the oscillator's dipole moment with the scattered magnetic-flux density from its neighbor. When the vector separation between the two induced currents is perpendicular to the electric field polarization of the incident field this force dominates all others in the scattered-field radiation zone. Again the induced moment and its time derivative stay constant for increasing separation while the radiation field for the magnetic-flux density falls off only inversely with the separation; hence the range is longer than might at first be expected.

For the perturbation treatment discussed above the time dependence of the forces comes from the product of a dipole moment or its time derivative and an electric field gradient or magnetic-flux density. Since both contain the harmonic time dependence of the incident external field, the induced forces between oscillators will have two Fourier time components, one at twice the frequency of the external fields and one at zero frequency. Although the high-frequency forces can have interesting consequences, it is the static (or time-average) part of the force with which the rest of this discussion will deal because it leads to optical binding.

Notice that the static forces we are considering are similar to those normally associated with the radiation pressure of light falling on a single isolated oscillator. In that case it is the phase shift in the internal oscillator motion ultimately connected to the reradiation of the incident field that gives rise to the static force of radiation pressure. In our case it is the phase shift associated with retardation between the oscillators which gives rise to the static internal forces between the oscillators (at least for the long-range part of the interaction). We should therefore not be surprised that spatially oscillatory forces of either sign should arise, because changing the distance between the oscillators results in all possible phase shifts between the induced moments and scattered fields.

Quantitatively the above effects are calculated by solving the coupled classical Maxwell-Lorentz equations self-consistently for the total fields and dynamical state of each oscillator. One then finds for the interaction energy, W, between the dipole oscillators

$$W = -\tfrac{1}{2}\alpha E^2[yf(x) + y^2g(x) + O(y^3)], \qquad (1)$$

with $y = \alpha k^3$ the expansion parameter, $x = kr$, $\beta = \sin^2\vartheta$, and

$$f(x) = [(\beta x^2 + 2 - 3\beta)\cos x + (2 - 3\beta)x\sin x]/x^3,$$

$$g(x) = [4(1 - \beta)(\cos x + x\sin x)^2$$
$$+ \beta(\cos x + x\sin x - x^2\cos x)^2]/x^6.$$

Here ϑ measures the angle between E and the direction connecting the oscillators, and $\alpha = e^2/m(\omega_0^2 - \omega^2)$ is the polarizability of the oscillator. When $kr \gtrsim 2\pi$ and $\beta = 1$ a satisfactory approximation for the interaction energy is

$$W = -\tfrac{1}{2}\alpha^2 E^2 k^2 \cos(kr)/r. \qquad (2)$$

Bound states are obtained at separations near multiple wavelengths, with well depths that fall off inversely with distance between the oscillators.

It is a remarkable fact that even if the oscillators are placed in a conducting fluid (as in our experiments) the static forces responsible for the binding described above are not screened out by free charges. Only plane-wave propagation at high frequencies is needed for the existence of the long-range static influences of interest.

For binding to occur at temperature T we require that $W \gtrsim k_B T$ at the first well ($kr \cong 2\pi$). This constrains the polarizability for which one can achieve significant binding to

$$\alpha > (4\pi k_B T/E^2 k^3)^{1/2}. \qquad (3)$$

To achieve binding experimentally an intense laser illuminates the oscillators which are realized as small dielectric spheres of radius a and relative index of refraction n for which the polarizability is given by

$$\alpha = a^3(n^2 - 1)/(n^2 + 2) \quad (ak < 1). \qquad (4)$$

The experimental setup is shown in Fig. 1. A 2.2-W argon-ion laser beam (0.387-μm wavelength in water) is

focused by two cylindrical lenses of different focal lengths to form a narrow line about 3 μm wide and 50 μm long at the common focal plane in the sample cell. The polarization of the beam is chosen so that E is perpendicular to the long axis of this trap. The sample (polystyrene spheres in water) is viewed either by diffraction of the incident beam on a screen 20 cm from the cell, or via an imaging system that collects the light from the incident beam in a projection-microscope arrangement onto a screen (see Fig. 1).

The sample cell is made of two fused-silica plates separated by 200 μm. A dilute concentration of monodisperse 1.43-μm-diam polystyrene spheres in water (relative index 1.20) with $10^{-4}M$ phosphite buffer is placed between the plates. The screening by the buffer (Debye length about 300 Å) eliminates all static monopole forces arising from any static charge on the spheres, but (as described above) has no effect on the interactions that we seek to explore here.

By manipulating the cell while observing the projected image it is possible to put two isolated spheres into the trap, initially separated by some distance—say 5 μm —and observe their interaction with time. Because of the strong gradient of the applied field in a direction perpendicular to the focal line, the spheres are constrained to move in one dimension along the line. Radiation pressure along the beam axis presses the two spheres against the upper cell surface, assuring the interaction geometry discussed earlier.

The relative distance separating the two spheres is determined from the fringe spacing on the diffraction screen. A video camera connected to a standard VCR records the time course of the fringes. Frame-by-frame measurement of the fringe pattern yields the separation of the two spheres at $\frac{1}{30}$-sec intervals. A plot of the relative separation of two spheres (measured in units of the wavelength of the light in water) for one 17-sec period is shown in Fig. 2, along with a histogram of the separations.

It is quite evident that as the spheres move together driven by diffusive forces and a weak optical gradient, there are discrete separations at which they are more likely to be found, and that these differ by distances approximately equal to the wavelength of the light, just as the above arguments indicated. The trajectory looks like that of a particle exhibiting fluctuating Brownian motion in a potential with periodic wells. The closest approach at about 3.8 wavelengths (1.43 μm) is in fact the separation for two spheres in contact. Higher laser power would mean deeper traps, and, assuming no additional thermal heating, longer dwell times near each minimum of the potential of Eq. (2). The laser power was chosen so the system hopped between many different wells during the time window of an experimental run.

For a number of technical reasons we have not yet been able to perform our experiments with spheres so small that $ka < 1$. The problems include thermal heat-

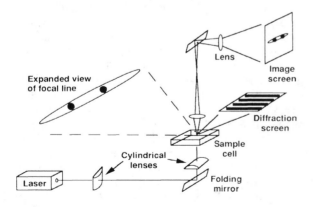

FIG. 1. Experimental setup for optical binding.

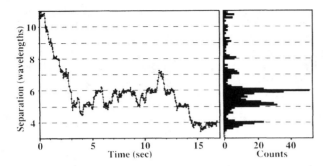

FIG. 2. Relative separation of two 1.43-μm-diam spheres measured in units of wavelength of illuminating light in water. The plot on the left shows the time course of the separation, sampled at $\frac{1}{30}$-sec intervals, and the plot on the right is the corresponding histogram.

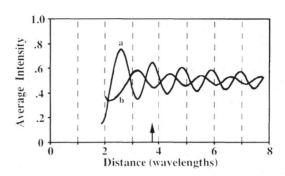

FIG. 3. Mie calculations of the time-average intensity of the near fields of (curve *a*) 1.43- and (curve *b*) 1.53-μm spheres. The arrow indicates the distance separating two identical spheres when touching.

ing of the spheres, poor image and diffraction signal-to-noise ratio, and thermal fluctuation times shorter than the sampling period. Thus while we retain the physical intuition contained in Eqs. (1)–(4), detailed numerical comparisons are inappropriate. An extension of the discrete-dipole approximation of Purcell and Pennypacker[6,7] is currently being implemented for detailed calculations. Here we take an intermediate, simplified approach to discuss our $ka \gtrsim 2\pi$ experimental case. The spheres are assumed to scatter light according to the exact (near-field) Mie equations[8] and the interference of this scattered light with the incident wave forms potential wells corresponding to those in the dipole calculation considered above. This simplified approach is similar to the introductory perturbation viewpoint in that the multiple-scattering fields's contributions are not considered. In actual fact these contributions are probably not negligible, but they affect the magnitude of the forces much more than the spatial location of the maxima and minima.

A plot of the time-average intensity near a single sphere calculated in this manner is shown in Fig. 3, curve *a*. It is expected that the spheres, reacting to the gradient force, will find potential minima at those places where peaks in the light intensity occur. Notice that the intensity peaks do asymptotically approach one-wavelength spacings at the larger distances, just as the dipole calculations predict. Note also that when the spheres are close together the peaks are separated by more than one wavelength, with the first physically accessible peak occurring at about 3.8 wavelengths—just the contact condition, and also the nearest approach noted in the experimental observations.

To appreciate the close correspondence between the data and calculations, curve *b* of Fig. 3 shows a plot of the intensity for a slightly different size sphere—here 1.53 μm. Notice that the intensity peaks occur at very different relative locations because of the changed phase shifts of the scattered partial waves. Notice also that the

intensity contrast, and hence potential well depth, is less as well. In fact, the 1.43-μm spheres are an optimum size; at any nearby size the binding potential would be less, and the first potential well would be at a distance other than the contact distance.

In conclusion, we have discussed and demonstrated the basic physics underlying the process of optical binding. This new interaction between bits of dielectric matter may well find application in the purposeful organization of both small and extended microscopic systems. Of course much more complicated structures than the one dealt with here can be envisioned; for example, increased dimensionality and number of spheres should yield a rich array of organized stable structures worthy of study in their own right. These organizations can be compared and contrasted with arrays of particles confined by mechanisms other than the one discussed here.[9,10] We mention that optically induced and organized structures formed in a liquid environment may be maintained in the absence of the applied field by freezing.

Finally we note that many readers of this journal view forces between elementary particles of nature as originating from the exchange of virtual quanta of fields to which they are coupled. The induced interaction discussed in this paper fits nicely into that scheme, but with real quanta being exchanged.[11] We wonder whether other particles and fields may be substituted for our dipoles and light, yielding analogous effects in other domains of physics.

We wish to thank Professor N. Bloembergen and Professor E. Purcell for useful discussions, Hang Xu for calculational assistance, and the Rowland Institute for Science for its continued support.

[1]A. Ashkin, J. M. Dziedzic, J. E. Bjorkholm, and S. Chu, Opt. Lett. **11**, 288 (1986).
[2]A. Ashkin, Phys. Rev. Lett. **24**, 156 (1970).

[3]A. Ashkin, Science **210**, 1081 (1980).

[4]A. Ashkin and J. M. Dziedzic, Science **187**, 1073 (1975).

[5]See references in S. Stenholm, Rev. Mod. Phys. **58**, 699 (1986).

[6]E. M. Purcell and C. R. Pennypacker, Astrophys. J. **186**, 705 (1973).

[7]B. T. Draine, Astrophys. J. **332(2)**, 848 (1988).

[8]H. C. van de Hulst, *Light Scattering by Small Particles* (Dover, New York, 1981).

[9]N. A. Clark, A. J. Hurd, and B. J. Ackerson, Nature (London) **281**, 57 (1979).

[10]C. A. Murray and D. H. Van Winkle, Phys. Rev. Lett. **58**, 1200 (1987).

[11]T. Thirunamachandran, Mol. Phys. **40**, 393 (1980).

One-Dimensional Optically Bound Arrays of Microscopic Particles

S. A. Tatarkova,* A. E. Carruthers, and K. Dholakia

School of Physics and Astronomy, University of St. Andrews, St. Andrews KY16 9SS, United Kingdom
(Received 9 August 2002; published 27 December 2002)

A one-dimensional optically coupled array of colloidal particles is created in a potential well formed by two counterpropagating Gaussian light beams. This array has analogies to linear chains of trapped atomic ions. Breathing modes and oscillations of the center of mass are observed. The stability of the array is in accordance with the Kramers model.

DOI: 10.1103/PhysRevLett.89.283901   PACS numbers: 42.60.–v, 45.50.Jf

Introduction.—The organization and manipulation of colloidal and biological matter at the microscopic level can be achieved using light forces. Gradients of the optical field can induce dielectric spheres of higher refractive index than their surrounding medium to be trapped in three dimensions in the light field maxima [1,2]. Such "optical tweezers" allow physicists to test several fundamental phenomena. Examples include thermally activated escape from a potential well [3], stochastic resonance phenomena [4], and various studies in colloid physics [5]. Recently the predetermined creation of arrays of microscopic particles, using light forces, has resulted in intense world-wide interest. Holographic methods [6], the phase contrast technique [7], the use of nonzero order light modes [8], and spatial light modulator technology [9] have successfully been used to create particle arrays in two and three dimensions. In 2D the light potential allows predesignated trap sites to be occupied by the particles of interest. Such tailored optical landscapes can give insights into mechanisms at the atomic level or, for example, the pinning of magnetic flux lines in type-II superconductivity [10].

Light forces may wholly dictate the assembly of a microscopic system and create analogs to atomic systems [10–12]. There have been a few observations of such "optical binding" notably by Golovchenko and co-workers [11,12]. They investigated systems where the interaction of coherently induced dipole moments of the spheres were said to interact to bind matter. Light forces may act to optically bind matter. These forces can organize microscopic particles with the prospects of studying "optical molecules" or systems in soft condensed matter physics. This topic has been controversial but potentially offers an important mechanism to self-assemble matter.

In this Letter we demonstrate the creation of one-dimensional coupled arrays of microscopic colloidal particles. An important distinction in our work is that the light forces that confine the particles also dictate the interparticle spacing due to light refocusing and may be deemed a form of optical binding [11,12]. Buican *et al.* [13] have studied optical guiding in a Gaussian beam and indicated the potential for creation of such regularly spaced particles. We provide a physical explanation of the creation and dynamics of the chain. We infer directly from our results quantitative information about the optical trap potential and trap oscillation frequency. Stability of the array is also investigated.

Atomic ion chains are deemed strong candidates for quantum computing. Chain dynamics can affect decoherence in these systems. Thus creating an analogous model with microscopic particles may offer valuable insights into similar dynamics at the atomic scale. Though our array separation is determined by light forces, it shows analogous behavior to a system of linear atomic ions where electrostatic forces dominate [14]. As particles in our optical trap can be considered harmonically bound, such systems can exhibit excitations, similar to those of atomic ions, including a center-of-mass motion and breathing modes. We observe such behavior in our experiments.

Experiment.—A continuous wave Ti-sapphire laser operating at 780 nm provides the trapping laser light. The beam was expanded and split into two equal (~ 25 mW) counterpropagating components which were then focused into a rectangular glass cell with their respective beam waists approximately 150 μm apart along the common axis far from walls. The focal length of the focusing lenses was 50 mm, the waist sizes were approximately 3.5 μm, and the cell outer dimensions were 5 mm $\times$ 5 mm $\times$ 20 mm. The cell was filled with uniform silica monodispersed colloidal microspheres in water of diameters 2.3 and 3 μm (Bangs Laboratories, Inc.). A microscope objective ($\times 20$, NA $= 0.4$, Newport) placed orthogonally to the laser beam propagation direction projected scattered light onto a charge-coupled device camera.

The separation of the beam waists in this counterpropagating geometry allows a single sphere to be trapped in the potential well between them (see Fig. 1); there is already tight confinement in directions transverse to the beam axis. Blocking one of the beams turns this geometry into one for optical guiding [1,13]. Theoretically, we consider the interaction of dielectric spheres of radius a with a Gaussian beam of wavelength λ when the relation $a \gg \lambda$ is satisfied. Our calculations of the axial force follow those elsewhere [15–17] with the sphere radius

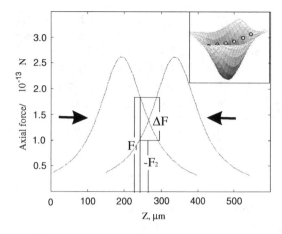

FIG. 1 (color online). Axial forces for the counterpropagating beam geometry. The peak point for each curve corresponds to the beam waist position. Forces from both beams (F_1 and F_2) are drawn as positive. The resultant axial force ΔF is the difference of two forces drawn. The inset shows how particles reside in the resulting potential well.

exceeding the trapping laser wavelength. The model allows us to calculate the variation of axial force along the beam propagation for our experimental parameters (Fig. 1). For collinear beams with slightly displaced waists (along the optic axis), a near parabolic optical potential results with the single equilibrium position at the minimum at the intersection of the curves in Fig. 1.

In the counterpropagating geometry, when a second sphere moves into the same trap the first (trapped) sphere (Fig. 2) is pushed from its equilibrium point. The new equilibrium position of two spheres in trap is above the bottom of the potential well with a separation between them as shown. Each sphere will experience the same value of differential force ΔF (Fig. 1) which can be calculated from the model. This process repeats each time a new sphere drifts into the trap region (Fig. 2). This behavior is observed with both 3 μm diameter and 2.3 μm diameter spheres. We stress at this point that the array length and importantly the interparticle spacing we observe is extremely long in comparison with typical distances over which electrostatic interactions occur in optical traps [5]. For two 3 μm particles we observe an interparticle separation of 48 μm [Fig. 2(a)].

For typical experimental conditions the maximum number of particles in a stable trapped array was seven. The overall array length was ~150 μm in this instance. Very occasionally, we observed trapping of eight or even nine spheres though this was rather unstable with the outer spheres of the array leaving the trap region very quickly. This instability is due to thermally activated loss as the outer particles within the chain have relatively weak optical potential barriers to overcome to escape the potential well. Thermal fluctuations also caused a variation in overall array length for greater than seven spheres for these trap parameters.

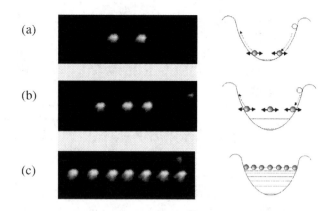

FIG. 2. Experimental data for arrays of (a) two, (b) three, and (c) seven spheres (each 3 μm in size). The diagrams on the right elucidate how we fill up the approximately harmonic potential well created by the two counterpropagating beams.

The mechanism for formation of this array is purely due to light refocusing and subsequent balancing of radiation pressure. Any given irradiated sphere refracts the majority of incident light thus acting like a lens. This creates a secondary light radiation pressure force on an adjacent sphere, that in combination with the light radiation pressure from the focused input beams, creates a new equilibrium in which the array of spheres can reside. Each sphere in the array is "optically" coupled to each of the other spheres. We have adapted a numerical ray optics ray model [17] to calculate the equilibrium position for two spheres subject to beam waists as in the experiments separated by 150 μm. The model takes into account the fraction of incident light focused by the sphere to calculate the interparticle forces. The optical force on the second sphere due to the first is calculated to be ~0.175 pN. The calculations of the model verify to within 10% the experimentally observed spacing (48 μm) and thus confirm the physical process responsible for the array generation. We also note that this explains earlier observations on two guided particles [13]. In terms of our model, spheres in the array reside in their own potential well. As an example, consider two spheres where each of the spheres focuses the light at a position approximately midway between the two spheres. In this instance this creates a situation where each of the spheres resides in one well of a double well light potential. For a greater number of spheres the idea is similar, with each sphere acting to create a new potential well within the system, forcing the particles farther up the well created by the counterpropagating beams in order to balance the forces and reach an equilibrium position. A detailed analysis will be presented elsewhere.

Electrostatic interactions could potentially play a role in the array formation. The interaction between charged spheres is based on the Derjaguin, Landau, Verwey, and Overbeek theory [5,18] which is limited by the screening length due to the atomic ions present in solution. For our experimental parameters the screening length is very short ($<$ 100 nm) and the electrostatic interaction restricted to a length scale of less than a micron, an order of magnitude lower than the interparticle spacing we observe. We dispersed the spheres in 1 M NaCl for some experimental runs and observed no discernible change in the interparticle spacing for our arrays, thus validating our premise that the array is created solely by light forces.

We have measured experimentally the equilibrium positions of the spheres for arrays from one to seven particles in length (Fig. 3). The form of these data is reminiscent of that for trapped atomic ions in a linear Paul trap [14]. We are able to extract detailed quantitative information from these experimental data such as the exact form of the trap potential and calculate the actual axial trap frequency. We have fitted the data in Fig. 3 in this way to a parabolic potential, determining the axial trap frequency to be approximately 300 Hz from our theory (Fig. 1) and our experimental parameters.

The lowest mode of oscillation corresponds to the center-of-mass motion of the particle chain. All of the particles move to and fro in unison in this mode. We observed center-of-mass motion of this array. When a chain of a given number of spheres was created, one of the trapping beams was blocked and the whole chain was observed to accelerate against the direction of propagation of the blocked light beam. Reintroducing the obstructed beam (within a few seconds) caused the chain to restore its initial position with a time scale determined by differential force at the current array position. Altering the laser power equivalently in both beams (no change in differential force; see Fig. 1) did not alter the sphere positions but did result in higher light scattering from the spheres. We also observed a breathing mode of

the trapped particle array. One of the lenses was mounted on the precision motorized z translation stage, to vary the focal position of one trapping beam. In Fig. 4 we see the corresponding breathing behavior of a long particle chain (of 2.3 μm particles) on the z displacement of one of the lenses. The reduced light pressure force on the one side of the potential well resulted in increases in interparticle spacing. As the lens returns to its original position, the chain self-restores. The time scale of the motion is very slow as the system is heavily overdamped. The dynamics are similar to center-of-mass and breathing modes in chains of linear trapped ions [14].

In the majority of research related to optical tweezing, little attention has been devoted to the temporal stability of the trap. Factors such as local temperature fluctuations caused by light heating and local convective microflows might affect the stability of the trap. Activated escape from such a trap underpins several physical and biological processes. We loaded spheres into our trap and measured their temporal stability. A quantitative description of diffusion activated escape from a one-dimensional potential well was given by Kramers [19]. It states that average residence time is a function of potential well parameters and obeys the equation $\langle \tau_K \rangle = \tau_0 \exp(U/k_B T)$, where U is potential well depth, T is temperature, and τ_0 is the time scale responsible for restoring relaxation dynamics within the well and can be expressed, for spheres far from walls and each other, as [3,4] $\tau_0 = 6\pi\rho a/m\omega^2$, where a is sphere radius and m is the mass, ρ is water viscosity, and ω is the associated frequency of the trap. Experimentally we measured the residence time as a function of number of spheres in the trap (Fig. 5). Each escape event is random and has a low probability in agreement with Poisson statistics.

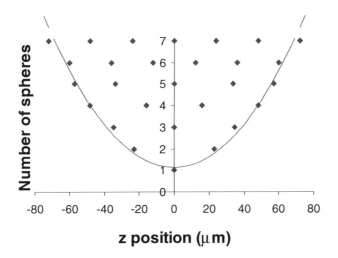

FIG. 3 (color online). Equilibrium positions for particles in the array. The parabolic fit shows the harmonic form (as expected) of the light beam potential. Notably, this allows us to extract important quantitative data about the trap (frequency, light potential).

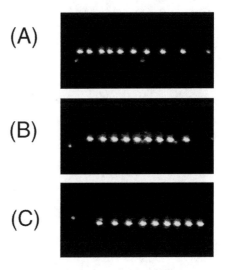

FIG. 4. Observation of a breathing mode. In (a) and (c) we see the displacement of the chain as a whole from the center with the interparticle spacing increasing as one goes farther from the center of the array.

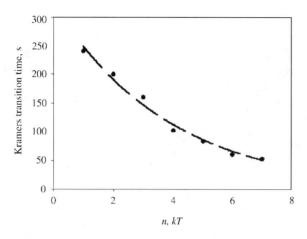

FIG. 5. Kramers transition time for the array. The main features of the observed residence times are supported by a general statistical theory showing exponential decay. The fit is appropriately described by an exponential function of the form $\langle \tau_K \rangle = \tau_0 \exp[\frac{U}{(n+c)k_B T}]$ and has the magnitude of the potential barrier $U/k_B T$ varied from 4.1 to 5.8 and $\tau_0 = 1$, $c = 16$ for particle number $n = 1, 2, \ldots 7$.

In conclusion, we have presented experimental observation and modes of oscillation of a one-dimensional optically bound system of colloidal particles that is akin to an extended line of coupled oscillators.

We thank D. Tatarkov for help with the figures. K. D. thanks R. C. Thompson, D. McGloin, and G. C. Spalding for useful discussions. This work is supported by the UK Engineering and Physical Sciences Research Council.

---

*Corresponding author.
Electronic addresses: kd1@st-andrews.ac.uk; sat3@st-andrews.ac.uk

[1] A. Ashkin, Phys. Rev. Lett. **24**, 156 (1970).
[2] A. Ashkin, J. M. Dziedzic, J. E. Bjorkholm, and S. Chu, Opt. Lett. **11**, 288 (1986).
[3] L. I. McCann, M. Dykman, and B. Golding, Nature (London) **402**, 785 (1999).
[4] A. Simon and A. Libchaber, Phys. Rev. Lett. **68**, 3375 (1992).
[5] For example, D. G. Grier, Curr. Opin. Colloid Interface Sci. **2**, 264 (1997); J. C. Crocker and D. G. Grier, Phys. Rev. Lett. **77**, 1897 (1996).
[6] E. R. Dufresne, G. C. Spalding, M. T. Dearing, S. A. Sheets, and D. G. Grier, Rev. Sci. Instrum. **72**, 1810 (2001).
[7] R. L. Eriksen, P. C. Mogensen, and J. Glückstad, Opt. Lett. **27**, 267 (2002).
[8] M. P. MacDonald *et al.*, Science **296**, 1101 (2002).
[9] J. E. Curtis, B. A. Koss, and D. G. Grier, Opt. Commun. **207**, 169 (2002); R. L. Eriksen, V. R. Daria, and J. Gluckstad, Opt. Express **10**, 597 (2002).
[10] P. Korda, G. C. Spalding, and D. Grier, Phys. Rev. B **66**, 024504 (2002).
[11] M. M. Burns, J.-M. Fournier, and J. A. Golovchenko, Phys. Rev. Lett. **63**, 1233 (1989).
[12] M. M. Burns, J.-M. Fournier, and J. A. Golovchenko, Science **249**, 749 (1990).
[13] T. N. Buican *et al.*, Appl. Opt. **26**, 5311 (1987).
[14] H. C. Nagerl *et al.*, Appl. Phys. B **66**, 603 (1998); A. M. Steane, Appl. Phys. B **64**, 623 (1998); D. F. V. James, Appl. Phys. B **66**, 181 (1998).
[15] T. C. B. Schut, G. Hesselink, B. G. de Grooth, and J. Greve, Cytometry **12**, 479 (1991).
[16] S. Nemoto and H. Togo, Appl. Opt. **37**, 6386 (1998).
[17] R. Gussgard, T. Lindmo, and T. Brevik, J. Opt. Soc. Am. B **9**, 1922 (1992).
[18] B. V. Derjaguin and L. Landau, Acta Physicochim. URSS **14**, 633 (1941); E. J. Verwey and J. Th. G. Overbeek, *Theory of the Stability of Lyophobic Colloids* (Elsevier, Amsterdam, 1948).
[19] H. A. Kramers, Physica (Amsterdam) **7**, 284 (1940).

6

Optical Spanners

When subjected to a spatial gradient in the intensity of the applied field, a dielectric particle experiences a force toward the high-field region. The polarizability of the particle means that the lowest overall energy of the system is reached when the dielectric is located at the position of maximum field. A tightly focused laser beam forms a field maximum at the focus, and hence a three-dimensional (3D) trap. In an equivalent view, the trapping force within optical tweezers arises from refraction of the transmitted light rays, and with this refraction a redirection of their momentum. The corresponding reaction force on the particle acts toward the highest intensity within the beam—potentially creating a 3D trap. Both of these views are useful as we consider the transformation of optical tweezers into an optical spanner.

In addition to its linear momentum, light can carry an angular momentum. Rather than understanding these momenta as photon properties, optical tweezers are best served by the recognition that at any point in the optical field, the energy and momentum flow are given by the direction and magnitude of the Poynting vector and local wave vector respectively; in an isotropic medium, these are always parallel [1].

Phase gradients across the beam—for example, arising from divergence—mean that at any one position the Poynting vector generally has transverse as well as axial components. Integrating the axial components of this vector over the beam cross section gives the linear momentum in the propagation direction, whereas integration of the transverse components, multiplied by a radius vector defined with respect to a defined axis, gives the angular momentum of the beam. Just as the transfer of linear momentum from light to a particle results in optical trapping, so the transfer of angular momentum from light to a particle results in rotation. As is now well recognized, light's angular momentum can be of a spin (manifest as the polarization state) or orbital (associated with phase cross section) type [2].

Although not uniquely a property of individual photons, it is convenient to express both the linear and angular components of light's momentum as a per-photon quantity. In free space, the linear momentum per photon is $\hbar k$. If this momentum is directed such that it has an azimuthal component, acting on a radius vector,

it exerts a torque. Applying this linear momentum in an azimuthal direction at the perimeter of an object maximizes this torque. Assuming an object radius r, the maximum torque, and hence angular momentum transfer, per photon is therefore $\hbar k r$. Whether the angular momentum is spin or orbital, this torque is the absolute maximum value that can be applied with a light beam. Beyond this upper limit, the torque is further reduced by the numerical aperture of the coupling optics, restricting the skew angle of the wave vector [3]. For a particle approximately one wavelength in diameter, the maximum angular momentum transfer is then of order $\hbar$, the spin angular momentum of the photon.

As mentioned above, light can carry an orbital angular momentum in addition to its spin component. Such light is typified by the Laguerre–Gaussian modes that have a helical phase structure due to an azimuthal phase term of the form $\exp(i\ell\theta)$. This results in a well-characterized azimuthal component to the Poynting vector, leading to an orbital angular momentum of $\ell\hbar$ per photon. However, this increased angular momentum comes only at an increase in the beam radius, which, if coupled to a particle, sets a lower limit to the particle size. The result is that the previous limit of angular momentum transfer to an object of $\hbar k r$ per photon still applies. Therefore, for larger particles, the maximum torque scales with the radius. However, larger particles also have a higher moment of inertia and are subject to higher drag forces, scaling with the fifth and third power of the radius respectively. Hence, optically induced rotation is only really effective in the micron-size range.

In his early pioneering work, Ashkin observed that rod-shaped bacteria would rotate upright, aligning vertically along the trapping axis of the beam [4]. This is best understood by recognizing that it is energetically favorable for the object, which is elongated, to align with the light beam. It was this phenomenon that was utilized in the first deliberate particle rotation within optical tweezers by Sato et al. [5], who used a high-order Hermite–Gaussian mode in optical tweezers to trap and rotate a red blood cell. In the transverse plane, the rectangular symmetry of both the mode and the cell meant that the long axis of the cells aligned with the long axis of the beam cross-section. An intracavity aperture was used to vary the orientation of the mode, causing a corresponding rotation of the cell.

Subsequently, a number of groups have successfully used various other techniques to create noncircular trapping beams that can be rotated. Interfering a Laguerre–Gaussian laser beam with its own mirror image creates spiral interference fringes [6], and shifting the frequency of either beam causes this pattern to translate. Paterson et al. [7] (see **PAPER 6.1**) used the same approach to create interference patterns within optical tweezers and rotate a Chinese hamster chromosome. Changing the indices of the Laguerre–Gaussian mode could set the rotational symmetry of the interference pattern, thereby optimizing its shape to that of the object. Another method of producing an asymmetric beam is to introduce a rectangular aperture into the optical path, between the laser and the tweezers, shaping the focused spot into an ellipse. A rotating rectangular aperture mounted in this way has been used by O'Neil and Padgett to trap and rotate assemblies of silica spheres [8] (see **PAPER 6.2**). This simple method for rotational control does not require high-order modes, interferometric precision, or computer-controlled optical modulators.

These torques and rotations arise directly from the gradient force generated by asymmetry trapping beams. However, rotation is also possible using circularly symmetric beams and a transfer of their optical angular momentum. In the 1930s, Beth used circularly polarized light and its inherent spin angular momentum to excite the torsional motion of a wave plate suspended by a quartz fiber. However, the detection of this motion required measurements of great precision [9].

In the mid 1990s, He's group [10], [11] (see **PAPER 6.3**) demonstrated that an absorbing particle could be set spinning, at several Hertz, by transfer of the orbital angular momentum carried by a Laguerre–Gaussian beam. The Laguerre–Gaussian beam was created using a computer-designed diffractive element, that is, a computer-generated hologram. In this early work, the holograms were produced photographically but more recently most groups tend to use a spatial light modulator (SLM; see Chapter 7). With respect to absorption of the light, both the orbital and spin angular momentum should be transferable to the trapped object. A circularly polarized Laguerre–Gaussian beam carries both spin and orbital angular momentum, which, depending upon the handedness of the two terms, can add or subtract. Changing the handedness of one component speeds up or slows down the rotation of the trapped object [12] or, in the case of a Laguerre–Gaussian beam with $\ell = 1$, causes the particle to stop; see Simpson et al. [13] (see **PAPER 6.4**).

The absorption of light by the trapped particle transfers both the spin and orbital angular momenta components with equal efficiency. However, rather that transferring

angular momentum by absorbing the light, as was the case in Beth's experiment, a torque can be applied simply by changing the angular momentum state of the transmitted light. Calcite and similar materials have a sufficiently high birefringence that a thickness of a few microns acts as a quarter-wave plate, changing the polarization state of the transmitted light from circular to linear. Consequently, as shown by Friese et al., a micron-sized birefringent particle trapped in a circularly polarized light beam experiences a torque comparable to the angular momentum flux in the beam. Rotation rates of hundreds of Hertz are possible [14].

For a particle trapped on the beam axis, the transfer of either spin or orbital angular momentum results in the particle spinning about the axis of the beam. A more interesting comparison occurs if the particle is small with respect to the diameter of the beam and is displaced from the beam axis. When a collection of metal particles was illuminated by a Laguerre–Gaussian beam with a high value of ℓ, the metal particles were observed to circulate around the beam axis in a direction set by the orbital angular momentum of the beam and, most importantly, independent of its spin angular momentum state [15]. In 2002, O'Neil et al. used the gradient force produced by a circularly polarized, $\ell = 8$ Laguerre–Gaussian beam to confine a small dielectric particle within its annular bright ring. The azimuthal component of the light's Poynting vector, arising from its orbital angular momentum, meant that light scattering forced the particle to orbit around the beam axis [16] (see **PAPER 6.5**). If the particle was itself birefringent, then the local transformation of the polarization state would result in a torque acting on the particle causing it to spin about its own axis. In these and similar experiments where the small particle is displaced from the axis of the laser beam, the transfer of spin and orbital angular momenta are distinguishable, producing a spinning and orbiting motion of the particle respectively. Equivalent results have also been obtained for high-order Bessel [17] and Mathieu [18] beams. In the former case, the multiple bright rings allowed quantitative confirmation of the rotation rate as a function of distance from the beam axis. Beyond trapping a single particle in the bright ring of a Laguerre–Gaussian mode that then orbits the beam axis due to an orbital angular momentum exchange, many particles can be added to the ring such that they all circulate around the beam axis [19]. In all these cases, the Laguerre–Gaussian (or similar helically phased beam) was obtained using a diffractive optical component to transform the phase structure of the trapping beam. This use of computer-generated holograms became easier still when the photographic processing could be replaced by the use of commercially available spatial light modulators for the generation of exotic beams. Modifying the hologram kinoform means that the annular ring of the beam can be distorted into petals and similar shapes around which the confined particles will travel; see Curtis and Grier [20] (**PAPER 6.6**). Combining several beams allows the precise angular momentum content of the beam and its effective torque to be finely tuned [21].

Following the holographic revolution of optical tweezers, one obvious application of spatial light modulators was to create two or more traps that, when positioned close together, could both trap and orientate asymmetric particles. Moving the traps around each other causes the particle to rotate. Because the SLM can introduce both axial and lateral displacement to the trap position, the circular trajectory and corresponding rotation can be about any axis, including, as shown by Bingelyte et al., an axis that is perpendicular to the optical axis of the tweezers [22] (see **PAPER 6.7**).

Having established methods for rotating particles based on the trapping beam's asymmetry or angular momentum, it is also possible to obtain rotation based upon the asymmetry of the particles: an optically driven propeller. The angular momentum exchange arises from the scattering of the light's linear momentum in an azimuthal direction. Alternatively, one can consider the object as being a simple form of mode converter, transforming the incident beam containing no angular momentum into a scattered beam that does. Higurashi et al. [23] fabricated four-armed silicon rotors approximately 10 μm in diameter. These were then trapped and set into rotation within optical tweezers. The sense of the rotation was determined by the handedness of the rotor construction. Whereas the bearings in conventional micromachines are subject to frictional wear, the use of optical trapping requires no mechanical constraint.

Asymmetric structures have been formed in other ways too, including partly silvered beads [24] and *in situ* formation of the objects using two-photon polymerization where the tweezers system itself has all the necessary optical control [25]. In this last example, multiple rotors were fabricated and positioned so that their vanes engaged, acting as gear teeth. Driving one of them caused both to rotate. Most recently, two-photon polymerization has been employed to create explicit mode converters based on miniature spiral phase plates where the helicity of the microcomponent has been designed to give transformation of the transmitted trapping beam into a Laguerre–Gaussian mode [26]. Two-photon polymerization has also been used to fabricate a micrometer-sized mechanical motor driven by light from an integrated waveguide within the sample [27].

Shaping the particle can lead to some bizarre effects. For example, if the particle has the form of a prism, refraction of the transmitted light can lead to a transverse or azimuthal force. This force can be greater than that due to the light scattering, leading to situations where a particle may appear to orbit the wrong way around an optical vortex. The apparent paradox is resolved by appreciating that in this special case, the transverse refraction force exceeds that of the scattering of the light's orbital angular momentum [28].

Beyond the insight provided to light's momentum properties and the general application to micromachines, rotational control within optical tweezers creates various other opportunities that are currently being investigated. Following their pioneering of the transfer of spin angular momentum to birefringent particles, Rubinsztein-Dunlop and colleagues developed a technique for producing near-spherical particles of vaterite. Vaterite is similar in structure to calcite and has a high birefringence, sufficient to give a quarter-wave thickness of only a few microns. By monitoring the change in the polarization state of the transmitted light, it is possible to calculate precisely the applied torque and the rotational speed [29,30]. By relating these measurements to the Stokes drag force, Bishop et al. were able to accurately measure the viscosity of the surrounding fluid [31] (see **PAPER 6.8**). Most recently, this rotational technique has been applied to measuring the viscosity of tear fluid, a measurement difficult by any other means but extremely important in the diagnosis of various eye conditions.

As an alternative to making the particle from a birefringent material, the particle can be made from an isotropic material but shaped in such a way that it has "form-birefringence." An early example of this was the trapping and orientation of long glass rods by a polarized beam [32]. Although this orientation is most easily understood as the action of the gradient force aligning the particle with the high-intensity region of the beam, an equivalent view is that the elongation of the object means that the polarization state of the scattered light is modified by the polarization of the dielectric. Effectively, the rods act as a primitive birefringent wave plate, resulting in an angular momentum transfer. This effect has recently been exploited in the manufacture of micron-sized gear wheels with grooves in the flat surfaces. The microscaled grooved surface is form-birefringent, such that the polarization state of the reflected light is again modified and angular momentum is exchanged [33]. These gear wheels can be both trapped and driven with light, and positioned to engage with other cogs to create the basis of a pump or other devices.

Like optical tweezers themselves, the rotational control within tweezers has been driven by developments within many different research groups. Since the initial curiosity-based demonstration of a rotating red blood cell by Sato et al. in 1992 [5], subsequent investigations using these optical spanners have targeted understanding the angular momentum properties of light. In parallel with this work has been the application of rotational control to micromachines, cogs, and others to form optically actuated pumps [34], [35] (see **PAPER 8.2**), or in the direct measurement of physical quantities like viscosity [31] (see **PAPER 6.8**).

Endnotes

1. L. Allen and M.J. Padgett, The Poynting vector in Laguerre-Gaussian beams and the interpretation of their angular momentum density, *Opt. Commun.* **184**, 67–71 (2000).

2. L. Allen, M.J. Padgett, and M. Babiker, The orbital angular momentum of light, *Prog. Opt.* **39** 291–372 (1999).

3. J. Courtial and M.J. Padgett, Limit to the orbital angular momentum per unit energy in a light beam that can be focussed onto a small particle, *Opt. Commun.* **173**, 269–274 (2000).

4. A. Ashkin, J.M. Dziedzic, and T. Yamane, Optical trapping and manipulation of single cells using infrared laser beams, *Nature* **330**, 769–771 (1987).

5. S. Sato, M. Ishigure, and H. Inaba, Optical trapping and rotational manipulation of microscopic particles and biological cells, *Electron. Lett.* **27**, 1831–1832 (1991).

6. M. Harris, C.A. Hill, and J.M. Vaughan, Optical helices spiral interference fringes, *Opt. Commun.* **106**, 161–166 (1994).

7. L. Paterson, M.P. MacDonald, J. Arlt, W. Sibbett, P.E. Bryant, and K. Dholakia, Controlled rotation of optically trapped microscopic particles, *Science* **292**, 912–914 (2001).

8. A.T. O'Neil and M.J. Padgett, Rotational control within optical tweezers by use of a rotating aperture, *Opt. Lett.* **27**, 743–745 (2002).

9. R.A. Beth, Mechanical detection and measurement of the angular momentum of light, *Phys. Rev.* **50**, 115–125 (1936).

10. H. He, N.R. Heckenberg, and H. Rubinsztein-Dunlop, Optical particle trapping with higher-order doughnut beams produced using high-efficiency computer generated holograms *J. Mod. Opt.* **42**, 217–223 (1995).

11. H. He, M.E.J. Friese, N.R. Heckenberg, and H. Rubinsztein-Dunlop, Direct observation of transfer of angular momentum to absorptive particles from a laser beam with a phase singularity, *Phys. Rev. Lett.* **75**, 826–829 (1995).

12. M.E.J. Friese, J. Enger, H. Rubinsztein-Dunlop, and N.R. Heckenberg, Optical angular-momentum transfer to trapped absorbing particles, *Phys. Rev. A* **54**, 1593–1596 (1996).

13. N.B. Simpson, K. Dholakia, L. Allen, and M.J. Padgett, Mechanical equivalence of spin and orbital angular momentum of light: An optical spanner, *Opt. Lett.* **22**, 52–54 (1997).

14. M.E.J. Friese, T.A. Nieminen, N.R. Heckenberg, and H. Rubinsztein-Dunlop, Optical alignment and spinning of laser-trapped microscopic particles, *Nature* **394**, 348–350 (1998).

15. A.T. O'Neil and M.J. Padgett, Three-dimensional optical confinement of micron-sized metal particles and the decoupling of the spin and orbital angular momentum within an optical spanner, *Opt. Commun.* **185**, 139–143 (2000).

16. A.T. O'Neil, I. MacVicar, L. Allen, and M.J. Padgett, Intrinsic and extrinsic nature of the orbital angular momentum of a light beam, *Phys. Rev. Lett.* **88**, 053601 (2002).

17. V. Garces-Chavez, D. McGloin, M.J. Padgett, W. Dultz, H. Schmitzer, and K. Dholakia, Observation of the transfer of the local angular momentum density of a multiringed light beam to an optically trapped particle, *Phys. Rev. Lett.* **91**, 093602 (2003).

18. C. Lopez-Mariscal, J.C. Gutierrez-Vega, G. Milne, and K. Dholakia, Orbital angular momentum transfer in helical Mathieu beams, *Opt. Express* **14**, 4182–4187 (2006).

19. J.E. Curtis and D.G. Grier, Structure of optical vortices, *Phys. Rev. Lett.* **90**, 133901 (2003).

20. J.E. Curtis and D.G. Grier, Modulated optical vortices, *Opt. Lett.* **28**, 872–874 (2003).

21. C.H.J. Schmitz, K. Uhrig, J.P. Spatz, and J.E. Curtis, Tuning the orbital angular momentum in optical vortex beams, *Opt. Express* **14**, 6604–6612 (2006).

22. V. Bingelyte, J. Leach, J. Courtial, and M.J. Padgett, Optically controlled three-dimensional rotation of microscopic objects, *Appl. Phys. Lett.* **82**, 829–831 (2003).

23. E. Higurashi, H. Ukita, H. Tanaka, and O. Ohguchi, Optically induced rotation of anisotropic micro-objects fabricated by surface micromachining, *Appl. Phys. Lett.* **64**, 2209–2210 (1994).

24. Z.P. Luo, Y.L. Sun, and K.N. An, An optical spin micromotor, *Appl. Phys. Lett.* **76**, 1779–1781 (2000).

25. P. Galajda and P. Ormos, Complex micromachines produced and driven by light, *Appl. Phys. Lett.* 78, 249–251 (2001).

26. G. Knoner, S. Parkin, T.A. Nieminen, V.L.Y. Loke, N.R. Heckenberg, and H. Rubinsztein-Dunlop, Integrated optomechanical microelements, *Opt. Express* **15**, 5521–5530 (2007).

27. L. Kelemen, S. Valkai, and P. Ormos, Integrated optical motor, *Appl. Opt.* **45**, 2777–2780 (2006).

28. A. Jesacher, S. Furhapter, C. Maurer, S. Bernet, and M. Ritsch-Marte, Reverse orbiting of microparticles in optical vortices, *Opt. Lett.* **31**, 2824–2826 (2006).

29. T.A. Nieminen, N.R. Heckenberg, and H. Rubinsztein-Dunlop, Optical measurement of microscopic torques, *J. Mod. Opt.* **48**, 405–413 (2001).

30. G. Knoner, S. Parkin, N.R. Heckenberg, and H. Rubinsztein-Dunlop, Characterization of optically driven fluid stress fields with optical tweezers, *Phys. Rev. E* **72**, 031507 (2005).

31. A.I. Bishop, T.A. Nieminen, N.R. Heckenberg, and H. Rubinsztein-Dunlop, Optical microrheology using rotating laser-trapped particles, *Phys. Rev. Lett.* **92**, 198104 (2004).

32. A.I. Bishop, T.A. Nieminen, N.R. Heckenberg, and H. Rubinsztein-Dunlop, Optical application and measurement of torque on microparticles of isotropic nonabsorbing material, *Phys. Rev. A* **68**, 033802 (2003).

33. S.L. Neale, M.P. Macdonald, K. Dholakia, and T.F. Krauss, All-optical control of microfluidic components using form birefringence, *Nat. Mater.* **4**, 530–533 (2005).

34. K. Ladavac and D.G. Grier, Microoptomechanical pumps assembled and driven by holographic optical vortex arrays. *Opt. Express* **12**, 1144–1149 (2004).

35. J. Leach, H. Mushfique, R. Di Leonardo, M. Padgett, and J. Cooper, An optically driven pump for microfluidics. *Lab Chip* **6**, 735–739 (2006).

Controlled Rotation of Optically Trapped Microscopic Particles

L. Paterson,[1] M. P. MacDonald,[1] J. Arlt,[1] W. Sibbett,[1] P. E. Bryant,[2] K. Dholakia[1]*

We demonstrate controlled rotation of optically trapped objects in a spiral interference pattern. This pattern is generated by interfering an annular shaped laser beam with a reference beam. Objects are trapped in the spiral arms of the pattern. Changing the optical path length causes this pattern, and thus the trapped objects, to rotate. Structures of silica microspheres, microscopic glass rods, and chromosomes are set into rotation at rates in excess of 5 hertz. This technique does not depend on intrinsic properties of the trapped particle and thus offers important applications in optical and biological micromachines.

Optical forces have been used to trap and manipulate micrometer-sized particles for more than a decade (*1*). Since it was shown that a single tightly focused laser beam could be used to hold, in three dimensions, a microscopic particle near the focus of the beam, this optical tweezers technique has now become an established tool in biology, enabling a whole host of studies. They can be used to manipulate and study whole cells such as bacterial, fungal, plant, and animal cells (*2*) or intracellular structures such as chromosomes (*3*). Optical tweezers make use of the optical gradient force. For particles of higher refractive index than their surrounding medium, the laser beam induces a force attracting the trapped particle into the region of highest light intensity.

The ability to rotate objects offers a new degree of control for microobjects and has important applications in optical micromachines and biotechnology. Various schemes have, therefore, been investigated recently to induce rotation of trapped particles within optical tweezers. This could be used to realize biological machines that could function within living cells or optically driven cogs to drive micromachines.

[1]School of Physics and Astronomy, St. Andrews University, North Haugh, St. Andrews, Fife KY16 9SS, Scotland. [2]School of Biology, Bute Building, St. Andrews University, St. Andrews, Fife KY16 9TS, Scotland.

*To whom correspondence should be addressed. E-mail: kd1@st-and.ac.uk

Besides the use of specially fabricated microobjects (*4*), two major schemes have successfully enabled trapped microobjects to be set into rotation. The first scheme uses Laguerre-Gaussian (LG) light beams (*5–7*). These beams have an on-axis phase singularity and are characterized by helical phase fronts (Fig. 1A). The Poynting vector in such beams follows a corkscrewlike path as the beam propagates, and this gives rise to an orbital angular momentum component in the light beam (*8*). This angular momentum is distinct from any angular momentum due to the polarization state of the light and has a magnitude of $l\hbar$ per photon. Specifically, l refers to the number of complete cycles of phase ($2\pi l$) upon going around the beam circumference. However, to transfer orbital angular momentum to a trapped particle with such a beam, the particle must typically absorb some of the laser light yet still be transparent enough to enable tweezing to occur. This in turn restricts the range of particles to which this method can be applied, and it also further limits this technique because any heating that arises from this absorption could damage the rotating particle. Furthermore, as the particle absorption can be difficult to quantify, controlled rotation of trapped objects in such a beam is very difficult to realize.

The other technique for rotation makes use of the change in polarization state of light upon passage through a birefringent particle (*9, 10*). For example, circularly polarized light has spin angular momentum that can be exchanged with a birefringent medium (e.g., calcite) upon propagation of the beam through the medium. This is analogous to Beth's famous experiment—where he measured the torque on a suspended half-wave plate as circularly polarized light passed through it (*11*)—but here we are working on a microscopic scale. This method has shown rotation rates of a few hundred hertz for irregular samples of crushed calcite, but it is difficult to control and is limited solely to birefringent media so it is not widely applicable. Although both of these methods have proven useful in specific applications, they do have serious shortcomings for general applications in rotating optical microcomponents and realizing optical micromachines.

We introduce a general scheme for rotating trapped microobjects. Specifically, we trap objects within the interference pattern of an LG beam and a plane wave (Fig. 1B) (*12*). By changing the path length of the interferometer, we are able to cause the spiral

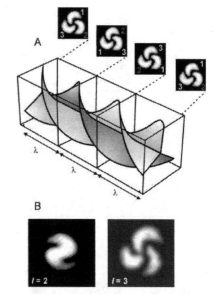

Fig. 1. (A) The phase fronts of an LG beam of azimuthal index $l = 3$ (helical structure) and intensity pattern when interfered with a plane wave. The phase fronts describe a triple start intertwined helix that repeats its shape every λ but only rotates fully after $l\lambda$. In **(B)**, we can see the experimental forms of the interference patterns of LG beams of index $l = 2$ and $l = 3$ with plane waves used in our experiments.

From L. Paterson, M. P. MacDonald, J. Arlt, W. Sibbett, P. E. Bryant, and K. Dholakia, "Controlled Rotation of Optically Trapped Microscopic Particles," Science. 292: 912–914 (2001). Reprinted with permission from AAAS.

pattern (and thus the trapped particles) to rotate in a controlled fashion about the axis of the spiral pattern. The rotation of the pattern occurs because of the helical nature of the phase fronts of an LG light beam. A single-ringed LG beam is described by its azimuthal mode index l, which denotes the number of complete cycles of phase upon going around the circumference of the mode. An $l = 2$ or $l = 3$ LG mode (beam) can be thought of as consisting of phase fronts that are double or triple start helices, respectively (see Fig. 1A). Interfering this beam with a plane wave will transform the azimuthal phase variation of the pattern into an azimuthal intensity variation, resulting in a pattern with l spiral arms. As we change the path length in one arm of the interferometer, these spiral arms will rotate around the beam axis. As an analogy, this is akin to considering what occurs along a

length of thick rope that consists of l intertwined cords. Now consider cutting this rope and viewing it end-on. As you move the position of the cut along the rope, any given cord rotates around the rope axis. This is analogous to altering the optical path length in the interferometer. With this technique, we rely solely on the optical gradient force to tweeze trapped particles in the spiral arms and then use the rotation of this spiral pattern under a variation of optical path length to induce particle rotation. The technique can therefore be applied in principle to any object (or group of objects) that can be optically tweezed, in contrast to the other methods listed above. This technique can be extended to the use of LG beams of differing azimuthal index, thus offering the prospect of trapping and rotating different shaped objects and groups of objects. Here, illustrative examples

of rotation with LG beams with azimuthal indices $l = 2$ and $l = 3$ are shown.

Figure 2 shows a schematic of the trapping arrangement (*13*). A change in the path length in one arm of the interferometer by $l \times \lambda$ will cause a full rotation of 360° of the pattern (and thus the trapped particle array) in the optical tweezers (*14*). We can readily change the sense of rotation by reducing the path length of one arm of the interferometer instead of increasing it. Thus, in contrast to other rotation methods, we have a very simple way of controlling both the sense and rate of rotation of our optically trapped structure.

The use of an LG $l = 2$ beam results in two spiral arms for our interference pattern, and we used this to rotate silica spheres and glass rods in our tweezers setup (using a ×100 microscope objective). In Fig. 3A, two 1-μm silica spheres are trapped and spun at a rate of 7 Hz. The minimum optical power required to rotate the 1-μm spheres (which is the minimum power required to rotate any of the structures) is 1 mW. Silica spheres coated with streptavidin can bind to biotinylated DNA, and thus one could rotationally orient DNA strands by extending this method. In Fig. 3B, a tweezed glass rod can be seen to rotate between the frames. This constitutes an all-optical microstirrer and has potential application for optically driven micromachines and motors. We also demonstrate rotation of a Chinese hamster chromosome in our tweezers using this same interference pattern (Fig. 3C), with the axis of our pattern placed over the centromere of the chromosome. This degree of flexibility could be used for suitably orienting the chromosome before, for example, the optical excision of sections for use in polymerase chain reactions. This latter demonstration shows the potential of our method for full rotational control of biological specimens.

The rotation of trapped particles in an interference pattern between an LG ($l = 3$) beam and a plane wave can be seen in Fig. 4. The number of spiral arms in the pattern is equivalent to the azimuthal index of the LG beam used. In this instance, we used a ×40 microscope objective to increase the overall size of the beam profile and thus tweeze and rotate larger structures. In Fig. 4, we see three trapped 5-μm silica spheres rotate in this pattern. One of the spheres has a slight deformity (denoted by the arrow), and the series of pictures charts the progress of this structure of spheres as the pattern is rotated. We typically achieved rotation rates in excess of 5 Hz in the above experiments, which were limited only by the amount of optical power (~13 mW) in our interference pattern at the sample plane. The use of optimized components would readily lead to rotation rates of tens to hundreds of hertz. One can envisage other fabricated microobjects being rotated in a similar fashion.

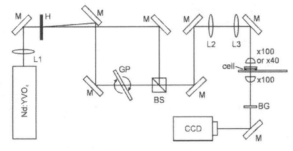

Fig. 2. The experimental arrangement for optical tweezing and subsequent particle rotation in the interference pattern. L, lens; M, mirror; H, hologram; GP, glass plate; BS, beam splitter; Nd:YVO₄, neodymium yttrium vanadate laser at 1064 nm; ×100 or ×40, microscope objectives; CCD, camera; and BG, infrared filter.

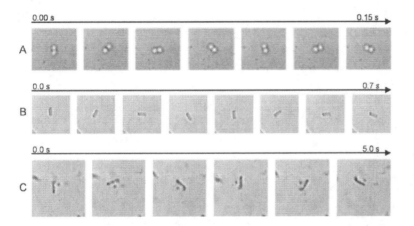

Fig. 3. Rotation of two-dimensionally trapped objects in an LG $l = 2$ interference pattern. **(A)** Rotation of two trapped 1-μm silica spheres. **(B)** Rotation of a 5-μm-long glass rod. In **(C)**, we see rotation of a Chinese hamster chromosome. The elapsed time t (in seconds) is indicated by the scale at the top of each sequence of images.

Fig. 4. Rotation of three trapped silica spheres each 5 μm in diameter. The slight deformity (indicated by arrow) on one of the spheres allows us to view the degree of rotation of the structure.

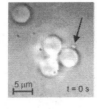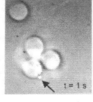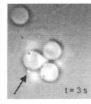

REPORTS

References and Notes

1. A. Ashkin, J. M. Dziedzic, J. E. Bjorkholm, S. Chu, *Opt. Lett.* **11**, 288 (1986).
2. A. Ashkin, J. M. Dziedzic, T. Yamane, *Nature* **330**, 769 (1987).
3. H. Liang, W. H. Wright, S. Cheng, W. He, M. W. Berns, *Exp. Cell Res.* **204**, 110 (1993).
4. P. Ormos, P. Galajda, *Appl. Phys. Lett.* **78**, 249 (2001).
5. H. He, M. E. J. Friese, N. R. Heckenberg, H. Rubinsztein-Dunlop, *Phys. Rev. Lett.* **75**, 826 (1995).
6. M. E. J. Friese, J. Enger, H. Rubinsztein-Dunlop, N. R. Heckenberg, *Phys. Rev. A* **54**, 1593 (1996).
7. N. B. Simpson, K. Dholakia, L. Allen, M. J. Padgett, *Opt. Lett.* **22**, 52 (1997).
8. L. Allen, M. W. Beijersbergen, R. J. C. Spreeuw, J. P. Woerdman, *Phys. Rev. A* **45**, 8185 (1992).
9. M. E. J. Friese, T. A. Nieminen, N. R. Heckenberg, H. Rubinsztein-Dunlop, *Nature* **394**, 348 (1998).
10. M. E. J. Friese, H. Rubinsztein-Dunlop, J. Gold, P. Hagberg, D. Hanstorp, *Appl. Phys. Lett.* **78**, 547 (2001).
11. R. A. Beth, *Phys. Rev.* **50**, 115 (1936).
12. M. Padgett, J. Arlt, N. Simpson, L. Allen, *Am. J. Phys.* **64**, 77 (1996).
13. The experimental setup consisted of a Nd:YVO$_4$ laser of 300-mW power at 1064 nm. This beam is then directed through an in-house manufactured holographic element (*15*) that yielded an LG beam in its first order with an efficiency of 30%. This LG beam is then interfered with the zeroth order beam from the hologram to generate our spiral interference pattern. This pattern propagates through our optical system and is directed through either a ×40 or a ×100 microscope objective in a standard optical tweezers geometry. Typically around 1 to 13 mW of laser light was incident on the trapped structure in our optical tweezers, with losses due to optical components and the holographic element. A charge-coupled device (CCD) camera was placed above the dielectric mirror for observation purposes (Fig. 2) when the ×100 objective was used. A similar setup was used when tweezing with a ×40 objective but with the CCD camera placed below the sample slide viewing through a ×100 objective. It is important to ensure exact overlap of the light beams to guarantee that spiral arms are observed in the interference pattern—at larger angles, linear fringe patterns (with some asymmetry) can result (*15*). To set trapped structures into rotation, the relative path length between the two arms of the interferometer must be altered. We achieved this by placing a glass plate on a tilt stage in one arm. Simply by tilting this plate, we can rotate accordingly the pattern in the tweezers.
14. The tilting of the glass plate to rotate the interference pattern has a limitation when the plate reaches its maximum angle. One can, however, envisage more advanced implementations for continuous rotation using, for example, a liquid crystal phase modulator in the arm of the interferometer containing the plane wave.
15. M. A. Clifford, J. Arlt, J. Courtial, K. Dholakia, *Opt. Commun.* **156**, 300 (1998).
16. We thank the UK Engineering and Physical Sciences Research Council for supporting our work.

26 December 2000; accepted 19 March 2001

Rotational control within optical tweezers by use of a rotating aperture

Anna T. O'Neil and Miles J. Padgett

Department of Physics and Astronomy, Kelvin Building, University of Glasgow, Glasgow, Scotland G12 8QQ

Received August 3, 2001

We demonstrate a simplified method of rotational control of objects trapped within optical tweezers that does not require high-order modes, interferometric precision, or computer-controlled optical modulators. Inserting a rectangular aperture into the optical beam results in a focused spot that also has rectangular symmetry. We show that an asymmetric object trapped in the beam has its angular orientation fixed such that rotation of the aperture results in a direct rotation of the particle. © 2002 Optical Society of America

OCIS codes: 140.7010, 350.0350.

Sixteen years ago Ashkin *et al.* demonstrated that tightly focused laser beams could trap and manipulate micrometer-sized particles.[1] So-called optical tweezers rely on the large gradient in the electric field generated by a tightly focused laser beam. Any dielectric particle in the vicinity of the focus is subject to a force, directed toward the region of highest intensity. For a tightly focused laser beam, the gradient force overcomes the gravitational force as well as light scattering; the particle becomes trapped in three dimensions. For particles a few micrometers in diameter suspended in water, a few milliwatts of laser power focused with an oil-immersion microscope objective is sufficient to form a robust trap. Optical tweezers are now commercially available[2,3] and are used widely in the biological community, e.g., for measuring the compliance of bacterial tails,[4] the forces exerted by single muscle fibers,[5] and the stretching of DNA.[6]

Physicists have used optical tweezers as a tool to study the transfer of angular momentum from light to particles. For particles trapped on the beam axis, the spin and orbital angular momentum have been shown to cause rotation of birefringent[7] and absorbing[8] particles, respectively. For absorbing particles, spin and orbital angular momenta can be transferred simultaneously so that the applied torque is proportional to the total angular momentum.[9] Most recently, optical tweezers have been used to explore the intrinsic and extrinsic nature of a light beam's angular momentum.[10] It should be emphasized, however, that the prime motivation behind all that work lay in the study of the optical properties of the beams rather than in the use of optical tweezers specifically as a technical tool. The linear momentum or radiation pressure of a single beam has also been used to impart rotation to suitably shaped micromachined objects[11] or assemblies of partly silvered beads.[12]

For practical applications, rotation within optical tweezers was to our knowledge first reported in 1991 when Sato *et al.* used a rotating high-order Hermite–Gaussian mode to induce rotation of red blood cells.[13] There the rotation relied on the rectangular symmetry of the beam's forcing the asymmetric cell to take up a particular orientation. Indeed, Ashkin

et al. observed that this inherent alignment of objects with the beam symmetry could cause rod-shaped bacteria to stand upright, aligning themselves vertically along the trapping axis of the beam.[14]

Optical tweezers have also been configured by use of multiple beams to trap more than one particle. These configurations have been implemented by rapid scanning of a single beam between two or more trap positions[15] or by use of computer-generated holograms to yield multiple beams simultaneously.[16] Dual beam traps have also been configured to create an interference pattern; the resulting intensity gradient gives rise to a trapping force.[17] Recently, by interfering beams with an azimuthal phase structure, Paterson *et al.* used this technique to induce rotation.[18]

In this work we demonstrate a simplified method of rotational control within optical tweezers. We do not require multiple beams, high-order modes, interferometric precision, or computer-controlled optical modulators. A rectangular aperture clipping each side of the beam results in a focused spot that also has rectangular symmetry. We show that an asymmetric object trapped in the beam has its angular orientation fixed such that rotation of the aperture results in direct rotation of the particle. The experimental configuration used is shown in Fig. 1 and is based around an inverted optical tweezers geometry, with the trapping beam directed upward. The inverted geometry is favored by biologists, as this allows easy access to the sample. The objective lens is an infinity-corrected, Zeiss Plan-Neofluar 1.3-N.A., 100× lens, widely used in the optical tweezers community. The trapping laser is a commercial Nd:YLF laser, frequency doubled to give 100 mW of power at 532 nm. A combination of relay lenses and beam-steering mirrors couples this beam into the objective via a highly reflective dielectric mirror, which separates the trapping and imaging optical paths. The sample cell is mounted on an *xyz* piezoelectric stage, which provides 100 μm of travel in each direction. Beam-steering mirrors allow the trapping beam to be positioned anywhere within the field of view of the microscope.

As discussed above, rotation in optical tweezers has already been achieved with a high-order Hermite–Gaussian mode.[13] In that work, the long axis

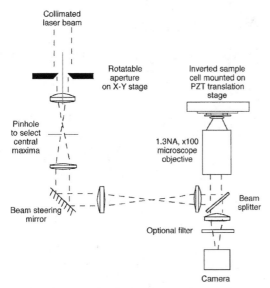

Fig. 1. Configuration of the inverted optical tweezers setup including the rotational aperture.

of the asymmetrically shaped cell aligned itself with the elongated direction of the beam cross section. The axis of the mode was then rotated by use of an aperture within the laser cavity, and the cell was observed to rotate, at all times maintaining its alignment with the mode. Clearly, that technique relied on a special laser and is therefore not readily adaptable. One alternative option for this approach, or indeed for achieving rotation by use of a laser within an elliptical output mode,[19] would be to create the asymmetric beam and then rotate the whole beam before the tweezers. Although simple in concept, the rotation of a beam exactly about its own axis is more difficult than it might appear. For example, a rotating Dove prism is known to rotate a transmitted beam, but such a prism requires precise angular and lateral alignment at a precision that is difficult to achieve at optical wavelengths.[20]

Perhaps the most obvious way to generate an asymmetric beam is to introduce an asymmetric aperture. Although almost any shape of aperture will suffice, and indeed some shapes would better match the symmetry of some objects, we chose a simple rectangular aperture, as it is readily engineered to the adjustable. Mounting the aperture within a rotation stage that is itself mounted on an xy translation stage means that its rotation axis is easily aligned with the beam axis. To minimize requirements on the mechanical stability and tolerance of the mounts, we insert the aperture into an expanded and collimated region of the trapping beam. The output of the trapping laser is expanded to give a collimated Gaussian beam with a $1/e$ diameter of 10 mm. Inserting a 4-mm-wide rectangular aperture at this point produces a diffraction pattern in the far field that is composed of a central maximum containing approximately 45% of the original power and additional diffraction orders containing approximately a further 10% of the power. The subsequent aperture removes these diffraction orders, such that only the central maximum is relayed to the sample plane.

The central maximum is itself elliptical in cross section, with an aspect ratio of 1:2, and it is this ellipticity that is responsible for causing the asymmetric objects to align with respect to the beam.

Although the use of an aperture in this way wastes laser power, it should be appreciated that it is the potential for optical damage[21] rather than the available laser power that typically sets the upper limit on the power that is incident in the sample plane. In any event, an efficiency of 50% compares favorably with the efficiency of many computer-generated holograms used for beam shaping in optical tweezers[22] and other applications.

To demonstrate this technique we prepared a sample of silica spheres with sizes of 1–5 μm suspended in water. Leaving the sample within the tweezers for 1 h invariably results in a number of the spheres' sticking together. The result is a collection of single spheres and asymmetric fused assemblies of silica spheres. Figure 2 shows the orientation of one such assembly that was trapped by the tweezers as the aperture was rotated. In this case the trapping power in the sample plane was set to be 30 mW. In

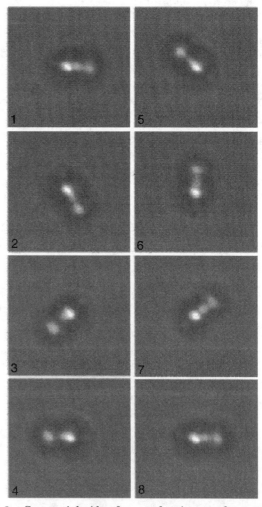

Fig. 2. Sequential video frames showing synchronous rotation (at $\approx$1 Hz) of a fused assembly of silica spheres with an aperture for rotational control.

our configuration the aperture is rotated by hand and therefore limited to a rotation rate of ~1 Hz. However, some particles in particular were held extremely tightly, such that even sudden movements of the aperture were unable to shake the object free, implying that significantly higher rotation speeds are possible. In practice, the maximum rotation speed is a complex function depending on the asymmetry of the focal spot, the viscosity of the fluid, and the shape of the object. The ideal object would be symmetrical in shape (thereby minimizing the drag) but asymmetrical in terms of its refractive index.

We also assessed the effect of the aperture on the performance of the lateral and axial trapping efficiencies of the tweezers by applying a ramp voltage to the translation stage and increasing the scan frequency until the particle fell out of the trap. At this point the trapping force could be equated to the drag force calculated from the fluid visocity and the scan speed. When the tweezers are used for trapping single spheres at a distance of $10-15$ μm above the bottom of the sample cell, we measure lateral (along the long axis of the elliptical spot) and axial Q values of 0.082 ± 0.012 and 0.079 ± 0.015 for 2-μm-diameter spheres and 0.173 ± 0.012 and 0.152 ± 0.018 for 5-μm-diameter spheres, respectively. The lateral trapping efficiencies are ~10% lower than those obtained without the aperture, which may be due to the elliptical beam's smaller intensity gradient. However, generally our results lie within the range of those reported by other groups that used conventional nonapertured Gaussian beams.[23]

In conclusion, we have demonstrated a simple, but to our knowledge unreported, method of achieving rotational control within optical tweezers. The inclusion of an aperture in the trapping beam results in a focused spot with the same cross-section symmetry as the aperture. This results in the preferential rotational alignment of an irregular object with the beam. Rotating the aperture rotates the focal spot within the tweezers, causing a synchronous rotation of the object. The technique could readily be applied to the experimental configurations of most existing tweezers, allowing its use in all areas of tweezers activity. However, it is most likely to be of interest in the manipulation and control of micromachines, where the potential use of rotational control is most apparent. We appreciate that

rotational control will perhaps be required only occasionally in most applications, and indeed the aperture is readily removed from our configuration, allowing reversion to a standard tweezers setup.

References

1. A. Ashkin, J. M. Dziedzic, J. E. Bjorkholm, and S. Chu, Opt. Lett. **11,** 288 (1986).
2. Cell Robotics International, Inc., Albuquerque, N.M.
3. P.A.L.M. GmbH, Bernried, Germany.
4. S. M. Block, H. C. Blair, and H. C. Berg, Nature **338,** 514 (1989).
5. J. T. Finer, R. M. Simmons, and J. A. Spudich, Nature **368,** 113 (1996).
6. W. D. Wang, H. Yin, R. Landick, J. Gelles, and S. M. Block, Biophys. J. **72,** 1335 (1997).
7. M. E. J. Friese, T. A. Nieminen, N. R. Heckenberg, and H. Rubinsztein-Dunlop, Nature **394,** 348 (1998).
8. H. He, M. E. J. Friese, N. R. Heckenberg, and H. Rubinsztein-Dunlop, Phys. Rev. Lett. **75,** 826 (1995).
9. N. B. Simpson, K. Dholakia, L. Allen, and M. J. Padgett, Opt. Lett. **22,** 52 (1997).
10. A. T. O'Neil, I. Mac Vicar, L. Allen, and M. J. Padgett, Phys. Rev. Lett. **88,** 053601 (2002).
11. E. Higurashi, H. Ukita, H. Tanka, and O. Ohguchi, Appl. Phys. Lett. **64,** 2209 (1994).
12. Z. P. Luo, Y. L. Sun, and K. N. An, Appl. Phys. Lett. **76,** 1779 (2000).
13. S. Sato, M. Ishigure, and H. Inaba, Electron. Lett. **27,** 1831 (1991).
14. A. Ashkin, J. M. Dziedzic, and T. Yamane, Nature **330,** 769 (1987).
15. K. Visscher, G. J. Brakenhoff, and J. J. Krol, Cytometry **14,** 105 (1993).
16. M. Reicherter, T. Haist, E. U. Wagemann, and H. J. Tiziani, Opt. Lett. **24,** 608 (1999).
17. A. E. Chiou, W. Wang, G. J. Sonek, J. Hong, and M. W. Berns, Opt. Commun. **133,** 7 (1997).
18. L. Paterson, M. P. MacDonald, J. Arlt, W. Sibbett, P. E. Bryant, and K. Dholakia, Science **292,** 912 (2001).
19. S. Sato and H. Inaba, Opt. Quantum Electron. **28,** 1 (1996).
20. J. Courtial, D. A. Robertson, K. Dholakia, L. Allen, and M. J. Padgett, Phys. Rev. Lett. **81,** 4828 (1998).
21. I. A. Vorobjev, L. Hong, W. H. Wright, and M. W. Berns, Biophys. J. **64,** 533 (1993).
22. K. T. Gahagan and G. A. Swartzlander, J. Opt. Soc. B **16,** 533 (1999).
23. H. Felgner, O. Mueller, and M. Schliwa, Appl. Opt. **34,** 977 (1995).

Direct Observation of Transfer of Angular Momentum to Absorptive Particles from a Laser Beam with a Phase Singularity

H. He, M. E. J. Friese, N. R. Heckenberg, and H. Rubinsztein-Dunlop

Department of Physics, The University of Queensland, Brisbane, Queensland, Australia Q4072
(Received 28 November 1994; revised manuscript received 4 April 1995)

Black or reflective particles can be trapped in the dark central minimum of a doughnut laser beam produced using a high efficiency computer generated hologram. Such beams carry angular momentum due to the helical wave-front structure associated with the central phase singularity even when linearly polarized. Trapped absorptive particles spin due to absorption of this angular momentum transferred from the singularity beam. The direction of spin can be reversed by changing the sign of the singularity.

PACS numbers: 42.25.Md, 42.40.My

It is well known that a circularly polarized beam carries angular momentum. Each photon of such a beam has an angular momentum of $\hbar$. The effects produced by the optical angular momentum are hard to observe in most circumstances as they represent extremely small quantities. The angular momentum flux carried by a circularly polarized 10 mW He-Ne laser beam is of the order of 10^{-18} mN. The first attempt to measure the torque produced by the optical angular momentum was made by Beth [1] 59 years ago. Beth reported that his results agreed with theory in sign and magnitude. More recently Santamato *et al.* [2] observed the light induced rotation of liquid crystal molecules, Chang and Lee [3] calculated the optical torque acting on a weakly absorbing optically levitated sphere, and Ashkin and Dziedzic [4] observed rotation of particles optically levitated in air.

A linearly polarized wave containing a central phase singularity also carries angular momentum associated with its helical structure. Figure 1 shows this structure for a TEM_{01}^* beam [5]. Each photon carries $l\hbar$ angular momentum where l is called the topological charge of the singularity. This contribution to the angular momentum is sometimes referred to as "orbital angular momentum" to distinguish it from "spin" angular momentum associated with circular polarization [6,7].

Allen *et al.* [6] have proposed to measure the angular momentum carried by a doughnut laser beam by measuring the torque acting on an optical device which reverses the chirality of the beam.

We have demonstrated in a previous paper [8] that black or reflective particles of sizes of $1-2$ μm can be trapped optically in a liquid by higher order doughnuts produced by high efficiency computer generated holograms. We also mentioned that some bigger absorptive particles were set into rotation by slightly defocused doughnuts. In this paper we report on further experiments concerned with the transfer of angular momentum to absorptive particles and their subsequent rotation. The detection is performed using a video camera. A large number of small particles have been trapped and have been recorded. Our results clearly show that the rotation directions of all these trapped particles agree with the sign of the doughnut.

In this Letter we analyze the results of the rotating particles in terms of possible torques acting on a trapped absorptive particle.

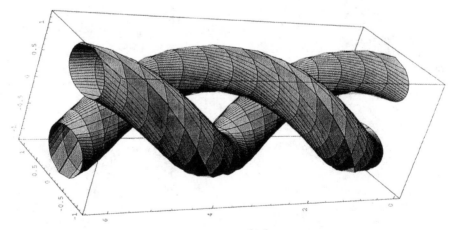

FIG. 1. Snapshot of the irradiance structure of a TEM_{01}^* $[(r/\omega)e^{-r^2/\omega^2}e^{i\theta}e^{ikz}]$ beam containing a first order phase singularity. The surface represented is that where the irradiance has half its peak value and we take $k = 1$, $\omega = 1$ for simplicity.

The angular momentum carried by light can be understood in two ways. Classically, electromagnetic radiation carries momentum, which can be both linear and angular. Allen *et al.* [6] have shown that for a Laguerre-Gaussian beam the angular momentum density is given as

$$M_z = \frac{l}{\omega} |u|^2 + \frac{\delta_z r}{2\omega} \frac{\partial |u|^2}{\partial r}, \qquad (1)$$

where u is the amplitude of the light field, l is the azimuthal index number of the Laguerre polynomial, $\delta_z = \mp 1$ for right-handed or left-handed circularly polarized light and $\delta_z = 0$ for linearly polarized light.

Quantum mechanically, one would say that each photon carries $(l \mp \delta_z)\hbar$ of angular momentum because of the well-known analogy between paraxial theory and quantum mechanics [9].

Although the above simple relationship is valid only within the paraxial approximation as shown recently by Barnett and Allen [7], for a linearly polarized light ($\delta_z = 0$), even when tightly focused as in an optical trapping experiment, the total angular momentum per second is still given by

$$\Gamma_z = \frac{P}{\omega} l, \qquad (2)$$

where P is the laser power.

A linearly polarized charge 3 singularity beam corresponds to $l = 3, \delta_z = 0$.

This suggests that an absorptive particle illuminated by such a singularity beam should be set into rotation in the same sense as the helical beam. It should reverse its rotation direction when the direction of rotation of the helical wave is reversed.

The experimental setup for the optical trap used in this work is similar to the one used in [10]. A linearly polarized TEM$_{00}$ laser beam from a 15 mW He-Ne laser (632.8 nm) illuminates a blazed charge 3 phase hologram. The hologram then produces a phase singular laser doughnut beam equivalent to a linearly polarized Gauss-Laguerre LG$_{03}$ mode with a power of approximately 7 mW. As the hologram used here is blazed, the sign of the doughnut can be simply reversed by turning the hologram around. We can also switch the diffracted beam between doughnut mode and Gaussian mode by moving the hologram sideways. The laser beam is now introduced into a microscope (Olympus CHT) through the aperture for the vertical illuminator. A dichroic mirror reflects the beam to fill the back aperture of an oil-immersion, high numerical aperture ($NA = 1.30$), 100× objective. The tightly focused doughnut beam had a diffraction limited beam waist within micrometers of the object plane of the objective.

The particles were illuminated from below by a lamp and green filters. A video camera placed vertically on the top of the microscope was used to record the motion of the trapped particles. We normally used black high-T_c superconductor ceramic powder as absorptive particles, dispersed in kerosene. Sizes of the particles that can be trapped are around 1–2 μm. We have performed similar experiments using CuO particles dispersed in water.

Using the charge 3 doughnut, we can trap a particle in the dark central spot. It can then be moved around relative to its surroundings by moving the microscope slide. The maximum measured trapping speed is around 5 μm/s. Particles adhering to the slide can often be set free by switching from the doughnut to the Gaussian mode and "kicking" them with the strong repulsive force.

Using a video camera, we clearly see a trapped particle rotating always in one direction determined by the helicity of the beam. The particle also keeps rotating in the same direction while being moved relative to its surroundings. The rotation of a trapped particle can be easily maintained for long periods of time. The particle can be set free and trapped again with the same rotation direction. In a session, more than 30 particles have been trapped, moved, set free, and trapped again. They all rotate in the same direction as the helicity of the beam. A few particles with very irregular shapes tumble wildly and occasionally rotate in the opposite direction for one or two turns but much more slowly than they rotate the other way. However, most of the time they rotate in the same direction as other particles. Those particles with close to spherical shapes always rotate in the same direction with high constant speed.

Turning the hologram back to front, that is reversing the charge of the singularity, we repeat the above procedures and we observe the same effects except that the rotation direction of all particles is reversed.

The trapped particle rotation speed varies from 1 to 10 Hz depending on shape and size.

Figure 2 shows six successive frames of a video recording. A fairly asymmetric particle, about 2 μm across, tumbling and rotating at about 1 Hz, has been chosen for illustration. It lies near the top of the frame and over a period of 0.24 s rotates through an angle of a little over $\pi/2$. Surrounding objects are stationary. More rapidly rotating symmetrical particles show up poorly in individual frames. However, the rotation is clear in continuous playback, in spite of motional blurring, very limited depth of field, and the fact that the microscope is operating near its limit of resolution.

When trapping CuO particles in water, if a little detergent is added, all the particles move very freely, and when one is trapped, almost immediately surrounding particles begin to move in radially and join a general circulation about it, in the same direction as the rotation of the trapped one.

Taking losses into consideration, the power incident on the focal plane is around 4 mW. According to Eq. (2), the angular momentum flux of such a doughnut beam produced by a charge 3 blazed hologram illuminated by a 633 nm HeNe laser is about 4×10^{-18} mN.

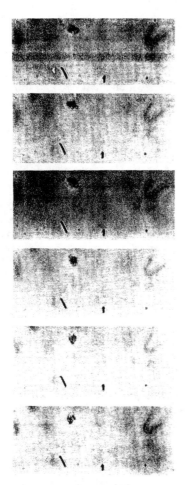

FIG. 2. Six successive frames of a video recording of a particle of black ceramic trapped in a charge three doughnut beam. The particle is near the top; other objects in the field are stationary. A rotation of just over $\pi/2$ occurs in the period shown.

Assume a particle absorbs 25% of the beam and hence 25% of the angular momentum. The rotation speed will become constant when the torque produced by the doughnut laser beam is balanced by the drag torque exerted by the surrounding liquid.

The drag torque acting on a spherical particle rotating with angular velocity ω_a is given by [11]

$$\tau_v = -8\pi\eta r^3 \omega_a, \tag{3}$$

where r is the radius of the particle, η is the viscosity of the liquid, taken as 1.58×10^{-3} N s m^{-2} for kerosene.

For our particles, with radius about 1 μm, this leads to a rotation speed of around 4 Hz, consistent with our observations, bearing in mind the actual form of our particles, and the fact that we have neglected the effects of the nearby slide surface. Our assumption that the particles absorb 25% of the incident power is based on the case where the sizes of particles and the central dark spot are

the same. We define the size of the dark spot to be the diameter of the maximum intensity ring of the doughnut beam. This estimate does not affect the relative sizes of the torques hypothetically present as they all would be proportional to the absorbed power.

Although absorbtion of angular momentum therefore accounts satisfactorily for the observed rotation, we briefly consider other explanations that might be advanced. It is possible that an unbalanced force, such as a thermal force, or a scattering force, may produce a torque acting upon a nonsymmetric particle. Among these forces, the scattering force is the biggest. However, since the direction of the scattering force is predominantly downward, the direction of the torque produced by the scattering force is therefore perpendicular to the beam axis. We believe that such torques are responsible for the irregular tumbling exhibited by asymmetric particles.

However, it is conceivable that reflections on the surface of an asymmetric particle may produce a torque in the same direction as that of the angular momentum from the beam.

Such a torque will be, at most,

$$\tau_a = k\gamma Fr, \tag{4}$$

where F is the force acting on the particle, r is an effective radius, γ is the reflection coefficient, and k is an asymmetry index, expressing the ratio of the area of a typical irregularity to the total area.

The scattering force (radiation pressure) cannot exceed

$$F_s = \frac{P}{c} = 1.3 \times 10^{-11} \text{ N}.$$

The value of k can vary from 0 to 1. However, only if the particle has a perfect "propeller" shape will k have a value approaching 1. Since our particles are mostly spheroidal and the effect of the trap is to center them in the beam, as the main asymmetry is due to the bumps on the surface of the particle, the value of k is approximately equal to the ratio of the bump area to the particle cross sectional area. Using a scanning electron microscopy image of these particles, we estimate that the value of k will normally be less than 1×10^{-2} as the irregularities are smaller than 10%. It is safe to assume the reflection coefficient to be smaller than 0.1 as our particles are highly absorptive. Hence, the torque due to asymmetric scattering forces is less than 10^{-20} N m, which is much smaller than the angular momentum from the doughnut laser beam. Obviously, a particle with an irregular shape may have a large value of k for a short period of time during which the scattering torque may dominate. However, the scattering torque is not constant, and hence the rotation direction changes. Furthermore, such a scattering force would be independent of the sign of the doughnut and therefore not able to explain our experimental results.

It is very difficult to estimate the magnitude of possible thermal effects like the photophoretic forces known to act on illuminated particles in air [12], but they could be expected to be smaller in our experiments because of the much higher thermal conductivities of the liquid media and the particles used. We have been unable to find any reports of such forces in liquids. Even if such forces are acting, considering the high thermal conductivity of the material and the fact that irregularities are small relative to the particle size and the wavelength, it is unlikely that they would exert torques leading to regular rotation with direction depending on the helicity of the wave.

Finally, if the rotation we observe was due to some thermally mediated torque, in order that angular momentum be conserved, it would be necessary that the surrounding liquid circulate in the opposite direction to the particle. However, our observations of the motion of nearby particles clearly eliminates this possibility. We see nearby particles swept around in the same direction as the rotating particle, consistent with a picture of an externally driven trapped particle stirring the surrounding liquid.

In conclusion, it has been demonstrated that absorptive particles trapped in the dark central minimum of a doughnut laser beam are set into rotation. The rotational motion of the particles is caused by the transfer of angular momentum carried by the photons. Since the laser beam is linearly polarized, this must originate in the "orbital" angular momentum associated with the helical wave-front structure and central phase singularity. We have shown that the direction of the rotational motion is determined by the chirality of the helical wave front. With a laser power of a few milliwatts, the rotation speed of the particles lies between 1 and 10 Hz depending on their sizes and shapes. This is in agreement with a simple model of absorption of the angular momentum of the radiation field.

We would like to thank Professor P. Drummond for useful discussions. We also want to thank R. McDuff and A. Noskoff for assistance with the images. This work was supported by the Australian Research Council.

[1] R. A. Beth, Phys. Rev. **50**, 115 (1936).

[2] E. Santamato, B. Daino, M. Romagnoli, M. Settemlre, and Y. R. Shen, Phys. Rev. Lett. **57**, 2433 (1986).

[3] S. Chang and S. S. Lee, J. Opt. Soc. Am. B **2**, 1853 (1985).

[4] A. Ashkin, Science **210**, 4474 (1980); **210**, 1081 (1980).

[5] N. R. Heckenberg, R. G. McDuff, C. P. Smith, H. Rubinsztein-Dunlop, and M. J. Wegerner, Opt. Quant. Electron. **24**, S951 (1992).

[6] L. Allen, M. W. Beijersbergen, R. J. C. Spreeuw, and J. P. Woerdman, Phys. Rev. A **45**, 8185 (1992).

[7] S. M. Barnett and L. Allen, Opt. Commun. **110**, 670 (1994).

[8] H. He, N. R. Heckenberg, and H. Rubinsztein-Dunlop, J. Mod. Opt. **42**, 217 (1995).

[9] D. Marcuse, *Light Transmission Optics* (Van Nostrand, New York, 1972).

[10] C. D'Helon, E. W. Dearden, H. Rubinsztein-Dunlop, and N. R. Heckenberg, J. Mod. Opt. **41**, 595 (1994).

[11] S. Oka, in *Rheology,* edited by F. R. Eirich (Academic Press, New York, 1960), Vol. 3.

[12] L. R. Eaton and S. L. Neste, AIAA J. **17**, 261 (1979).

Mechanical equivalence of spin and orbital angular momentum of light: an optical spanner

N. B. Simpson, K. Dholakia, L. Allen, and M. J. Padgett

J. F. Allen Physics Research Laboratories, Department of Physics and Astronomy, University of St. Andrews, North Haugh, St. Andrews, Fife KY16 9SS, Scotland

Received August 27, 1996

We use a Laguerre–Gaussian laser mode within an optical tweezers arrangement to demonstrate the transfer of the orbital angular momentum of a laser mode to a trapped particle. The particle is optically confined in three dimensions and can be made to rotate; thus the apparatus is an optical spanner. We show that the spin angular momentum of $\pm \hbar$ per photon associated with circularly polarized light can add to, or subtract from, the orbital angular momentum to give a total angular momentum. The observed cancellation of the spin and orbital angular momentum shows that, as predicted, a Laguerre–Gaussian mode with an azimuthal mode index $l = 1$ has a well-defined orbital angular momentum corresponding to $\hbar$ per photon. © 1997 Optical Society of America

The circularly symmetric Laguerre–Gaussian (LG) modes form a complete basis set for paraxial light beams.[1] Two indices identify a given mode, and the modes are normally denoted $LG_p{}^l$, where l is the number of 2π cycles in phase around the circumference and ($p + 1$) is the number of radial nodes. In contrast to the planar wave fronts of the Hermite–Gaussian (HG) mode, for $l \neq 0$ the azimuthal phase term $\exp(il\phi)$ in the LG modes results in helical wave fronts.[2] These helical wave fronts are predicted by Allen *et al.* to give rise to an orbital angular momentum for linearly polarized light of $l\hbar$ per photon.[3] This orbital angular momentum is distinct from the angular momentum associated with the photon spin manifested in circularly polarized light. For a collimated beam of circularly polarized light the photon spin is well known to be $\pm\hbar$; however, for a tightly focused beam the polarization state is no longer well defined.[4] The total angular momentum of a beam, found by a rigorous solution of Maxwell's equations that reduces to a LG beam in the paraxial approximation, is given by[5]

$$\left[l + \sigma_z + \sigma_z \left(\frac{2kz_r}{2p + l + 1} + 1 \right)^{-1} \right] \hbar \qquad (1)$$

per photon, where $\sigma_z = 0, \pm 1$ for linearly and circularly polarized light, respectively, and where z_r is the Rayleigh range and k is the wave number of the light. Note that for a collimated beam $kz_r \gg 1$, and expression (2) reduces to

$$(l + \sigma_z)\hbar \qquad (2)$$

per photon, whereas for linearly polarized light $l\hbar$ per photon is strictly correct even in the absence of any approximation.

Lasers have been made that operate in LG modes,[6] but it is easier to obtain them from the conversion of HG modes. Three different classes of mode converter have been demonstrated. Two of these, spiral phase plates[7] and computer-generated holographic converters,[8,9] introduce the azimuthal phase term to a $HG_{0,0}$ Gaussian beam to produce a LG mode with $p = 0$ and specific values of l. In all these devices a screw phase dislocation, produced on axis, causes destructive interference, leading to the characteristic ring intensity pattern in the far field. The other class of converter is the cylindrical-lens mode converter,[10] which employs the change in Gouy phase in the region of an elliptically focused beam to convert higher-order HG modes, of indices m and n, into the corresponding LG modes (or vice versa) with 100% efficiency. The transformation is such that $l = |m - n|$ and $p = \min(m, n)$.

Consider the various macroscopic mechanisms whereby angular momentum can be transferred between a well-defined transverse laser mode and matter. One can transfer spin angular momentum by using a birefringent optical component, resulting in a change in the polarization state of the light. One can transfer orbital angular momentum by using a component that possesses an azimuthal dependence of its optical thickness, as with a cylindrical lens or a spiral phase plate. In contrast, absorption of the light allows both spin and orbital angular momentum to be transferred.

The prediction that LG laser modes possess well-defined orbital angular momentum has led to considerable research activity.[3,6-14] The transfer of orbital angular momentum from a light beam to small particles held at the focus has been modeled,[12] but to date only one group of researchers has demonstrated that these modes do indeed possess orbital angular momentum. He *et al.*[13] demonstrated that orbital angular momentum could be transferred from the laser mode to micrometer-sized ceramic and metal-oxide particles near the focus of a LG beam with an azimuthal mode index $l = 3$. More recently the same authors reported the use of a circularly polarized mode and the corresponding speeding up or slowing down of the rotation, depending on the relative sign of the spin and orbital angular-momentum terms.[14] In each case the particles were trapped with radiation pressure, first forcing the totally absorbing particle into the dark region on the beam axis (x–y trapping) and then forcing the particle against the microscope slide (z trapping). LG modes were recently also used in an optical tweezers geometry for trapping hollow glass spheres[15]; however, no rotation was observed because these spheres were totally transparent.

Here we use weakly absorbing dielectric particles, larger than the dimensions of the tightly focused LG mode. The particle experiences a net force within the electric field gradient toward the focus of the beam and is thus held in an all-optical $x-y-z$ trap. This clearly distinguishes our experiments from previous ones,[13,14] which relied on mechanical restraint in the z direction. Such all-optical traps, using zero-order Gaussian modes, were demonstrated by Ashkin *et al.*[16] and are commonly referred to as optical tweezers. The transfer of angular momentum and the subsequent rotation of the trapped particle through the use of a LG mode leads us to refer to our modified optical geometry as an optical spanner.

Figure 1 shows our experimental arrangement. We use a diode-pumped Nd:YLF laser at 1047 nm operating in a high-order HG mode. This mode is converted into the corresponding LG mode by the use of the previously mentioned cylindrical-lens mode converter; a quarter-wave plate sets the polarization state. A telescope arrangement optimizes the mode size to fill the back aperture of the objective lens, and an adjustable mirror enables the focused beam to be translated within the field of view of the objective lens. A dichroic beam splitter couples the beam into the optical spanner. The optical spanner is based on a 1.3-N.A., 100× oil-immersion microscope objective. The associated tube lens forms an image of the trapped particle on a CCD array. The sample cell comprises a microscope slide, an ≈60-μm-thick vinyl spacer, and a cover slip and is backlit with a standard tungsten bulb. The LG mode is focused such that the trapped object is held in the focal plane of the objective lens, and a colored glass filter can be inserted before the CCD array to attenuate the reflected laser light. We independently confirmed the polarization state of the light in the plane of the trapped particle by monitoring the power transmitted through a linear polarizer at various orientations.

When the $l = 1$ LG mode was focused by the objective lens, its beam waist was approximately 0.8 μm, which corresponds to a high-intensity ring 1.1 μm in diameter. For dielectric particles smaller than ~1 μm we observe that the gradient force results in the particles' being trapped off axis, centered within the high-intensity region of the ring itself. However, particles 1 μm or larger are trapped on the beam axis and for incident power above a few milliwatts are held stably in three dimensions. If the particle is partially absorbing, the transfer of angular momentum from the laser beam to the particle causes the particle to rotate. The strong z-trapping force enables us to lift the particle optically off the button of the sample cell, allowing it to rotate more freely. We have successfully rotated a number of a different particle types, including particles of absorbing glass (e.g., Schott BG38 glass) and Teflon spheres. The particles were dispersed in various fluids, including water, methanol, and ethanol.

For the linearly polarized $l = 1$ LG mode and an incident power of ≈25 mW, typical rotation speeds for

the Teflon particles were ~1 Hz. The relationship between applied torque τ and limiting angular velocity $\omega_{\lim}$ of a sphere of radius r in a viscous medium of viscosity η is[17]

$$\omega_{\lim} = \tau/(8\pi\eta r^3). \tag{3}$$

This implies that the absorption of the Teflon particle is of the order of 2%.

The principal motivation of this study was to demonstrate that, with respect to absorption and the resulting mechanical rotation of a particle, the spin and the orbital angular momenta of light are equivalent. As discussed above, for the spin and orbital angular-momentum content of the laser mode to interact in an equivalent manner with the trapped particle it is essential that the dominant mechanism for transfer of the angular momentum be absorption. The birefringence or astigmatism of the trapped particle will preferentially transfer spin or orbital angular momentum, respectively. For comparison of spin and orbital angular momentum we selected Teflon particles, or amalgamated groups of Teflon particles, larger than the focused beam waist. Trapped on the beam axis, these particles interact uniformly with the whole of the beam, can be optically levitated off the bottom of the sample cell, and are observed to rotate smoothly with a constant angular velocity. This ability to lift the particles optically away from the cell boundary differentiates our research from that previously reported. Less regular particles could be lifted and made to rotate in the same sense as the more regular particles but often failed to stop when the spin and orbital angular-momentum terms were subtractive. This imperfect cancellation is due to additional transfer of orbital angular momentum owing to the asymmetry of the particle.

For our experimental configuration and an $l = 1$ mode, expression (1) shows that the total angular

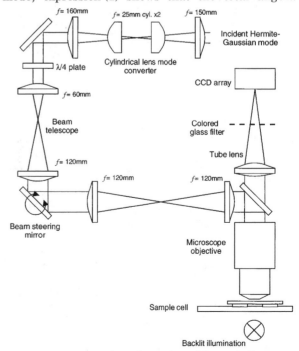

Fig. 1. Experimental configuration of the optical spanner.

time →

| | orbital - spin | orbital | orbital + spin |

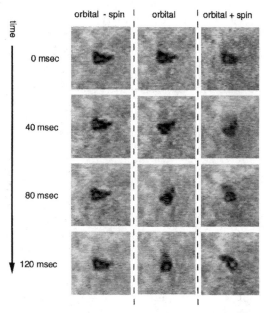

0 msec

40 msec

80 msec

120 msec

Fig. 2. Successive frames of the video image showing the stop–start behavior of a 2-μm-diameter Teflon particle held with the optical spanner.

momentum is ~2.06$\hbar$ per photon when the spin and orbital terms are additive and ~0.06$\hbar$ per photon when the spin and orbital terms are subtractive.

Figure 2 shows successive frames of the video image of a 2-μm-diameter Teflon particle held within the optical spanner, trapped with an $l = 1$ mode of various polarization states. The slight asymmetry in the particle geometry allows the particle rotation to be assessed. In more than 80% of cases, once a smoothly rotating particle had been selected, orienting the wave plate to circularly polarize the light would cause the particle to speed up or stop completely, depending on whether the spin and orbital angular-momentum terms were added or subtracted. In the other 20% of cases, although the particles would speed up or slow down, they could not be stopped completely. We believe that this was because these particles were insufficiently uniform, resulting in an unwanted mode transformation and giving an additional exchange of orbital angular momentum. With the well-behaved particles, when the spin and orbital terms are subtracted the low value of the torque, owing to a total angular momentum of no more than 0.06$\hbar$ per absorbed photon, is insufficient to overcome the stiction present within the system, and the particle ceases to rotate. Stiction can arise from slight particle asymmetry coupled with residual astigmatism in the laser mode, which will favor the particles' being trapped in particular orientations. When the spin and orbital terms are additive the rotation speed increases significantly, but in only few cases is it seen to double. We attribute this to a nonlinear relationship between the applied torque and the terminal rotation speed, and the particle asymmetry. We confirmed this nonlinearity in the relationship between torque and rotation speed by deliberately changing the optical power while maintaining the same laser mode.

We observe, in agreement with our interpretation, that reversing the sense of the cylindrical-lens mode converter caused the particles to rotate in the opposite direction; similarly, by use of a HG$_{0,0}$ mode, particles could be rotated in either direction when circularly polarized light of the appropriate handedness was used.

Our experiment uses a LG mode within an all-optical $x-y-z$ trap to form an optical spanner for rotating micrometer-sized particles. By controlling the polarization state of a LG mode with $l = 1$ we can arrange for the angular momentum associated with the photon spin to add to the orbital angular momentum, giving a total angular momentum of 2$\hbar$ per photon, in which case the Teflon particle spins more quickly. Alternatively, that momentum can subtract, resulting in no overall angular momentum, and the particle comes to a halt. We observe the cancellation of the spin and orbital angular momentum in a macroscopic system, which verifies that for an $l = 1$ LG mode the orbital angular momentum is indeed well defined and corresponds to $\hbar$ per photon.

This research is supported by Engineering and Physical Sciences Research Council grant GR/K11536. M. I. Padgett is a Royal Society Research Fellow.

References

1. A. E. Siegman, *Lasers* (University Science, Mill Valley, Calif., 1986), Sec. 17.5, pp. 685–695.
2. J. M. Vaughan and D. V. Willetts, Opt. Commun. **30,** 263 (1979).
3. L. Allen, M. W. Beijersbergen, R. J. C. Spreeuw, and J. P. Woerdman, Phys. Rev. A **45,** 8185 (1992).
4. D. N. Pattanayak and G. P. Agrawal, Phys. Rev. A **22,** 1159 (1980).
5. S. M. Barnett and L. Allen, Opt. Commun. **110,** 670 (1994).
6. M. Harris, C. A. Hill, and J. M. Vaughan, Opt. Commun. **106,** 161 (1994).
7. M. W. Beijersbergen, R. P. C. Coerwinkel, M. Kristensen, and J. P. Woerdman, Opt. Commun. **112,** 321 (1994).
8. N. R. Heckenberg, R. McDuff, C. P. Smith, and A. G. White, Opt. Lett. **17,** 221 (1992).
9. N. R. Heckenberg, R. McDuff, C. P. Smith, H. Rubinsztein-Dunlop, and M. J. Wegener, Opt. Quantum Electron. **24,** S951 (1992).
10. M. W. Beijersbergen, L. Allen, H. E. L. O. van der Veen, and J. P. Woerdman, Opt. Commun. **96,** 123 (1993).
11. M. Babiker, W. L. Power, and L. Allen, Phys. Rev. Lett. **73,** 1239 (1994); S. J. van Enk and G. Nienhuis, Opt. Commun. **94,** 147 (1992).
12. N. B. Simpson, L. Allen, and M. J. Padgett, J. Mod. Opt. **43,** 2485 (1996).
13. H. He, M. E. J. Friese, N. R. Heckenberg, and H. Rubinsztein-Dunlop, Phys. Rev. Lett. **75,** 826 (1995).
14. M. E. J. Friese, J. Enger, H. Rubinsztein-Dunlop, and N. R. Heckenberg, Phys. Rev. A **54,** 1593 (1996).
15. K. T. Gahagan and G. A. Swartzlander, Jr., Opt. Lett. **21,** 827 (1996).
16. A. Ashkin, J. M. Dziedzic, J. E. Bjorkholm, and S. Chu, Opt. Lett. **11,** 288 (1986).
17. S. Oka, in *Rheology,* F. R. Eirich, ed. (Academic, New York, 1960), Vol. 3.

Intrinsic and Extrinsic Nature of the Orbital Angular Momentum of a Light Beam

A. T. O'Neil, I. MacVicar, L. Allen, and M. J. Padgett

Department of Physics and Astronomy, University of Glasgow, Glasgow, G12 8QQ, Scotland
(Received 28 June 2001; published 16 January 2002)

We explain that, unlike the spin angular momentum of a light beam which is always intrinsic, the orbital angular momentum may be either extrinsic or intrinsic. Numerical calculations of both spin and orbital angular momentum are confirmed by means of experiments with particles trapped off axis in optical tweezers, where the size of the particle means it interacts with only a fraction of the beam profile. Orbital angular momentum is intrinsic only when the interaction with matter is about an axis where there is no net transverse momentum.

DOI: 10.1103/PhysRevLett.88.053601   PACS numbers: 42.50.Ct

Introduction.—Some 65 years ago Beth [1] demonstrated that circularly polarized light could exert a torque upon a birefringent wave plate suspended in the beam by the transfer of angular momentum. The angular momentum associated with circular polarization arises from the spin of individual photons and is termed spin angular momentum.

More recently, Allen *et al.* [2] showed that for beams with helical phase fronts, characterized by an $\exp(il\phi)$ azimuthal phase dependence, the orbital angular momentum in the propagation direction has the discrete value of $l\hbar$ per photon. Such beams have a phase dislocation on the beam axis that in related literature is sometimes referred to as an optical vortex [3]. In general, any beam with inclined phase fronts carries orbital angular momentum about the beam axis which, when integrated over the beam, can be an integer or noninteger [4,5] multiple of $\hbar$.

In this paper, we experimentally examine the motion of particles trapped off axis in an optical tweezers and are able to associate specific aspects of the motion with the distinct contributions of spin and orbital angular momentum of the light beam. The interpretation of the experiments, when combined with a numerical calculation of the spin and orbital contributions derived from established theory, allows a distinction to be made between the intrinsic and extrinsic aspects of the angular momentum of light.

Angular momentum of a light beam.—The cycle-averaged linear momentum density, **p**, and the angular momentum density, **j**, of a light beam may be calculated from the electric, **E**, and magnetic, **B**, fields [6]:

$$\mathbf{p} = \varepsilon_0 \langle \mathbf{E} \times \mathbf{B} \rangle, \tag{1}$$

$$\mathbf{j} = \varepsilon_0 (\mathbf{r} \times \langle \mathbf{E} \times \mathbf{B} \rangle) = \mathbf{r} \times \mathbf{p}. \tag{2}$$

Equation (2) encompasses both the spin and orbital angular momentum density of a light beam.

Within the paraxial approximation, the local value of the linear momentum density of a light beam is given by [2]

$$\mathbf{p} = i\omega \frac{\varepsilon_0}{2} (u^* \nabla u - u \nabla u^*) + \omega k \varepsilon_0 |u|^2 \mathbf{z}$$

$$+ \omega \sigma \frac{\varepsilon_0}{2} \frac{\partial |u|^2}{\partial r} \Phi, \tag{3}$$

where $u \equiv u(r, \phi, z)$ is the complex scalar function describing the distribution of the field amplitude. Here σ describes the degree of polarization of the light; $\sigma = \pm 1$ for right- and left-hand circularly polarized light, respectively, and $\sigma = 0$ for linearly polarized.

The cross product of this momentum density with the radius vector $\mathbf{r} = (r, 0, z)$ yields an angular momentum density. The angular momentum density in the z direction depends upon the Φ component of **p**, such that

$$j_z = r p_\phi. \tag{4}$$

The final term in Eq. (3) depends upon the polarization but is independent of the azimuthal phase and, consequently, this term may be linked directly to the spin angular momentum. The first term in Eq. (3) depends upon the phase gradient and not the polarization, and so gives rise to the orbital angular momentum.

For many mode functions, u, such as for circularly polarized Laguerre-Gaussian modes, Eqs. (3) and (4) can be evaluated analytically such that the local angular momentum density in the direction of propagation is given by [2]

$$j_z = \varepsilon_0 \left\{ \omega l |u|^2 - \frac{1}{2} \omega \sigma r \frac{\partial |u|^2}{\partial r} \right\}. \tag{5}$$

The angular momentum integrated over the beam is readily shown to be equivalent to $\sigma\hbar$ per photon for the spin and $l\hbar$ per photon for the orbital angular momentum [2], that is

$$J_z = (l + \sigma)\hbar. \tag{6}$$

A theoretical discussion of the behavior of local momentum densities has been published elsewhere [7], and it should be noted that the local spin and orbital angular momentum do not have the same functional form.

As is well known, spin angular momentum does not depend upon the choice of axis and so is said to be *intrinsic*. The angular momentum which arises for any light beam from the product of the z component of linear momentum about a radius vector, may be said to be an *extrinsic* because its value depends upon the choice of calculation axis.

Berry showed [8] that the orbital angular momentum of a light beam does not depend upon the lateral position of the axis and can therefore also be said to be intrinsic, provided the direction of the axis is chosen so that the transverse momentum is zero. When integrated over the whole beam the angular momentum in the z direction is

$$\mathbf{J}_z = \varepsilon_0 \iint dx\, dy\, \mathbf{r} \times \langle \mathbf{E} \times \mathbf{B} \rangle. \tag{7}$$

If the axis is laterally displaced by $\mathbf{r}_0 \equiv (r_{0x}, r_{0y})$ it is easy to show that the change in the z component of angular momentum is given by

$$\begin{aligned} \Delta J_z &= (r_{0x} \times P_y) + (r_{0y} \times P_x) \\ &= r_{0x}\varepsilon_0 \iint dx\, dy\, \langle \mathbf{E} \times \mathbf{B} \rangle_y \\ &\quad + r_{0y}\varepsilon_0 \iint dx\, dy\, \langle \mathbf{E} \times \mathbf{B} \rangle_x. \end{aligned} \tag{8}$$

The angular momentum is intrinsic only if ΔJ_z equals zero for all values of r_{0x} and r_{0y}. This condition is satisfied only if z is stipulated as the direction for which the transverse momenta $\varepsilon_0 \iint dx\, dy\, \langle \mathbf{E} \times \mathbf{B} \rangle_x$ and $\varepsilon_0 \iint dx\, dy\, \langle \mathbf{E} \times \mathbf{B} \rangle_y$ are exactly zero.

For Laguerre-Gaussian light beams truncated by apertures, Eqs. (3) and (8) can only be evaluated numerically. Nevertheless, for all apertures, of whatever size or position, the spin angular momentum remains $\sigma\hbar$ irrespective of the choice of calculation axis and so is, as expected, intrinsic; see Fig. 1. Any beam with a helical phase front apertured symmetrically about the beam axis has zero transverse momentum and, consequently, an orbital angular momentum of $l\hbar$ per photon, independent of the axis of calculation. The orbital angular momentum of the light beam may therefore be described as intrinsic. However, when the beam is passed through an off-axis aperture, its transverse momentum is nonzero and the orbital angular momentum depends upon the choice of calculation axis and so must be described as extrinsic; see Fig. 1. An interesting result occurs when the orbital angular momentum of the apertured beam is calculated about the original beam axis. Even though the transverse momentum is nonzero, the orbital angular momentum remains $l\hbar$ per photon because r_{0x} and r_{0y} are both zero. However, it does not follow that the angular momentum of the apertured beam is intrinsic as the result does depend upon the choice of calculation axis. When any beam is apertured off axis, it is simpler and more accurate to

understand its interaction with particles by considering the components of $\mathbf{p}$ in the x-y plane. For beams with helical phase fronts, these transverse components are in the Φ direction with respect to the beam axis. It is this distinction between spin and orbital angular momentum which gives rise to differences in behavior for the interaction of light with matter.

The transfer of spin and orbital angular momentum to small particles. — The interaction of small particle with the angular momentum of a light beam has been investigated by a number of groups with the use of optical tweezers. Usually implemented by use of a high numerical-aperture microscope, optical tweezers rely on the gradient force to confine a dielectric particle near the point of highest light intensity [9]. For particles trapped on the beam axis, both the spin and orbital angular momentum have been shown to cause rotation of birefringent [10] and absorptive [11] particles, respectively. For absorbing particles, both spin and orbital angular momenta are transferred with the same efficiency so that the applied torque is proportional to the total angular momentum [12], that is $(\sigma + l)\hbar$ per photon.

In this present work we also use optical tweezers, but in this instance the particles are trapped away from the beam axis. This allows us to demonstrate the difference between particle interactions with spin and orbital angular momentum. The experimental configuration is shown in Fig. 2. Our optical tweezers are based on a 1.3 numerical aperture, $\times 100$ objective lens, configured with the trapping beam directed upwards, which allows easier access to the sample plane. This beam is generated from the 100 mW output of a commercial Nd:YLF laser transformed, using a computer generated hologram, to give a Laguerre-Gaussian mode of approximately 30 mW. The beam is circularly polarized, $\sigma = \pm 1$ with a high azimuthal mode index, $l = \pm 8$. The sign of the spin or the orbital angular momentum may be reversed by the insertion of a half-wave plate or a Dove prism, respectively [13]. The radius of maximum intensity, $r_{\max}$ of a Laguerre-Gaussian mode is given by [14],

$$r_{\max} = \sqrt{\frac{z_R l}{k}}, \tag{9}$$

where z_R is the Rayleigh range of the beam. Even under the tight focusing associated with optical tweezers, the peak intensity ring of a Laguerre-Gaussian mode of high index l may be made several μm in diameter and, consequently, be much larger than the particles it is attempting to trap.

It is not surprising, for such conditions, that we observe the particles to be confined by the gradient force at the radius of maximum light intensity and not on the beam axis. When a birefringent particle such as a calcite fragment is trapped, and circularly polarized light is converted to linear, we observe that the particle spins about its own axis. The sense of rotation is governed by the handedness of the circular polarization.

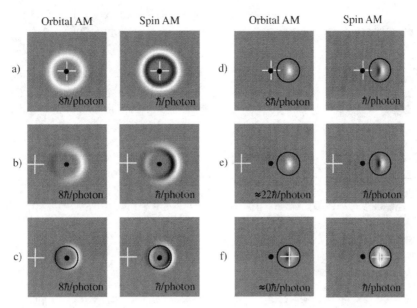

FIG. 1. Numerically calculated local spin and orbital angular momentum densities in the direction of propagation for a $l = 8$ and $\sigma = 1$ Laguerre-Gaussian mode. A positive contribution is shown in white, gray represents zero, and black a negative contribution; the black spot marks the axis of the original beam, the white cross marks the axis about which the angular momenta are calculated and, where appropriate, the black circle marks the position of a soft edged aperture. Note that the spin angular momentum is equivalent to $\sigma \hbar$ per photon irrespective of the choice of aperture or calculation axis, whereas the orbital angular momentum is only $l \hbar$ per photon if the aperture or calculation axes coincide with the axis of the original beam.

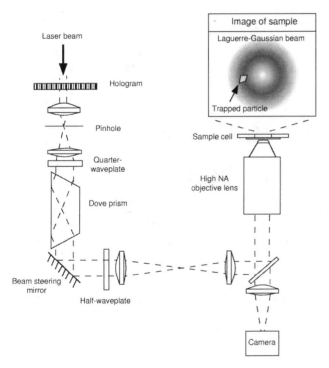

FIG. 2. The optical tweezers use a high-order Laguerre-Gaussian beam to trap the particle in the region of maximum light intensity, away from the beam axis. The circularly polarized Laguerre-Gaussian beam is generated using a quarter-wave plate and a hologram. The sense of the orbital and spin angular momentum can be reversed by the inclusion of a half-wave plate or a Dove prism, respectively.

For small particles the force arising from the light scattering, the momentum recoil force, becomes important. For a tightly focused Laguerre-Gaussian mode, the dominant component of the scattering force lies in the direction of beam propagation. The gradient force again constrains the particle to the annulus of maximum beam intensity. However, as the intensity distribution is cylindrically symmetric, the particle is not constrained azimuthally. Because the particle is trapped off the beam axis, the inclination of the helical phase fronts and the corresponding momentum result in a tangental force on the particle. We observe that a small particle, while still contained within the annular ring of light, orbits the beam axis in a direction determined by the handedness of the helical phase fronts; see Fig. 3. We conclude that the larger calcite and small particles are interacting with intrinsic spin and extrinsic orbital angular momentum, respectively. In principle, it should be possible to observe both the orbital and spin angular momenta acting simultaneously upon the same small birefringent particle. However, our observations have been inconclusive as birefringent particles small enough for the scattering force to induce a rotation about the beam axis are typically too small to see whether they are spinning about their own axis.

This orbital and spin behavior is entirely consistent with the formulation summarized in Eqs. (7) and (8). If one considers the cross section of an off-axis trapped particle to play the rôle of an aperture, then we see that the intrinsic spin of the angular momentum creates a torque about the

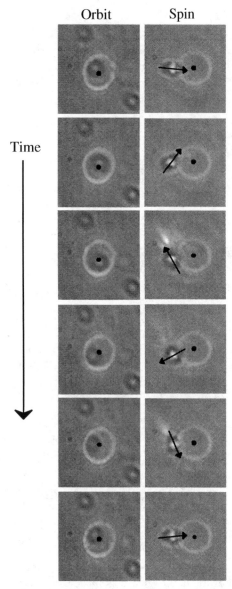

Orbit   Spin

Time

FIG. 3. Successive video frames showing particles trapped near the focus of an $l = 8$ and $\sigma = 1$ Laguerre-Gaussian mode. The left column shows particles of ≈ 1 μm diam. These particles are sufficiently small to be subject to a well-defined scattering force, allowing them to interact with the orbital angular momentum of the beam. They are set in motion, orbiting the beam axis at a frequency of ≈ 1 Hz. The right column shows a calcite fragment with a length of ≈ 3 μm and a width of about ≈ 1.5 μm, which is large enough not to interact detectably with the beam's orbital angular momentum. However, due to its birefringence it interacts with the spin angular momentum of the beam and is set spinning about its own axis at ≈ 0.3 Hz.

particle's own axis, causing it to spin. A calculation of the particle's angular momentum about an arbitrary axis shows a clear distinction between the intrinsic angular momentum associated with its spinning motion and the extrinsic angular momentum associated with its orbital motion. In this situation, orbital angular momentum is better described as the result of a linear momentum component directed at a tangent to the radius vector.

Unlike the spin angular momentum of a light beam which is always intrinsic, the z component of the orbital angular momentum can be described as intrinsic only if the z direction can be stipulated such that the transverse momentum integrated over the whole beam is zero. If an interaction is with only a fraction of the beam cross section, then the orbital angular as measured about the original axis is extrinsic.

L. Allen is pleased to thank the Leverhulme Trust for its support. M. J. Padgett thanks the Royal Society for its support. A. T. O'Neil is supported by the EPSRC.

[1] R. E. Beth, Phys. Rev. **50**, 115 (1936).

[2] L. Allen, M. W. Beijersbergen, R. J. C. Spreeuw, and J. P. Woerdman, Phys. Rev. A **45**, 8185 (1992).

[3] I. V. Basistiy, M. S. Soskin, and M. V. Vasnetsov, Opt. Commun. **119**, 604 (1995).

[4] M. Soljačić and M. Segev, Phys. Rev. Lett. **86**, 420 (2001).

[5] J. Courtial *et al.*, Opt. Commun. **144**, 210 (1997).

[6] J. D. Jackson, *Classical Electrodynamics* (Wiley, New York, 1962).

[7] L. Allen and M. J. Padgett, Opt. Commun. **184**, 67 (2000).

[8] M. V. Berry, in *International Conference on Singular Optics*, edited by M. S. Soskin, SPIE Proceedings Vol. 3487 (SPIE–International Society for Optical Engineering, Bellingham, WA, 1998), p. 6.

[9] A. Ashkin, J. M. Dziedzic, J. E. Bjorkholm, and S. Chu, Opt. Lett. **11**, 288 (1986).

[10] M. E. J. Friese, T. A. Nieminen, N. R. Heckenberg, and H. Rubinsztein-Dunlop, Nature (London) **394**, 348 (1998).

[11] H. He, M. E. J. Friese, N. R. Heckenberg, and H. Rubinsztein-Dunlop, Phys. Rev. Lett. **75**, 826 (1995).

[12] N. B. Simpson, K. Dholakia, L. Allen, and M. J. Padgett, Opt. Lett. **22**, 52 (1997).

[13] M. J. Padgett and J. Courtial, Opt. Lett. **24**, 430 (1999).

[14] M. J. Padgett and L. Allen, Opt. Commun. **121**, 36 (1995).

Modulated optical vortices

Jennifer E. Curtis* and David G. Grier

*Department of Physics, James Franck Institute and Institute for Biophysical Dynamics,
University of Chicago, Chicago, Illinois 60637*

Received January 9, 2003

Single-beam optical gradient force traps created by focusing helical modes of light are known as optical vortices. Modulating the helical pitch of such a mode's wave front yields a new class of optical traps whose dynamically reconfigurable intensity distributions provide new opportunities for controlling motion in mesoscopic systems. An implementation of modulated optical vortices based on the dynamic holographic optical tweezer technique is described.   © 2003 Optical Society of America

OCIS codes: 140.7010, 090.1760, 350.5030, 350.3850.

A single-beam optical gradient force trap, known as an optical tweezer, is created by focusing a beam of light with a strongly converging high-numerical-aperture (NA) lens.[1] Optical tweezers can trap and move materials noninvasively at length scales ranging from tens of nanometers to tens of micrometers and so have provided unprecedented access to physical, chemical, and biological processes in the mesoscopic domain.[2] Variants of optical tweezers based on specially crafted modes of light have demonstrated additional useful and interesting properties: optical vortices created from helical modes of light exert torques on trapped objects,[3-11] traps based on Bessel beams facilitate controlled transport over long distances,[12] and optical rotators provide fine orientation control.[13] These specialized traps have potentially widespread applications in biotechnology[14] and micromechanics,[15] particularly when they are created as integrated optical systems by use of holographic techniques.[16] This Letter introduces a generalized class of optical vortices with novel properties and describes an implementation based on computed holography.[16]

A vortex-forming helical mode is distinguished from a plane wave by an overall phase factor, $\exp(il\theta)$, where θ is the azimuthal angle around the optical axis and l is the integral winding number characterizing an l-fold helix. Semiclassical theory suggests that each photon in a helical beam carries an orbital angular momentum $l\hbar$ that is distinct from the photon's intrinsic spin angular momentum and yet quantized in units of Planck's constant.[10,17] This has been confirmed through measurements of particles' motions in optical vortices.[4,6,8,18,19] The topological charge l also determines the annular intensity distribution characteristic of an optical vortex,[17,19,20] where the intensity is distributed into a ring at radius R_l from the focal point. Recently, we demonstrated[19,21] that an optical vortex's radius scales linearly with l:

$$R_l \approx a\frac{\lambda}{\text{NA}}\left(1 + \frac{l}{l_0}\right), \qquad (1)$$

where λ is the wavelength of light, NA is the objective lens's numerical aperture, and the constants a and l_0 depend on the beam's radial amplitude profile. This observation suggests the generalization

$$R(\theta) = a\frac{\lambda}{\text{NA}}\left[1 + \frac{1}{l_0}\frac{\mathrm{d}\varphi(\theta)}{\mathrm{d}\theta}\right], \qquad (2)$$

in which the local radius of maximum intensity at angle θ in the plane of the optical vortex depends on the local winding number. For example, Eq. (2) predicts that

$$\varphi(\theta) = l[\theta + \alpha\sin(m\theta + \beta)] \qquad (3)$$

should produce an m-fold symmetric Lissajous pattern whose depth of modulation is controlled by α and whose orientation depends on β. Direct visualization with the dynamic holographic optical tweezer technique[16] confirms this prediction.

Our optical trapping system, shown in Fig. 1(a), has been described in detail elsewhere.[16,19] We use a

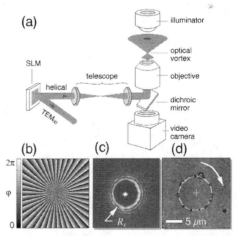

Fig. 1.   Creating optical vortices with dynamic holographic optical tweezers. (a) Schematic diagram of the experimental apparatus. A reflective spatial light modulator (SLM) imprints the phase modulation $\varphi(\mathbf{r})$ onto the wave front of a TEM$_{00}$ laser beam. The transformed beam is relayed by a telescope to the back aperture of a microscope objective lens that focuses it into an optical trap. A conventional illuminator and video camera create images of objects in the trap. (b) Phase modulations encoding an $l = 40$ optical vortex. (c) Resulting optical vortex's intensity in the focal plane. (d) Trajectory of a single 800-nm-diameter silica sphere traveling around the optical vortex's circumference, measured at 1/6-s intervals over 5 s.

reflective liquid-crystal SLM[22] to imprint a desired phase profile $\varphi(\mathbf{r})$ onto the wave front of a collimated TEM$_{00}$ beam ($\lambda = 532$ nm). The modified beam is relayed to the input pupil of a high-NA objective lens mounted in an inverted light microscope. Figure 1(b) shows a typical phase mask encoding an optical vortex with $l = 40$, and Fig. 1(c) shows the resulting intensity distribution imaged by replacement of the sample cell with a mirrored microscope slide. The SLM has a diffraction efficiency of roughly 50%, and the central spot in Fig. 1(c) is a conventional optical tweezer centered on the optical axis formed from the undiffracted portion of the input beam. Because the SLM can impose phase shifts only in the range 0 to 2π rad, the projected phase function wraps around at $\varphi = 2\pi$ to create a scalloped appearance.

When an optical vortex is projected into a sample of colloidal microspheres dispersed in water, optical gradient forces draw spheres onto the ring of light, and the beam's orbital angular momentum drives them around the circumference, as shown in Fig. 1(d). The resulting motion entrains flows that have yet to be studied systematically but whose qualitative features suggest opportunities for pumping and mixing extremely small sample volumes.

Figure 2 shows how periodically modulating an optical vortex's phase affects its geometry. The phase mask in Fig. 2(a) includes an $m = 5$ fold modulation of amplitude $\alpha = 0.1$ superimposed on an $l = 60$ helical pitch. The radial profile predicted with Eq. (2) appears in Fig. 2(b) and agrees well with the observed intensity distribution in Fig. 2(c). Comparably good agreement is obtained with our apparatus for modulated helical phases up to $m = 12$ and $\alpha = 1$ and $l = 60$. Figure 3 shows typical intensity patterns obtained by variation of m with fixed α and by variation of α with fixed m. Increasing the modulation beyond $\alpha_c = (l_0/l + 1)/m$ causes the locus of maximum intensity to pass through the origin and to create lobes of negative parity, as shown in the right-hand images in Fig. 3.

Just as uniform optical vortices exert torques on trapped particles, modulated optical vortices exert tangential forces. These forces can drive particles through quite complicated trajectories, as demonstrated in Fig. 4. Here, two 800-nm-diameter polystyrene spheres are driven through water by a threefold modulated optical vortex, each completing one circuit in ~2 s. Because the tangential driving force depends on intensity, spheres tend to circulate most rapidly where $R(\theta)$ is smallest and the light is most intense. Spheres drawn into deeply modulated patterns such as those in Fig. 3 must be supported in the axial direction by a bounding glass wall. Adjusting the microscope's focus thus allows the particle to explore different axial slices. In some planes, the particles circulate freely. In others, they become localized at the smallest radii. In still others, they are projected outward from the tips of the extended lobes rather than circulating around them. Such optically mediated distribution could be useful for manipulating samples in microfluidic devices. Unlike distribution methods based on translating discrete optical tweez-

ers,[23] the present approach can be implemented with a single static diffractive optical element.

In addition to translating particles, the forces exerted by modulated optical vortices can be used to distinguish particles on the basis of their size, shape, and optical properties. Consequently, modulated optical vortices may also provide a basis for sorting and fractionation,[24] applications that will be described in the future.

One can rotate a modulated optical vortex to any angle by varying β in Eq. (3). An asymmetric

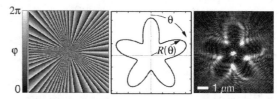

Fig. 2. Modulated optical vortex with $m = 5$, $\alpha = 0.1$: (a) phase modulation, (b) predicted radial profile $R(\theta)$, (c) experimental intensity distribution.

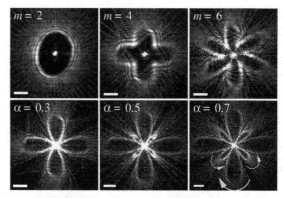

Fig. 3. Modulated optical vortices at (top row) $\alpha = 0.1$ with different m values and (bottom row) $m = 4$ with different α values. Additional lobes appear in the bottom patterns for $\alpha > \alpha_c \approx 0.25$, with the direction of tangential forces indicated by arrows. All patterns were created with $l = 60$. Scale bars indicate 1 μm.

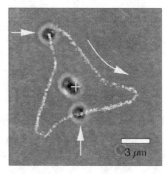

Fig. 4. Two particles' transit around a modulated optical vortex. Data points show the positions of two 800-nm-diameter polystyrene spheres measured at 1/10-s intervals over 10 s. The two spheres, indicated by arrows, move along a trap with $l = 60$, $m = 3$, and $\alpha = 0.1$ at 300 mW, in the direction indicated by the curved arrow. Two additional spheres are trapped motionlessly in the undiffracted central spot.

object that is comparable in extent to the trapping pattern can be immobilized on the pattern's asperities, and its orientation can be controlled by variation of the phase angle. The negative-parity lobes of deeply modulated optical vortices exert retrograde tangential forces that are useful for canceling the overall torque on large illuminated objects. Researchers have implemented comparable controlled rotation by interfering an optical vortex with a conventional optical tweezer[13] and by creating optical traps with elliptically polarized light.[25] The present approach offers several advantages: The trapped object can be oriented by a single beam of light without mechanical adjustments, the intensity distribution can be tailored to the targeted sample's shape through Eq. (2), and the same apparatus can create multiple independent rotators simultaneously.[16] These enhanced capabilities suggest applications for modulated optical vortices in actuating microelectromechanical systems such as pumps and valves in microfluidic and lab-on-a-chip devices.

This work was supported by the Materials Research and Engineering Centers program of the National Science Foundation through grant DMR-9880595. D. G. Grier's e-mail address is d-grier@uchicago.edu.

*Present address, Institut für Physikalische Chemie, Universität Heidelberg, Heidelberg, Germany.

References

1. A. Ashkin, J. M. Dziedzic, J. E. Bjorkholm, and S. Chu, Opt. Lett. **11,** 288 (1986).
2. A. Ashkin, IEEE J. Sel. Top. Quantum Electron. **6,** 841 (2000).
3. H. He, N. R. Heckenberg, and H. Rubinsztein-Dunlop, J. Mod. Opt. **42,** 217 (1995).
4. H. He, M. E. J. Friese, N. R. Heckenberg, and H. Rubinsztein-Dunlop, Phys. Rev. Lett. **75,** 826 (1995).
5. N. B. Simpson, L. Allen, and M. J. Padgett, J. Mod. Opt. **43,** 2485 (1996).
6. M. E. J. Friese, J. Enger, H. Rubinsztein-Dunlop, and N. R. Heckenberg, Phys. Rev. A **54,** 1593 (1996).
7. K. T. Gahagan and G. A. Swartzlander, Jr., Opt. Lett. **21,** 827 (1996).
8. H. Rubinsztein-Dunlop, T. A. Nieminen, M. E. J. Friese, and N. R. Heckenberg, Adv. Quantum Chem. **30,** 469 (1998).
9. K. T. Gahagan and G. A. Swartzlander, J. Opt. Soc. Am. B **16,** 533 (1999).
10. L. Allen, M. J. Padgett, and M. Babiker, Prog. Opt. **39,** 291 (1999).
11. A. T. O'Neil and M. J. Padgett, Opt. Commun. **185,** 139 (2000).
12. J. Arlt, V. Garces-Chavez, W. Sibbett, and K. Dholakia, Opt. Commun. **197,** 239 (2001).
13. L. Paterson, M. P. MacDonald, J. Arlt, W. Sibbett, P. E. Bryant, and K. Dholakia, Science **292,** 912 (2001).
14. W. He, Y. G. Liu, M. Smith, and M. W. Berns, Microsc. Microanal. **3,** 47 (1997).
15. M. E. J. Friese, H. Rubinsztein-Dunlop, J. Gold, P. Hagberg, and D. Hanstorp, Appl. Phys. Lett. **78,** 547 (2001).
16. J. E. Curtis, B. A. Koss, and D. G. Grier, Opt. Commun. **207,** 169 (2002).
17. L. Allen, M. W. Beijersbergen, R. J. C. Spreeuw, and J. P. Woerdman, Phys. Rev. A **45,** 8185 (1992).
18. A. T. O'Neil, I. MacVicar, L. Allen, and M. J. Padgett, Phys. Rev. Lett. **88,** 053601 (2002).
19. J. E. Curtis and D. G. Grier, Phys. Rev. Lett. **90,** 133901 (2003).
20. M. J. Padgett and L. Allen, Opt. Commun. **121,** 36 (1995).
21. Standard results for Laguerre–Gaussian modes suggest that $R_l \propto \sqrt{l}$. However, they do not account for diffraction by the objective lens's aperture. See, for example, Refs. 10 and 17.
22. Hamamatsu Model X7550 parallel-aligned nematic liquid-crystal SLM.
23. S. C. Grover, A. G. Skirtach, R. C. Gauthier, and C. P. Grover, J. Biomed. Opt. **6,** 14 (2001).
24. P. T. Korda, M. B. Taylor, and D. G. Grier, Phys. Rev. Lett. **89,** 128301 (2002).
25. M. E. J. Friese, T. A. Nieminen, N. R. Heckenberg, and H. Rubinsztein-Dunlop, Nature **394,** 348 (1998).

Optically controlled three-dimensional rotation of microscopic objects

V. Bingelyte, J. Leach, J. Courtial, and M. J. Padgett[a)]
Department of Physics and Astronomy, University of Glasgow, Glasgow G12 8QQ, Scotland,
United Kingdom

(Received 30 September 2002; accepted 16 December 2002)

We demonstrate that microscopic objects held in optical tweezers can be set into controlled rotation about any axis of choice. Our approach relies on the use of a spatial light modulator to create a pair of closely separated optical traps holding different parts of the same object. The pair of traps can be made to revolve around each other in any plane, rotating the trapped object with them. This technique overcomes the previous restriction on the orientation of the rotation axis to be parallel to the beam axis, and extends the versatility of optical tweezers as micromanipulation tools. © *2003 American Institute of Physics.* [DOI: 10.1063/1.1544067]

In 1986, Ashkin *et al.* demonstrated a technique now referred to as optical tweezing.[1] Their original paper has received hundreds of citations, roughly half of which occurred during the last five years. Nowadays, optical tweezers can be bought commercially,[2] and have found numerous applications, particularly in the biological sciences; for example, measuring the compliance of bacterial tails[3] and the forces exerted by motor proteins,[4] and stretching DNA molecules.[5]

The trapping mechanism in optical tweezers relies on the high intensity near the focus of a laser beam: any dielectric, that is, transparent, object in the beam experiences a force in the direction of the intensity gradient. In optical tweezers, microscopic dielectric objects become trapped near the focus and can be moved with it. In addition to this ability to translate objects, some optical tweezers are also able to rotate them.[6–11] However, until now it has only been possible to rotate objects about an axis parallel to the axis of the laser beam, which is approximately parallel to the axis of the objective lens. Furthermore, the short working distance (about 100 μm) means that the objective lens, and with it the rotation axis in all previous work, cannot be angled significantly with respect to the sample cell.

Here, we demonstrate that objects can be set into a controlled rotation about any axis, including axes perpendicular to the beam axis. This generalization of rotation control in optical tweezers has important applications. For example, it further enhances the capabilities of optical tweezers for assembling complex micromachines.

During the last decade, optical tweezers have been used to trap more than one object simultaneously. At first, this was achieved with a single laser beam rapidly switching between two or more trap positions,[12,13] or by splitting an initial beam early in the optical circuit to produce multiple beams that are later recombined before entering the microscope objective used to focus the beams.[14] Multiple-beam optical tweezers have also been implemented using computer-controlled spatial light modulators (SLMs), which can shape a beam into patterns of bright spots, which in turn can then act as independent traps.[15,16] Most recently, an SLM was used to create reconfigurable patterns of traps, where each individual trap

could be positioned and moved in three dimensions.[17] Closely related to these techniques, in which multiple traps hold different objects, are those in which multiple traps hold different parts of the *same* object. Such traps are often formed by the intensity maxima of more complex beams, for example, the lobes of higher-order Hermite–Gaussian beams,[6] or the maxima of interference patterns involving Laguerre–Gaussian (LG) beams.[9] In both cases, rotation of the beam about its axis leads to a rotation of the trapped object. Our approach for rotating trapped objects about an arbitrary axis relies on the use of a spatial light modulator to create a pair of closely separated optical traps revolving around the rotation axis (Fig. 1).

Rotation of trapped objects in optical tweezers has long been studied with great interest. The transfer of the two contributions to the angular momentum, spin[18] and orbital,[8] from a light beam to a microscopic object held within optical tweezers caused the object to rotate. Recently, the spinning and orbiting of optically trapped objects has provided deeper insights into the respective nature of these two kinds of angular momentum of light.[19] In parallel with these approaches, a completely different mechanism was investigated in which the linear momentum of the trapping beam drives the rotation of propeller-shaped objects, in much the same way as wind drives a windmill.[7] As a further example of the exciting prospects for micro-assembly and optically driven micromachines opened up by rotational control, microscopic

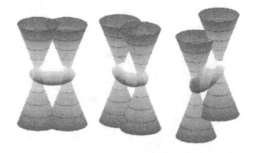

FIG. 1. Schematic representation of a pair of laser beam foci holding different parts of the same trapped object. Common translation or rotation of the foci around each other respectively translates or rotates the trapped object.

[a)]Electronic mail: m.padgett@physics.gla.ac.uk

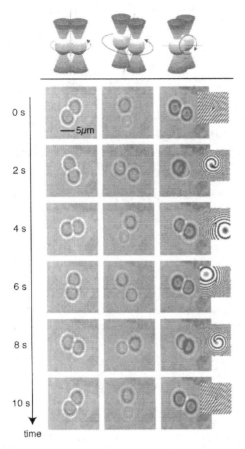

FIG. 2. Video sequences showing controlled rotations of an asymmetric object comprising two spheres of silica, each with a diameter of about 5 μm, fused together. The three columns demonstrate controlled rotation, from left to right, about the beam axis, an axis slightly inclined with respect to the beam axis, and an axis perpendicular to the beam axis. The further an object is out of the image plane, the more blurred is its image. The insets in the pictures of the right column show gray-scale representations of the corresponding phase patterns displayed on the SLM.

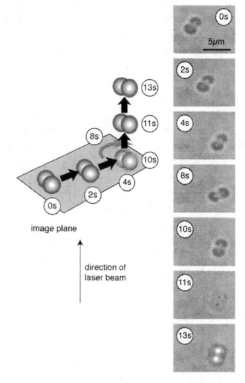

FIG. 3. 3D schematic and video sequence demonstrating controlled 3D translation and rotation of an object in optical tweezers. This sequence was controlled interactively through a computer interface, which in turn controlled a spatial light modulator used for shaping the laser beam. After being translated in the transverse direction, the object was rotated in the same plane through almost 360° and then lifted upwards. The relative times at which the frames were recorded are shown in circles.

cogs have recently been driven by light.[10] Controlled three-dimensional (3D) rotation literally adds one dimension to this body of work.

The numerical aperture (NA) of the objective sets a minimum size for a beam focus, which in turn places a lower limit on the separation between two foci, for which they function as two separate traps. Curtis *et al.*,[17] for example, state that "this precludes forming a three-dimensional cubic optical tweezer array with a lattice constant much smaller than 10 μm" in their setup. The problem is most pronounced in the direction of the beam axis, in which the bright focal region is almost always largest; it leads to the tendency of optical tweezers to align oblong objects with the beam axis[20] or even stack up multiple objects along the beam axis. For our application, this is a problem, because a rotation about an axis perpendicular to the beam axis passes through a position in which the two foci that hold different parts of the trapped object are directly above one another. This can lead to alignment of the object with the axis of one of the focused beams instead of the direction defined by the two foci. Our experimental observations show that this problem can be eased by using, instead of the more commonly used Gaussian beams,

LG laser beams, which are known to provide better axial trapping,[21,22] and which can easily be generated with phase holograms.[23,24] This allowed us to use our pair of foci as independent traps for two objects (silica spheres) as small as 2 μm in diameter, to place them on top of one another in the direction of the beam axis such that they were touching, and to subsequently separate them in a lateral or axial direction without either trap "capturing" the other object.

In our experiment, we use a commercial Nd:YVO$_4$ laser, frequency-doubled to give a power of 100 mW at 532 nm (Casix). After expansion and collimation, the beam from this laser is reflected off a computer-controlled SLM (based on Hamamatsu PAL-SLM X7665). The plane of the SLM is then imaged to the rear aperture of a microscope objective lens, which has a NA of 1.3 and 100× magnification (Zeiss Plan-Neofluar, oil-immersed, infinity-corrected), and which focuses the beam. The resulting laser beam is directed upwards ("inverted optical tweezers geometry"), allowing access to the sample from above. The sample cell comprises a 3-mm-thick aluminum slide with a 10-mm-diameter hole in the middle, which is sealed on the bottom using a standard glass cover slip. The resulting sample chamber can be filled from the top with a solution (in our case water) containing the objects to be manipulated in suspension.

An SLM can be thought of as a phase diffractive optical element (DOE) that can be configured into any desired form. As with all DOEs, the resulting beam can comprise a number of diffraction orders, the relative brightness of which depends upon the form and depth of the phase profile. Usually, DOEs are designed such that as much of the energy as possible is directed into the +1st diffraction order, with the other orders removed using a spatial filter. We can use our SLM in this fashion, combining the phase profiles of a Fresnel lens and a simple grating to produce a first-order diffracted beam that is then tightly focused by the microscope objective lens. Changing the focal length of the Fresnel lens or the pitch of the grating translates the focus in the axial or lateral direction, respectively. This focus can then form a trap whose 3D position can be controlled by computer. We can also add an azimuthal phase ramp to the SLM, converting the incident Gaussian beam into an approximation to a LG beam with an annular intensity distribution. Applying a suitable mapping function of the phase of the composite optical element to the phase level of the SLM allows us to direct upwards of 60% of the incident light into this +1st order. The SLM is then effectively the phase hologram of a combination of a wedge, a lens, and a spiral phase plate.

However, within our specific application, we apply a symmetric mapping function, such that equal amounts of energy are diffracted into the +1st and −1st diffraction orders, which focus to diametrically opposed positions on either side of the focus of the zero-order (nondiffracted) beam. The zero-order beam is removed by a spatial filter positioned in a Fourier plane prior to the objective lens. This gives two optical traps, positioned on either side of the zero-order focus position. Continually adjusting both the grating period and the focal length of the Fresnel lens allows us to rotate these two traps around a circular trajectory about any axis of choice. We use this pair of traps to hold different parts of the same object, which can then be rotated with the foci. Examples of the rotation of an asymmetric object, two fused, 5-μm-diameter, silica spheres, about various axes are shown in Fig. 2. In this case the spheres become fused simply by allowing the sample to rest for some time. Waiting for 10–30 minutes after loading the cell ensures that a number of spheres will join to produce larger groups. The fused pair of spheres shown in Fig. 2 were picked at random; the ability to rotate or otherwise manipulate them was representative of the sample as a whole. They could be trapped for several hours while such parameters as rotation axis and rotation speed could be adjusted interactively and in real time, simply by changing the parameters of the hologram sequence. In our case, the maximum rotation rate was set by the update rate of the spatial light modulator. For smooth rotation, we typically calculate 50 hologram steps for each revolution. In practice, we find the maximum update rate of the spatial light modulator to be about 10 Hz, giving a rotation rate of 0.2 Hz.

We can easily add an additional phase element to the same SLM such that the two traps can be laterally and axially translated to offset the rotation center in three dimensions. The resulting fully 3D translation and rotation is illustrated in Fig. 3.

To conclude, we have succeeded in inducing a controlled rotation of an object about any axis of choice; this contrasts with all previous work, in which the controlled rotation was constrained to be about the beam axis alone. Such techniques have clear implications in the positioning and control of miniature components within optically driven micromachines.

Two of the authors (J.C. and M.P.) acknowledge support from the Royal Society through University Research Fellowships.

[1] A. Ashkin, J. M. Dziedzic, J. E. Bjorkholm, and S. Chu, Opt. Lett. **11**, 288 (1986).

[2] Cell Robotics International Inc. (Albuquerque, NM); P.A.L.M. GmbH (Bernried, Germany).

[3] S. M. Block, D. F. Blair, and H. C. Berg, Nature (London) **338**, 514 (1989).

[4] J. T. Finer, R. M. Simmons, and J. A. Spudich, Nature (London) **368**, 113 (1994).

[5] M. D. Wang, H. Yin, R. Landick, J. Gelles, and S. M. Block, Biophys. J. **72**, 1335 (1997).

[6] S. Sato, M. Ishigure, and H. Inaba, Electron. Lett. **27**, 1831 (1991).

[7] E. Higurashi, H. Ukita, H. Tanaka, and O. Ohguchi, Appl. Phys. Lett. **64**, 2209 (1994).

[8] H. He, M. E. J. Friese, N. R. Heckenberg, and H. Rubinsztein-Dunlop, Phys. Rev. Lett. **75**, 826 (1995).

[9] L. Paterson, M. P. MacDonald, J. Arlt, W. Sibbett, P. E. Bryant, and K. Dholakia, Science **292**, 912 (2001).

[10] P. Galajda and P. Ormos, Appl. Phys. Lett. **78**, 249 (2001).

[11] A. T. O'Neil and M. J. Padgett, Opt. Lett. **27**, 743 (2002).

[12] K. Visscher, G. J. Brakenhoff, and J. J. Krol, Cytometry **14**, 105 (1993).

[13] J. E. Molloy, J. E. Burns, J. C. Sparrow, R. T. Tregear, J. Kendrickjones, and D. C. S. White, Biophys. J. **68**, S298 (1995).

[14] H. M. Warrick, R. M. Simmons, J. T. Finer, T. Q. P. Uyeda, S. Chu, and J. A. Spudich, Methods Cell Biol. **39**, 1 (1993).

[15] M. Reicherter, T. Haist, E. U. Wagemann, and H. J. Tiziani, Opt. Lett. **24**, 608 (1999).

[16] E. R. Dufresne, G. C. Spalding, M. T. Dearing, S. A. Sheets, and D. G. Grier, Rev. Sci. Instrum. **72**, 1810 (2001).

[17] J. E. Curtis, B. A. Koss, and D. G. Grier, Opt. Commun. **207**, 169 (2002).

[18] M. E. J. Friese, T. A. Nieminen, N. R. Heckenberg, and H. Rubinsztein-Dunlop, Nature (London) **394**, 348 (1998).

[19] A. T. O'Neil, I. MacVicar, L. Allen, and M. J. Padgett, Phys. Rev. Lett. **88**, 053601 (2002).

[20] A. Ashkin, J. M. Dziedzic, and T. Yamane, Nature (London) **330**, 769 (1987).

[21] A. Ashkin, Biophys. J. **61**, 569 (1992).

[22] N. B. Simpson, D. McGloin, K. Dholakia, L. Allen, and M. J. Padgett, J. Mod. Opt. **45**, 1943 (1998).

[23] V. Y. Bazhenov, M. V. Vasnetsov, and M. S. Soskin, JETP Lett. **52**, 429 (1990).

[24] N. R. Heckenberg, R. McDuff, C. P. Smith, and A. G. White, Opt. Lett. **17**, 221 (1992).

Optical Microrheology Using Rotating Laser-Trapped Particles

Alexis I. Bishop,* Timo A. Nieminen, Norman R. Heckenberg, and Halina Rubinsztein-Dunlop

Centre for Biophotonics and Laser Science, Department of Physics, The University of Queensland, Brisbane QLD 4072, Australia
(Received 29 July 2003; published 14 May 2004)

We demonstrate an optical system that can apply and accurately measure the torque exerted by the trapping beam on a rotating birefringent probe particle. This allows the viscosity and surface effects within liquid media to be measured quantitatively on a micron-size scale using a trapped rotating spherical probe particle. We use the system to measure the viscosity inside a prototype cellular structure.

DOI: 10.1103/PhysRevLett.92.198104   PACS numbers: 87.80.Cc, 83.85.Cg, 83.85.Ei

It is well known that light can transport and transfer angular momentum as well as linear momentum [1]. In recent years, this has been widely applied to rotate microparticles in optical tweezers. Such rotational micromanipulation has been achieved using absorbing particles [2,3], birefringent particles [4,5], specially fabricated particles [6,7], and nonspherical particles using linearly polarized light [8–13] or nonaxisymmetric beams [14,15]. Notably, particles rotated in rotationally symmetric beams [4,12] spin at a constant speed, indicating that the optical torque driving the particle is balanced by viscous drag due to the surrounding medium. It has been noted that the optical torque can be measured optically [16], allowing direct independent measurement of the drag torque.

For a pure Gaussian beam that is circularly polarized, each photon also has an angular momentum of $\pm\hbar$ about the beam axis, equivalent to an angular momentum flux of $\pm P/\omega$ for the beam, where P is the power of an incident beam, and ω is the angular frequency of the optical field, which can be transferred to an object that absorbs the photon, or changes the polarization state [1]. The reaction torque on a transparent object that changes the degree of circular polarization of the incident light is given by [16]

$$\tau_R = \Delta\sigma P/\omega, \tag{1}$$

where $\Delta\sigma$ is the change in circular polarization. Thus, by monitoring the change in circular polarization of light passing through an object, the reaction torque on the object is found. We exploit this simple principle as the basis for an optically driven viscometer. The viscosity of a liquid can be determined by measuring the torque required to rotate a sphere immersed in the liquid at a constant angular velocity—a concept we have implemented at the micrometer scale using a system based on optical tweezers. We have rotated birefringent spheres of synthetically grown vaterite that were trapped three dimensionally in a circularly polarized optical tweezers trap and measured the frequency of rotation, and the polarization change of light passing through the particle, to determine the viscosity of the surrounding fluid.

At least four distinct techniques have been used previously for microrheology: tracking the diffusion of tracer particles [17,18], tracking the rotational diffusion of disk-shaped tracer particles [19], measurement of correlations in the Brownian motion of two optically trapped probe particles [20], and the measurement of viscous drag acting on an optically trapped probe particle in linear motion [21]. The method we use here, direct optical measurement of the rotation of a spherical probe particle in a stationary position, allows highly localized measurements to be made, since the probe particle does not move in the surrounding medium, and, compared with the rotational diffusion measurements by Cheng and Mason using microdisks [19], the use of spherical probe particles greatly simplifies the theoretical analysis of the fluid flow. Also, since the probe particle is rotationally driven by the optical trap, the rotation rate can be readily controlled.

The expression for the drag torque, τ_D, on a sphere rotating in a fluid for low Reynolds number flows is well known [22] and is given by

$$\tau_D = 8\pi\mu a^3\Omega, \tag{2}$$

where μ is the viscosity of the fluid, a is the radius of the sphere, and Ω is the angular frequency of the rotation. Equating the expressions for torques τ_R and τ_D gives

$$\mu = \Delta\sigma P/(8\pi a^3\Omega\omega). \tag{3}$$

This equation suggests that the viscosity can be determined using knowledge of only a few elementary parameters—the power and the polarization change of the incident beam passing through the particle, and the size and rotation rate of the particle. The power, polarization state, and rotation rate can all be measured optically.

A significant obstacle to implementing this approach has been obtaining spherical, transparent, birefringent particles of suitable size for trapping (approximately 1–10 μm in diameter). We have developed a novel technique to grow nearly perfectly spherical crystals of the calcium carbonate mineral vaterite, which has similar birefringence properties to calcite, within the required size regime. We produce the vaterite crystals by adding six drops of $0.1M$ K_2CO_3 to a solution of 1.5 ml of $0.1\dot{M}$

CaCl$_2$ plus four drops of 0.1M MgSO$_4$. The solution is strongly agitated by pipetting. The solution initially appears milky, becoming clearer as the crystals form within a few minutes of mixing. The typical mean size is found to be 3 μm with a spread of approximately $\pm$1.5 μm. Figure 1 shows both an optical image and an electron micrograph of typical crystals used for viscosity measurements. Vaterite is a positive uniaxial birefringent material, with $n_o = 1.55$ and $n_e = 1.65$.

A typical optical tweezers arrangement was used as the basis for the experiment, which delivered approximately 300 mW of light at 1064 nm into a focal spot with a diameter of around 0.85 μm using a 100$\times$ oil-immersion objective of high numerical aperture (NA = 1.3). A $\lambda/4$ plate located immediately before the objective converts the linearly polarized light from the laser into a circularly polarized beam which is able to rotate birefringent objects [4]. Light passing through the probe particle is collected by an oil coupled condenser lens (NA = 1.4) and then sent to a series of polarization measuring devices. The degree of circular polarization remaining in the beam is determined by passing it through a $\lambda/4$ plate with the fast axis oriented at 45° to the axes of a polarizing beam splitter cube. Photodiodes monitoring the orthogonal outputs of the beam splitter provide a measure of the degree of circular polarization remaining in the beam, which allows the change in angular momentum, and hence the torque applied to the particle to be known. The photodiodes are calibrated such that they provide a measure of the power at the trap focus. The sum of the two signals gives the total trapping power, and the difference between the signals gives the degree of circular polarization of the beam measured in units of power at the trap focus. The torque applied to the particle is determined by measuring the difference between the initial degree of circular polarization of the beam in the absence of the particle and the final polarization, in accordance with Eq. (1). The frequency of rotation of the particle can be accurately determined by monitoring the transmitted light through a linear polarizer, as suggested by Nieminen *et al.* [16]. A small amount of light is

diverted from the main beam and is passed through a polarizing beam splitter cube, and the forward direction is monitored using a photodiode. The signal is modulated at twice the rotation rate of the particle, and the depth of the modulation is proportional to the amount of linearly polarized light in the beam. If the particle is thick enough to act as a $\lambda/2$ plate, then circularly polarized light passing through the particle is reversed in handedness and the frequency of the rotation cannot be measured as there is no linear component to the transmitted light. However, as the particles are spherical and are of similar diameter to the beam, the polarization at different radial distances from the rotation axis varies, and there will almost always be some linear component after transmission.

The calculated viscosity depends on the cube of the radius of the particle, and thus it is important to determine the size of the particles accurately. The particles used for viscosity measurements are in the range of 1.5 to 3.5 μm in diameter and consequently to obtain the viscosity within 10% requires the diameter to be measured with an accuracy on the order of 90 nm. It was found that obtaining an accurate measurement of the diameter by direct visualization was problematic. The method of measurement preferred by the authors is to put two spheres of nearly identical size in contact and measure the distance between centers, which can be found accurately due to the spherical symmetry. The diameter is inferred using elementary geometry, and the error in the measurement by this method is estimated as being $\pm$40 nm. The degree of asphericity for typical particles was estimated to be less than 3%.

The viscosity of water was measured using the system described. A range of vaterite particles was suspended in distilled water and trapped three dimensionally in a sealed slide/coverslip cell, with a depth of 50 μm. Particles that are trapped begin to rotate immediately. However, the rotation may be stopped by aligning the $\lambda/4$ plate located before the objective so that the polarization is made linear. The size of the particles ranged from 1.5 to 3.5 μm in diameter, and rotation rates up to 400 Hz (24 000 rpm) were observed for powers up to 350 mW at the trap focus. The Reynolds number of the fluid flow around the rotating spheres is quite low, on the order of 1×10^{-3} for rotations up to 1000 Hz, and hence the flow is well within the creeping flow regime required by (2). The signals from the circular polarization and linear polarization monitoring photodiodes were recorded using a 16 bit analog to digital converter sampling at 10 kHz for a period of 5 s. Typical signals recorded during an experiment are reproduced in Fig. 2. The strong uniform modulation of the linear polarization signal [Fig. 2(a)] shows that the particle rotates very uniformly, taking a large fraction of the angular momentum carried by the beam. Using a particle of diameter 2.41 μm, the viscosity was found to be $\mu = (9.6 \pm 0.9) \times 10^{-4}$ Pa s, which is in

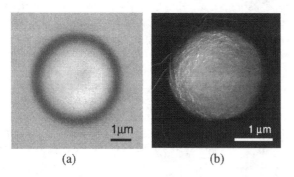

(a)   (b)

FIG. 1.   (a) Optical microscope image of a typical vaterite crystal used for viscosity measurements. (b) Scanning electron microscope image of a vaterite crystal.

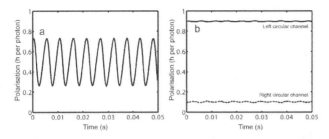

FIG. 2. (a) Signal recorded by the linear polarization measurement apparatus during rotation of a vaterite crystal. The frequency of rotation is (94.0 ± 0.7) Hz. (b) Signal traces from the circular polarization measurement apparatus during rotation of a vaterite crystal.

excellent agreement with the established value for the viscosity of water at 23 °C of 9.325×10^{-4} Pa s [23]. A series of measurements was made with a range of different sized particles at a range of rotation rates and trapping powers. The variation in viscosity over a range of powers below 100 mW was found to be on the order of 0.7% using a single particle. The variation in inferred viscosity using 11 particles, spanning a range of sizes, was approximately 4.7%. The most significant sources of error arise from the calibration of the trapping power at the focus of the trap (7%), the asphericity of the particles (3%), resulting in a 6% error in the viscosity, and the uncertainty in the measurement of the particle size (1.4%), which contributes to a 4% error in the viscosity. The total error for the viscosity measurement from all error sources is estimated as being 10%.

The inferred viscosity was found to be independent of the rotation rate, although for trapping powers above approximately 100 mW, the viscosity was observed to decrease with increasing power. This is expected, due to heating of the particle which results from the absorption of light through the sample, and also heating of the surrounding liquid due to absorption of the trapping beam by water, which is on the order of 10 K per W at the focus [24]. Figure 3 demonstrates the behavior of the measured viscosity as a function of the power at the trapping focus. It was observed that higher trapping powers could be used with smaller particles before the onset of the decrease in viscosity, which is consistent with volume absorption effects. The observed decrease in viscosity with power was initially approximately linear and was approximately 0.13% per mW, which is equivalent to a temperature rise of around 0.06 °C per mW. The total power loss for the trapping beam passing through a particle was measured for a sample of vaterite particles and was found to be within the range of 0.6% to 1.2%. The reflection loss due to refractive index mismatches at the particle surfaces is approximately 0.6%.

We have also demonstrated the ability to measure the viscosity inside a small confined region, such as a cellular membrane. Hexane filled vesicles were formed by emul-

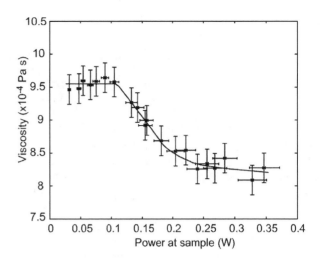

FIG. 3. Variation in viscosity with trapping power for a vaterite crystal of diameter 2.41 μm. The viscosity is independent of power for powers at the trap focus less than approximately 100 mW.

sifying hexane, containing a small amount of soy lecithin (1 g/l), with water containing a quantity of vaterite spheres. Spherical membrane structures were formed with diameters up to approximately 20 μm which occasionally contained single vaterite crystals. The viscosity inside a 16.7 μm diameter vesicle was measured with a vaterite particle of diameter 3.00 μm, using the same procedure as outlined earlier, and was found to be $(2.7 \pm 0.5) \times 10^{-4}$ Pa s for powers less than 100 mW, which is in good agreement with the established value of viscosity for hexane at 23 °C of 3.07×10^{-4} Pa s (extrapolated) [25]. This value is significantly less than that of water $(9.325 \times 10^{-4}$ Pa s at 23 °C [23]. Figure 4 shows the hexane filled vesicle with the vaterite probe particle inside.

The viscous drag torque acting on a spherical probe particle of radius a rotating with an angular frequency Ω

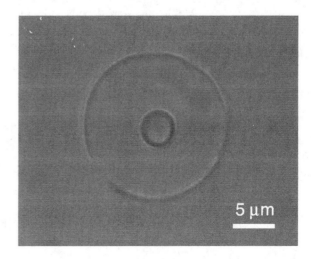

FIG. 4. Free vaterite crystal located inside a hexane-filled lipid-walled vesicle of 16.7 μm in diameter.

at the center of a sphere of radius R of fluid of viscosity μ_1, surrounded by a fluid of viscosity μ_2 is [22]

$$\tau_D = 8\pi\mu a^3 \Omega R^3/(R^3 - a^3 + a^3\mu_1/\mu_2). \qquad (4)$$

This can be used to determine the viscosity within a vesicle if the viscosity of the surrounding fluid is known or to estimate the error if Eq. (3) is used instead. For other geometries, such as if the probe particle is not in the center of the vesicle or is near a plane interface [26], the effect will be more complicated. However, in most cases the effect of the boundary can be ignored if the probe particle is at least one diameter away from the interface. The effect of nearby boundaries was determined experimentally by measuring the viscous drag torque at varying distances from a solid interface. For rotation about an axis parallel to the interface, no change in the drag torque was observed until the probe particle was within one diameter of the interface. For rotation about an axis normal to the interface, no change in the drag torque was observed until the probe particle was very close to the interface. For the above measurement of the viscosity of hexane, the drag in the vesicle is only 0.4% greater than in pure hexane. The rotation rate of the vesicle is $\Omega_R = \Omega/[1 + \mu_2(R^3 - a^3)/(\mu_1 a^3)]$ [22], equal to $\Omega/522$ for the case above. Small particles in the surrounding fluid near the vesicle were observed to orbit the vesicle at approximately this rate.

The most significant contribution to the error in these measurements is the uncertainty in accurately determining the diameter of the spherical particle. It was not found to be possible to drag a particle through the membrane, and hence the method of size measurement using two spheres in contact could not be used. The diameter was instead directly estimated from an image recorded of the sphere. The error in determining the diameter is estimated as being 5%, which is the main contribution to the total measurement error of 19%.

It is feasible to produce smaller vaterite particles, and the use of these would allow the viscosity to be probed on an even smaller size scale—potentially probing volumes with a capacity of a cubic micron. The ability to functionalize the probe particle surface would enable the selective attachment of the particles to various biological structures, which would allow the torsional response of these structures to be investigated quantitatively. We note that the torque and angular deflection can still be measured accurately in the absence of continuous rotation. The demonstrated ability to measure viscosity within micronsized volumes makes the system developed here of great value for probing the microrheology of liquid-based materials, such as colloids.

This work was partially supported by the Australian Research Council. We wish to thank Professor Rane Curl, University of Michigan for valuable discussions.

*Present address: Department of Physics, Heriot-Watt University, Riccarton, Edinburgh, EH14 4AS, Scotland, United Kingdom.
Electronic address: A.I.Bishop@hw.ac.uk

[1] R. A. Beth, Phys. Rev. **50**, 115 (1936).

[2] M. E. J. Friese, J. Enger, H. Rubinsztein-Dunlop, and N. R. Heckenberg, Phys. Rev. A **54**, 1593 (1996).

[3] M. E. J. Friese, T. A. Nieminen, N. R. Heckenberg, and H. Rubinsztein-Dunlop, Opt. Lett. **23**, 1 (1998).

[4] M. E. J. Friese, T. A. Nieminen, N. R. Heckenberg, and H. Rubinsztein-Dunlop, Nature (London) **394**, 348 (1998); **395**, 621(E) (1998).

[5] E. Higurashi, R. Sawada, and T. Ito, Phys. Rev. E **59**, 3676 (1999).

[6] E. Higurashi, H. Ukita, H. Tanaka, and O. Ohguchi, Appl. Phys. Lett. **64**, 2209 (1994).

[7] P. Galajda and P. Ormos, Appl. Phys. Lett. **78**, 249 (2001).

[8] S. Bayoudh, T. A. Nieminen, N. R. Heckenberg, and H. Rubinsztein-Dunlop, J. Mod. Opt. **50**, 1581 (2003).

[9] K. D. Bonin, B. Kourmanov, and T. G. Walker, Opt. Express **10**, 984 (2002).

[10] E. Santamato, A. Sasso, B. Piccirillo, and A. Vella, Opt. Express **10**, 871 (2002).

[11] Z. Cheng, P. M. Chaikin, and T. G. Mason, Phys. Rev. Lett. **89**, 108303 (2002).

[12] A. I. Bishop, T. A. Nieminen, N. R. Heckenberg, and H. Rubinsztein-Dunlop, Phys. Rev. A **68**, 033802 (2003).

[13] P. Galajda and P. Ormos, Opt. Express **11**, 446 (2003).

[14] L. Paterson, M. P. MacDonald, J. Arlt, W. Sibbett, P. E. Bryant, and K. Dholakia, Science **292**, 912 (2001).

[15] A. T. O'Neil and M. J. Padgett, Opt. Lett. **27**, 743 (2002).

[16] T. A. Nieminen, N. R. Heckenberg, and H. Rubinsztein-Dunlop, J. Mod. Opt. **48**, 405 (2001).

[17] D. T. Chen, E. R. Weeks, J. C. Crocker, M. F. Islam, R. Verma, J. Gruber, A. J. Levine, T. C. Lubensky, and A. G. Yodh, Phys. Rev. Lett. **90**, 108301 (2003).

[18] A. Mukhopadhyay and S. Granick, Curr. Opin. Colloid Interface Sci. **6**, 423 (2001).

[19] Z. Cheng and T. G. Mason, Phys. Rev. Lett. **90**, 018304 (2003).

[20] L. Starrs and P. Bartlett, Faraday Discuss. **123**, 323 (2003).

[21] B. K. Rafał Ługowski and Y. Kawata, Opt. Commun. **202**, 1 (2002).

[22] L. D. Landau and E. M. Lifshitz, *Fluid Mechanics*, Course of Theoretical Physics Vol. 6 (Butterworth-Heinemann, Oxford, 1987), 2nd ed.

[23] *CRC Handbook of Chemistry and Physics, 1974–1975* (CRC Press, Boca Raton, FL, 1974), 55th ed.

[24] Y. Liu, G. J. Sonek, M. W. Berns, and B. J. Tromberg, Biophys. J. **71**, 2158 (1996).

[25] *CRC Handbook of Chemistry and Physics, 2002–2003* (CRC Press, Boca Raton, FL, 2002), 83rd ed.

[26] K. D. Danov, T. D. Gurkov, H. Raszillier, and F. Durst, Chem. Eng. Sci. **53**, 3413 (1998).

7

Multitrap and Holographic Optical Tweezers

Since the first demonstration of optical tweezers, various applications have required the simultaneous trapping of not one but many particles. Lateral control over a single trap is readily accomplished by having a beam-steering mirror positioned in a Fourier plane to that of the sample, with an angular deviation of the beam reflected from the mirror giving a lateral displacement of the trap. In early systems, rapid control of the mirror allowed a single laser beam to be time-shared between different positions, effectively forming multiple traps [1,2]; see Visscher et al. [3] (**PAPER 7.1**), [4].

The dynamics of all optical traps are that of a damped harmonic oscillator, where the restoring force is created by the focused laser beam and the damping is provided by the fluid medium in which the trapped particles are suspended. For a liquid medium, a typical trap has a resonance frequency in the range 100 Hz–1 kHz but is overdamped by an order of magnitude or more. Time-sharing a single laser beam between many traps is indistinguishable from a true multibeam trap provided that the refresh rate of each trap is much greater than the trap's resonant frequency. In practice, multitraps can be maintained at lower refresh rates, but the modulation in trap strength is then detectable in the motion of the individual particles. Hence the speed at which a single laser beam can be switched between traps limits the number of traps that can be maintained. For beam switching based upon mirrors, mechanical considerations limit this to about tens of kHz, restricting the number of traps to ≈ 10.

Instead of a mirror, an angular deviation of the trapping beam can also be introduced using an acousto-optic modulator. These use a piezoelectric element to create an acoustic compression wave along the length of a suitable crystal. This wave acts as a Bragg grating such that an incident laser beam is diffracted. Changing the frequency of the piezo element changes the period of the grating such that the angular deflection of the transmitted laser beam can be controlled. Such modulators typically operate at a center frequency of 100 MHz but with a bandwidth of tens of MHz. As shown by Visscher et al., this means that a single laser beam can be switched between positions at MHz rates—two to three orders of magnitude faster than by using a mirror [5].

This switching is sufficiently rapid that several hundred traps can be maintained and independently positioned; see Vossen et al. [6]. Since each trap requires a few milliwatts of average laser power to hold the particle against its thermal motion, the ultimate limit to the number of traps that can be maintained is set by the available laser power, or at least that which can be handled, without damage, by the modulators and objective lens.

Both the scanning mirrors and acousto-optic modulators can introduce angular deflections of the trapping beam, resulting in a lateral motion of the trap. The precise lateral motion is given by the product of the focal length of the objective lens and the angular deflection of the beam, as measured at the back aperture of the lens. The lateral range over which the traps can be positioned is typically 10–100 μm. By contrast, axial motion of the trap requires control of the beam divergence at the back aperture of the lens, a parameter that cannot be set by either the mirror or modulator system. One method is to replace the beam-steering mirror with a deformable one [7]. The limited dynamic range of such devices means that the amount of axial trap motion is limited to a couple of microns. However, the high mechanical bandwidth of the deformable mirror means that the trap can be modulated quickly.

Optical tweezers can readily be adapted to have two independent traps by introducing multiple beam paths, each with its own steering mirror [8]. Inserting an adjustable lens into one or both paths can set different axial positions for the traps, but the speed at which such lenses can be adjusted means that they cannot be used in a switching configuration [9]. In a slight variant of this approach, the incorporation of a Pockels cell or other similar device into the optical path allows setting of the polarization state to route the beam though one or two different lenses so that it is possible to switch any number of traps at high speed to fall in one of two possible trapping planes [6]. An alternative to splitting the laser beam into two or more paths is to use multiple laser sources. Although this approach may not be applicable if the lasers are bulky, the possibility of using an array of semiconductor lasers means the individual traps positions can simply be switched on or off [10] (see **PAPER 7.2**), [11].

In the mid to late 1990s, the possibility of an alternative to using either mirrors or acousto-optic modulators to create multiple traps was realized. This alternative was to use diffractive optical components. When illuminated with a plane wave, a simple grating produces multiple diffraction orders, giving a line of spots in the far field. Usually, the far-field condition is obtained by inserting a lens into the diffracted beams and imaging the intensity distribution at its rear focal plane. Within optical tweezers, it is the microscope objective that acts as the transform lens. Hence, if the beam-steering mirror in an optical tweezers is replaced with a grating, the result is a linear array of traps. Blazing the grating to emphasize various orders changes the intensity balance between the traps. If the grating were replaced with a Fresnel lens, then the diffraction order would be shifted out of the far field, moving the trap axially.

Gratings and Fresnel lenses are just two simple examples of diffractive optical components. In general, there is no limit to the complexity of elements that can be designed, thereby producing arbitrary arrays of spots in the far field. Most importantly, these spots can be displaced in both the lateral and axial directions. The axial control over trap position is a clear advantage of the diffractive optics approach over scanning mirrors or acousto-optic modulators. Since diffractive optics is usually designed to convert a plane-wave, spatially coherent, illumination into a specific intensity distribution in the far field, such techniques are frequently called computer-designed holograms. Such holograms were used in the mid 1990s to control the phase structure of single optical traps, converting a normal Gaussian beam into an annular beam with helical phase fronts and associated orbital angular momentum [12].

Perhaps the first use of diffractive optics, albeit not positioned in the Fourier-plane, for multibeam optical tweezers was described by Fournier [13]. Another very early use of holograms within multibeam optical tweezers was by Dufresne and Grier [14] (see **PAPER 7.3**), who used a bespoke diffractive optical element designed to produce a 4×4 array of optical traps. They also foresaw many possible applications centered in colloidal science, ranging from fundamental studies, including those associated with phase transitions, through to the assembly of complex structures, including the fabrication of photonic materials.

A number of groups across the world recognized this holographic approach to tweezing could be made more versatile by replacing the static diffractive optical component with an addressable spatial light modulator. As with all diffractive optics, the diffraction efficiency is maximized by ensuring the device modulates the phase rather than the intensity of the incident light. Originally these spatial light modulators were adapted from miniature liquid crystal display devices and consequently had low diffraction efficiencies of only a few percent. However, subsequently a number of phase-only modulators became commercially available, and diffraction efficiencies now exceed 50%. Early successes with spatial light modulators include Refs. [15] and [16], but it was Liesener et al. [17] (see **PAPER 7.4**) that demonstrated the full lateral and axial positioning capability of an optical tweezers based upon a spatial light modulator. The wider community woke up to the potential of spatial light modulators in 2002 when Curtis et al. [18] (see **PAPER 7.5**) demonstrated the combination of multitrapping and the modification of the individual trap types, establishing holographic optical tweezers as a tool for other applications.

Key to the interactive use of spatial light modulators is the algorithm used to calculate the phase hologram required to produce the desired distribution of traps. For conventional traps, this can be accomplished by the appropriate additions of the kinoforms for simple lateral (i.e., a blazed grating) and axial (i.e., a Fresnel lens) designs for single traps [17]. Alternatively, adaptation of the Gerchberg–Saxton algorithm [19], direct binary search [20], or kinoform sectioning [21,22] also allows complicated patterns of modified optical traps to be established in three dimensions throughout the sample volume. Further work has been undertaken to understand [23] and improve the intensity uniformity [24] between traps.

The exact range over which the traps can be positioned is largely determined by the spatial resolution of the spatial light modulator. Increasing the number of pixels relayed to the back aperture of the objective lens increases the maximum angle through which the beams can be diffracted. For a spatial light modulator with video resolution, the working field of view within the sample volume is typically several tens of microns in both diameter and depth [25]. The precision to which individual traps can be positioned within the field of view of the microscope is complicated. However, it is clear that commercial designs of spatial light modulators have sufficient resolution to allow trap positions to be potentially defined with nanometer precision [26].

In addition to using the spatial light modulator to create multiple and/or modified traps, one can also use them to correct for aberrations in the optical system. These aberrations may arise from using the microscope objective away from its design wavelength, or its optimum conjugate ratio. The dominant aberration in such cases is spherical and can been corrected using a deformable mirror [27]. Aberrations can also be corrected by modifying the hologram kinoform applied to the spatial light modulator [28,29]. In these latter cases, the main source of aberration is most likely the spatial light modulator itself, often involving astigmatism. Most recently, Ritsch-Marte and co-workers [30] have demonstrated a powerful technique for correcting for all the aberrations in an optical system, using a modified Gerchberg–Saxton algorithm to create the required hologram kinoform based on an uncorrected image of the diffracted beam.

The increase in speed due to the advent of multiprocessor, multicore desktop computers means that many algorithms for kinoform design can be implemented in real time, potentially redefining the positions of tens of optical traps many times per second. Coupled with ever increasing sophistication in the user interface, access to cheap computing power has enabled holographic tweezers to be controlled by joystick or hand-driven interfaces; see Whyte et al. [31] (**PAPER 7.6**), and also with haptic feedback [32]. Ultimately, the update rate of the kinoform is restricted by the speed of the spatial light modulator. Most spatial light modulators used in holographic tweezers are a liquid crystal, twisted nematic design, allowing different phase levels to be set but being intrinsically restricted in their speed of response to video-frame rates. However, if holographic tweezers are to be used in, for example, a closed-loop force measuring configuration, then a much faster response will be needed. With that in mind, it has also been demonstrated that ferroelectric configurations of liquid crystal can be used [33]. Although having only a binary phase response, which limits their diffraction efficiency, these devices can be updated at kHz rates.

One of the surprising things about optical tweezers is that the extreme numerical aperture of the focused laser beams means that even though all the trapping beams emerge from the same objective lens, a particle in one of the traps only slightly perturbs the beams forming the other traps. As shown by Bingelyte et al., this is best exemplified by the fact that it is possible to trap two particles in the same lateral position but displaced from each other axially [34] (see **PAPER 6.7**). It is therefore possible to use optical tweezers to assemble both three-dimensional (3D) crystalline (see Sinclair et al. [35] **PAPER 7.7**) and quasi-crystalline structures [36]. These structures can be made permanent by using a gel as the fluidic medium that, when set, holds the particles (see Mio and Marr [37]) or even laser manipulated cells [38] into position.

Extended structures can be formed within optical tweezers in other ways too. Burns et al. [39] showed how interference between optical beams, and the resulting standing waves, could bind microscopic beads to align with the local intensity maxima. This interference approach was exploited within a microscope geometry, more representative of optical tweezers, using two plane waves to create a sinusoidal interference pattern in which particles were trapped [40]. Shifting the frequency of one of the beams caused the fringes and the trapped particles to move laterally. By using helically phased, annular beams, Paterson et al. showed that a frequency

shift between the beams causes a rotation of the fringes and a corresponding rotation of the particles [41] (see **PAPER 6.1**). The interference patterns themselves can be used to create 3D trapped structures, either as the interference between two phase-structured beams [42] or in a multibeam configuration, created either explicitly [43] or through the Talbot effect [13]. However, it is clear that the use of a fully programmable spatial light modulator offers a control over the positions of the optical traps that cannot be matched by interference between fixed beams.

In parallel with the development of holographic optical tweezers, where the spatial light modulator is positioned in the Fourier plane of the sample, has been the approach of Glückstad and co-workers where the spatial light modulator is placed in the image plane [44]. By configuring the system in phase-contrast mode, it is possible to direct virtually all the incident light into the optical traps [45]. Since the spatial light modulator is placed in the image plane, the kinoform is simply the pattern of desired optical traps, greatly simplifying the computational requirement. In this case, the size of the optical trap is set by the imaged size of the pixel onto the sample plane. Consequently, these systems tend to be run at a lower numerical aperture, meaning that they trap in 2D only. However, the longer working distance of the lower magnification objective lens enables easy coupling of additional optical systems, including an opposing objective lens, so that a 3D stable trap can be formed by two counterpropagating optical beams [46] (see **PAPER 7.8**), [47]. Although requiring a more complicated optical system than conventional tweezers, the increased optical access to the sample volume [48] made possible by this phase-contrast approach offers clear advantages in a number of potential applications.

As with the first use of diffractive optics in multiple tweezers, one is not restricted to placing the spatial light modulator in either the Fourier or image plane of the sample. As shown by Polin et al., moving the spatial light modulator slightly away from the Fourier plane means that the unwanted zero-order diffraction spot is focused to a plane with the sample volume, thereby eliminating the unwanted trap [49] (see **PAPER 7.9**). Moving the spatial light modulator yet further out of the Fourier plane means that traps formed at different lateral positions, as back-projected onto the modulator, fill only part of its aperture and are displaced from the optical axis. Similarly, a smaller aperture kinoform falling within the overall aperture of the spatial light modulator is sufficient to create a single optical trap. While still illuminating the whole aperture of the modulator, translating the same kinoform within the aperture translates the trap. These Fresnel optical tweezers [50] significantly reduce the computational power required to move multiple traps around the sample volume, something that is potentially important for the ultra-high-resolution or high-speed spatial light modulator technology that is being developed.

In conclusion, there are a number of approaches to achieving multitrap optical tweezers. If only two traps are required, it is tempting simply to use two optical paths, each with its own beam-steering mirror and adjustable lens. Larger numbers of traps require either a time-shared configuration based on galvo mirrors, acousto-optic modulators, or a holographic approach. Modern acousto-optic modulators allow many hundreds of traps to be defined, each of which can be precisely positioned in the lateral plane and the intensity well controlled. The holographic approach comes into its own if the required distribution of traps is 3D, requiring both lateral and axial interactive control. Precisely in which plane one places the spatial light modulator depends upon the detailed requirements or preferences of the user. In all cases, it is likely that the continued advances in low-cost computing power and the development of new spatial light modulator technologies (largely driven by the display market) are likely to lead to improved performance in all embodiments.

Endnotes

1. K. Sasaki, M. Koshioka, H. Misawa, N. Kitamura, and H. Masuhara, Laser scanning micromanipulation and spatial patterning of fine particles, *Jpn. J. Appl. Phys. 2 Lett.* **30**, L907–L909 (1991).

2. K. Sasaki, M. Koshioka, H. Misawa, N. Kitamura, and H. Masuhara, Pattern formation and flow control of fine particles by laser scanning, *Opt. Lett.* **16**, 1463–1465 (1991).

3. K. Visscher, G.J. Brakenhoff, and J.J. Krol, Micromanipulation by multiple optical traps created by a single fast scanning trap integrated with the bilateral confocal microscope, *Cytometry* **14**, 105–114 (1993).

4. P.J.H. Bronkhorst, G.J. Streekstra, J. Grimbergen, E.J. Nijhof, J.J. Sixma, and G.J. Brakenhoff, A new method to study shape recovery of red blood cells using multiple optical trapping, *Biophys. J.* **69** 1666–1673 (1995).

5. K. Visscher, S.P. Gross, and S.M. Block, Construction of multiple-beam optical traps with nanometer-resolution position sensing, *IEEE J. Sel. Top. Quantum Electron.* **2**, 1066–1076 (1996).

6. D.L.J. Vossen, A. van der Horst, M. Dogterom, and A. van Blaaderen, Optical tweezers and confocal microscopy for simultaneous three-dimensional manipulation and imaging in concentrated colloidal dispersions, *Rev. Sci. Instrum.* **75**, 2960–2970 (2004).

7. T. Ota, S. Kawata, T. Sugiura, M.J. Booth, M.A.A. Neil, R. Juskaitis, and T. Wilson, Dynamic axial-position control of a laser-trapped particle by wave-front modification, *Opt. Lett.* **28**, 465–467 (2003).

8. H. Misawa, K. Sasaki, M. Koshioka, N. Kitamura, and H. Masuhara, Multibeam laser manipulation and fixation of microparticles, *Appl. Phys. Lett.* **60**, 310–312 (1992).

9. E. Fällman and O. Axner, Design for fully steerable dual-trap optical tweezers, *Appl. Opt.* **36**, 2107–2113 (1997).

10. Y. Ogura, K. Kagawa, and J. Tanida, Optical manipulation of microscopic objects by means of vertical-cavity surface-emitting laser array sources, *Appl. Opt.* **40**, 5430–5435 (2001).

11. R.A. Flynn, A.L. Birkbeck, M. Gross, M. Ozkan, B. Shao, M.M. Wang, and S.C. Esener, Parallel transport of biological cells using individually addressable VCSEL arrays as optical tweezers, *Sensors Actuators B* **87**, 239–243 (2002).

12. H. He, N.R. Heckenberg, and H. Rubinsztein-Dunlop, Optical particle trapping with higher order doughnut beams using high-efficiency computer-generated holograms, *J. Mod. Opt.* **42**, 217–223 (1995).

13. J.M.R. Fournier, M.M. Burns, and J.A. Golovchenko, Writing diffractive structures by optical trapping, in *Practical Holography IX*, ed. S. A. Benton, Bellingham, WA, SPIE (1995), 101–111.

14. E.R. Dufresne and D.G. Grier, Optical tweezer arrays and optical substrates created with diffractive optics, *Rev. Sci. Instrum.* **69**, 1974–1977 (1998).

15. Y. Hayasaki, M. Itoh, T. Yatagai, and N. Nishida, Nonmechanical optical manipulation of microparticle using spatial light modulator, *Opt. Rev.* **6** 24–27 (1999).

16. M. Reicherter, T. Haist, E.U. Wagemann, and H.J. Tiziani, Optical particle trapping with computer-generated holograms written on a liquid-crystal display, *Opt. Lett.* **24** 608–610 (1999).

17. J. Liesener, M. Reicherter, T. Haist, and H.J. Tiziani, Multifunctional optical tweezers using computer-generated holograms, *Opt. Commun.* **185**, 77–82 (2000).

18. J.E. Curtis, B.A. Koss, and D.G. Grier, Dynamic holographic optical tweezers, *Opt. Commun.* **207**, 169–175 (2002).

19. E.R. Dufresne, G.C. Spalding, M.T. Dearing, S.A. Sheets, and D.G. Grier, Computer-generated holographic optical tweezer arrays, *Rev. Sci. Instrum.* **72**, 1810–1816 (2001).

20. J. Leach, G. Sinclair, P. Jordan, J. Courtial, M.J. Padgett, J. Cooper, and Z.J. Laczik, 3D manipulation of particles into crystal structures using holographic optical tweezers, *Opt. Express* **13**, 5434–5439 (2005).

21. M. Montes-Usategui, E. Pleguezuelos, J. Andilla, and E. Martin-Badosa, Fast generation of holographic optical tweezers by random mask encoding of Fourier components, *Opt. Express* **14**, 2101–2107 (2006).

22. F. Belloni and S. Monneret, Quadrant kinoform: An approach to multiplane dynamic three-dimensional holographic trapping, *Appl. Opt.* **46**, 4587–4593 (2007).

23. J.E. Curtis, C.H.J. Schmitz, and J.P. Spatz, Symmetry dependence of holograms for optical trapping, *Opt. Lett.* **30**, 2086–2088 (2005).

24. R. Di Leonardo, F. Ianni, and G. Ruocco, Computer generation of optimal holograms for optical trap arrays, *Opt. Express* **15**, 1913–1922 (2007).

25. G. Sinclair, P. Jordan, J. Leach, M.J. Padgett, and J. Cooper, Defining the trapping limits of holographical optical tweezers, *J. Mod. Opt.* **51**, 409–414 (2004).

26. C. Schmitz, J. Spatz, and J. Curtis, High-precision steering of multiple holographic optical traps, *Opt. Express* **13** 8678–8685 (2005).

27. E. Theofanidou, L. Wilson, W.J. Hossack, and J. Arlt, Spherical aberration correction for optical tweezers, *Opt. Commun.* **236** 145–150 (2004).

28. Y. Roichman, A. Waldron, E. Gardel, and D.G. Grier, Optical traps with geometric aberrations, *Appl. Opt.* **45**, 3425–3429 (2006).

29. K.D. Wulff, D.G. Cole, R.L. Clark, R. Di Leonardo, J. Leach, J. Cooper, G. Gibson, and M.J. Padgett, Aberration correction in holographic optical tweezers, *Opt. Express* **14**, 4169–4174 (2006).

30. A. Jesacher, A. Schwaighofer, S. Furhapter, C. Maurer, S. Bernet, and M. Ritsch-Marte, Wavefront correction of spatial light modulators using an optical vortex image, *Opt. Express* **15**, 5801–5808 (2007).

31. G. Whyte, G. Gibson, J. Leach, M. Padgett, D. Robert, and M. Miles, An optical trapped microhand for manipulating micron-sized objects, *Opt. Express* **14**, 12497–12502 (2006).

32. C. Basdogan, A. Kiraz, I. Bukusoglu, A. Varol, and S. Doganay, Haptic guidance for improved task performance in steering microparticles with optical tweezers, *Opt. Express* **15**, 11616–11621 (2007).

33. W.J. Hossack, E. Theofanidou, J. Crain, K. Heggarty, and M. Birch, High-speed holographic optical tweezers using a ferroelectric liquid crystal microdisplay, *Opt. Express* **11**, 2053–2059 (2003).

34. V. Bingelyte, J. Leach, J. Courtial, and M.J. Padgett, Optically controlled three-dimensional rotation of microscopic objects, *Appl. Phys. Lett.* **82**, 829–831 (2003).

35. G. Sinclair, P. Jordan, J. Courtial, M. Padgett, J. Cooper, and Z.J. Laczik, Assembly of 3-dimensional structures using programmable holographic optical tweezers, *Opt. Express* **12**, 5475–5480 (2004).

36. Y. Roichman and D.G. Grier, Holographic assembly of quasicrystalline photonic heterostructures, *Opt. Express* **13**, 5434–5439 (2005).

37. C. Mio and D.W.M. Marr, Tailored surfaces using optically manipulated colloidal particles, *Langmuir* **15**, 8565–8568 (1999).

38. P. Jordan, J. Leach, M. Padgett, P. Blackburn, N. Isaacs, M. Goksor, D. Hanstorp, A. Wright, J. Girkin, and J. Cooper, Creating permanent 3D arrangements of isolated cells using holographic optical tweezers, *Lab Chip* **5**, 1224–1228 (2005).

39. M.M. Burns, J.M. Fournier, and J.A. Golovchenko, Optical matter: Crystalisation and finding in intense optical fields, *Science* **249**, 749–754 (1990).

40. A.E. Chiou, W. Wang, G.J. Sonek, J. Hong, and M.W. Berns, Interferometric optical tweezers, *Opt. Commun.* **133**, 7–10 (1997).

41. L. Paterson, M.P. MacDonald, J. Arlt, W. Sibbett, P.E. Bryant, and K. Dholakia, Controlled rotation of optically trapped microscopic particles, *Science* **292**, 912–914 (2001).

42. M.P. MacDonald, L. Paterson, K. Volke-Sepulveda, J. Arlt, W. Sibbett, and K. Dholakia, Creation and manipulation of three-dimensional optically trapped structures, *Science* **296**, 1101–1103 (2002).

43. E. Schonbrun, R. Piestun, P. Jordan, J. Cooper, K.D. Wulff, J. Courtial, and M. Padgett, 3D interferometric optical tweezers using a single spatial light modulator, *Opt. Express* **13**, 3777–3786 (2005).

44. P.C. Mogensen and J. Glückstad, Dynamic away generation and pattern formation for optical tweezers, *Opt. Commun.* **175**, 75–81 (2000).

45. R.L. Eriksen, P.C. Mogensen, and J. Glückstad, Multiple-beam optical tweezers generated by the generalized phase-contrast method, *Opt. Lett.* **27**, 267–269 (2002).

46. P.J. Rodrigo, V.R. Daria, and J. Glückstad, Four-dimensional optical manipulation of colloidal particles, *Appl. Phys. Lett.* **86**, 074103 (2005).

47. P.J. Rodrigo, I.R. Perch-Nielsen, C.A. Alonzo, and J. Glückstad, GPC-based optical micromanipulation in 3D real-time using a single spatial light modulator, *Opt. Express* **14**, 13107–13112 (2006).

48. I.R. Perch-Nielsen, P.J. Rodrigo, and J. Glückstad, Real-time interactive 3D manipulation of particles viewed in two orthogonal observation planes, *Opt. Express* **13**, 2852–2857 (2005).

49. M. Polin, K. Ladavac, S.H. Lee, Y. Roichman, and D.G. Grier, Optimized holographic optical traps, *Opt. Express* **13**, 5831–5845 (2005).

50. A. Jesacher, S. Furhapter, S. Bernet, and M. Ritsch-Marte, Diffractive optical tweezers in the Fresnel regime, *Opt. Express* **12**, 2243–2250 (2004).

Micromanipulation by "Multiple" Optical Traps Created by a Single Fast Scanning Trap Integrated With the Bilateral Confocal Scanning Laser Microscope

K. Visscher, G.J. Brakenhoff, and J.J. Krol

Department of Molecular Cell Biology, Section Molecular Cytology, University of Amsterdam, Plantage Muidergracht 14, NL-1018 TV Amsterdam, The Netherlands

Received for publication December 30, 1991; accepted July 22, 1992

We have developed a novel micromanipulator consisting of *multiple* optical traps created by scanning one single beam trap along a variable number of positions. Among other things, this enables the orientation of irregularly shaped and relatively large structures which could not be oriented by just one trap as is demonstrated on long *Escherichia coli* bacteria filaments. We expect that the multiple trap manipulator will broaden the field of applications of optical trapping as a micromanipulation technique. For example, it facilitates the study of mechanical properties of extended structures as illustrated by a "bending"-experiment using *E. coli* bacterium filaments. A special application of the multiple trap manipulator is the "indirect trapping" of objects which we did by keeping them held between other optically trapped particles. Indirect trapping makes it possible to trap particles which either cannot be trapped directly due to their optical properties (refractive index) or for which exposure to the laser radiation is undesirable. The multiple optical trap manipulator is controlled interactively by a UNIX workstation coupled to a VME instrumentation bus. This provides great flexibility in the control of the position and the orientation of the optical traps.

Micromanipulation makes it desirable to have real time 3D microscopy for imaging and guidance of the optical traps. Therefore we integrated optical micromanipulation and a specially developed real-time confocal microscope. This so called bilateral confocal scanning laser microscope (bilateral CSLM) [Brakenhoff and Visscher, J Microsc 165:139–146, 1992] produces images at video rate.

Key terms: Optical trapping, multiple optical traps, micromanipulation, bilateral optical scanning laser microscopy

---

Micromanipulation is becoming an important technique in the biomedical or biological sciences. It can be used to study, for instance, the motility of bacteria (9) and sperm cells (11,22) or intra-cellular phenomena (6, 8). Micromanipulation by mechanical means where needles or micro-pipettes are used to remove or displace internal cell structures is an invasive technique which causes damage and effects the experiment itself. The use of a *non-contact* micromanipulator utilizing an optical trap, also called an "optical tweezer" was first demonstrated by Ashkin (3). This technique minimizes the invasive character of micromanipulation. For intra-cellular manipulation it is no longer necessary to penetrate the cell wall or membrane. Optical traps are created using a high numerical aperture lens to sharply focus the trap laser beam. The focal spot forms the tip of the "tweezer." An object illuminated by such a focused laser beam experiences a scattering force based on radiation pressure which pushes it forward into the direction of the energy flux. In addition a so called gradient force is exerted which pulls the object towards regions with a high electric energy density (generally the focal region) (4,15). The latter force originates from the large gradients in the time-averaged electric field density near the focal plane due to focusing. In regions where the two forces have opposite signs and cancel each other particles may be trapped at a stationary three-dimensional (3D) position. For dielectric particles with a refractive index larger than the surrounding medium this generally is in the laser beam on the optical axis near the focal plane (4,24,25). By displacing the focal position of the laser beam the trapped particle may be moved around in 3D space.

Most optical micromanipulators only use one optical single beam gradient trap, which results in only one "tweezer" or point of action in the specimen. Contrary to the self-orientation of 'smaller' particles, which have no spherical optical symmetry such as red blood cells (7) or the algae *Chlamydomonas* (as we saw in our laboratory), the orientation of, e.g., long rod-like particles is not possible using a single trap. We saw, while trapping a filamentous *Escherichia coli* bacterium using a single trap, that it was possible to move the filament around, but not to orient it. This encouraged us to develop an optical micromanipulator equipped with a variable number of optical traps.

The orientation of large, rod-like objects such as the *E. Coli* filament requires at least two points of action at the object. Several approaches have been presented to create such multiple traps. Ashkin has mentioned the possibility to use two laser beams for the levitation and orientation of nonspherical particles (2) or for optical trapping and orientation of rod-like bacteria (5). In this way the two beams may be positioned independently within the specimen. Nevertheless the number of traps is limited to the amount of laser beams, which in this case is two. A larger number of traps can be created using interference, as has been demonstrated by Burns (13). He simultaneously trapped multiple particles at the positions of maximal intensity of a standing wave field created by interference. Two- or three-dimensional periodic structures comprised of microscopical particles can be created in this way. Contrary to Ashkins solution these traps cannot be positioned independently. Interactively changing the interference pattern looks hard to do in this set-up. Just recently, Sasaki and Misawa (16,18,19) reported the trapping of several particles using a computer controlled scanning laser beam. A number of particles are then trapped along the scanned path as long as the repetition rate of scanning is set higher than the cut-off frequency for the mechanical response of the particles (18). This means that the scanning rate is high enough to prevent drifting of the particles. We, independently of their work, succeeded in the creation of a multiple trap manipulator based on the scanning of one single beam gradient trap together with a periodical blocking of the laser beam.

Our multiple trap manipulator created by scanning only one single beam gradient trap is based on the fact that the forces exerted at several positions at the object do not have to be applied simultaneously to trap it as a whole. From single trap experiments we saw that it was possible to keep the object trapped at a stationary position by just periodically illuminating it: the laser beam could be switched off for a short time interval (milliseconds—seconds, depending on the mobility of the object) without losing the object from the trap. To maintain effective trapping, the time interval during which the beam is switched off should be less than the time in which the object may drift away from the trap location. A multiple trap manipulator can now be obtained by rapidly illuminating different spots within the specimen after each other. The laser beam is switched off when it is scanned from one position to the other, so that no particles are trapped on a scanned line between the two trap positions. So, in order to create a multiple trap manipulator, a single beam trap is sequentially scanned along a desired number of trap positions in the specimen.

Because (optical) micromanipulation can be done in three dimensions this calls for a 3D imaging system for both visualization and control. The confocal microscope, with its optical sectioning capabilities is an excellent imaging tool for 3D micromanipulation. We integrated a fast CCD-based bilateral CSLM (12) and an optical micromanipulator based on a commercially available inverted microscope.

MATERIALS AND METHODS

Optical trapping and its imaging are often done using one and the same objective of the microscope (3,8–10,21). This is achieved using one of the extra photo ports of the microscope to couple the laser beam. This approach has several advantages such as maintaining the complete imaging capabilities of the original microscope, e.g., bright field, phase contrast, Nomarski differential contrast, dark-field, and epi-fluorescence microscopy. However by optical trapping the object may also be vertically displaced. This is achieved in two ways: by either moving the object stage of the microscope (Fig. 1b) or by moving the focal position of the trap laser beam through the displacement of a relay lens (Fig. 1c). Using conventional microscopy this results in a dilemma: either the trapped particle stays in focus and the former surroundings go into defocus (Fig. 1b) or vice versa (Fig. 1c).

The problem stated above may be solved using a real-time 3D imaging device. We made a first step towards this goal using the real-time bilateral CSLM (12). In this microscope, the movement of the confocal imaging probe along the optical axis (necessary to make the optical sections) is done by displacing the imaging objective. In this case when imaging and trapping are done using the same objective there is a conflict between 3D confocal imaging and vertical positioning of the trap. For instance, when rapidly scanning the object along the optical axis to acquire a 3D image, fast moving correctional optics would be necessary to keep the optical trap at its intended vertical position. Although this problem may be solved technically, we found this an impractical solution. Therefore in our apparatus the optical paths of the optical manipulator and the (confocal) microscope are completely separated and the two devices can operate independently.

Separation of the optical paths is achieved by replacing the original condenser of the microscope (Olympus IMT2, Olympus Optical Co., Ltd., Tokyo, Japan) by the

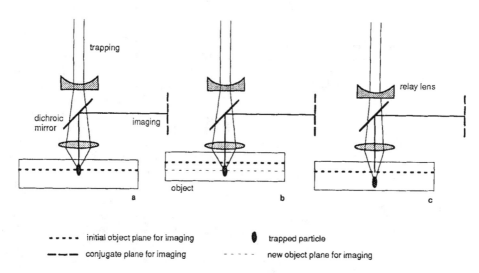

FIG. 1. Optical trapping and imaging utilizing the same objective: consequences of vertical trap displacements for imaging. **a)** initial situation, **b)** displacement of object stage, **c)** displacement of relay lens.

objective ($100 \times$, NA $= 1.3$) which focuses the trap laser beam. The original illumination set-up which was for bright-field and phase-contrast is adapted so that we can use bright-field, phase-contrast together with epi-fluorescence 2D-microscopy, as well as fluorescence confocal microscopy. Note that the epi-illumination for fluorescence 2D-imaging is not affected by introducing the optical trap.

Optical Trapping

The multiple trap micromanipulator is based on the fast scanning of one single beam gradient trap along a desired number of points in the specimen. The optical scheme of the trap set-up is shown in Figure 2. The laser used for trapping is a Nd:YAG laser (Spectron SL503, Rugby, England) emitting at 1,064 nm. An Acoustic-Optic Modulator (ACM, IntraAction Corp., Bellwood, IL) functions as a fast shutter to block the beam when it is scanned from one point of action to the other. Two galvanometric scan mirrors, M_1 (General Scanning Inc. G120DT, Watertown, MA), are used to deflect the beam into the x and y direction while the z positioning of the trap is done by a piezo device (PIFOC, Physik Intrumente GmbH & Co., Waldbronn, Germany), which moves the objective. The beam is expanded to a diameter of 2.5 mm ($1/e^2$) to fill the pupil of the objective L_1. A variable attenuator (Newport, M930, Fountain Valley, CA) is used to control the laser power. To maintain classical 2D phase-contrast imaging the phase ring is imaged, through the dichroic mirror by L_1, L_3, and L_4 onto the phase plate inside L_4. Note that for proper phase contrast imaging L_3 should move with L_1, which in practice we do not. Nevertheless the imaging quality in phase contrast is sufficient to visualize and control trapping.

Control of the Single and Multiple Trap Manipulator

The electronic control system was initially designed for *single* trap manipulation. As a basis for our control system design we formulated three requirements: 1) The system should be straightforward both in hard- and soft-ware organization. This can be achieved by using analog voltages to drive commercially available mirrors and piezo-actuators operating in closed-loop with position sensors. 2) The system has to be integrated in the local computer network to optimize further processing of collected data obtained with the confocal microscope. 3) The control system should run as independently as possible of the CPU load of the workstation to which it is connected in order to achieve real-time performance on a multi-user and multi-tasking computer. This can be realized by implementing a VME sub-system as described below.

Complete electronic control of trap positioning (Fig. 3a) is accomplished by an HP/Apollo 9000/425s UNIX workstation equipped with an AT-bus interfaced with a VME instrumentation bus by a bus-translator (IBM PC/AT VME Adaptor, model 403, BIT3 Computer Corporation, Minneapolis, MN), which consists of 32 Kb Dual Port Ram (DPR). The VME based system consists of a CPU 68030 (SYS68K/CPU-32, Force Computers, Inc., Campbell, CA), a fast Digital-Analog Converter board (DAC, model MPV954, Pentland Systems Ltd., Livingston, Scotland), and a dialbox for manual control. For *single* trap operation a software program for translating the dialbox knob settings into its x,y,z coordinates is loaded from the UNIX workstation through the DPR into the 68030 CPU. The coordinates are directed to the DAC which drives the closed loop control electronics of the mirrors and the piezo. This program operates completely independently of the load

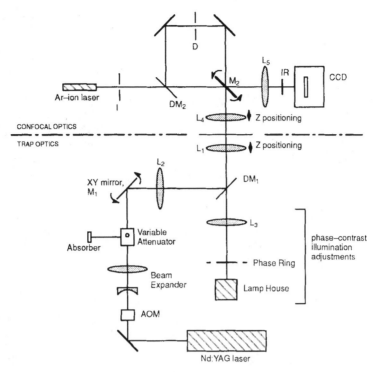

FIG. 2. Optical scheme of both the optical trap and the CCD-based confocal microscope. Trap Optics: for a detailed description of the optical trap set-up and phase-contrast adjustments see the text. (L_1, objective, L_2, relay lens imaging the XY mirror M_1 into the pupil plane of L_1; L_3 relay lens to image the Phase ring onto the phase plate of L_4. DM_1 is a dichroic mirror transparent for visible light and reflecting at 1,064 nm. The AOM is the Acousto Optic Modulator which acts as a fast shutter.) Confocal Optics: The light source I, which may be a pinhole or an other illuminated pattern, is scanned over the specimen by the front surface of the double-sided mirror M_2. The fluorescence light emitted by the specimen is descanned by the same surface of M_2 and reflected by the dichroic mirror DM_2 (transparent for the excitation). It passes the detection slit D, which is placed into a conjugate plane of I to create the confocal effect. The basic feature of the bilateral scan technique is that the detection pattern D is imaged onto the CCD (by L_5) and the final image is generated by scanning D over the CCD, with the back surface of the non-transparent mirror M_2. The respective surfaces of M_2 are located in the pupil of L_4 and L_5. A perfect geometrical registration of the specimen data is assured by this coupled scan technique. In front of the CCD an IR-blocking filter (IR) is incorporated, to partially block the trap beam, so that its position can be imaged properly.

level of UNIX workstation. By this approach we fully satisfy our requirements for the case of *single* trap operation.

For *multiple* trap operation, some adjustments and new features are necessary such as a user-interface which is required to change the trap pattern interactively. This user interface (Fig. 3b) is displayed at the monitor of the UNIX workstation. We used a 2D representation in which the actual x and y positions are represented by circles in the z = 0 reference plane. The reference plane for trapping is the plane where the z-actuator is initially; it is maneuvered within the imaged volume interactively. The actual z distance of each trap relative to the reference plane is represented by the radii of the circles displayed (Fig. 3b). The further away from the reference plane the larger the radius is. A cross is added to the circle when the trap is below the reference plane. Each trap can be selected and positioned independently by the mouse. Furthermore the entire trap pattern can be translated and rotated (using the dialbox) around the x, y, and z axis. The location of these axis is represented in the by the cross inside the rotation marker (RM), which, momentarily, can be positioned only within the reference plane (z = 0).

For *multiple* trap operation the control software had to be changed to include the user interface. In the adjusted software the 68030 VME processor still performs the translations of the entire trap pattern independently of the load level of the UNIX workstation. However as soon as operations are done which require the calculation of new coordinates, like changing the individual trap position through the user interface, or rotations of the entire trap pattern, the UNIX workstation is used to calculate the offsets on the initial coordinates. The offset table is refreshed by the workstation every time such an operation is carried out. The offsets are written to the DPR of the bus-translator from which they are read by the 68030 CPU and which adds them to the coordinates corresponding to the actual x,y,z knob settings of the dialbox. The resulting coordinates are again directed to the DAC.

These extension, to a certain extent, make the position control system dependent of the CPU load level of

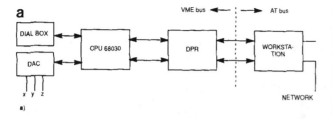

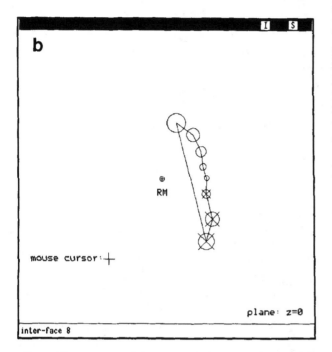

FIG. 3. Electronic control diagram (**a**) and user interface (**b**) of the multiple trap manipulator. a) DPR is the Dual Port Ram of the AT-VME bustranslator which interfaces the AT bus of the UNIX workstation and the VME instrumentation bus. The CPU is a VME based 68030 processor board. DAC, Digital to Analog converter board (VME). With the dialbox, translations (trackerball) and rotations (knobs) of the entire trap pattern are performed. A further description is given in the text. b) The circles represent the 3D position of the traps within the specimen. RM denotes the center of rotation (rotation marker). For a detailed explanation see the text.

the multi user, multi-tasking UNIX workstation (the job running the user interface and calculating the offsets on the old coordinates is time-shared with other jobs). In future developments, the software (originally designed for single trap operation) will be reorganized so that the necessary calculations of new trap coordinates are done entirely by the 68030 VME processor and the UNIX workstation only is used as an user interface environment. Also the integration of the user-interface into the X window environment then will be taken to hand.

Confocal Imaging

We now briefly discuss the realization of the fast bilateral confocal scanning laser microscope which provides us with real-time 3D images. The CCD based confocal microscope is extensively discussed by Brakenhoff and Visscher (12) and is schematically shown in Figure 2. Here we restrict ourselves to a short description. The optics necessary for realizing confocal imaging are coupled to one of the additional photo ports of the microscope. The key elements in the design are the use of a CCD detector for confocal data acquisition and the use of a double sided mirror (M_2), which bilaterally scans the illumination pattern over the object and its image over the detecting element, the CCD. The double-sided mirror can be slid in and out of the optical path of the microscope so that 2D, non-confocal imaging capabilities can be maintained (not shown in Fig. 2). The confocally collected image is visualized on a monitor and can be stored in computer memory using a frame grabber. The confocal microscope is controlled by the same UNIX workstation used for the optical trap control.

The confocal microscope in principle can either be used with illumination and detection slits or with pinholes. The use of slits results in a somewhat reduced optical sectioning effect and lateral resolution (in one direction) compared to the use of pinholes. The advantages of using slits, as we do at the moment, are in the first place the high image repetition rate that can be achieved (up to video rate) and second, a considerable gain in the strength of the detected signal, which is especially important in fast fluorescence imaging.

Experimental Results

First we tested the 3D imaging of optically trapped fluorescent polystyrene beads with a diameter of 6.51 μm using a *single* trap. Such a bead could be easily lifted out of the confocal plane using the trap. Figure 4 shows the results for a sphere which is both lateral and axially displaced within the 3D image. After contrast enhancement we used the Simulated Fluorescence Process (SFP) algorithm, which is described by Van der Voort [26], for visualization. This SFP algorithm generates a "shadow image" so that a feeling for the 3D spatial organization is obtained. In Figure 4a the two spheres lie in approximately the same xy-plane as can be seen from the shadows which are approximately at the same distance from the beads, while in Figure 4b the bead at the right has been displaced both laterally and vertically (its shadow is closer because of the opposite top-bottom orientation of trapping and imaging). During confocal imaging the slit was opened further than is required for true confocal imaging to enhance the detected signal. Therefore the images look a little fuzzy, but nevertheless the background suppression was sufficient to make these 3D images.

When we tried to displace an *E. coli* filament (ftsZ84Ts mutant, LMC509) with a *single* trap we were only able to move it around but by no means could we rotate it into a desired orientation or lift it as a whole. Although the orientation of smaller filaments in an optical trap was observed, it could not be done in a controlled way. This lack of controlled orientation of

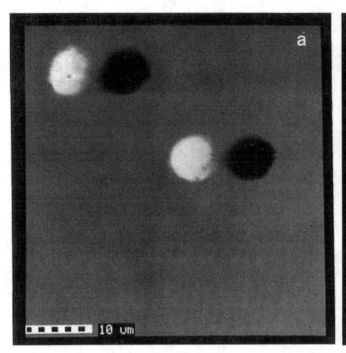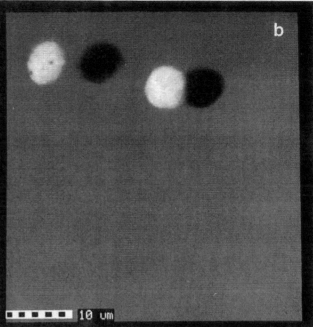

FIG. 4. Three-dimensional representation using the Simulated Fluorescence Process algorithm (26) of an optical trapped and in three dimensions displaced 5.61 μm polystyrene sphere. Data are collected using the confocal set-up taking optical sections approximately 1 μm apart. **a)** Initial situation, **b)** displaced (lifted) sphere.

certain objects is one of the limitations of the availability of just one trap for micromanipulation.

With the *multiple* trap set-up, using the scan approach outlined above, we have shown that the orientation of extended objects such as the *E. coli* filament is indeed possible. In Figure 5 the rotation of a filamentous *E. coli* bacterium is demonstrated. The optical traps are imaged as bright spots. Orientation of the bacterium is performed within the focal plane by rotating the entire trap pattern using the dialbox. In another experiment (not shown), using confocal imaging we were able to lift and to rotate an entire *E. coli* filament out of the plane of focus using multiple traps. Confocal fluorescent imaging of the bacteria still was of poor quality due to the lack of enough excitation laser power at the moment.

To explore the further possibilities of multiple traps for manipulation of filamentous structures we tried to bend the *E. coli* bacterium by gently displacing the trap at the right end with respect to the two left traps which were kept at fixed positions (Fig. 6). In this figure two images, which are drawn from a video recording, are superimposed: one in the initial situation and one with the right trap displaced. The filament is somewhat curved and therefore parts of it are out of focus. From the resulting image it can be distinguished that the left part of the bacterium stays in place while the right part is moved over a small distance, showing that the bacterium is bent (we ensured that the bending was not

confused with a defocusing effect due to a vertical displacement caused by trapping). We were only able to displace the end part over about 1–2 μm. When the left trap was displaced further the bacterium acted like a spring and its right part jumped back to its initial position. By no means we were able to fold or even break the *E. coli* filament. These experiments demonstrate that the multiple trap manipulator may be a useful tool in the study of mechanical properties (such as the spring constant in this case) of biological objects.

In Figure 7 we demonstrate another important application of multiple traps: the *indirect* manipulation, in this case of a spherical oil particle. The specimen is an emulsion of sunflower oil and water. The spherical oil particle is held between three other optically trapped oil particles. The center particle is not illuminated by the laser beam, so no direct optically induced forces are exerted to it. We were able to move this non-illuminated "trapped" particle in the xy plane by displacing the three optical traps positioned at the three other particles simultaneously. The center particle could easily be set free again by removing the three trapped particles.

DISCUSSION

As mentioned above, a multiple trap is created by rapidly scanning a single beam gradient trap along the desired positions in the specimen. The trap efficiency at each position therefore depends on the part of the time

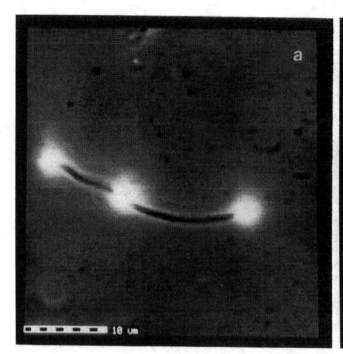
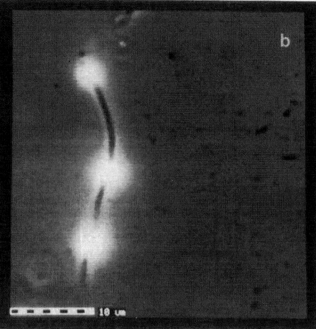

FIG. 5. Rotation of an *E. coli* bacterium using the *multiple* trap manipulator, imaged by the adjusted phase-contrast mode. The three points of action the optical trap is scanned along are clearly imaged as bright spots. **a)** Initial situation, **b)** after rotation through the dialbox.

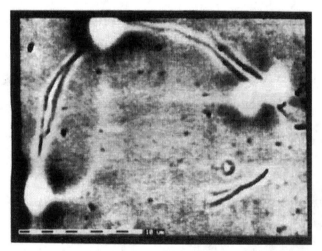

FIG. 6. Bending of a filamentous *E. coli* bacterium using *multiple* trap manipulator. Two video images have been super-imposed into the presented image: one in the initial situation and one with the right most trap displaced showing the two different positions of the right trap. The traps are imaged as bright spots. We tried to improve quality of the initially poor images by rigorous contrast enhancement.

the particle is illuminated and the laser power. For instance when we scan along four points at a typical frequency of 50 Hz the particles are illuminated for about 4 ms (mind the time lost for scanning between

two positions) with a period of 20 ms. So they are illuminated for 1/5 of the total time. This results in considerably higher laser power levels required for multiple traps compared to a single trap which is effective for 100% of the time. In practice we try to minimize the optical power for a chosen amount of traps. This is done by adjusting the scan frequency interactively so that trapping of the object looks efficient as is checked by eye on the monitor. The frequency a trap pattern is scanned is limited by the maximum operating frequencies of the two galvanometric mirrors and the z piezo-actuator, respectively, 300 Hz and 160 Hz.

As illustrated by this paper we mainly use phase-contrast microscopy to control manipulation. In our experience 2D imaging methods are sufficient in many applications, especially when imaging is optically *independent* of trapping. When imaging and trapping are dependent, i.e., when they both use the same objective, the adjustments to the microscope to integrate an optical trap are minor and very simple (4). However in our apparatus we separated both optical paths so that 3D optical micromanipulation and 3D imaging could be combined in the most practical way. It also has the advantage that the user can decide what to hold in focus when using 2D imaging microscopy during micromanipulation: either the trapped object or its surroundings. This separation of the optical paths implies a somewhat more complex optical set up and has some implications on the preparation of the specimen of which we already reported in a previous paper (23):

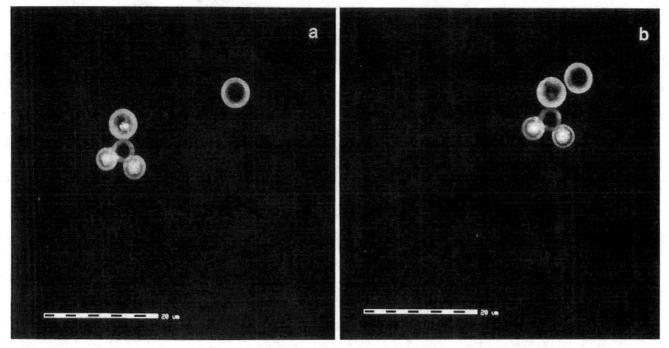

FIG. 7. *Indirect trapping* and lateral displacement of a spherical oil particle in water held between three other optically trapped oil particles. The optical traps are shown as bright spots on the trapped spheres. **a)** Initial situation, **b)** after displacement.

The maximal thickness of the specimen should preferably not exceed 50 μm. The object should be immersed between two cover slides. When using a water-immersion objective for trapping, the top cover slide can be omitted which can be advantageous in specimen preparation.

In our first optical trap manipulator the positioning of the traps into *all* three dimensions was done using a so called "scan objective" which was originally designed in our laboratory for confocal scan purposes but was used with good results in the single trap manipulator version (23). This scan objective had good performance in a multiple manipulator set-up when operating at low frequencies up to 10 Hz. Its limited response at higher frequencies made it unpractical for the creation of multiple traps, so that galvanometric scan mirrors are used for beam positioning. Nevertheless the scan objective enables a relatively easy realization of a controllable single trap or slowly scanning trap, 2D imaged micromanipulator, because it is compatible with general microscope objectives. Only a stationary laser beam has to be coupled to one of the photo ports of the microscope.

Operation of the *multiple* trap, individually trap positioning as well as translations and rotations of the entire trap pattern, were very easily done using the user-interface. As long as we could prohibit use of the workstation CPU by other uses or processes, manipulation could be done real-time and interactively. As soon as the load of the CPU increased due to the ap-

pearance of other users, the translation of individual traps or the rotation of the entire pattern of the computer was delayed. At the moment we are busy to re-design the software to cope with this limitation.

Before optical trapping and biological material is discussed, the wavelength of the laser used for the optical trap is discussed. The choice of this wavelength is governed by the damage caused to the biological objects by the laser radiation. To prevent this damage, the laser wavelength of the trap laser beam must be chosen such that minimal absorption will take place. At absorption, the production of heat or photo-chemical effects may kill the live specimen. For instance, Ashkin reported of damages to biological objects caused by using 514.5 nm radiation for trapping (2). Later on when he used IR (1,064 nm) radiation instead, no damage was induced and (5) the reproduction of yeast cells and *E. Coli* bacteria in a single beam gradient traps was observed. Because of these experiments we also chose this wavelength for trapping and we saw that no damage was caused to yeast cells which were trapped, isolated, and grown on a culturing plate. Although this wavelength is successful for these specific biological objects, it cannot be assumed that micromanipulation at 1,064 nm is free of damage for any object. This is because absorption is a process which depends on the wavelength as well as the object itself. Biological systems often have absorption in the uv and blue regions of the spectrum and it therefore looks sensible to use longer wavelengths. Possibly one does not have to go as far as 1,064 nm; some groups use diode lasers which emit around 840 nm (1) [or even at 1,300 nm (20)]. But in practice for each object the viability of the optical trap working at a specific wavelength can only be assured if it is checked experimentally.

The experiments with the *E. coli* filament show some of the possibilities of multiple traps. With the multiple trap we could lift the entire filament instead of only a part of it. So, controlled 3D translations of extended objects becomes possible. In addition we could easily rotate the filament within the focal plane as demonstrated in Figure 5. Rotation in three dimensions in principle is possible and we succeeded in the rotation of an about 10 μm long filament over almost 90° out of the focal plane (around the x-axis). But at this first experiment (with poor quality confocal imaging) the orientation looked not stable in the sense that it vibrated around the optical axis. Further experiments are necessary to investigate the possibilities of rotations into the axial direction, i.e., into the direction perpendicular to the object stage of the microscope. It should be noted that the orientation of particles by multiple traps, as a supplement of the earlier mentioned self-orientation of smaller particles in a single trap, in both cases depends on the optical spherical symmetry of the particles. The orientation is only possible for non-symmetrical particles while the orientation of optical spherical symmetric particles both by multiple traps and by self-orientation in a single trap remains impossible!

The bending of the *E. coli* filament showed that in principle we can determine the spring constant of such a bacterium. We expect that multiple traps can be used to determine several mechanical properties of biological objects. We also did an experiment in which we separated two mating algae *Chlamydomonas* to determine the position where their flagella stick to each other during the mating process. This application demonstrates the possibilities of using multiple traps in cell-cell interaction studies.

Probably the most promising property of our manipulator is the principle of indirect trapping using multiple optically trapped particles to hold and move a fragile object. *Indirect trapping* using *multiple* traps makes it possible to manipulate particles that cannot be trapped directly because of their unsuitable optical properties (i.e., low refractive index) or because exposure to intense laser illumination may be undesirable because of potential damage that even at 1,064 nm may still occur. We demonstrated indirect trapping only within a 2D plane. When the friction between the beads and the particle is sufficient the particle may evenly be lifted together with the trapped beads. We do not expect that indirect trapping in 3D can be obtained by placing trapped beads on top of and below the particle because it then will be illuminated by the (defocused) laser beam which we did not want in the first place. In addition when a bead is placed on top of the particle, the particle itself is influenced by the laser

light scattered at this bead: it may be pushed away when its refraction index is below that of the surrounding medium. Another way of *indirect* trapping of biological objects is the attachment of polystyrene or latex bead coated with specific binding monoclonal antibodies to them (10,14). Using this technique even very small particles attached to the bead may be positioned within the specimen. It further has the advantage that 3D manipulation is possible by lifting the attached bead. On the other hand the bead will stay attached to the particle during the entire experiment. Our way of indirect trapping enables it to remove the beads when the particle has reached its intended position within the specimen.

The bilateral CSLM is a real-time confocal microscope in the sense that the confocal plane is imaged immediately. We used this to image the vertical lifting of a *E. coli* filament using multiple traps. As soon as it was vertically displaced it disappeared out of the confocal plane and was not imaged anymore. The filament could be imaged again after re-focusing. Three dimensional images of dimensions $256 \times 256 \times 16$ can be acquired with a speed in principle sufficient for real-time 3D imaging. At the moment the 3D representation of the data is the bottleneck. The SFP algorithm, at this moment, takes too much time, but the group of van der Voort in our lab is working on this problem. A solution may be the calculation of stereoscopic views, which can be done very rapidly. But not everyone (only 10–30%) can see "3D" from stereoscopic views. Therefore, for the time being, we use the confocal microscope only for imaging and presentation purposes rather than for controlling the traps. Control of the positioning of the traps is temporarily being done using phase contrast microscopy.

In this paper we mentioned some experiments of trapping *E. coli* filaments which were not supported with photographs. These experiments concern the use of the confocal microscope for imaging. In the prototype of the bilateral CSLM we use a low powered laser source, so that image intensifier is needed to image the fluorescent labeled filaments. The resulting images were very noisy and worse the traps could not be visualized because these intensifiers generally are not sensitive for 1,064 nm radiation, so that there is no visual feedback of the positions of the traps within the specimen. We improvised by quickly alternating between 2D phase-contrast and confocal fluorescence imaging. A more rigorous solution is under development. We intend to make an overlay image of the microscopes image and the new (X window based) user-interface, so that the traps do not need to be imaged anymore but can be represented symbolically in the image.

CONCLUSIONS

We presented an approach to 3D *multiple* optical trapping visualized by a confocal microscope. We expect that in many applications it is possible to use a *multiple* trap manipulator in which the individual traps are not simultaneously but periodically illuminated. In all these applications one may benefit from the flexible positioning of the individual traps in this kind of manipulator. We emphasize that in this optical manipulator it is not only possible to move and rotate the entire trap pattern but it is also possible to create any desired trap pattern by positioning the individual traps independently. Great flexibility is assured through individual control of each position by mouse input and the user-interface. The multiple trap manipulator may greatly enhance the possibilities of optical trapping within biomedical and biological sciences. The study of mechanical properties of microbiological objects is only one possible application of this technique. The mechanics and 3D spatial structure of coiled chromosomes in the anaphase may be a suitable object for further studies in this laboratory (17). Also cell fusion experiments as presented by Steubing (21) may benefit from using multiple traps by holding the cells together or keeping them oriented into desired directions. Another promising application is the *indirect trapping* by which a particle is "trapped" using other optically trapped particles to hold it, thereby broadening the field of application to particles that otherwise could not be optically trapped.

The use of the bilateral confocal scanning laser microscope appears to be very promising for visualization of micromanipulation by optical trapping. The developed bilateral CSLM, in principle, is able to produce 3D data sets at image repetition rates sufficient for proper control and visualization. Regretfully real time 3D representation software is not available at present. The SFP algorithm in principle looks very suitable to visualize 3D optical trapping. But at the moment it is not fast enough to produce real time 3D representations. New software developments concerning the SFP algorithm in the group of Van der Voort look very promising to obtain real time performance. For the time being we use conventional 2D microscopy for control.

ACKNOWLEDGMENTS

We like to thank Taco Visser for reviewing this paper. This work has been supported by the Technical Foundation (STW) of The Netherlands, grant ANS90.1917.

LITERATURE CITED

1. Afzal RS, Treacy EB: Optical Tweezers using a diode laser. Rev Sci Instr 64:2157–2163, 1992.
2. Ashkin A, Dziedzic JM: Observation of light scattering from non-spherical particles using optical levitation. Appl Optics 19:660–668, 1980.
3. Ashkin A, Dziedzic JM: Optical trapping and Manipulation of viruses and bacteria. Science 235:1517–1520, 1987.
4. Ashkin A, Dziedzic JM, Bjorkholm JE, Chu S: Observation of a single-beam gradient force optical trap for dielectric particles. Optic Lett 11:288–290, 1986.
5. Ashkin A, Dziedzic JM, Yamane T: Optical trapping and Manipulation of single cells using infrared laser beam. Nature 330: 769–771, 1987.
6. Ashkin A, Schuetze K, Dziedzic JM, Euteneuer U, Schliwa M: Force generation of organelle transport measured in vivo by an infrared laser trap. Nature 348:346–348, 1990.
7. Ashkin A: The study of cells by optical trapping and manipulation of living cells using infrared laser beams. ASGSB Bull 4: 133–146, 1991.
8. Berns MW, Wright WH, Tromberg BJ, Profeta GA, Andrews JJ, Walter RJ: Use of laser-induced optical force trap to study chromosome movement on the mitotic spindle. Proc Natl Acad Sci USA 86:4539–4543, 1989.
9. Block SM, Blair DF, Berg HC: Compliance of bacterial flagella measured with optical tweezers. Nature 338:514–517, 1989.
10. Block SM, Goldstein LSB, Schnapp BJ: Bead movement by single kinesin molecules studied with optical tweezers. Nature 348:348–352, 1989.
11. Bonder EM, Colon J, Dziedzic JM, Ashkin A: Force production of by swimming sperm-analysis using optical tweezers. J Cell Biol 111:421A, 1990.
12. Brakenhoff GJ, Visscher K: Confocal imaging with bilateral scanning and array detectors. J Microsc 165:139–146, 1992.
13. Burns MM, Fournier J, Golovchenko JA: Optical matter: Crystallization and binding in intense optical fields. Science 249:749–754, 1990.
14. Chu S: Laser manipulation of atoms and particles. Science 253: 861–866, 1991.
15. Gordon JP: Radiation forces and momenta in dielectric media. Phys Rev A 8:14–21, 1973.
16. Misawa H, Sasaki K, Koshioka M, Kitamura N: Multibeam laser manipulation and fixation of microparticles. Appl Phys Lett 60: 310–312, 1992.
17. Oud JL, Mans A, Brakenhoff GJ, Van der Voort HTM, Van Spronsen EA, Nanninga N: Three-dimensional chromosome arrangement of *Crepis capillaris* in mitotic prophase and anaphase as studied by confocal scanning laser microscopy. J Cell Sci 92:329–339, 1989.
18. Sasaki K, Koshioka M, Misawa H, Kitamura N, Masuhara H: Pattern formation and flow control of fine particles by laser-scanning micromanipulation. Optic Lett 16:1463–1465, 1991.
19. Sasaki K, Koshioka M, Misawa H, Kitamura N: Optical trapping of a metal particle and a water droplet by a scanning laser beam. Appl Phys Lett 60:807–809, 1992.
20. Sato S, Ohyumi M, Shibata H, Inaba H: Optical trapping of small particles using 1.3 μm compact InGaAsp diode laser. Optics Lett 16:282–284, 1991.
21. Steubing RW, Cheng S, Wright WH, Numajiri Y, Berns MW: Laser induced cell fusion in combination with optical tweezers: The laser cell fusion trap. Cytometry 12:505–510, 1991.
22. Tadir Y, Wright WH, Vafa O, Ord T, Ash RH, Berns MW: Micromanipulation of sperm by a laser generated optical trap. Fertil Steril 52:870–873, 1989.
23. Visscher K, Brakenhoff GJ: Single beam optical trapping integrated in a confocal microscope for biological applications. Cytometry 12:486–491, 1991.
24. Visscher K, Brakenhoff GJ: Theoretical study of optically induced force on spherical particles in a single beam trap I: Rayleigh scatterers. Optik 89:174–180, 1992.
25. Visscher K, Brakenhoff GJ: Theoretical study of optically induced force on spherical particles in a single beam trap II: Mie scatterers. Optik 90:57–60, 1992.
26. Van der Voort HTM, Brakenhoff GJ, Baarslag MW: Three-dimensional visualization methods for confocal microscopy. J Microsc 153:123–132, 1988.

Optical manipulation of microscopic objects by means of vertical-cavity surface-emitting laser array sources

Yusuke Ogura, Keiichiro Kagawa, and Jun Tanida

We report on experimental verification of optical trapping using multiple beams generated by a vertical-cavity surface-emitting laser (VCSEL) array. Control of the spatial and temporal emission of a VCSEL array provides flexibility for manipulation of microscopic objects with compact hardware. Simultaneous capture of multiple objects and translation of an object without mechanical movement are demonstrated by an experimental system equipped with 8×8 VCSEL array sources. Features and applicability of the method are also discussed. © 2001 Optical Society of America

OCIS codes: 140.7010, 250.7260, 140.3290, 120.4640, 350.4990.

1. Introduction

Since Ashkin *et al.* demonstrated trapping a microscopic dielectric particle by a three-dimensional optical gradient force in 1986,[1] optical beam trapping has been used for a wide range of applications, such as manipulation of biological particles,[2,3] measurement of piconewton force produced by a single kinesin molecule,[4] breakage of an actin filament,[5] alignment and spin of a birefringent object,[6] and an optical spin micromotor.[7] The optical trap technique is useful as a noncontact manipulation method for microscopic objects, and its importance is unquestionable. In addition, by introducing beam modulation, we can extend the variety of manipulations. The examples are trapping and translation of an object by use of two-beam interferometric fringes,[8] arrangement of microparticles according to intensity distribution by multiple-beam interference,[9] rotation by a beam with a helical wave-front structure,[10] and manipulation of low-index particles by a single dark optical vortex laser beam.[11]

In conventional systems, including the above examples, optical manipulation is usually performed with a single light source. Optical components to generate desired light patterns for a specific manipulation are incorporated into the system with the light source and a reduction imaging system. When passive devices are used, complex manipulation is difficult because of poor flexibility in pattern generation. On the other hand, modulation of the light distribution by a rewritable spatial light modulator is an alternative method for flexible pattern generation.[12] However, high computational cost is required to calculate proper display patterns on the spatial light modulator, and unavoidable errors in the design and in the displayed pattern often disturb delicate manipulation for multiple objects. As a result, to achieve the desired function, the system tends to be complex and difficult to control. To overcome these problems, we present an optical manipulation method for microscopic objects that uses a vertical-cavity surface-emitting laser (VCSEL) array, which we refer to as VCSEL array trapping.

Emission intensities of individual VCSELs can be controlled independently with high frequency. Therefore, flexible manipulation for micro-objects is achieved by control of the spatial and temporal intensity distribution generated by the VCSEL array sources. VCSEL array trapping has many advantages. The VCSEL array can be easily combined with micro-optics. For example, a board-to-board free-space optical interconnect is implemented by the VCSEL array and microlenses without external relay optics.[13,14] These facts indicate that VCSEL array

When this study was done, Y. Ogura (ogura@gauss.ap.eng.osaka-u.ac.jp), K. Kagawa, and J. Tanida were with the Department of Material and Life Science, Graduate School of Engineering, Osaka University, 2-1 Yamadaoka, Suita, Osaka 565-0871, Japan. K. Kagawa is now with the Graduate School of Material Science, Nara Institute of Science and Technology, 8916-5 Takayamacho, Ikoma, Nara 630-0101, Japan.

Received 13 March 2001; revised manuscript received 25 June 2001.

0003-6935/01/305430-06$15.00/0

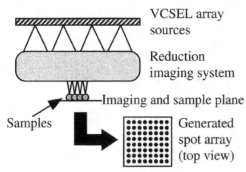

Fig. 1. Schematic diagram of VCSEL array trapping.

trapping has sufficient capability to reduce hardware complexity. Because no controls other than switching of the VCSEL array are required, additional devices for a specific function are not necessary and a troublesome control method is avoided. Various kinds of manipulation can be implemented by the same system configuration. Therefore, we expect to explore new applications for VCSEL array trapping.

The purpose of our study was to verify the capabilities of VCSEL array trapping and to clarify its features. In Section 2 we describe the procedure and the experimental system of VCSEL array trapping. In Section 3 we use VCSEL array trapping to demonstrate simultaneous manipulation of multiple particles, translation of a particle without mechanical movement, and position control of a particle by changing the emission intensities of multiple-beam trapping. In Section 4 we discuss the capability and applicability of the method for practical applications, and we provide our conclusions in Section 5.

2. Vertical-Cavity Surface-Emitting Laser Array Trapping

A tightly focused laser beam on a microscopic particle can be used to trap an object optically.[15] The particle receives a piconewton force induced by interaction with light such as absorption, reflection, and refraction. When the refractive index of the target particle is higher than that of the surrounding medium, the force pushes the particle toward the brighter illuminated region. As a result, the particle is drawn to the focused spot and can be translated if the spot is moved.

The trapping force depends on the illumination distribution on the particle. An example of the trapping capability enhancement is achieved by use of the Laguerre–Gaussian laser mode for axial trapping.[16] Another example is a single dark optical vortex laser beam to capture low-index particles.[11] As reported in these papers, flexibility in light pattern generation is an important requirement to enhance the performance and the functionality of the optical trap technique.

Use of parallel controllable beams is also effective as another solution for this problem. A schematic diagram of VCSEL array trapping is shown in Fig. 1. The VCSEL array is an array of semiconductor laser

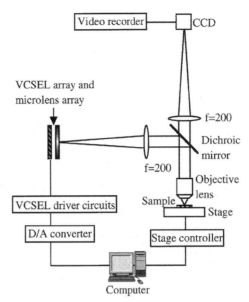

Fig. 2. Experimental setup of VCSEL array trapping. D/A, digital-to-analog.

sources arranged with high density on a substrate. Emission intensities of the individual pixels can be independently controlled by electronics. The maximum intensity of each pixel is several milliwatts. With the VCSEL array one can achieve a faster than megahertz intensity modulation. Therefore, in terms of scanning rate, enough performance can be obtained that is comparable with that from a galvanomirror or an acousto-optic deflector.

The VCSEL array sources are useful for the manipulation of multiple microscopic objects and are effective for function extension and simple system configuration, because special devices are not required to generate arbitrary spot array patterns, and the pattern can be switched by control of the emission intensities of the individual VCSELs.

Figure 2 illustrates the experimental setup of the VCSEL array trapping system. Because we aim to verify the basic capability of VCSEL array trapping, we used bulk lenses instead of micro-optics. Emission distribution of the VCSEL array and the position of the sample stage are controlled with a personal computer (Dell 800-MHz PentiumIII processor). The VCSEL array from NTT Photonics Laboratory has 8×8 pixels, an 854- $\pm$ 5-nm wavelength, greater than 3-mW maximum output, a 15-μm ϕ aperture, and a 250-μm pixel pitch. The micro lens array of the system has a 720-μm focal length and a 250-μm lens pitch and was set to increase the light efficiency. Each VCSEL is modulated independently with an analog signal by the driver circuit of our own composition. A simple voltage-to-current conversion circuit was used to drive the VCSEL. The voltage signal input to the driver circuit was generated by a digital-to-analog converter. The emission intensity can be modulated to 100 kHz by the circuit. The VCSEL is driven by a 500-Hz rectangular wave sig-

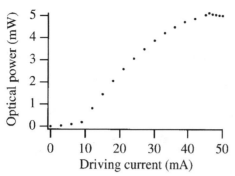

Fig. 3. Dependence of the emission intensity of the VCSEL on the driving current.

nal to prevent system overload. Dependence of the emission intensity measured by a photodetector placed close to the VCSEL on the driving current is shown in Fig. 3. The threshold current was approximately 10 mA, and the emission intensity increased to 5.1 mW with a driving current of 46 mA.

We used an immersible, long-distance objective lens (Olympus LUMPlan) with 60× focal length infrared radiation and 0.90 NA as the focusing lens. The VCSEL pixels were imaged onto the sample plane with a magnification of 1/67. Thus the optical spot pitch on the sample plane is 3.75 μm, and the maximum light intensity is approximately 1.1 mW. The sample plane was observed and recorded by an 8-mm video recorder. The sample objects were 6- and 10-μm-diameter polystyrene particles (Polysciences, Inc., Polybead Polystyrene Microspheres)

with 1.60 refractive index and 1.05-g/ml density that were mixed and dispersed in water. The sample was positioned on the sample stage that was controlled in 1-μm steps by a simple control method.

3. Experimental Results

The desired spot array patterns were generated by a combination of emitting pixels of the VCSEL array. When different pixels of the VCSEL array were assigned to individual objects, multiple objects were captured and translated simultaneously by control of the spot pattern. Two kinds of particle were two-dimensionally captured with two VCSEL pixels. Unfortunately, with this experimental configuration, we cannot observe three-dimensional trapping. We translated the particles by moving the stage. A series of six pictures at 10-s interval is shown in Fig. 4. The target objects were polystyrene particles, 6 μm (lower right) and 10 μm (upper left) in diameter, as indicated by circles. The ● designates the origin of the stage. The stage was moved to the left and then upward. Since the observation field is stationary, the trapped particles remain at the same position while the other particles move with the stage. The result indicates that the multiple objects are translated, with an average of 0.48-μm/s velocity for simultaneous translation.

One can achieve various kinds of manipulation by switching the emission pattern of the VCSEL array. As an example, the translation and the position of a stationary object are verified experimentally. Seven VCSEL pixels arranged in an L shape on the 8 × 8 VCSEL array were used for nonmechanical transla-

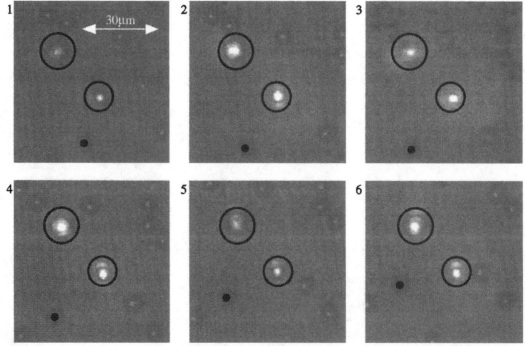

Fig. 4. Series of observed pictures at 10-s intervals during simultaneous capture and translation of two particles. The target particles are indicated by circles; the ● designates the origin of the stage.

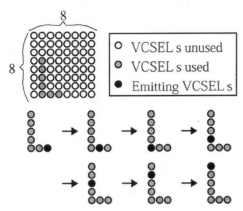

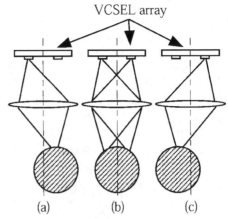

Fig. 5. VCSEL pixels that were used for nonmechanical translation (upper) and an emission sequence of the VCSELs (lower).

Fig. 7. Position control of particles by two VCSELs: (a) left, (b) both, (c) right. The crosshatched circles designate the target particle.

tion. The VCSEL pixels were turned on sequentially from the lower right to the upper left of the L shape as shown in Fig. 5. Figure 6 shows a series of six pictures at 10-s intervals. The target object was a 6-μm-diameter polystyrene particle, which is indicated by a circle; the ● designates the initial position of the target. The target particle shifts on the glass slide in response to the emission pixels of the VCSEL array. The average velocity for nonmechanical translation is 0.45 μm/s. No operation other than switching the emission pixels was carried out during the manipulation. We have verified that the method is capable of translating an object without mechanical movement.

Position control is based on the phenomenon that

an object is captured at the position where total radiation pressure is balanced, as shown in Fig. 7. The target particle is trapped at the bright spot when a single VCSEL array is emitted [Figs. 7(a) and 7(c)]. On the other hand, when both VCSEL arrays are emitted at equivalent intensities, the particle is captured at the midpoint of the spots [Fig. 7(b)]. If the intensities of the VCSEL arrays are varied, the particle moves according to the intensity ratio. A 10-μm-diameter polystyrene particle was illuminated by two adjacent pixels on the VCSEL array during the experiment. Figure 8 shows the response time of the horizontal position of the target particle (●) when

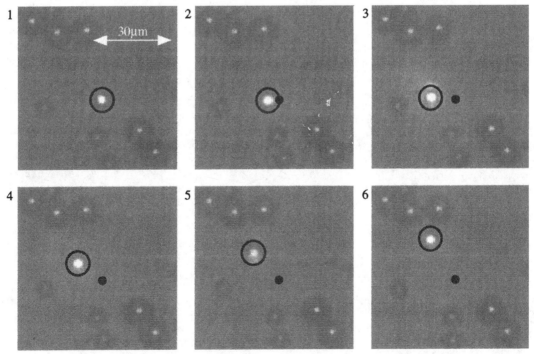

Fig. 6. Series of observed pictures at 10-s intervals in nonmechanical translation. The target particles are indicated by circles; the ● designates the initial position of the target.

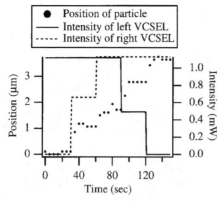

Fig. 8.   Time response of the particle position for modulation of two VCSELs.

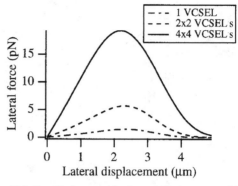

Fig. 9.   Relationship between the lateral displacement of the particle and the lateral force.   Coincidental illumination of the particles by solid curve, a single VCSEL; dashed curve, 2×2 VCSELs; dot–dash curve, 4×4 VCSELs.

the intensities of the pixels (solid and broken curves) were stepwise modulated.   The particle responds to the emission pattern of the VCSEL array, and the position to be trapped is changed according to the intensity ratio between the two pixels.   The result demonstrates that, with this method, the object position is controlled to within an accuracy that is equal to or less than 1 μm.

4.   Discussion

We used a high-power laser to obtain a strong trapping force in conventional optical trapping systems.   In contrast, the maximum emission intensity of a single VCSEL pixel is less than 10 mW at this stage.   In spite of weak intensity, our experiments demonstrate that an object of several or tens of micrometers diameter can be captured and translated two dimensionally with a single pixel on the VCSEL array.   This is a significant result for the proposed method because it suggests the potential capability for flexible manipulation with the setup of simple hardware.   Also, VCSEL array trapping is useful for simultaneous manipulation.   Note that the averaged trapping force given to the particle does not depend on the light intensity but on the total light energy.   Trapping forces for multiple particles can be applied continuously because scanning the light beam is not required.   Therefore, the light distribution for individual particles is not restricted by the scanning rate, and more flexible manipulation can be implemented than use of the time-averaged light distribution by fast beam scanning.[17]

However, the current experimental system has problems with translation speed and position accuracy, which are addressed at low light power and a coarse pitch of the optical spots at the sample plane.   We expect to obtain better performance of the VCSEL array trapping by improving the configuration of the optical system.

To solve the above problems, several techniques can be considered.   First, the assignment of multiple VCSEL pixels to an individual particle is a direct

method to increase the translation velocity.   When the particle is illuminated by N VCSELs, the total power of the trapping beams is N times as much as for a single VCSEL.   We can now estimate the trapping force.   We consider three cases of coincidental illumination of the particle: (1) a single VCSEL (current setup), (2) 2×2 VCSELs, and (3) 4×4 VCSELs.   The trapping force for the lateral direction is estimated by the calculation method based on a ray-optic model developed by Gauthier *et al.*[18]   The parameters used for the computer simulation were determined according to the experimental conditions; the target particle is 6 μm in diameter with 1.6 refractive index, the light power is 1.1 mW for a single VCSEL with a wavelength of 850 nm, and the individual beams have a Gaussian intensity profile with a beam waist of 2.6 μm (value measured with the current system).   The spot pitch was assumed to be 0.5 μm at the sample plane.   Figure 9 shows the calculation result of the relationship between the lateral displacement of a particle and the induced lateral force for all three cases.   The maximum forces increased 3.8 times in case (2) and 13.1 times in case (3) compared with that in case (1).   According to Stokes law, the required force for translating a particle in a fluid is proportional to the translation velocity.   We obtained a 0.45-μm/s translation velocity for nonmechanical translation, which corresponds to case (1).   We expect to achieve a translation velocity of 1.7 μm/s for case (2) and 5.9 μm/s for case (3).   In addition, by shrinking the spot pitch, we reduced the amount of wasted illumination power and consistently obtained a strong force.   Therefore, the average translation velocity can be significantly increased.

Based on the above results, we have determined that translation velocity can be increased to greater than several tens of micrometers per second by assigning multiple VCSELs to an individual particle.   If we use the same technique, the accuracy of the position control method can be improved.   For that reason, the illumination pattern can be switched based on position and motion of the target particle. An increase of the total illumination power is obviously effective for accurate control. If we combine the method for two-dimensional position control of a particle with the nonmechanical translation technique, the particle can travel over a wide area. Although improvement of the optical system is required, VCSEL array trapping can be applied to biological and other applications. VCSEL array trapping has several features that would make it useful for three-dimensional trapping: advanced rotation and oscillation control of an object, and manipulation of objects with complicated shapes.

5. Conclusion

We have presented optical manipulation of microscopic objects by means of VCSEL array sources. Simultaneous manipulation of multiple objects as well as nonmechanical translation and position control of an object have been experimentally demonstrated. Use of the VCSEL array provides an effective implementation technique for flexible manipulation with compact hardware and a simple control method. In addition, VCSEL array trapping opens the door to simultaneous manipulation for multiple objects because trapping forces for the objects can be applied continuously. Owing to versatility and simplicity, VCSEL array trapping can be used as a manipulation method for microscopic objects.

The VCSEL array was supplied by the NTT Photonics Laboratory through the System Photo-Electronics Consortium in Japan. The authors appreciate their support of our studies on optoelectronic application systems.

References

1. A. Ashkin, J. M. Dziedzic, J. E. Bjorkholm, and S. Chu, "Observation of a single-beam gradient force optical trap for dielectric particles," Opt. Lett. **11**, 288–290 (1986).
2. A. Ashkin and J. M. Dziedzic, "Optical trapping and manipulation of viruses and bacteria," Science **235**, 1517–1520 (1987).
3. A. Ashkin, J. M. Dziedzic, and T. Yamane, "Optical trapping and manipulation of single cells using infrared laser beams," Nature (London) **330**, 769–771 (1987).
4. S. C. Kuo and M. P. Sheetz, "Force of single kinesin molecules measured with optical tweezers," Science **260**, 232–234 (1993).
5. Y. Arai, R. Yasuda, K. Akashi, Y. Harada, H. Miyata, K. Kinosita, Jr., and H. Itoh, "Tying a molecular knot with optical tweezers," Nature (London) **399**, 446–448 (1999).
6. M. E. J. Friese, T. A. Nieminen, N. R. Heckenberg, and H. Rubinsztein-Dunlop, "Optical alignment and spinning of laser-trapped microscopic particles," Nature (London) **394**, 348–350 (1998).
7. Z. P. Luo, Y. L. Sun, and K. N. An, "An optical spin micromotor," Appl. Phys. Lett. **76**, 1779–1781 (2000).
8. A. E. Chiou, W. Wang, G. J. Sonek, J. Hong, and M. W. Berns, "Interferometric optical tweezers," Opt. Commun. **133**, 7–10 (1997).
9. M. M. Burns, J. M. Fournier, and J. A. Golovchenko, "Optical matter: crystallization and binding in intense optical fields," Science **249**, 749–754 (1990).
10. H. He, M. E. J. Friese, N. R. Heckenberg, and H. Rubinsztein-Dunlop, "Direct observation of transfer of angular momentum to absorptive particles from a laser beam with a phase singularity," Phys. Rev. Lett. **75**, 826–829 (1995).
11. K. T. Gahagan and G. A. Swartzlander, Jr., "Optical vortex trapping of particles," Opt. Lett. **21**, 827–829 (1996).
12. Y. Hayasaki, M. Itoh, T. Yatagai, and N. Nishida, "Nonmechanical optical manipulation of microparticle using spatial light modulator," Opt. Rev. **6**, 24–27 (1999).
13. F. B. McCormick, F. A. P. Tooley, T. J. Cloonan, J. M. Sasian, and H. S. Hinton, "Optical interconnections using microlens arrays," Opt. Quantum Electron. **24**, S465–S477 (1992).
14. E. M. Strzelecka, D. A. Louderback, B. J. Thibeault, G. B. Thompson, K. Bertilsson, and L. A. Coldren, "Parallel free-space optical interconnect based on arrays of vertical-cavity lasers and detectors with monolithic microlenses," Appl. Opt. **37**, 2811–2821 (1998).
15. S. M. Block, "Making light work with optical tweezers," Nature (London) **360**, 493–495 (1992).
16. N. B. Simpson, D. McGloin, K. Dholakia, L. Allen, and M. J. Padgett, "Optical tweezers with increased axial trapping efficiency," J. Mod. Opt. **45**, 1943–1949 (1998).
17. K. Sasaki, M. Koshioka, H. Misawa, N. Kitamura, and H. Masuhara, "Laser-scanning micromanipulation and spatial patterning of fine particles," Jpn. J. Appl. Phys. **30**, L907–L909 (1991).
18. R. C. Gauthier and S. Wallace, "Optical levitation of spheres: analytical development and numerical computations of the force equations," J. Opt. Soc. Am. B **12**, 1680–1686 (1995).

Optical tweezer arrays and optical substrates created with diffractive optics

Eric R. Dufresne and David G. Grier[a)]

The James Franck Institute and Department of Physics, The University of Chicago, 5640 S. Ellis Ave., Chicago, Illinois 60637

(Received 9 December 1997; accepted for publication 10 February 1998)

We describe a simple method for creating multiple optical tweezers from a single laser beam using diffractive optical elements. As a demonstration of this technique, we have implemented a 4×4 square array of optical tweezers—the hexadeca tweezer. Not only will diffractively generated optical tweezers facilitate many new experiments in pure and applied physics, but they also will be useful for fabricating nanocomposite materials and devices, including photonic bandgap materials and optical circuit elements. © *1998 American Institute of Physics.* [S0034-6748(98)02905-0]

I. INTRODUCTION

Since their introduction a decade ago,[1,2] optical tweezers have become indispensable tools for physical studies of macromolecular[3] and biological[4] systems. Formed by bringing a single laser beam to a tight focus, an optical tweezer exploits optical gradient forces to manipulate micrometer-sized particles. Optical tweezers have allowed scientists to probe the fantastically small forces which characterize the interactions of colloids,[5-9] polymers,[10-16] and membranes,[17-20] and to assemble small numbers of colloidal particles into mesoscopic structures.[21,22] These pioneering studies each required only one or two optical tweezers. Extending their techniques to larger and more complex systems will require larger and more complex arrays of optical tweezers.

We describe a simple and effective means to create multiple optical tweezers in arbitrary patterns from a single laser beam using diffractive optical elements. These tweezer arrays and their variants should have immediate applications for probing phenomena in biological systems and complex fluids, and in organizing soft matter into mesoscopically textured composite materials.

II. OPTICAL TWEEZER ARRAYS

The optical forces generated by a milliwatt of visible light are more than enough to overwhelm the random thermal forces which drive the dynamics of microparticles. The goal in creating an optical tweezer is to direct the optical forces from a single laser beam to trap a particle in all three dimensions. While quite general formulations of this problem have been developed,[23-26] a simplified discussion suffices to motivate the design of optical tweezer arrays. We will consider the forces exerted by monochromatic light of wavenumber k on a dielectric sphere of radius a in the Rayleigh limit, where $a \ll 2\pi/k$. The total optical force, $\mathbf{F}$, is the sum of two contributions:[27]

$$\mathbf{F} = \mathbf{F}_\nabla + \mathbf{F}_s, \qquad (1)$$

the first of which arises from gradients in the light's intensity and the second of which is due to scattering of light by the particle. The gradient force on a particle of dielectric constant ϵ immersed in a medium of dielectric constant ϵ_0 and subjected to an optical field with Poynting vector $\mathbf{S}$,

$$\mathbf{F}_\nabla = 2\pi a^3 \frac{\sqrt{\epsilon_0}}{c} \left(\frac{\epsilon - \epsilon_0}{\epsilon + 2\epsilon_0} \right) \nabla |\mathbf{S}|, \qquad (2)$$

tends to draw the particle toward the region of highest intensity. The scattering force,

$$\mathbf{F}_s = \frac{8}{3} \pi (ka)^4 a^2 \frac{\sqrt{\epsilon_0}}{c} \left(\frac{\epsilon - \epsilon_0}{\epsilon + 2\epsilon_0} \right)^2 \mathbf{S}, \qquad (3)$$

drives the particle along the direction of propagation of the light.

An optical tweezer can be formed by focusing a laser beam to a diffraction limited spot with a high numerical aperture lens. The gradient force attracts the particle to the beam's focal point, while the scattering force drives the particle along the beam's axis. In order to form a full three dimensional trap, the axial intensity gradient must be large enough to overcome the scattering force.

In a typical experimental setup, a microscope objective lens focuses a Gaussian TEM_{00} laser beam into a tweezer while simultaneously imaging the trapped particles. In order to maximize the axial intensity gradient, the incident beam should be expanded to fill the objective's back aperture. An optical tweezer can be translated across the microscope's focal plane by adjusting the beam's angle of incidence at the back aperture and can be displaced along the optical axis by changing the curvature of its wavefront.

Similarly, if several collimated beams pass through the back aperture at different angles, they form separate tweezers at different locations in the focal plane. For instance, dual optical tweezer systems have been created by splitting and recombining a single laser beam with beamsplitters and refractive optics.[21,28] This approach, however, becomes cumbersome for more than a few traps. An elegant alternative for creating two-dimensional arrays of traps involves scanning a single tweezer rapidly among a number of positions to create

[a)]Electronic mail: grier@fafnir.uchicago.edu

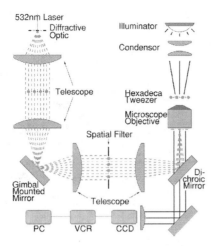

FIG. 1. Schematic diagram of a practical diffractively generated optical tweezer array.

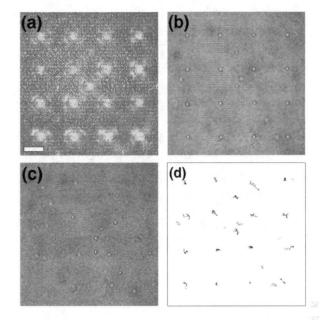

FIG. 2. 4×4 Optical tweezer array created from a single laser beam using a holographic array generator. (a) The tweezer array illuminates the 70 ×70 μm^2 field of view with light backscattered from trapped silica particles. The scale bar represents 10 μm. (b) The particle array 1/30 s after being released. (c) The same field of view 3.1 s later. (d) Trajectories of the particles in the field of view after being released. Dark traces indicate particles initially trapped in the array. Shorter tracks indicate particles which diffused out of the ±200 nm depth of focus.

a time-averaged extended trapping pattern.[21,29–31] This approach has proved highly effective for shaping small colloidal assemblies.

Yet another alternative for creating two dimensional arrays of traps is to split and steer the light from a single beam with diffractive optical elements. As demonstrated in the next section, inexpensive commercially available diffractive pattern generators (e.g. MEMS Optical Inc., Huntsville AL) are ideally suited to this task. In addition, diffractive optics can be used to change the curvature of beam wavefronts, thereby facilitating the creation of three-dimensional arrays of traps. Additionally, diffractive optics can be used to modify beam profiles. For instance, computed holograms can convert a TEM$_{00}$ Gaussian beam into multiple Gauss–Laguerre beams, otherwise known as optical vortices.[32–34] Tightly focused optical vortices can be used to trap low dielectric constant and reflective microparticles. Finally, the use of addressable liquid crystal phase shifting arrays allows for the dynamic reconfiguration of a tweezer array for active particle manipulation and assembly. Thus, diffractively generated optical fields can be used to configure arbitrary numbers of microscopic particles into useful and interesting arrangements. We will discuss some applications of these techniques after describing a practical demonstration of diffractively generated optical tweezer arrays.

III. THE HEXADECA TWEEZER

The schematic diagram in Fig. 1 shows one implementation of diffractively generated optical tweezer arrays. This design was used to create a 4×4 square array of tweezers—the hexadeca tweezer. The optical tweezer array is powered by a 100 mW diode-pumped frequency-doubled Nd:YAG laser operating at 532 nm. Two Keplerian telescopes arranged in series produce two planes conjugate to the back aperture of the microscope's objective lens (100×, N.A. 1.40). The laser beam passes through eyepoints in both planes to create a conventional single optical tweezer. Introducing a diffractive 4×4 square array generator (Edmund Scientific No. P53191, angular divergence 25 mrad at λ = 532 nm) at the first eyepoint creates the pattern of rays

desired for the hexadeca tweezer. A gimbal mounted mirror centered at the second eyepoint allows us to translate the entire tweezer array in the microscope's field of view and to dynamically stiffen the tweezers. A spatial filter placed in the inner focal plane of either telescope removes spurious rays created by imperfections in the low-cost diffractive optic. Conventional white-light illumination is used to form an image of the trapped particles. The image passes through a dichroic beam splitter and is captured with a video camera (NEC TI-324A) and recorded with a VCR (NEC PC-VCR) for later analysis.

For our demonstration, we used the hexadeca tweezer to trap silica spheres (a = 0.50±0.03 μm, ϵ = 2.3, Duke Scientific, Palo Alto CA, Cat. No. 8100) suspended in deionized water (ϵ_0 = 1.7) at room temperature. The suspension was sandwiched between a microscope slide and cover glass separated by 40 μm. The tweezer array was focused 8 μm above the lower glass wall and left in place to acquire particles. Figure 2(a) shows all 16 of the primary optical traps filled with particles. Since the focal plane is several sphere diameters from the nearest wall, each sphere is trapped stably in three dimensions. Residence times in excess of 100 s, suggest a trapping potential deeper than $6k_BT$ per particle given the free spheres' self-diffusion coefficient of 0.4 μm^2/s. Comparable trapping efficiency would be expected for a single conventional optical tweezer operating at equivalent light intensity. Two additional spheres are trapped less strongly at spurious peaks in the diffraction pattern.

Figure 2(b) shows the same field of view 1/30 s after the laser was interrupted and before the spheres have had time to diffuse away. After 3.1 s [Fig. 2(c)], the pattern has completely dispersed with some of the spheres wandering out of the imaging volume altogether. Figure 2(d) shows two-dimensional projections of the particles' trajectories over this period.[6] Studying the dynamic relaxation of artificially structured colloidal crystals is just one application we foresee for manipulating soft matter with holographically patterned light.

IV. OPTICAL SUBSTRATES AND APPLICATIONS

Not all holographically generated optical fields will create arrays of traps. A large uniformly illuminated volume, for example, will not have the intensity gradients required for three-dimensional trapping. Such nontrapping patterns also have applications, however, and can be created in the same manner as the hexadeca tweezer, with diffractive optics. The resulting optical fields may be useful for forcing particles against a surface and into desired configurations.

Optically induced ordering has been demonstrated by intersecting several discrete beams.[35–42] The resulting periodic intensity patterns are observed to induce order in colloidal monolayers. In this respect, the pattern of light plays the role of a modulated substrate potential and affects the phase behavior of the illuminated two-dimensional system. This effect is believed to be closely analogous to the influence of atomic corrugation on the phase transitions of adsorbed atomic and molecular overlayers.[43,44] The optical substrate's symmetry, periodicity, and depth of modulation all can be adjusted experimentally. This system therefore has great promise for exploring the mechanisms of surface phase transitions. Diffractively generated optical substrates, moreover, need not be periodic. Quasiperiodic and aperiodic patterns will be useful for studies of the effect of pinning on monolayer dynamics, a potentially powerful analog for dynamics in superconducting vortex lattices and sliding charge density waves.

Beyond these applications to studies in fundamental condensed matter physics, optical substrates should be useful for assembling composite systems textured on the micron scale. In this case, the patterned illumination acts as a template for depositing particles directly or assisting self-assembly. The resulting mesoscopically arranged structures could be gelled in place and combined to create extended structures and devices. The fabrication of optical circuit elements might be based on such an approach.

Both tweezing arrays and optical substrates can be used to direct the self-assembly of three-dimensionally ordered colloidal crystals. These crystals recently have been shown to have a promising future as photonic circuit elements[45,46] and as quantitative chemical sensors.[47,48] Their full commercial exploitation will require the ability to fabricate large single crystals with desirable symmetry and lattice constants. Diffractive optical arrays can be used to organize the first layer of a growing colloidal crystal and register it with a desired substrate. Subsequent layers then grow epitaxially on the first, leading to much longer-ranged three-dimensional order than might be possible otherwise. This principle has been demonstrated with lithographically patterned substrates.[49] Optical substrates could be adjusted *in situ* to achieve optimal ordering.

Multilayer tweezer arrays could be used to directly form multilayer photonic crystals such as photonic bandgap materials,[50] with tuned defect states in the bandgap created with planned interstitials. The high-dielectric-constant materials required for realizing a full photonic bandgap most likely cannot be trapped with conventional tweezer arrays. Arrays of optical vortex tweezers, however, should be amenable to the task.

Finally, scanned and otherwise time-dependent optical substrates can provide the time-modulated spatially asymmetric potentials required for practical particle size fractionation through directed diffusion.[51] The optical ratchet principle required for practical directed diffusion has been demonstrated with a single scanned tweezer.[52–54] Extended optical substrates may turn this demonstration into a practical technique.

ACKNOWLEDGMENTS

This work was supported primarily by the MRSEC Program of the National Science Foundation under Award No. DMR-9400379. Additional support was provided by the National Science Foundation under Grant No. DMR-9320378. Mr. Dufresne was supported by a GAANN Fellowship of the Department of Education under Award No. P200A-10052.

[1] A. Ashkin, J. M. Dziedzic, J. E. Bjorkholm, and S. Chu, Opt. Lett. **11**, 288 (1986).

[2] A. Ashkin, Proc. Natl. Acad. Sci. USA **94**, 4853 (1997).

[3] D. G. Grier, Curr. Opin. Colloid Interface Sci. **2**, 264 (1997).

[4] K. Svoboda and S. M. Block, Annu. Rev. Biophys. Biomol. Struct. **23**, 247 (1994).

[5] J. C. Crocker and D. G. Grier, Phys. Rev. Lett. **73**, 352 (1994).

[6] J. C. Crocker and D. G. Grier, J. Colloid Interface Sci. **179**, 298 (1996).

[7] A. D. Dinsmore, A. G. Yodh, and D. J. Pine, Nature (London) **383**, 239 (1996).

[8] A. E. Larsen and D. G. Grier, Nature (London) **385**, 230 (1997).

[9] T. Sugimoto *et al.*, Langmuir **13**, 5528 (1997).

[10] S. B. Smith, L. Finzi, and C. Bustamante, Science **258**, 1122 (1992).

[11] C. Bustamante, J. F. Marko, E. D. Siggia, and S. Smith, Science **265**, 1599 (1994).

[12] H. Yin *et al.*, Science **270**, 1653 (1995).

[13] T. T. Perkins, D. E. Smith, R. G. Larson, and S. Chu, Science **268**, 83 (1995).

[14] S. B. Smith, Y. Cui, and C. Bustamante, Science **271**, 795 (1996).

[15] H. Felgner, R. Frank, and M. Schliwa, J. Cell. Sci. **109**, 509 (1996).

[16] M. D. Wang *et al.*, Biophys. J. **72**, 1335 (1997).

[17] R. Bar-Ziv, Phys. Rev. Lett. **73**, 1392 (1994).

[18] R. Bar-Ziv, R. Menes, E. Moses, and S. A. Safran, Phys. Rev. Lett. **75**, 3356 (1995).

[19] R. Bar-Ziv, T. Frisch, and E. Moses, Phys. Rev. Lett. **75**, 3481 (1995).

[20] J. D. Moroz, P. Nelson, R. Bar-Ziv, and E. Moses, Phys. Rev. Lett. **78**, 386 (1997).

[21] K. Sasaki *et al.*, Opt. Lett. **16**, 1463 (1991).

[22] H. Misawa *et al.*, Appl. Phys. Lett. **60**, 310 (1992).

[23] G. Gouesbet, B. Maheu, and G. Gréhan, J. Opt. Soc. Am. A **5**, 1427 (1988).

[24] A. Ashkin, Biophys. J. **61**, 569 (1992).

[25] K. F. Ren, G. Gréhan, and G. Gouesbet, Appl. Opt. **35**, 2702 (1996).

[26] Ø. Farsund and B. U. Felderhof, Physica A **227**, 108 (1996).

[27] Y. Harada and T. Asakura, Opt. Commun. **124**, 529 (1996).

[28] E. Fällman and O. Axner, Appl. Opt. **36**, 2107 (1997).

[29] L. P. Faucheux and A. J. Libchaber, Phys. Rev. E **49**, 5158 (1994).

[30] K. Visscher, S. P. Gross, and S. M. Block, IEEE J. Sel. Top. Quantum Electron. **2**, 1066 (1996).

[31] K. Sasaki *et al.*, Jpn. J. Appl. Phys. **30**, L907 (1997).

[32] N. R. Heckenberg *et al.*, Opt. Quantum Electron. **24**, S951 (1992).

[33] H. He, N. R. Heckenberg, and H. Rubinsztein-Dunlop, J. Mod. Opt. **42**, 217 (1995).

[34] K. T. Gahagan and J. G. A. Swartzlander, Opt. Lett. **21**, 827 (1996).

[35] A. Chowdhury, B. J. Ackerson, and N. A. Clark, Phys. Rev. Lett. **55**, 833 (1985).

[36] B. J. Ackerson and A. H. Chowdhury, Faraday Discuss. Chem. Soc. **83**, 1 (1987).

[37] K. Loudiyi and B. J. Ackerson, Physica A **184**, 1 (1992).

[38] K. Loudiyi and B. J. Ackerson, Physica A **184**, 26 (1992).

[39] M. M. Burns, J.-M. Fournier, and J. A. Golovchenko, Science **249**, 749 (1990).

[40] A. Chowdhury, B. J. Ackerson, and N. A. Clark, Science **252**, 1049 (1991).

[41] M. M. Burns, J.-M. Fournier, and J. A. Golovchenko, Science **252**, 1049 (1991).

[42] A. E. Chiou *et al.*, Opt. Commun. **133**, 7 (1997).

[43] J. Chakrabarti, H. R. Krishnamurthy, and A. K. Sood, Phys. Rev. Lett. **73**, 2923 (1994).

[44] J. Chakrabarti, H. R. Krishnamurthy, A. K. Sood, and S. Sengupta, Phys. Rev. Lett. **75**, 2232 (1995).

[45] G. Pan, R. Kesavamoorthy, and S. A. Asher, Phys. Rev. Lett. **78**, 3860 (1997).

[46] D. G. Grier, Phys. World **10**, 24 (1997).

[47] J. H. Holtz and S. A. Asher, Nature (London) **389**, 829 (1997).

[48] D. G. Grier, Nature (London) **389**, 784 (1997).

[49] A. van Blaaderen, R. Ruel, and P. Wiltzius, Nature (London) **385**, 321 (1997).

[50] J. D. Joannopoulos, R. D. Meade, and J. N. Winn, *Photonic Crystals* (Princeton University Press, Princeton, 1995).

[51] M. Bier and R. D. Astumian, Phys. Rev. Lett. **76**, 4277 (1996).

[52] L. P. Faucheux, L. S. Bourdieu, P. D. Kaplan, and A. J. Libchaber, Phys. Rev. Lett. **74**, 1504 (1995).

[53] L. P. Faucheux, G. Stolovitzky, and A. Libchaber, Phys. Rev. E **51**, 5239 (1995).

[54] R. C. Gauthier, Appl. Phys. Lett. **67**, 2269 (1995).

Multi-functional optical tweezers using computer-generated holograms

J. Liesener *, M. Reicherter, T. Haist, H.J. Tiziani

Institut für Technische Optik, Universität Stuttgart, Pfaffenwaldring 9, D-70569 Stuttgart, Germany

Received 17 July 2000; accepted 6 September 2000

Abstract

Optical tweezers are capable of trapping microscopic particles by photon momentum transfer. The use of dynamic computer-generated holograms for beam shaping allows a high flexibility in terms of trap characteristics and features. We use a liquid crystal display (LCD) to display the holograms. Efficiency losses caused by the periodic electrode structure of the LCD have been clearly reduced by use of an optically addressed spatial light modulator. We realized multiple traps, which can hold and move at least seven silica spheres independently in real time. We also demonstrate the controllability of trapped particles in three dimensions without the need for mechanical elements in the setup. © 2000 Elsevier Science B.V. All rights reserved.

1. Introduction

The ability to manipulate micrometer-sized particles with laser beams has led to applications, mainly in biological fields. Early works of Ashkin et al. in 1970 [1] and in the 1980s [2] served as seeds for this technique now known as optical tweezers. Therein ray optics and photon momentum considerations were used to show that, roughly speaking, a high-index particle ($n_{particle} > n_{medium}$) is attracted by the beam focus if the parameters are chosen appropriately.

In experimental works of several authors [3–8] different setups have been realized showing the usability of optical tweezers for manipulation and investigation purposes. Normally a trapped particle is moved by motion of a microscope stage. Alternatively adjustable mirrors or accousto-optical modulators can be used to steer the beam and thus the trapped particle [9]. Such setups become quite complicated if three-dimensional steering or multiple trapping is desired [10].

In Ref. [11] we have presented a holographic tweezer setup in which computer-generated holograms written on a liquid crystal display (LCD) have been used to control the number, positions, and shapes of optical traps in two dimensions. The basic setup is briefly explained in Section 2. We extend this method to the full three-dimensional manipulation of multiple objects as shown in Section 3. In Section 4 we present a modified setup with an optically and an electrically addressed LCD in order to improve the diffraction efficiency considerably and hence much more laser power is available for the hologram reconstruction.

---

* Corresponding author. Fax: +49-711-685-6586.

E-mail address: liesener@ito.uni-stuttgart.de (J. Liesener).

2. Basic tweezer setup

The basic experimental setup of the tweezer is shown in Fig. 1. Its central element is the LCD which is controlled by a personal computer. It displays the Fourier holograms which are read out by a collimated 1 W Ar$^+$-laser (Spectra Physics 165) at a wavelength of 488 nm. The beam is then coupled into the microscope setup by a dichroic mirror (DM). The beam diameter is reduced to fit the 2 mm aperture of the water immersion microscope objective, MO (Zeiss Achroplan 100×/1.0 W). The hologram reconstruction in the focal plane of the MO forms the trap. This is where the particles are trapped for manipulation. The illumination from below with a white light source is used for imaging the particles onto the CCD camera (Sony XC–55). The strong backscattered laser light is filtered out by an interference filter.

The twisted nematic LCD (Epson) with VGA resolution (640 × 480 pixels) was removed from a data projector (InFocus LitePro 580). It has a fill factor of 44% and a pixel pitch of 42 μm. The active area has a size of 26.9 × 20.2 mm^2. By removing the polarizers on both sides of the LCD panel the LCD no longer acts as an amplitude modulator, as originally desired. The phase shifting properties can be optimized for maximum diffraction efficiency by the choice of the input polarization. A phase shift of 1.6π can be accomplished at 488 nm by the LCD used.

The computation of Fourier holograms is straightforward for manipulation in two dimensions. In this case the reconstruction is just the Fourier transform of the hologram. Therefore a central spot, capable of trapping a high index particle, is generated if the hologram consists of a uniform phase Φ. A lateral displacement of this trap can be achieved if the hologram is chosen to be a blazed grating:

$$\Phi(x,y) = \left(\frac{2\pi}{\Lambda_x}x + \frac{2\pi}{\Lambda_y}y \right) \bmod 2\pi \qquad (1)$$

Λ_x and Λ_y are the fringe periods in the x and y direction. For light modulators with a maximum phase shift below 2π, as in our case, the efficiency is reduced and part of the intensity remains in the central (zeroth order) spot. Multiple traps can be realized by summing up the complex functions $e^{i\Phi_j(x,y)}$ and then calculating the argument of that complex function:

$$\Phi(x,y) = \arg\left(\sum_j e^{i\Phi_j(x,y)} \right) \qquad (2)$$

By considering only the phase of the complex sum, trap intensity variations can occur. The behavior is complex and not yet studied thoroughly by us. However, these variations did not appear hindering in our experiments.

We also employed more complex fields like doughnuts [11] which have advantages in some applications. The keyboard is used to control the tweezer features like the number of traps, their positions, and their shapes.

Fig. 2 shows the simultaneous trapping of seven 1 μm silica spheres which have been precisely arranged in a V-shape. Each of the trapped particles can be moved independently by the user. Among the spheres two appear larger because the corresponding traps hold two particles on top of each other. The computation of one hologram with a personal computer (Pentium III 600) takes less than 1 s for seven simultaneous traps.

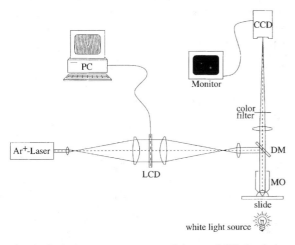

Fig. 1. Optical tweezer setup containing an LCD for holographic beam shaping.

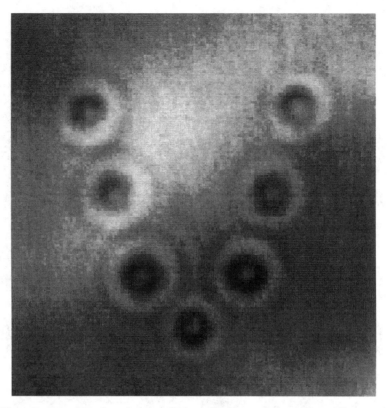

Fig. 2. Seven silica spheres, positioned precisely in a V-shape. The 1-μm-sized particles can be moved independently.

3. Three-dimensional trapping

By no means is one limited to a manipulation in just one plane. By adding a lens term to the blazed gratings the beam focus can be shifted up and down parallel to the optical axis:

$$\Phi(x,y) = \left(\frac{2\pi}{\Lambda_x}x + \frac{2\pi}{\Lambda_y}y + \Gamma(x^2 + y^2) \right) \bmod 2\pi \quad (3)$$

Γ controls the axial position of the trap. The appearance of a typical hologram is depicted in Fig. 3. It is a hologram for three laterally and axially displaced traps which fills the entire LCD panel.

Fig. 4 shows an experiment that demonstrates the controllability in three dimensions. To our knowledge this is the first time that full three-dimensional particle manipulation with a single beam optical trap was accomplished without moving parts.

Two equivalent silica spheres, again of 1 μm diameter, are trapped. By changing the hologram, one of the spheres is pushed out of the observation plane and therefore appears blurred and bigger. After being moved to the left the sphere is shifted back into the observation plane on the other side of the stationary sphere.

4. Improved setup using an optically addressed liquid crystal display

For the practical use of optical tweezers and holographic applications in general it is often desirable to have maximum laser power available in the hologram reconstruction. The setup of Fig. 1, however, leads to some loss of laser power caused by the LCD structure.

The loss can be traced back to two major limiting factors: First, about 56% of the laser power is

Fig. 3. Phase hologram for three laterally and axially displaced traps.

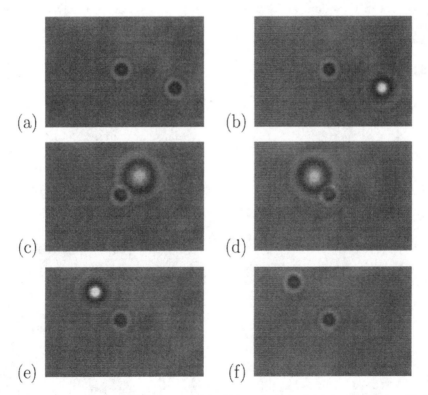

(a)

(b)

(c)

(d)

(e)

(f)

Fig. 4. Manipulation in three dimensions. One of the 1 μm particles (a) is pushed out of the observation plane (b). It is passed over to the other side of the the stationary particle (c–e) and shifted back into the observation plane (f).

blocked by the mask structure surrounding each pixel which also defines the fill factor of 44%. Second, the periodic structure of the LCD acts as an optical grating which deflects 56% of the remaining power into side orders. Only the central diffraction order of the LCD can be used for the hologram reconstruction. Considering further reflection and absorption losses, only 15% of the original laser power is available for the hologram reconstruction. Those are general limitations whenever LCDs are used in holography.

We overcome the described losses for the most part by use of an additional optically addressed liquid crystal display (OALCD). The basic optical tweezer setup (Fig. 1) is modified by replacing the LCD by a low-pass filter setup, as shown in Fig. 5.

The OALCD (Jenoptik SLM-O30) has one planar-nematic liquid crystal layer of 30 mm in diameter. A dielectric mirror between the liquid crystal and the photoconductor layer prevents the read light from addressing the photoconductor and allows a read out in reflection. A 2π phase shift of the reading light at 488 nm has been achieved with a writing light intensity of 0.5 mW/cm^2 at 633 nm. A resolution of 25 line pairs/mm can be accomplished with the element. As before, an Ar$^+$-laser at 488 nm was used to read out the hologram information, this time however in reflection. A HeNe laser at 633 nm illuminates the LCD.

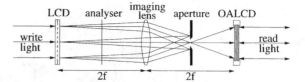

Fig. 6. The hologram information is transferred from the LCD to the OALCD. The phase disribution is first displayed by the LCD, encoded as an intensity distribution. After a low-pass filtering process, the hologram information is converted back into a phase distribution by the OALCD.

Fig. 6 illustrates how the hologram information is transferred from the LCD to the OALCD in a 4f setup. The coherent imaging system contains an aperture for low-pass filtering purposes, located in the Fourier plane of the imaging lens.

The diffraction orders due to the LCD pixel structure appear as equally spaced copies of the Fourier-transformed hologram in the Fourier plane of the imaging lens. By blocking those side orders, the information about the LCD pixel structure gets completely lost, leaving the hologram information only.

Fig. 7 demonstrates how the LCD pixel structure of the write light gradually vanishes when the aperture diameter is reduced. In this example a blazed grating has been written on the LCD. Diffraction loss is reduced because the resulting hologram – displayed on the OALCD – is not pixelated any more and additionally the effective fill factor now is 100%.

Minor sources of losses are the limited reflectivity of the broad-band dielectric mirror in the OALCD and some absorption loss in the ITO layer [12]. The OALCD reflects 85% of the reading light at 488 nm.

In order to quantify the improvement obtained by the OALCD, we measured the diffraction efficiency, defined as the ratio of the power in the first diffraction order of a blazed grating to the input laser power. For the unmodified setup it is 9.5%. This reduced value as opposed to the 15% from above is due to the imperfect polarization and phase shift properties of the LCD. With the combination of the electrically and the optically addressed LCD we considerably improved diffraction efficiency by a factor of nearly six to 53%.

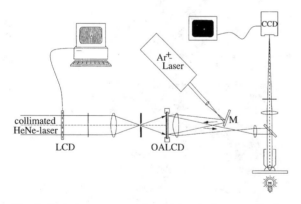

Fig. 5. Optical tweezer setup containing the low-pass filtering setup.

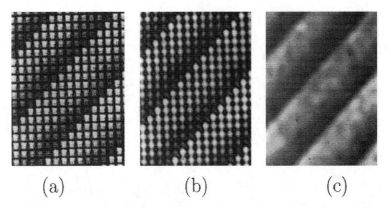

Fig. 7. The effect of low-pass filtering. (a) The aperture is wide open, the LCD pixel structure is clearly visible. (b) The central five diffraction orders pass the partly closed aperture, the pixel structure appears blurred. (c) Only the central diffraction order passes the aperture, the pixel structure completely vanishes.

5. Conclusion

The versatility of optical tweezers using computer-generated holograms displayed on an LCD has been demonstrated. The technique qualifies for a wide range of applications. We demonstrated the ability to trap and manipulate multiple particles independently in three dimensions without mechanical elements. In terms of hologram reconstruction efficiency a clear improvement has been accomplished by use of an additional, OA-LCD. The efficiency was raised by a factor of almost 6 compared to a single electrically addressed LCD.

Acknowledgements

We thank the Deutsche Forschungsgemeinschaft for the financial support.

References

[1] A. Ashkin, Phys. Rev. Lett. 24 (4) (1970) 156–159.
[2] A. Ashkin, J.M. Dziedzic, J.E. Bjorkholm, S. Chu, Opt. Lett. 11 (1986) 288–290.
[3] Y. Tadir, W.H. Wright, O. Vafa, T. Ord, R.H. Asch, M.W. Berns, Fertil. Steril. 53 (1990) 944–947.
[4] M.D. Wang, H. Yin, R. Landick, J. Gelles, S.M. Block, Biophys. J. 72 (1997) 1335–1346.
[5] S.C. Kuo, M.P. Sheetz, Science 260 (1993) 232–234.
[6] P.C. Mogensen, J. Glückstad, Opt. Commun. 175 (2000) 75–81.
[7] P.J.H. Bronkhorst, G.J. Steekstra, J. Grimbergen, E.J. Nijhof, J.J. Sixma, G.J. Brakenhoff, Biophys. J. 69 (1995) 1666–1673.
[8] K.T. Gahagan, G.A. Swartzlander, Opt. Lett. 21 (1996) 827–829.
[9] R.M. Simmons, J.T. Finer, S. Chu, J. Spudich, Biophys. J. 70 (4) (1996) 1813–1822.
[10] E. Fällman, O. Axner, Appl. Opt. 36 (10) (1997) 2107–2113.
[11] M. Reicherter, T. Haist, E.U. Wagemann, H.J. Tiziani, Opt. Lett. 24 (1999) 608–610.
[12] M. Gilo, R. Dahan, N. Croitoru, Opt. Eng. 38 (1999) 953–957.

Dynamic holographic optical tweezers

Jennifer E. Curtis, Brian A. Koss, David G. Grier *

James Franck Institute and Institute for Biophysical Dynamics, University of Chicago, 5640 S. Ellis Ave., Chicago, IL 60637, USA

Received 29 March 2002; received in revised form 18 April 2002; accepted 29 April 2002

Abstract

Optical trapping is an increasingly important technique for controlling and probing matter at length scales ranging from nanometers to millimeters. This paper describes methods for creating large numbers of high-quality optical traps in arbitrary three-dimensional configurations and for dynamically reconfiguring them under computer control. In addition to forming conventional optical tweezers, these methods also can sculpt the wavefront of each trap individually, allowing for mixed arrays of traps based on different modes of light, including optical vortices, axial line traps, optical bottles and optical rotators. The ability to establish large numbers of individually structured optical traps and to move them independently in three dimensions promises exciting new opportunities for research, engineering, diagnostics, and manufacturing at mesoscopic lengthscales. © 2002 Elsevier Science B.V. All rights reserved.

PACS: 42.40.Jv; 87.80.Cc

An optical tweezer uses forces exerted by intensity gradients in a strongly focused beam of light to trap and move a microscopic volume of matter [1]. Optical tweezers' unique ability to manipulate matter at mesoscopic scales has led to widespread applications in biology [2,3], and the physical sciences [4]. This paper describes how computer-generated holograms can transform a single laser beam into hundreds of independent optical traps, each with individually specified characteristics, arranged in arbitrary three-dimensional configurations. The enhanced capabilities of such dynamic holographic optical trapping

systems offer new opportunities for research and engineering, as well as new applications in biotechnology, nanotechnology, and manufacturing.

Holographic optical tweezers (HOT) use a computer-designed diffractive optical element (DOE) to split a single collimated laser beam into several separate beams, each of which is focused into an optical tweezer by a strongly converging lens [5–7]. Originally demonstrated with microfabricated DOEs [8], holographic optical tweezers have since been implemented with computer-addressed liquid crystal spatial light modulators [9,10]. Projecting a sequence of computer-designed holograms reconfigures the resulting pattern of traps. Unfortunately, calculating the phase hologram for a desired pattern of traps is not straightforward, and the lack of appropriate algorithms has prevented dynamic holographic op-

---

* Corresponding author. Tel.: +1-773-702-9176; fax: +1-773-702-5863.

E-mail address: d-grier@uchicago.edu (D.G. Grier).

tical tweezers from achieving their potential. This paper introduces new methods for computing phase holograms for optical trapping and demonstrates their use in a practical dynamic holographic optical trapping system.

The same optical gradient forces exploited in conventional optical tweezers [1] also operate in holographic optical tweezers. A dielectric particle approaching a focused beam of light is polarized by the light's electric field and then drawn up intensity gradients toward the focal point. Radiation pressure competes with this optical gradient force and tends to displace the trapped particle along the beam's axis. For this reason, optical tweezers usually are designed around microscope objective lenses whose large numerical apertures and minimal aberrations optimize axial intensity gradients.

An optical trap can be placed anywhere within the objective lens' focal volume by appropriately selecting the input beam's propagation direction and degree of collimation. For example, a collimated beam passing straight into an infinity-corrected objective lens comes to a focus in the center of the lens' focal plane, while another beam entering at an angle comes to a focus proportionately off-center. A slightly diverging beam focuses downstream of the focal plane while a converging beam focuses upstream. By the same token, multiple beams simultaneously entering the lens' input pupil each form optical traps in the focal volume, each at a location determined by its angle of incidence and degree of collimation. This is the principle behind holographic optical tweezers.

Our implementation, shown schematically in Fig. 1, uses a Hamamatsu X7550 parallel-aligned nematic spatial light modulator (SLM) [11] to reshape the beam from a frequency-doubled Nd:YVO$_4$ laser (Coherent Verdi) into a designated pattern of beams. Each is transferred by relay optics to the entrance pupil of a $100\times$ NA 1.4 oil immersion objective mounted in a Zeiss Axiovert S100TV inverted optical microscope and then focused into a trap. A dichroic mirror reflects the laser light into the objective while allowing images of the trapped particles to pass through to a video camera. When combined with a $0.63\times$ widefield video eyepiece, this optical train offers a 85×63 μm^2 field of view.

The Hamamatsu SLM can impose selected phase shifts on the incident beam's wavefront at each 40 μm wide pixel in a 480×480 array. The SLM's calibrated phase transfer function offers 150 distinct phase shifts ranging from 0 to 2π at the operating wavelength of $\lambda = 532$ nm. The phase shift imposed at each pixel is specified through a computer interface with an effective refresh rate of 5 Hz for the entire array. Quite sophisticated trapping patterns are possible despite the SLM's inherently limited spatial bandwidth. The array of 400 functional optical traps shown in Fig. 1 is the largest created by any means. Improvements in the number and density of effective phase pixels, in their diffraction efficiency, in the resolution of the available phase modulation, and in the refresh rate for projecting new phase patterns will correspondingly improve the performance of dynamic holographic optical tweezer systems. Other phase modulating technologies, such as micromirror arrays could offer the additional benefit of creating optical traps in multiple wavelengths simultaneously.

Modulating only the phase and not the amplitude of the input beam is enough to establish any desired intensity pattern in the objective's focal volume and thus any pattern of traps [7]. Such intensity-shaping phase gratings are often referred to as kinoforms. Previously reported algorithms for computing optical trapping kinoforms produced only two-dimensional distributions of traps [7,9] or patterns on just two planes [10]. Moreover, the resulting traps were suitable only for dielectric particles in low-dielectric media, and could not be adapted to handle metallic particles or samples made of absorbing, reflecting, or low-dielectric-constant materials. A more general approach relaxes all of these restrictions.

We begin by modeling the incident laser beam's electric field $E_0(\vec{\rho}) = A_0(\vec{\rho}) \exp(i\psi)$, as having constant phase, $\psi = 0$ in the DOE plane, and unit intensity: $\int_\Omega |A_0(\vec{\rho})|^2 \, d^2\rho = 1$. Here $\vec{\rho}$ denotes a position in the DOEs aperture Ω, $A_0(\vec{\rho})$ is the real-valued amplitude profile of the input beam. The DOE then imposes a phase modulation $\varphi(\vec{\rho})$ onto the input beam's wavefront which, in principle, encodes the desired pattern of outgoing beams.

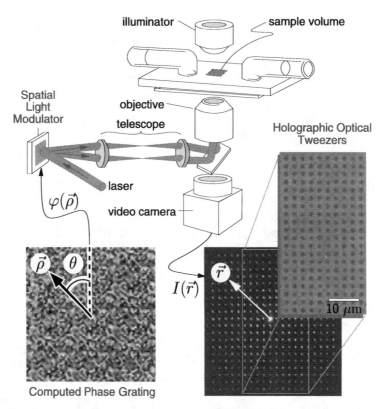

Fig. 1. Schematic implementation of dynamic holographic optical tweezers using a reflective liquid crystal spatial light modulator. The inset phase grating is 1/25 of the hologram $\varphi(\vec{\rho})$ encoding a 20×20 array of traps, with white regions corresponding to local phase shifts of 2π radians and black to 0. A telescope relays the diffracted beams to a high-numerical-aperture objective which focuses them into optical traps. The sample, enclosed in a glass flow cell, can be imaged through conventional video microscopy. The inset video micrograph shows the intensity $I(\vec{r}) = \sum_j |\epsilon_j|^2 \delta(\vec{r} - \vec{r}_j)$ of light from the traps reflected off a mirror temporarily placed in the object plane. The smaller inset shows these traps in action, holding two hundred colloidal polystyrene spheres, each 800 nm in diameter.

The electric field $\epsilon_j = \alpha_j \exp(\mathrm{i}\phi_j)$ at each of the discrete traps is related to the electric field $E(\vec{\rho})$ in the plane of the DOE by a generalized Fourier transform

$$E(\vec{\rho}) = \sum_{j=1}^{N} \int \epsilon_j \delta(\vec{r} - \vec{r}_j) K_j^{-1}(\vec{r}, \vec{\rho})$$

$$\times \exp\left(\mathrm{i}\frac{2\pi\vec{r} \cdot \vec{\rho}}{\lambda f}\right) \mathrm{d}^3 r$$

$$= \sum_{j=1}^{N} \epsilon_j K_j^{-1}(\vec{r}_j, \vec{\rho}) \exp\left(\mathrm{i}\frac{2\pi\vec{r}_j \cdot \vec{\rho}}{\lambda f}\right), \qquad (1)$$

$$\equiv A(\vec{\rho}) \exp(\mathrm{i}\varphi(\vec{\rho}), \qquad (2)$$

where f is the effective focal length of the optical train, including the relay optics and objective lens.

The kernel $K_j(\vec{r}, \vec{\rho})$ can be used to transform the jth trap from a conventional tweezer into another type of trap, and K_j^{-1} is its inverse. For conventional optical tweezers in the focal plane, $K_j = 1$.

If the calculated amplitude, $A(\vec{\rho})$, were identical to the laser beam's profile, $A_0(\vec{\rho})$, then $\varphi(\vec{\rho})$ would be the kinoform encoding the desired array of traps. Unfortunately, this is rarely the case. More generally, the spatially varying discrepancies between $A(\vec{\rho})$ and $A_0(\vec{\rho})$ direct light away from the desired traps and into ghosts and other undesirable artifacts. Despite these shortcomings, combining kinoforms with Eq. (1) is expedient and can produce useful trapping patterns [10]. Still better and more general results can be obtained by using Eqs. (1) and (2) as the basis for an iterative search for the ideal kinoform.

Following the approach pioneered by Gerchberg and Saxton (GS) [12], we treat the phase $\varphi(\vec{\rho})$ calculated with Eqs. (1) and (2) as an estimate, $\varphi^{(n)}(\vec{\rho})$, for the desired kinoform and use this to calculate the fields at the trap positions given the laser's actual profile $A_0(\vec{\rho})$

$$\epsilon_j^{(n)} = \int_\Omega A_0(\vec{\rho}) \exp(\mathrm{i}\varphi^{(n)}(\vec{\rho})) K_j(\vec{r}_j, \vec{\rho})$$

$$\times \exp\left(-\mathrm{i}\frac{2\pi\vec{r}_j \cdot \vec{\rho}}{\lambda f}\right) \mathrm{d}^2\rho. \qquad (3)$$

The index n refers to the nth iterative approximation to $\varphi(\vec{\rho})$.

The classic GS algorithm replaces the amplitude $\alpha_j^{(n)}$ in this estimate with the desired amplitude α_j, leaving the corresponding phase $\phi_j^{(n)}$ unchanged, and solves for the next estimate $\varphi^{(n+1)}(\vec{\rho})$ using Eqs. (1) and (2). The fraction $\sum_j |\alpha_j^{(n)}|^2$ of the incident power actually delivered to the traps by the nth approximation is useful for tracking the algorithm's convergence.

For the present application, the simple GS substitution leads to slow and non-monotonic convergence. We find that an alternate replacement scheme

$$\alpha_j^{(n+1)} = \left[(1-\xi) + \xi\frac{\alpha_j}{\alpha_j^{(n)}}\right]\alpha_j \qquad (4)$$

leads to rapid monotonic convergence for $\xi \approx 0.5$. The resulting estimate for $\varphi(\vec{\rho})$ then can be discretized and transmitted to the SLM to establish a trapping pattern. In cases where the SLM offers only a few distinct phase levels, discretization can be incorporated into each iteration to minimize the associated error. In all of the examples discussed below, this algorithm yields kinoforms with theoretical efficiencies exceeding 80% in two or three iterations starting from a random choice for the traps' initial phases ϕ_j and often converges to solutions with better than 90% efficiency. Iterative optimization with Eqs. (2) and (3) is computationally efficient because discrete transforms are calculated only at the actual trap locations.

Fig. 2(a) shows 26 colloidal silica spheres 0.99 μm in diameter suspended in water and trapped in a planar fivefold pattern of optical tweezers created with Eqs. (2) and (3). Replacing this kinoform with another whose traps are slightly displaced moves the spheres into the new configuration. Projecting a sequence of trapping patterns translates the spheres deterministically into an entirely new configuration. Fig. 2(b) shows the same spheres after 16 such hops, and Fig. 2(c) after 38. Powering each trap with 1 mW of light traps the particles stably against thermal forces. Increasing the trapping power to 10 mW and updating the trapping pattern in 2 μm steps allows us to translate particles at up to 10 μm/s.

Comparable planar motions also have been implemented by rapidly scanning a single tweezer through a sequence of discrete locations, thereby creating a time-shared trapping pattern [13]. The continuous illumination of holographic optical traps offer several advantages, however. HOT patterns can be more extensive both spatially and

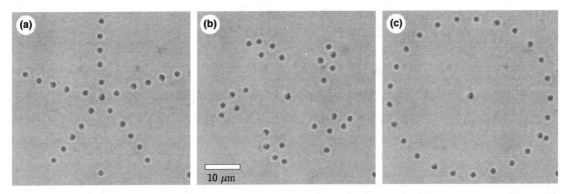

Fig. 2. A fivefold pattern of 26 colloidal silica spheres 0.99 μm in diameter is transformed into a circle using dynamic holographic optical tweezers. (a) The original configuration; (b) after 16 steps; (c) the final configuration after 38 steps.

in number of traps than time-shared arrays which must periodically release and retrieve each trapped particle. Additionally, the lower peak intensities required for continuously illuminated traps are less damaging to sensitive samples [14].

Similar rearrangements also would be possible with previous dynamic HOT implementations [10]. These studies used fast Fourier transforms to optimize the projected intensity over the entire trapping plane, and routinely achieved theoretical efficiencies exceeding 95% [7]. However, the discrete transforms adopted here allow us to encode more general patterns of traps.

Dynamic holographic optical tweezers need not be limited to planar configurations. If the laser beam illuminating the SLM were slightly diverging, then the entire pattern of traps would come to a focus downstream of the focal plane. Such divergence can be introduced with a Fresnel lens, encoded as a phase grating with

$$\varphi_z(\vec{\rho}) = \frac{2\pi\rho^2 z}{\lambda f^2} \quad \text{mod}\, 2\pi, \tag{5}$$

where z is the desired displacement of the optical traps relative to the focal plane in an optical train with effective focal length f. Rather than placing a separate Fresnel lens into the input beam, the same functionality can be obtained by adding the lens' phase modulation to the existing kinoform: $[(\varphi(\vec{\rho}) + \varphi_z(\vec{\rho})] \,\text{mod}\, 2\pi$. Fig. 3(a) shows a typical array of optical tweezers collectively displaced out of the plane in this manner. The accessible range of

out-of-plane motion in our system is approximately ± 10 μm.

Instead of being applied to the entire trapping pattern, separate lens functions can be applied to each trap individually with kernels

$$K_j^z(\vec{r}_j, \vec{\rho}) = \exp\left(i\frac{2\pi\rho^2 z_j}{\lambda f^2}\right) \tag{6}$$

in Eqs. (1) and (3). Fig. 3(b) shows spheres being moved independently through multiple planes in this way.

Other phase modifications implement additional functionality. For example, the phase profile

$$\varphi_\ell(\vec{\rho}) = \ell\theta \quad \text{mod}\, 2\pi \tag{7}$$

converts an ordinary Gaussian laser beam into a Laguerre–Gaussian mode [15], and its corresponding optical tweezer into a so-called optical vortex [16–18]. Here θ is the polar coordinate in the DOE plane (see Fig. 1) and the integer ℓ is the beam's topological charge [15].

Because all phases are present along the circumference of a Laguerre–Gaussian beam, destructive interference cancels the beam's intensity along its axis. Optical vortices thus appear as bright rings surrounding dark centers. Such dark traps have been demonstrated to be useful for trapping reflecting, absorbing [19] or low-dielectric particles [18] not otherwise compatible with conventional optical tweezers.

Adding $\varphi_\ell(\vec{\rho})$ to a kinoform encoding an array of optical tweezers yields an array of identical

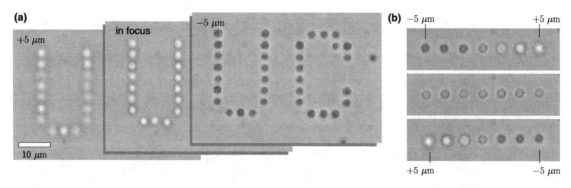

Fig. 3. Three-dimensional motion with holographic optical tweezers. The images in (a) show 34 silica spheres 0.99 μm in diameter trapped in a single plane and then displaced by ±5 μm using Eq. (5). The spheres' images change as they move relative to the focal plane. (b) Seven spheres trapped and moved independently through seven different planes using kinoforms calculated with Eq. (6).

optical vortices, as shown in Fig. 4(a). Here, the light from the array of traps is imaged by reflection off a front-surface mirror placed in the microscope's focal plane. The vortex-forming phase function also can be applied to individual traps through

$$K_j^l(\vec{r}_j, \vec{\rho}) = \exp(i\ell_j \theta) \qquad (8)$$

as demonstrated in the mixed array of optical tweezers and optical vortices shown in Fig. 4(b).

Previous reports of optical vortex trapping have considered Laguerre–Gaussian modes with relatively small topological charges, $\ell \leqslant 5$. The $\ell = 30$ examples in Fig. 4(b) are thus the most highly charged optical vortices so far reported, and traps with $\ell > 100$ are easily created with the present system.

Fig. 4(c) shows multiple colloidal particles trapped on the bright circumferences of a 3×3 array of $\ell = 15$ vortices. Because Laguerre–Gaussian modes have helical wavefronts, particles trapped on optical vortices experience tangential forces [19]. Optical vortices are useful, therefore, for driving motion at small length scales, for example in microelectromechanical systems (MEMS). Particles trapped on a vortex's bright circumference, such as the examples in Fig. 4(c) circulate rapidly around the ring, entraining cir-

culating fluid flows as they move. The resulting hydrodynamic coupling influences particles' motions on single vortices and leads to cooperative motion in particles trapped on neighboring vortices. The resulting fluid flows can be reconfigured dynamically by changing the topological charges, intensities and positions of optical vortices in an array, and may be useful for microfluidics and lab-on-a-chip applications.

The vortex-forming kernel K^ℓ can be combined with K^z to produce three-dimensional arrays of vortices. Such heterogeneous trapping patterns are useful for organizing disparate materials into hierarchical three-dimensional structures and for exerting controlled forces and torques on extended dynamical systems.

While the present study has demonstrated how a single Gaussian laser beam can be modified to create three-dimensional arrays of optical tweezers and optical vortices, other generalizations follow naturally, with virtually any mode of light having potential applications. For example, the axicon phase profile $\varphi_\gamma(\vec{\rho}) = \gamma\rho$ creates an approximation of a Bessel mode which focuses to an axial line trap whose length is controlled by γ [20]. These and other generalized trapping modes will be discussed elsewhere. Linear combinations of optical vortices and conventional tweezers have been shown to

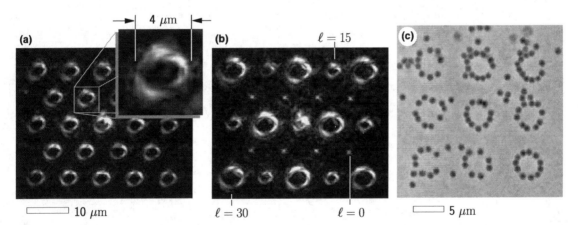

Fig. 4. (a) Triangular array of optical vortices with topological charge $\ell = 20$ created from an equivalent array of tweezers using Eq. (7). Light from the focused optical vortices is imaged by reflection from a mirror in the sample plane. The inset shows a more detailed view of one vortex's structure. (b) Mixed array of optical tweezers ($\ell = 0$) and optical vortices with $\ell = 15$ and $\ell = 30$ in the same configuration as (a), calculated with Eq. (8). The traps' amplitudes α_j were adjusted for uniform brightness. (c) Colloidal polystyrene spheres 800 nm in diameter trapped in 3×3 square array of $\ell = 15$ optical vortices.

operate as optical bottles [21] and controlled rotators [22]. All such trapping modalities can be combined dynamically using the techniques described above.

The complexity of realizable trapping patterns is limited in practice by the need to maintain three-dimensional intensity gradients for each trap, and by the maximum information content that can be encoded accurately in the SLM. For example, the former consideration precludes forming a three-dimensional cubic optical tweezer array with a lattice constant much smaller than 10 μm, while the latter limits our optical vortices to $\ell \approx 200$.

Within such practical bounds, dynamic holographic optical tweezers are highly reconfigurable, operate non-invasively in both open and sealed environments, and can be coupled with computer vision technology to create fully automated systems. A single apparatus thus can be adapted to a wide range of applications without modification. Dynamic holographic optical tweezers have a plethora of potential biotechnological applications including massively parallel high throughput screening, sub-cellular engineering, and macromolecular sorting. In materials science, the ability to organize materials into hierarchical three-dimensional structures constitutes an entirely new category of fabrication techniques. As research tools, dynamic holographic optical tweezers combine the demonstrated utility of optical tweezers with unprecedented flexibility and adaptability.

Acknowledgements

This work was funded by a sponsored research grant from Arryx Inc., using equipment purchased under Grant Number 991705 from the W.M. Keck Foundation. The spatial light modulator used in this study was made available by Hamamatsu Corp., as a loan to The University of Chicago. Additional funding was provided by the National Science Foundation through Grant Number DMR-9730189, and by the MRSEC program of the National Science Foundation through Grant Number DMR-980595.

References

[1] A. Ashkin, J.M. Dziedzic, J.E. Bjorkholm, S. Chu, Opt. Lett. 11 (1986) 288.

[2] K. Svoboda, S.M. Block, Annu. Rev. Biophys. Biomol. Struct. 23 (1994) 247.

[3] A. Ashkin, IEEE J. Sel. Top. Quantum Electron. 6 (2000) 841.

[4] D.G. Grier, Curr. Opin. Colloid Interface Sci. 2 (1997) 264.

[5] E.R. Dufresne, D.G. Grier, Rev. Sci. Instrum. 69 (1998) 1974.

[6] D.G. Grier, E.R. Dufresne, US Patent 6,055,106, The University of Chicago (2000).

[7] E.R. Dufresne et al., Rev. Sci. Instrum. 72 (2001) 1810.

[8] D.G. Grier, Nature 393 (1998) 621.

[9] M. Reicherter, T. Haist, E.U. Wagemann, H.J. Tiziani, Opt. Lett. 24 (1999) 608.

[10] J. Liesener, M. Reicherter, T. Haist, H.J. Tiziani, Opt. Commun. 185 (2000) 77.

[11] Y. Igasaki et al., Opt. Rev. 6 (1999) 339.

[12] R.W. Gerchberg, W.O. Saxton, Optik 35 (1972) 237.

[13] K. Sasaki et al., Opt. Lett. 16 (1991) 1463.

[14] K.C. Neuman et al., Biophys. J. 77 (1999) 2856.

[15] N.R. Heckenberg et al., Opt. Quantum Electron. 24 (1992) S951.

[16] H. He, N.R. Heckenberg, H. Rubinsztein-Dunlop, J. Mod. Opt. 42 (1995) 217.

[17] N.B. Simpson, L. Allen, M.J. Padgett, J. Mod. Opt. 43 (1996) 2485.

[18] K.T. Gahagan, G.A. Swartzlander Jr., Opt. Lett. 21 (1996) 827.

[19] H. He, M.E.J. Friese, N.R. Heckenberg, H. Rubinsztein-Dunlop, Phys. Rev. Lett. 75 (1995) 826.

[20] J. Arlt, V. Garces-Chavez, W. Sibbett, K. Dholakia, Opt. Commun. 197 (2001) 239.

[21] J. Arlt, M.J. Padgett, Opt. Lett. 25 (2000) 191.

[22] L. Paterson et al., Science 292 (2001) 912.

An optical trapped microhand for manipulating micron-sized objects

Graeme Whyte, Graham Gibson, Jonathan Leach, and Miles Padgett.

Department of Physics and Astronomy, University of Glasgow, Glasgow G12 8QQ, UK

m.padgett@physics.gla.ac.uk

http://www.physics.gla.ac.uk/Optics

Daniel Robert

School of Biological Sciences, Woodland Road, University of Bristol, Bristol BS8 1UG, UK

Mervyn Miles

H. H. Wills Physics Laboratory, Tyndall Avenue, University of Bristol, Bristol BS8 1TL, UK

Abstract: We have developed a real time interface for holographic optical tweezers where the operator's fingertips are mapped to the positions of silica beads captured in optical traps. The beads act as the fingertips of a microhand which can be used to manipulate objects that otherwise do not lend themselves to tweezers control, e.g. objects that are strongly scattering or highly light sensitive. We illustrate the use of the microhand for the real time manipulation of micron sized chrome particles.

OCIS codes: (090.1970) Diffractive optics; (140.7010) Trapping; (230.6120) Spatial light modulators

---

References and links

1. A. Ashkin, J. M. Dziedzic, J. E. Bjorkman, and S. Chu, "Observation of a single-beam gradient force optical trap for dielectric particles," Opt. Lett. **11**, 288–290 (1986).
2. D. G. Grier, "A revolution in optical manipulation," Nature **424**, 810-816 (2003).
3. E. R. Dufresne and D. G. Grier, "Optical tweezers arrays and optical substrates created with diffractive optics," Rev. Sci. Instr. **69**, 1974-1977 (1998).
4. V. Bingelyte, J. Leach, J. Courtial, and M. J. Padgett, "Optically controlled three-dimensional rotation of microscopic objects," Appl. Phys. Lett. **82**, 829-831 (2003).
5. M. Reicherter, T. Haist, E. U. Wagemann, and H. J. Tiziani, "Optical particle trapping with computer generated holograms written on a liquid crystal display," Opt. Lett. **24**, 608-610 (1999).
6. J. E. Curtis, B. A. Koss, and D. G. Grier, "Dynamic holographic optical tweezers," Opt. Commun. **207**, 169-175 (2002).
7. G. Sinclair, P. Jordan, J. Courtial, M. J. Padgett, J. Cooper, and Z. J. Laczik, "Assembly of 3-dimensional structures using programmable holographic optical tweezers," Opt. Express **12**, 5475-5480 (2004).
8. J. Leach, G. Sinclair, P. Jordan, J. Courtial, M. J. Padgett, J. Cooper, and Z. J. Laczik, "3D manipulation of particles into crystal structures using holographic optical tweezers," Opt. Express **12**, 220-226 (2004).
9. V. Emiliani, D. Cojoc, E. Ferrari, V. Garbin, C. Durieux, M. Coppey-Moisan, and E. Di Fabrizio, "Wave front engineering for microscopy of living cells," Opt. Express **13**, 1395-1405 (2005).
10. E. Ferrari, V. Emiliani, D. Cojoc, V. Garbin, M. Zahid, C. Durieux, M. Coppey-Moisan, and E. Di Fabrizio, "Biological samples micro-manipulation by means of optical tweezers," Microelectron. Eng. **78-79**, 575-581 (2005).
11. K. Svoboda and S. M. Block, "Optical trapping of metallic rayleigh particles," Opt. Lett. **19**, 930-932 (1994).
12. K. Sasaki, M. Koshioka, H. Misawa, N. Kitamura, and H. Masuhara, "Optical trapping of a metal-particle and a water droplet by a scanning laser-beam," Appl. Phys. Lett. **60**, 807-809 (1992).

13. S. Sato, Y. Harada, and Y. Waseda, "Optical Trapping of microscopic metal particles," Opt. Lett. **19,** 1807-1809 (1994).

14. J. Liesener, M. Reicherter, T. Haist, and H. J. Tiziani, "Multi-functional optical tweezers using computer-generated holograms," Opt. Commun. **185,** 77-82 (2000).

15. J. Leach, K. Wulff, G. Sinclair, P. Jordan, J. Courtial, L. Thomson, G. Gibson, K. Karunwi, J. Cooper, Z. J. Laczik, and M. Padgett, "Interactive approach to optical tweezers control," Appl. Opt. **45,** 897-903 (2006).

16. J. Curtis, C. Schmitz, and J. Spatz, "Symmetry dependence of holograms for optical trapping," Opt. Lett. **30,** 2086-2088 (2005).

17. C. Schmitz, J. Spatz, and J. Curtis, "High-precision steering of multiple holographic optical traps," Opt. Express **13,** 8678-8685 (2005).

1. Introduction

Optical tweezers [1] have been revolutionised by the incorporation of spatial light modulators to split the laser beam into many individual traps that can be independently positioned; a technique called holographic optical tweezers [2]. Here we report on an interface using the operator's fingers to simultaneously determine the position and motion of several optical traps. In effect we use the beads captured in optical traps as the fingertips of a manipulating microhand. One traditional limitation of optical tweezers is their reliance on the gradient-force generated by the interaction of transparent objects with the focused laser light. Consequently, the trapping of opaque particles has remained problematic. Circumventing that fundamental problem, the present dynamic holographic system is demonstrated to use the digits of the microhand to accurately manipulate micron-sized metallic objects, expanding the range of particles that can be manipulated.

When placed in the tight focus of a laser beam, a micron-sized object is subject to a gradient-force and a light scattering force, both adding to gravity and forces associated with Brownian motion. For dielectric (i.e. transparent) particles the gradient force can correspond to a substantial fraction of the optical momentum, dominating over all other forces to create an optical trap centred on the beam focus.

For positioning a single optical trap, it is normal to place a beam steering mirror in a Fourier-plane of the trap itself, where an angular deviation of the collimated beam gives a lateral shift of the focused trap. This arrangement lends itself to replacement of the mirror with a diffractive optical element [3] that gives the additional freedom of introducing focal power and hence axial control of the trap position. One point of particular interest is the extremely high numerical aperture of the microscope systems typically employed. One important consequence is that two or more objects can be trapped at different heights along the microscope's optical axis, and then separated again [4]. Positioning the diffractive element in the Fourier plane of the trap means it is operating as a hologram, hence the term holographic optical tweezers. Using a spatial light modulator as the diffractive element allows the positioning of the traps at video frame-rates [5, 6]. Thus multiple optical traps can be controlled independently and simultaneously. Various algorithms have been employed for the design of the hologram allowing 3D arrangements of multiple traps, such as those used for creating complex 3D structures [7, 8] and performing simultaneously 3D optical manipulation and optical sectioning [9, 10]. In this latter work a 3D arrangement of beads was created in order to hold a biological sample indirectly. However, this 3D arrangement relied on a sequence of pre calculated holograms and therefore offered limited real time control.

For metallic particles between 10 and 100nm in diameter, the excitation of surface plasmons makes trapping at the beam focus possible [11]. However, for larger, micron-sized metallic particles, scattering forces dominate, repelling the particle from the beam. Although rapidly scanning a single beam around the particle [12] or using beams with annular intensity cross sections [13] allows trapping of metal particles, trapping of multiple particles at differing axial position can be problematic. Similarly, the direct trapping of biological samples such as cells can also be problematic. The cell can be damaged by intense illumination, or simply the low contrast in the refractive index between the cell and the surrounding fluid generates only very small trapping forces.

2. Hologram design

Whilst holographic optical tweezers offer much flexibility and control in the positioning of multiple traps, they can be computationally demanding. A simple and computationally efficient algorithm for hologram design relies upon the fact that a lateral displacement of a single trap requires a hologram corresponding to a blazed diffraction phase-grating with a period proportional to the displacement of the trap away from the beam axis. For an axial shift the hologram corresponds to a Fresnel lens. The modulo 2π addition of multiple holograms produces a hologram giving a single trap shifted from the original focus by the vectorial sum of the shifts produced by the individual holograms. Multiple traps are produced by the complex addition of the holograms defining individual traps. The magnitude of each input hologram corresponds to the individual trap strength and the argument of their sum gives the desired multi-trap hologram [14, 15]. The time-averaged contrast of the resulting traps can be maximized by introducing a random phase-shift between the individual holograms prior to their addition [16]. Although the optical resolution of the microscope system is 100's nm, the precision to which each of the beads can be positioned is better than 10nm [17], providing control at the nanoscale.

Within our application, we define the position of several traps of equal strength and map the $0 - 2\pi$ phase of the hologram onto an 8-bit grey-scale image, relayed onto the spatial light modulator via a video card interface. The 8-bit grey-scale lends itself to the use of inherently modulo 8-bit arithmetic. This computational efficiency means that a twin-processor, desktop computer can readily calculate and display 512x512 pixel holograms at 8 frames per second. This rate is sufficient for a real-time interface.

3. Experimental configuration

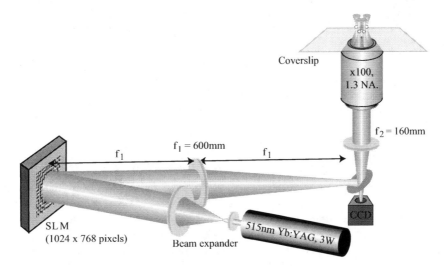

Fig. 1. The experimental configuration of the holographic optical tweezers. The plane of the SLM is positioned in the Fourier plane of the sample.

Our optical tweezers are configured around an inverted microscope with a 1.3NA, x100, Plan

Neo-fluar objective, as illustrated in Fig. 1. The trapping laser is a diode-pumped, frequency-doubled solid-state laser emitting up to 3 watts of 515nm light that is expanded to fill the aperture of an electrically addressed spatial light modulator (Holoeye LC-R 2500), the plane of which is imaged onto the back aperture of the objective lens. The novelty of our system is that the trap locations are controlled by the positions of the operator's finger tips.

A single fire-wire interface camera positioned above the working area images the x and y positions of white beads attached to the fingers of a black glove, as illustrated in Fig.2. The z positions of the beads are unambiguously inferred from their apparent size in the image. These positions are scaled and fed to the hologram calculation algorithm to produce optical traps at the corresponding positions within the 3D space accessible to the holographic tweezers. The position of each fingertip is directly mapped to the position of each trapped bead within the tweezers. For trapping robustness and stability, the interface limits the maximum translation velocity of the traps ($< 5\mu$m/sec) so that too rapid a movement of the fingers results in a locking of the trap positions. This also allows a static arrangement of traps to be created by rapidly moving the fingers away from the camera, allowing further experiments on the trapped object without requiring the fingers to be held in place. To regain control on the traps, the fingers have to be re-positioned in coincidence with the locked traps (as shown in the first two frames of Fig. 4).

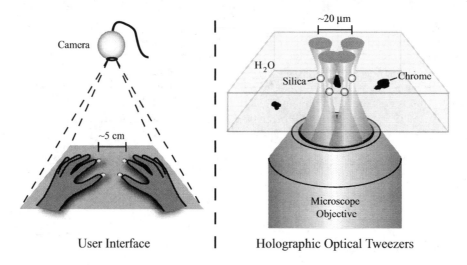

Fig. 2. A new approach to controlling optical tweezers. A single camera images the coordinates of white beads attached to fingertips. The position of each fingertip is mapped to the position of optical traps, providing a direct, visually controlled manipulation of microscale objects.

The selection of silica bead size is an important parameter in the operation of the microhand. Although the trapping beams are tightly focused, the comparatively high light intensity means that some light is still present around the edge of the beads. If the object to be manipulated is particularly sensitive to light or is highly scattering then this residual light may damage the object or expel it from the microhand. Whereas 2-micron diameter beads result in higher applied forces, 5-micron diameter beads better isolate the object from the trapping light. Optical tweezers typically can create a force of order 1pN per 1mW, meaning that the maximum force that can be exerted by the hand is limited by the power of the laser, and efficiency of the SLM and relay optics, to order of a few 100pN.

The holographic technique presented here allows for the fine control of multiple independent

Fig. 3. Split screen video sequence showing control of the microhand in the lateral and axial directions. The axial position of each silica bead (5μm diameter) is determined by the size of it's corresponding white marker in the image. See accompanying movie, 3beads.avi (1.64MB).

traps, and provides the possibility for sophisticated interactions with micro- and nano-scale objects. In the present scheme, the operator's fingers can be adjusted to grasp and manipulate objects with complex geometries. Each of the beams acts as a digit of an optically controlled hand.

The spatial and temporal correspondence between fingertips and optical traps can be monitored by a split screen video (Fig. 3). Fingertips with white markers attached, can control the position and velocity of beads captured in the optical traps. In addition to controlling the lateral positions of the silica beads, determined by the lateral positions of the corresponding white markers, their axial positions are determined by the size of the white markers in the image.

We have demonstrated the microhand as a tool for manipulating micron sized chrome particles (Fig. 4). In this example, a chrome particle is selected and moved using four 100 mW power optical traps. In these conditions, the trapped five micron diameter beads can be independently translated at up to 5μm/s over a field of view approximately 80μm in diameter. The system conveniently enables grabbing, displacing, positioning, orientating, lifting and gentle releasing of irregular objects.

4. Discussion and conclusions

We believe that the microhand interface will make optical tweezers a more accessible tool within the multidisciplinary workplace. Removing the need to trap objects directly will greatly extend the range of applications to include both strongly scattering and highly light sensitive

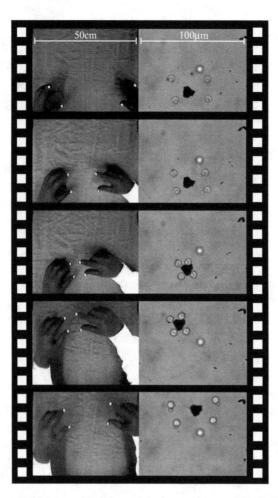

Fig. 4. Split screen video sequence of user interface and trapped beads (5μm diameter). Here, the microhand is used to select and move an irregularly shaped and opaque chrome particle. See accompanying movie, 4beads.avi (2.29MB).

objects. For example, multiple traps guided by the microhand interface could be put to use in cell sorting tasks requiring mechanically soft intervention with low light levels. The accurate and soft handling of cells, bacteria, or even subcellular organelles could be automated; enticingly, learning algorithms could acquire the operator's manual demonstration of the proper manipulative procedure, and generalize it to entire cell populations. More generally, the capacity of manipulating matter in three dimensions at the micro and nanoscale has great potential for future technologies in areas such as biomedicine and material research.

Acknowledgments

This work is supported by the RCUK.

Assembly of 3-dimensional structures using programmable holographic optical tweezers

Gavin Sinclair, Pamela Jordan, Johannes Courtial, Miles Padgett

Department of Physics and Astronomy, University of Glasgow, Glasgow, G12 8QQ, UK
g.sinclair@physics.gla.ac.uk

Jon Cooper

Department of Electronics, University of Glasgow, Glasgow, G12 8LT, UK

Zsolt John Laczik

Department of Engineering Science, University of Oxford, Oxford, OX1 3PJ, UK

Abstract: The micromanipulation of objects into 3-dimensional geometries within holographic optical tweezers is carried out using modified Gerchberg-Saxton (GS) and direct binary search (DBS) algorithms to produce the hologram designs. The algorithms calculate sequences of phase holograms, which are implemented using a spatial light modulator, to reconfigure the geometries of optical traps in many planes simultaneously. The GS algorithm is able to calculate holograms quickly from the initial, intermediate and final trap positions. In contrast, the DBS algorithm is slower and therefore used to pre-calculate the holograms, which are then displayed in sequence. Assembly of objects in a variety of 3-D configurations is semi-automated, once the traps in their initial positions are loaded.

OCIS Codes: (020.7010) Trapping, (170.4520) Optical confinement and manipulation, (090.1760) Computer holography, (230.6120) Spatial light modulators.

References and Links

1. A. Ashkin, J. M. Dziedzic, J. E. Bjorkman and S. Chu, "Observation of a single-beam gradient force optical trap for dielectric particles" Opt. Lett. **11,** 288-290, (1986).
2. J. E. Molloy and M. J. Padgett, "Lights, action: optical tweezers" Contemp. Phys. **43,** 241-258 (2002).
3. D. G. Grier, "A revolution in optical manipulation" Nature, **424,** 810-816 (2003).
4. J. E. Curtis, B. A. Koss and D. G. Grier, "Dynamic holographic optical tweezers" Opt. Commun. **207,** 169-175 (2002).
5. N. R. Heckenberg, R. McDuff, C. P. Smith, H. Rubinsztein-Dunlop and M. J. Wegner, "Laser-beams with phase singularities" Opt. Quantum Electron. **24,** S951-S962 (1992).
6. J. Glückstad, "Phase contrast image synthesis" Opt. Commun. **130,** 225-230 (1996).
7. M. Reicherter, T. Haist, E. U. Wagemann and H. J. Tiziani, "Optical particle trapping with computer generated holograms written on a liquid crystal display" Opt. Lett. **24,** 608-610 (1999).
8. G. Gibson, J. Courtial, M. Vasnetsov, S. Barnett, S. Franke-Arnold and M. Padgett, "Increasing the data density of free-space optical communications using orbital angular momentum" Proc. SPIE **5550,** In Press.
9. G. Sinclair, J. Leach, P. Jordan, G. Gibson, E. Yao, Z. J. Laczik, M. J. Padgett and J. Courtial, "Interactive application in holographic optical tweezers of a multi-plane Gerchberg-Saxton algorithm for three-dimensional light shaping" Opt. Express **12,** 1665-1670 (2004).
   http://www.opticsexpress.org/abstract.cfm?URI=OPEX-12-8-1665
10. M. A. Seldowitz, J. P. Allebach and D. W. Sweeney, "Synthesis of digital holograms by direct binary search" Appl. Opt. **26,** 2788-2798 (1987).
11. R. A. Gabel and B. Liu, "Minimization of reconstruction errors with computer generated binary holograms" Appl. Opt. **9,** 1180-1190 (1970).
12. K. Nagashima, "3D computer-generated holograms using 1D Fourier transform operations" Opt. Laser Technol. **30,** 361-366 (1998).

13. R. W. Gerchberg and W. O. Saxton, "A practical algorithm for the determination of the from image and diffraction plane pictures" Optik **35**, 237-246 (1972).

14. Z. J. Laczik, "3D beam shaping using diffractive optical elements" Proc. SPIE **4770**, 104-111 (2002)

15. J. Leach, G. Sinclair, P. Jordan, J. Courtial, M. J. Padgett, J. Copper and Z. J. Laczik, "3D mainpulationn of particles into crystal structures using holographic optical tweezers" Opt. Express **12**, 220-226 (2004). http://www.opticsexpress.org/abstract.cfm?URI=OPEX-12-1-220

16. P. Jordan, H. Clare, L. Flendrig, J. Leach, J. Cooper and M. Padgett, "Permanent 3D structures in a polymeric host created using holographic optical tweezers" J. Mod. Opt. **51**, 627-632 (2004).

1. Introduction

Optical tweezers are used for manipulating both single and multiple, micron-sized, particles suspended in solution [1]. Various types of particles can be trapped including transparent silica or polymer spheres, metallic particles and biological specimens [2]. Recent advances in spatial light modulators (SLMs) enable a single laser beam to be split into many beams, enabling the simultaneous trapping of many objects [3]. Such experimental arrangements are termed holographic optical tweezers (HOTs) [4]. The SLM is usually positioned in the Fourier-plane with respect to the sample such that angular deflection of the beam at the SLM gives a lateral translation of the trap, and a change in the wavefront curvature gives an axial shift of the trap. The spatial resolution of the SLM, and the aberrations within the system, limit the maximum displacements to a few 10's µm.

To date, many different algorithms have been applied to design computer-generated holograms for use with HOTs [4-9]. In this work we use both a modified Gerchberg-Saxton (GS) algorithm [9] developed to specify arbitrary geometries of traps in multiple planes, and a direct binary search (DBS) algorithm developed to specify trap positions in continuous 3-dimensional space [10]. We provide examples in which sequences of holograms are used to trap objects initially in simple geometries and then transform their configuration into complicated 3-D structures.

2. Algorithms for designing holograms

In addition to photographic techniques, that produce holograms directly from the light scattered by an object, numerous computer algorithms have been developed to design the hologram pattern required to produce the desired target image [11-12]. Such holograms are called computer-generated holograms. Within optical tweezers, rather than requiring an arbitrary light field, all that is needed is to specify the positions of high intensity within a 3-D volume, which significantly simplifies the computational problem.

The GS algorithm was initially developed to produce a target intensity distribution in one plane [13]. It is based upon the Fourier-transform relationship between the hologram and target planes. Repeated transformation between the planes, accompanied by repeated substitution of the target and input intensities whilst maintaining the phase distribution, gives rapid convergence of the hologram design. By incorporating additional beam-propagation steps into the algorithm, intensities can be specified in multiple planes [7]. However, propagation between the planes requires further Fourier-transform operations, slowing both the iteration cycle and convergence to final hologram design.

Another approach makes random changes to a hologram design. The changes are either kept or discarded depending on whether or not it improves the match between the resulting beam and target intensity distribution, where the latter can be specified in either two or three dimensions. These direct binary search strategies [10,14] are particularly useful because they are not restricted to defining the trap positions on a discrete grid. Further considerations can be targeted, such as positions of zero intensity separating two traps.

3. Hologram display interfaces

We have previously designed holograms to create multiple traps distributed in 3-D, allowing the corresponding structures to be assembled [15]. However, loading the traps in these complex structures can be problematic since some of the traps are embedded within the structure, i.e. are surrounded by many other traps. Filling all the traps with only a single object can be difficult and the assembly process may require many attempts. Alternatively, complex 3-D structures can be morphed into such configurations starting from a simply geometry. This requires a sequence of holograms to be designed that gradually transforms the initial simple trap arrangement into the final complex structure.

To facilitate the morphing process, an interface was developed using the LabVIEW programming environment, to display a sequence of holograms. The starting point for both the GS and DBS algorithms is a spreadsheet file specifying the x, y, z positions for all the traps. This can be written explicitly or interpolated from a few key positions. In either case, it is important that the sequential displacement of individual traps does not exceed a fraction of the object diameter, typically 25%. For one, two or three planes, we find that the GS algorithm converges quickly enough using a desktop computer (Pentium4 2x2.8Ghz) to calculate ≈ 2 holograms per second at a resolution of 512x512. The convergence of the algorithm is improved by the fact that each particular hologram is, itself, used as the starting point for the next hologram in the sequence. For more than a few ten's of traps in more than three planes, the convergence of both the GS and DBS algorithms are too slow for real time calculation. In these cases we pre-calculate sequences of holograms and save them to file. These can then be displayed on the SLM in a real time.

Currently, commercially available SLMs have diffraction efficiencies in the region of 40%. Much of the remaining light appears in the zero order, although some of the incident power also appears in higher diffraction orders. Consequently, all our hologram designs are modified to create the desired configuration of traps typically displaced by 20μm from the zero order. This means that most of the ghost traps produced by unwanted diffraction orders can be blocked by a spatial filter.

4. Experimental arrangement

The optical tweezers are based on a NA1.3, x100, Zeiss Plan Neofluar oil immersion microscope objective used in an inverted geometry. A sample cell was mounted on a 3-axis piezo stage. The trapping laser was a frequency-doubled Nd:YVO$_4$ laser, with a maximum output power of 1.5 W at 532nm. The laser beam was expanded to slightly over-fill the active area of a Holoeye LC-R 2500 spatial light modulator. The SLM was imaged with a magnification of about 1/3 to fill the pupil plane of the microscope objective. The losses in the optical system coupled with the diffraction efficiency of the SLM results in a laser power in the trapping plane of the order of 100-200mW, distributed between each of the traps.

5. Results

To demonstrate the versatility of the techniques, we show a number of example structures morphed using both the GS and DBS algorithms (Figs. 1-3). For both the GS and DBS algorithms, all holograms were displayed at 512x512 pixels in size, utilising the full resolution of the SLM.

The first sequence of video frames (Fig. 1) shows nine, 2μm diameter, silica spheres morphing into three triangles stacked in planes above each other, separated by 6μm. In this case, once the structure is assembled, the hologram sequence is reversed to return the line configuration. The ability to return to this starting configuration proves that the traps remain distinct throughout the manipulation despite periods where one object is immediately in line with another, along the optic axis. The holograms were derived using the multi-plane GS algorithm to calculate holograms in real time, as determined from the spreadsheet file specifying the trap trajectories.

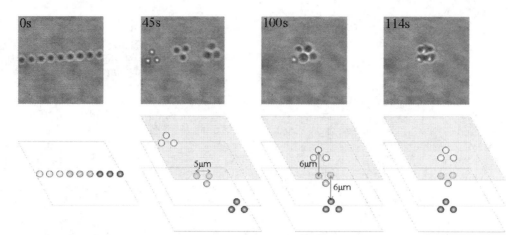

Fig. 1. [2.3MB] Sequence of video frames showing nine 2μm silica spheres morphed to form three triangles with 5μm side lengths stacked in three planes, with 6μm between planes. The times at which each frame was taken is shown in the top corner.

The second sequence of video frames (Fig. 2) shows an initial geometry of three lines of nine 1μm diameter silica spheres manipulated into three grids of nine spheres and each layer separated axially by 8μm. This geometry of spheres would be particularly difficult to trap by filling each trap from a static hologram since the central trap in the middle layer is completely surrounded by other traps. By automating the manipulation process, the user only loads the initial arrangement of traps. Again the ability to reverse the sequence shows that the individual traps remain distinct despite being in close proximity to each other both laterally and axially. Note, during the morphing sequence the sample stage was translated in respect to the traps. The translation shows the individual traps are robust enough to maintain their integrity against the resulting Stokes drag force.

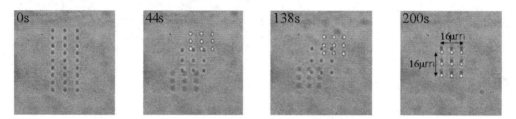

Fig. 2. [0.85MB] Sequence of video frames showing twenty-seven 1μm silica spheres initially trapped in three lines of nine spheres and morphed to form three grids of 3 by 3 in three different planes, separated by 8μm. The times at which each frame was taken is shown in the top corner.

The third sequence of frames (Fig. 3) shows eighteen 1μm diameter silica spheres, initially positioned in two lines, and then morphed into a configuration corresponding to the unit cell of a diamond lattice. In the initial orientation with the [100] direction parallel to the optical axis, this configuration comprises five separate planes. The assembly sequence of holograms was pre-calculated using the DBS algorithm. An additional sequence of holograms rotates the unit cell by 360 degrees, about a high crystal-index axis before reversing the procedure, returning to the original dual-line configuration. In the rotation phase we are effectively calculating trap positions in up to 18 different planes. Using the GS algorithm, we were unable to reproduce this manipulation and in addition, the large number of planes in the rotation stage results in a slow iteration of the GS algorithm.

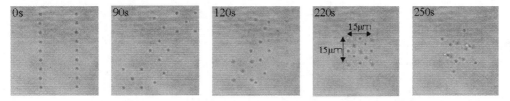

Fig. 3. [2.6MB] Sequence of video frames showing the morphing of 15μm diamond unit cell made from eighteen 1μm silica spheres. The whole sequence took 5 minutes to complete.

The final frame sequence (Fig. 4) shows a similar geometry of trapped spheres to the diamond unit cell (Fig. 3), except four 1μm silica spheres are replaced with 2μm spheres to give a mixed unit cell corresponding to a zincblende structure. Again the formation of the structure in five planes is followed by a rotation about a high order axis. The holograms were pre-calculated using the DBS algorithm.

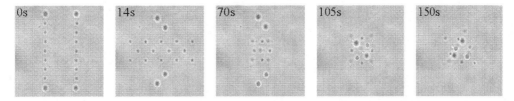

Fig. 4. [2.2MB] Sequence of video frames showing the morphing of a 15μm zincblendee unit cell made from fourteen 1μm and four 2μm silica spheres. The whole sequence took 5 minutes to complete.

6. Discussion

The methods used to calculate the hologram patterns reflect the different strengths and weaknesses of the design algorithms. The GS algorithm converges fast enough for holograms to be displayed in real-time on the SLM, although its main limitation is that it slows with increasing numbers of planes (making, for example, rotation of a multi-plane 3-D structure around arbitrary axes problematic). The DBS algorithm is slower than the GS algorithm, preventing it from being used interactively. However, the DBS algorithm allows control of the intensity in an arbitrary number of planes, making it more suited for the generation of more complex structures.

All the examples we show within this work involve sequences of holograms, displayed over many tens of seconds, to assemble structures with dimensions of tens of microns. The time required for the manipulation of the particles corresponds to speeds of up to 1 micron per second and is limited by a number of factors. As each hologram corresponds to a discrete pattern of traps, individual movements of each trap between successive holograms must not exceed the capture range of the trap. We restrict these steps to be a maximum of 0.5μm. When using the GS algorithm for the real time calculation of holograms, this limits the update rate to 2 holograms per second, hence the observed movement rates. For pre-calculated holograms, we are in principle limited only by the update rate of the SLM, which is in the region of 10Hz for the SLM used in our experiments. However we find that we are still limited to similar movement rates. Additional factors limiting the maximum translation speeds of individual traps in a complex 3-D structure are degradation of the trap quality due to aberrations in the optical systems, scattering from neighbouring objects within the structure and reduced contrast in the trapping light field.

7. Conclusion

We show particles initially trapped in lines can be manipulated into 3-D geometries using sequences of holograms. The holograms can be calculated in real-time using a Gerchberg-Saxton algorithm or pre-calculated using a direct binary search algorithm. The use of hologram sequences means that multiple objects can be loaded into traps whilst arranged in a single plane, and then subsequently transformed into the chosen geometry. For both algorithms we observe that two objects within a structure can be displaced both laterally and axially with respect to each other. The ability to displace two objects in the axial direction is, at first sight, surprising since one might think that traps further away would be degraded by the shadow of those positioned between them and the objective lens. However, this is not the case since the high NA of the trapping beams means that the obscuration of one trap by another is minimised.

In addition to the intrinsic excitement of forming microscopic 3-D structures, we believe that such structures will have applications in measuring mechanical properties of materials, photonic and biological crystal growth, tissue engineering, permanent extended 3-D structures, and manipulation within microfluidic devices. Some of these latter applications are aided by the fact that 3-D structures formed within optical tweezers can be made permanent by using a gel solution in which to trap the objects which are then locked in place when the gel sets [16].

Four-dimensional optical manipulation of colloidal particles

Peter John Rodrigo, Vincent Ricardo Daria, and Jesper Glückstad[a)]

Optics and Plasma Research Department, Risø National Laboratory, DK-4000 Roskilde, Denmark

(Received 8 October 2004; accepted 12 January 2005; published online 11 February 2005)

We transform a TEM_{00} laser mode into multiple counterpropagating optical traps to achieve four-dimensional simultaneous manipulation of multiple particles. Efficient synthesis and dynamic control of the counterpropagating-beam traps is carried out via the generalized phase contrast method, and a spatial polarization-encoding scheme. Our experiments genuinely demonstrate real-time, interactive particle-position control for forming arbitrary volumetric constellations and complex three-dimensional trajectories of multiple particles. This opens up doors for cross-disciplinary cutting-edge research in various fields. © 2005 *American Institute of Physics.* [DOI: 10.1063/1.1866646]

Sculpted light fields have advanced our ability to manipulate colloidal aggregates and brought unique opportunities for fundamental and applied science.[1] From a physical stance, tailored optical potentials were used to investigate underlying mechanisms of optically bound matter,[2,3] stochastic resonance in Brownian particles,[4] and light-matter angular momentum transfer.[5] Of equivalent significance are the applications of optical manipulation for powering colloidal microfluidic machines[6] and fractionating mixtures through optical lattices.[7,8] Tailored optical landscapes have also been applied to organize matter. Arrays of microscopic objects have been patterned in one dimension,[9] two dimensions (2D),[10,11] and three dimensions (3D).[12] In particular, material engineers have seen the potentials of optical manipulation in synthesizing advanced materials. This is substantiated in recent works showing the complementary role of optical trapping with laser-initiated photopolymerization for the construction and gelling of permanent particle arrays from linear to crystal-like structures.[13,14] Aside from the ability to form predefined structures in 3D, however, a number of applications, including those of biological relevance, would significantly gain from the power of being able to arbitrarily adjust the relative positions of particle aggregates over a full volume and in a manner that satisfies real human response time.

With the advent of spatial light modulators (SLM), rapid developments boosting the degree of control for optically trapped particles have been seen in two techniques based on diffractive optics (DO)[12,13] and the generalized phase contrast (GPC) method.[11,15–17] Recently, we have shown the suitability of the GPC approach for direct and user-interactive manipulation of a colony of particles and living cells simultaneously trapped and manipulated in a two-dimensional plane with no surface contact.[18] Here, we implement the GPC method to demonstrate optical micromanipulation of colloidal particles independently trapped and real-time manipulated throughout a large volume. This is achieved using a light-efficient configuration based on multiple counterpropagating-beam traps. Using a lossless polarization scheme,[19] the axial positions of each individually trapped particle are varied by changing the relative powers of orthogonally polarized counterpropagating beams. This is accomplished by using a spatially addressable polarization

modulator, which effectively separates the command on the relative strengths of individual pairs of opposing beams forming each trap. Since the transverse intensity patterns used to construct the trapping configurations are synthesized by the GPC method, transverse trap positions and dynamics are inherently reconfigurable in real time. The combined result enables, four-dimensional (4D) optical manipulation of a plurality of simultaneously trapped microscopic particles.

The instrumentation of our 4D optical manipulation system is shown in Fig. 1. An expanded TEM_{00} mode from a near-infrared laser is illuminating a phase-only SLM. Upon reflection, the approximately planar incident wave front acquires a 2D phase distribution encoded on the SLM front-panel surface. The programmable phase function, described by $\phi(u,v)$, is represented as a spatial pattern of up to 256 grey levels on a designated computer graphic monitor. A reasonably linear relation is set between the SLM phase and monitor grey level. For reconfigurable patterns of phase $\phi(u,v)$ in our system implementation, binary grey levels corresponding to phases $\phi=0$ and $\phi=\pi$, suffice. This implies that faster types of SLMs based on ferroelectric liquid crystals, quantum well devices, or microelectromechanical systems-mirrors can be applied. By utilizing a GPC-based light projection setup, the binary phase $\phi(u,v)$ is efficiently

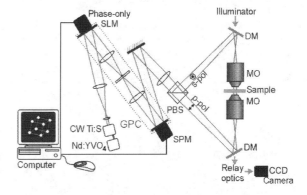

FIG. 1. (Color online) Experimental setup for implementing 4D optical manipulation. cw Ti:S, continuous-wave titanium:sapphire (wavelength =830 nm; maximum power=1.5 W); Nd:YVO$_4$, neodymium:yttrium-vanadate (wavelength=532 nm); GPC, generalized phase contrast system; SLM, spatial light modulator; SPM, spatial polarization modulator; PBS, polarizing beam splitter; MO, microscope objective [×60; numerical aperature (NA)=0.85]; DM, dichroic mirror.

[a)]Electronic mail: jesper.gluckstad@risoe.dk

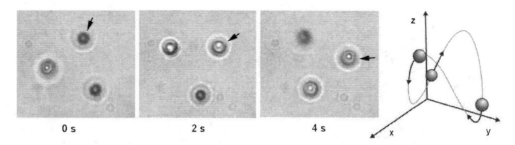

FIG. 2. (Color online) Defining 3D trajectories for multiple ($N=3$) optically trapped polystyrene spheres (diameter$=3$ μm). The trace of the 3D path is shown, right (see Ref. 23).

converted into a high-contrast intensity pattern $I(u',v')$.[20–22] The intensity pattern is then projected onto the front-panel of a spatial polarization modulator (SPM). Here, the process is described by

$$\phi(u,v) = \pi \sum_{n=1}^{N} f_n(u-u_n, v-v_n) \xrightarrow{\text{GPC}} I(u',v')$$

$$\approx I_0 \sum_{n=1}^{N} f_n(u'-u_n', v'-v_n'), \qquad (1)$$

where $f_n(u,v)=f_n(u'/s,v'/s)$ (i.e., s magnification 4-f imaging) denotes the desired geometrical feature of the nth trap centered at (u_n',v_n') with the condition that no two traps overlap. For brevity, we also assume a Dirac-delta type imaging point-spread function in our description so that the same functional form of $I(u',v')$ acceptably represents the projected pattern to our sample plane. Thus, for the traps used (in general, $f_n(u,v)$ sets an arbitrary pattern for each trap) with top-hat cross-sectional intensity profiles, we can write $f_n(u,v)=\text{circ}(\sqrt{u^2+v^2}/a_n)$, where a_n defines the cross-sectional radius, which is individually adjustable for each trap. Standard optimization of the GPC process, described in our previous work,[22] is followed to attain high photon efficiency and optimal visibility in generating $I(u',v')$, i.e., I_0 in Eq. (1) is maximized.

At this point, it is worth to discuss the various modes under which the system can operate. To enumerate these modes, we need to describe the reflection that occurs on the SPM. The SPM modulates the incident linearly polarized field $E(u',v')\hat{e}_v$ whose dot product with its complex conjugate is proportional to $I(u',v')$. A computer encodes a unique

phase function $\psi(u',v')$ onto this SPM. Since the director axis of this birefringent device is oriented at 45° with $\hat{e}_v$, the time-dependent reflected intensity pattern $I_r(u',v';t)$ is proportional to the squared modulus of the complex reflected field given by[19]

$$\vec{E}_r(u',v';t) = E(u',v';t)\{-i\sin[\psi(u',v';t)/2]\hat{e}_u$$
$$+ \cos[\psi(u',v';t)/2]\hat{e}_v\}. \qquad (2)$$

The bracketed terms of Eq. (2) denotes a polarization landscape that, in general, varies both in space and time. Depending on how we choose the space and the time dependence of $\phi(u',v';t) \Rightarrow I(u',v';t)$ and $\psi(u',v';t)$, a set of operating modes for 4D optical manipulation can take place. Using a polarizing beam splitter (PBS), we have losslessly decomposed the orthogonally polarized components (s-pol and p-pol) of Eq. (2). The two components are finally projected to the sample along a common optical axis but with opposite $\boldsymbol{k}$ vectors. The system prealignment procedure requires the two projections to be superimposed transversely and focused onto respective planes that have a slight axial separation. The control over the individual relative strengths of polarization components associated with each trap is totally decoupled from each other, as seen from Eq. (2). In other words, using N traps, each particle in a colloidal system under manipulation can be given a generally time-dependent position vector $\vec{r}_n(t)=x_n(t)\hat{e}_x+y_n(t)\hat{e}_y+z_n(t)\hat{e}_z$ or its higher time derivatives (velocity, acceleration, and so on), as well. Functions $x_n(t)$ and $y_n(t)$ respectively possess direct proportionality to the SLM-controlled variables $u_n(t)$ and $v_n(t)$ in Eq. (1), and $z_n(t)$ is separately controlled by $\psi(u,v;t)$ in Eq. (2). In Fig. 2, we illustrate one operation mode for creating full-space trajectories of simultaneously trapped polystyrene microbeads.[23] The user-controlled positions $|(u_n,v_n)|_{n=1,2,3}$ of three traps enable us to interactively position three particles at the vertices of an imaginary triangle. We set $|(u_n,v_n)|_{n=1,2,3}$ to progressively vary taking coordinates that lie on a circle with $\psi=\psi(u',v')$, i.e., a temporally fixed polarization landscape. This effectively results in identical but out-of-phase time-varying position vectors for the particles, i.e.: $\vec{r}_1(t)=\vec{r}_2(t-T/3)=\vec{r}_3(t-2T/3)$, which trace out the 3D path shown in Fig. 2. T denotes the time it takes for a sphere to traverse one cycle of the path.

The most general mode of operation is achieved by making the axial trap position an active control parameter like the transverse trap position rather than applying a fixed polarization landscape. To do so, we simply encode $\psi(u',v';t)$ of the following form:

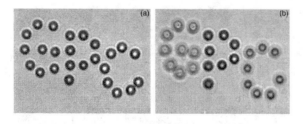

FIG. 3. (Color online) User-coordinated patterning of commercially dyed polystyrene spheres (diameter$=3$ μm) in (a) 2D; and (b) 3D, ($N=25$), forming the text "GPC," using a distinct sphere-color for each character. In (b) virtual planes $z=-5$ μm, $z=0$, and $z=+5$ μm intersect the centers-of-mass of the red, blue, and yellow-dyed spheres, respectively.

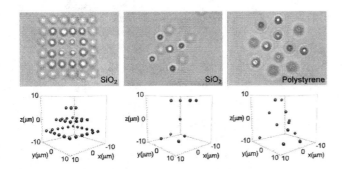

FIG. 4. (Color online) Optical trapping ($N_{max}=36$) and manipulation into arbitrary constellations of colloidal microspheres (SiO$_2$, diameter $=2.25$ μm; polystyrene, diameter$=3$ μm), top. 3D rendered view of the spheres' relative positions, bottom (see Ref. 23).

$$\psi(u',v';t) = \pi \sum_{n=1}^{N} \gamma_n(t) f_n[u' - u'_n(t), v' - v'_n(t)]. \tag{3}$$

Note that this is similar to $\phi(u,v)$ in Eq. (1), differing only by a factor $0 \le \gamma_n(t) \le 1$ inside the summation. By recalling that in Eq. (2) $\psi(u',v')$ determines the relative strengths of the orthogonal polarization components, one finds that the indexed weighting term $\gamma_n(t)$ enables axial control to be completely independent for each trap. The application of various operation modes has enabled us to perform genuinely real-time interactive optical manipulation of large arrays of particles (Figs. 3 and 4).[23] We believe that these illustrations of interactively formed colloidal constellations and their user-defined dynamics have profound implications. It demonstrates the possibility of our system to allow access into exciting experiments involving arbitrarily patterned or dynamically driven systems of colloidal particles, including colloidal arrays created as analogs for modeling systems that are irrepressible or inaccessible within atomic and molecular domains.[2,3] For microbiologists, it serves as a noninvasive tool for manipulating cells into spatial configurations that may trigger variations in developmental features.

Over other alternatives, the GPC-based system offers several advantages. Further scaling of the real-time user-interactive manipulation to larger particle arrays is feasible since the GPC setup can in theory reconstruct fully dynamic arbitrary arrays with more than 90% efficiency requiring no computational overhead, which is compulsory in a DO-based approach.[12] In the GPC-based system, trap resolution may be set equal to the SLM resolution and the lateral intensity distribution that form the traps stretch along the extent of the microscope field of view. Photon efficiency in a DO-based system depends on the spatial trap locations, and is compromised by the limited space-bandwidth product (SBWP) of the SLM, by the unwanted zero order and higher orders accompanying any desired diffraction pattern and by the strong spherical aberrations along the depth dimension associated with the use of high numerical aperture objective lens focusing. The GPC-based 4D manipulation system described here uses two key devices: a SLM to modulate phase and a SPM to modulate polarization of light. Nevertheless, unlike the

complex task imposed by a DO-based method to synthesize a 3D light distribution on a single 2D phase-encoding device, separated transverse and axial trap control, which are not constrained by the SBWP, are assigned to the two devices, respectively. Finally, we emphasize that the simplicity of the binary phase modulation and the spatial polarization modulation scheme implies the future use of low-cost liquid crystal display technology.

Focusing on two attributes of an electromagnetic field, namely phase and polarization, we have formulated an energy-efficient way for crafting a single laser beam into real-time adjustable arrays of counterpropagating light fields, whose confining optical potentials enable the manipulation of multiple particles in 4D. The proposed system can, in due course, become an essential tool for unraveling issues that are yet to be understood in the physics of colloidal systems and the biology of microorganisms. It also has the potential for a range of technological applications where current optical manipulation schemes fall short.

This work has been funded by the European Science Foundation through the Eurocores-SONS program and partially by an internal grant awarded by Risø National Laboratory. The authors also acknowledge T. Hara and Y. Kobayashi of Hamamatsu Photonics for valuable discussions.

[1]D. G. Grier, Nature (London) **424**, 810 (2003).
[2]M. M. Burns, J. M. Fournier, and J. A. Golovchenko, Phys. Rev. Lett. **63**, 1233 (1989).
[3]M. M. Burns, J. M. Fournier, and J. A. Golovchenko, Science **249**, 749 (1990).
[4]A. Simon and A. Libchaber, Phys. Rev. Lett. **68**, 3375 (1992).
[5]A. T. O'Neil, I. MacVicar, L. Allen, and M. J. Padgett, Phys. Rev. Lett. **88**, 053601 (2002).
[6]A. Terray, J. Oakley, and D. W. M. Marr, Science **296**, 1841 (2002).
[7]M. P. MacDonald, G. C. Spalding, and K. Dholakia, Nature (London) **426**, 421 (2003).
[8]J. Glückstad, Nature Mater. **3**, 9 (2004).
[9]S. A. Tatarkova, A. E. Carruthers, and K. Dholakia, Phys. Rev. Lett. **89**, 283901 (2002).
[10]J. P. Hoogenboom, D. L. J. Vossen, C. Faivre-Moskalenko, M. Dogerom, and A. van Blaaderen, Appl. Phys. Lett. **80**, 4828 (2002).
[11]P. J. Rodrigo, R. L. Eriksen, V. R. Daria, and J. Glückstad, Opt. Express **10**, 1550 (2002).
[12]J. Leach, G. Sinclair, P. Jordan, J. Courtial, M. J. Padgett, J. Cooper, and Z. J. Laczik, Opt. Express **12**, 220 (2004).
[13]P. Jordan, H. Clare, L. Flendrig, J. Leach, J. Cooper, and M. Padgett, J. Mod. Opt. **51**, 627 (2004).
[14]A. Terray, J. Oakley, and D. W. M. Marr, Appl. Phys. Lett. **81**, 1555 (2002).
[15]R. L. Eriksen, V. R. Daria, and J. Glückstad, Opt. Express **10**, 597 (2002).
[16]V. R. Daria, R. L. Eriksen, and J. Glückstad, J. Mod. Opt. **50**, 1601 (2003).
[17]V. R. Daria, P. J. Rodrigo, and J. Glückstad, Appl. Phys. Lett. **84**, 323 (2004).
[18]P. J. Rodrigo, V. R. Daria, and J. Glückstad, Opt. Lett. **19**, 2270 (2004).
[19]R. L. Eriksen, P. C. Mogensen, and J. Glückstad, Opt. Commun. **187**, 325 (2001).
[20]J. Glückstad, Opt. Commun. **130**, 225–230 (1996).
[21]J. Glückstad, US Patent No. 6011874 (2000).
[22]J. Glückstad and P. C. Mogensen, Appl. Opt. **40**, 268 (2001).
[23]Associated movie files can be found at http://www.ppo.dk/files/APL-supplementary.

Optimized holographic optical traps

Marco Polin,[1]**, Kosta Ladavac,**[2]**, Sang-Hyuk Lee,**[1] **Yael Roichman,**[1]
and David G. Grier[1]

[1]*Dept. of Physics and Center for Soft Matter Research, New York University, New York, NY 10003*
[2]*Dept. of Physics, James Franck Institute and Institute for Biophysical Dynamics, The University of Chicago, Chicago, IL 60637*

david.grier@nyu.edu

Abstract: Holographic optical traps use the forces exerted by computer-generated holograms to trap, move and otherwise transform mesoscopically textured materials. This article introduces methods for optimizing holographic optical traps' efficiency and accuracy, and an optimal statistical approach for characterizing their performance. This combination makes possible real-time adaptive optimization.

OCIS codes: (140.7010) Trapping; (090.1760) Computer holography; (120.4610) Optical fabrication

References and links

1. A. Ashkin, J. M. Dziedzic, J. E. Bjorkholm and S. Chu. "Observation of a single-beam gradient force optical trap for dielectric particles." Opt. Lett. **11**, 288–290 (1986).
2. E. R. Dufresne and D. G. Grier. "Optical tweezer arrays and optical substrates created with diffractive optical elements." Rev. Sci. Instr. **69**, 1974–1977 (1998).
3. M. Reicherter, T. Haist, E. U. Wagemann and H. J. Tiziani. "Optical particle trapping with computer-generated holograms written on a liquid-crystal display." Opt. Lett. **24**, 608–610 (1999).
4. J. Liesener, M. Reicherter, T. Haist and H. J. Tiziani. "Multi-functional optical tweezers using computer-generated holograms." Opt. Commun. **185**, 77–82 (2000).
5. E. R. Dufresne, G. C. Spalding, M. T. Dearing, S. A. Sheets and D. G. Grier. "Computer-generated holographic optical tweezer arrays." Rev. Sci. Instr. **72**, 1810–1816 (2001).
6. J. E. Curtis, B. A. Koss and D. G. Grier. "Dynamic holographic optical tweezers." Opt. Commun. **207**, 169–175 (2002).
7. D. G. Grier. "A revolution in optical manipulation." Nature **424**, 810–816 (2003).
8. P. T. Korda, G. C. Spalding, E. R. Dufresne and D. G. Grier. "Nanofabrication with holographic optical tweezers." Rev. Sci. Instr. **73**, 1956–1957 (2002).
9. P. T. Korda, M. B. Taylor and D. G. Grier. "Kinetically locked-in colloidal transport in an array of optical tweezers." Phys. Rev. Lett. **89**, 128301 (2002).
10. A. Jesacher, S. Furhpater, S. Bernet and M. Ritsch-Marte. "Size selective trapping with optical "cogwheel" tweezers." Opt. Express **12**, 4129–4135 (2004).
11. K. Ladavac, K. Kasza and D. G. Grier. "Sorting by periodic potential energy landscapes: Optical fractionation." Phys. Rev. E **70**, 010901(R) (2004).
12. M. Pelton, K. Ladavac and D. G. Grier. "Transport and fractionation in periodic potential-energy landscapes." Phys. Rev. E **70**, 031108 (2004).
13. S.-H. Lee, K. Ladavac, M. Polin and D. G. Grier. "Observation of flux reversal in a symmetric optical thermal ratchet." Phys. Rev. Lett. **94**, 110601 (2005).
14. S.-H. Lee and D. G. Grier. "Flux reversal in a two-state symmetric optical thermal ratchet." Phys. Rev. E **71**, 060102(R) (2005).
15. A. Jesacher, S. Fürhapter, S. Bernet and M. Ritsch-Marte. "Size selective trapping with optical "cogwheel" tweezers." Opt. Express **12**, 4129–4135 (2004).
16. V. Soifer, V. Kotlyar and L. Doskolovich. *Iterative Methods for Diffractive Optical Elements Computation* (Taylor & Francis, Bristol, PA, 1997).

17. H. He, N. R. Heckenberg and H. Rubinsztein-Dunlop. "Optical particle trapping with higher-order doughnut beams produced using high efficiency computer generated holograms." J. Mod. Opt. **42**, 217–223 (1995).

18. H. He, M. E. J. Friese, N. R. Heckenberg and H. Rubinsztein-Dunlop. "Direct observation of transfer of angular momentum to absorptive particles from a laser beam with a phase singularity." Phys. Rev. Lett. **75**, 826–829 (1995).

19. K. T. Gahagan and G. A. Swartzlander. "Optical vortex trapping of particles." Opt. Lett. **21**, 827–829 (1996).

20. N. B. Simpson, L. Allen and M. J. Padgett. "Optical tweezers and optical spanners with Laguerre-Gaussian modes." J. Mod. Opt. **43**, 2485–2491 (1996).

21. M. Meister and R. J. Winfield. "Novel approaches to direct search algorithms for the design of diffractive optical elements." Opt. Commun. **203**, 39–49 (2002).

22. J. L. Aragón, G. G. Naumis and M. Torres. "A multigrid approach to the average lattices of quasicrystals." Acta Cryst. **A58**, 352–360 (2002).

23. K. Svoboda, C. F. Schmidt, B. J. Schnapp and S. M. Block. "Direct observation of kinesin stepping by optical trapping interferometry." Nature **365**, 721–727 (1993).

24. L. P. Ghislain, N. A. Switz and W. W. Webb. "Measurement of small forces using an optical trap." Rev. Sci. Instr. **65**, 2762–2768 (1994).

25. F. Gittes, B. Schnurr, P. D. Olmsted, F. C. MacKintosh and C. F. Schmidt. "Microscopic viscoelasticity: Shear moduli of soft materials determined from thermal fluctuations." Phys. Rev. Lett. **79**, 3286–3289 (1997).

26. E.-L. Florin, A. Pralle, E. H. K. Stelzer and J. K. H. Hörber. "Photonic force microscope calibration by thermal noise analysis." Appl. Phys. A **66**, S75–S78 (1998).

27. K. Berg-Sørensen and H. Flyvbjerg. "Power spectrum analysis for optical tweezers." Rev. Sci. Instr. **75**, 594–612 (2004).

28. F. Gittes and C. F. Schmidt. "Interference model for back-focal-plane displacement detection in optical tweezers." Opt. Lett. **23**, 7–9 (1998).

29. J. C. Crocker and D. G. Grier. "Methods of digital video microscopy for colloidal studies." J. Colloid Interface Sci. **179**, 298–310 (1996).

30. G. E. P. Box and G. M. Jenkins. *Time Series Analysis: Forecasting and Control* (Holden-Day, San Francisco, 1976).

31. H. Risken. *The Fokker-Planck Equation* (Springer-Verlag, Berlin, 1989), 2nd ed.

32. M. Polin, D. G. Grier and S. Quake. "Anomalous vibrational dispersion in holographically trapped colloidal arrays." Phys. Rev. Lett. submitted for publication (2005).

33. P. T. Korda, G. C. Spalding and D. G. Grier. "Evolution of a colloidal critical state in an optical pinning potential." Phys. Rev. B **66**, 024504 (2002).

1. Introduction

A single laser beam brought to a focus with a strongly converging lens forms a type of optical trap widely known as an optical tweezer [1]. Multiple beams of light passing simultaneously through the lens' input pupil focus to multiple optical tweezers, each at a location determined by the associated beam's angle of incidence and degree of collimation as it enters the lens. Their intersection at the input pupil yields an interference pattern whose amplitude and phase corrugations characterize the downstream trapping pattern. Imposing the same modulations on a single incident beam at the input pupil would yield the same pattern of traps. Such wavefront modification can be performed by a computer-designed diffractive optical element (DOE), or hologram.

Holographic optical trapping (HOT) uses computer-generated holograms (CGHs) to project arbitrary configurations of optical traps [2, 3, 4, 5, 6], and so provides exceptional control over microscopic materials dispersed in fluid media. Holographic micromanipulation provides the basis for a rapidly growing field of applications in the physical and biological sciences as well as in industry [7].

This article describes refinements to the HOT technique that help to optimize the traps' performance. It also introduces self-consistent and statistically optimal methods for characterizing their performance. Section 2 describes modifications to the basic HOT optical train that compensate for practical limitations of dynamic holography. Section 3 discusses a direct search algorithm for HOT CGH computation that is both faster and more accurate than commonly used iterative refinement algorithms. Together, these modifications yield marked improvements in the holographic traps' performance that can be quantified rapidly using techniques introduced in Section 4. These techniques are based on optimal statistical analysis of trapped colloidal spheres' thermally-driven motions, and lend themselves to simultaneous real-time characterization and optimization of entire arrays of traps through digital video microscopy. Such adaptive optimization is demonstrated experimentally in Section 5.

2. Improved optical train

Figure 1(a) depicts a conventional HOT implementation in which a collimated laser beam is modified by a computer-designed DOE, and thereafter propagates as a superposition of independent beams, each with individually specified wavefront characteristics [4, 6]. These beams are relayed to the input pupil of a high-numerical-aperture lens, typically a microscope objective, which focuses them into optical traps. Although a transmissive DOE is shown in Fig. 1, comparable results are obtained with reflective DOEs. The same objective lens used to form the optical traps also can be used to create images of trapped objects. The associated illumination and image-forming optics are omitted from Fig. 1 for clarity.

Practical DOEs only diffract a portion of the incident light into the intended modes and directions. Some of the incident beam may not be diffracted at all, and the undiffracted portion typically forms an unwanted trap in the middle of the field of view [8]. This "central spot" has been removed in previous implementations by spatially filtering the diffracted beam [8, 9]. Practical DOEs also tend to project spurious "ghost" traps into symmetry-dictated positions within the sample. Spatially filtering a large number of ghost traps generally is not practical, particularly in the case of dynamic holographic optical tweezers whose traps move freely in three dimensions. Projecting holographic traps in the off-axis Fresnel geometry automatically

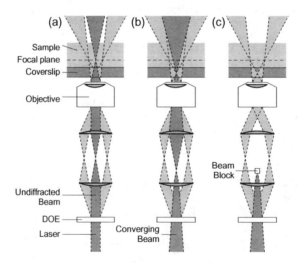

Fig. 1. Simplified schematic of a holographic optical tweezer optical train before and after modification. (a) A collimated beam is split into multiple beams by the DOE, each of which is shown here as being collimated. The diffracted beams pass through the input pupil of an objective lens and are focused into optical traps in the objective's focal plane. The undiffracted portion of the beam, shown here with the darkest shading, also focuses into the focal plane. (b) The input beam is converging as it passes through the DOE. The DOE collimates the diffracted beams, so that they focus into the focal plane, as in (a). The undiffracted beam comes to a focus within the coverslip bounding the sample. (c) A beam block can eliminate the undiffracted beam without substantially degrading the optical traps.

eliminates the central spot [10], but limits the number of traps that can be projected, and also does not address the formation of ghost traps.

Figure 1(b) shows a basic improvement that minimizes the central spot's influence and effectively eliminates ghost traps. Rather than illuminating the DOE with a collimated laser beam, a converging beam is used. This moves the undiffracted central spot upstream of the objective's normal focal plane. The intended traps can be moved back to the focal plane by incorporating wavefront-shaping phase functions into the hologram's design [6]. Deliberately decollimating the input beam allows the central spot to be projected outside of the sample volume, thereby ensuring that the undiffracted beam lacks both the intensity and the intensity gradients needed to influence the sample's dynamics.

An additional consequence of the traps' displacement relative to the modified optical train's focal plane is that most ghost traps also are projected out of the sample volume. This is a substantial improvement for processes such as optical fractionation [11, 12] and optical ratchets [13, 14], which require defect-free intensity distributions.

Even though the undiffracted beam may not create an actual trap in this modified optical train, it still can exert radiation pressure on parts of the sample near the center of the field of view. This is a particular problem for large arrays of optical traps in that the central spot, which typically receives a fixed proportion of the input beam, can be brighter than the intended traps. Illuminating the DOE with a diverging beam [15] reduces the undiffracted beam's influence by projecting some of its light out of the optical train. In a thick sample, however, this has the deleterious effect of projecting both the weakened central spot and the undiminished ghost traps into the sample.

These problems all can be mitigated by placing a beam block as shown in Fig. 1(c) in the intermediate focal plane within the relay optics to spatially filter the undiffracted portion of the beam. The trap-forming beams focus downstream of the beam block and therefore are only partially occluded, even if they pass directly along the optical axis. This has little effect on the performance of conventional optical tweezers and can be compensated by increasing the occluded traps' relative brightness.

3. Algorithms for HOT CGH Calculation

Holographic optical tweezers' efficacy is determined by the quality of the trap-forming DOE, which in turn reflects the performance of the algorithms used in their computation. Previous studies have applied holograms calculated by simple linear superposition of the input fields [3], with best results being obtained with random relative phases [4, 6], or with variations [4, 5, 6] on the classic Gerchberg-Saxton and adaptive-additive algorithms [16]. Despite their generality, these algorithms yield traps whose relative intensities can differ greatly from their design values, and often project an unacceptably large fraction of the input power into ghost traps. These problems can become acute for complicated three-dimensional trapping patterns, particularly when the same hologram also is used as a mode converter to project multifunctional arrays of optical traps [4, 6]. This section describes faster and more effective algorithms for HOT DOE calculation based on direct search optimization.

The holograms used for holographic optical trapping typically operate only on the phase of the incident beam, and not its amplitude. Such phase-only holograms, also known as kinoforms, are far more efficient than amplitude-modulating holograms, which necessarily divert light away from the traps. Quite general trapping patterns can be achieved with kinoforms because optical tweezers rely for their operation on intensity gradients and not on phase variations. The challenge is to find a phase pattern in the input plane of the objective lens that encodes the desired intensity pattern in the focal volume.

Most approaches to designing phase-only holograms are based on scalar diffraction theory, in which the complex field $E(r)$, in the focal plane of a lens of focal length f is related to the field, $u(\rho) \exp(i\varphi(\rho))$, in its input plane by a Fraunhofer transform,

$$E(r) = \int u(\rho) \exp(i\varphi(\rho)) \exp\left(-i\frac{kr \cdot \rho}{f}\right) d^2\rho. \qquad (1)$$

Here, $u(\rho)$ and $\varphi(\rho)$ are the real-valued amplitude and phase, respectively, at position ρ in the input pupil, and $k = 2\pi/\lambda$ is the wavenumber of light of wavelength λ. If $u(\rho)$ is taken to be the amplitude profile of the input laser beam, then $\varphi(\rho)$ is the kinoform encoding the intensity distribution $I(r) = |E(r)|^2$. Finding the kinoform $\varphi(\rho)$ to project a particular pattern, $I(r)$, is nontrivial because the inherent nonlinearity of Eq. (1) defies straightforward inversion. Even so, kinoforms may be estimated through indirect search algorithms. The particular requirements of holographic trapping lend themselves to especially fast and effective computation.

Simply computing the superposition of input beams required to create a desired trapping pattern, disregarding the resulting amplitude variations, and retaining the phase as an estimate for $\varphi(\rho)$ turns out to be remarkably effective [3, 4, 6]. Fidelity to design is particularly good if the input beams are assumed to have random relative phases. Not surprisingly, such randomly phased superpositions yield traps with widely varying intensities as well as a great many ghost traps.

The process of refining such an initial estimate begins by noting that most practical DOEs, including those projected with SLMs, consist of an array of discrete phase pixels, $\varphi_j = \varphi(\rho_j)$, centered at locations ρ_j, each of which can impose any of P possible discrete phase shifts on the incident beam. The field in the focal plane due to such an N-pixel DOE is, therefore,

$$E(r) = \sum_{j=1}^{N} u_j \exp(i\varphi_j) T_j(r), \qquad (2)$$

where the transfer function describing the light's propagation from input plane to output plane is

$$T_j(r) = \exp\left(-i\frac{kr \cdot \rho_j}{f}\right). \qquad (3)$$

Unlike more general holograms, the desired field in the output plane of a holographic optical trapping system consists of M discrete bright spots located at r_m:

$$E(r) = \sum_{m=1}^{M} E_m \delta(r - r_m), \quad \text{with} \qquad (4)$$

$$E_m = \alpha_m \exp(i\xi_m), \qquad (5)$$

where α_m is the relative amplitude of the m-th trap, normalized by $\sum_{m=1}^{M} |\alpha_m|^2 = 1$, and ξ_m is its (arbitrary) phase. Here $\delta(r)$ represents the amplitude profile of the focused beam of light, and may be approximated by a two-dimensional Dirac delta function. For simplicity, we may also approximate the input beam's amplitude profile by a top-hat function with $u_j = 1$ within the input pupil's aperture, and $u_j = 0$ elsewhere. In these approximations, the field at the m-th trap is [6]

$$E_m = \sum_{j=1}^{N} \mathbf{K}_{j,m}^{-1} \mathbf{T}_{j,m} \exp(i\varphi_j), \qquad (6)$$

with $\mathbf{T}_{j,m} = T_j(r_m)$. We introduce the inverse operator $\mathbf{K}_{j,m}^{-1}$ because the hologram φ_j may modify the wavefronts of each of the diffracted beams it creates in addition to establishing its direction of propagation. Such wavefront distortions are useful for creating three-dimensional arrays of multifunctional traps. However, they also distort the traps' otherwise sharply peaked profiles in the focal plane, which were assumed in Eq. (4). The inverse operators correct for these distortions so that even generalized traps can be treated discretely.

For example, a trap can be displaced a distance z away from the focal plane by curving the input beam's wavefronts into a parabolic profile

$$\varphi_z(\rho, z) = \frac{k\rho^2 z}{2f^2}. \tag{7}$$

The operator that displaces the m-th trap to z_m is [4, 6]

$$\mathbf{K}_{j,m}^z = \exp(i\varphi_z(\rho_j, z_m)). \tag{8}$$

Its inverse, $\mathbf{K}_{j,m}^{z}{}^{-1} = \mathbf{K}_{j,m}^{z}{}^{*}$ returns the m-th trap to best focus in the focal plane.

Similarly, a conventional TEM$_{00}$ beam can be converted into a helical mode with the phase profile

$$\varphi_\ell(\rho, \ell) = \ell\theta, \tag{9}$$

where θ is the azimuthal angle around the optical axis and ℓ is a winding number known as the topological charge. Such corkscrew-like beams focus to ring-like optical traps known as optical vortices, which can exert torques as well as forces [17, 18, 19, 20]. The topology-transforming kernel [6]

$$\mathbf{K}_{j,m} = \exp(i\varphi_\ell(\rho_j, \ell_m)) \tag{10}$$

can be composed with $\mathbf{T}_{j,m}$ in the same manner as the displacement-inducing $\mathbf{K}_{j,m}^z$ to convert the m-th trap into an optical vortex. A variety of analogous phase-based mode transformations have been described, each with applications to single-beam optical trapping [7], all of which can be applied to each trap independently in this manner.

Calculating the fields only at the traps' positions greatly reduces the computational burden of HOT CGH refinement. It also eliminates the need to account for the beams' propagation through intermediate trapping planes when designing three-dimensional patterns [4]. Unlike more general FFT-based algorithms [5], this restricted approach does not directly optimize the field between the traps. If the converged amplitudes match the design values, however, no light is left over to create ghost traps.

Applying Eq. (6) directly in an iterative refinement algorithm [6] also has drawbacks. In particular, only the $M - 1$ relative phases ξ_m in Eq. (5) can be adjusted when inverting Eq. (6) to solve for φ_j. Having so few free parameters severely limits the improvement over simple superposition that can be obtained. Equation (6) suggests an alternative approach that not only is far more effective, but also is substantially more efficient.

The operator $\mathbf{K}_{j,m}^{-1}\mathbf{T}_{j,m}$ describes how light in the mode of the m-th trap propagates from the j-th phase pixel on the DOE to the trap's projected position r_m. Changing the pixel's value by $\Delta\varphi_j$ therefore changes each trap's field by

$$\Delta E_m = \mathbf{K}_{j,m}^{-1}\mathbf{T}_{j,m}\exp(i\varphi_j)\left[\exp(i\Delta\varphi_j) - 1\right]. \tag{11}$$

If such a change were to improve the overall pattern, we would be inclined to retain it, and to seek other such improvements. This is the basis for direct search algorithms. The simplest involves selecting a pixel at random from a trial phase pattern, changing its value to any of the $P - 1$ alternatives, and computing the effect on the projected pattern. Quite clearly, there is a considerable computational advantage in calculating changes only at the M traps' positions, rather than over the entire focal plane. The updated trial amplitudes then are compared with their design values and the proposed change is accepted if the overall error is reduced. The process is repeated until the result converges to the design or the acceptance rate for proposed changes dwindles.

Effective and efficient refinement by the direct search algorithm depends on the choice of metric for quantifying convergence. The standard cost function, $\chi^2 = \sum_{m=1}^{M}(I_m - \varepsilon I_m^{(D)})^2$, assesses the mean-squared deviations of the m-th trap's projected intensity $I_m = |\alpha_m|^2$ from its design value $I_m^{(D)}$, assuming an overall diffraction efficiency of ε. It requires an accurate estimate for ε and places no emphasis on uniformity in the projected traps' intensities. An alternative proposed by Meister and Winfield [21],

$$C = -\langle I \rangle + f\sigma, \tag{12}$$

avoids both shortcomings. Here, $\langle I \rangle$ is the mean intensity at the traps and

$$\sigma = \sqrt{\frac{1}{M}\sum_{m=1}^{M}(I_m - \gamma I_m^{(D)})^2} \tag{13}$$

measures the deviation from uniform convergence to the design intensities. Selecting

$$\gamma = \frac{\sum_{m=1}^{M} I_m I_m^{(D)}}{\sum_{m=1}^{M}\left(I_m^{(D)}\right)^2} \tag{14}$$

minimizes the total error and optimally accounts for non-ideal diffraction efficiency [21]. The weighting fraction f sets the relative importance attached to diffraction efficiency versus uniformity, with $f = 0.5$ providing a generally useful balance.

A direct binary search proceeds with any candidate change that reduces C being accepted, and all others being rejected. In a worst-case implementation, the number of trials required for convergence should scale as NP, the product of the number of phase pixels and the number of possible phase values. In practice, this estimate is accurate if P and N are comparatively small and if the starting phase function is either uniform or purely random. Much faster convergence can be obtained by starting from the a randomly phased superposition of input beams. In this case, convergence typically is obtained within N trials, even for fairly complex trapping patterns, and thus requires a computational effort comparable to the initial superposition.

As a practical demonstration, we have implemented a quasiperiodic array of optical traps, which is challenging because it has no translational symmetries. The traps are focused with a $100\times$ NA 1.4 S-Plan Apo oil immersion objective lens mounted in a Nikon TE-2000U inverted optical microscope. The traps are powered by a Coherent Verdi frequency-doubled diode-pumped solid state laser operating at a wavelength of 532 nm. Computer-generated phase holograms are imprinted on the beam with a Hamamatsu X8267-16 parallel-aligned nematic liquid crystal spatial light modulator (SLM). This SLM can impose phase shifts up to 2π radians at each pixel in a 768×768 array. The face of the SLM is imaged onto the objective's 5 mm diameter input pupil using relay optics designed to minimize aberrations. The beam is directed into the objective with a dichroic beamsplitter, which allows images to pass through to a low-noise charge-coupled device (CCD) camera (NEC TI-324AII). The video stream is recorded as uncompressed digital video with a Pioneer 520H digital video recorder (DVR) for processing.

Figure 2(a) shows the intended planar arrangement of 119 holographic optical traps designed by the dual generalized method for generating quasiperiodic lattices [22]. Even after adaptive-additive refinement, the hologram resulting from simple superposition with random phases fares poorly for this aperiodic pattern. Figure 2(b) shows the intensity of light reflected by a front-surface mirror placed in the sample plane. This image reveals extraneous ghost traps, an exceptionally bright central spot, and large variability in the intended traps' intensities. Imaging photometry on this and equivalent images produced with different random relative phases for the beams yields a typical root-mean-square (RMS) variation of more than 50 percent in the projected traps' brightness. The image in Fig. 2(c) was produced using the modified optical train described in Sec. 2 and the direct search algorithm described in Sec. 3, and suffers from none of these defects. Both the ghost traps and the central spot are suppressed, and the apparent relative brightness variations are smaller than 5 percent, a factor of ten improvement. Figure 2(d) shows 119 colloidal silica spheres, $2a = 1.5 \pm 0.3$ μm in diameter (Bangs Labs, lot 5238), dispersed in water at $T = 27°C$ and trapped in the quasiperiodic array.

To place the benefits of the direct search algorithm on a more quantitative basis, we augment standard figures of merit with those introduced in Ref. [21]. In particular, the DOE's theoretical diffraction efficiency is commonly defined as

$$\mathscr{Q} = \frac{1}{M} \sum_{m=1}^{M} \frac{I_m}{I_m^{(D)}},$$ (15)

and its root-mean-square (RMS) error as

$$e_{\text{rms}} = \frac{\sigma}{\max(I_m)}.$$ (16)

The resulting pattern's departure from uniformity is usefully gauged as [21]

$$u = \frac{\max(I_m/I_m^{(D)}) - \min(I_m/I_m^{(D)})}{\max(I_m/I_m^{(D)}) + \min(I_m/I_m^{(D)})}.$$ (17)

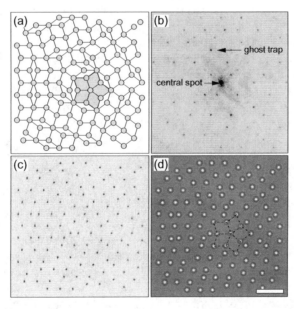

Fig. 2. (a) Design for 119 identical optical traps in a two-dimensional quasiperiodic array. (b) Trapping pattern projected without optimizations using the adaptive-additive algorithm. (c) Trapping pattern projected with optimized optics and adaptively corrected direct search algorithm. (d) Bright-field image of colloidal silica spheres 1.53 μm in diameter dispersed in water and organized in the optical trap array. The scale bar indicates 10 μm

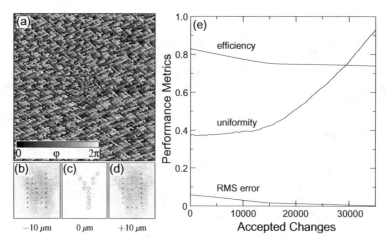

Fig. 3. A three-dimensional multifunctional holographic optical trap array created with the direct search algorithm. (a) Refined DOE phase pattern. (b), (c) and (d) The projected optical trap array at $z = -10$ μm, 0 μm and $+10$ μm. Traps are spaced by 1.2 μm in the plane, and the 12 traps in the middle plane consist of $\ell = 8$ optical vortices. (e) Performance metrics for the hologram in (a) as a function of the number of accepted single-pixel changes. Data include the DOE's overall diffraction efficiency as defined by Eq. (15), the projected pattern's RMS error from Eq. (16), and its uniformity, $1 - u$, where u is defined in Eq. (17).

Figure 3 shows results for a HOT DOE encoding 51 traps, including 12 optical vortices of topological charge $\ell = 8$, arrayed in three planes relative to the focal plane. The excellent results in Fig. 3 were obtained with a *single* pass of direct-search refinement. The resulting traps, shown in the bottom three images, again vary from their planned relative intensities by less than 5 percent. In this case, the spatially extended vortices were made as bright as the point-like optical tweezers by increasing their requested relative brightness by a factor of 15. This single hologram, therefore, demonstrates independent control over three-dimensional position, wavefront topology, and brightness of all the traps. Performance metrics for the calculation are plotted in Fig. 3(b) as a function of the number of accepted single-pixel changes, with an overall acceptance rate of 16 percent. Direct search refinement achieves greatly improved fidelity to design over randomly phase superposition at the cost of a small fraction of the diffraction efficiency and roughly doubled computation time. The entire calculation can be completed in the refresh interval of a typical liquid crystal spatial light modulator.

4.   Optimal characterization

Gauging a HOT system's performance numerically and by characterizing the projected intensity pattern does not provide a complete picture. The real test is in the projected traps' ability to localize particles. A variety of approaches have been developed for measuring the forces exerted by optical traps. The earliest involved estimating the hydrodynamic drag required to dislodge a trapped particle [23]. This has several disadvantages, most notably that it identifies only the marginal escape force in a given direction and not the trap's actual three-dimensional potential. Complementary information can be obtained by measuring a particle's thermally driven motions in the trap's potential well [24, 25, 26]. For instance, the measured probability density $P(r)$ for displacements r is related to the trap's potential $V(r)$ through the Boltzmann distribution

$$P(r) \propto \exp(-\beta V(r)), \tag{18}$$

where $\beta^{-1} = k_B T$ is the thermal energy scale at temperature T. Similarly, the power spectrum of $r(t)$ for a harmonically bound particle is a Lorentzian whose width is the viscous relaxation time of the particle in the well [24, 27].

Both of these approaches require amassing enough data to characterize the trapped particle's least probable displacements, and therefore oversample the trajectories. Oversampling is acceptable when data from a single optical trap can be collected rapidly, for example with a quadrant photodiode [24, 25, 26, 28]. Tracking multiple particles in holographic optical traps, however, requires the area detection capabilities of digital video microscopy [29], which yields data much more slowly. Analyzing video data with optimal statistics [30] offers the benefits of thermal calibration by avoiding oversampling.

An optical trap is accurately modeled as a harmonic potential energy well [25, 26, 27, 28],

$$V(r) = \frac{1}{2} \sum_{i=1}^{3} \kappa_i r_i^2, \tag{19}$$

with a different characteristic curvature κ_i along each axis. This form also is convenient because it is separable into one-dimensional contributions. The trajectory of a colloidal particle localized in a viscous fluid by a harmonic well is described by the one-dimensional Langevin equation [31]

$$\dot{x}(t) = -\frac{x(t)}{\tau} + \xi(t), \tag{20}$$

where the autocorrelation time $\tau = \gamma/\kappa$, is set by the particle's viscous drag coefficient γ and by the curvature of the well, κ. The Gaussian random thermal force, $\xi(t)$, has zero mean, $\langle \xi(t) \rangle = 0$, and variance

$$\langle \xi(t) \xi(s) \rangle = \frac{2k_B T}{\gamma} \delta(t - s). \tag{21}$$

If the particle is at position x_0 at time $t = 0$, its trajectory at later times is given by

$$x(t) = x_0 \exp\left(-\frac{t}{\tau}\right) + \int_0^t \xi(s) \exp\left(-\frac{t-s}{\tau}\right) ds. \tag{22}$$

Sampling such a trajectory at discrete times $t_j = j\Delta t$, yields

$$x_{j+1} = \phi x_j + a_{j+1}, \tag{23}$$

where $x_j = x(t_j)$,

$$\phi = \exp\left(-\frac{\Delta t}{\tau}\right), \tag{24}$$

and where a_{j+1} is a Gaussian random variable with zero mean and variance

$$\sigma_a^2 = \frac{k_B T}{\kappa} \left[1 - \exp\left(-\frac{2\Delta t}{\tau}\right) \right]. \tag{25}$$

Because $\phi < 1$, Eq. (23) is an example of an autoregressive process [30], which is readily invertible.

In principle, the particle's trajectory $\{x_j\}$ can be analyzed to extract ϕ and σ_a^2, and thus the trap's stiffness, κ, and the particle's viscous drag coefficient γ. In practice, however, the experimentally measured particle positions y_j differ from the actual positions x_j by random errors b_j, which we assume to be taken from a Gaussian distribution with zero mean and variance σ_b^2. The measurement then is described by the coupled equations

$$x_j = \phi x_{j-1} + a_j \quad \text{and} \quad y_j = x_j + b_j, \tag{26}$$

where b_j is independent of a_j. We still can estimate ϕ and σ_a^2 from a set of measurements $\{y_j\}$ by first constructing the joint probability

$$p(\{x_i\}, \{y_i\} | \phi, \sigma_a^2, \sigma_b^2) = \prod_{j=2}^{N} \left[\frac{\exp\left(-\frac{a_j^2}{2\sigma_a^2}\right)}{\sqrt{2\pi\sigma_a^2}} \right] \prod_{j=1}^{N} \left[\frac{\exp\left(-\frac{b_j^2}{2\sigma_b^2}\right)}{\sqrt{2\pi\sigma_b^2}} \right]. \tag{27}$$

The probability density for measuring the trajectory $\{y_j\}$, is then the marginal [30]

$$p(\{y_j\} | \phi, \sigma_a^2, \sigma_b^2) = \int p(\{x_j\}, \{y_j\} | \phi, \sigma_a^2, \sigma_b^2) \, dx_1 \cdots dx_N \tag{28}$$

$$= \frac{(2\pi\sigma_a^2\sigma_b^2)^{-\frac{N-1}{2}}}{\sqrt{\sigma_b^2 \det(\mathbf{A}_\phi)}} \exp\left(-\frac{1}{2\sigma_b^2} (y)^T \left[\mathbf{I} - \frac{\mathbf{A}_\phi^{-1}}{\sigma_b^2} \right] y \right), \tag{29}$$

where $y = (y_1, \ldots, y_N)$ with transpose $(y)^T$, $\mathbf{I}$ is the $N \times N$ identity matrix, and

$$\mathbf{A}_\phi = \frac{\mathbf{I}}{\sigma_b^2} + \frac{\mathbf{M}_\phi}{\sigma_a^2}, \tag{30}$$

with the tridiagonal memory tensor

$$\mathbf{M}_\phi = \begin{pmatrix} \phi^2 & -\phi & 0 & 0 & \cdots & 0 \\ -\phi & 1+\phi^2 & -\phi & 0 & \cdots & \vdots \\ 0 & -\phi & 1+\phi^2 & -\phi & \cdots & \vdots \\ 0 & 0 & -\phi & \ddots & \cdots & \vdots \\ \vdots & \vdots & \cdots & -\phi & 1+\phi^2 & -\phi \\ 0 & 0 & \cdots & 0 & -\phi & 1 \end{pmatrix}. \tag{31}$$

Calculating the determinant, $\det(\mathbf{A}_\phi)$, and inverse, $\mathbf{A}_\phi^{-1}$, of $\mathbf{A}_\phi$ is greatly facilitated if we artificially impose time translation invariance by replacing M_ϕ with the $(N+1) \times (N+1)$ matrix that identifies time step $N+1$ with time step 1. Physically, this involves imparting an impulse, a_{N+1}, that translates the particle from its last position, x_N, to its first, x_1. Because diffusion in a potential well is a stationary process, the effect of this change is inversely proportional to the number of measurements, N. With this approximation,

$$\det(\mathbf{A}_\phi) = \prod_{n=1}^{N} \left\{ \frac{1}{\sigma_b^2} + \frac{1}{\sigma_a^2} \left[1 + \phi^2 - 2\phi \cos\left(\frac{2\pi n}{N}\right) \right] \right\} \tag{32}$$

and

$$(\mathbf{A}_\phi^{-1})_{\alpha\beta} = \frac{1}{N} \sum_{n=1}^{N} \frac{\sigma_a^2 \sigma_b^2 \exp\left(i\frac{2\pi}{N} n(\alpha - \beta)\right)}{\sigma_a^2 + \sigma_b^2 \left[1 + \phi^2 - 2\phi \cos\left(\frac{2\pi n}{N}\right)\right]}, \tag{33}$$

so that the conditional probability for the measured trajectory, $\{y_j\}$, is

$$p(\{y_j\} | \phi, \sigma_a^2, \sigma_b^2) = (2\pi)^{-\frac{N}{2}} \prod_{n=1}^{N} \left\{ \sigma_a^2 + \sigma_b^2 \left[1 + \phi^2 - 2\phi \cos\left(\frac{2\pi n}{N}\right) \right] \right\}^{-\frac{1}{2}}$$

$$\times \exp\left(-\frac{1}{2\sigma_b^2} \sum_{n=1}^{N} y_n^2\right) \exp\left(\frac{1}{2\sigma_b^2} \frac{1}{N} \sum_{m=1}^{N} \frac{\tilde{y}_m^2 \sigma_a^2}{\sigma_a^2 + \sigma_b^2 \left[1 + \phi^2 - 2\phi \cos\left(\frac{2\pi m}{N}\right)\right]}\right), \tag{34}$$

where $\tilde{y}_m$ is the m-th component of the discrete Fourier transform of $\{y_n\}$. This can be inverted to obtain the likelihood function for ϕ, σ_a^2, and σ_b^2:

$$L(\phi, \sigma_a^2, \sigma_b^2 | \{y_i\}) = -\frac{N}{2} \ln 2\pi - \frac{1}{2\sigma_b^2} \sum_{n=1}^{N} y_n^2 + \frac{\sigma_a^2}{2\sigma_b^2} \frac{1}{N} \sum_{n=1}^{N} \frac{\tilde{y}_n^2 \sigma_a^2}{\sigma_a^2 + \sigma_b^2 \left[1 + \phi^2 - 2\phi \cos\left(\frac{2\pi n}{N}\right) \right]}$$
$$-\frac{1}{2} \sum_{n=1}^{N} \ln \left(\sigma_a^2 + \sigma_b^2 \left[1 + \phi^2 - 2\phi \cos\left(\frac{2\pi n}{N}\right) \right] \right). \quad (35)$$

Best estimates $(\hat{\phi}, \hat{\sigma_a^2}, \hat{\sigma_b^2})$ for the parameters $(\phi, \sigma_a^2, \sigma_b^2)$ are solutions of the coupled equations

$$\frac{\partial L}{\partial \phi} = \frac{\partial L}{\partial \sigma_a^2} = \frac{\partial L}{\partial \sigma_b^2} = 0. \quad (36)$$

4.1. Case 1: No measurement errors ($\sigma_b^2 = 0$)

Equations (36) can be solved in closed form if $\sigma_b^2 = 0$. In this case,

$$\hat{\phi}_0 = \frac{c_1}{c_0} \quad \text{and} \quad \hat{\sigma}_{a0}^2 = c_0 \left[1 - \left(\frac{c_1}{c_0} \right)^2 \right], \quad (37)$$

where

$$c_m = \frac{1}{N} \sum_{j=1}^{N} y_j y_{(j+m) \bmod N} \quad (38)$$

is the barrel autocorrelation of $\{y_j\}$ at lag m. The associated statistical uncertainties are

$$\Delta \hat{\phi}_0 = \sqrt{\frac{\hat{\sigma}_{a0}^2}{N c_0}} \quad \text{and} \quad \Delta \hat{\sigma}_{a0}^2 = \hat{\sigma}_{a0}^2 \sqrt{\frac{2}{N}}. \quad (39)$$

In the absence of measurement errors, c_0 and c_1 constitute *sufficient statistics* for the time series [30] and thus embody all of the information that can be extracted.

4.2. Case 2: Small measurement errors ($\sigma_b^2 \ll \sigma_a^2$)

The analysis is less straightforward when $\sigma_b^2 \neq 0$ because Eqs. (36) no longer are simply separable. The system of equations can be solved approximately if $\sigma_b^2 \ll \sigma_a^2$. In this case, the best estimates for the parameters can be expressed in terms of the error-free estimates as

$$\hat{\phi} \approx \hat{\phi}_0 \left\{ 1 + \frac{\sigma_b^2}{\hat{\sigma}_{a0}^2} \left[1 - \hat{\phi}_0^2 + \frac{c_2}{c_0} \right] \right\} \quad \text{and} \quad \hat{\sigma}_a^2 \approx \hat{\sigma}_{a0}^2 - \frac{\sigma_b^2}{\hat{\sigma}_{a0}^2} c_0 \left[1 - 5\hat{\phi}_0^4 + 4\hat{\phi}_0^2 \frac{c_2}{c_0} \right], \quad (40)$$

to first order in σ_b^2 / σ_a^2, with statistical uncertainties propagated in the conventional manner. Expansions to higher order in σ_b^2 / σ_a^2 involve additional correlations, and the exact solution involves correlations at all lags m. If measurement errors are small enough for Eq. (40) to apply, the computational savings relative to other approaches can be substantial, and the amount of data required to achieve a desired level of accuracy in the physically relevant quantities, κ and γ, can be reduced dramatically.

The errors in locating colloidal particles' centroids can be calculated from knowledge of the images' signal to noise ratio and the optical train's magnification [29]. Centroid resolutions of 10 nm or better can be attained routinely for micrometer-scale colloidal particles in water using conventional bright-field imaging. In practice, however, mechanical vibrations, video jitter and other processes may increase the measurement error. Quite often, the overall measurement error is most easily assessed by increasing the laser power to the optical traps to minimize the particles' thermally driven motions. In this case, $y_j \approx b_j$, and σ_b^2 can be estimated directly.

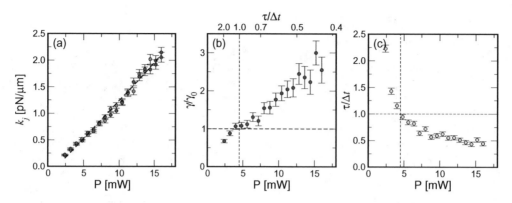

Fig. 4. Power dependence of (a) the trap stiffness, (b) the viscous drag coefficient and (c) the viscous relaxation time for a 1.53 μm diameter silica sphere trapped by an optical tweezer in water.

4.3. Trap characterization

The stiffness and viscous drag coefficient can be estimated simultaneously as

$$\frac{\kappa}{k_B T} = \frac{1 - \hat{\phi}^2}{\hat{\sigma}_a^2} \quad \text{and} \quad \frac{\gamma}{k_B T \Delta t} = -\frac{1 - \hat{\phi}^2}{\hat{\sigma}_a^2 \ln \hat{\phi}}, \tag{41}$$

with error estimates, $\Delta\kappa$ and $\Delta\gamma$, given by

$$\left(\frac{\Delta\kappa}{\kappa}\right)^2 = \left(\frac{\Delta\hat{\sigma}_a^2}{\hat{\sigma}_a^2}\right)^2 + \left(\frac{2\hat{\phi}^2}{1 - \hat{\phi}^2}\right)^2 \left(\frac{\Delta\hat{\phi}}{\hat{\phi}}\right)^2 \quad \text{and} \tag{42}$$

$$\left(\frac{\Delta\gamma}{\gamma}\right)^2 = \left(\frac{\Delta\hat{\sigma}_a^2}{\hat{\sigma}_a^2}\right)^2 + \left(\frac{2\hat{\phi}^2}{1 - \hat{\phi}^2} + \frac{1}{\ln \hat{\phi}}\right)^2 \left(\frac{\Delta\hat{\phi}}{\hat{\phi}}\right)^2. \tag{43}$$

If the measurement interval, Δt, is much longer than the viscous relaxation time $\tau = \gamma/\kappa$, then ϕ vanishes and the error in the drag coefficient diverges. Conversely, if Δt is much smaller than τ, then ϕ approaches unity and both errors diverge. Consequently, this approach does not benefit from excessively fast sampling. Rather, it relies on accurate particle tracking to minimize $\Delta\hat{\phi}$ and $\Delta\hat{\sigma}_a^2$. For trap-particle combinations with viscous relaxation times exceeding a few milliseconds and typical particle excursions of at least 10 nm, digital video microscopy provides the resolution needed to simultaneously characterize multiple optical traps [29].

In the event that measurement errors can be ignored ($\sigma_b^2 \ll \sigma_a^2$),

$$\frac{\kappa_0}{k_B T} = \frac{1}{c_0} \left[1 \pm \sqrt{\frac{2}{N}\left(1 + \frac{2c_1^2}{c_0^2 - c_1^2}\right)}\right] \quad \text{and} \quad \frac{\gamma_0}{k_B T \Delta t} = \frac{1}{c_0 \ln\left(\frac{c_0}{c_1}\right)}\left(1 \pm \frac{\Delta\gamma_0}{\gamma_0}\right) \tag{44}$$

where

$$N\left(\frac{\Delta\gamma_0}{\gamma_0}\right)^2 = 2 + \frac{1}{c_0^2 - c_1^2}\left[\frac{c_0^2 + 2c_1^2 \ln\left(\frac{c_1}{c_0}\right) - c_1^2}{c_1 \ln\left(\frac{c_1}{c_0}\right)}\right]^2. \tag{45}$$

These results are not reliable if $c_1 \lesssim \sigma_b^2$, which arises when the sampling interval Δt is much longer or much shorter than the viscous relaxation time, τ. Accurate estimates for κ and γ still may be obtained in this case by applying Eq. (40).

As a practical demonstration, we analyzed the thermally driven motions of a single silica sphere of diameter 1.53 μm (Bangs Labs lot number 5328) dispersed in water and trapped in a conventional optical tweezer. With the trajectory resolved to within 6 nm at 1/30 sec intervals, 1 minute of data suffices to measure both κ and γ to within 1 percent error using Eqs. (41). The results plotted in Fig. 4(a) indicate trapping efficiencies of $\kappa_x/P = \kappa_y/P = 142 \pm 3$ pN/μmW. Unlike κ, which depends principally on c_0, γ also depends on c_1, which is accurately measured only for $\tau \gtrsim 1$. Over the range of laser powers for which this condition holds, we obtain the expected $\gamma_x/\gamma_0 = \gamma_y/\gamma_0 = 1.0 \pm 0.1$, as shown in Fig. 4(b). The viscous relaxation time becomes substantially shorter than our sampling time for higher powers, so that estimates for γ and its error both become unreliable, as expected.

5. Adaptive optimization

Optimal statistical analysis offers insights not only into the traps' properties, but also into the properties of the trapped particles and the surrounding medium. For example, if a spherical probe particle is immersed in a medium of viscosity η far from any bounding surfaces, its hydrodynamic radius a can be assessed from the measured drag coefficient using the Stokes result $\gamma = 6\pi\eta a$. The viscous drag coefficients, moreover, provide insights into the particles' coupling to each other and to their environment. The independently assessed values of the traps' stiffnesses then can serve as a self-calibration in microrheological measurements and in measurements of colloidal many-body hydrodynamic coupling [32]. In cases where the traps themselves must be calibrated accurately, knowledge of the probe particles' differing properties gauged from measurements of γ can be used to distinguish variations in the traps' intrinsic properties from variations due to differences among the probe particles.

These measurements, moreover, can be performed rapidly enough, even at conventional video sampling rates, to permit real-time adaptive optimization of the traps' properties. Each trap's stiffness is roughly proportional to its brightness. So, if the m-th trap in an array is intended to receive a fraction $|\alpha_m|^2$ of the projected light, then instrumental deviations can be corrected by recalculating the CGH with modified amplitudes:

$$\alpha_m \rightarrow \alpha_m \sqrt{\frac{\sum_{i=1}^N \kappa_i}{\kappa_m}}. \tag{46}$$

Analogous results can be derived for optimization on the basis of other performance metrics. A quasiperiodic pattern similar to that in Fig. 2(c) was adaptively optimized for uniform brightness, with a single optimization cycle yielding better than 12 percent variance from the mean. Applying Eqs. (41) to data from images such as Fig. 2(d) allows us to correlate the traps' appearance to their actual performance.

With each trap powered by 3.4 mW, the mean viscous relaxation time is found to be $\tau/\Delta t = 1.14 \pm 0.11$. We expect reliable estimates for the viscous drag coefficient under these conditions, and the result $\gamma/\gamma_0 = 0.95 \pm 0.10$ with an overall measurement error of 0.01, is consistent with the manufacturer's rated 10 percent polydispersity in particle radius. Variations in the measured stiffnesses, $\langle \kappa_x \rangle = 0.38 \pm 0.06$ pN/μm and $\langle \kappa_y \rangle = 0.35 \pm 0.10$ pN/μm, can be ascribed to a combination of the particles' polydispersity and the traps' inherent brightness variations. This demonstrates that adaptive optimization based on the traps' measured intensities also optimizes their performance in trapping particles.

6. Summary

The quality and uniformity of the holographic optical traps projected with the methods described in the previous sections represent a substantial advance over previously reported results. We have demonstrated that optimized and adaptively optimized HOT arrays can be used to craft highly structured potential energy landscapes with excellent fidelity to design. These optimized landscapes have potentially wide-ranging applications in sorting mesoscopic fluid-borne objects through optical fractionation [11, 12], in fundamental studies of transport [33, 9], dynamics [13, 14] and phase transitions in macromolecular systems, and also in precision holographic manufacturing.

We have benefitted from extensive discussion with Alan Sokal. This work was supported by the National Science Foundation under Grant Number DBI-0233971 with additional support from Grant Number DMR-0451589. S.L. acknowledges support from the Kessler Family Foundation.

8

Applications in Microfluidics

Microfluidics is one of a number of topical burgeoning research fields. Allowing the study of very fundamental physical processes as well as the development of highly applied technologies, the field sits comfortably at the junction of physics, chemistry, materials science, life sciences, and engineering. The highly interdisciplinary nature of microfluidics has made it incredibly attractive for a huge range of researchers to develop new techniques and to integrate old ones into new technology.

Optical manipulation techniques share a number of similarities with microfluidics in that they have a highly interdisciplinary nature, even if their overall scope for applications is more limited. It seems natural then to fuse the two techniques, but there are difficulties with this approach. The field of view of conventional optical tweezers is limited compared to microfluidic systems, and integration of high numerical aperture (NA) optical systems into microfluidic devices is not necessarily straightforward.

One solution to this problem is to make use of counterpropagating beam traps, first demonstrated by Ashkin [1] (see **PAPER 1.2**). The free-space beams used by Ashkin are not suitable for integration into a device but the idea was advanced by Constable et al. [2] (see **PAPER 8.1**), who made use of optical fibers to deliver the light into the trapping region. The dual-beam technique not only allows a wider trapping field of view but also produces larger forces enabling much larger particles to be trapped than in conventional optical tweezers. For example, in Ref. [3], Jess et al. make use of a dual-beam fiber trap to trap a 100-μm diameter polymer sphere, an object way beyond the reach of gradient-force trapping alone. Also of note in Ref. [3] is the fact that the authors show that dual-beam tweezers can be integrated with Raman spectroscopy and microfluidics.

The dual-beam technique has been successfully integrated into a number of microfluidic devices. The most prominent has been the work by Guck et al. and their "optical stretcher" [4] (see **PAPER 2.2**), [5,6]. An object placed in the dual-beam trap becomes stretched: while the total force acting on the object is zero, there is a momentum transfer from the light to the surface as the light passes through the interface acting away from the beam propagation direction. Hence, a suitably elastic object will become stretched in a dual-beam trap. The

technique allows one to probe the viscoelastic properties of the trapped particles, which is particularly useful for cellular material where the deformability provides information about the cytoskeleton. This has been shown to be useful as a cell marker for detecting cancerous cells [7]. By integrating the device into a microfluidic system, cells can be analyzed in a high-throughput manner [8,9].

In recent years, one of the most interesting developments in merging optical manipulation and microfluidics has been the work in which diode lasers are directly integrated onto the microfluidic channel, producing a miniature dual-beam optical trap [10,11]. This holds great promise for making true lab-on-chip devices incorporating light sources for trapping, sample excitation, and light detection or for stretching-type applications.

One of the traditional areas in which optical tweezers have sparked significant research interest is in their ability to allow the properties of the trapping beam to be probed via the particle being trapped. A key example of this is in the area of the angular momentum of light (see Chapter 6). One of the long-standing applications of this work is to rotate micromachines or to drive pumps or mixers in microfluidic devices. Friese et al. demonstrated the basic principal of driving micromachines making use of spin angular momentum [12], while others [13] have made use of shape birefringence to produce optically driven rotating cogs. Optically driven devices can be created directly using light through the two-photon polymerization technique [14–16].

A simple but powerful demonstration of this type of pumping was demonstrated by Leach et al. [17] (see **PAPER 8.2**) in which they make use of two trapped birefringent particles that can be rotated using circularly polarized light. By trapping the particles within a microfluidic channel and then rotating them in opposite directions, fluid can be pumped within the channel at rates of up to 200 fL^{-1}. Moreover, by making use of a probe particle that is trapped, released, and trapped again, they are able to map out the magnitude and direction of fluid flow moving around the rotating particles.

The type of work in which devices are driven by light is largely proof of concept, and little in the way of practical devices has yet been reported. More promising perhaps is the work in which colloidal particles are trapped and used to carry out useful microfluidic functions. One of the earliest examples of this type of work was carried out by Terray et al. [18] (see **PAPER 8.3**), [19] in which they use a scanning optical tweezers to trap multiple particles that they then can optically drive to create rudimentary pumps and multiway valves. The use of holographic optical tweezers has also allowed multiple particles to be trapped and turned into, in particular, microfluidic pumps. Laguerre–Gaussian beams carrying optical angular momentum can be used to rotate a large number of particles quickly enough to drive fluid past the rotating spheres [20]. Similar effects have been demonstrated at an air–water interface [21].

The use of a shaped optical field is the key to one of the seminal papers in the area of optical manipulation and microfluidics. Particles flowing through an optical lattice, an interference pattern of two or more beams, are deflected through their interaction with the light. If the deflection is somehow dependent on the properties of the particles, then this effect can be used for sorting, which was shown by MacDonald et al. [22] (see **PAPER 2.1**). As optical gradient forces are dependent on the polarizability of the particle, the size, refractive index, and even morphology of the particle become sorting parameters, as the polarizability is dependent on these properties. In Ref. [22], sorting of two species is demonstrated based on both size and refractive index (separating polymer beads from silica beads). The concept clearly has a huge potential in biological high-throughput separation with a simple proof of concept experiment separating mouse erythrocytes and lymphocytes demonstrated in Ref. [23]. The use of holographically generated fields can also be used to carry out similar sorting or fractionation processes [24].

There are various problems associated with the optical lattice method of sorting. One problem is that it becomes difficult to scale the process up to sort more than a couple of types of particles. This issue is addressed in Ref. [25] in which, rather than use a static interference pattern, a scanning optical field created by an acousto-optical deflector is employed. A pattern can be "drawn" out by the user to effectively sort multiple particles into multiple steams (four are shown in the paper).

Other problems are that as the particles flow through the sorting region they are affected by colloidal jamming and also by particle–particle interactions. These problems may be eased by making use of translating or flashing optical potentials [26] in which the optical interaction with the particles is reduced, which in turn reduces the density of colloidal motion within the lattice site. A more conventional approach to optical sorting is shown in Ref. [27] in which a traditional fluorescent activated cell sorting (FACS) setup is incorporated

with optical fields that act to push a particle into one flow steam or another (i.e., sorting them) based on fluorescence.

These techniques are akin to other optical manipulation techniques for moving particles, such as optical thermal ratchets that rely on the interaction of a particle undergoing Brownian motion with an optical field. The key paper in this area is by Faucheux et al. [28] (see **PAPER 8.4**). Ratcheting behaviour is designed to allow particles to only move one way through the optical potential and allows directed transport of particles. Such techniques can be used to produce an optical "peristaltic" pump [29] as well as more generally study thermal ratchets produced using holographic optical tweezers [30] and to observe flux reversal particles, in which the particles move against the direction of the moving optical potential [31].

There are various other types of optical sorting methods. The optical lattice technique relies on an interaction between particles propelled in a fluid flow and the optical interaction. It is possible to carry out such sorting techniques in a more passive manner in the absence of any fluid flow. This can be achieved by making use of a shaped static optical potential that drives the particles from one point to another. An excellent example of such an optical potential is a Bessel beam [32] in which the electric field of the beam is described by a Bessel function, with a much higher intensity at the core of the beam surrounded by concentric rings of decreasing intensity. Particles placed within the rings will be preferentially moved toward the center of the beam at a rate that depends on their polarizability and Brownian motion. This has been demonstrated for both colloidal particles [33] and cells, separating erythrocytes and lymphocytes [34], and blood cells [35].

Other types of flowless sortings are based on moving optical patterns, in which a fringe pattern is swept over a sample. In Ref. [36], a fringe pattern is created in a Mach–Zender type of setup. A mirror in one of the arms is driven by a piezoelectric signal, causing the pattern to move in one direction. Sorting of polystyrene beads of different sizes and beads of silica and polystyrene of the same size is demonstrated. Sorting using moving potentials produced using interfering counterpropagating beams on a prism surface has also been shown [37]. The time-varying potential is produced by introducing a continuous phase shift between the beams. Using this technique, sorting of submicron-diameter particles has been demonstrated and holds promise for large area sorting (as it is carried out on a surface).

Optical chromatography is a technique analogous to chemical chromatography in which particles with different properties are separated out. In this type of experiment, a laser is directed along a channel in which fluid is flowing. Particles with different properties feel a different radiation pressure force and as such reach a different balancing point in the channel depending on how much the light holds them back. First demonstrated by Imasaka et al. [38] (see **PAPER 8.5**), the work has been significantly advanced recently [39,40] and it holds promise for the detection of biological agents such as anthrax and other airborne spores [41]. Samples can be enriched using the same type of technique [42], while the optical concentration of nanoparticles in fluid flow has also been demonstrated [43].

Also worth mentioning in the context of optical sorting and microfluidics techniques is the ability to control electrical fields using optical ones. This idea is called light-induced electrophoresis and is used to create opto-electronic tweezers in which a dielectropheric force (the electric analogue of optical tweezers) is induced by illuminating a photoconductor built into a microfluidic chip. This idea has been demonstrated most impressively by Chiou et al. [44], who created thousands of trap sites by illuminating an amorphous silicon photoconductor with a patterned optical field using an LED and a digital micromirror device. The trap sites require far less power than optical tweezers (10 nW/μm^2) and may offer one of the most powerful tools for integrating optical techniques into microfluidic systems in the near future. A recent comprehensive review of optical sorting techniques can be found in Ref. [35].

One of the most promising uses of optical manipulation in soft condensed matter physics is in microrheology [45]. Microrheological techniques have developed in various different ways, employing particle tracking [46,47], magnetic beads [48], laser tracking [49], and optical manipulation [50]. The techniques are very powerful and have broad applications in colloidal science looking at properties of gels [51], for example. Many of the applications are based on the idea of measuring the properties of biological systems such as cells [49, 52–56]. Combining optical trapping with optical rotation is the work of Bishop et al. [57] (see **PAPER 6.8**), in which the rotational drag force can be used to calculate the torque applied by the trapping laser and hence the liquid viscosity can be calculated. One of the key aspects of this work is the development of a birefringent spherical particle (vaterite) that allows standard equations for torque to be applied and significantly advanced rotational studies, which traditionally made use of irregular chunks of calcite. The measurement error in the viscosity is claimed to be ~19% and is mainly due difficulty in sizing the vaterite particle accurately. The same authors have now applied their technique to measuring intracellular viscosity by pushing a trapped particle through a hole in the cell membrane [58].

So far we have discussed microfluidics as a continuous flow process, but this need not be the case. Powerful microfluidic systems can also be produced making use of droplets, a technique called droplet or digital microfluidics [59]. This allows for more flexible system architectures and may enhance such processes as mixing and on-chip storage. As we are now dealing with discrete droplets of liquid, this enables the direct manipulation of both the liquid itself and the contents of a droplet using optical tweezers.

There is growing interest in this area and work has been demonstrated in two distinct size regimes, in large droplets of the order of 100 μm in diameter and in much smaller droplets in the sub 10-μm regime. The larger droplets are not easily trapped by optical tweezers but can be manipulated indirectly by light making use of thermal forces. This optothermal manipulation [60] works as the light is absorbed at the water–oil interface (typically digital microfluidics makes use of water droplets in oil) where the water contains a small amount of absorbing material. This type of effect has been used by Ref. [60] to create valves and droplet routing (a precursor of sorting) on a microfluidic chip. Optothermal manipulation has a couple of important advantages over direct optical trapping: firstly, it is a large field-of-view effect and does not rely on high NA microscope objectives, and secondly, it produces much larger forces than optical tweezers. However, as it makes use of a thermal force, it is not clear if the droplets can be loaded as robustly (say with cells) as smaller optically trapped droplets.

The direct optical manipulation of droplets in an emulsion (coined "hydrosomes" in Ref. [61] and particularly referring to single trapped droplets) is a more mature field and one that shows great promise for applications. Normally, water droplets in oil have a lower refractive index than their surrounds and so cannot be trapped using conventional optical tweezers. It has been shown [62] that droplets can be directly trapped by Gaussian beam tweezers if one chooses an appropriate oil—in this case, a fluorocarbon oil. Single-molecule encapsulation and excitation within the droplet, one of the key drivers of this type of work, can then be achieved—in Ref. [62], by looking at single dye molecules. This, however, may not be the best oil for more general encapsulation of biological material and so one must use vortex beams with a dark core in which the droplet is trapped. Development of a platform to produce femtoliter droplets and encapsulate particles of choice within them has been demonstrated [63], which makes use of optical tweezers to transport the particles both outside the droplet and within it once encapsulated. This work was extended to allow direct optical manipulation of the droplets, using vortex beams, by Lorenz et al. [64]. Simple droplet fusion is also shown and a more detailed study is found in Ref. [65]. One of the possible drawbacks of this type of technique is that high laser powers are needed to move the water droplets in the viscous oil medium and this may make it difficult to carry out operations on multiple droplets simultaneously (hundreds of mW at 1064 nm). This technique can be used to carry out cellular nanosurgery [66] and chemical concentration [67].

The integration of droplet microfluidics can also be used to carry out intriguing functions such as droplet shaping. Ward et al. [68] (see **PAPER 8.6**) have shown that, using oil droplets in water with very low interfacial tensions, the droplets can be shaped into objects such as triangles and squares. This could be used to create new types of optofluidics devices. They have also shown that the droplets can be pulled apart so that they are attached by only a thin strand of liquid, which throws up the possibility of creating two linked droplet test tubes into which material could be placed in one and then allowed to move into the other chamber.

A final area worth mentioning with regard to digital microfluidics is the growing interest in optical manipulation of airborne particles, in particular aerosols. Ashkin carried out some of his original work on airborne particles [1] (see **PAPER 1.2**), and there has been subsequent work making use of radiation pressure to trap and probe aerosols [69,70] but little work has been carried out on the direct gradient-force trapping of such particles. This is mainly due to the more difficult nature of trapping in air and also because of the lack of research drivers. However, optical trapping of aerosols gives an exquisite amount of control over both single and multiple particles. The first airborne tweezing was carried out by Ref. [71], but the first indication of applications for atmospheric science were outlined by Hopkins et al. [72] (see **PAPER 8.7**), in which the power of cavity-enhanced Raman spectroscopy (CERS) was used to accurately size (to ~2 nm) aerosols and determine their composition. The paper also shows two droplets being trapped using a dual-beam tweezers and coagulated, while CERS is used to show that no volume is lost in the fusion process. This technique has undoubted power in future probing and characterizing of aerosols and their chemical and thermodynamic properties [73–78].

Airborne trapping can be combined with other advanced optical trapping techniques. Burnham and McGloin [79] (see **PAPER 8.8**) have shown that holographic optical trapping can be used to combine and fuse multiple droplets and that spatial light modulators are fast enough to carry out useful coagulation studies. Moreover, they have illustrated that there is interesting physics to study when trapping in air, highlighting subtle differences between trapping in air and liquid.

The trapping of particles in air also leads to new dynamical regimes, allowing optical tweezers to trap particles in the underdamped regime for the first time (normally all particles in a liquid-based trap are heavily overdamped). This has already led to the observation of parametric resonance in optical tweezers [80] and will undoubtedly lead to more exciting observations in future, as the inclusion of mass into the optical tweezers situation begins to open up more analogues to condensed matter and chaotic systems that cannot currently be probed.

The integration of optical tweezers into microfluidic systems seems a natural extension of the technique and is sure to open up new and exciting research topics. Running hand in hand with this integration is the new field of "optofluidics" [81] that merges light and liquid into new types of devices, and is already incorporating optical manipulation [82]. This is clearly a field of study with a bright future.

Endnotes

1. A. Ashkin, Acceleration and trapping of particles by radiation pressure, *Phys. Rev. Lett.* **24**, 156–159 (1970).
2. A. Constable, J. Kim, J. Mervis, F. Zarinetchi, and M. Prentiss, Demonstration of a fiberoptic light-force trap, *Opt. Lett.* **18**, 1867–1869 (1993).
3. P.R.T. Jess, V. Garces-Chavez, D. Smith, M. Mazilu, L. Paterson, A. Riches, C.S. Herrington, W. Sibbett, and K. Dholakia, Dual beam fibre trap for Raman microspectroscopy of single cells, *Opt. Express* **14**, 5779–5791 (2006).
4. J. Guck, R. Ananthakrishnan, H. Mahmood, T.J. Moon, C.C. Cunningham, and J. Kas, The optical stretcher: A novel laser tool to micromanipulate cells, *Biophys. J.* **81**, 767–784 (2001).
5. J. Guck, R. Ananthakrishnan, C.C. Cunningham, and J. Kas, Stretching biological cells with light, *J. Phys.-Condes. Matter* **14**, 4843–4856 (2002).
6. J. Guck, R. Ananthakrishnan, T.J. Moon, C.C. Cunningham, and J. Kas, Optical deformability of soft biological dielectrics, *Phys. Rev. Lett.* **84**, 5451–5454 (2000).
7. J. Guck, S. Schinkinger, B. Lincoln, F. Wottawah, S. Ebert, M. Romeyke, D. Lenz, H.M. Erickson, R. Ananthakrishnan, D. Mitchell, J. Kas, S. Ulvick, and C. Bilby, Optical deformability as an inherent cell marker for testing malignant transformation and metastatic competence, *Biophys. J.* **88**, 3689–3698 (2005).
8. B. Lincoln, S. Schinkinger, K. Travis, F. Wottawah, S. Ebert, F. Sauer, and J. Guck, Reconfigurable microfluidic integration of a dual-beam laser trap with biomedical applications. *Biomed. Microdevices* **9**, 703–710 (2007).
9. B. Lincoln, F. Wottawah, S. Schinkinger, S. Ebert, and J. Guck, High-throughput rheological measurements with an optical stretcher, *Cell Mechanics* **83**, 397–423 (2007).
10. S. Cran-McGreehin, T.F. Krauss, and K. Dholakia, Integrated monolithic optical manipulation, *Lab Chip* **6**, 1122–1124 (2006).
11. S.J. Cran-McGreehin, K. Dholakia, and T.F. Krauss, Monolithic integration of microfluidic channels and semiconductor lasers, *Opt. Express* **14**, 7723–7729 (2006).
12. M.E.J. Friese, H. Rubinsztein-Dunlop, J. Gold, P. Hagberg, and D. Hanstorp, Optically driven micromachine elements, *Appl. Phys. Lett.* **78**, 547–549 (2001).

13. S.L. Neale, M.P. Macdonald, K. Dholakia, and T.F. Krauss, All-optical control of microfluidic components using form birefringence, *Nat. Mater.* **4**, 530–533 (2005).

14. P. Galajda and P. Ormos, Complex micromachines produced and driven by light, *Appl. Phys. Lett.* **78**, 249–251 (2001).

15. P. Galajda and P. Ormos, Rotors produced and driven in laser tweezers with reversed direction of rotation, *Appl. Phys. Lett.* **80**, 4653–4655 (2002).

16. L. Kelemen, S. Valkai, and P. Ormos, Integrated optical motor, *Appl. Opt.* **45**, 2777–2780 (2006).

17. J. Leach, H. Mushfique, R. Di Leonardo, M. Padgett, and J. Cooper, An optically driven pump for microfluidics, *Lab Chip* **6**, 735–739 (2006).

18. A. Terray, J. Oakey, and D.W.M. Marr, Microfluidic control using colloidal devices, *Science* **296**, 1841–1844 (2002).

19. A. Terray, J. Oakey, and D.W.M. Marr, Fabrication of linear colloidal structures for microfluidic applications, *Appl. Phys. Lett.* **81**, 1555–1557 (2002).

20. K. Ladavac and D.G. Grier, Microoptomechanical pumps assembled and driven by holographic optical vortex arrays, *Opt. Express* **12**, 1144–1149 (2004).

21. A. Jesacher, S. Furhapter, C. Maurer, S. Bernet, and M. Ritsch-Marte, Holographic optical tweezers for object manipulations at an air-liquid surface, *Opt. Express* **14**, 6342–6352 (2006).

22. M.P. MacDonald, G.C. Spalding, and K. Dholakia, Microfluidic sorting in an optical lattice, *Nature* **426**, 421–424 (2003).

23. M.P. MacDonald, S. Neale, L. Paterson, A. Richies, K. Dholakia, and G.C. Spalding, Cell cytometry with a light touch: Sorting microscopic matter with an optical lattice, *J. Biol. Regul. Homeost. Agents* **18**, 200–205 (2004).

24. K. Ladavac, K. Kasza, and D.G. Grier, Sorting mesoscopic objects with periodic potential landscapes: Optical fractionation, *Phys. Rev. E* **70**, 010901 (2004).

25. G. Milne, D. Rhodes, M. MacDonald, and K. Dholakia, Fractionation of polydisperse colloid with acousto-optically generated potential energy landscapes, *Opt. Lett.* **32**, 1144–1146 (2007).

26. R.L. Smith, G.C. Spalding, K. Dholakia, and M.P. MacDonald, Colloidal sorting in dynamic optical lattices, *J. Opt. A—Pure Appl. Opt.* **9**, S134–S138 (2007).

27. M.M. Wang, E. Tu, D.E. Raymond, J.M. Yang, H.C. Zhang, N. Hagen, B. Dees, E.M. Mercer, A.H. Forster, I. Kariv, P.J. Marchand, and W.F. Butler, Microfluidic sorting of mammalian cells by optical force switching, *Nat. Biotech.* **23**, 83–87 (2005).

28. L.P. Faucheux, L.S. Bourdieu, P.D. Kaplan, and A.J. Libchaber, Optical thermal ratchet, *Phys. Rev. Lett.* **74**, 1504–1507 (1995).

29. B.A. Koss and D.G. Grier, Optical peristalsis, *Appl. Phys. Lett.* **82**, 3985–3987 (2003).

30. S.H. Lee and D.G. Grier, One-dimensional optical thermal ratchets, *J. Phys. Condens. Matter* **17**, S3685–S3695 (2005).

31. S.H. Lee, K. Ladavac, M. Polin, and D.G. Grier, Observation of flux reversal in a symmetric optical thermal ratchet, *Phys. Rev. Lett.* **94**, 110601 (2005).

32. D. McGloin, V. Garces-Chavez, and K. Dholakia, Interfering Bessel beams for optical micromanipulation, *Opt. Lett.* **28**, 657–659 (2003).

33. G. Milne, K. Dholakia, D. McGloin, K. Volke-Sepulveda, and P. Zemanek, Transverse particle dynamics in a Bessel beam, *Opt. Express* **15**, 13972–13987 (2007).

34. L. Paterson, E. Papagiakoumou, G. Milne, V. Garces-Chavez, S.A. Tatarkova, W. Sibbett, F.J. Gunn-Moore, P.E. Bryant, A.C. Riches, and K. Dholakia, Light-induced cell separation in a tailored optical landscape, *Appl. Phys. Lett.* **87**, 123901 (2005).

35. K. Dholakia, M.P. MacDonald, P. Zemanek, and T. Cizmar, Cellular and colloidal separation using optical forces, *Methods Cell Biol.* **82**, 467–495 (2007).

36. I. Ricardez-Vargas, P. Rodriguez-Montero, R. Ramos-Garcia, and K. Volke-Sepulveda, Modulated optical sieve for sorting of polydisperse microparticles, *Appl. Phys. Lett.* **88**, 121116 (2006).

37. T. Cizmar, M. Siler, M. Sery, P. Zemanek, V. Garces-Chavez, and K. Dholakia, Optical sorting and detection of submicrometer objects in a motional standing wave, *Phys. Rev. B* **74**, 035105 (2006).

38. T. Imasaka, Y. Kawabata, T. Kaneta, and Y. Ishidzu, Optical chromatography, *Anal. Chem.* **67**, 1763–1765 (1995).

39. S.J. Hart, A. Terray, K.L. Kuhn, J. Arnold, and T.A. Leski, Optical chromatography of biological particles, *Am. Lab.* **36**, 13–17 (2004).

40. A. Terray, J. Arnold, and S.J. Hart, Enhanced optical chromatography in a PDMS microfluidic system, *Opt. Express* **13**, 10406–10415 (2005).

41. S.J. Hart, A. Terray, T.A. Leski, J. Arnold, and R. Stroud, Discovery of a significant optical chromatographic difference between spores of Bacillus anthracis and its close relative, Bacillus thuringiensis, *Anal. Chem.* **78**, 3221–3225 (2006).

42. S.J. Hart, A. Terray, J. Arnold, and T.A. Leski, Sample concentration using optical chromatography, *Opt. Express* **15**, 2724–2731 (2007).

43. Y.Q. Zhao, B.S. Fujimoto, G.D.M. Jeffries, P.G. Schiro, and D.T. Chiu, Optical gradient flow focusing, *Opt. Express* **15**, 6167–6176 (2007).

44. P.Y. Chiou, A.T. Ohta, and M.C. Wu, Massively parallel manipulation of single cells and microparticles using optical images, *Nature* **436**, 370–372 (2005).

45. F.C. MacKintosh and C.F. Schmidt, Microrheology, *Curr. Opin. Colloid Interface Sci.* **4**, 300–307 (1999).

46. T. Gisler and D.A. Weitz, Tracer microrheology in complex fluids, *Curr. Opin. Colloid Interface Sci.* **3**, 586–592 (1998).

47. R. Di Leonardo, J. Leach, H. Mushfique, J.M. Cooper, G. Ruocco, and M.J. Padgett, Multipoint holographic optical velocimetry in microfluidic systems, *Phys. Rev. Lett.* **96**, 134502 (2006).

48. F.G. Schmidt, F. Ziemann, and E. Sackmann, Shear field mapping in actin networks by using magnetic tweezers, *Eur. Biophys. J. Biophys. Lett.* **24**, 348–353 (1996).

49. S. Yamada, D. Wirtz, and S.C. Kuo, Mechanics of living cells measured by laser tracking microrheology, *Biophys. J.* **78**, 1736–1747 (2000).

50. E.M. Furst, Applications of laser tweezers in complex fluid rheology, *Curr. Opin. Colloid Interface Sci.* **10**, 79–86 (2005).

51. F.K. Oppong, L. Rubatat, B.J. Frisken, A.E. Bailey, and J.R. de Bruyn, Microrheology and structure of a yield-stress polymer gel, *Phys. Rev. E* **73**, 041405 (2006).

52. E. Helfer, S. Harlepp, L. Bourdieu, J. Robert, F.C. MacKintosh, and D. Chatenay, Viscoelastic properties of actin-coated membranes, *Phys. Rev. E* **63**, 021904 (2001).

53. A.C. De Luca, G. Volpe, A.M. Drets, M.I. Geli, G. Pesce, G. Rusciano, A. Sasso, and D. Petrov, Real-time actin-cytoskeleton depolymerization detection in a single cell using optical tweezers, *Opt. Express* **15**, 7922–7932 (2007).

54. P. Panorchan, J.S.H. Lee, B.R. Daniels, T.P. Kole, Y. Tseng, and D. Wirtz, Probing cellular mechanical responses to stimuli using ballistic intracellular nanorheology, *Cell Mech.* **83**, 115–140 (2007).

55. J.C. Crocker and B. Hoffman, Multiple-particle tracking and two-point microrheology in cells, *Cell Mech.* **83**, 141–178 (2007).

56. S. Parkin, G. Knoner, W. Singer, T.A. Nieminen, N.R. Heckenberg, and H. Rubinsztein-Dunlop, Optical torque on microscopic objects, *Methods Cell Biol.* **82**, 525–561 (2007).

57. A.I. Bishop, T.A. Nieminen, N.R. Heckenberg, and H. Rubinsztein-Dunlop, Optical microrheology using rotating laser-trapped particles, *Phys. Rev. Lett.* **92**, 198104 (2004).

58. S.J. Parkin, G. Knöner, T.A. Nieminen, N.R. Heckenberg, and H. Rubinsztein-Dunlop, Picoliter viscometry using optically rotated particles, *Phys. Rev. E* **76**, 041507 (2007).

59. R.B. Fair, Digital microfluidics: Is a true lab-on-a-chip possible? *Microfluid. Nanofluid.* **3**, 245–281 (2007).

60. C.N. Baroud, M.R. de Saint Vincent, and J.P. Delville, An optical toolbox for total control of droplet microfluidics, *Lab Chip* **7**, 1029–1033 (2007).

61. G.M. Lowman, A.M. Jofre, M.E. Greene, J.Y. Tang, J.P. Marino, K. Helmerson, and L.S. Goldner, Single molecule FRET: Comparing surface immobilization and hydrosome encapsulation, *Biophys. J. Sup. S* 655A–656A (2007).

62. J.E. Reiner, A.M. Crawford, R.B. Kishore, L.S. Goldner, K. Helmerson, and M.K. Gilson, Optically trapped aqueous droplets for single molecule studies, *Appl. Phys. Lett.* **89**, 013904 (2006).

63. M.Y. He, J.S. Edgar, G.D.M. Jeffries, R.M. Lorenz, J.P. Shelby, and D.T. Chiu, Selective encapsulation of single cells and subcellular organelles into picoliter- and femto-liter-volume droplets, *Anal. Chem.* **77**, 1539–1544 (2005).

64. R.M. Lorenz, J.S. Edgar, G.D.M. Jeffries, and D.T. Chiu, Microfluidic and optical systems for the on-demand generation and manipulation of single femtoliter-volume aqueous droplets, *Anal. Chem.* **78**, 6433–6439 (2006).

65. R.M. Lorenz, J.S. Edgar, G.D.M. Jeffries, Y.Q. Zhao, D. McGloin, and D.T. Chiu, Vortex-trap-induced fusion of femtoliter-volume aqueous droplets, *Anal. Chem.* **79**, 224–228 (2007).

66. G.D.M. Jeffries, J.S. Edgar, Y.Q. Zhao, J.P. Shelby, C. Fong, and D.T. Chiu, Using polarization-shaped optical vortex traps for single-cell nanosurgery, *Nano Lett.* **7**, 415–420 (2007).

67. G.D.M. Jeffries, J.S. Kuo, and D.T. Chiu, Dynamic modulation of chemical concentration in an aqueous droplet, *Angew. Chem.-Int. Edit.* **46**, 1326–1328 (2007).

68. A.D. Ward, M.G. Berry, C.D. Mellor, and C.D. Bain, Optical sculpture: Controlled deformation of emulsion droplets with ultralow interfacial tensions using optical tweezers, *Chem. Commun.* 4515–4517 (2006).

69. N. Jordanov and R. Zellner, Investigations of the hygroscopic properties of ammonium sulfate and mixed ammonium sulfate and glutaric acid micro droplets by means of optical levitation and Raman spectroscopy, *Phys. Chem. Chem. Phys.* **8**, 2759–2764 (2006).

70. D. McGloin, Optical tweezers: 20 years on. *Philos. Trans. R. Soc. A—Math. Phys. Eng. Sci.* **364**, 3521–3537 (2006).

71. N. Magome, M.I. Kohira, E. Hayata, S. Mukai, and K. Yoshikawa, Optical trapping of a growing water droplet in air, *J. Phys. Chem. B* **107**, 3988–3990 (2003).

72. R.J. Hopkins, L. Mitchem, A.D. Ward, and J.P. Reid, Control and characterisation of a single aerosol droplet in a single-beam gradient-force optical trap, *Phys. Chem. Chem. Phys.* **6**, 4924–4927 (2004).

73. J. Buajarern, L. Mitchem, and J.P. Reid, Characterizing multiphase organic/inorganic/aqueous aerosol droplets, *J. Phys. Chem. A* **111**, 9054–9061 (2007).

74. J. Buajarern, L. Mitchem, A.D. Ward, N.H. Nahler, D. McGloin, and J.P. Reid, Controlling and characterizing the coagulation of liquid aerosol droplets, *J. Chem. Phys.* **125** (2006).

75. L. Mitchem, J. Buajarern, R.J. Hopkins, A.D. Ward, R.J.J. Gilham, R.L. Johnston, and J.P. Reid, Spectroscopy of growing and evaporating water droplets: Exploring the variation in equilibrium droplet size with relative humidity, *J. Phys. Chem. A* **110**, 8116–8125 (2006).

76. L. Mitchem, J. Buajarern, A.D. Ward, and J.P. Reid, A strategy for characterizing the mixing state of immiscible aerosol components and the formation of multiphase aerosol particles through coagulation, *J. Phys. Chem. B* **110**, 13700–13703 (2006).

77. L. Mitchem, R.J. Hopkins, J. Buajarern, A.D. Ward, and J.P. Reid, Comparative measurements of aerosol droplet growth, *Chem. Phys. Lett.* **432**, 362–366 (2006).

78. J.P. Reid and L. Mitchem, Laser probing of single-aerosol droplet dynamics, *Annu. Rev. Phys. Chem.* **57**, 245–271 (2006).

79. D.R. Burnham and D. McGloin, Holographic optical trapping of aerosol droplets, *Opt. Express* **14**, 4175–4181 (2006).

80. R. Di Leonardo, G. Ruocco, J. Leach, M.J. Padgett, A.J. Wright, J.M. Girkin, D.R. Burnham, and D. McGloin, Parametric resonance of optically trapped aerosols, *Phys. Rev. Lett.* **99**, 010601 (2007).

81. C. Monat, P. Domachuk, and B.J. Eggleton, Integrated optofluidics: A new river of light, *Nat. Phot.* **1**, 106–114 (2007).

82. P. Domachuk, F.G. Omenetto, B.J. Eggleton, and M. Cronin-Golomb, Optofluidic sensing and actuation with optical tweezers, *J. Opt. A—Pure Appl. Opt.* **9**, S129–S133 (2007).

Demonstration of a fiber-optical light-force trap

A. Constable, Jinha Kim, J. Mervis, F. Zarinetchi, and M. Prentiss

Department of Physics, Harvard University, Cambridge, Massachusetts 02138

Received April 26, 1993

We demonstrate a fiber-optical version of a stable three-dimensional light-force trap, which we have used to hold and manipulate small dielectric spheres and living yeast. We show that the trap can be constructed by use of infrared diode lasers with fiber pigtails, without any external optics.

In 1969, Ashkin demonstrated that two counter-propagating mildly focused laser fields form a stable three-dimensional trap for small dielectric spheres.[1] A similar trap was later demonstrated for atoms.[2] In this Letter we demonstrate a fiber-optical implementation of this trap, whereby the light is introduced directly into the sample from the side by two easily aligned single-mode optical fibers [see Figs. 1(a) and 1(b)]. We have used the trap to capture, hold, and manipulate polystyrene spheres with diameters between 0.1 and 10 μm, as well as living yeast.

The trap is not a direct replacement for the single-beam gradient trap, commonly referred to as optical tweezers,[3,4] but it does present a variety of advantages. In the limit where the trapped particles are <1 μm, the capture volume can be approximately 5 orders of magnitude larger than that of most existing optical tweezers, so the trap can significantly enhance sample densities. The trap's fibers can be widely separated from the trapped particle, so several other optical or mechanical probes can simultaneously investigate it. The trap is decoupled from the microscope, which permits greater freedom in viewing and manipulating samples and permits a variety of measurements of absorption and elasticity that are not possible with optical tweezers.

The fiber-optical implementation is simple, robust, and inexpensive. A conventional optical microscope can be used; no parallel focusing, additional filters, telescopes, or beam splitters are needed. In addition, there are no requirements on the mode quality of the source because the trapping light will be spatially filtered by the fiber. The apparatus that uses the diode lasers with fiber pigtails is particularly simple since no external optics are required. The diode lasers can also be rapidly switched and are tunable, so a wide variety of experiments can be done simply and easily.

We will consider only the theory of trap operation for the case in which the two trapping beams have a large frequency difference. In this case there is no time-averaged force associated with interference terms between the two fields, and the effective trapping force is simply the sum of the light forces that are due to the two individual fields acting separately. We will confine ourselves to the case shown in Fig. 1, in which each focus is at the face of a single-mode fiber and the numerical aperture out of each fiber is

of the order of 0.1 (the beam divergence shown in Fig. 1 is greatly exaggerated).

The trapping forces can be resolved into two components: the gradient force, which pushes particles in the direction of increasing intensity, and the scattering force, which pushes particles in the direction of propagation. The gradient force provides the restoring force in the directions $\hat{x}$ and $\hat{y}$, whereas the scattering force provides the restoring force along the direction of propagation ($\hat{z}$).

Since the gradient component of the force is well known from optical tweezers, we will discuss only the scattering force. Assume that one trapping beam, designated red by the subscript r, is propagating in the $+\hat{z}$ direction and that the other trapping beam, designated green in the same manner, is propagating in the $-\hat{z}$ direction. For either color, if P is the power transmitted by the fiber, and the light from the fiber can be represented by a Gaussian beam with waist w_0 at the fiber face, then the intensity at the fiber face is given by $I_0 = 2P/(\pi w_0^2)$. For a plane wave with intensity I, the scattering force on a sphere of radius R is given by $(r\pi^2)IQ_{pr}/c$, where c is the speed of light an Q_{pr} is defined in Ref. 5.

Let $z = 0$ at the center point between the fibers. Using an approximation in which the intensity and

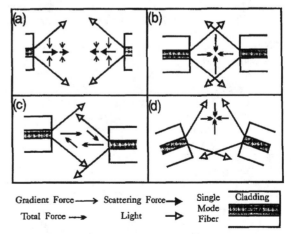

Fig. 1. (a) Schematic of the gradient and scattering forces for each of the two fibers that compose the trap. (b)–(d) Directions of the total forces when the fibers are (b) perfectly aligned, (c) translationally misaligned, and (d) rotationally misaligned.

the phase of the trapping beams are constant over the area of a sphere with radius R, we can express the total scattering force as

$$F_s = \frac{aP_rQ_{\mathrm{pr},r}/w_{0r}^2}{1 + d_r^{-2}(S/2 + z)^2} - \frac{aP_gQ_{\mathrm{pr},g}/w_{0g}^2}{1 + d_g^{-2}(S/2 - z)^2}, \quad (1)$$

where $a = 2R^2/c$, $d^{-1} = \lambda/(\pi w_0^2)$, S is the separation between the two fiber faces, and λ and Q are the wavelength and radiation pressure coefficient, respectively, for the designated color.

Let z_{eq} be the value of z for which the force given in Eq. (1) is zero. When a particle is displaced from z_{eq}, a restoring force results from the increase in intensity with decreasing distance from the fiber face. This restoring force is simply a manifestation of the scattering force described in Eq. (1) and can be expanded to first order in $\epsilon = z - z_{\mathrm{eq}}$, resulting an equation of the form $F = -k\epsilon$, where k is given by

$$k = 16\pi^2 aS\left[\frac{P_gQ_{\mathrm{pr},g}w_{0g}^2}{\lambda_g^2(S^2 + 4d_g^2)^2} + \frac{P_rQ_{\mathrm{pr},r}w_{0r}^2}{\lambda_r^2(S^2 + 4d_r^2)^2}\right]. \quad (2)$$

k, as a function of S, is a maximum when S is approximately twice the Rayleigh range.

The discussion above assumes that the two trapping beams are exactly counterpropagating. However, there are two possible types of fiber misalignment: a positional misalignment, in which the beams are propagating in the $\pm\hat{z}$ direction but the two fibers are translationally displaced [Fig. 1(c)], and a rotational misalignment, in which both fiber faces still have their centers on the $\hat{z}$ axis but are at skewed angles to each others' faces and therefore the two light beams are not counterpropagating [Fig. 1(d)]. Both types of misalignment may occur at the same time. The alignment of the two counterpropagating beams to within a fraction of the beam waist is critical for good trap operation. For example, if the fibers are translationally misaligned, then the particle can oscillate back and forth between the two fiber faces instead of finding a stable trap position [Fig. 1(c)].

We formed one version of the trap using two diode lasers with fiber pigtails. One diode had a wavelength of 1.3 μm and a single-mode pigtail with a 9-μm core diameter. The other diode had a wavelength of 0.831 μm and a single-mode pigtail with a 5-μm core diameter. We also did a second series of experiments, which used an Ar laser at 514 nm and a He–Ne laser at 633 nm as sources of trapping light. Each laser was coupled into a separate single-mode fiber by use of external optics. In both experimental cases, we positioned the fibers so as to form a trap using the sample cell described in Fig. 2.

We made the cell, which was fabricated on a microscope slide, by sandwiching a piece of 180-μm OD glass capillary tubing between the slide and a glass coverslip [see Fig. 2(a) for a top view]. One edge of the coverslip was glued to the slide with Krazy glue. This pressed the capillary tubing tightly against the slide, forming a smooth water-tight seal. The three other sides of the cell were left open. The fibers were

introduced into the cell through two short pieces of larger glass capillary (1-mm OD, also Krazy glued), which held them in place. This alignment technique is much quicker and simpler than the bulk-optical version.[1]

We aligned the fibers by pressing them against the cylindrical capillary tubing, which forms a backstop for the fibers and provides a V groove in which they sit [see Fig. 2(b)]. We could easily charge the separation of the fibers by translating them in this groove, where they were held firmly by their own springlike properties. The cell was also fairly quiet, as the watertight seal formed between the capillary and the microscope slide prevented large fluid currents near the trap center.

We successfully trapped polystyrene spheres with diameters from 2 to 10 μm using the diode laser at approximately 7 mW and at fiber separations of 20–280 μm. At similar fiber separations, we were able to trap particles with diameters from 6 to 0.1 μm by using ion lasers at a range of 3–100 mW. In the ion-laser experiments, the typical peak-to-peak motion of trapped particles in the longitudinal direction, with a fiber separation of 100 μm and 5 mW of power in each beam, was approximately 3 μm for 4-μm spheres and approximately 4 μm for 2-μm spheres. Stability in the transverse direction was much greater—the peak-to-peak motion for all balls was typically less than 1 μm. We could enhance the stability in both directions by increasing the power and optimizing the fiber separation. The stability might also be enhanced by an increase of the divergence of the light by use of lenses on the ends of the fibers.[6]

We measured the restoring longitudinal spring constant k by displacing a particle from z_{eq} and then permitting the trap to force the particle back to z_{eq} while monitoring the particle's position as a function of time. In order to displace the trapped particle from z_{eq}, we briefly decreased the intensity of one of the trapping beams so that the particle was translated to a new $z \neq z_{\mathrm{eq}}$. We then restored the original power so that the particle was pushed back toward z_{eq}. We measured k with a CCD camera and a VCR to track the displaced particle's motion as a function of time as it returned to z_{eq}. We then digitized the information with a frame grabber. Since the system is strongly overdamped, we determined k by fitting the position as a function of time to

$$\epsilon = c_1 \exp(-kt/6\pi\mu R) + c_2, \quad (3)$$

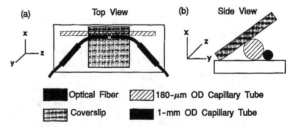

Fig. 2. Schematics of the trap design: (a) top view of the sample cell, which is constructed on a microscope slide; (b) side view showing the V groove that aligns fibers.

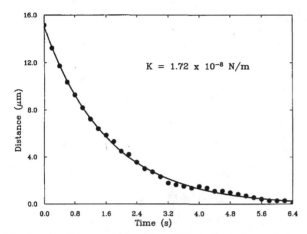

Fig. 3. Position as a function of time for a 3-μm sphere that was initially displaced from equilibrium. The curve shows a least-squares fit to a decaying exponential.

where c_1 and c_2 are constants and μ is the viscosity of water.[7] A typical graph of position versus time for a 2.98-μm-diameter sphere is shown in Fig. 3, in which the filled circles represent coordinates obtained from the frame grabber. The trapping light was provided by the two diode pigtails, both of which had output powers of approximately 7 mW. The fiber separation for these data was approximately 200 μm. A least-squares fit of Eq. (3) to the data is shown by the solid curve, which gives a measured spring constant k of 1.72×10^{-8} N/m. This value agrees well with the theoretical value of 1.75×10^{-8} N/m obtained from Eq. (2).

In this Letter we have concentrated on measurements under conditions in which $z_{eq} \approx 0$. However, there are many important applications of the trap for which z_{eq} is nearer one of the fiber surfaces. We were able to adjust the ratio of the powers of the trapping beams so that $z_{eq} \approx S/2$. This permitted us to place one or more balls directly on the center of the core of one of the optical fibers in a stacked formation. The stacked balls formed compound lenses.

Similarly, to expose the particle to a large force, let $z_{eq} \approx S/2$, which means that the particle is trapped near the surface of one of the fibers. If P, the power transmitted by that fiber, is then suddenly increased to a much greater value, the particle will experience a large force. For polystyrene spheres in water, the maximum force is approximately equal to $PQ_{pr}/c \approx P$ nN, where P is expressed in watts. Higher index differences between the trapped particle and the medium can greatly increase Q_{pr}.

Viscosity measurements of the medium in which the particles are suspended could be made by measurements of the trapped particle's velocity as a function of time under the influence of the applied force. We could make elasticity measurements by attaching one end of an elastic object to one of the fiber faces, the other end to a particle, and then trapping the particle. The tension on the object could be adjusted, and the corresponding length easily and exactly measured.[8] A trap made with light of different wavelengths could also be used to measure differential absorption since the equilibrium position of a particle in the trap depends on the degree to which it scatters the trapping light.

In contrast, if both trapping beams have the same frequency and polarization, then there will be a force that is due to the standing-wave interference pattern between the two fields. This force will tend to localize particles in the intensity maxima,[9] thereby reducing the fluctuations of trapped particles.

We have demonstrated a simple, inexpensive, and robust fiber-optical implementation of a three-dimensional light-force trap. We have measured the trapping force along the direction of propagation and found it to be consistent with theory. Finally, we have suggested several applications for this trap, including measurement of differential absorption and elasticity.

We acknowledge the contributions of Lisa Mills, Steve Smith, and Howard Berg and funding by the U.S. Office of Naval Research (ONR-N0014-91-J-1808) and the National Science Foundation (PYI-9157 485).

References

1. A. Ashkin, Phys. Rev. Lett. **24,** 156 (1970).
2. P. L. Gould, P. D. Lett, P. S. Julienne, and W. D. Phillips, Phys. Rev. Lett. **60,** 788 (1988).
3. A. Ashkin and J. M. Dziedzic, Science **235,** 1517 (1987).
4. A. Ashkin, Biophys. J. **61,** 569 (1992).
5. H. C. van de Hulst, *Light Scattering by Small Particles* (Dover, New York, 1981), p. 14.
6. J. Mervis, A. H. Bloom, G. Bravo, L. Mills, F. Zarinetchi, S. P. Smith, and M. Prentiss, Opt. Lett. **18,** 325 (1993).
7. G. K. Batchelor, *An Introduction to Fluid Dynamics* (Cambridge U. Press, Cambridge, 1991).
8. R. M. Simmons, in *Mechanism of Myfilament Sliding in Muscle Contraction,* H. Sugi and G. H. Pollack, eds. (Plenum, New York, 1993).
9. M. M. Burns, J. Fournier, and J. A. Golovchenko, Science **249,** 749 (1990).

An optically driven pump for microfluidics†

Jonathan Leach,[ab] **Hasan Mushfique,**[ab] **Roberto di Leonardo,**[ac] **Miles Padgett**[a] **and Jon Cooper**[b]

Received 7th February 2006, Accepted 7th April 2006
First published as an Advance Article on the web 28th April 2006
DOI: 10.1039/b601886f

We demonstrate a method for generating flow within a microfluidic channel using an optically driven pump. The pump consists of two counter rotating birefringent vaterite particles trapped within a microfluidic channel and driven using optical tweezers. The transfer of spin angular momentum from a circularly polarised laser beam rotates the particles at up to 10 Hz. We show the that the pump is able to displace fluid in microchannels, with flow rates of up to 200 $\mu m^3 s^{-1}$ (200 fL s^{-1}). The direction of fluid pumping can be reversed by altering the sense of the rotation of the vaterite beads. We also incorporate a novel optical sensing method, based upon an additional probe particle, trapped within separate optical tweezers, enabling us to map the magnitude and direction of fluid flow within the channel. The techniques described in the paper have potential to be extended to drive an integrated lab-on-chip device, where pumping, flow measurement and optical sensing could all be achieved by structuring a single laser beam.

Introduction

The miniaturisation of analytical processing as part of the development of lab-on-a-chip devices promises to revolutionise biological and chemical measurement. Fluid manipulation on such devices not only gives the possibility of automated analysis systems but has the advantages of providing faster and higher throughput, reducing the volume of the materials used and providing the ability to perform multiple processes in parallel.[1] Central to the success of microfluidic systems has been the development of innovative methods for the manipulation of fluids within microchannels. Techniques based on electrokinetics[2] and hydraulic control[3] have been shown to be effective in developing well controlled fluid flow, but require the implementation of external transducers.

Optical tweezers, originally developed by Ashkin[4] in the late 1980's, have been an enabling factor behind investigations into the interaction of light and matter on the micron scale. The technique is based on the principle that dielectric objects are drawn to regions of high optical intensity through a gradient force which, if the light is focussed tightly enough, is sufficient to confine objects in three dimensions. The technique can trap objects ranging in size from 100's of nanometers to 10's of microns.[5,6] The addition of a spatial light modulator to split a single laser beam to create multiple traps is termed holographic optical tweezers[7] and have been applied to the

assembly of complex 3D geometries of trapped objects, including particles[8,9] and cells.[10]

Optical tweezers have also been widely used for the study of light's angular momentum.[11] It is now appreciated that light carries both a spin and orbital angular momentum, macroscopically linked with the polarization and phase structure respectively.[12] When circularly polarized light is used to trap a micron-sized birefringent particle, the transfer of spin angular momentum from the light to the particle can result in a rotation rate of up to a few hundred Hertz.[13] These spinning particles can be used as micro probes to measure the rheological properties of systems.[14,15]

In recent years, a promising new method for fluidic control has been developed using optical tweezers to drive a pumping process. The method allows the possibility of incorporating programmable holographic optical tweezers to automate functions within microdevices using a single light source. Recent studies have described optical tweezers to drive fluid flow. Terray *et al.*[16] reported peristaltic pumps and valves using time-shared optical traps, in which multiple silica particles were moved within a microchannel by a scanning mirror to steer the tweezing beam and trapped particle to induce fluid motion. Subsequently, Ladavac *et al.*[17] used holographic optical tweezers to transfer orbital angular momentum to rings of trapped particles resulting in a displacement of colloids through a liquid medium. Most recently, Neale *et al.*[18] reported the fabrication and rotation of form-birefrigent microgears which could, in future, be used as a micropump.

We report on a new system for producing an optically driven pump which does not require specific design of microfluidic channels or scanning mirrors. Our approach uses the transfer of spin angular momentum from a circularly polarised beam to trapped birefringent particles. The transfer of the angular momentum causes the particles to rotate and results in a flow of the surrounding fluid. Positioning the particles close to the channel edge results in a net fluid flow along the channel. We also incorporate a novel optical sensing method to map the

[a] *Physics and Astronomy, University of Glasgow, Glasgow, UK.*
E-mail: j.leach@physics.gla.ac.uk; Fax: +44 (0) 141 330 2893;
Tel: +44 (0) 141 330 6432
[b] *Electronics and Electrical Engineering, University of Glasgow,*
Glasgow, UK. E-mail: j.cooper@elec.gla.ac.uk;
Fax: +44 (0) 141 330 6010; Tel: +44 (0) 141 330 4931
[c] *INFM-CRS SOFT, Dipartimento di Fisica, Roma, Italy.*
E-mail: roberto.dileonardo@phys.uniroma1.it; Fax: +39 06 4463158;
Tel: +39 06 49913548

† Electronic supplementary information (ESI) available: Movie showing optically driven pump and probe particle. See DOI: 10.1039/b601886f

magnitude and direction of fluid flow in a microfluidic channel. Due to the simplicity of the design, it is possible to integrate the pump into a lab-on-chip device, and use the same system, to perform a sensory function.

Method

Fig. 1 shows the schematic of the optical tweezers setup, based on an inverted microscope (Zeiss Axiovert 200) housing a $100 \times$ high numerical aperture (NA = 1.3) objective. The optical traps used in the pump are created with a continuous-wave, $\lambda = 1064$ nm, laser (Ventus 1064, Laser Quantum). The laser light was collimated and coupled into the microscope using an IR mirror. The beam waist of the light was selected so that the light fills the entrance pupil of the microscope and the circularly polarised beams are brought to a focus, forming two optical traps with opposite angular momentum in the image plane of the lens. The illumination light was coupled *via* the condensing lens and the resulting image viewed using a mounted camera (Pulnix TM9701). A half-wave plate sets the linearly polarised output beam at 45° to the axis of a Wollaston prism which then introduces an angular deviation to the two orthogonal polarisations. The Wollaston prism is positioned near the Fourier plane of the traps such that the angular deviation between the beams corresponds to a lateral separation of the traps. Immediately prior to the objective lens a quarter-wave plate transforms the orthogonal polarisations to circular polarisations, one left-, the other right.

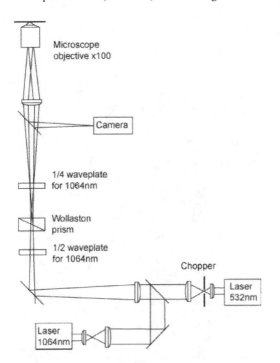

Fig. 1 The optical set up required for the optical pump. The polarisation of the 1064 nm laser is aligned appropriately so that on transmission through the Wollaston prism it is split into two orthogonally polarised components. The two traps are formed at the focus of the high magnification microscope objective. The second laser, at 532 nm, is used to trap the probe particle used for the flow measurements.

The microfluidic channels were made using standard soft lithographic techniques,[19] namely pattern transfer, deep dry etch and replica moulding of the elastomeric polymer, poly(dimethylsiloxane), PDMS, Sylgard 184. A positive photoresist was spin-coated onto a silicon wafer, and the channel pattern was transferred into the resist layer by photolithography. The patterned wafer was then dry etched using an anisotropic deep silicon etch (STS, Cardiff). Smooth, high aspect ratio masters, 15 μm deep were produced and subsequently used to cast the channels in PDMS. A mixture of de-gassed poly(dimethylsiloxane), PDMS, Sylgard 184, pre-polymer and curing agent was poured over the silicon master mold, which, after curing, revealed channel structures. Finally, the PDMS channels were sealed against a 150 μm thick coverglass producing the microchannels.

To transfer the spin angular momentum from the circularly polarized laser beam, the particles must be birefringent. Vaterite particles were produced using the procedure report by Bishop.[13] The torque τ, exerted by a circularly polarised light beam of wavelength λ upon an object of birefringence Δn, and thickness d is proportional to,[20]

$$\tau \alpha 1 - \cos\left(\frac{2\pi d \Delta n}{\lambda}\right). \qquad (1)$$

The maximum torque occurs when the particle acts as half-wave plate, reversing the handedness of the light as it passes through. For vaterite, which has a birefringence of 0.10, the optimum thickness for maximum torque is 5.3 μm, eqn (1). Hence, although it is possible to grow vaterite crystals ranging in diameter from 1–10 μm and shaped from cubic to spherical, we chose to produce spherical particles of 5–7 μm in diameter.

The particles were introduced into the sample cell and the 1064 nm beam was used to trap two particles in the microfluidic channel. The opposite handedness of the trapping beams caused the two particles to rotate in opposite directions; one clockwise, the other anti-clockwise. The distance between the two traps could be changed by adjusting the position of the Wollaston prism along the optical train. The particles were placed as close as possible to the channel walls to reduce any possible back-flow that would otherwise reduce the overall pumping efficiency. The position and height of the two particles, within the channels could be readily modified either through translation of the motorised microscope stage or by using the optical tweezers, *per se*. For the results that we report, the rotating particles were held mid-height within a 15 μm deep channel, *i.e.* approximately 5 μm, from the channel floor. The fluid could be pumped in either direction by setting the orientation of the quarter-wave plate, thereby interchanging the circular polarisation states.

A third probe particle (diameter 1 μm) was trapped within the channel using a 532 nm (Coherent VersaDisk) coupled into the optical train before entering the microscope objective. The probe particle was trapped in the same plane as the rotating particles by slight adjustments to the collimation of the 532 nm laser. In the simplest experiments, the bead was released and the flow was tracked by using video imaging. A more detailed study of the fluid flow was made using the bead as a flow sensor. In these experiments the bead was repeatedly trapped and then released from a specific position within the channel,

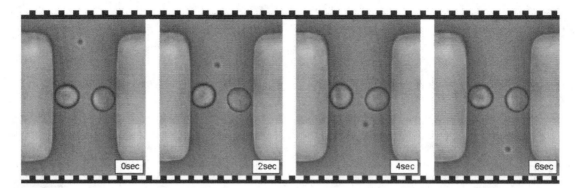

Fig. 2 Frames from a movie showing the optically driven pump and probe particle. The channel is 15 μm (height) by 15 μm (width) by 27 μm (length) and the vaterite particles are 6 μm in diameter. The probe particle is driven from one end of the channel to the other by the fluid flow.

using a beam steering mirror and a mechanical chopper running at 40 Hz.[21] When released the particle travelled a given distance and direction, depending upon the microfluidic flow, prior to the trap being re-established, and the bead being drawn back into the optical trap. Images of the particle were taken on consecutive trigger signals with a relative delay of a few ms, so that the magnitude and direction of the flow could be calculated at that point.

Results

In the example we report here, the microfluidic channel measured 15 μm (height) by 15 μm (width) by 27 μm (length). The two vaterite particles of the pump were measured to spin at constant speeds of 8.7 Hz and 9.2 Hz respectively. To measure the fluid flow induced by the optically driven pump, we positioned and trapped a third silica bead, 1 μm in diameter at the entrance of the channel. When the trapping beam for the probe particle was blocked the particle was released into the fluid flow. For example, Fig. 2 shows a sequence of images taken at 2 s intervals showing the probe particle being pumped through the channel. Fig. 3 shows a trace of the path taken by the silica bead (for video, see ESI†). The colour of the line indicates the speed of the probe particle (as deduced from measurements of particle position in subsequent frames) as it moves along the channel, it is clear to see the increase in speed as the probe passes in-between the rotating particles. Measuring the speed of the pumped bead though video analysis indicated a maximum flow rate of 8.3 μm s^{-1} as the bead passed through the middle of the pump. Further from the rotating beads, the flow occurs over a larger cross-section and is therefore reduced to 1 μm s^{-1} at the entrance and exit of the channel.

The vector plots in Fig. 4a, were obtained using methods previously reported by ourselves[21] and show a complete flow map within the channel. The pump generates a significant flow within a microchannel, the magnitude of which increases near the vaterite particles. The direction of flow at the probed positions is always found to be consistent with that anticipated. Fig. 4b shows the measured flow produced at the centre of the pump as a function of both forward and reverse flow. The forward and reverse rotation rate of the vaterite was controlled by adjusting the power of the laser and measured by

examining the periodicity of the small fluctuations in the power of the transmitted IR light as it passed through the sample. Care was taken so that pressure at each channel entrance was equal so that there was no residual flow. In addition, we did not observe any effects due to possible heating of the fluid from absorption of the IR light.

The observed variation in the fluid flow speed along the channel is consistent with uniform flux. Across the entrance and exit of the channel, the flow rate is approximately uniform at 1 μm s^{-1}. The extremely low Reynolds number of the fluid system, coupled with distance from the rotating spheres,[22] implies that the direction of the flow is largely constant with depth. This enables us to estimate the flux at the entrance and exit of the channel of 200 μm^3 s^{-1}. Between the spheres, the flow is higher, $\approx$ 8 μm s^{-1}, but the effective cross-section of the flow is less. Measurements of the local fluid flow near a

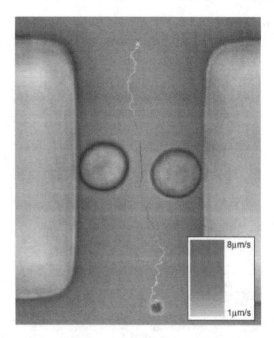

Fig. 3 The traced path of a 1 μm silica particle being pumped through a 15 μm wide PDMS channel. The colour of the trace changes as the particle accelerates and then decelerates through the channel due to the pumping effect. The probe is held in a third optical trap and then released after a few seconds.

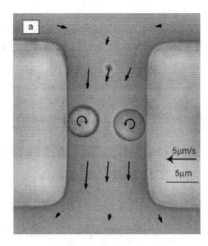
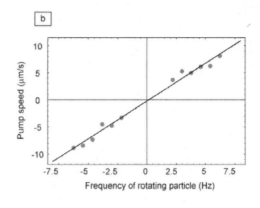

Fig. 4 (a) Measured flow field mapped while fluid is being pumped. (b) Fluid speed measured at the centre of the pump as a function of the rotation rate of the vaterite particles.

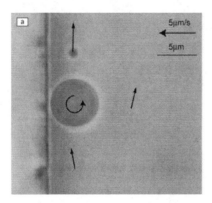
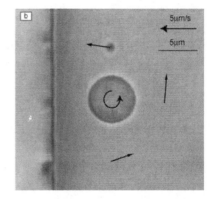

Fig. 5 Measured flow field around a spinning 5 μm vaterite particle placed at two positions from a wall. (a) In the 1st case, the back-flow is suppressed as the particle is placed as close as possible to the wall. The vectors are parallel to the wall. (b) At the 2nd position, the back-flow results in a circulation of the fluid around the vaterite particle.

rotating sphere suggest that the vertical extent of the moving fluid corresponds to 1.5 particle diameters.[21] This implies that the fluid flow flux is also of order 200 μm³ s⁻¹, consistent with a stable flux along the channel length.

Further experiments were performed to investigate the effect of the wall on the motion of the fluid. The drag forces induced by a spinning spherical object in an infinite fluid would cause the surrounding medium to circulate freely,[21] resulting in no net translational flow. In our case we deliberately bring the rotating particles as close as possible to a wall in order to minimize any unwanted backflow. Fig. 5 shows a 5 μm vaterite particle spinning anticlockwise at constant frequency while a 1 μm diameter silica probe was used to measure the change in flow around the particle as it is moved away from the wall. It clearly shows that in order to pump effectively in a channel the vaterite particles need to be positioned close enough to the walls such that the back flow around each is blocked.

Conclusions

In conclusion we have demonstrated an optically driven pump in a microfluidic channel. The pump was formed by placing two counter-rotating birefringent particles in a micro channel. The flux caused by the pump was measured to be of order 200 μm³ s⁻¹ (200 fL s⁻¹). This method of pumping can be used to accurately transport small quantities of liquid. The technique could be extended to include a chain of pumps so that pumping in a more complex microfluidic system is achieved. Whilst the present system requires large lasers coupled into a microscope, there is a possibility to integrate the optical tweezers system into a lab-on-a-chip device.

References

1 V. R. Daria, P. J. Rodrigo and J. Gluckstad, *Proceedings of the 49th SPIE Annual Meeting on Nanotechnology and Organic Materials*, Denver, Colorado, 2004.
2 S. R. Quake and T. M. Squires, Microfluidics: Fluid physics at the nanoliter scale, *Rev. Mod. Phys.*, 2005, **77**, 977–1026.
3 M. A. Unger, P. Chou, T. Thorsen, A. Scherer and S. R. Quake, Monolithic Microfabricated Valves and Pumps by Multilayer Soft Lithography, *Science*, 2000, **288**, 113–116.
4 A. Ashkin, J. M. Dziedzic, J. E. Bjorkholm and S. Chu, Observation Of A Single-Beam Gradient Force Optical Trap For Dielectric Particles, *Opt. Lett.*, 1986, **11**, 288–290.
5 K. C. Neuman and S. M. Block, Optical trapping, *Rev. Sci. Instrum.*, 2004, **2787**, 75–9.

6　J. E. Molloy and M. J. Padgett, Lights, action: optical tweezers, *Contemp. Phys.*, 2002, **43–4**, 241–258.

7　J. E. Curtis, B. A. Koss and D. G. Grier, Dynamic holographic optical tweezers, *Opt. Commun.*, 2004, **207**, 169–175.

8　J. Leach, G. Sinclair, P. Jordan, J. Courtial, M. J. Padgett, J. Cooper and Z. J. Laczik, 3D manipulation of particles into crystal structures using holographic optical tweezers, *Opt. Express*, 2004, **12**, 220–226.

9　G. Sinclair, P. Jordan, J. Courtial, M. Padgett, J. Cooper and Z. J. Laczik, Assembly of 3-dimensional structures using programmable holographic optical tweezers, *Opt. Express*, 2004, **12**, 5475.

10　P. Jordan, J. Leach, M. Padgett, P. Blackburn, N. Isaacs, M. Goksör, D. Hanstorp, A. Wright, J. Girkin and J. Cooper, Creating permanent 3D arrangements of isolated cells using holographic optical tweezers, *Lab Chip*, 2005, **5**, 11, 1224–1228.

11　A. T. O'Neil, I. MacVicar, L. Allen and M. J. Padgett, Intrinsic and extrinsic nature of the orbital angular momentum of a light beam, *Phys. Rev. Lett.*, 2002, **88**, 053601.

12　L. Allen, M. J. Padgett and M. Babiker, The orbital angular momentum of light, *Prog. Opt.*, 1999, **39**, 291–372.

13　M. E. J. Friese, T. A. Nieminen, N. R. Heckenberg and H. Rubinsztein-Dunlop, Optical alignment and spinning of laser-trapped microscopic particles, *Nature*, 1998, **16**, 348–350.

14　A. I. Bishop, T. A. Nieminen, N. R. Heckenberg and H. Rubinsztein Dunlop, Optical Microrheology Using Rotating Laser-Trapped Particles, *Phys. Rev. Lett.*, 2004, **92**, 19, 198104.

15　G. Knöner, S. Parkin, N. R. Heckenberg and H. Rubinsztein-Dunlop, Characterization of optically driven fluid stress fields with optical tweezers, *Phys. Rev. E*, 2005, **72**, 031507.

16　A. Terray, J. Oakey and D. W. M. Marr, Microfluidic control using colloidal devices, *Science*, 2002, **296**, 1841–1844.

17　K. Ladavac and G. Grier David, Microoptomechanical pumps assembled and driven by holographic optical vortex arrays, *Opt. Express*, 2004, **12**, 1144–1149.

18　S. Neale, M. MacDonald, K. Dholakia and T. Krauss, All-optical control of microfluidic components using form birefringence, *Nat. Mater.*, 2005, **4**, 530–533.

19　L. Ceriotti, K. Weible, N. F. de Rooij and E. Verpoorte, Rectangular channels for lab-on-a-chip applications, *Microelectron. Eng.*, 2003, **67–68**, 865–871.

20　M. E. J. Friese, T. A. Nieminen, N. R. Heckenberg and H. Rubinsztein-Dunlop, Optical alignment and spinning of laser-trapped microscopic particles, *Nature*, 1998, **394**, 348–350.

21　R. di Leonardo, J. Leach, H. Mushfique, J. Cooper, G. Ruocco and M. J. Padgett, *Phys. Rev. Lett.*, 2006, **96**, 134502.

22　L. Landau and E. Lifshitz, in *Fluid Mechanics,* 2nd edn, Elsevier, Massachusetts, 2004, vol. 6, ch. 2, p. 65.

Microfluidic Control Using Colloidal Devices

Alex Terray, John Oakey,* David W. M. Marr*

By manipulating colloidal microspheres within customized channels, we have created micrometer-scale fluid pumps and particulate valves. We describe two positive-displacement designs, a gear and a peristaltic pump, both of which are about the size of a human red blood cell. Two colloidal valve designs are also demonstrated, one actuated and one passive, for the direction of cells or small particles. The use of colloids as both valves and pumps will allow device integration at a density far beyond what is currently achievable by other approaches and may provide a link between fluid manipulation at the macro- and nanoscale.

Microscale devices designed to accomplish specific tasks have repeatedly demonstrated superiority over their macroscale analogs (*1, 2*). The advantages of such devices are due largely to unique transport properties resulting from laminar flows and vastly increased surface-to-volume ratios (*3*) and have enabled microscale sensors (*4*) and fabrication schemes (*5*) not possible on the macroscale. Additionally, microfluidic processes may be easily parallelized for high throughput (*6*) and require vastly smaller sample volumes, a major benefit for applications in which reagents or analytes are either hazardous or expensive (*7*). The utility, speed,

Chemical Engineering Department, Colorado School of Mines, Golden, CO 80401, USA.

*To whom correspondence should be addressed. E-mail: joakey@mines.edu (J.O.); dmarr@mines.edu (D.W.M.M.)

REPORTS

and performance of microsystems typically increase as the overall device size decreases. The need to pump and direct fluids at very small length scales, however, has long been the limiting factor in the development of microscale systems, thus generating a tremendous amount of interest in microfluidics development (8). As improved actuation techniques have become available, conventional valving and pumping schemes have been miniaturized, yet continue to dwarf microchannels and other chip-top features (9).

Recently, several approaches conceived explicitly for the microscale have been developed, including platforms based upon electrohydrodynamics (10), electroosmosis (11), interfacial phenomena (12, 13), conjugated materials (14), magnetic materials (15), and multilayer soft lithography (16). Although these microfluid handling techniques enable functional devices on microscopic length scales, they also impose unique constraints upon potential device capability, flexibility, and performance. To fully integrate multiple fluidic processes within a single microsystem, methods for microfluid handling must be developed that can accommodate fluids of complex and dynamic composition and are of comparable scale to that of the processes into which they are embedded.

Development of devices that can function at these length scales has focused on complex fabrication schemes for intricate components such as gears (17, 18), cantilevers (19), and other microscale objects (20, 21). The fabrication and actuation of these devices, however, has been limited to bulk environments external to the microfluidic geometries. Because no practical implementation scheme has been developed for their incorporation into functioning microfluidic systems, they have not realized their suggested potential as microfluidic pumps and valves. We avoid the complexities inherent in other approaches by using colloidal microspheres as the active flow-control element. These materials are an excellent choice because they can be synthesized at length scales easily transported through microfluidic networks and their surface chemistry may be readily altered. Colloidal silica, for example, is easily modified for dispersion in both aqueous and nonaqueous solvents and is well known for its biocompatibility, allowing its use in a host of microfluidic applications.

It has also been shown that colloids can be directly manipulated through the application of external fields. The processes of electrophoresis, dielectrophoresis, and magnetophoresis have all been used to control and influence the motion of small particles in solution. Particularly useful for our preliminary studies, however, is the technique of optical trapping, which has become popular because it allows the direct manipulation of individual colloids. This noncontact, noninvasive technique eliminates the need to physically interface to the macroscopic

Fig. 1. Pump design illustrating lobe movement (the top pair rotate clockwise, the bottom counterclockwise). Also shown is 3-μm colloidal silica undergoing rotation at 2 Hz within a 6-μm channel. Frames are separated by two cycles to show movement of the 1.5-μm colloidal silica tracer particles. To create the pump, we first

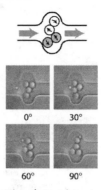

cured PDMS over a negative photoresist pattern created with conventional photolithography. The resulting elastomeric layer (~1 mm thick) was then placed upon a base fabricated from a block of Plexiglas designed to allow easy colloidal fluid introduction to the channel network through a syringe pump (Kd Scientific, KDS200, New Hope, PA). Holes through the PDMS membrane were then aligned with the fluid ports in the base with which the PDMS layer seals tightly and reversibly. A cover slip placed on top of the channel network created a tight, reversible seal and a closed microfluidic system. The optical trap consisted of a laser (Spectra Physics Millennia V, Mountain View, CA; 532-nm wavelength, typically operated at 0.8 W for actuation) and a microscope (Optiphot 150, Nikon) with an oil immersion 100× objective, numerical aperture = 1.3 (CFN plan fluor, Nikon) used to create a tight laser focus of ~1 μm and to observe the experiments. The piezoelectric mirror (Physik Instrumente, Karlsruhe, Germany) used to create the scanning trap was controlled by computer (Macintosh G3, Apple) with routines written in LabView (National Instruments, Austin, TX) and scanned at a rate of 50 to 100 Hz.

world and allows for the manipulation of complex asymmetric objects or multiple objects at once, as would be required for the actuation of a microfluidic valve or pump.

For multiple trap generation, we used a scanning approach in which a piezoelectric mirror is translated to rapidly reflect a laser beam in a desired pattern. If the piezoelectric mirror is scanned over the desired pattern at a frequency greater than that associated with Brownian time scales, a time-averaged trapping pattern is created. Through the application of this technique, known as scanning laser optical trapping (SLOT) (22, 23), we are able to rearrange microspheres into functional structures and subsequently actuate these structures to generate microfluidic pumping and valving. However, using an optical actuation scheme and transmission microscopy to monitor device performance requires a method for the creation of channels at single-micrometer length scales in a transparent housing. Because of this requirement, we applied soft lithography techniques, pioneered by the Whitesides group, which allow for the inexpensive fabrication of microfluidic networks in poly(dimethylsiloxane) (PDMS), an optically transparent elastomer (3, 24–26).

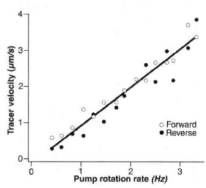

Fig. 2. Tracer particle velocity as a function of gear rotation rate for both forward and reverse directions. The line is a fit to guide the eye.

Working at microscopic length scales offers unique challenges for colloidal pump design. This is illustrated through calculation of the Reynolds number, $Re = \rho v D/\eta$, where for colloidal length scales in aqueous solutions, $\rho = 1$ g/cm^3, $\eta = 0.01$ g/cm·s, $D \approx 5$ μm, and $v \approx 5$ μm/s, giving $Re \approx 10^{-5} \ll 1$ and corresponding to laminar flow. Under these circumstances, fluid flow is fully reversible, and pump designs that rely on centrifugal action, such as impeller-type approaches, are inappropriate. For this reason, our designs are based on positive-displacement pumping techniques that operate by imparting forward motion to individual plugs of fluid.

Our first design is a two-lobe gear pump in which small, trapped pockets of fluid are directed through a specially designed cavity fabricated in a microchannel by rotating two colloidal dumbbells or "lobes" in opposite directions. Over repeated and rapid rotations, the accumulated effect of displacing these fluid pockets is sufficient to induce a net flow. This motion is illustrated in Fig. 1, where clockwise rotation of the top lobe combined with counterclockwise rotation of the bottom lobe induces flow from left to right. In these experiments, each of the lobes consisted of two independent 3-μm silica spheres (Bangs Laboratories, Fishers, IN). To create these structures, we first maneuvered the colloids with the optical trap to a 3-μm-deep section of channel designed with a region of wider gap to accommodate lobe rotation. Once the particles were properly positioned, the laser was scanned in a manner such that a time-averaged pattern of four independent optical traps was created, one for each microsphere composing the two-lobe pump. By rotating the two traps in the upper part of the channel and the two traps in the lower part of the channel in opposite directions and offset by 90°, the overall pump and the corresponding fluid movement were achieved. Flow direction was easily and quickly reversed by changing the rotation direction of both top and bottom lobes.

To aid visualization of flow and provide a means of estimating flow rate, we added tracer

Fig. 3. Three-micrometer colloidal silica used as a peristaltic pump, operating at 2 Hz, to induce flow from right to left within a 6-μm channel. Frames are separated by four cycles to show movement of the 1.5-μm tracer particle.

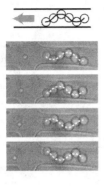

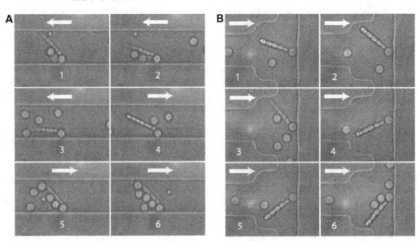

Fig. 4. (A) A passive colloidal valve, where arrows indicate the direction of fluid flow. In images 1 to 3, a flow of ~2 nl/hour pushes the valve arm against the channel wall, allowing particulates in the flow to pass. In images 4 to 6, the flow is reversed, swinging the valve structure across the channel, restricting the flow of the 3-μm colloids but not the smaller 1.5-μm tracer particles. The aqueous solution used to photopolymerize the valve structure consisted of an aqueous monomer, 1.99 M acrylamide (Sigma-Aldrich) and cross-linker solution, 0.048 M bis-acrylamide (Sigma-Aldrich), containing both a photoinitiator, 0.0006 M triethanolamine (J. T. Baker, Phillipsburg, NJ), and a coinitiator, 0.0002 M riboflavin 5′ phosphate sodium salt (Sigma-Aldrich). This solution was pumped into the microfluidic channel network through a syringe pump. Initially fabricated in a deeper region of the network, the linear structures were optically maneuvered into an 11-μm-wide and 3.2-μm-deep straight section of the PDMS microfluidic channel. **(B)** An actuated, three-way colloidal valve. Images 1 to 3 show the valve structure in a position to direct 3-μm spheres to the lower channel with flow rates of ~2 nl/hour. In images 4 to 6, the valve is moved downward to direct particulates to the upper channel.

particles consisting of 1.5-μm silica spheres to the aqueous solution (Movie S1). The measured tracer particle velocities were determined as a function of the gear rotation rate both in the forward and reverse directions and indicate a maximum flow rate of 1 nl/hour. The dependence appears to be linear and independent of direction, as expected from the laminar nature of the flow (Fig. 2). From the measured tracer particle velocities, we estimate that pressure drops of ~30 μm H_2O are achievable at our fastest rotation rates.

The gear pump design illustrates the success of positive-displacement pumping through the use of colloidal microspheres; however, its design may prove particularly harsh to certain solutions. Although we have been able to pump individual cells using the gear pump (Movie S2), concentrated cellular suspensions may be damaged by the aggressive motion of the meshing "gears" of the pump. We have therefore investigated a second approach that incorporates a peristaltic design also based on the concept of positive fluid displacement, effectively a pseudo two-dimensional analog of a three-dimensional, macroscopic screw pump. If instead of rotating the particles as in the gear pump, they are translated back and forth across the channel in a cooperative manner, fluid propagation can be achieved.

The colloidal movement required to direct flow with this approach is illustrated in Fig. 3. The optical trap moves the colloids in a propagating sine wave within which a plug of fluid is encased. Direction of the flow can be reversed by changing the direction of colloidal wave movement. Again, these experiments were performed with independent, 3-μm silica spheres; however, a larger number of particles was used in the experiments shown in Fig. 3 to represent a complete wavelength. Fabrication of these pumps required first maneuvering the colloids into the channel section. Once in place, the optical trap was scanned such that multiple independent traps were created, one for each colloid composing the peristaltic pump. Tracer particles were also used in these experiments and indicate that comparable flow rates could be achieved with this approach. The "snake-like" motion of this pump is best viewed as movie clips (Movies S3 and S4).

The physical, colloid-based, in situ positive-displacement pumping scheme of these two pumps has a number of advantages in addition to its diminutive size. Because colloidal particles are used, and depending on the design, the actuation scheme could be electrophoretic, magnetophoretic, or optical based. This range of actuation schemes will allow pumping of complex suspensions and nonpolar organic solvents, two fluid classes in which electrophoretic pumping techniques falter or fail.

To restrict and direct the flow of cells or colloids within microfluidic networks, we created two types of valve (Fig. 4) by using laser-initiated photopolymerization to first lock colloids into specific geometries. Once polymerized, these structures are positioned and, in some cases, actuated by the same laser used for their construction. Each valve consists of a 3-μm silica sphere photopolymerized to several 0.64-μm silica spheres that form a linear structure. For passive applications, the device was maneuvered into a straight channel and the 3-μm sphere was held next to the wall, allowing the arm to rotate freely in the microchannel. As the flow direction was changed (Fig. 4A), the valve selectively restricted the flow of large particles in one direction while allowing passage of all particles in the other. To actively direct particulates to one of two exit channels, the passive valve was maneuvered into a confining "T" geometry. As the valve structure was rotated about its swivel point by using the optical

trap, the top or bottom channel was sealed, directing flow of particulates toward the open channel (Fig. 4B) (Movies S5 and S6).

We have shown here that colloidal particles can be used to fabricate true micrometer-scale microfluidic pumps and valves that are much smaller than those constructed with current approaches (*16, 27, 28*). Although the use of an optical trap provides a number of advantages, including the elimination of physical connection to macroscopic hardware and the ability to instantly alter device design or location in situ, actuation of these devices via other applied fields is certainly feasible. As discussed previously, appropriately selected colloids will also translate in applied electric and magnetic fields. Because of its versatility, a colloid-based approach to microfluidic flow generation and control may indeed prove a powerful technique for the creation of complex, highly integrated, microscale total analysis systems.

References and Notes
1. C. S. Effenhauser, A. Manz, H. M. Widmer, *Anal. Chem.* **65**, 2637 (1993).
2. P. Wilding, M. A. Shoffner, L. J. Kricka, *Clin. Chem.* **40**, 1815 (1994).
3. G. M. Whitesides, A. D. Stroock, *Phys. Today* **54**, 42 (June 2001).
4. A. E. Kamholz, B. H. Weigl, B. A. Finlayson, P. Yager, *Anal. Chem.* **71**, 5340 (1999).
5. P. J. A. Kenis, R. F. Ismagilov, G. M. Whitesides, *Science* **285**, 83 (1999).
6. R. P. Hertzberg, A. J. Pope, *Curr. Opin. Chem. Biol.* **4**, 445 (2000).
7. R. F. Service, *Science* **282**, 400 (1998).
8. A. van den Berg, W. Olthuis, P. Bergveld, *Micro Total*

REPORTS

Analysis Systems 2000 (Kluwer Academic, Dordrecht, Netherlands, 2000).

9. S. Shoji, M. Esashi, *J. Micromech. Microeng.* **4**, 157 (1994).
10. S. F. Bart, L. S. Tarrow, M. Mehregany, J. H. Lang, *Sensors Actuators A* **21–23**, 193 (1990).
11. A. Manz *et al.*, *J. Micromech. Microeng.* **4**, 257 (1994).
12. D. E. Kataoka, S. M. Troian, *Nature* **402**, 794 (1999).
13. B. Zhao, J. S. Moore, D. J. Beebe, *Science* **291**, 1023 (2001).
14. E. W. H. Jager, O. Inganas, I. Lundstrom, *Science* **288**, 2335 (2000).
15. A. Hatch, A. E. Kamholz, G. Holman, P. Yager, K. F. Boehringer, *J. Microelectromech. Syst.* **10**, 215 (2001).
16. M. A. Unger, H.-P. Chou, T. Thorsen, A. Scherer, S. R. Quake, *Science* **288**, 113 (2000).
17. R. C. Gauthier, R. N. Tait, H. Mende, C. Pawlowicz, *Appl. Opt.* **40**, 930 (2001).

18. S. Maruo, K. Ikuta, K. Hayato, in *Proceedings of the 14th Annual IEEE International Conference on MEMS* (IEEE, Piscataway, NJ, 2001), pp. 594–597.
19. S. M. Kuebler *et al.*, in *Micro- and Nano-Photonic Materials and Devices*, J. W. Perry, A. Scherer, Eds. (International Society for Optical Engineering, Bellingham, WA, 2000), pp. 97–105.
20. P. Galajda, P. Ormos, *Appl. Phys. Lett.* **78**, 249 (2001).
21. S. Kawata, H.-B. Sun, T. Tanaka, K. Takada, *Nature* **412**, 697 (2001).
22. C. Mio, T. Gong, A. Terray, D. W. M. Marr, *Rev. Sci. Instrum.* **71**, 2196 (2000).
23. C. Mio, D. W. M. Marr, *Adv. Mater.* **12**, 917 (2000).
24. D. C. Duffy, O. J. A. Schueller, S. T. Brittain, G. M. Whitesides, *J. Micromech. Microeng.* **9**, 211 (1999).
25. J. C. McDonald *et al.*, *Electrophoresis* **21**, 27 (2000).

26. D. C. Duffy, J. C. McDonald, O. J. A. Schueller, G. M. Whitesides, *Anal. Chem.* **70**, 4974 (1998).
27. D. J. Beebe *et al.*, *Nature* **404**, 588 (2000).
28. Q. Yu, J. M. Bauer, J. S. Moore, D. J. Beebe, *Appl. Phys. Lett.* **78**, 2589 (2001).
29. Supported by the NSF (grants CTS-9734136 and CTS-0097841) and the Office of Life and Microgravity Sciences and Applications, NASA (grant NAG9-1364). We thank T. Vestad and N. Flannery for their assistance with device fabrication and Y. Munakata for useful discussions.

Supporting Online Material
www.sciencemag.org/cgi/content/full/296/5574/1841/DC1
Movies S1 to S6

22 March 2002; accepted 30 April 2002

Optical Thermal Ratchet

L. P. Faucheux, L. S. Bourdieu, P. D. Kaplan, and A. J. Libchaber

*NEC Research Institute, 4 Independence Way, Princeton, New Jersey 08540
and Princeton University, Physics Department, Princeton, New Jersey 08544*
(Received 31 August 1994)

We present an optical realization of a thermal ratchet. Directed motion of Brownian particles in water is induced by modulating in time a spatially periodic but asymmetric optical potential. The net drift shows a maximum as a function of the modulation period. The experimental results agree with a simple theoretical model based on diffusion.

PACS numbers: 05.40.+j

Let us consider a Brownian particle diffusing in a one-dimensional periodic well-shaped potential. If the potential height is much larger than the thermal noise, the particle is localized in a minimum. Suppose that this potential is asymmetric and characterized by two length scales λ_f and λ_b (forward and backward) and assume that λ_b is larger than λ_f (time $\tau = 0$ in Fig. 1). In an equilibrium situation, not net motion of particles can be induced by a periodic potential, since there is no large scale gradients. However, a time modulation of such a potential, when asymmetric, can induce motion in the following way: Turn the potential off; the particle diffuses freely (time $\tau \leq \tau_{off}$ in Fig. 1). We call P_f the probability that the particle diffuses forward by more than λ_f during the time τ_{off} (and similarly P_b for the backward probability). Switching the potential on again after a time τ_{off} forces the particle to the forward well with a probability P_f and to the backward one with a probability P_b (time $\tau = \tau_{off}$ in Fig. 1). We define as $J = P_f - P_b$, the probability current for a particle to advance one step in the periodic potential. Because λ_b is larger than λ_f, P_b is smaller than P_f and the drift is nonzero. As proposed earlier, the time modulation of a periodic asymmetric potential creates directed motion of thermally fluctuating particles [1]. Similar models of engines that extract work from random noise have been recently proposed under the denomination of "thermal ratchets" [2–6]. These models may have some connection with biological motor proteins [7–14].

How does one experimentally realize such a spatially periodic but asymmetric forcing of Brownian particles? One way is to deposit two metallic films on a glass substrate in a periodic but asymmetric fashion, so that applying an ac electric field through these electrodes creates the desired potential for colloidal particles in an aqueous solution. Recent experiments using such a setup confirmed the induced drift [15,16]. However, hydrodynamic interactions and the complicated electrical response of charges in water limited these experiments to only qualitative agreement with theory.

In this Letter, to avoid hydrodynamic interactions we study only one particle (a 1.5 μm diameter polystyrene sphere in room temperature water). To avoid electrolytic effects, the potential is created optically by strongly focusing an infrared laser beam to form an optical tweezer [17,18]. Two oscillating mirrors move the optical trap along a circle at a frequency too high for the particle to feel any net azimuthal force. The radial force, however, does not average to zero, and the particle is confined to diffuse along the circle. We then produce a periodic, asymmetric spatial modulation of the tweezing strength along the circle by synchronizing the rotation of a neutral density filter wheel with the rotation of the optical tweezer. Modulating in time this spatial intensity profile induces a net drift of the Brownian particle. The experiment is in agreement with the theory. We also observe a maximum of the induced motion as a function of the modulation time, related to stochastic resonance [19,20] (see Ref. [21] for an experimental case of resonance using a setup similar to ours). We then discuss the feasibility of particle separation using thermal ratchets.

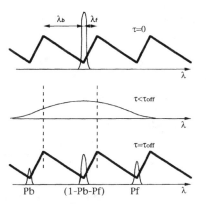

FIG. 1. The asymmetric potential is drawn as the thick line. The forward and backward length scales defining the asymmetry are λ_f and λ_b. The particle probability densities are drawn as thin lines. At time $\tau = 0$, the particle is localized and the probability density is sharply peaked. For times $\tau \leq \tau_{off}$, the potential is off and the particle diffuses freely. At time $\tau = \tau_{off}$, the potential is back on and the particle is forced to the forward and backward minimum with probabilities P_f and P_b.

Let us first describe the experimental setup. The samples are prepared by diluting suspensions of 1.5 μm diameter polystyrene spheres in pure water to a volume fraction of 10^{-4}, so that typically only a few beads are seen in the microscope field of view ($50 \times 50\ \mu\text{m}^2$). Mylar sheets of 50 μm thickness are cut and used as spacers between a microscope slide and a coverslip previously cleaned and dried using a nitrogen gas ionizing gun. The cells are then filled with the spheres in suspension and sealed with fast epoxy. The sample, placed on the translation stage of an upright microscope, is observed under bright-field illumination.

The TEM$_{00}$ linearly polarized output of a 1 W power Nd:YAG laser (wavelength = 1064 nm) is inserted into the microscope's optical path via the beam splitter B (see Fig. 2). Strongly focused by a 100$\times$ oil coupled objective (OBJ), the beam converges inside the sample to a sharp focal point that acts as an optical trap for the polystyrene sphere. Because of optical losses, the laser power at the sample level is of order 30 mW, and the trapping force is of order a piconewton. Two mirrors $M1$ and $M2$ mounted on galvanometers oscillate around two perpendicular axes with a $\pi/2$ phase difference. Two telescopes $T1$ and $T2$ allow the beam to pivot about the center of the iris

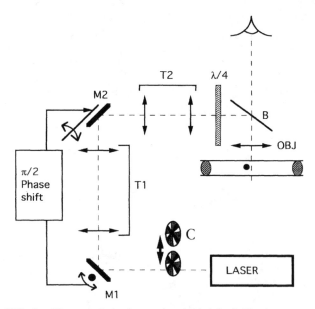

FIG. 2. The sample is observed under bright-field microscopy. The beam splitter B adds the infrared laser beam to the microscope's optical path. The microscope objective OBJ focuses the beam to a focal point, defining an optical trap. Two mirrors $M1$ and $M2$ oscillate around two perpendicular axes with a $\pi/2$ phase difference. They move the trap along a circle (see Fig. 3). Two telescopes $T1$ and $T2$ pivot the beam at the iris diaphragm of the objective OBJ. Circular polarization of the beam is achieved using a quarter wave plate ($\lambda/4$). The filter wheel is mounted on a chopper (C) synchronized with the mirrors. The chopper itself is put on a translation stage which is pulled in and out of the laser path.

diaphragm of the microscope objective as the mirrors oscillate (thus preserving the Gaussian beam profile) [22]. This setup moves the optical trap along a 7 μm diameter circle. A quarter wave plate $\lambda/4$ transforms the linear beam polarization into a circular one: This is necessary to keep the beam intensity constant along the circle. The two galvanometers are synchronously driven at 100 Hz. This frequency is such that the optical tweezer moves too fast for the sphere to follow the trap; the azimuthal force on the particle averages out to zero. The sphere diffuses then freely along the circle *but is still confined in the radial direction.* We checked experimentally that the bead moves diffusively along the circle. We also checked that the trap intensity is constant to within a few percent.

We now spatially modulate the beam intensity along the circle in a periodic but asymmetric way. A neutral density filter wheel is mounted on a chopper, the rotation speed of which is synchronized with the rotation of the optical trap (Fig. 2). The shape of the transmission coefficient of the wheel as a function of the rotation angle θ is an asymmetric triangle repeated 4 times as θ goes from 0 to 2π. The maximum attenuation factor is 10^{-2}. Figure 3 shows the beam intensity profile along the circle described by the optical trap. Note that the particle is not trapped and does not rotate at 100 Hz; the spatial modulation localizes, however, the particle in the regions of maximum intensity (see Fig. 3, modulation on). The shape of the effective potential experienced by the sphere along the circle is then indeed an asymmetric triangle characterized by the two length scales λ_f and λ_b shown in Fig. 1. The smaller length scale λ_f is limited by both the laser wavelength and the particle diameter; λ_f is around 2 μm. Experimentally, it takes a few seconds for the particle to fall to the bottom of the potential. Once there, it stays

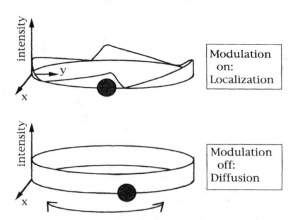

FIG. 3. Modulation on: the spatially asymmetric modulation of the beam intensity along the circle described by the optical trap is shown as the thin solid line (4 modulations per optical trap cycle). The 1.5 μm diameter particle is shown localized in a region of maximum beam intensity. Modulation off: the beam intensity is constant along the circle. The particle diffuses freely in one dimension.

localized: The height of the potential is much larger than the thermal noise k_BT.

We now modulate the beam intensity in time. The chopper is mounted on a translation stage. We manually pull the chopper in and out of the laser path in order to switch on and off the asymmetric spatial modulation of the potential along the circle. When on, the particle gets azimuthally localized in a potential minimum (time $\tau = 0$ in Fig. 1, modulation on in Fig. 3). When off, the particle is still confined on the circle but now diffuses freely along it (time $\tau \le \tau_{\text{off}}$ in Fig. 1, modulation off in Fig. 3). Turning the spatial asymmetry on again after a time τ_{off} localizes the particle to the minimum where the local gradient of light leads to (time $\tau = \tau_{\text{off}}$ in Fig. 1, modulation on in Fig. 3). The experimental time scales are small enough that the particle does not diffuse by more than the spatial extent of one well. The uncertainty in τ_{off} is of order 1 s. We repeat this experiment N times (N is of order 80) for different τ_{off} and count the total number of forward (N_f) and backward (N_b) bead motions. The forward probability is $P_f = N_f/N$ and the backward one $P_b = N_b/N$. In Fig. 4, the measured probabilities P_f and P_b are plotted as a function of τ_{off}. The vertical error bars on P_f (P_b) represent the statistical uncertainty $\sqrt{N_f}/N$ ($\sqrt{N_b}/N$). As expected for small τ_{off}, the particle does not have enough time to diffuse to the next minimum and stays in the same well; both probabilities tend to zero. For large τ_{off} the particle diffuses to the forward or backward minimum with equal probability $1/2$. (This assumes that the particle does not diffuse by more than the spatial extension of one well, which is the case for the range of experimental time scales, it is a short time scale approximation.) The forward probability P_f is then given by [23]

$$P_f = \frac{1}{2}\,\text{Erfc}\left[\frac{\lambda_f}{\sqrt{4D\tau_{\text{off}}}}\right] = \frac{1}{2}\,\text{Erfc}\left[\sqrt{\frac{\tau_f}{2\tau_{\text{off}}}}\right], \quad (1)$$

where Erfc is the complement of the error function and τ_f the average time for the particle to diffuse a distance λ_f, $\tau_f = \lambda_f^2/2D$. D is the diffusion coefficient of a

1.5 μm diameter sphere in water at room temperature ($D = 0.3\ \mu\text{m}^2/\text{s}$). This estimate agrees with the value we find by observing the bead diffusion along the circle when the modulation is switched off. We approximate Eq. (1) by [16,23]

$$P_f = \frac{1}{2}\exp(-\tau_f/\tau_{\text{off}}). \quad (2a)$$

In a similar way the backward probability is approximated by

$$P_b = \frac{1}{2}\exp(-\tau_b/\tau_{\text{off}}), \quad (2b)$$

where the characteristic time τ_b is equal to $\tau_b = \lambda_b^2/2D$. The solid lines in Fig. 4 are fitted to the experimental points using the simple exponentials (2a) and (2b). The results are $\tau_f = 5.3 \pm 0.3$ s and $\tau_b = 12.8 \pm 0.5$ s. From these times we estimate the two length scales, $\lambda_f = 1.8 \pm 0.2\ \mu$m and $\lambda_b = 2.8 \pm 0.2\ \mu$m. The forward characteristic length scale λ_f is close to the bead diameter as expected. These estimates agree with direct observation of the fall of a particle into the potential well.

The net drift ($P_f - P_b$) is shown in Fig. 5 as a function of τ_{off}. The vertical error bars represent the statistical uncertainty $\sqrt{N_f + N_b}/N$. The observed drift increases sharply from zero as τ_{off} approaches τ_f (of order 5 s), reaches a maximum for a time $\tau_{\text{max}} = (\tau_b - \tau_f)/\ln[\tau_b/\tau_f]$ of order 8 s, and then decreases slowly back to zero as τ_{off} becomes larger than τ_b (13 s). The solid line in Fig. 5 is the difference of the two fits in Fig. 4. The value of the drift at resonance is quite small (0.15). It could reach as much as 0.5 in the limit of large (τ_b/τ_f).

A particle with a different diffusion coefficient will show a similar curve with different characteristic time scales. This could be used in principle to sort particles according to their sizes [24]. This is, however, rather unpractical in this experiment. The first limitation comes from the broad width of the resonance (Fig. 5). Making it sharper by reducing the difference between τ_f and τ_b will decrease the strength of the resonance. There are no values of τ_f and τ_b that maximize the quality

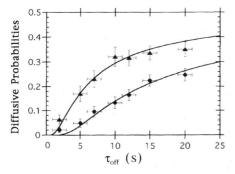

FIG. 4. The probabilities P_f and P_b for a particle to move forward (triangles) and backward (circles) as a function of τ_{off}. The two solid lines are fitted to the data following Eqs. (2a) and (2b).

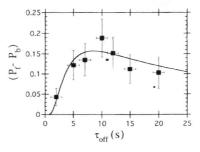

FIG. 5. The probability current ($P_f - P_b$) as a function of τ_{off}. The solid line is the difference between the two fits from Fig. 4.

factor (height over width) of the resonant curve. Another limitation comes from the statistical character of the induced drift: It is only *on average* that the particle motion follows the drift. The on and off switching of the potential then has to be repeated a large enough number of times so that a net particle motion emerges from the diffusive noise.

In conclusion, we have experimentally demonstrated the principle of a thermal ratchet: Broken spatial symmetry and time modulation are indeed enough to induce directed motion from random noise, with a maximum for a characteristic modulation time. Thermal noise can be a tool rather than a physical limit to the efficiency of motors [25].

We gratefully acknowledge the help of A. Ott and A. Schweitzer in designing and building the experimental setup. We thank D. Chatenay, D. Muraki, J. Prost, and M. Shelley for enlightening discussions.

---

[1] A. Ajdari and J. Prost, C.R. Acad. Sci. Paris **315**, 1635 (1992).

[2] R.P. Feynman, R.B. Leighton, and M. Sands, *The Feynman Lectures in Physics* (Addison-Wesley, Reading, 1966).

[3] J. Prost, J.F. Chauvin, L. Peliti, and A. Ajdari, Phys. Rev. Lett. **72**, 2652 (1994).

[4] M. Magnasco, Phys. Rev. Lett. **71**, 1477 (1993).

[5] C.R. Doering, W. Horsthemke, and J. Riordan, Phys. Rev. Lett. **72**, 2984 (1994).

[6] R.D. Astumian and M. Bier, Phys. Rev. Lett. **72**, 1766 (1994).

[7] B. Alberts, D. Bray, J. Lewis, M. Raff, K. Roberts, and J.D. Watson, *The Molecular Biology of the Cell* (Garland, New York, 1989).

[8] M. Magnasco, Phys. Rev. Lett. **72**, 2656 (1994).

[9] C.S. Peskin, G.M. Odell, and G.F. Oster, Biophys. J. **65**, 316 (1993).

[10] S.M. Simon, C.S. Peskin, and G.F. Oster, Proc. Natl. Acad. Sci. U.S.A. **89**, 3770 (1992).

[11] R. Vale and F. Oosawa, Adv. Biophys. **26**, 97 (1990).

[12] K. Svoboda, C.F. Schmidt, B.J. Schnapp, and S.M. Block, Nature (London) **365**, 721 (1993).

[13] J.T. Finer, R.M. Simmons, and J.A. Spudich, Nature (London) **368**, 113 (1994).

[14] J.M. Scholey, *Motility Assays for Motor Proteins* (Academic, New York, 1993).

[15] S. Leibler, Nature (London) **370**, 412 (1994).

[16] J. Rousselet, L. Salome, A. Ajdari, and J. Prost, Nature (London) **370**, 447 (1994).

[17] A. Ashkin, Phys. Rev. Lett. **24**, 156 (1970).

[18] A. Ashkin, J.M. Dziedic, J.E. Bjorkholm, and S. Chu, Opt. Lett. **11**, 288 (1986).

[19] H.A. Kramers, Physica (Utrecht) **7**, 284 (1940).

[20] R. Benzi, G. Parisi, A. Sutera, and A. Vulpiani, SIAM J. Appl. Math. **43**, 565 (1983).

[21] A. Simon and A. Libchaber, Phys. Rev. Lett. **68**, 3375 (1992).

[22] K. Svoboda and S. Block, A. Rev. Biophys. Biomol. Str. **23**, 247 (1994).

[23] A. Ajdari, Ph.D. thesis, Université Paris 6, Chap. VI.

[24] A. Ajdari and J. Prost, Proc. Natl. Acad. Sci. U.S.A. **88**, 4468 (1991).

[25] H.S. Leff and A.F. Rex, *Maxwell's Demon; Entropy, Information, Computing* (Princeton University Press, Princeton, 1990).

AC Research

Accelerated Articles

Anal. Chem. **1995**, *67*, 1763–1765

Optical Chromatography

Totaro Imasaka,* Yuji Kawabata, Takashi Kaneta, and Yasunori Ishidzu

Department of Chemical Science and Technology, Faculty of Engineering, Kyushu University, Hakozaki, Fukuoka 812, Japan

A new and potentially useful method for separation of particles by optical radiation pressure is described and demonstrated in this study. A laser beam is focused into the solution, which contains particles counterflowing coaxially in a capillary. The particle is focused into the center line of the laser beam by radiation pressure. The particle is turned around, accelerated, passed through a beam waist, decelerated by a liquid flow, and drifts, at which point the radiation pressure is identical to the force induced by the liquid flow, resulting in separation of particles as a function of size.

In 1906 Tswett reported on a separation technique that is now generally called chromatography.[1] Classical adsorption chromatography involves the adsorption of a substance (for example, a dye) to the solid support or solid phase. The adsorbed compound can then be eluted (desorbed) from the solid support by passing a liquid solvent or mixture of solvents over the solid phase which desorbs the compound from the support, thereby affecting a separation and purification. Chromatography has been studied by many researchers in the interim years and is now used almost universally in chemical analysis and separation science. Chromatography, when used in the classical way, however, has serious limitations that have not yet been (and may never be) overcome. Some of the limitations of chromatography are as follows: (1) Columns must be replaced frequently in order to optimize separation, making the technique time consuming. (2) Retention times must be measured for a series of compounds prior to the analysis of an unknown sample, since it cannot be precisely calculated from its physical properties. (3) Resolution is limited by diffusion in the column and cannot be greatly improved by extending the separation time. (4) Biological cells and large molecules that are of biochemical interest are poorly separated by chromatography as well as by other techniques such as field-

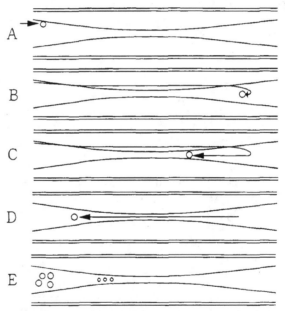

Figure 1. Schematic motion of particle. The laser beam is introduced from the right-hand side and the liquid from the left-hand side: (A) particle introduction; (B) focus of particle into beam center by gradient force; (C) acceleration of particle; (D) deceleration of particle; (E) particles drifting at equilibrium position.

flow fractionation or flow cytometry. (5) Detection efficiencies are well below unity. (6) The concentration detection limit is rather poor, even in state-of-the-art chromatographic techniques, such as capillary electrophoresis combined with laser fluorometric detection, thus making preconcentration a necessity, requiring additional instrument and time. (7) It is desirable to recover the sample at the same concentration that was used in the starting run, so that further experiments can be performed with the sample. (8) The sample is obtained in an eluent, usually in fractions, and is in a diluted state. (9) It is usually difficult or

(1) Pecsok, R. L.; Shields, L. D.; Cairns, T.; McWilliam, I. G. *Modern Methods of Chemical Analysis*, 2nd ed.; John Wiley & Sons: New York, 1968.

Figure 2. Photograph of particles separated by radiation pressure. The 1-μm particles are localized in a line at around the mark ($\times$) in the right-hand side, and the 3-μm particles at around the mark ($\times$) in the left-hand side. This experiment used a multiline argon laser (454.5–514.5 nm) at 2 W and a 200-μm-i.d. capillary at a 10 μm/s flow rate. Refractive index: polymer beads, 1.59; water, 1.33.

impossible to perform in situ reactions on the compound as it exists in the chromatographic column. (10) Column dimensions, optimally, should be minimized when used as a preseparator in conjunction with a chemical sensor, but the columns are far too long for this to be achieved.

In 1970, Ashkin reported an optical trapping technique.[2] By tightly focusing a laser beam, a particle with a higher refractive index than the medium is trapped at the focal point with radiation pressure. This technique is useful in the sense that a particle can be "held" in place and then subjected to chemical analysis in the micrometer region, e.g., fluorescence measurement of dyes that are adsorbed on a particle.[3] This technique allows the manipulation of a single particle by complete holding like "adsorption", though it is possible to control many particles sequentially by scanning the laser beam.[4]

In this study, we demonstrate a new approach for the separation of particles or molecules using radiation pressure by incomplete holding under a liquid flow. The overall technique affects a chromatographic-like separation but involves principles that are different from classical chromatography. The approach is quite simple; a laser beam is focused into the solution, which contains substances counterflowing coaxially in a capillary. The behavior of the substances is recorded by a video camera equipped with a microscope objective.

RESULTS AND DISCUSSION

The motion of the substance, a particle in this study, under radiation pressure is schematically shown in Figure 1: (A) A particle is introduced into a capillary by a liquid flow. (B) The

particle is focused into the center line of the laser beam by the gradient force.[5] (C) The particle is turned around and accelerated by the scattering force.[5] (D) The particle is decelerated by a liquid flow as the particle moves away from the beam waist. (E) The particle drifts when the radiation pressure becomes identical to the force induced by the liquid flow. A photograph of the system at equilibrium is shown in Figure 2. The 1- and 3-μm particles are clearly separated from each other; the 6-μm particles injected are also separated but are located beyond the frame of this picture and are not shown. These data clearly demonstrate that particles can be separated using this technique as a function of particle size.

The present approach has unique characteristics: (1) The particle separation can be easily controlled by altering the beam focusing condition, which corresponds to a replacement of the column in classical chromatography. (2) The position of the particle can be calculated provided its physical properties, e.g., size and refractive index, are known. (3) The separation resolution can be improved by measuring the particle position more accurately by extending the time for measurement. (4) Particles (or large molecules) are more easily separated, although smaller molecules can also be separated by using a stronger laser that is emitting at shorter wavelengths. (5) The separated particle is recorded at all times by a video camera and is detected with an efficiency of 1.0. (6) Separation and concentration are performed simultaneously with no additional instrument and time required. (7) The concentration of the sample recovered by interrupting the laser beam is identical to the original solution. (8) The particles are collected in order of size by decreasing the laser power, and the concentration can be increased. (9) A particle separated in the capillary can be trapped by introducing a second, perpendicular laser beam, thus allowing in situ chemical reactions

(2) Ashkin, A. *Phys. Rev. Lett.* **1970**, *24*, 156–159.

(3) Misawa, H.; Koshioka, M.; Sasaki, K.; Kitamura, N.; Masuhara, H. *Chem. Lett.* **1990**, 1479–1482.

(4) Sasaki, K.; Koshioka, M.; Misawa, H.; Kitamura, N.; Masuhara, H. *Opt. Lett.* **1991**, *16*, 1463–1465.

(5) Ashkin, A. *Biophys. J.* **1992**, *61*, 569–582.

to be performed in succeeding stages. (10) The column size can be reduced to micrometers. This technique addresses most of the problems and limitations of classical chromatography that are described above.

There are several advanced technologies for particle separation, such as field-flow fractionation invented by Giddings.[6] However, optical chromatography as demonstrated herein has several advantages over other separation methods: for example, the driving force, i.e., radiation pressure and liquid flow, can be changed immediately and independently from outside for modification of separation conditions. Moreover, concentration and separation can be performed simultaneously, and then the present method can be applied to diluted samples. The collection efficiency can, in theory, be improved to 100% by increasing the laser output power and by matching the laser beam diameter at the inlet port to the capillary inner diameter, though the efficiency is presently much less than 1%. The separation resolution, which is affected by the beam focusing conditions and the stabilities of the laser and liquid-flow parameters, can be improved by optimizing conditions, though unavoidable limiting factors, e.g., based on a Brownian motion of the molecule in the liquid, exist. It is noted that an irregular or opaque particle can be separated similarly but a light-absorbing, i.e., black, particle is beyond the application of this technique.

The present method may be used in a variety of fields. A straightforward application might be separation of polystyrene beads, biological cells, or biopolymers such as protein, DNA, and RNA. For example, antibody-bound polymer beads are coagulated in the presence of antigen, so that a specific protein can be detected.[7] Since the present method allows the detection of particles at extremely low levels, in theory, a single protein molecule could be detected.

The aim of the present study is not only the demonstration of a new separation technique but also the proposal of a new "field" for chemical reactions and analysis in the micrometer region. For example, ion cyclotron resonance (ICR) mass spectrometry (MS) first isolates a specific ion, which is then dissociated, e.g., by electron impact, to generate daughter ions that are further analyzed by MS. In an analogous manner, a specific molecule could be isolated by optical chromatography and then be optically trapped by a second laser. The molecule can be further reacted, e.g., with an enzyme introduced by an electroosmotic flow, and the products may be further separated and analyzed by optical chromatography. Thus, the present method provides for an ultrasensitive means for structural analysis of large molecules or even particles that cannot be applied to a vapor-phase experiment such as ICR-MS.

Received for review January 17, 1995. Accepted March 19, 1995.[⊗]

AC950052K

(6) Giddings, J. C. *Anal. Chem.* **1981**, *53*, 1170A–1178A.
(7) Rosenzweig, Z.; Yeung, E. S. *Anal. Chem.* **1994**, *66*, 1771–1776.

[⊗] Abstract published in *Advance ACS Abstracts*, April 15, 1995.

Optical sculpture: controlled deformation of emulsion droplets with ultralow interfacial tensions using optical tweezers

Andrew D. Ward,*[a] Mark G. Berry,[a] Christopher D. Mellor[b] and Colin D. Bain*[c]

Received (in Cambridge, UK) 14th July 2006, Accepted 31st August 2006
First published as an Advance Article on the web 28th September 2006
DOI: 10.1039/b610060k

We report a technique for deforming micron-sized emulsion droplets that have ultralow interfacial tensions, by the manipulation of multiple optical trapping sites within the droplets.

Laser-based optical trapping has become a widespread technique for manipulation and control of colloidal and biological particles.[1] The advantages of the technique lie in the precise and non-destructive control of microscopic particles without the requirement for sample perturbation by mechanical means. Optical trapping has been primarily used to measure forces exerted externally on a trapped object. Here we describe the deformation of a single emulsion droplet using multiple optical tweezers to provide a means for applying internal forces.

In general, small objects trapped with optical tweezers do not deform because the radiation pressure of the continuous wave (CW) lasers used in optical trapping is feeble compared to the Young's modulus of most solids or the Laplace pressure within a micron-sized liquid droplet. Thus, droplets form and retain their perfect spherical geometry to minimise the surface area, and hence the Gibbs free energy, for a given volume. Deformation studies have been performed on liposomes and red blood cells,[2] in which a thin lipid bilayer membrane encapsulates the cell. Distortions in the membrane are then possible because of the low values of the elastic shear and bending moduli of the bilayer.

It has been known, however, since the experiments of Ashkin and Dziedzic 30 years ago[3] that focused (and especially pulsed) laser beams can cause measurable deformations in a planar or pseudo-planar (*i.e.* large radius of curvature) surface. Recently, there has been a flourish of interest in deforming surfaces this way. For instance, Zhang and Chang[4] studied the effect of pulsed lasers on macrodroplets of water, Casner and Delville[5] showed that giant deformations could be induced with continuous wave lasers if the interfacial tension was low, and Mitani and Sakai[6] developed laser interface manipulation as a means of measuring interfacial tensions.

In this paper we describe how large deformations can be achieved in droplets of an oil-in-water emulsion with a low-power CW laser. For this phenomenon to occur the interfacial tension at the oil–water surface has to be reduced to a value comparable with the force constant of the optical trap, which is typically 10^{-5} to 10^{-6} N m^{-1}. Since oil–water interfacial tensions are around 0.05 N m^{-1} for an alkane, the interfacial tension has to be reduced by four orders of magnitude. Such ultralow tensions are achieved by adsorption of surfactants under conditions where the emulsion is close to the microemulsion phase boundary (where the interfacial tension vanishes). The system we chose comprises AOT–hexane–brine. Temperature was used to fine-tune the interfacial tension and large deformations were observed with a laser power of 10^{-2} W.

The optical trapping experiments were carried out in the Laser Microscopy Laboratory of the Central Laser Facility at the Rutherford Appleton Laboratory, Oxfordshire, UK. For these initial experiments we prepared an oil-in-water emulsion by blending together heptane, water, salt (0.05 M NaCl) and the surfactant Aerosol OT (1 mM AOT). Heptane (Sigma-Aldrich, >99%) was used after passing through an alumina column three times and AOT (Sigma-Aldrich, >99%) was used as received. The oil was added to the surfactant solution and the two briefly agitated to begin emulsification. At this point, the emulsion is in a dynamic state where the droplets are a few microns in diameter and the heptane has not yet microemulsified. The interfacial tension of this emulsion system has been characterised by Aveyard[7] using spinning drop tensiometry. It was also used by Mitani to demonstrate laser induced deformation of a planar oil–water interface.[6] A region of ultralow interfacial tension ($<3 \times 10^{-6}$ N m^{-1}) is known to exist between 20 °C and 23 °C. In our study, the emulsion was injected into a 100 μm deep flow cell as shown in Fig. 1. Single droplets were captured near the upper surface as droplets in the emulsion started to cream. The trapped droplet was dragged vertically down into the bulk solution to be isolated from other droplets. All deformation experiments were performed at 20.0 ± 0.5 °C.

The experimental apparatus has been reported previously.[8] In brief, a 1 W Nd:YAG laser (Laser 2000) at 1064 nm was passed through a pair of perpendicular acousto-optic deflectors (AOD), whose function was to steer the laser beam in the *x*–*y* plane. The first-order diffracted beam was expanded and then directed into an inverted microscope (Leica, DM-IRB) *via* a dichroic mirror and microscope objective (Leica, ×63 water immersion, NA = 1.2). The signals applied to the AOD were multiplexed using computer software to generate up to four optical traps with individually controllable trapping position. Laser powers at the focus were 11–27 mW; at these values laser heating is expected to be minimal (<1 °C).[9] Dividing the optical traps into multiple positions is assumed to lower the average laser power at each position in the same ratio. The images were recorded using a CCD camera and

[a] *Lasers for Science Facility, CCLRC, Rutherford Appleton Laboratory, Chilton, Didcot, UK OX11 OQX. E-mail: a.d.ward@rl.ac.uk; Fax: +44 (0)1235 445693; Tel: +44 (0) 1235 446751*
[b] *National Institute for Medical Research, Mill Hill, London, UK NW7 1AA*
[c] *Department of Chemistry, Durham University, South Road, Durham, UK DH1 3LE. E-mail: c.d.bain@durham.ac.uk; Fax: +44 (0)191 3342051; Tel: +44 (0) 191 3342138*

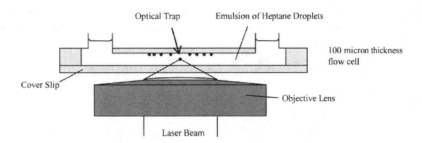

Fig. 1 Schematic diagram of the flow cell illustrating the position of optical trapping and droplet deformation within the sample.

show the shape of the droplet as viewed in plan from beneath the sample using brightfield illumination. A side-on view of the droplet is not possible in the current optical configuration. To prevent the CCD from saturating with 1064 nm laser light, an optical filter was placed before the camera.

The maximum force exerted by a single optical trap was determined using an escape force technique, where a viscous drag force[10] was applied to a heptane droplet in the absence of salt, thus the oil–water interface was not in the ultralow region. As the solution flowed past the trapped droplet the laser power was reduced until the droplet just escaped. A linear relationship between laser power and the applied drag force was recorded, yielding an escape force of 0.69 pN mW^{-1} for a single optical trap.

Fig. 2 shows an image of an oil droplet in a single optical trap (Fig. 2a) and a series of droplets deformed symmetrically by 2, 3 and 4 traps (Fig. 2b, c and d). With a single optical trap, the interfacial tension at the oil–water surface is sufficiently low such that the droplet deforms into an ellipse along the axial direction of the laser beam. This deformation is inferred from the observation that increasing laser power decreases the radius of the droplet and *vice versa*. The measured droplet radius as a function of laser power was extrapolated to zero power to obtain the spherical volume of the droplet. Across the range of laser powers used in this

experiment the radius decreases by approximately 10%. Although we do not know the 3-dimensional shape of the structures formed with multiple optical traps, the average depth in the axial direction can be calculated on the reasonable assumption that the volume of the droplet remains constant throughout the deformation process. For example, the volume of the spherical droplet in a single trap with corrected radius, r, can be compared to that of an ellipsoid in the two trap deformation. The semi-major, a, and semi-minor, b, dimensions are estimated to the nearest 0.1 μm from the microscope images (1 pixel is 0.095 μm). From this assumption the semi-minor depth value, c, is equivalent to b within the experimental error of measurement. Thus, the droplet is deformed to a prolate spheroid with a small increase in surface area. The triangle and square droplet shapes appear to have a thicker central region that thins in the "corners". The contrast in the images does not suggest that there is a change from positive to negative curvature. In qualitative terms, the optical sculptures can be understood as photonic forces acting on the oil towards the focal points of the laser beam, balanced by a weak interfacial tension. The droplet assumes a shape of minimum surface area between the positions of the laser traps.

A series of experiments were then performed in which a droplet was deformed into an ellipsoid, with two optical traps at a set laser power, until the surface area could not be extended further. If the trap positions were separated beyond this point, the interfacial elasticity overcame the optical trapping force and the stretched droplet dislodged from one of the traps. The droplet remained in the other trap and the traps could be reset to repeat the deformation. In this way, the maximum deformation could be measured as a function of the laser power (Fig. 3). Extending our geometry calculations enables a first order approximation of the interfacial tension of the droplets. The Laplace pressure arising from the curvature of the prolate spheroid is balanced by the applied force, F, over the droplet cross-section (eqn 1).

$$\frac{F}{\pi a^2} = \gamma \left(\frac{2}{R_1} - \frac{1}{a} - \frac{1}{b} \right) \qquad (1)$$

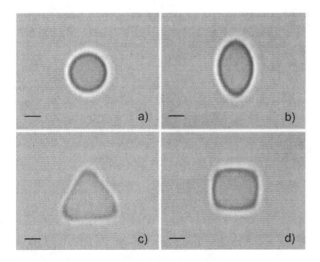

Fig. 2 Shapes formed by manipulation of multiple optical traps inside an emulsion droplet: a) droplet in a single trap, b) ellipsoid formed by stretching the droplet with two traps, c) triangular shape formed using three traps, d) square shaped droplet formed with four traps. The images are in plan view and the scale bar represents 2 μm. Laser power is 24 mW divided between respective number of trapping positions.

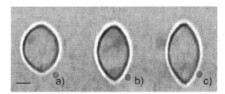

Fig. 3 Limiting droplet extensions with two optical traps as a function of laser power: a) 11 mW, b) 19 mW, c) 22 mW. Scale bar represents 2 μm.

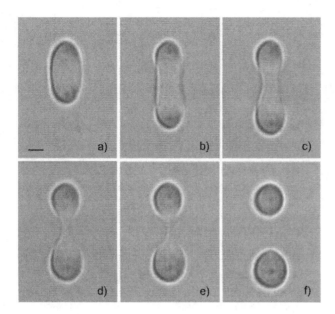

Fig. 4 With very low tensions the droplets form dumbbell shapes on extension and can be separated and rejoined. Time elapsed between each image is 1 second. Careful inspection of image f) reveals that a thin thread connects the separated droplets. Laser power is 27 mW.

where: R_1 is the radius of curvature at the end of the droplet. From the escape force calibration, the optical force exerted at each node at the maximum extension was 8 pN; this value is only an estimate, since the shape of the interface through which the laser passes is no longer spherical. The values of R_1, a and b are approximately 1 μm, 2 μm and 5 μm, from which we estimate a value of the interfacial tension of $\sim 1 \times 10^{-6}$ N m^{-1}, which is in good agreement with the results obtained by Aveyard[7] and Mitani[6] using spinning drop interfacial tensiometry. The value of the interfacial tension depends sensitively on temperature. At present, we have fixed both temperature and salt concentration during the deformation experiments. However, varying these parameters will allow a better comparison of the interfacial properties determined from tensiometry and optical deformation techniques.

Under higher powers (27 mW), some droplets continued to stretch into a dumbbell shape (Fig. 4) which then divided into two smaller droplets as the trap separation was increased. These droplets appeared to remain connected by a thin oil thread barely discernable in the microscopic images and reminiscent of the ligament that connects a droplet to a nozzle in an inkjet printer. Closer inspection of the two droplets showed that they formed a point where the threads were attached, indicating the presence of a small tethering force. The two parts could readily be joined again, with no barrier to coalescence, to reform the initial droplet.

We have shown that it is possible to deform spherical emulsion droplets into a variety of shapes such as triangles and squares using multiple optical traps, provided that the interfacial tension, which is of the order of 10^{-6} N m^{-1}, is comparable to the force constant of the trap. Such low interfacial tensions can be achieved by working with oil–surfactant systems, such as heptane, Aerosol OT and brine, that are known to form microemulsions.

Notes and references

1 *Methods in Cell Biology*, ed. M. P. Sheetz, Academic Press, San Diego, CA, USA, 1998, vol. 55.
2 P. J. H. Bronkhorst, G. J. Streekstra, J. Grimbergen, E. J. Nijhof, J. J. Sixma and G. J. Brakenhoff, *Biophys. J.*, 1995, **69**, 1666; J. Guck, R. Ananthakrishnan, T. J. Moon, C. C. Cunningham and J. Käs, *Phys. Rev. Lett.*, 2000, **84**, 5451.
3 A. Ashkin and J. M. Dziedzic, *Phys. Rev. Lett.*, 1973, **30**, 139.
4 J.-Z. Zhang and R. K. Chang, *Opt. Lett.*, 1988, **13**, 916.
5 A. Casner and J.-P. Delville, *Phys. Rev. Lett.*, 2001, **87**, 054503.
6 S. Mitani and K. Sakai, *Phys. Rev. E: Stat. Phys., Plasmas, Fluids, Relat. Interdiscip. Top.*, 2002, **66**, 031604.
7 R. Aveyard, B. P. Binks, S. Clark and J. Mead, *J. Chem. Soc., Faraday Trans.*, 1986, **82**, 125.
8 C. D. Mellor, M. A. Sharp, C. D. Bain and A. D. Ward, *J. Appl. Phys.*, 2005, **97**, 103114.
9 E. J. G. Peterman, F. Gittes and C. F. Schmidt, *Biophys. J.*, 2003, **84**, 1308.
10 R. Aveyard, B. P. Binks, J. H. Clint, P. D. I. Fletcher, T. S. Horozov, B. Neumann, V. N. Paunov, J. Annesley, S. W. Botchway, D. Nees, A. W. Parker, A. D. Ward and A. N. Burgess, *Phys. Rev. Lett.*, 2002, **88**, 246102.

COMMUNICATION

www.rsc.org/pccp

PCCP

Control and characterisation of a single aerosol droplet in a single-beam gradient-force optical trap

Rebecca J. Hopkins, Laura Mitchem, Andrew D. Ward† and Jonathan P. Reid*

School of Chemistry, University of Bristol, Bristol, UK BS8 1TS

Received 20th September 2004, Accepted 6th October 2004
First published as an Advance Article on the web 12th October 2004

Optical tweezers are used to control aerosol droplets, 4–14 μm in diameter, over time frames of hours at trapping powers of less than 10 mW. When coupled with cavity enhanced Raman scattering (CERS), the evolution of the size of a single droplet can be examined with nanometre accuracy. Trapping efficiencies for water and decane droplets are reported and the possible impact of droplet heating is discussed. We demonstrate that the unique combination of optical tweezing and CERS can enable the fundamental factors governing the coagulation of two liquid droplets to be studied.

Heterogeneous atmospheric aerosol chemistry is an experimentally challenging area of research. Measurements often must be made on an ensemble of particles, characterised by distributions of particle size and composition, and complicated by indeterminacy of mixing state, phase and morphology.[1] With heterogeneous chemistry often governed by the surface-to-volume ratio, interfacial composition and phase, the fundamental factors governing the chemical and physical transformation of aerosols may be obscured or at best averaged. To complement measurements made on ensembles of particles, measurements made on a single particle may provide detailed information on the mechanisms of heterogeneous chemistry, the factors governing mixing state, and the processes that lead to the physical transformation of the particle through phase transformation and growth.

In this Communication, we demonstrate for the first time that single-beam gradient force optical trapping,[2,3] or optical tweezing, can be used to trap a single water or decane droplet, 4–14 μm in diameter, for timescales of hours, affording the prospect of characterising the mechanisms of chemical and physical transformations of a single droplet. When coupled with cavity enhanced Raman scattering (CERS), the size of the droplet can be determined with nanometre precision.[4,5] By creating two optical traps, two droplets can be manipulated simultaneously, providing the unique opportunity to study the coagulation of two droplets directly.

The interplay of scattering and gradient forces must be considered when a dielectric sphere is illuminated with a Gaussian laser beam.[3,6] The scattering force acts in the direction of propagation of the laser and is proportional to the light intensity, ultimately acting to displace the particle from the beam focus. Conversely, the gradient force acts to draw the dielectric sphere towards the beam focus and is proportional to the gradient of the light intensity. When the gradient force dominates the scattering force, a restoring force with a magnitude of piconewtons acts to confine the particle to the most intense region in the laser beam and a stable trap is created.[3] Such a scenario can be achieved with a tightly focussed laser

beam using a microscope objective, forming a three dimensional trap with the particle position constrained both in the horizontal plane and on the vertical axis. Tweezers have found widespread application in the biological and colloidal sciences, but have not been developed to address key issues in aerosol science.[3] Tweezing has been applied to trapping solid glass spheres in air[6] and scattering forces have been used to levitate particles,[7–13] but there has been little success in tweezing liquid aerosols and no reported entrapment of water droplets.[14]

A schematic diagram of the experiment is shown in Fig. 1(a). Trapping is achieved with a cw argon ion laser operating at 514.5 nm. Not only is the complex refractive index of water near a minimum at this wavelength,[1] minimising droplet heating, but the size parameter, x, is larger than it would be at the more usual longer wavelengths chosen for tweezing.[3] This is extremely important for recording a CERS fingerprint from the trapped droplet.[4,5] The size parameter is defined as the ratio of the circumference of the droplet, radius a, to the wavelength of the light, λ $(x = 2\pi a/\lambda)$.[15]

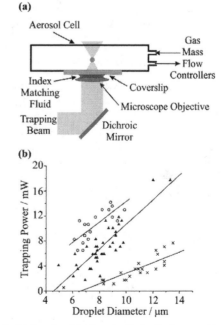

Fig. 1 (a) Experimental schematic, illustrating the aerosol trapping cell. The working distance is ~120 μm for the oil-immersion 60× objective used in this work; (b) Measurement of the minimum trapping power for a range of aerosol droplet sizes (circles: water droplets, water immersion objective; triangles: water droplets, oil immersion objective; crosses: decane droplets, oil immersion objective).

† Lasers for Science Facility, CCLRC, Rutherford Appleton Laboratory, Chilton, Didcot, UK OX11 0QX.

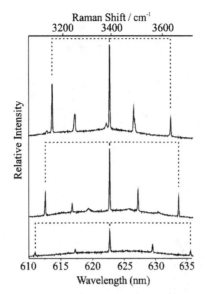

Fig. 2 Three CERS signatures from trapped water droplets, showing the spontaneous OH stretching band and the stimulated Raman scattering at whispering gallery mode wavelengths. Droplet radii are 5.103, 4.547 and 3.918 μm (top to bottom, respectively). The trapping power was 4.7 mW and the spectral integration time 1 s.

The tweezing instrument has been described in a previous publication and will be outlined only briefly here.[16] The trapping beam is passed through two sets of beam expansion optics, reflected from a holographic notch filter (HSPF-514.5, Kaiser Optical Systems) and directed into a Leica DM IRB microscope. Two microscope objectives are compared in this work: a 63× water immersion objective (NA of 1.2), and a 60× oil immersion objective (NA of 1.4). The backscattered Raman light from the trapped droplet is collimated by the objective lens, passed through the filter, and focussed onto the entrance slit of a 0.5 m spectrograph (1200 g/mm grating) coupled with a CCD, with a spectral resolution of 0.05 nm. The trapped droplet is imaged onto a CCD camera using conventional brightfield microscopy.

The aerosol was generated with an Omron NE-U07 Ultrasonic Nebuliser, introduced into an aerosol cell mounted on the translation stage of the microscope (Fig. 1(a)), along with a flow of humidified nitrogen using a mass flow controller in order to regulate the humidity in the aerosol cell. To achieve stable trapping in the aerosol cell, which is always at a relative humidity of less than 100%, the water aerosol was doped with sodium chloride at a concentration of ~40 mM. Köhler theory enables the equilibrium droplet size to be estimated at a particular relative humidity and for a given dry particle diameter.[1] Doping the water droplets with sodium chloride reduces the vapour pressure of the droplet so that an equilibrium size is rapidly attained, and enables the droplets to be trapped for timescales of hours. The aerosol trap was loaded by providing a brief dose of aerosol from the nebuliser. A stable trap was achieved for water droplets and decane droplets 4–14 μm in diameter, with laser trapping powers of 5–15 mW (Fig. 1(b)). The axial trapping efficiency of a droplet, Q, can be determined by measuring the minimum incident laser power, P, required to counterbalance the gravitational force experienced by the particle, F.[6,14,17]

$$Q = \frac{\text{radiation pressure exerted on the particle}}{\text{radiation pressure of the incident beam}}$$

$$= \frac{F}{nP/c} \text{ where } F = \frac{4}{3}\pi a^3 \rho g$$

The refractive index of the surrounding medium is denoted by n, the speed of light by c, the particle density by ρ and the gravitational acceleration by g.[17] The minimum power required to trap a droplet is estimated by systematically reducing the laser power until a droplet falls out of the trap. Q was measured as 0.07 ± 0.02 and 0.1 ± 0.04 for water droplets trapped using the water and oil immersion objectives, respectively, and 0.45 ± 0.12 for decane droplets with the oil immersion objective. The trapping efficiencies are comparable to those measured for particles in liquids, allowing the prolonged trapping and manipulation of droplets for timescales of hours.[3]

In previous publications we have shown that CERS can be used to probe dynamic changes in the size and the composition of aerosol droplets.[4,5,18] A spherical droplet behaves as a high finesse optical cavity at wavelengths commensurate with whispering gallery modes (WGM).[19] Such modes occur when an integer number of wavelengths, n, form a standing wave around the circumference of the droplet, providing a mechanism for optical feedback. A CERS fingerprint can result, with amplification of the Raman scattering only occurring at wavelengths commensurate with WGMs. This leads to stimulated Raman scattering, which is apparent as structure superimposed on the spontaneous Raman bands. The fingerprint of resonant wavelengths can be used to determine the droplet size with nanometre precision.[4,5]

Fig. 2 illustrates three CERS spectral fingerprints obtained from optically trapped water droplets of different sizes. The spectra are comprised of the spontaneous Raman scattering envelope arising from the O–H stretch of water, with resonance structure superimposed from stimulated Raman scattering at wavelengths matching WGM wavelengths. Successive modes of the same scattered polarisation, as indicated by the brackets, correspond to sequential integer values of the mode number, n. The trend in resonance wavelengths illustrates the increase in mode spacing with diminishing droplet size. Mie scattering calculations are performed to determine the droplet radius with an estimated error of ±2 nm, limited primarily by the spectral resolution of the spectrograph.[5] All sizes should be assumed to have this associated error. As a further illustration of the sizing capability of CERS, Fig. 3 illustrates the variation in the trapped droplet size when equilibrating with the local relative humidity in the aerosol cell, each spectrum corresponding to 1 s of signal integration. This demonstrates that dynamic size changes at the nanometre level can be investigated by CERS.

Perhaps the most important issue to address is the influence of the laser on the droplet temperature through absorption.[20] Fig. 4(a) illustrates that varying the laser power by a factor of three over a period of ~10 min leads to droplet size fluctuations of <±2 nm. Indeed, the invariance of the resonant wavelengths with power also suggests that the refractive index of the droplet is constant, reinforcing the conclusion that there is no significant change in temperature.[5] A shift to shorter wavelength of 0.047 nm/K would be apparent if heating were

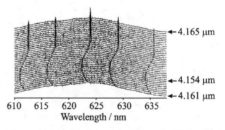

Fig. 3 Illustration of the sizing capability of CERS, showing the evolution in the size of a trapped water droplet over a ~40 s period. The trapping power was 4.7 mW and the spectral integration time 1 s.

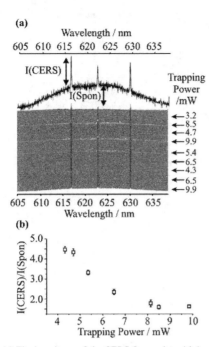

Fig. 4 (a) The invariance of the CERS fingerprint with laser power; (b) Inverse correlation between the intensity of the CERS signal and laser power when compared with the growth in the spontaneous scatter.

occurring, reflecting the change in refractive index with temperature.[5] The magnitude of this shift is comparable to the experimental resolution of the spectrograph and it can be concluded that any droplet heating must correspond to a temperature change of <1 K.

Assuming an absorbance of 0.005, the upper limit of the absorbance in the visible of the HPLC grade water used in this work, the complex refractive index can be calculated[15] to be less than 1×10^{-8}. Mie scattering calculations of the absorption efficiency[4] indicate that $Q_{abs} \sim 2 \times 10^{-7}$. By applying the procedure adopted by Popp *et al.*, this correlates with an increase in droplet temperature due to heating of 0.01 K under steady state conditions, consistent with the temperature invariance inferred from the experimental data.[21] It should be noted that a droplet resonant with the incident laser field will show

enhanced heating and this would lead to an enhancement in Q_{abs} of two orders of magnitude and a temperature rise of the order of 1–10 K. The low density of WGMs for the droplet sizes considered in this work and the invariance of the refractive index and resonant wavelengths suggest this does not occur.[21] However, a final caveat should be added. The intensity of the CERS signal shows a strong inverse correlation with laser power (Fig. 4(b)), suggesting that the droplet cavity losses increase with increasing laser power, and this may be symptomatic of enhanced absorption losses in the droplet.[4,19] Studies of this will be undertaken to examine further any role that laser induced heating may play.

The nature of the optical trap enables 3D manipulation of the aerosol droplets, an advantage over purely levitating the droplet with the scattering force. A beam splitter was inserted into the laser path between the two sets of expansion optics, creating a second optical trap and allowing two aerosol droplets to be trapped simultaneously. The position of each droplet could be controlled independently enabling the coagulation of the two droplets to be studied. Two water droplets were trapped simultaneously and their radii monitored until stable by examining the CERS fingerprint from each droplet (radii of 3.014 μm and 4.038 μm), as illustrated in Fig. 5. Following controlled coagulation, a final CERS fingerprint was recorded to determine the size of the coagulated droplet (4.533 μm). The results from the coagulation measurement shown in Fig. 5 illustrate that the additive volumes of the two individual droplets and the volume of the coagulated droplet are in agreement to within $\pm 3 \times 10^{-19}$ m³.

By coupling optical tweezers with CERS, a wide range of aerosol dynamics can now be explored. We have performed preliminary measurements of the growth of water droplets through the uptake of ethanol and water and are exploring further the fundamental factors that govern aerosol coagulation. Not only can accurate measurements be made of droplet size, but the evolution of the chemical composition can be monitored by spontaneous and stimulated Raman scattering.

Acknowledgements

The authors acknowledge the support of an EPSRC grant (GR/S52261/01) and the award of CCLRC facility time at the LSF (CM8C1/04). They also acknowledge helpful discussions with Dr D. McGloin at the University of St Andrews, Dr M. King at RHBNC and Dr P. Bartlett at the University of Bristol.

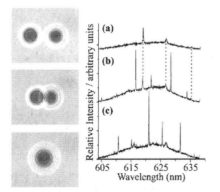

Fig. 5 The images show the simultaneous control that can be exercised over two aerosol droplets, manipulating them to the point of coagulation. CERS fingerprints: (a) right droplet in the image, (b) left and right droplets (combined volume 3.905×10^{-16} m³) and (c) coagulated droplet (volume 3.902×10^{-16} m³).

References

1 J. H. Seinfeld and S. N. Pandis, *Atmospheric Chemistry and Physics: From Air Pollution to Climate Change*, John Wiley & Sons, New York City, 1998.
2 A. Ashkin, J. M. Dziedzic, J. E. Bjorkholm and S. Chu. *Opt. Lett.*, 1986, **11**, 288.
3 J. E. Molloy and M. J. Padgett, *Contemp. Phys.*, 2002, **43**, 241.
4 R. Symes, R. M. Sayer and J. P. Reid, *Phys. Chem. Chem. Phys.*, 2004, **6**, 474.
5 R. M. Sayer, R. D. B. Gatherer, R. Gilham and J. P. Reid, *Phys. Chem. Chem. Phys.*, 2003, **5**, 3732.
6 R. Omori, T. Kobayashi and A. Suzuki, *Opt. Lett.*, 1997, **22**, 816.
7 E. J. Davis, *Aerosol Sci. Technol.*, 1997, **26**, 212.
8 A. Ashkin and J. M. Dziedzic, *Science*, 1975, **187**, 1073.
9 T. R. Lettieri, W. D. Jenkins and D. A. Swyt, *Appl. Opt.*, 1981, **20**, 2799.
10 R. Thurn and W. Kiefer, *Appl. Opt.*, 1985, **24**, 1515.
11 J. C. Carls, G. Moncivais and J. R. Brock, *Appl. Opt.*, 1990, **29**, 2913.
12 C. Esen, T. Kaiser and G. Schweiger, *Appl. Spectrosc.*, 1996, **50**, 823.

13   J. Popp, M. Lankers, M. Trunk, I. Hartmann, E. Urlaub and W. Kiefer, *Appl. Spectrosc.*, 1998, **52**, 284.

14   N. Magome, M. I. Kohira, E. Hayata, S. Mukai and K. Yoshikawa, *J. Phys. Chem. B*, 2003, **107**, 3988.

15   C. F. Bohren and D. R. Huffman, *Absorption and Scattering of Light by Small Particles*, John Wiley and Sons, Inc., 1983.

16   J. M. Sanderson and A. D. Ward, *Chem. Commun.*, 2004, **9**, 1120.

17   W. H. Wright, G. J. Sonek and M. W. Berns, *Appl. Phys. Lett.*, 1993, **63**, 715.

18   R. J. Hopkins, R. Symes, R. M. Sayer and J. P. Reid, *Chem. Phys. Lett.*, 2003, **380**, 665.

19   S. C. Hill and R. E. Benner, in *Optical Effects Associated with Small Particles*, ed. P. W. Barber and R. K. Chang, World Scientific, Singapore, 1988.

20   E. J. G. Peterman, F. Gittes and C. F. Schmidt, *Biophys. J.*, 2003, **84**, 1308.

21   J. Popp, M. Lankers, K. Schaschek, W. Kiefer and J. T. Hodges, *Appl. Opt.*, 1995, **34**, 2380.

Holographic optical trapping of aerosol droplets

D. R. Burnham and D. McGloin

SUPA, School of Physics and Astronomy, University of St. Andrews, North Haugh, St. Andrews, Fife KY16 9SS, UK
dm11@st-and.ac.uk

Abstract: We demonstrate the use of holographic optical tweezers for trapping particles in air, specifically aerosol droplets. We show the trapping and manipulation of arrays of liquid aerosols as well as the controlled coagulation of two or more droplets. We discuss the ability of spatial light modulators to manipulate airborne droplets in real time as well as highlight the difficulties associated with loading and trapping particles in such an environment. We conclude with a discussion of some of the applications of such a technique.

OCIS codes: (010.1110) Aerosols, (010.7340) Water, (090.2890) Holographic optical elements, (120.5060) Phase modulation, (140.7010) Trapping, (230.6120) Spatial light modulators

---

References and Links

1.   A. Ashkin, "Acceleration and trapping of particles by radiation pressure," Phys. Rev. Lett. **24**, 156 (1970).
2.   A. Ashkin, J. M. Dziedzic, J. E. Bjorkhom and S. Chu, "Observation of a single beam gradient force trap for dielectric particles," Opt. Lett. **11**, 288-290 (1986).
3.   E. A. Abbondanzieri, W. J. Greenleaf, J. W. Shaevitz, R. Landick and S. M. Block, "Direct observation of base-pair stepping by RNA polymerase," Nature **438**, 460-465 (2005).
4.   S. Dumont, W. Cheng, V. Serebrov, R. K. Beran, I. Tinoco Jr., A. M. Pyle and C. Bustamante "RNA translocation and unwinding mechanism of HCV NS3 helicase and its coordination by ATP," Nature **439**, 105-108 (2006).
5.   P. T. Korda, G. C. Spalding and D. G. Grier, "Evolution of a colloidal critical state in an optical pinning potential landscape," Physical Review B **66**, 024504 (2002).
6.   P. M. Hansen, J. K. Dreyer, J. Ferkinghoff-Borg, and L. Oddershede "Novel optical and statistical methods reveal colloid-wall interactions inconsistent with DLVO and Lifshitz theories,", Journal of Colloid and Interface Science, **287** 561-571 (2005).
7.   V. Garcés-Chávez, D. McGloin, M. J. Padgett, W. Dultz, H. Schmitzer and K. Dholakia, "Observation of the transfer of the local angular momentum density of a multi-ringed light beam to an optically trapped particle," Phys. Rev. Lett. **91**, 093602 (2003).
8.   N. B. Simpson, K. Dholakia, L. Allen and M. J. Padgett, "The mechanical equivalence of the spin and orbital angular momentum of light: an optical spanner," Opt. Lett. **22**, 52-54 (1997).
9.   A. Ashkin and J.M. Dziedzic, "Optical levitation of liquid drops by radiation pressure," Science **187** 1073-1075 (1975).
10. A. Biswas, H. Lati, R. L. Armstrong and R.G. Pinnick, "Double-resonance stimulated Raman scattering from optically levitated glycerol droplets," Phys. Rev. A **40**, 7413-7416 (1989).
11. R. Thurn and W. Kiefer, "Raman microsampling technique applying optical levitation by radiation pressure," Appl. Spectrosc. **38**, 78-83 (1984).
12. R. Omori, T. Kobayashi and A. Suzuki, "Observation of a single beam gradient-force optical trap for dielectric particles in air," Opt. Lett. **22**, 816-818 (1997).
13. N. Magome, M.I. Kohira, E. Hayata, S. Mukai and K. Yoshikawa, "Optical Trapping of a Growing Water Droplet in Air" J. Phys. Chem. B **107**, 3988-3990 (2003).
14. M.D. King, K.C. Thompson and A.D. Ward, "Laser Tweezers Raman Study of Optically Trapped Aerosol Droplets of Seawater and Oleic Acid Reacting with Ozone: Implications for Cloud-Droplet Properties," J. Am. Chem. Soc. **126**, 16710-16711 (2004).
15. R. J. Hopkins, L. Mitchem, A. D. Ward and J. P. Reid, "Control and characterisation of a single aerosol droplet in a single-beam gradient-force optical trap," Phys. Chem. Chem. Phys. **6**, 4924-4927 (2004).
16. M. Reicherter, T. Haist, E. U. Wagemann and H. J. Tiziani "Optical particle trapping with computer generated holograms written on a liquid crystal display," Opt. Lett. **24**, 608-610 (1999).
17. J. A. Curtis, B. A. Koss and D. G. Grier, "Dynamic Holographic Optical Trapping," Opt. Commun. **207**, 169-175 (2002).
18. J. Enger, M. Goksor, K. Ramser, P. Hagberg, and D. Hanstorp, "Optical tweezers applied to a microfluidic system," Lab on a Chip **4**, 196-200 (2004).

19. K. Ramser, J. Enger, M. Goksör, D. Hanstorp, K. Logg, M. Käll, "A microfluidic system enabling Raman measurements of the oxygenation cycle in single optically trapped red blood cells," Lab on a Chip **5**, 431-436 (2005).

20. C. Creely, G. Volpe, M. Soler and D. Petrov, "Raman imaging of floating cells," Opt. Express **13**, 6105-6110 (2005).

21. Y. Roichman and D. G. Grier, "Holographic assembly of quasicrystalline photonic heterostructures," Opt. Express **13**, 5434-5439 (2005).

22. V. Soifer, V. Kotlyar, and L. Doskolovich, *Iterative Methods for Diffractive Optical Elements Computation* (Taylor and Francis, London, 1997).

23. J. H. Dennis, C. A. Pieron and K. Asai, "Aerosol Output and Size from Omron NE-U22 nebulizer," in *Proceedings of the 14th International Congress International Society for Aerosols in Medicines*, Baltimore June 14-18 2003. Journal of Aerosol Medicine 16:2 213, 2003.

24. J. H. Seinfeld and S. N. Pandis, *Atmospheric Chemistry and Physics: Air Pollution to Climate Change* (John Wiley and Sons Inc., 1997).

25. Laura Mitchem, Particle Dynamics Group, School of Chemistry, University of Bristol, Bristol, BS8 1TS (personal communication, 2006).

26. M. Polin, K. Ladavac, S. -H. Lee, Y. Roichman, and D. Grier, "Optimized holographic optical traps," Opt. Express **13**, 5831-5845 (2005).

27. P. Kaye, W. R. Stanley, E. Hirst, E. V. Foot, K. L. Baxter, and S. J. Barrington, "Single particle multichannel bio-aerosol fluorescence sensor," Opt. Express **13**, 3583-3593 (2005).

28. K. Davitt, Y. -K. Song, W. Patterson III, A. Nurmikko, M. Gherasimova, J. Han, Y. -L. Pan, and R. Chang, "290 and 340 nm UV LED arrays for fluorescence detection from single airborne particles," Opt. Express **13**, 9548-9555 (2005).

29. N. J. Beeching, D. A. B. Dance, A. R. O. Miller, and R. C. Spencer, "Biological warfare and bioterrorism," BMJ **324**, 336-339 (2002).

30. E. F. Mikhailov, S. S. Vlasenko, Lutz Krämer and Reinhard Niessner, "Interaction of soot aerosol particles with water droplets: influence of surface hydrophilicity," J. Aero. Sci. **32**, 697-711 (2001).

31. P.F. Smith, "Direct detection of weakly interacting massive particles using non-cryogenic techniques," Phil. Trans. R. Soc. A **361**, 2591-2606 (2003).

1. Introduction

Optical manipulation techniques have matured considerably in the 35 years since Ashkin first demonstrated the use of radiation pressure to guide particles [1]. The field was firmly established by the demonstration of the optical tweezers technique by Ashkin *et al* [2], and these have developed into a tool that is routinely used to probe biological function [3,4], colloidal dynamics [5,6], properties of light beams [7,8], and to facilitate the stable trapping and manipulation of particles at the micron scale in a wide range of disciplines. Almost the entire body of work on optical manipulation is carried out on particles suspended in a liquid medium. This medium acts to damp out the motion of particles as they are trapped and this fact combined with the buoyancy of particles in such samples makes their trapping relatively straightforward. To trap particles in the absence of such a damping medium (such as air) is more difficult and less relevant for studies to date. Ashkin studied the optical levitation of airborne droplets [9] and others have used these techniques to probe the size and composition of aerosols [10,11] but it is only relatively recently that their optical tweezing (that is gradient force trapping) has been demonstrated [12,13].

The ability to trap and interrogate airborne particles in a controlled manner offers much to those who wish to study aerosol properties, composition and dynamics. The localization of droplets in this way has recently been used to probe the Raman spectra of a trapped seawater droplet [14] as well as the controlled oxidation of oleic acid within the droplet. The stable trapping of the droplet also allowed the reaction and growth dynamics of the process to be followed. In a related experiment [15] the cavity enhanced Raman signal from a trapped droplet acting as a cavity was used to size droplet diameters to +/- 2nm and using a dual beam system two droplets could be controllably coagulated. Such experiments should allow systematic studies of processes on and within aerosols to be carried out with the atmospheric sciences being the main beneficiaries.

In this paper we outline the first use of holographic optical tweezers [16,17] for the manipulation of airborne particles, in this case aerosol droplets. Making use of such devices allows for arrays of particles to be trapped simultaneously, as well as allowing their controlled *xyz* translation. We show that controlled coagulation can be achieved and discuss the performance of the spatial light modulator in this context as well as outlining future applications of such a technique.

2. Experimental setup

Holographic tweezers are now a well established technique for the simultaneous manipulation of many particles. Recent work using these tools has explored applications in microfluidics [18,19], Raman imaging [20] and the creation of complex colloidal structures [21]. Our system makes use of a Holoeye LC-R 2500 spatial light modulator (SLM) to modulate the phase of 532nm cw light from a Laser Quantum Finesse laser (maximum output 4W). A schematic of our system is shown in Fig. 1. We first rotate the laser's plane of polarisation to optimize the efficiency of the phase modulation, then having expanded the beam via a telescope (L1 and L2) it is incident on, and covers, the SLM. Two 4f (L3 and L4, and L5 and L6) imaging systems are placed directly after the SLM to reduce the beam size to slightly overfill the back aperture of the microscope objective (Nikon CFI E Plan Achromat 100X oil, N.A. 1.25). The 4f imaging systems also make the SLM and back aperture planes conjugate. Having focused the light into the trapping plane the same objective is used for imaging.

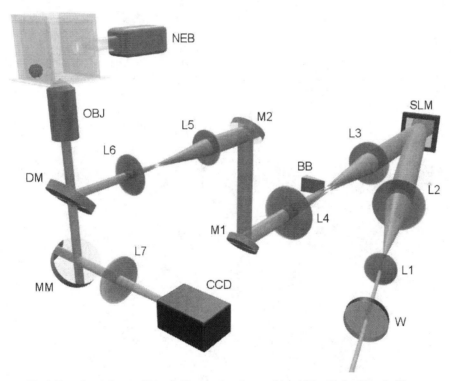

Fig. 1. Experimental setup. W is a half-wave plate. Lenses L1 and L2, with focal lengths 75mm and 750mm respectively act to expand the laser source. Lenses L3, L4, L5, and L6 form the two 4f imaging systems with focal lengths 400mm, 250mm, 200mm and 100mm respectively. Finally lens L7, with a focal length of 200mm, is the tube lens used in conjuction with the microscope objective, OBJ, which focuses the desired trapping pattern into the trapping chamber containing a water soaked tissue. Mirrors M1 and M2 are broadband dielectric mirrors. Mirror DM is a green dichroic mirror and MM is a metallic mirror for imaging into the camera CCD. A beam block, BB, is used to remove the SLM's zeroth diffraction order. The aerosols are produced using an ultrasonic nebuliser, NEB.

We generate our holograms making use of a custom LabVIEW program implementing an adaptive-additive algorithm [22] which treats the SLM as a secondary microdisplay.

The aerosol droplets are created using an ultrasonic nebuliser (Omron U22(NE-U22-E)), which produces aerosols with a mass median aerodynamic diameter (MMAD) between approximately 3 and 5 microns [23]. Custom glassware is attached to the nebuliser allowing the droplets to be fed accurately into a sealed glass cube, with an inlet and outlet hole in either side, which acts as our trapping chamber. The bottom of the cube is formed from a glass coverslip having been soaked in a 2% weight by volume solution of sodium dodecyl sulphate, for reasons explained in the discussion. To help control the humidity inside the cell we place a small water soaked paper tissue in the cube. In the experiments making use of one type of aerosol we block the second hole to avoid unnecessary air currents.

For the work outlined here we make use of water aerosols. However, water droplets generally only exist in oversaturated atmospheres (that is with a relative humidity greater than 100%) so we use aqueous solutions of NaCl to decrease the vapour pressure of the droplets which allows them to form in relative humidity <100% [24]. The majority of the work has been carried out with 0.34817 Molar concentrations [25] but others have been tested, for example 40mM and seawater, which provide similar results, although they do alter the droplet dynamics. The ambient laboratory temperature is $22.5 \pm 1°C$.

3. Experimental results

We are able to trap arrays of aerosol droplets as shown in Fig. 2, which also outlines one of the difficulties of trapping in air. In a liquid medium it can be hard to reliably load a large array of trap sites with colloidal particles. One has to actively seek out the particles, but this can result in particles falling out of the trap or in multiple particles being trapped at a single site. Alternatively the trap can be static with particles flown in to try and fill the sites and this can lead to jamming, or particles knocking each other out. In air we have no choice but to wait for particles to fall into the traps. In the experimental geometry trap loading is far from optimized, and the images shown illustrate that although we have a relatively simple trap pattern, which would be easy to fill in a liquid, we may be unable to fill the sites at all if no aerosol flies into the trapping region of that site.

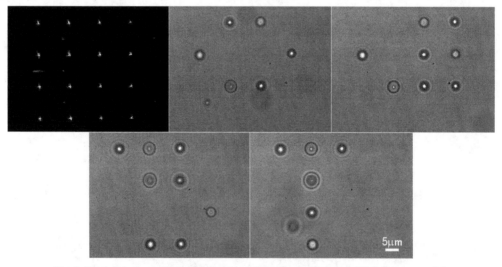

Fig. 2. Top left image shows the backscattered light from the bottom of the microscope slide revealing the holographic trapping pattern. The remaining images show the resulting trapped water droplets after multiple uses of the nebuliser in attempts to fill all trap sites. As indicated in the text, although we have a relatively simple trapping pattern it is hard to fill all the sites.

The current system allows xy translation of droplets in single jumps of up to $3.40 \pm 0.02\mu m$, which combined with a high refresh rate of ~25 frames per second allows xy translation at speeds up to $85\mu ms^{-1}$. The precision to which we can translate droplets in x and y is 227 ± 1nm with control of droplets over the entire field of view; 45 by 60μm. The apparatus also allows z control of the droplets as shown in Fig. 3.

Fig. 3. A series of microscope images demonstrating the z control of water droplets from -10 to +10 microns. The white bar indicates a scale of 5 microns.

We are able to trap a range of droplet sizes from ~2.5 to 12μm in diameter and determine that the minimum power required to trap an aerosol is 0.38 ± 0.02mW and an axial Q value of 0.22 ± 0.01, which are all comparable to results in conventional tweezers [15]. The droplet size is dependent on power with larger droplets requiring greater powers, as shown in Fig. 4.

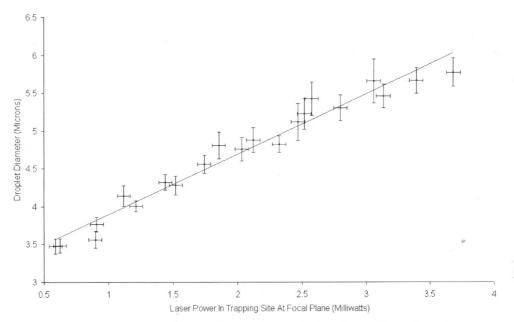

Fig. 4. Graph showing the variation of droplet size as a function of power. The vertical error has standard error bars increasing in size with power indicating that not only can higher powers trap, on average, larger droplets but also a greater distribution of sizes. The horizontal error bars mostly arise from the non perfect intensity uniformity of the trapping sites. However, the graph cannot convey the difficulty of achieving initial trapping at low powers.

This stable trapping and precise manipulation of water droplets allows not only their individual control but also the ability to coagulate multiple droplets both of which are demonstrated in Fig. 5.

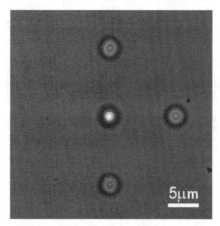

Fig. 5. Movie showing control and coagulation of multiple water droplets.

One may imagine the Brownian motion of aerosol particles is too high, and the SLM refresh rate too slow to enable such precise control of water droplets. However, by calculating the displacement due to Brownian diffusivity using Eq. (1):

$$\langle x \rangle^2 = \frac{2kTC_c t}{3\pi\mu D_p},$$

(1)

where k is Boltzmann's constant, T is the temperature in Kelvin, C_c is the slip correction factor, μ is the viscosity of the surrounding fluid (air), D_p is the droplet diameter, and t is the time. Plotting the results, Fig 6, it is clear that for the droplet sizes studied here the displacement due to Brownian motion after the SLM response time of 16ms is significantly below a single droplet radius [24].

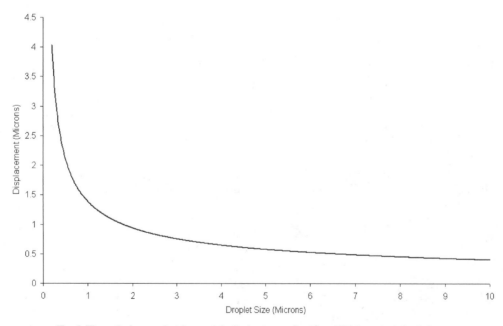

Fig. 6. Theoretical curve showing particle displacement after 16ms (SLM response time) due to Brownian diffusion.

4. Discussion and conclusion

Clearly from Figs. 3, 4, and 6 a major hurdle in quantitative analysis is the inability to accurately size droplets from video images alone, due not only to their dynamic nature but also to the poor definition of their edges on the video output. However, precise droplet size measurement techniques have already been demonstrated [15] allowing for future accuracy improvement and more precise quantitative results.

The relationship between laser power and droplet size combined with the ability of holographic optical tweezers to control each individual trap site's intensity will eventually allow some degree of size selective trapping.

One problem encountered was the formation of water 'puddles' from parts of the aerosol cloud settling onto the microscope slide. Measuring tens of microns across they create sufficient aberrations in the beam to cause the loss of droplets from traps unless they are sited directly over the puddle's centre where curvature is at a minimum. As a solution we applied a detergent surfactant to the slide thus increasing the wetting and reducing the formation of puddles.

The quantitative results shown above are specific to one concentration of sodium chloride solution, with any alteration in the aerosol composition significantly altering the droplet's properties. This obviously increases the complexity over simple monodispersed or even polydispersed microspheres as they have fixed dimensions and properties where as droplets can condense, evaporate, absorb, and constantly interact with their surroundings.

The experiments described above show for the first time the optical trapping and control of arrays of aerosol particles. Further they demonstrate the applicability of using the slow (in comparison to acousto-optical deflection techniques) phase only SLMs to manipulate airborne particles and to coagulate them. However issues remain with the techniques as described for some applications. Better kinoform generation algorithms need to be implemented [26] in order to reduce the effects of ghost traps and increase the intensity uniformity of the pattern. As we are using a flow of aerosols from a nebuliser source we mimic how atmospheric aerosols might behave, and for real sampling strategies we must make robust devices that can cope with a range of particle sizes and velocities. To some extent this is what we have demonstrated. For other applications, such as airborne digital microfluidics, where we wish to have less random choices about which particles we trap and carry out droplet reactions with we must come up with better loading strategies for the traps. We envisage using single droplet maker devices (such as ink jet printer heads) to position the drops ready for accurate loading. Combining this sort of technique with the random process would also allow better implementation of digital microfluidics for real time sampling.

We envisage many applications for airborne optical manipulation and the initial type of work in this area is outlined in [14] and [15] with regard to atmospheric chemistry. Other application areas include the sampling and characterisation of aerosols [27,28] such as bioaerosol sampling for anti-bioterrorism related work [29] and combustion studies [30]. There also exist many opportunities in more traditional optical manipulation studies, for example studying particle dynamics in the absence of a damping medium, optical rotation studies and even esoteric suggestions such as building neutrino detectors [31].

Acknowledgments

We would like to thank both Jonathan Reid and Laura Mitchem from Bristol University for helpful discussions and Kishan Dholakia (St. Andrews) for the use of the SLM. DRB would like to thank EPSRC for support, while DM is a Royal Society University Research Fellow. The work was partly funded by NERC and EPSRC.

Index